RECENT ADVANCES IN NUCLEOSIDES:

CHEMISTRY AND CHEMOTHERAPY

RECENT ADVANCES IN NUCLEOSIDES:

CHEMISTRY AND CHEMOTHERAPY

Edited by

C.K. CHU

2002

ELSEVIER

Amsterdam – Boston – London – New York – Oxford – Paris
San Diego – San Francisco – Singapore – Sydney – Tokyo

ELSEVIER SCIENCE B.V.
Sara Burgerhartstraat 25
P.O. Box 211, 1000 AE Amsterdam, The Netherlands

First edition 2002

Library of Congress Cataloging in Publication Data
A catalog record from the Library of Congress has been applied for.

British Library Cataloguing in Publication Data
A catalogue record from the British Library has been applied for.

ISBN: 0-444-50951-8

∞ The paper used in this publication meets the requirements of ANSI/NISO Z39.48-1992 (Permanence of Paper).
Printed in The Netherlands.

CONTENTS

PREFACE

The chemistry and biology of nucleosides and nucleotides has undoubtedly been a very exciting field for the past fifty years, particularly to those investigators who have been closely involved in the area. If we compare the field of nucleosides and nucleotides to an apple tree, the early fifties and mid-sixties were the main root-, trunk-, and branch-growing periods, and the next thirty years from the middle sixties to mid-nineties have been the fruit harvesting period. As we all know from our experience, without strong roots, a sturdy trunk and copious branches, we would not expect to harvest abundant and delicious fruits in the fall. Fortunately, we have been harvesting abundant and delicious fruits of nucleosides for the past thirty years. Particularly, during the last twenty years, we have witnessed a number of clinically useful nucleosides being developed as antiviral and anticancer agents. Furthermore, it doesn't appear likely that this rate of development of nucleosides as chemotherapeutic agents will be slowing down for the foreseeable future. However, those of us who are still in the field and enjoy the dividend of the fruit tree planted a half century ago should not forget those who nurtured the nucleoside tree at the early stage of the field. Thus, this book is dedicated to Dr. Jack Fox, who is one of the pioneers in nucleoside chemistry. He dedicated his entire half-century career to the chemistry of nucleosides at Memorial Sloan-Kettering Institute Cancer Center, New York, until his retirement in 1987. We, the contemporary nucleoside chemists and biochemists, are greatly indebted to his contribution in the chemistry and biology of nucleosides and nucleotides.

The chapters in this book are mainly based on the symposium honoring Dr. Fox, which was held as a satellite symposium of the Round Table in September of 2000. I would like to express my sincere appreciation to contributing authors for this book, and I would also like to thank those who helped the symposium as chairs (Drs. Bergstrom, Meyer, Lopez, Secrist, Bischofberger, Kalman, Herdewijn, Colacino, Eriksson, Broom, Rabi, Cook, Seela, Tanaka, and L. Townsend). I would like to recognize those pharmaceutical firms (Lilly Research Laboratories, Microbiologica, Bukwang Pharmaceuticals, Noviro, Pharmasset, Yamasa, Tiho Pharmaceuticals, and Gilead Sciences) who contributed their financial assistance for the symposium. My special thanks goes to Drs. Carlos Lopez and Joseph Colacino of Lilly Research Laboratories, and Dr. Jaime Rabi of Microbiologica, who arranged significant financial contributions for the symposium. Without their help, we would not have had such a high quality symposium that was enjoyed by all who attended. Finally, my special thanks goes to my assistant, Ms. Wendy Nix, for her dedication and hard work to complete the difficult task, from beginning to end, of the symposium as well as this book.

David Chu
College of Pharmacy
The University of Georgia
Athens, Georgia
June, 2001

MISSION ORIENTED RESEARCH: AN EXPERIENCE IN DR. JACK J. FOX'S LABORATORY

KYOICHI A. WATANABE

Pharmasset, Atlanta, GA, USA

1.   Introduction

In the early 1960s, Professor Yoshihisa Mizuno told me as a part of my thesis work to chemically synthesize 6-azauridine, which had been prepared by the Sorm's group and studied extensively in Czechoslovakia. As a young new professor, Dr. Mizuno might have had a grand plan to create a large library of biologically active nucleosides in his laboratory for future medicinal developments. However, I was not able to understand his plan and did not like the given project. There were many excellent chemists in Sorm's group, such as Prystas, Zemlicka, Piskala, Farkas, etc., and, I thought, they must have already undertaken the chemical synthesis of the nucleoside that they had developed. I rather wanted to develop our own nucleosides. In those days, there were two rather ineffective methods to synthesize pyrimidine nucleosides; the Hilbert-Johnson reaction and the Fox' "mercuri" procedure.

Handschumacher reported (Handschumacher, 1960) the condensation of 6-azathymine with ribose by the "mercuri" procedure, and obtained a mixture of variously ribosylated products, 1,3-bisribosyl-6-azathymine being the major product. In a similar manner, when I condensed chloromercuri-6-azauracil with acetobromoglucose, 1,3-bisglucosyl and 3-glucosyl-6-azauracil nucleoside derivatives were obtained in crystalline form, but the desired 1-glucosyl-6-azauracil was only detected on paper chromatography. In order to synthesize 6-azauridine, a method to avoid glycosylation at $N3$ and force the sugar to react selectively at N1 had to be developed. One possible way I thought was to fix the nitrogen at the 3 position in the azomethine structure by finding conditions for monothiation of 6-azauracil followed by S-methylation. The S-methyl group could be displaced by reaction with various nucleophiles. Thus, this method would lead to a new chemistry for the synthesis of a variety of nucleosides. I was excited with this idea. At that time, selective monothiation of uracil was considered not possible. However, I was able to find conditions for selective thiation at the 4 carbonyl of uracil and 6-azauracil (1, Scheme 1) simply by using a considerably small amount of phosphorus pentasulfide (Mizuno Y. *et al.*, 1962). S-Methylation of 4-thio-6-azauracil (2) proceeded smoothly, and the product 3 was converted into the "mercuri" derivative and condensed with tri-*O*-benzoyl-D-ribofuranosyl chloride by the Fox procedure. After the reaction, protected nucleoside 4 was obtained in crystalline form. Ammonia treatment of 4 gave 6-azacytidine (5) in good yield (Mizuno Y., 1962). After hydrolysis of the S-methyl group to 6, followed by saponification

1

Recent Advances in Nucleosides: Chemistry and Chemotherapy, Ed. by C.K. Chu. 1 — 20

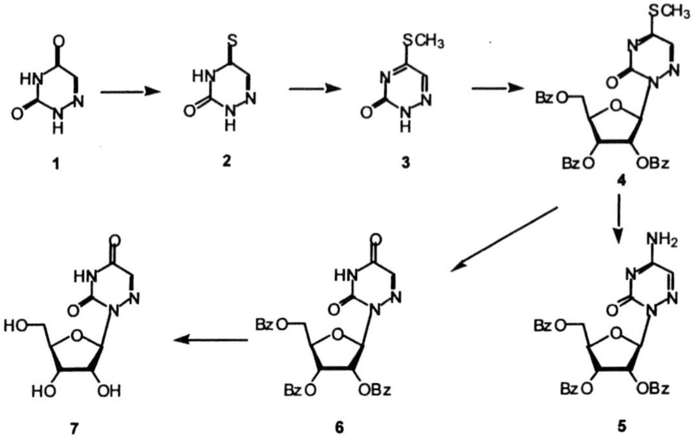

Scheme 1.

gave 6-azauridine (7) (Mizuno Y. *et al.*, 1963). This was, however, not the first synthesis of 6-azacytidine. Dr. Beranek and his group in Sorm's laboratory synthesized this nucleoside by a completely different procedure a few months before my synthesis. The *S*-methyl chemistry was later reinvented for the synthesis of certain 6-methylpyrimidine nucleosides (Winkley R., 1968). When I was working on these compounds, Dr. Mizuno took a sabbatical to Rockefeller University, and he met Dr. Fox at Sloan-Kettering Institute for Cancer Research across the street. Dr. Fox was interested in 6-azacytidine, and I received letter from Dr. Mizuno asking me to synthesize one gram (!) of the nucleoside and send it to Dr. Fox.

It was a difficult task. Since 6-azauracil was not commercially available, I had to synthesize it from mesoxalic acid and thiosemicarbazide. Large-scale synthesis of 6-azauracil was not possible as there was a decarboxylation step in this process. More troublesome was the preparation of ribose. I had to go to the Sapporo Beer factory where I obtained several kilograms of brewery yeast, from which RNA was extracted, and hydrolyzed to obtain nucleosides. Guanosine was the easiest nucleoside to isolate. This nucleoside was converted into 1-*O*-acetyl-2,3,5-tri-*O*-benzoyl-D-ribose by the method developed in Mizuno's laboratory. I was able to prepare about 100 mg of 6-azacytidine at a time. Just after sending the last batch (Watanabe K. A. *et al.*, 1981) to Dr. Fox, Professor Mizuno came back to Sapporo and asked me what I had done during his absence. He was rather unhappy with my response that I had not done anything but repeat the same sequence of reactions again and again to prepare one gram of 6-azacytidine. I was awfully unhappy with Dr. Mizuno's remark, but because my first priority was to obtain my degree, I just followed his direction and finished my thesis work. Probably due to 6-azacytidine, Dr. Mizuno proposed Dr. Fox's laboratory for my postdoctoral training. To assure this Dr. Mizuno did all the necessary paperwork for me. After graduation, I had a faculty position at the newly founded Faculty of Science and Engineering at Jochi (Sophia) University in Tokyo. I worked for one semester to save money for the trip to New York, and joined Dr. Fox's group as his Research Associate in 1963. Drs. Tohru Ueda and Jiri Farkas were about to leave the laboratory after their contracts were over. I was fortunate to have been able to meet them there. Tohru, who was my former research supervisor, was Dr. Fox's first postdoctoral fellow. Later he became one of the leading nucleic acid chemists. The Fox's laboratory equipment, however, was disappointingly insufficient. There was a Cary 15 UV spectrometer, but everything else belonged in a museum item. There was no NMR. Later, I had to drive about 20 miles to the Tarrytown Union Carbide Research Center to take NMR spectra to determine the structure of gougerotin (Fox J. J. *et al.*, 1965).

By the time I finished my Ph.D. work, I had dreams of synthesizing ψ-uridine and two complicated nucleoside antibiotics, gougerotin and blasticidin S (Figure 1) starting from a simple sugar molecule. I told Dr. Fox that I wanted to work on the synthesis of these natural products in his laboratory. The ψ-uridine project was out of the question.

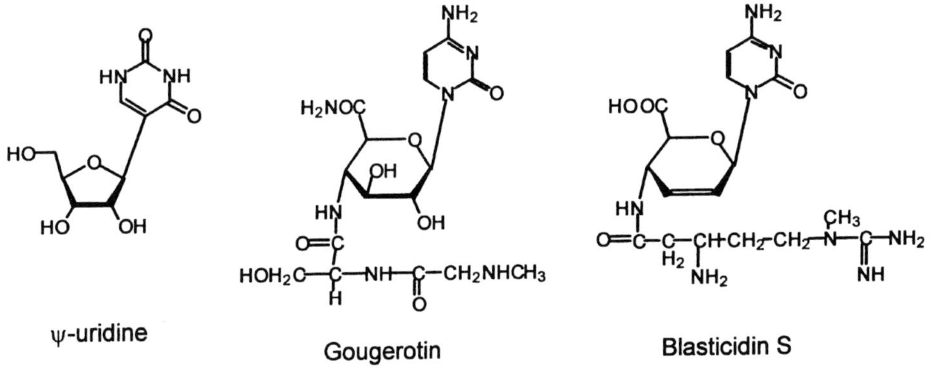

ψ-uridine

Gougerotin

Blasticidin S

Figure 1.

Already at that time, the compound was known to be the catabolic end product of tRNA to be excreted into urine and the nucleoside has no biological activity whatsoever. Dr. Fox said: "The Institute is mission oriented. You have to work on a cancer-related project. *Good science is not good enough for me*". According to Dr. Fox, the total synthesis of natural products would have been a long term projects, and the probability of anticancer drug development out of total synthesis was quite dubious. Thus, I had to do something practical. This first conversation with Dr. Fox haunted me ever since, for I went to Dr. Mizuno's laboratory and then joined Dr. Fox's group just to learn the *chemistry* of nucleic acids. In order to be practical, one has to be good at biology and biochemistry, and I was uninterested in these subjects. Dr. Fox offered me to choose from one of three projects. Two of them looked rather easy but the third one seemed challenging. The project was to apply the Fischer-Baer reaction to nucleosides.

H. O. L. Fischer, the youngest son of Emil Fischer had developed a cyclization method of sugar dialdehyde with nitromethane to a nitro-sugar, which was then reduced to an amino-sugar. Hans Baer, Professor at the University of Ottawa, had used the reaction he had developed with Fischer to synthesize kanosamine, 3-amino-3-deoxy-D-glucose (Baer H. H., 1968). I chose this project because the originally reported structure of gougerotin contained a 3-amino-sugar, thus eventually I would have had a chance to use the reaction for the total synthesis of this antibiotic.

The project was certainly difficult. According to Dr. Fox's suggestion, uridine (8, Scheme 2) was oxidized with metaperiodate to the dialdehyde 9 and then treated with nitromethane under various conditions to cyclize to 3'-nitro-D-hexopyranosyluracil. I worked very hard day and night for three months without any visible results. At that time Dr. John F. Codington, Naishun Miller, Iris Wempen and Iris Doerr were the residents of the laboratory. Lloyd Stempel was a graduate student. Every day at around 4 o'clock, Dr. Fox came to the laboratory and asked everyone about the day's results. He called everybody by first name except me. Unfortunately for a long time I had nothing to report. The reaction did not proceed or gave me a mess. I purified the solvents and reagents all by myself. One day my purified ethanol ran out, so I took some commercial solvent and ran the reaction. Something happened. I was able to isolate three crystalline compounds after ion-exchange chromatography. However, I was not sure the products I had obtained had the desired structures.

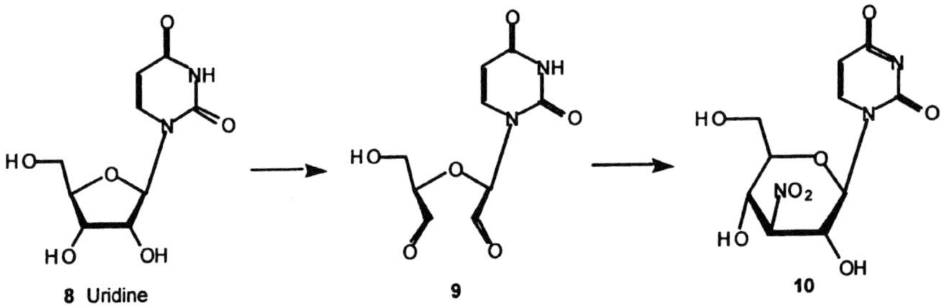

Scheme 2.

So, when Dr. Fox came to me that day, I still had no good news for him. Dr. Fox then asked when I would go back to Japan. When I responded that my contract would give me nine more months to work in the laboratory, he advised me go back earlier.

In Dr. Fox's laboratory, water occasionally and unexpectedly plays tricks. For example, his famous nucleoside thiation reactions (Fox, J. J. *et al.*, 1958, 1959) did not go without very small amounts of added water. Therefore, I added a small amount of water to the supposedly anhydrous reaction next time, and I was able to isolate a small amount of crystalline product without the help of chromatography. I then used water alone, and the result was even better. A high yield of shiny crystalline product was obtained which was later found to have the 3-nitro-3-deoxy-D-glucopyranose configuration (10) (Watanabe K. A. *et al.*, 1964, 1965). When I showed the crystals to Dr. Fox, he patted my shoulder and said "Kyo that's good" with a big smile in his face. For the first time he had called me by my first name. Since then I felt that I belonged to the laboratory. He then worked for my family to come to join me, and my one year contract was extended for another year. It was an interesting coincidence that Professor Lichtenthaler at Darmstadt, Germany, synthesized the same compound 10 almost simultaneously by uridine dialdehyde-nitromethane condensation.

During the second year Dr. Herbert A. Friedman and Dr. Jiri Beranek joined the group. We worked together and extended the dialdehyde-nitromethane reactions to other nucleosides. (Beranek J. *et al.*,1965; Friedman H. A. *et al.*, 1967) After almost 40 years, Herb and I still keep in contact. Jiri organized "Bechnye" conference on nucleic acid chemistry. For a long time this conference was the only chance for East and West scientists to meet, especially Germans from both sides of the wall. Unfortunately, he passed away.

Toward the end of my contract, and while I was packing up my belongings to go back to Tokyo, my apartment in New York burned down. The fire started in the basement around midnight, and when the building was in flame we did not have a chance to salvage anything including our passports but barely escaped from the building in pajamas. There was a firehouse next to our apartment building and we had been disturbed many times by the fire alarms and fire engines. But this time, the firehouse was silent. When we came to the street, a police car patrolling stopped by us, then the officers took us to the police station. I called up Dr. Fox at about 2 o'clock in the morning and explained what had happened. He came to pick us up and took us to his home where we stayed a couple of days. I lost not only my personal belongings but also lost the job in Tokyo, because I was not able to come back in time for the new semester. In this unexpected and dramatic situation, I accepted Dr. Fox's offer and took a permanent position. However, I felt needed carbohydrate chemistry experience and requested further training in Professor Lemieux's laboratory. Dr. Fox immediately called up Professor Lemieux in Canada and arranged everything for me. Just before I left for Edmonton, Dr. Brian A. Otter and Dr. Robert J. Cushley joined Dr. Fox's laboratory. Brian came from England with a solid carbohydrate background. Bob was Lemieux's student, but became more obsessed with NMR spectroscopy rather than synthetic carbohydrate chemistry. He came to the laboratory at the most opportune time, as just before his arrival the Institute had finally purchased a Varian A-60 spectrometer. With this instrument, we quickly solved a problem (Watanabe K. A. *et al.*, 1966) with a strange "sulfur containing product" Tohru had obtained treating 3-methyl-4-thiouracil with dimethylamine (Ueda T. *et al.*, 1963). Instead of affording the desired 3-methyl-N^4,N^4-dimethylaminocytosine, the uracil ring opened and a thioacrylamide derivative was formed. Bob published many papers from Dr. Fox's laboratory collaborating with almost everybody. Later he analyzed moon soil brought by astronauts by NMR and found no organic materials in the moon.

Two postdoctoral years in Lemieux's laboratory were very fruitful in the long run. The laboratory was equipped with state-of-the-art instruments. There were many professors of various fields. I always found someone to talk to whenever I encountered any scientific problem. There were many outstanding departmental seminars. Lectures by Dr. Nakanishi on the structure of gingkolide, by Dr. Breslow on the synthesis of tetracycline antibiotics, and the progress toward a total synthesis of vitamin B_{12} by Dr. Eschenmoser were especially impressive. The chemistry library had all the necessary journals and books. I read many papers related to nucleic acid chemistry, and slowly an idea of unifying mechanisms of nucleoside synthesis by condensation came to me (Watanabe K. A. *et al.*, 1974). In Dr. Lemieux' laboratory, I was involved in the synthesis of many deoxy and unsaturated sugars with Dr. Andre Pavia (Lemieux R. U. *et al.*, 1968; Lemieux R. U. *et al.*, 1969) . Andre is currently Professor at the University of Montpellier, France. Even though Dr. Lemieux was a very busy person, I was able to talk to him almost every day since the location of my laboratory was just next to his office. Thus, quite often regardless of his excuse: "Kyo, I have to go. Jeannine is waiting", he spent some time discussing on the results of the day and then several other topics other than chemistry. He introduced me to many outstanding chemists visiting his laboratory. Later, I frequently found him in the audience when I gave a talk at national and international conferences.

Once we discussed the qualification for good scientific publications. According to his opinion, there were three types of good publications: (1) useful publications, (2) papers reporting the results that respond to current interest, and (3) publications containing something that is very new although its practical value is quite dubious. Drug development research belongs to the first category and total synthesis of complicated natural products such as chlorophyll or vitamin B_{12} is considered to be in the second category. However, it is the third type of research that advances the science. Many young scientists in the field do not know the names of Friedrich Miescher, the discoverer of nucleic acid, Albert Kossel, the discoverer of thymine and adenine in nucleic acid, and even P. A. Levene who identified the sugar components of both RNA and DNA, and isolated adenosine and guanosine from nucleic acid.[1] These pioneers really pushed the chemistry of nucleic acids forward, but they did not know how their work would lead to a useful contribution to the public. Unlike sending a rocket to the moon, the chemistry of nucleic acids did not appeal to the public especially at its infancy. In college I first learned the term "nucleic acid" originally introduced by Altmann. Miescher had hard time publishing his papers. We also encountered similar difficulties when we tried to publish something very new. Without any precedents, reviewers are either very reluctant or incapable of evaluating manuscripts that belong to the third category. Many of the manuscripts incorporating our best work were rejected by the so-called good journals, but later an alternative or improved approach by others appeared in the same journal that had rejected our original work. On the other hand, we experienced little trouble publishing less innovative work. I will present a few such examples later.

In the meantime, Dr. Fox worked on solving my visa problem, and two years later, I came back to his laboratory. Many researchers in Dr. Fox' laboratory had left while I was in Canada, but Brian and Iris Wempen were still there. Brian discovered interesting pyrimidine ring transformations of 5-substituted pyrimidine nucleosides. He was a meticulous chemist, not publishing anything unsure. In addition, there were two young postdoctoral fellows, Dr. Robert S. Klein and Dr. Michael P. Kotick working in the laboratory. Dr. Fox was promoted to Vice President of Research in the Institute, and became ever more enthusiastic in developing anticancer drugs. The major activity was to synthesize a number of analogues of active compounds. Such studies are definitely important, but I also wanted to be challenged with something more chemically exciting. I started to synthesize gougerotin from D-galactose with Mike.[2] We found that the carbohydrate moiety of grougerotin was not 3-amino-D-allopyranuronamide (Iwasaki H. *et al.*, 1962) but a 4-amino sugar 6 which turned out to be 4-amino-D-glucopyranuronamide (Fox J. J. *et al.*, 1968). Mike came from Professor Thomas Bardos' laboratory, very intelligent, skillful and somewhat shy. We worked together only for 6 months, but successfully completed the synthesis of 4-amino-D-glucose (Kotick M. P. *et al.*, 1969) and 4-amino-D-glucuronic acid (Watanabe, K. A. *et al.*, 1969) and the nucleosides out of them, which were found to be identical with the nucleosides derived from the antibiotic (Watanabe K. A. *et al.*, 1970). A couple of years later, we were able to complete the total synthesis of gougerotin (Watanabe K. A., Falco E. A., Fox J. J., 1972) with the help of Iris Wempen (Watanabe K. A., Wempen I., Fox J. J., 1972) and Elvira Falco. We were surprised by the fact that Professor Lichtenthaler's laboratory was also involved in the total synthesis of gougerotin at the time we were working on it. We also worked on blasticidin S total synthesis with Dr. Roger Goody and synthesized 4-amino-D-glycerohex-2-enuronic acid, a new type of carbohydrate and its nucleoside, (Goody R. S. *et al.*, 1970; Watanabe K. A., Goody R. S., Fox J. J., 1970) completing a formal total synthesis of this antibiotic (Watanabe K. A., Wempen I. Fox J. J., 1970). We learned later that Professor Goto's laboratory in Nagoya was also achieved a formal total synthesis of blasticidin S about the same time. Roger is currently Professor at Max Planck Institut fur medizinische Forschung. One of the nucleosides, 1-(4-amino-3,4-dideoxy-β-D-ribohexopyranuronosyl)cytosine, (Chiu T. M. K. *et al.*, 1973; Watanabe K. A. *et al.*, 1976) synthesized with Dr. Tony M. K. Chiu was later found in nature. We also synthesized a number of natural nucleosides, pentopyranins, elaborated by *Streptomyces griseochromogenes* and discovered by Professor Seto of Tokyo University. My Canadian experience in sugar chemistry was essential for the successful total synthesis of these natural products.

I have to emphasize here that the antibiotic syntheses were not our first priority. Our major efforts were focused on more mission oriented projects, which eventually led to the development of Dr. Fox's "masked precursor" concept in drug development (Scheme 3). The antileukemic nucleoside, 1-(β-D-arabinofuranosyl)cytosine (ara-C or cytarabin) is one of the most effective drug for the treatment of adult human myeloblastic leukemia. The trouble with this drug is its short half-life in plasma, because it is enzymatically deaminated rapidly to the inactive uracil derivative. We found that modification of the carbohydrate moiety of cytosine nucleoside affects the rate of deaminase action (Kreis W. *et al.*, 1978). Thus, if a compound were so designed that it is not a good substrate of deaminase but can undergo chemical rearrangement in the plasma to ara-C, such a compound might become a better anti-leukemic agent than the parent ara-C. Indeed, unmodified xylosylcytosine was completely inactive but 3'-bromo-3'-deoxyxylosyl-cytosine (15, X = Br) was an active compound, and its activity is reversed, like ara-C, by addition of deoxycytidine. 2'-Bromo-2'-deoxycytidine (11, X = Br) and 2'-bromo-2'-deoxyara-C (13, X = Br) were found to be active and their activity

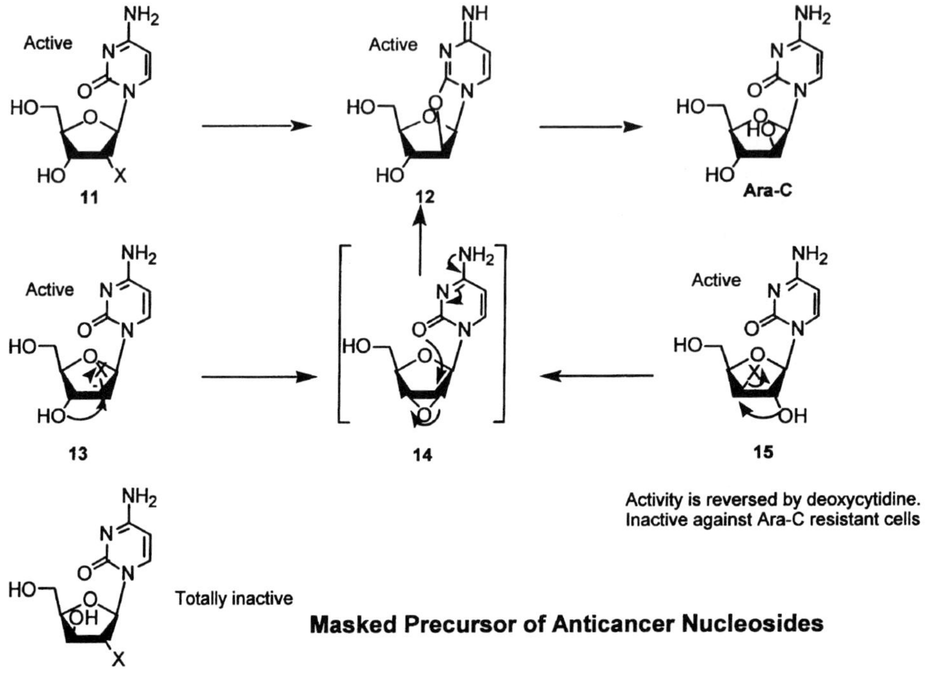

Scheme 3.

was reversed as in the case of ara-C by deoxycytidine. The intermediate 2,2'-anhydro-C (12) was not a substrate of deaminase (Hoshi A. *et al.*, 1973). Introduction of a fluorine substituent at C-5 in these sugar modified nucleosides may act by dual mechanisms; they may first act as ara-C, and after deamination, and after glycosyl cleavage as 5-fluorouracil (Watanabe K. A. *et al.*, 1980). We made a number of cytosine nucleosides, and Dr. Fox's hypothesis worked well. Much later, the "pro-drugs" approach became rather popular, but Dr. Fox was one of the pioneers if not *the* pioneer of this type of drug design.

Chemical synthesis of ψ-uridine was one of my dreams and wanted to work on the chemistry of *C*-nucleosides. I found a paper by Jardetzky in the Journal of Biological Chemistry stating that the molecular shape of chloramphenicol was similar to uridine (Jardetsky C. D., 1963). Together with Dr. Klein and Dr. Kotick, we synthesized 1-β-D-ribofuranosyl-4-nitrobenzene (Klein R. S. *et al.*, 1971), which should be more similar to uridine than the antibiotic itself. I was not serious about the rationale which was just used for an excuse to enter *C*-nucleoside chemistry. Somehow, we received an unexpectedly large number of reprint requests for this work. More seriously, I tried to condense diethyl malonate or ethyl formylacetate or ethyl α,α-dimethoxyacetate with acetobromoglucose whenever I found time, without success. Even in Sapporo, Akihiro

Yamazaki and myself tried the reaction. Akihiro, later, developed the famous synthesis of guanosine from 4-aminoimidazole-5-carboxamide riboside (AICAR) (Yamazaki A. *et al.*, 1971). I tried several times the malonic ester reaction under different conditions in Canada, and with Mike in New York. It was, therefore, complete a shock to me when Stephen Hanessian reported his successful condensation of malonate and a sugar at a Gordon Conference. I immediately asked the special secret for the success. The answer was the solvent, 1,2-dimethoxyethane. Malonate is not a good starting material for ψ-uridine synthesis because cyclization with urea gives barbiturate derivatives and removal of the 6-oxo group from the product is not straightforward. I had an idea of making α-ribosylacetate (17, Scheme 4) which has an active methylene group and should be amenable to formylation. The only problem was that I did not have a good excuse to work on it.

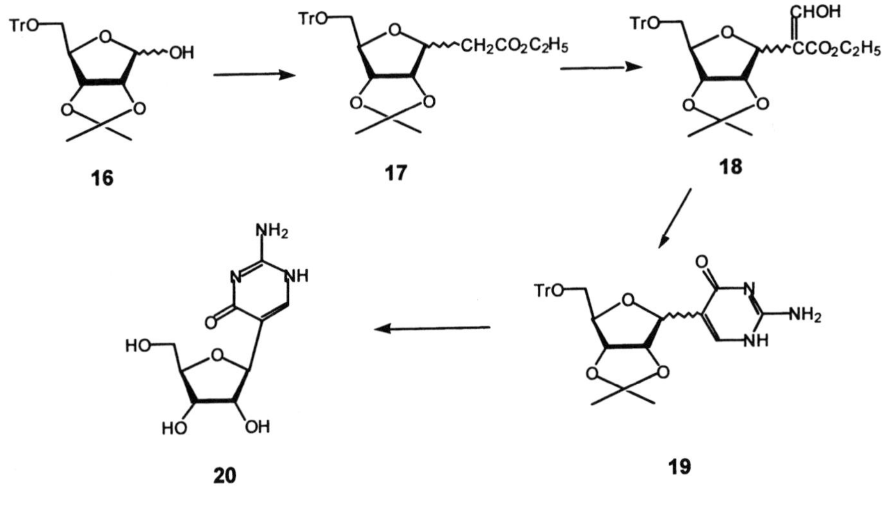

Scheme 4.

In 1975, Dr. David C. K. Chu joined our group. He was trained as a medicinal chemist in Bardos' laboratory. When I talked to David about the synthesis of ψ-uridine and my difficulty of cooking up the rationale, he immediately responded by saying to cyclize the formylacetate derivative with guanidine, the product, ψ-isocytidine, would be considered as an analogue of an antitumor antibiotic, 5-azacytidine. The latter is known to be a good anticancer agent, but rather unstable. In contrast, our ψ-isocytidine is not an unstable *s*-triazine but a stable pyrimidine. David solved my problem of necessary justification to work on the *C*-nucleosides which troubled me more than a decade. David synthesized ψ-isocytidine (Chu C. K., Watanabe, K. A., Fox, J. J., 1975; Reichman U. *et al.*, 1977; Chu C. K., Wempen I. *et al.*, 1976) in a few weeks according to the procedure of Scheme 4. Later, a number of publications by others dealing with the synthesis of various *C*-nucleosides, but most of them used the same principle; *i.e.*, preparation of α-glycosyl-acetate or –acetonitrile, formylation of the active methylene group, followed by construction of a heterocyclic aglycone. We also synthesized several pyrimidine or purine-like *C*-nucleosides using the same concept. Dr. Fox, of course, was very helpful in arranging biological tests. The *C*-nucleoside was, as expected, a very potent inhibitor of various leukemic cells (Burchenal, J. H. *et al.*, 1976), which naturally pleased both David and Dr. Fox, but I was pleased more by the successful synthesis of ψ-uridine[42] (Chu, C. K., et al., 1976) using the procedure that had been planned and desired for many years. Our first paper, which opened a new avenue for synthesis of many types of biologically active *C*-nucleosides, was not accepted by ACS journals.

The excellent biological activity of ψ-isocytidine[3] (Burchenal, J. H., et al., 1976) gave me nightmares. We had to synthesize much larger amounts of the *C*-nucleoside for further biological studies. However, as it happened to me at the beginning of my carrier (see 6-azacytidine synthesis for Dr. Fox) I could not ask my colleagues to synthesize large amounts of the known compound repeatedly.

Development of a new method of synthesis was necessary. I remembered a chapter written by Aaron Bendich in Chargaff and Davidson's "The Nucleic Acids" which I had read as a student. In the chapter he discussed Levene and Bass' discovery of transformation of uracil into pyrazolone by treatment with hydrazine.[4] In this reaction, the urea portion (N-C-N) of the molecule is displaced by hydrazine (N-N). If guanidine had been used instead of hydrazine in this uracil transformation reaction, the urea portion might have been displaced by guanidine forming isocytosine. At that time, fortunately, Dr. Kosaku Hirota joined our group. He had also been trained as a medicinal chemist and at the same time a heterocyclic chemist. I explained my idea and asked him to treat uracil with guanidine. The reaction did not occur. Probably in the presence of a strong base like guanidine, uracil dissociates and the formed anion would repel the approaching nucleophile. Therefore if one could alkylate both nitrogens of the uracil, no dissociation would occur and approach of nucleophile should be uninhibited. Kosaku made 1,3-dimethyluracil, treated it with guanidine and isolated isocytosine in good yield (Hirota K. *et al.*, 1977; Hirota K. *et al.*, 1978). Fortunately for us, Kowa Hakko in Tokyo had a patent on the production of ψ-uridine by fermentation, and they had about a kilogram on their shelf. The company kindly gave all the ψ-uridine they had to us to use as a starting material for the synthesis of new *C*-nucleosides with potential anticancer activity. We converted it to the corresponding 1,3-dimethyl derivative 22 (Scheme 5) and then treated the product with guanidine to obtain isocytosine. Again, our first paper of this subject was not accepted by ACS journals, although it contained a true sense of new chemistry, which triggered the later discoveries of new heterocyclic ring transformation reactions. This two-step process was certainly better than the original synthesis and amenable to scale-up. Its only limitation was the size of the flask we could handle. Usually we started with 60 grams of ψ-uridine, which we converted in two days into about 56-58 grams of crystalline ψ-isocytidine, isolated as the hydrochloride salt.

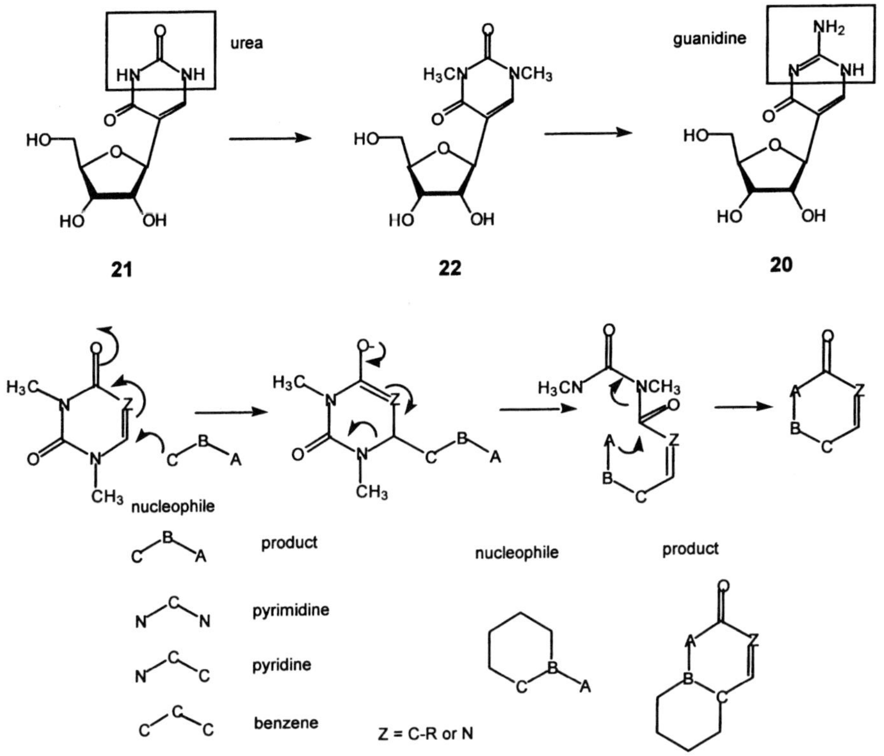

Scheme 5.

We believed that we were the first ones to invent this pyrimidine to pyrimidine transformation. However, quite independently and almost simultaneously Professor van der Plas achieved displacement of the N_1-C_2-N_3 fragment of pyrimidine with other N-C-N fragments. His paper appeared prior to our publication in a Dutch journal (Oostveen, E. A., et al., 1976). Kosaku later developed some useful ring transformations from 1,3-dimethyluracil derivatives into other pyrimidines, pyridines (Hirota K. *et al.*, 1979; Hirota K. *et al.*, 1981), pyridopyrimidines (Hirota K. *et al.*, 1981). He is currently president of Gifu Pharmaceutical University. Also, ring transformations from an s-triazine to another s-triazine and s-triazine to pyrimidine were developed by Dr. Won Keun Chung and Dr. Moon Woo Chun in our laboratory (Chung W. K. *et al.*, 1979). Later Dr. Tsann-Long Su from Professor Vorbruggen's laboratory came to our laboratory, and developed another new type of pyrimidine to pyridopyrimidine (Su T-L., Watanabe K. A., 1982; Su T-L. *et al.*, 1984) transformation (Scheme 6) as well as pyrimidine to benzene transformation (Su T-L. *et al.*, 1982). Actually, Dr. Fox's

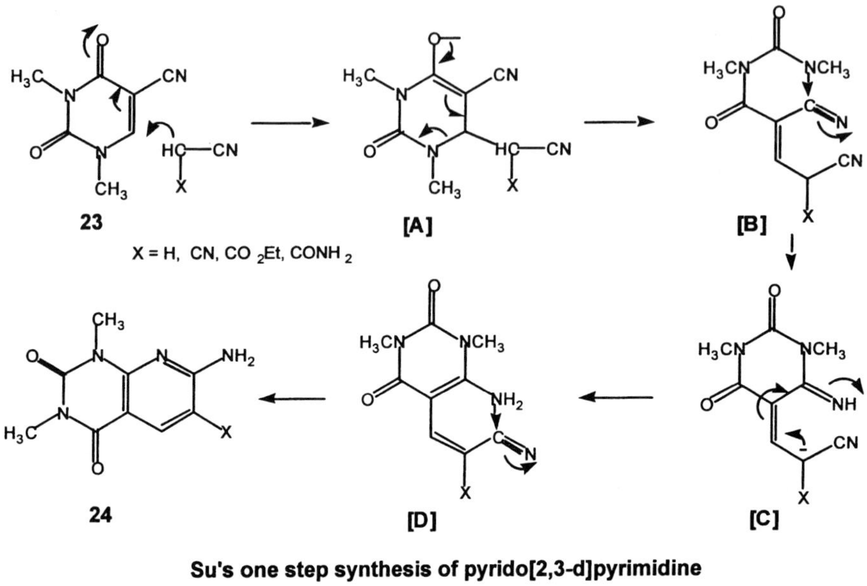

Su's one step synthesis of pyrido[2,3-d]pyrimidine

Scheme 6.

graduate student, Lloyd Stempel was the first one who discovered pyrimidine to benzene ring transformation reaction in 1963 when he added a base in an acetone solution of 5-nitropyrimidin-2-one. A yellow color immediately developed, which disappeared upon neutralization, but no 5-nitropyrimidin-2-one was recovered. He isolated *p*-nitrophenol instead (Fox J. J. *et al.*, 1982). In his case, N_1-C_2-N_3 of the pyrimidine was displaced by C-C-C of acetone in the presence of base. The pyrimidine to pyridopyrimidine transformation was later utilized in the synthesis of folic acid analogues by Dr. Su in our laboratory (Su T-L. *et al.*, 1986; Su T-L. *et al.*, 1988).

About that time all of my student time dreams came true Dr. Fox told me to prepare 2'-fluoro-ara-C, which had been synthesized in his laboratory by Dr. John Wright (Wright J. A. *et al.*, 1970) and was found to show good cytotoxicity in tissue culture. He needed the compound for animal studies. John was from Dr. Norman Taylor's laboratory in England and had experience with fluorinated sugars.

His method was perfect to make the compound, but looked too complicated to me. Almost every step produced a mixture of close isomers and required separation. I would not be able to prepare the nucleoside in an amount sufficient for animal studies by John's method. Here again, my carbohydrate experience in Dr. Lemieux' laboratory helped me. Nucleophilic substitution in methyl glycosides on C-2 is difficult, especially by a poor nucleophile as fluoride but on C-3 is rather easy. In general, nucleophilic substitution at a carbon atom is difficult when the adjacent (vicinal) carbon bears an electron withdrawing group. I explained this simply in the following way: in order for substitution to take place, the leaving group leaves as an anion by pulling out an electron pair from the carbon. If the leaving group is attached to a carbon with lower electron density (due to the presence of electronegative substituent on the adjacent carbon) it should be difficult to dissociate from carbon by extracting electrons. Although my explanation is simple and does not deal with molecular orbital theory, I have not found any exception to my conjecture for more than 40 years. Thus, ribo and arabino nucleosides are more resistant to hydrolysis than 2'-deoxynucleosides. I thought of a procedure for the synthesis of 2-fluoro-D-arabinofuranoside via introduction of a fluorine on C-3 of hexose as shown in Scheme 7. This was to me a novel and exciting carbohydrate chemistry. The procedure should afford only the desired furanose and each step should produce only single product. Key to the success of this method was to prepare 3-fluoro-D-glucose effectively and economically. Dr. Uri Reichman just joined our group, and he undertook the project. Uri obtained his Ph.D. under Professor Felix Bergmann with no carbohydrate chemistry background. It turned out, however, we could not find a better person to perform the chemistry outlined in Scheme 7. He introduced the inexpensive KF-acetamide combination for fluorination and success-fully obtained 3-fluoro-glucofuranose, which he converted into the desired 2-fluoro-D-arabinofuranose. He prepared a few grams of 2'-F-ara-C. Unfortunately, the compound did not show good activity in animal studies (Reichman U. *et al.*, 1975). Again, this paper was not accepted by an ACS journal.[5] Uri's bench was always messy in contrast to David's. Some times Uri used even laboratory floor and occasionally invaded David's

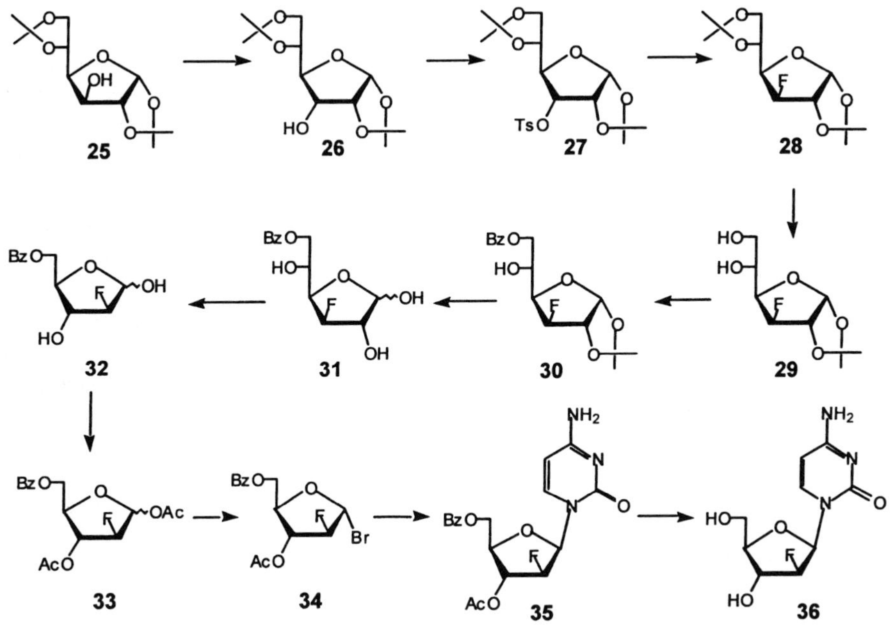

Scheme 7.

bench. David complained but cleaned the mess. However, Uri's experiments were always very accurate and reproducible. Fortunately they are very good friends, and created a comfortable and productive atmosphere in our group.

I have always remembered a conversation with Dr. Morio Ikehara which took place a number of years ago. He was the supervisor of Eiko Ohtsuka and Tohru was mine, but we all were close and freely discussed many things. One day, Morio showed me a short article in C & E News mentioning that 5-iodo-2'-deoxyuridine (IdU, Figure 2) synthesized by Dr. William Prusoff[6] of Yale University exhibited remarkable activity against herpes keratitis. Later I read somewhere that IdU was readily decomposed by nucleoside phosphorylase. This glycosyl instability is a common problem of deoxy nucleosides including BVdU and 5-ethyl-2'-deoxyuridine. Now we had a method of synthesis of 2-fluoro-D-arabinofuranose which contains the very electronegative fluorine substituent at C-2. We expected that 2'-fluoro analogues of these antiviral nucleosides should have been stable. We synthesized a number of 2'-fluorinated nucleosides and tested them for their activity against herpes simplex viruses (Watanabe K. A., Reichman U., Hirota K., Lopez C., Fox J. J. , 1979). The timing was good but not perfect, because those days herpes simplex type 2 infection was a big social concern. We were able to attract the NIH funding for many years. During Phase II clinical trials (Young C.W. *et al.*, 1983; Leyland-Jones B. *et al.*, 1986) of one of our nucleosides FIAC for treatment of herpes infection in cancer patients, acyclovir (acycloguanosine) became available,

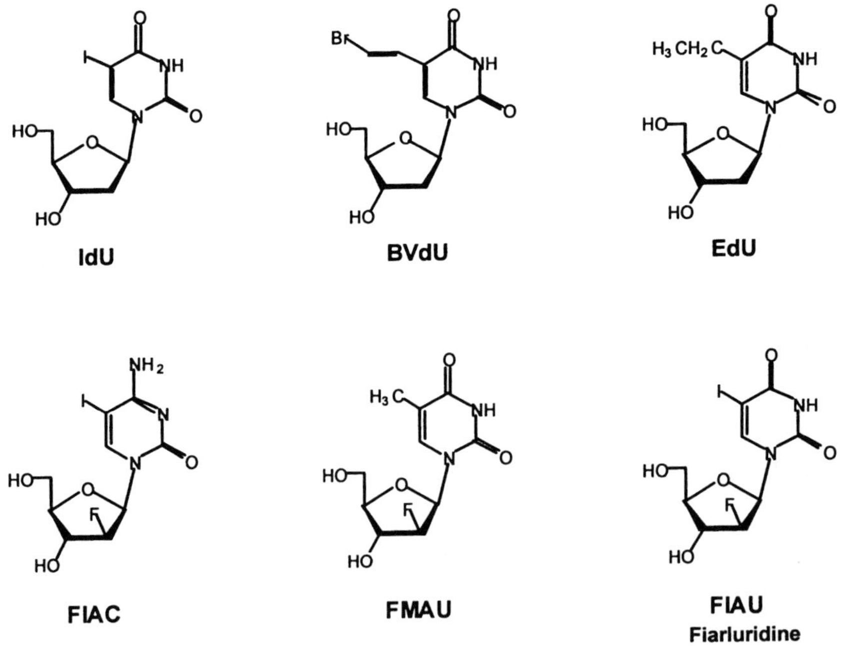

Figure 2.

and the Memorial Sloan-Kettering Cancer Center discontinued further development of FIAC. Many people participated in this project; including Dr. Michael E. Perlman, Dr. Jasenka Matulic-Adamic, Dr. Akira Matsuda in addition to David, Uri, Tsann-long and Moon Woo. Akira and David are the organizers of the Fox symposium and editors of this book. Moon Woo is currently the Dean of College of Pharmacy, Seoul National University. Tsann-long is Deputy Director of the Institute of Biomedical Sciences, Academia Sinica in Taiwan and Jasenka is at Ribozyme, Inc., publishing a number of papers. Those 2'-fluoro-nucleosides we synthesized are indeed resistant to chemical and enzymatic hydrolysis.

In 1980 Dr. Krzysztof W. Pankiewicz and Dr. Akira Matsuda joined my group. Kris was from Professor Wojciech Stec's laboratory of Polish Academy of Science in Lodz[7] Although Kris did not have any experience in nucleosides, he learned everything very quickly and became very productive. Kris' discovery of triflyl migration (Pankiewicz K. W. *et al.*, 1986) and his idea of introducing fluorine in the C-2' position of preformed nucleosides (Pankiewicz K. W., Kim J. H., Watanabe K. A., 1985; Pankiewicz K. W., Watanabe K. A., Takayanagi H., Itoh T., Ogura H., 1985; Pankiewicz K. W., Krzeminski J., Ciszewsk L. A., Ren W-Y., Watanabe K. A., 1992; Pankiewicz K. W., Krzeminski J., Watanabe K. A., 1992) attest to his excellent chemical sense. Later we started nicotinamide *C*-nucleoside synthesis with Dr. Marek Kabat from the Institute of Organic Chemistry, Polish Academy of Sciences in Warsaw (Kabat M. *et al.*, 1987). Kris is now Director of Chemistry at Pharmasset, Inc., (see his own chapter in this book). Akira was trained by Tohru Ueda, my former supervisor, and succeeded his laboratory which was originally founded by Dr. Mizuno.

In the early 1980's the public interest shifted from herpes to AIDS. We continued working on the discovery of antiviral agents but the main target became the treatment of AIDS patients. Many colleagues were involved in this program. Three Polish woman scientists, Dr. Joanna Zeidler, Dr. Barbara Nawrot and Dr. Elzbieta Sochacka were the early participants of this program. We synthesized many nucleosides and tested them in-house by Dr. Bruce Polsky of the Infectious Diseases Department. Some of them showed significant activity. A good candidate for clinical development was 3'-deoxy-3'-fluorothymidine 5'-hydrogenphosphonate (Matulic-Adamic J. *et al.*, 1987). However because of the severe toxicity of the parent nucleoside no efforts were made for further development, although this compound was remarkably non-toxic to mice.

After Dr. Fox's retirement, I became more interested in the biological mechanism of action of active nucleosides. My biological colleagues claimed with evidence that these nucleosides were incorporated into nucleic acids, disrupting their functions. (Grant A. J., *et al.*, 1982; Fox J. J. *et al.*, 1982; Chou T.-C. *et al.*, 1983; Lewis W. *et al.*, 1996) My simple curiosity is the reason for the disruption of nucleic acid function caused by incorporated "artificial" nucleoside. For example, in FMAU – a very potent anti-HSV and anti-HBV compound but also very toxic – the aglycon is natural thymine. The only difference between FMAU and natural thymidine is a fluorine that replaces the 2'-β hydrogen. The 2'-fluoro-arabino nucleosides apparently are conformationally very similar to corresponding the 2'-deoxynucleosides, as analysis of FIAC by X-ray crystallography indicated (Birnbaum G. I. *et al.*, 1982) that surprisingly the fluorine substituent causes little conformational change. When FMAU displaces a few thymidines in DNA, the backbone is natural phosphodiester linkage, the aglycon is natural thymine and the sugar conformation is quite similar to that of "deoxyribose". For us organic chemists, the simplest approach to this question would be to synthesize oligonucleotides containing biologically active nucleosides and compare their biochemical and biophysical properties with their natural oligonucleotide counterparts. We synthesized various modified oligomers using home-made synthesizers (Rosenberg I. *et al.*, 1993). Commercial synthesizers did not work well for many of our purposes. At the beginning we worked on a manual synthesizer made by Dr. Ivan Rosenberg. Later, Ivan with Dr. Jaime Farras Soler and Dr. Wu-Yun Ren constructed a fully automated unique synthesizer. Dr. Zdenek Tocik and Dr. Pavol Kois then joined the synthesis of oligomers with our automated synthesizer. Interesting results started to emerge, but the progress was painfully slow due to lack of funds (Kois P. *et al.*, 1993). I was unable to make up the rationale to persuade Study Section members of the significance of such research.[8] While synthesizing many modified oligonucleotides, I also became interested in "gene repair" using triplex. I was fortunate to some extent that I was later able to be involved in initial stage of attempts to convert mutated gene back to normal using modified oligonucleotides as the third strand (Majumdar A. *et al.*, 1998). The major planners for the chemistry part were Dr. Alexander Khorlin and his wife Dr. Natalie Dyatkina both from Engelhardt Institute of Molecular Biology, Russian Academy of Sciences. Both joined my group by recommendation of Natalie's mother, Professor Maria Preobrazhenskaya. I met Maria for the first time in 1964 when she visited Dr. Fox at Sloan-Kettering Institute.

A large part of my scientific career was spent in Dr. Fox's laboratory at Sloan-Kettering Institute for Cancer Research. During my tenure with Dr. Fox, I was able to concentrate in and enjoyed only science with many open-minded colleagues. More than one third of my publications are coauthored with Dr. Fox, and a handful of compounds from Dr. Fox's laboratory underwent Phase I and Phase II clinical trials. I am happy that I was involved in the development of some of these compounds.

References

Baer, H. H. *J. Org. Chem.* 1968, *32*, 2822.

Beranek, J.; Friedman, H. A.; Watanabe, K. A.; Fox, J. J. *J. Heterocycl. Chem.* 1965, 2, 188-191.

Birnbaum, G. I.; Cygler, M.; Watanabe, K. A.; Fox, J. J. *J. Am. Chem. Soc.* 1982, *104*, 7626-7630.

Burchenal, J. H.; Ciovacco, K.; Kalaher, K.; O'Toole, T.; Kiefner, R.; Dowling, M. D.; Chu, C. K.; Watanabe, K. A.; Wempen, I.; Fox, J. J. *Cancer Res.* 1976, 36, 1520-1523.

Chou, T.-C.; Lopez, C.; Colacino, J. M.; Fox, J. J. *Cancer Res.* 1983, *24*, 305.

Chiu, T. M. K.; Warnock, D. H.; Watanabe, K. A.; Fox, J. J. J. Heterocycl. Chem. 1973, 10, 607.

Chu, C. K.; Reichman, U.; Watanabe, K. A.; Fox, J. J. *J. Heterocycl. Chem.* 1977, 14, 1119-1121.

Chu, C. K.; Wempen, I.; Watanabe, K. A.; Fox, J. J. *J. Org. Chem.* 1976, 41, 2793-2797.

Chu, C. K.; Watanabe, K. A.; Fox, J. J. J. Heterocycl. Chem. 1975, 12, 817.

Chung, W. K.; Chu, C. K.; Watanabe, K. A.; Fox, J. J. *J. Org. Chem.* 1979, 44, 3982.

Fox, J. J.; Kuwada, Y.; Watanabe, K. A.; Ueda, T.; Whipple, E. B. *Anitmicrob. Agents Chemother.* 1965, 518-529.

Fox, J. J.; Kuwada, Y.; Watanabe, K. A. *Tetrahedron Lett.* 1968, 6029-6032.

Fox, J. J.; Stempel, L. M.; Su, T-L.; Watanabe, K. A. *J. Org. Chem* 1982, *47*, 1081-1084.

Fox, J. J.; Van Praag, D.; Wempen, I.; Doerr, I. L.; Cheong, L.; Knoll, J. E.; Eidinoff, M. L.; Bendich, A.; Brown, G. B. *J. Am. Chem. Soc.* 1959, *81*, 178.

Fox, J. J.; Watanabe, K. A.; Lopez, C.; Philips, F. S.; Leyland-Jones, B. In "Herpesvirus. Clinical, Pharmacological and Basic Aspects." Shiota, N.; Cheng, Y-C.; Prusoff, W. H., Eds., Excerpta Medica, Amsterdam, 1982, pp. 135-147.

Fox, J. J.; Wempen, I.; Hampton, A.; Doerr, I. *J. Am. Chem. Soc.* 1958, *80*, 1669.

Friedman, H. A.; Watanabe, K. A.; Fox, J. J. *J. Org. Chem.* 1967, *32*, 3775-3780.

Goody, R. S.; Watanabe, K. A.; Fox, J. J. *Tetrahedron Lett.* 1970, 293-296.

Grant, A. J.; Feinberg, A.; Chou, T-C.; Watanabe, K. A.; Fox, J. J.; Philips, F. S. *Biochem. Pharmacol.* 1982, *31*, 1103-1108.

Handschumacher, R. E. *J. Biol Chem.* 1960, *235*, 764.

Hirota, K.; Watanabe, K. A.; Fox, J. J. *J. Heterocycl. Chem.* 1977, *14*, 537.

Hirota, K.; Watanabe, K. A.; Fox, J. J. *J. Org. Chem.* 1978, *43*, 1193-1196.

Hirota, K.; Kitade, Y.; Senda, S.; Halat, M. J.; Watanabe, K. A.; Fox, J. J. *J. Am. Chem. Soc.* 1979, *101*, 4423.

Hirota, K.; Kitade, Y.; Senda, S.; Halat, M. J.; Watanabe, K. A.; Fox, J. J. *J. Org. Chem.* 1981, *46*, 846.

Hoshi, A.; Kanzawa, F.; Kuretani, K.; Saneyoshi, M.; Arai, Y. *GANN*, 1973, *64*, 519.

Iwasaki, H. *Yakugaku Zasshi*, 1962, *82*, 1393.

Jardetsky, C. D. *J. Biol Chem.* 1963, *238*, 2498.

Kabat, M. M.; Pankiewicz, K. W.; Watanabe, K. A. *J. Med. Chem* 1987, *30*, 924-927.

Klein, R. S.; Kotick, M. P.; Watanabe, K. A.; Fox, J. J. *J. Org. Chem.* 1971, *36*, 4113-4116.

Kois, P.; Tocik, Z.; Ren, W-Y.; Spassova, M.; Rosenberg, I.; Farras Soler, J.; Watanabe, K. A. *Nucleosides Nucleotides* 1993, *12*, 1093-1109.

Kotick, M. P.; Klein, R. S.; Watanabe, K. A.; Fox, J. J. *Carbohydr. Res.* 1969, *11*, 369-377.

Kreis, W.; Watanabe, K. A.; Fox, J. J. *Helv. Chim. Acta* 1978, *61*, 1011-1016.

Lemieux, R. U.; Fraga, E.; Watanabe, K. A. *Can. J. Chem.* 1968, *46*, 61-69.

Lemieux, R. U.; Watanabe, K. A.; Pavia, A. A. *Can. J. Chem.* 1969, *47*, 4413-4426.

Lemieux, R. U.; Pavia, A. A.; Martin, J. C.; Watanabe, K. A. *Can. J. Chem.* 1969, *47*, 4427-4439.

Lewis, W.; Levine, E. S.; Griniuviene, B.; Tankersley, K. O.; Colacino, J. M.; Sommadossi, J-P.; Watanabe, K. A.; Perrino, F. W. *Proc. Nat. Acad. Sci. USA* 1996, *93*, 3592-3597.

Matulic-Adamic, J.; Rosenberg, I.; Krayevsky, A. A.; Watanabe, K. A.; Arzumanov, A. A.; Dyatkina, N. B.; Shirokova, E. A. *Nucleosides Nucleotides* 1993, *12*, 1085-1092.

Majumdar, A.; Khorlin, A.; Dyatkina, N.; Lin, M.; Powell, J.; Liu, J.; Fei, Z.; Khripine, Y.; Watanabe, K. A.; George, J.; Glazer, P. M.; Seidman, M. M. *Nature Genetics* 1998, *20*, 212-214.

Mizuno, Y.; Ikehara, M.; Watanabe, K. A. *Chem. Pharm. Bull.* 1962, *10, 647-652. The* amount of *phosphorus* pentasulfide used for selective thiation was about one fifth of that used by Elion and Hitchings (J. Am. Chem. Soc. 1947, 69, 2138). When I came to Sloan-Kettering Institute, both Elion and Hitchings were at Burroughs Wellcome laboratory nearby Eastchester, New York. I had the privilege to meet them frequently. In 1988, we had invited Dr. Hitchings for Institutional seminar at Sloan-Kettering. The date which had been set for the seminar abo*ut half a year ea*rlier *was* the day that the announcement was made for his Nobel *Prize award.*

Mizuno, Y.*; Ik*ehara, M.; Watanabe, K. A. Chem. Pharm. Bull. 1962, 10, 653-659.

Mizuno, Y.; Ikehara, M.; Watanabe, K. A. Chem. Pharm. Bull. 1963, 11, 293-296.

Oostveen, E. A.; van der Plas, H. C.; Jongejan, H. Recl. Trav. Chim. Pays-Bas, 1976, 95, 209.

Pankiewicz, K. W.; Nawrot, B.; Watanabe, K. A. *J. Org. Chem.* 1986, *51*, 1525.

Pankiewicz, K. W.; Kim, J. H.; Watanabe, K. A. *J. Org. Chem.* 1985, *50*, 3319-3322.

Pankiewicz, K. W.; Watanabe, K. A.; Takayanagi, H.; Itoh, T.; Ogura, H. *J. Heterocycl. Chem.* 1985, *22*, 1703-1710.

Pankiewicz, K. W.; Krzeminski, J.; Ciszewski, L. A.; Ren, W-Y.; Watanabe, K. A. *J. Org. Chem.* 1992, *57*, 553-559.

Pankiewicz, K. W.; Krzeminski, J.; Watanabe, K. A. *J. Org. Chem.* 1992, *57*, 7315-7321.

Reichman, U.; Watanabe, K. A.; Fox, J. J. *Carbohydr. Res.* 1975, 42, 233-240.

Rosenberg, I.; Farras Soler, J.; Tocik, Z.; Ren, W-Y.; Ciszewski, L. A.; Kois, P.; Pankiewicz, K. W.; Spassova, M.; Watanabe, K. A. *Nucleosides Nucleotides* 1993, *12*, 381-401.

Su, T-L.; Watanabe, K. A. *J. Heterocycl. Chem.* 1982, *19*, 1261-1262.

Su, T-L.; Watanabe, K. A.. *Heterocycl. Chem.* 1984, *21*, 1543-1547.

Su, T-L.; Watanabe, K. A.; Fox, J. J. *Tetrahedron* 1982, *32*, 1405-1408.

Su, T-L.; Huang, J-T.; Burchenal, J. H.; Watanabe, K. A.; Fox, J. J. *J. Med. Chem.* 1986, *29*, 709-715.

Su, T-L.; Huang, J-T.; Chou, T-C.; Otter, G. M.; Sirotnak, F. M.; Watanabe, K. A. *J. Med. Chem.* 1988, *31*, 1209.

Ueda, T.; Fox, J. J. *J. Am. Chem. Soc.* 1963, *85*, 4024.

Watanabe, K. A.; Beranek, J.; Friedman, H. A.; Fox, J. J. *J. Org. Chem.* 1965, *30*, 2735-2739.

Watanabe, K. A.; Chiu, T. M. K.; Reichman, U.; Chu, C. K.; Fox, J. J. *Tetrahedron*, 1976, *32*, 1493.

Watanabe, K. A.; Falco, E. A.; Fox, J. J. *J. Am. Chem. Soc.* 1972, *94*, 3272-3274.

Watanabe, K. A.; Fox, J. J. *Chem. Pharm. Bull.* 1964, *12*, 975-976.

Watanabe, K. A.; Friedman, H. A.; Cushley, R. J.; Fox, J. J. *J. Org. Chem.* 1966, *31*, 2942-2845.

Watanabe, K. A.; Goody, R. S.; Fox, J. J. *Tetrahedron* 1970, *26*, 3883.

Watanabe, K. A.; Hollenberg, D. H.; Fox, J. J.. *Carbohydr. Nucleosides Nucleotides.* 1974, *1*, 1.

Watanabe, K. A.; Kotick, M. P.; Fox, J. J. *Chem. Pharm. Bull.* 1969, *17*, 416-418.

Watanabe, K. A.; Kotick, M. P.; Fox, J. J. *J. Org. Chem.* 1970, *35*, 231-236.

Watanabe, K. A.; Reichman, U.; Fox, J. J.; Chou, T-C. Chem.-Biol. Interactions 1981, 37, 41. Almost 20 years later, when we worked on the mechanism of enzymatic deamination, this sample, found in the Sloan-Kettering sample room in still a nice crystalline condition, was most useful to identify the site of enzyme attack.

Watanabe, K. A.; Reichman, U.; Chu, C. K.; Hollenberg, D. H.; Fox, J. J. *J. Med. Chem.* 1980, *23*, 1088.

Watanabe, K. A.; Reichman, U.; Hirota, K.; Lopez, C.; Fox, J. J. *J. Med. Chem.* 1979, 22, 21-24.

Watanabe, K. A.; Wempen, I.; Fox, J. J. *Carbohydr. Res.,* 1972, *21*, 148-153.

Watanabe, K. A.; Wempen, I.; Fox, J. J. *Chem. Pharm. Bull.* 1970, *18*, 2368.

Winkley, Robins, R. K. J. Org. Chem. 1968, 32, 2822.

Wright, J. A.; Wilson, D. P.; Fox, J. J. *J. Med. Chem.* 1970, *13*, 269.

Yamazaki, A.; Kumashiro, I.; Takenishi, T. *J. Org. Chem.* 1967, *32*, 1825.

Young, C.W.; Schneider, R.; Reyland-Jones,B.; Armstrong, D.; Tan, C.T.C.; Lopez, C.; Watanabe, K.A.; Fox, J. J.; Philips, F. S. *Cancer Res.* 1983, *43*, 5006-5009.

Leyland-Jones, B.; Donnelly, H.; Groshen, S.; Myskowski, P.; Donner, A. L.; Fanucchi, M.; Fox, J.; and the Memorial Sloan-Kettering Antiviral Work Group. *J. Infect. Dis.* 1986, *154*, 430-436.

1. In the middle of 1970s at the carbohydrate chemistry section of the Gordon Research Conference, I happened to meet Dr. R. Stuart Tipson who was Levene's colleague for nine years, and made many important contributions to nucleic acid chemistry, especially in the area of carbohydrate and nucleoside

components. Levene and Tipson introduced 2',3'-*O*-isopropylidenation and 5'-*O*-tritylation to nucleoside chemistry and established the furanose structure of the sugar in nucleosides in early 1930s.

2. Methyl galactopyranoside was not commercially available. I was fortunate because while I was in Edmonton, I learned a facile method to synthesize both anomers of methyl galactosides from Dr. Mills, a visiting Australian scientist at Lemieux' laboratory.

3. Almost simultaneously, Wise and Townsend ("The Chemistry and Biological Activity of C-Nucleosides related to ψ-Uridine". In "Chemistry and Biology of Nucleosides and Nucleotides", Eds. Harmon, R. E.; Robins, R. K.; Townsend, L. B., Academic Press, New York, 1978, pp. 109-120) successfully executed multi-step conversion of ψ-uridine into ψ-isocytidine. Once Wise argued that our nomenclature is based on the base isocytosine, but the *C*-nucleoside is not an analogue of isocytidine but cytidine. Also it acts biologically as analogue of cytidine but not as isocytidine. Therefore, ψ-cytidine would be a better name for this *C*-nucleoside. Either nomenclature has sound basis, but we agreed that we should not confuse literature and keep the name ψ-isocytidine for this *C*-nucleoside, since we had already synthesized 5-ribosyl-cytosine and named it ψ-cytidine, which actually is an isocytidine analogue.

4. Recently, I tried to find this reaction in Bendich's chapter. It was not there but in the next chapter J. Baddiley discussed this transformation as a part of determination of the point of attachment of the sugar to the base. This type of reactions were later used to prepare apurinic and apyrimidinic acid synthesis for biochemical purposes, but surprisingly never explored for organic synthetic means. I have a very fond memory of Professor Bendich. When I met this chairman of Sloan-Kettering Division of Cornell University Graduate School of Medical Sciences first almost accidentally, he asked me the reason for my choice of chemistry for my career. I was completely unprepared and simply answered because I liked it. He gazed at me for a while, and said "I've been a professor of this medical school for a long time, and I always asked the same question to young people. Everybody gave me more or less the same answer. He or she wanted to contribute to the society or help struggling patients or something the like, but you are different." I felt I was arrogant, but I insisted that I did not think Beethoven had ever thought about the society or audience when he composed his symphonies. Apparently he liked my unsophisticated attitude and a few months later he appointed me to a faculty position in the graduate school.

5. Many years later, I found a very similar scheme in *J. Org. Chem.* 1991, *56*, 3608. The authors needed a 2'-fluoro-arabino nucleoside as a starting material for other synthesis. They described the preparation of this sugar in details in the experimental section, however, they did not cite our work properly. As the consequence many ordinary readers thought the compound was synthesized by these authors and impressed by their work.

6. More than 30 years later, I met Professor Prusoff at Waldorf Astolia Hotel by accident at a conference dinner. Surprisingly he recognized my name, and ever since he treated me as one of his old students.

7. In the early 1970's I attended an ACS meeting in Chicago with several members of my group. One of the lectures was impressive, which was given by a Polish professor who did not use any sophisticated equipment or fancy reagents but with brain and muscle he constructed several sugars from simple molecules. That evening when we were about to leave the hotel for Chinatown for dinner, I saw this Polish chemist at the door. I am rather shy and rarely talk to anyone not acquainted well, but somehow I asked him if he was interested in joining us for Chinese dinner. He was very pleased. We had a very pleasant time that evening. Later I learned that he was Professor Aleksander Zamoyski and a member of the Polish Academy of Sciences. Several years later in 1979 I was surprised by an invitation to Poland by the Academy with Professor Zamoyski being the sponsor. I gave lectures at Lodz, Poznan and Warsaw. I was very warmly welcome everywhere I visited, and later I enjoyed working with young colleagues from Poland recommended by my Polish friends.

8. It is certainly easier for the Study Section members to support for the study of a clearly delineated objective than for a plan to forage in ill-understood areas of knowledge in the hope of true discovery. As the results, scientists are forced to behave more like surveyors than explorers, and true discoveries became solely dependent upon serendipity than sincere search motivated by curiosity.

DEVELOPMENT OF NEW RADICAL REACTIONS WITH A VINYLSILYL GROUP AND THEIR APPLICATION TO THE SYNTHESIS OF BRANCHED-CHAIN SUGAR NUCLEOSIDES

SATOSHI SHUTO, MAKIKO KANAZAKI,
ISAMU SUGIMOTO, SATOSHI ICHIKAWA, YUKI NAGASAWA,
YOSHIHITO UENO, HIROSHI ABE, NORIAKI MINAKAWA,
MAKOTO SUKEDA, TETSUYA KODAMA,
MAKOTO NOMURA and AKIRA MATSUDA

Graduate School of Pharmaceutical Sciences, Hokkaido University,
Kita-12, Nishi-6, Kita-ku, Sapporo 060-0812, JAPAN

In recent years, we have been engaged in the synthesis of biologically active branched-chain sugar nucleosides. Among them, we have found that 1-(2-deoxy-2-methylene-β-D-*erythro*-pentofuranosyl)cytosine (DMDC) (Takenuki *et al.*, 1988; Matsuda *et al.*, 1991a; Yamagami *et al.*, 1991; Matsuda *et al.*, 1992; Ono *et al.*, 1996; Miwa *et al.*, 1998; Eda *et al.*, 1998), 1-(2-*C*-cyano-2-deoxy-β-D-*arabino*-pentofuranosyl)cytosine (CNDAC) (Matsuda *et al.*, 1991b; Tanaka *et al.*, 1992; Matsuda *et al.*, 1993; Azuma *et al.*, 1993; Matsuda and Azuma, 1995; Azuma *et al.*, 1995; Obata *et al.*, 1998; Hayakawa *et al.*, 1998; Hanaoka *et al.*, 1999), and 1-(3-*C*-ethynyl-β-D-*ribo*-pentofuranosyl)cytosine (ECyd) (Matsuda *et al.*, 1996; Tabata *et al.*, 1996; Hattori *et al.*, 1996; Tabata *et al.*, 1997; Takatori *et al.*, 1998; Hattori *et al.*, 1998; Matsuda *et al.*, 1999; Takatori *et al.*, 1999) are potent antitumor antimetabolites, which significantly inhibit the growth of various human solid tumor cells both *in vitro* and *in vivo*. These nucleosides are being examined in clinical studies against solid tumors. Although a number of procedures for preparing branched-chain sugar nucleosides have been developed, examples of synthesis of 1'- and 4'-branched-chain sugar nucleosides are rare. Furthermore, because of the lack of efficient synthetic methods for their preparation, the biological activities of 1'- and 4'-branched-chain sugar nucleosides have not been systematically investigated.

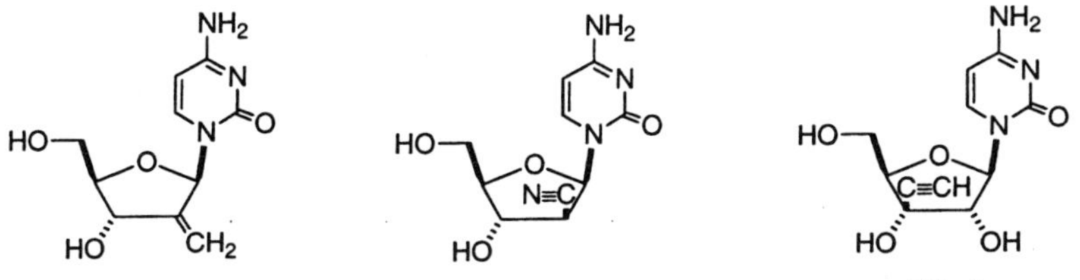

Figure 1.

Recent Advances in Nucleosides: Chemistry and Chemotherapy, Ed. by C.K. Chu. 21 — 55

Therefore, we decided to develop new efficient methods for preparing such branched-chain sugar nucleosides. A radical cyclization reaction has been known as a highly versatile method for forming C-C bonds. Silicon-containing groups are very useful for the regio- and stereoselective introduction of a carbon substituent based on a temporary silicon connection and there is a growing interest in their use in intramolecular radical cyclization reactions. In this review, we describe our recent progress in evaluation of new types of radical cyclization reactions and their application to the synthesis of modified nucleosides branched at 1'-, 2'- and 4'-positions.

1.   Development of a new radical cyclization reaction using a 2-bromo-1-indanol as a model system

We hypothesized that if a radical intermediate **b** generated from vinylsilyl ethers of halohydrins or α-phenylselenoalkanols is cyclized to 6-*endo*-product **c**, then stereoselective introduction of a 2-hydroxyethyl group at the β-position of the hydroxyl can be achieved to give **d**, after an oxidative ring-cleavage reaction (Scheme 1) (Shuto *et al.*, 1997).

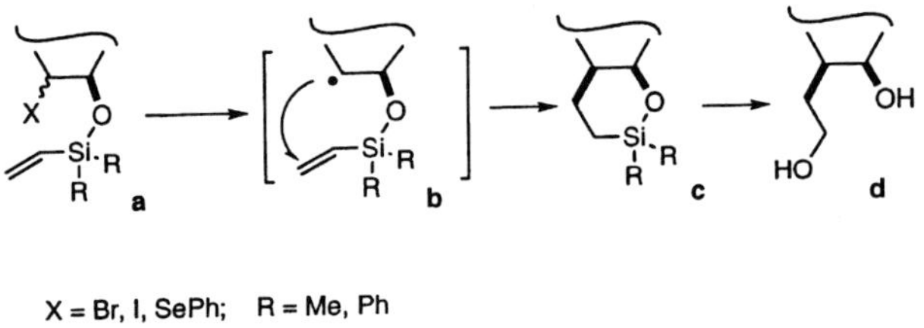

Scheme 1.

We selected commercially available (±)-*trans*-2-bromo-1-indanol (**1**) as a starting material and prepared the diphenyl- and dimethylvinylsilyl ethers 2a and 2b (Sieburth and Fensterbank, 1992) as model compounds (Scheme 2). Radical reactions were performed with Bu_3SnH and either AIBN or Et_3B in benzene, followed by Tamao oxidation (Tamao *et al.*, 1983), to give a mixture of diols **5** and **6**, and the results are summarized in the Table 1. First, a mixture of Bu_3SnH (1.1 equiv) and AIBN in benzene was added slowly over 4 h to a solution of **2a** in benzene (0.01 M) under reflux, to give the desired 2-hydroxyethyl derivative **5** *via* 6-*endo* cyclization product **3a**, as a major product, along with **6** *via* 5-*exo* cyclization product **4a** (entry 1). The selectivity for the formation of **5** increased significantly when a lower concentration of Bu_3SnH was employed (entry 2). Interestingly, when the reaction was performed at room temperature, the regioselectivity was almost completely reversed to give **6** preferentially (entry 3). Furthermore, the radical reaction of **2a** in the presence of excess of Bu_3SnH at 80 °C also gave **6** with high selectivity (entry 4). Similar results were obtained when dimethylvinylsilyl derivative **2b** was used as a substrate (entries 5-7). These results suggest that the formation of the 6-*endo* product **3** may not be kinetic but thermodynamic, since the ratio of the *endo* and *exo* products should be independent of the concentration of Bu_3SnH if the reaction is controlled kinetically. These results conflict with well-known Baldwin-Beckwith rule that the cyclization reactions of hexenyl radicals and their equivalents are controlled kinetically to give 5-*exo* cyclization products preferably over 6-*endo* cyclization products (Baldwin, 1976; Beckwith, 1981; Beckwith and Schiesser, 1985; Spellmeyer and Houk, 1987). Two pathways may explain the selective formation of 6-*endo* cyclization product 3: 1) the cyclization reaction is reversible, or 2) 5-*exo* cyclized radical B, which is initially formed, is rearranged to give C (Scheme 3). However, it is unlikely that the cyclization is reversible, since reversible radical cyclizations of hexenyl radical or their equivalents have been observed only when radical centers are attached to radical-stabilizing groups, such as carbonyl groups (Julia, 1967; Beckwith *et al.*, 1972; Curran and Chang, 1989).

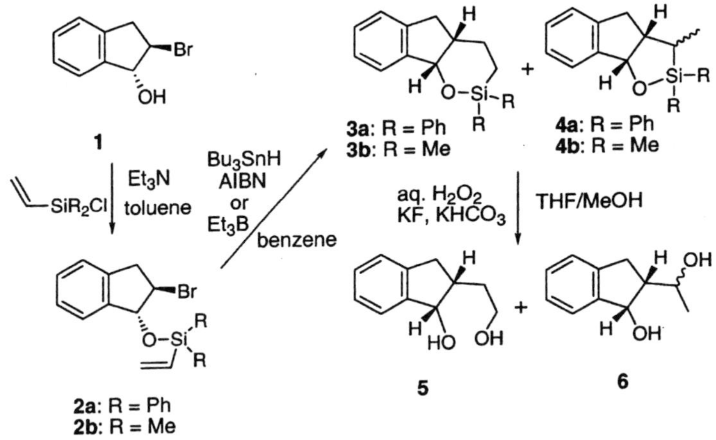

Scheme 2.

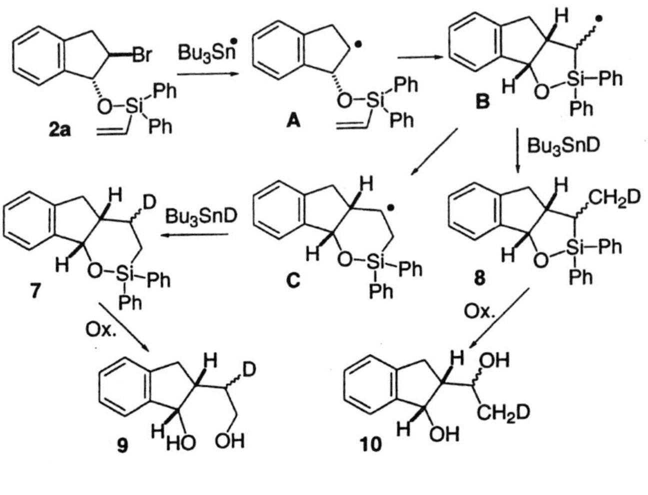

Scheme 3.

Table 1.   Synthesis of **5** and **6** *via* radical cyclization reaction of **2a** or **2b**.[a]

entry	substrate	method[b]	temp	yield (5 + 6)	ratio[d] (5 : 6)
1	**2a**	A	80 °C	71%[c]	6 : 1
2	**2a**	B	80 °C	72%[d]	15 : 1
3	**2a**	C	26 °C	91%[d]	1 : 11
4	**2a**	D	80 °C	84%[d]	1 : 17
5	**2b**	A	80 °C	70%[d]	3 : 1
6	**2b**	C	26 °C	72%[d]	1 : 23
7	**2b**	D	80 °C	81%[d]	1 : 31

[a]   Compounds 5 and 6 were obtained after treating the crude reaction mixture of the radical reaction under Tamao oxidation conditions.

[b]   A: To a solution of substrate (0.01 M) in benzene, a mixture of Bu_3SnH (1.1 equiv) and AIBN (0.6 equiv) in benzene was added slowly over 4 h. B: To a solution of substrate (0.002 M) in benzene, a mixture of Bu_3SnH (1.1 equiv) and AIBN (0.6 equiv) in benzene was added slowly over 7 h. C: To a solution of substrate (0.01 M) in benzene, a solution of Bu_3SnH (1.1 equiv) in benzene and a solution of Et_3B (0.6 equiv) in benzene were simultaneously added over 4 h. D: To a mixture of substrate (0.01M) and Bu_3SnH (3.0 equiv) in benzene, AIBN (0.6 equiv) in benzene was added over 2 h.

[c]   Isolated yield.

[d]   Determined by HPLC.

To examine the reaction mechanism, the reaction was performed with Bu_3SnD under the same conditions as for entry 1. After Tamao oxidation, the protons β to the primary hydroxyl were exclusively replaced by deuterium only in product 9. On the other hand, product 10 was deuterated exclusively at the methyl group. These results suggest that this cyclization would be irreversible and that the 5-*exo* cyclized radical B would be formed first and is mainly trapped when the concentration of $Bu_3SnH(D)$ is high enough or the reaction is done at room temperature; under a low $Bu_3SnH(D)$ concentration at a higher reaction temperature, radical B is rearranged into the ring-enlarged radical C, which is then trapped with $Bu_3SnH(D)$ (Scheme 3). To the best of our knowledge, such a ring-enlarging 1,2-radical rearrangement of β-silyl carbon-centered radicals has not been previously reported (Dowd and Zhang, 1993; Johnson *et al.*, 1979; Harris *et al.*, 1991; Tsai and Cherng, 1991; Kulicle *et al.*, 1992).

2.   Mechanistic study of the ring-enlargement reaction of (3-oxa-2-silacyclopentyl)methyl radicals into 4-oxa-3-silacyclohexyl radicals

In the above reaction, two possible pathways for the ring-enlargement reaction may be postulated (Scheme 4); one via a transition state (or an intermediate) x, in which the silicon atom expands its valence shell to five (path x), the other via β-elimination to give the ring-opened silyl radical y (path y), which subsequently undergoes 6-*endo*-cyclization to give f. The mechanism of this radical rearrangement reaction, which may be related mechanistically to the known radical 1,2-silicon shifts in nitrogen- (West and Boudjouk, 1973; Harris *et al.*, 1991; Harris *et al.*, 1993; Roberts and Vazquez-Persaud, 1995), oxygen- (Tsai and Cherng, 1991), and sulfur- (Pitt and Fowler, 1968) centered radicals, has been of great interest to us. Therefore, we try to understand whether or not the ring-enlargement reaction occurs via a pentavalent silicon-bridging radical transition state (or an intermediate) x.

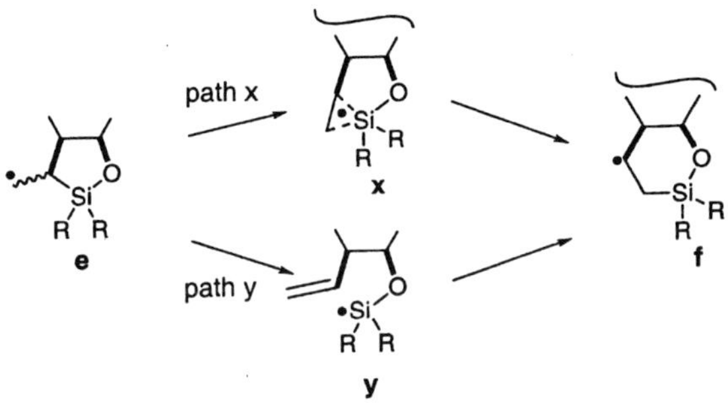

Scheme 4.

First, we investigated the reaction of 11 and 12, which are precursors for a (3-oxa-2-silacyclopentyl)methyl radical B' and a 4-oxa-3-silacyclohexyl radical C', respectively (Scheme 5) (Sugimoto *et al.*, 1999b, Shuto *et al.* 2000a). Such experiments would clearly confirm that radical C' is produced from radical B' via a novel ring-enlargement reaction, as previously suggested by deuterium-labeling experiments, and also would clarify whether the corresponding reverse reaction, i.e., ring-contraction of radical C' into radical B', actually occurs.

A solution of Bu$_3$SnH (1.2 equiv) and AIBN (0.6 equiv) in benzene was added slowly over 4 h to a solution of 11 in refluxing benzene. The reaction gave the ring-enlargement product 14 as a major product along with the directly reduced product 13 (yield 70%, 13:14 = 9:91). On the other hand, when 12 was treated under conditions identical to those for 11, only the direct reduction of radical C' occurred affording 14 in 72% yield as the sole product; the corresponding ring-contracted product 13 was not obtained. These results clearly demonstrate that the radical B' readily rearranged into the radical C' (Scheme 5), which is consistent with the previous results suggested by deuterium-labeling experiments. The results also suggest that the corresponding reverse reaction, i.e. ring-contraction of C' into B', did not occur or was very slow.

We next investigated the reaction mechanism of the radical rearrangement by using 15a and 15b as substrates. Both have a methyl and a phenyl group on the silicon

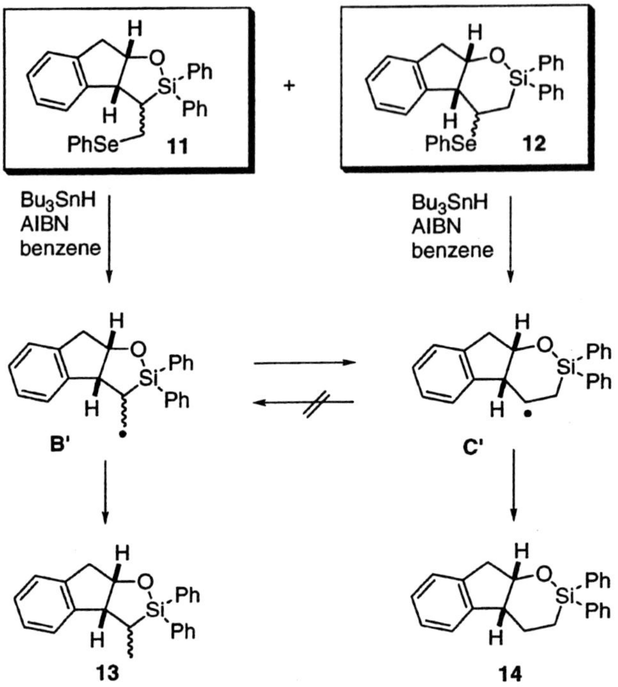

Scheme 5.

atom and are therefore stereoisomers. The configurations at the silicon atom in the ring-enlargement reaction products derived from the radical reactions of 15a, b should be dominated by the reaction mechanism shown in Scheme 6. Treatment of 15a (*endo*-SiMe isomer) or 15b (*exo*-SiMe isomer) with Bu₃SnH/AIBN would produce radical ia or ib, respectively. Direct reduction of ia or ib by Bu₃SnH gives 16a or 16b, respectively. If the radical ring-enlargement reaction of ia or ib proceeds via ring-opened silyl radical iiia or iiib (path y or y' in Scheme 6), isomerization at the silicon atom of iiia or iiib should occur, at least to some extent, before re-cyclization to give a mixture of 6-*endo*-cyclized radicals iva and ivb. Consequently, a mixture of 17a (*endo*-SiMe product) and 17b (*exo*-SiMe product) would be obtained. Alternatively, the configuration at the silicon atom of 15a or 15b should be retained during the rearrangement process to give the ring enlargement product 17a or 17b, respectively, when the radical rearrangement proceeds via the pentavalent silicon-bridging radical transition state iia or iib (path x or x' in Scheme 6).

When 15a was treated with BuSn₃H/AIBN in benzene, the ring-enlargement product 17a and the directly reduced product 16a were obtained without producing 16b and 17b. Similarly, the radical reaction of 15b gave 16b and 17b. These results demonstrate that the configuration at the silicon atom is retained during the ring-enlargement

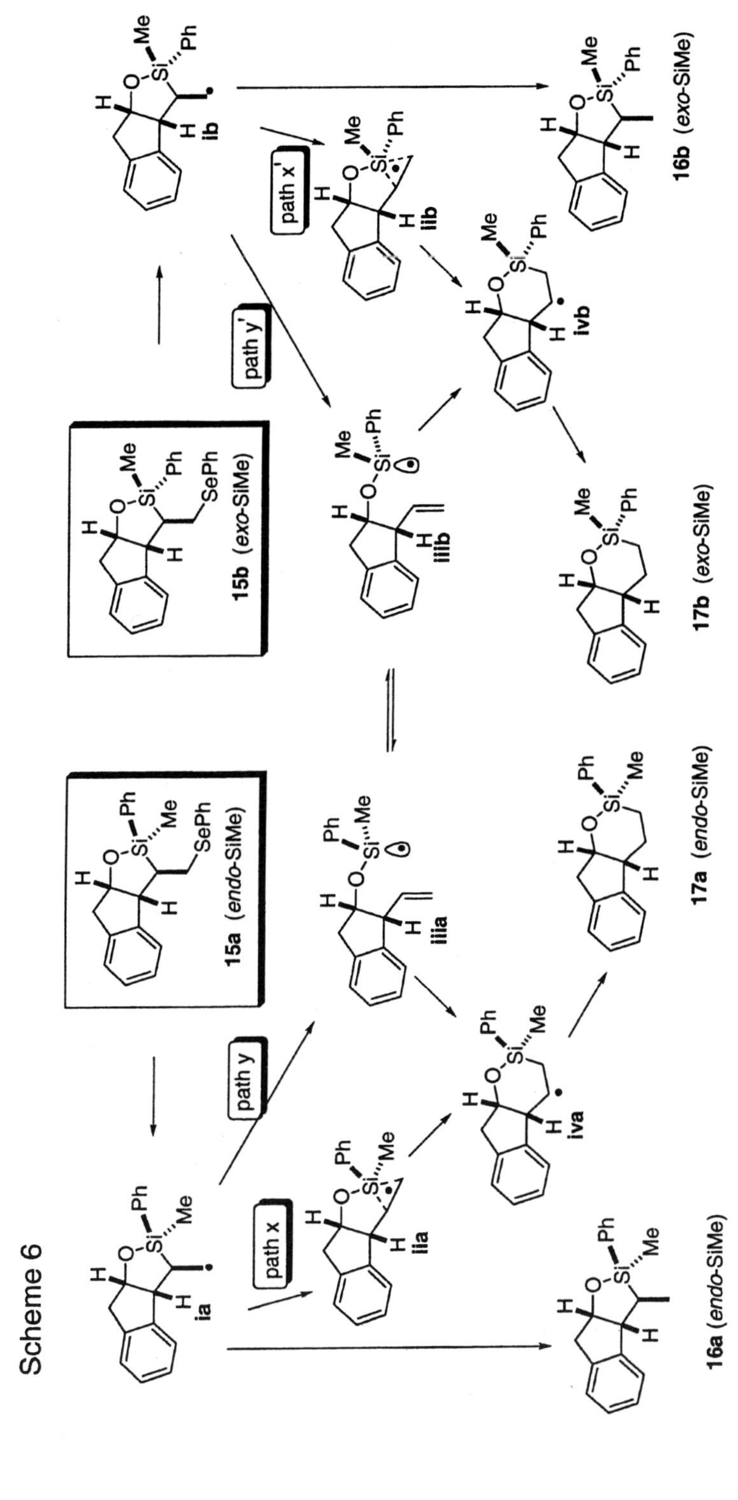
Scheme 6
ia
ib
path x
path y
path x'
path y'
15a (endo-SiMe)
15b (exo-SiMe)
iia
iib
iiia
iiib
iva
ivb
16a (endo-SiMe)
16b (exo-SiMe)
17a (endo-SiMe)
17b (exo-SiMe)

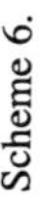

Scheme 6.

reaction and that a ring-opened silicon radical like iiia or iiib is not produced. The most likely explanation for the configuration-retaining reaction pathway of the radical ring-enlargement is that it proceeds via pentavalent silicon-bridging radical transition states iia and iib (path x and x').

We next designed a deuterium-labeled substrate 18D to further elucidate the reaction pathway of the ring-enlargement reaction. Our strategy is summarized in Scheme 7. Reaction of 18D with Bu$_3$SnH/AIBN would generate the *exo*-cyclized radical vi via the allylic radical v. If the ring-enlargement reaction of vi proceeds via a pentavalent silicon-bridging radical transition state vii, it would produce the ring-enlargement product 20D, with the deuterium-label at the terminal methylene carbon (path x). On the other hand, if β-elimination of vi occurs, a mixture of ring-enlargement products 19D and 20D should be obtained, since the resulting silyl radical viii can cyclize to both the labeled and unlabeled terminal methylenes which are regiochemically equivalent.

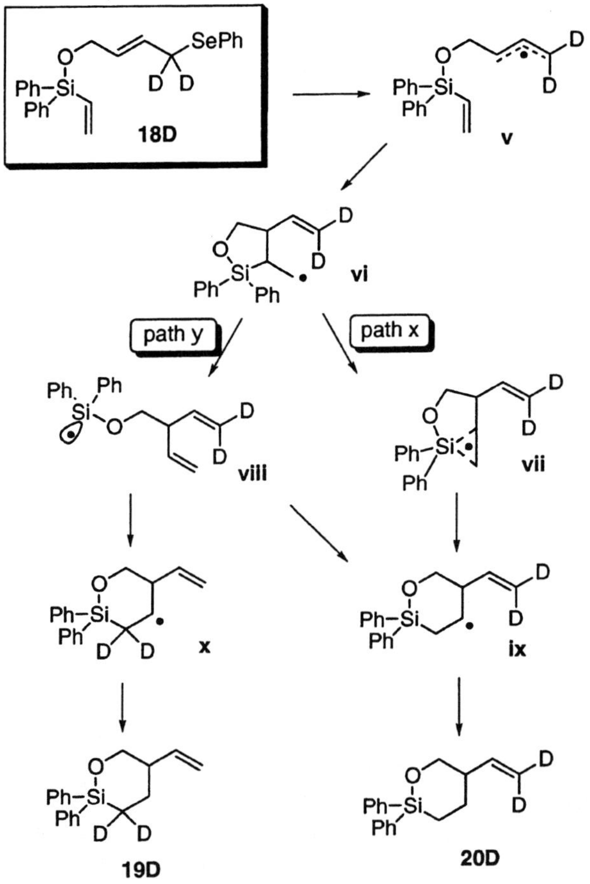

Scheme 7.

The radical reaction of 18 was performed under thermodynamic conditions. The reaction mixture was subsequently treated under Tamao oxidation conditions, and the resulting 3-hydroxymethyl-4-pentenol was isolated as the corresponding dibenzoate 22 (Scheme 8). A similar reaction of 18 with Bu_3SnD gave the corresponding deuterium-labeled product 23. These results confirmed that the radical ring-enlargement reaction occurred in this system as expected. Therefore, we next investigated the radical reaction of deuterium-labeled substrate 18D. The product, after purification by HPLC, was analyzed by ^1H NMR. The spectrum clearly shows that the protons at the terminal methylene in 24D were exclusively replaced by deuteriums and that the regioisomerically labeled 25D was not detected at all (Scheme 8). Accordingly, these results suggest that the ring-enlargement reaction of radical vi is not likely to occur via a ring-opened silicon radical viii but rather via the pentavalent silicon-bridging radical transition state (or intermediate) vii in Scheme 7.

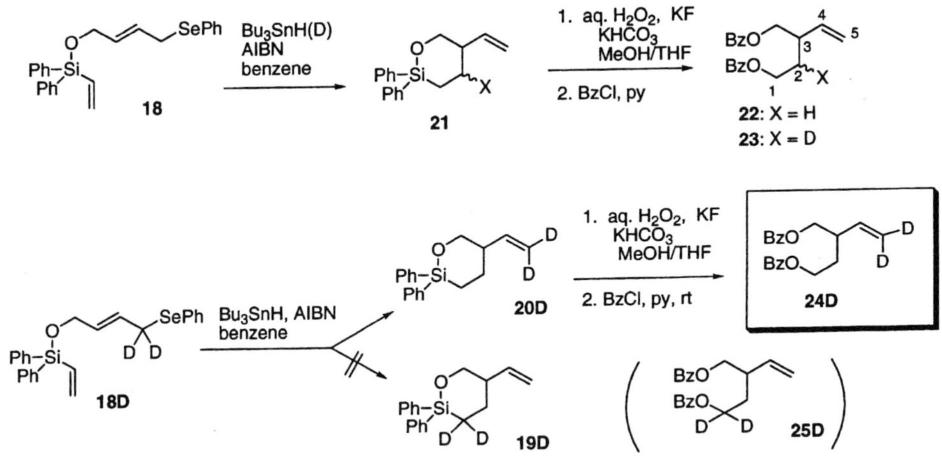

Scheme 8.

The rearrangement reaction appears to be irreversible based on the study with 12. We undertook a theoretical investigation of the radical B' and the C' by computational methods to compare their stabilities. Geometry optimizations and single-point energy calculations were performed using PM3 and *ab initio* calculations at the UHF/STO-3G level, respectively. Two stereoisomers, i.e. *endo*-Me radical B'$_1$ and *exo*-Me radical B'$_2$ should be considered for (3-oxa-2-silacyclopentyl)methyl radical B'. The ring-enlarged radical C' is 11.5 kcal/mol and 6.3 kcal/mol more stable than B'$_1$ and B'$_2$, respectively, based on the heat of formation (Scheme 9). Similar results were obtained from the calculations on radicals ia, ib, iva, and ivb with an asymmetric silicon center (Scheme 9). These computational results clearly support the experimental data indicating that the ring-enlargement rearrangement is irreversible.

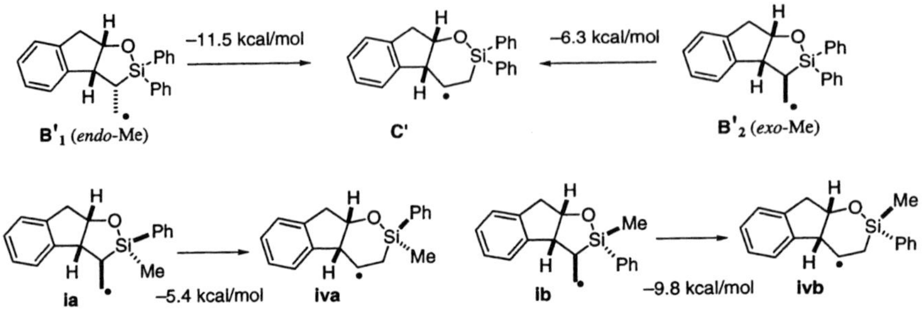

-11.5 kcal/mol

-6.3 kcal/mol

B'₁ (*endo*-Me)   **C'**   **B'₂** (*exo*-Me)

ia -5.4 kcal/mol   **iva**   **ib** -9.8 kcal/mol   **ivb**

Scheme 9

The reactions with substrates 15a and 15b show that the configuration at the silicon atom is retained during the rearrangement reaction. This suggests that the radical ring-enlargement proceeds via pentavalent silicon-bridging radical transition states iia and iib (path x and x'), without producing a ring-opened silicon radical like iiia or iiib. The rates of inversion at the silicon center of asymmetric silyl radicals should be considered, since asymmetric silyl radicals have been reported to be configurationally stable compared with asymmetric carbon radicals (Sakurai *et al.*, 1969; Brook and Duff, 1969; Sommer and Ulland, 1972; Sakurai *et al.*, 1977; Chatgilialoglu *et al.*, 1982). The rates of inversion at the silicon center of asymmetric silyl radicals were determined to be $(3-12) \times 10^9$ s^{-1} at temperatures from 0 to 80 °C (Chatgilialoglu *et al.*, 1982). On the other hand, 6-*endo*-cyclization of pent-4-enylsilyl radicals has been studied (Chatgilialoglu *et al.*, 1983; Barton and Revis, 1984; Sarasa *et al.*, 1986; Cai and Roberts, 1998). The rate constant for 6-*endo*-cyclization of 3,3-dimethylpent-4-enylsilyl radical was also measured to be $< 10^9$ s^{-1} at -100 °C and $>10^7$ s^{-1} at room temperature (Chatgilialoglu *et al.*, 1983). Considering these results, isomerization at the silicon atom should occur, at least to some extent, if the ring-opened silicon radical iiia or iiib is involved in the reaction process, since 6-*endo*-cyclization of pent-4-enylsilyl radicals has been shown to be slower than isomerization at the silicon atom.

The results with the asymmetric silicon substrates 15a and 15b were further confirmed by the study with deuterium-labeled substrate 18D. Therefore, we conclude that the ring-enlargement reaction is an irreversible process that occurs via a transition state (or an intermediate) x in which the silicon atom expands its valence shell to five (path x in Scheme 4).

This study should be of interest, since it presents the first experimental evidence for the pentavalent silicon transition state (or intermediate) in radical reactions of organic silicon compounds.

3. Synthesis of 4'-branched-chain sugar nucleosides and their introduction into oligodeoxyribonucleotides. Preparation and properties of nuclease-resistance phosphodiester oligodeoxyribonucleotides.

As described in previous sections, we have successfully developed the new radical cyclization reaction. Consequently, we describe herein an application of the reaction for synthesis of 4'-branched-chain sugar nucleosides. Such nucleosides when incorporated into oligodeoxynucleotides (ODNs) are expected to hydrolysis of phosphodiester (PO) bonds by endo- and exonucleases.

ODNs and their analogs have been shown to specifically inhibit gene expression (Uhlmann and Peyman, 1990; Milligan *et al.*, 1993; Crooke and Lebleu, 1993; Thuong and Helene, 1993; Agrawal, 1996a). Because of their potential to control diseases of known genetic etiology, development of these compounds as therapeutic agents is of great interest. Antisense ODNs bind to mRNAs by Watson-Crick base-pairing and inhibit translation of mRNAs in a sequence-specific manner. One of the major problems encountered when naturally occurring PO-ODNs are used as antisense or antigene molecules is their rapid degradation by nucleases found in cell culture media and inside cells. Therefore, many types of backbone-modified ODNs such as methylphosphonates (PM), phosphoramidates, and phosphorothioates (PS) have been synthesized and used for antisense and antigene studies (Beaucage and Iyer, 1993a,b). However, the benefits of such stabilization against enzymatic degradations are sometimes counteracted by the loss of other properties that are important for antisense activity. PS-ODNs tend to have lower binding affinity for their complementary RNA targets than unmodified PO-ODNs, presumably because they are diastereomeric mixtures at the PS linkages (Cosstick and Eckstein, 1985; LaPlanche *et al.*, 1986; Latimer *et al.*, 1989; Hacia *et al.*, 1994). Although RNase H cleavage is important in antisense strategy, RNA is not a substrate for the enzyme when PM-ODN is the complementary strand (Tidd *et al.*, 1988; Walder and Walder, 1988). Furthermore, PS-ODNs have been reported to exhibit non-sequence-specific activity (Stein *et al.*, 1991; Stein and Narayanan, 1996; Agrawal, 1996b).

ODNs having natural PO-linkages have also been studied as antisense molecules. These ODNs form thermally stable duplexes with their complementary RNAs and also elicit RNase H activity, but are often rapidly hydrolyzed by nucleases. We reasoned that the PO-ODNs with basic aminoalkyl chains near to their PO moieties might be resistant to nucleases. Nucleases hydrolyze PO-linkages by a general acid-base catalysis mechanism, including acidic and/or basic amino acid residues at their active sites. The presence of a basic amino group very near to the PO moiety of ODNs may prevent nucleolytic hydrolysis by forming an ionic bond with the acidic PO moiety of ODNs. It is also possible that the amino group attached to ODNs interrupts the catalytic system of nucleases by bonding with an acidic amino acid residue or by repulsing a basic amino acid residue at the enzyme active sites.

On the other hand, naturally occurring polyamines, such as spermidine and spermine, are known to bind strongly to DNAs and to stabilize duplex and triplex formation (Tabor and Tabor, 1976, 1984; Etter, 1990). The enhanced thermal stability of duplexes and triplexes is explained by the reduction of the anionic electrostatic repulsion between the phosphate moieties by the cationic amino groups. Several groups reported that the ODNs carrying the polyamines or the aminoalkyl groups thermally stabilize the duplexes and the triplexes with the complementary DNAs and the target duplexes (Dan *et al.*, 1993; Ono *et al.*, 1993; Ono *et al.*, 1994; Haginoya *et al.*, 1997; Nomura *et al.*, 1997; Ueno *et al.*, 1997a,b, Ueno *et al.*, 1998a, Tung *et al.*, 1993; Prakash *et al.*, 1994; Barawkar *et al.*, 1994, 1996; Schmid and Behr, 1995; Sund *et al.*, 1996; Hashimoto *et al.*, 1993a,b, Ozaki *et al.*, 1995; Shinozuka *et al.*, 1998; Griffey *et al.*, 1996; Cuenoud *et al.*, 1998). Thus, we envisioned that the aminoalkyl groups at the 4'-position of the nucleosides would impart both the nuclease-resistant and the duplex-stabilizing properties to the ODNs.

3.1.   Synthesis of 2'-deoxy-4'-*C*-(2-hydroxyethyl)adenosine

We used 2'-deoxyadenosine as a starting material, since a method for introducing a phenylseleno group at the 4'-position of 3'-*O*-acetyl-N^6,N^6-dibenzoyl-2'-deoxyadenosine has been developed (Giese *et al.*, 1994). We used N^6,N^6,3'-tri-*O*-benzoyl-2'-deoxyadenosine (26) as a protected nucleoside for further derivatization (Scheme 10) (Shuto *et al.*, 1998). Thus, 26 was treated under Swern oxidation conditions followed by treatment with PhSeCl and Et_3N in CH_2Cl_2, which afforded the 4'-phenylseleno derivative 27 as a diastereomeric mixture at the 4'-position. When the formyl group was reduced with Bu_4NBH_3CN in THF, the resulting diastereomeric mixture was successfully separated by silica gel column chromatography to give the 4'-*C*-phenylseleno-2'-deoxyadenosine derivative 29 and its 4'-diastereomer 28 in yields of 21% and 72%, respectively. The compound 28 was treated with diphenylchlorovinylsilane and Et_3N in the presence of DMAP in toluene to give vinylsilyl derivative 30.

The reaction of 30 with Bu_3SnH and AIBN in benzene, followed by Tamao oxidation, gave cyclonucleoside 32 as the major product. Although the radical reaction of 30 was investigated under various conditions, the desired 4'-branched nucleosides were not obtained. This result suggests a tandem radical cyclization mechanism (Scheme 11): a 5-*exo*-cyclized radical intermediate was first produced from the 4'-radical which did not react with Bu_3SnH to give the desired 33 or 34 but rapidly added to the 8-position of the adenine moiety. The 8-hydrogen was subsequently abstracted by a phenylseleno radical to afford 31. Similar formations of the cyclonucleoside 36 *via* intramolecular radical additions at the adenine 8-position have previously been reported by our laboratory (Matsuda *et al.*, 1976; Matsuda *et al.*, 1978; Usui and Ueda, 1986a,b; Usui *et al.*, 1986).

We subsequently introduced the silicon tether at the 3'-hydroxyl group of the 4'-phenylseleno adenine nucleoside derivative and investigated its radical reactions (Scheme 12) (Shuto *et al.*, 1998a). The primary hydroxyl group of 4'-*C*-phenylseleno derivative 28 was selectively protected by a dimethoxytrityl (DMTr) group to give 38, which was treated with diphenylchlorovinylsilane to give 39. Treatment of 39 with Bu_3SnH/AIBN in benzene, followed by Tamao oxidation, gave a diastereomeric mixture of 4'-*C*-(1-hydroxyethyl) derivatives 43 which were derived from a 5-*exo*-cyclized product 42, in almost quantitative yield. When a solution of Bu_3SnH/AIBN in toluene was added slowly over 4 h to a solution of 39 in toluene at 110 °C, the regioselectivity was completely reversed. The reaction did not give 42 at all, but rather 6-*endo*-cyclized 40. Tamao oxidation of 40 gave 4'-*C*-(2-hydroxyethyl) derivative 41.

Scheme 10.

Scheme 11.

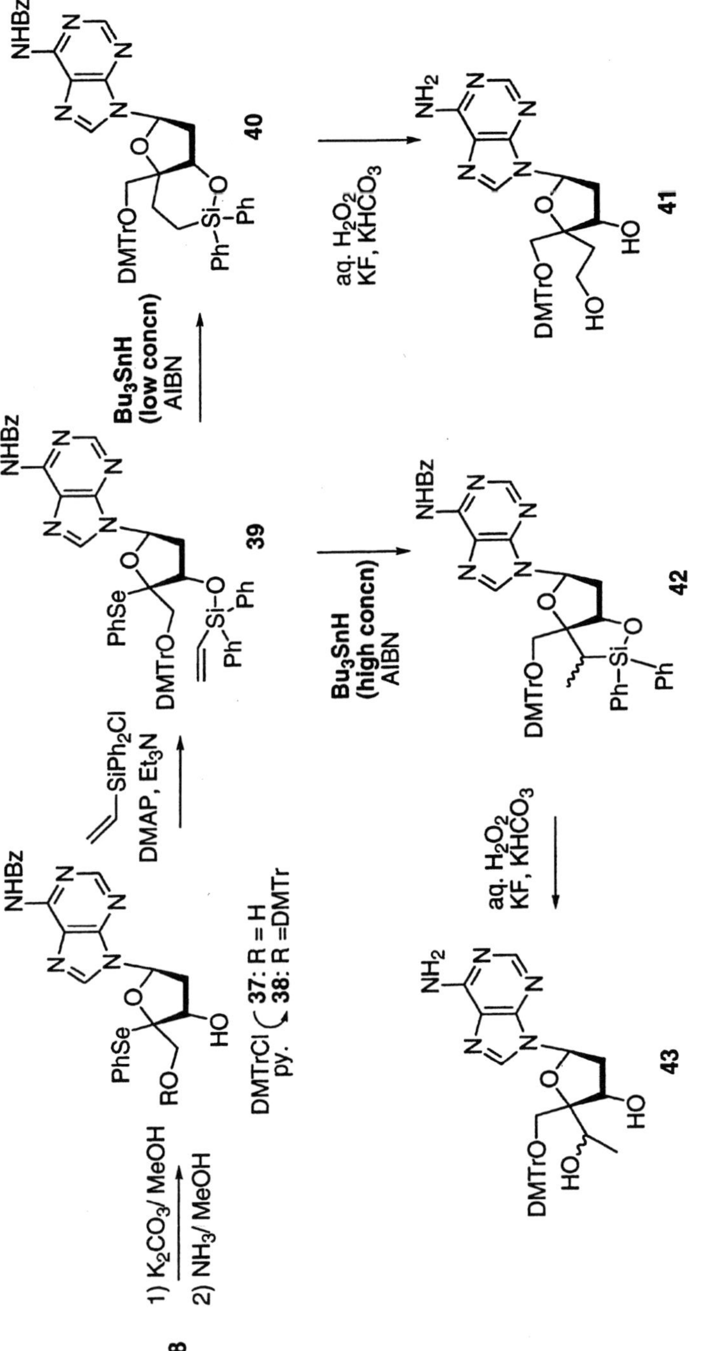

Scheme 12.

3.2. Synthesis of 4'-*C*-(2-aminoethyl)- and 4'-*C*-(3-aminopropyl)thymidines and their incorporation into ODNs

Previously, ODNs containing **44** were reported to be more resistant to nucleolytic hydrolysis by snake venom phosphodiesterase (a 3'-exonuclease) although endonuclease resistance was not examined (Wang and Seifert, 1996). Therefore, we designed 4'-*C*-aminoalkylthymidines **45-48** (Figure 2) to compare the ability to prevent hydrolysis by both endo- and exonucleases with those of **44** (Kanazaki *et al.* 2000). To synthesize these nucleoside units, we used our newly developed radical chemistry using a vinylsilyl (Shuto *et al.*, 1997; Ueno *et al.*, 1998b, Shuto *et al.*, 1998; Sugimoto *et al.*, 1999a,b, Shuto *et al.* 2000a) or an allylsilyl (Xi *et al.*, 1992; Shuto *et al.* 2000b) group as a temporary radical acceptor.

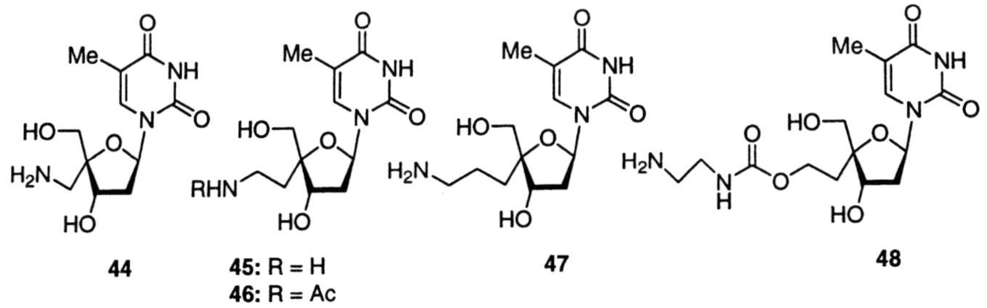

Figure 2.

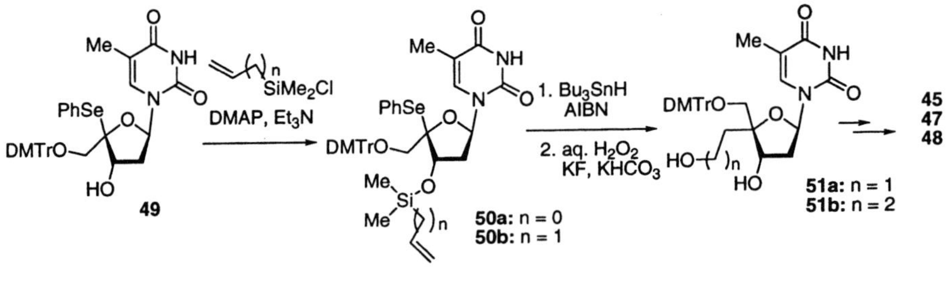

Scheme 13.

The 4'-*C*-phenylselenothymidine derivative **49** (Giese *et al.*, 1994) was silylated with dimethylchlorovinylsilane and allylchlorodimethylsilane to give **50a** and **50b**, respectively (Scheme 13). An intramolecular radical cyclization reaction of **50a** with Bu$_3$SnH/AIBN and subsequent Tamao oxidation readily provided **51a**. When a solution of Bu$_3$SnH/AIBN in benzene was added slowly to a solution of 50b in benzene, followed by Tamao oxidation, the desired 51b, which was derived from the 7-*endo*-cyclized product, was isolated. Then, these 4'-*C*-hydroxyalkylthymidines 51a and 51b were converted to the 4'-*C*-aminoalkylthymidines 45, 47, and 48.

The ODNs were synthesized on a DNA synthesizer by the phosphoramidite method (Beaucage and Caruthers, 1981; Gait, 1984). The nucleosides 45, 47, and 48 were incorporated into the 18-mer (see Table 2), instead of T at various positions. We also prepared a 21-mer ODN-15 with a mixed sequence [5'-d[ACETGATEGCAEAΛΛECTTAE]-3', where E is 45]. The ODNs containing 44 were also synthesized according to the previous method (Wang and Seifert, 1996). The ODN-8E, which contains five residues of 45 was treated with Ac$_2$O in 0.2 M HEPES buffer to give the ODN-9.

Table 2.　　Sequences of ODNs.

ODNs	sequences
1	5'-MTMTMTMTMTMTMTMTMT-3'
2-Y, E, P or Z	5'-MTMTMTMTMTMTMTMTMT-3'
3-Y, E, P or Z	5'-MTMTMTMTMTMTMTMTMT-3'
4-Y, E, P or Z	5'-MTMTMTMTMTMTMTMTMT-3'
5-Y, E, P or Z	5'-MTMTMTMTMTMTMTMTMT-3'
6-Y, E, P or Z	5'-MTMTMTMTMTMTMTMTMT-3'
7-Y, E, P or Z	5'-MTMTMTMTMTMTMTMTMT-3'
8-Y, E, P or Z	5'-MTMTMTMTMTMTMTMTMT-3'
9	5'-MTMTMTMTMTMTMTMTMT-3'
10	5'-d[TG(GA)$_9$GGT]-3'
11	5'-r[(AG)$_9$A]-3'
12	5'-d[ACTTGATTGCATAAATCTTAT]-3'
13	5'-d[ATAAGATTTATGCAATCAAGT]-3'
14	5'-r[AUAAGAUUUAUGCAAUCAAGU]-3'
15	5'-d[ACETGATEGCAEAAAECTTAE]-3'

Y : (T = 44); E : (T = 45); P : (T = 47); Z : (T = 48); ODN-9 : (T = 46); M = 5-methyl-2'-deoxycytidine

3.2.1. *Thermal stability*

Thermal stability of duplexes formed by these ODNs and the complementary DNA-10 or RNA-11 was studied by thermal denaturation in a buffer of 0.01 M sodium phosphate (pH 7.0) containing 0.01 M or 0.1 M NaCl. Melting temperatures (T_ms) are listed in Table 3 with DNA-10 as a complementary strand. All of the ODNs containing 44, 45, or 47 stabilized the ODN/DNA duplexes. The duplexes became more stable as the number of 44, 45, or 47 increased. The ΔT_m values for the ODNs containing five residues of 44, 45, or 47 were +4.5, +5.7, or +4.1 °C, respectively. The ODNs containing one or two residues of 48 destabilized the duplexes, whereas the ODNs containing three, four, or five residues of 48 stabilized the duplexes. The ΔT_m value for the ODNs containing five residues of 48 was +2.4 °C. The ΔT_m value for the ODNs containing 45 was greater than those for the ODNs containing the same numbers of 44 or 47 except for the ODNs-3P and 7P. Additionally, the ΔT_m value for the ODN-15 with a mixed sequence, which has five residues of 45, was +6.8 °C. Therefore, analogs containing 45 seemed to efficiently stabilize the ODN/DNA duplexes.

On the other hand, the ODN-9 containing five residues of 46, which has an acetamidoethyl group, destabilized the ODN/DNA duplex as compared with the control duplex ($\Delta T_m = -6.1$ °C). Thus, the enhanced thermal stability of the duplexes containing 45 was likely due to the effect of the terminal ammonium ion.

When RNA-11 was used as a complementary strand, all of the ODNs containing 44, 45, 47, or 48 slightly destabilized the ODN/RNA duplexes. The duplexes became less stable as the number of the modified nucleosides increased. However, even when five residues of 44, 45, 47, or 48 were incorporated into the 18-mers, the ΔT_m values for these ODNs were –2.2, –2.7, –2.7, and –2.5 °C, respectively. The ΔT_m value for the ODN-15 was only –0.4 °C. Therefore, DNA/RNA duplexes formed by the ODNs containing 44, 45, 47, or 48, are stable enough for use in antisense studies.

3.2.2. *Nuclease resistance*

The susceptibility of the ODNs to nucleolytic digestion was examined with two kinds of nucleases, snake venom phosphodiesterase (a 3'-exonuclease) and DNase I (an endonuclease). The stability of the ODNs in human serum was also investigated.

The ODNs-8Y, 8E, 8P, and 8Z (labeled at the 5'-end with ^{32}P) containing five residues of each nucleoside analog were incubated with snake venom phosphodiesterase or DNase I. When the ODNs were incubated with snake venom phosphodiesterase, the half-lives of ODNs-1, 8Y, 8E, 8P, and 8Z were about 2 min, 2.9 h, 14.4 h, 17.8 h, and 5.2 h, respectively. Among the ODNs, the ODNs-8E and 8P were highly resistant to snake venom phosphodiesterase.

The phosphodiester linkages around the modified nucleosides were also highly resistant to the endo-hydrolysis by DNase I. The half-lives of the ODNs-8Y, 8E, 8P, and 8Z were about 27, 29, 28, and 13 h, respectively, while that of the control ODN-1 was 20 min. Interestingly, the ODNs-8Y, 8E, and 8P having relatively shorter aminoalkyl chains were more resistant to the nuclease than the ODN-8Z with a longer aminoalkyl chain.

Next, we examined the effects of the terminal ammonium cations of the aminoalkyl chains on the resistance of the ODNs to an endonuclease. We compared the susceptibility of ODN-9 containing 46 to nucleolytic digestion by DNase I with that of ODN-8E containing 45. The half-lives of the ODNs-1, 8E, and 9 were 35 min, 30 h, and 3.4 h, respectively. The ODN-9 was much less resistant to the nuclease than the ODN containing 45. Therefore, this suggested that the effects of the terminal ammonium ions of the aminoalkyl groups play an important role in nuclease resistance of the ODNs.

Table 3. Hybridization data.

ODN	ODN/DNA[a]		ODN/RNA[b]	
	0.01 M NaCl T_m (°C)	ΔT_m[c] (°C)	0.1 M NaCl T_m (°C)	ΔT_m[c] (°C)
1	54.5	-	73.9	-
2Y	54.4	+0.0	73.9	+0.0
3Y	54.4	+0.0	73.7	-0.2
4Y	54.5	+0.1	73.3	-0.6
5Y	54.4	+1.0	72.9	-1.0
6Y	56.8	+2.4	73.0	-0.9
7Y	57.5	+3.1	72.3	-1.6
8Y	58.9	+4.5	71.7	-2.2
2E	55.0	+0.6	73.7	-0.2
3E	54.5	+0.1	73.8	-0.1
4E	55.5	+1.1	73.5	-0.4
5E	55.7	+1.3	72.1	-1.8
6E	56.9	+2.5	72.3	-1.6
7E	58.1	+3.7	71.5	-2.4
8E	60.1	+5.7	71.2	-2.7
2P	54.4	+0.0	73.3	-0.6
3P	54.6	+0.2	73.4	-0.5
4P	55.1	+0.7	73.1	-0.8
5P	55.6	+1.2	72.5	-1.4
6P	56.4	+2.0	72.5	-1.4
7P	58.3	+3.9	72.0	-1.9
8P	58.5	+4.1	71.2	-2.7
2Z	52.1	-2.3	73.2	-0.7
3Z	52.7	-1.7	73.5	-0.4
4Z	50.8	-3.6	72.8	-1.1
5Z	53.9	-0.5	72.3	-1.6
6Z	54.9	+0.5	72.6	-1.3
7Z	55.6	+1.2	71.8	-2.1
8Z	56.8	+2.4	71.4	-2.5
9	48.3	-6.1	-	-
12	42.3	-	49.8	-
15	49.1	+6.8	49.4	-0.4

[a] The complementary DNA: DNA-10 for ODNs-1–9; DNA-13 for ODNs-12 and -15. [b]The complementary RNA: RNA-11 for ODNs-1–9; RNA-14 for ODNs-12 and -15. [c]ΔT_m = [T_m (each ODN)-T_m (the control ODN 1)].

We also investigated the susceptibility of the ODN-15 with a mixed sequence against DNase I and human serum. While the half-life of the control ODN-12 was 7 min, the ODN-15 was not hydrolyzed after 48 h incubation with DNase I. The ODN-15 was also about 21-fold more stable than the control ODN-12 during incubation in human serum ($t_{1/2}$ of ODN-12 and 15; 37 min and 13 h, respectively).

From these results, the ODNs containing nucleosides with the aminoalkyl chains at the 4'-position, especially those containing nucleosides with the aminoethyl and aminopropyl chains, were found to be resistant enough to enzymatic degradations to be used as antisense molecules.

3.2.3. *Degradation by RNase H*

It has been postulated that antisense activity of antisense ODNs is due, at least in part, to cleavage of the RNA strand of a DNA/RNA duplex by RNase H (Crooke and Lebleu, 1993; Chiang *et al.*, 1991; Neckers *et al.*, 1991; Monia *et al.*, 1992). We therefore examined whether the ODN/RNA heteroduplex between an ODN containing three residues (ODN-6Y, 6E, 6P, or 6Z, respectively) and five residues (ODN-8Y, 8E, 8P, or 8Z, respectively) of 44, 45, 47, or 48 and its complementary RNA-11 could elicit RNase H activity. The RNAs in the duplexes with the ODNs containing three residues of the modified nucleosides were completely hydrolyzed after 20 min, and the rates were similar to those of a control experiment with ODN-1 using *E. coli* RNase H. When the ODNs contained five residues of the modified nucleosides, the complementary RNAs in the duplexes were also degraded, although the rates were slightly decreased compared with those in the unmodified duplex. The ODN-15 with a mixed sequence also elicited effective cleavage of the complementary RNA-14.

For ODN therapeutics, antisense ODNs are often required to elicit RNase H activity in human cells. Therefore, we next examined cleavage of RNAs by HeLa cell nuclear extracts as a source of human RNase H (Agrawal *et al.*, 1990; Monia *et al.*, 1993). The experiments were carried out using the same heteroduplexes described above. All the RNA strands in the duplexes were found to be effectively cleaved by HeLa cell nuclear extracts.

E. coli RNase H requires at least four contiguous unmodified 2'-deoxyribonucleotide residues to elicit cleavage of the RNA strand (Inoue *et al.*, 1987; Furdon *et al.*, 1989), and a minimum of five contiguous unmodified 2'-deoxyribonucleotides was required for efficient activation of HeLa RNase H (Monia *et al.*, 1993). Therefore, it is noteworthy that ODNs-8 elicit efficient cleavage of the RNA strand by RNase H, although they have four regions of three contiguous unmodified 2'-deoxyribonucleotide residues in the antisense strand.

These results demonstrated that the natural phosphodiester ODNs containing the 4'-*C*-aminoalkyl thymidines, especially 45 with the aminoethyl group, possess the desired properties for an antisense molecule: 1) formation of the thermally stable duplex with complementary DNA and RNA, 2) resistance to nucleases, and 3) RNase H activity in the duplex with RNA. Thus, the ODNs are good candidates for antisense molecules. Applications of the ODN containing 45 as an antisense ODN are currently being studied.

4. Stereoselective introduction of a vinyl group via an atom-transfer radical cyclization

As described in previous sections, we successfully developed a new radical cyclization reaction, in which a 1-hydroxyethyl or 2-hydroxyethyl group can be introduced at the position β to a hydroxyl group. These 1-hydroxyethyl or 2-hydroxyethyl group can further be transformed to other carbon-linked functionalities. Using this idea, we prepared several 4'-carbon-substituted thymidine derivatives 52 and 53 (Figure 3) (Sugimoto *et al.*, 1999a). We also prepared 54-57 starting from 2'-deoxy-4'-*C*-hydroxymethylthymidine or -cytidine (Nomura *et al.*, 1999a,b). Some of these nucleosides showed interesting biological activities such as anti-HIV-1. However, the synthetic methods of these nucleosides were rather lengthy. Therefore, we developed an efficient one-pot method for introducing a vinyl group *via* an atom-transfer radical cyclization reaction with a diphenylvinylsilyl group followed by a fluoride ion-promoted elimination reaction (Sugimoto *et al.*, 1999c).

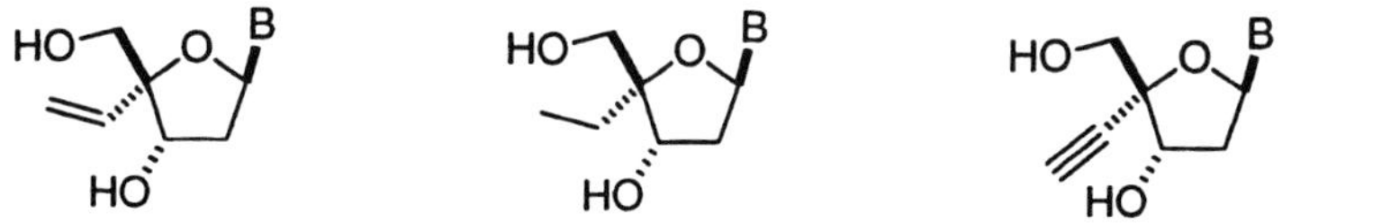

52: B = thymin-1-yl
55: B = cytosin-1-yl

53: B = thymin-1-yl
56: B = cytosin-1-yl

54: B = thymin-1-yl
57: B = cytosin-1-yl

Figure 3.

Our strategy is shown in Scheme 14. We presumed that when the reaction with the vinylsilyl ether a is performed under radical atom-transfer conditions, 5-exo-cyclized radical e would be trapped with a halogeno or a seleno group to give g. If this indeed occurred, subsequent treatment of g with fluoride ion would promote an elimination reaction to give the desired vinyl derivative h.

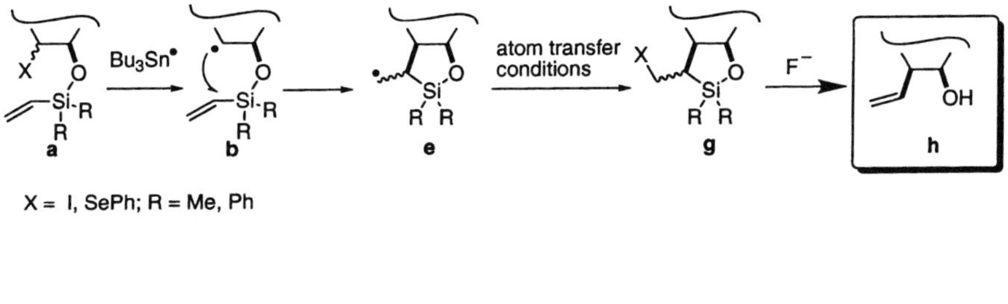

X = I, SePh; R = Me, Ph

Scheme 14.

the yield of 72 (entries 2-4). Treatment of 71 under conditions identical to entry 2 also gave 72 (entry 5). Thus, the radical atom-transfer reaction with a vinylsilyl group effectively occurred under thermal conditions with $(Me_3Sn)_2$ and AIBN, which is more convenient than the previous method by irradiation (Sugimoto *et al.*, 1999c), especially for large-scale experiments.

Table 5.   Synthesis of 2'α-*C*-vinyluridine derivative **72** by the radical atom-transfer reaction.[a]

entry	substrate	reagents (equiv)	solvent	temperature, °C	yield, %[b]
1	**70**	$(Bu_3Sn)_2$ (0.5), AIBN (0.6)	benzene	80	29
2	**70**	$(Me_3Sn)_2$ (0.5), AIBN (0.6)	benzene	80	62
3	**70**	$(Me_3Sn)_2$ (0.7), AIBN (0.6)	benzene	80	64
4	**70**	$(Me_3Sn)_2$ (1.0), AIBN (0.6)	benzene	80	66
5	**71**	$(Me_3Sn)_2$ (0.5), AIBN (0.6)	benzene	80	69

[a] A mixture of the substrate, $(Me_3Sn)_2$ or $(Bu_3Sn)_2$, and AIBN in benzene or toluene was heated under reflux until the substrate disappeared on TLC. [b]The product was isolated as the 3'-*O*-TBS derivative **72**.

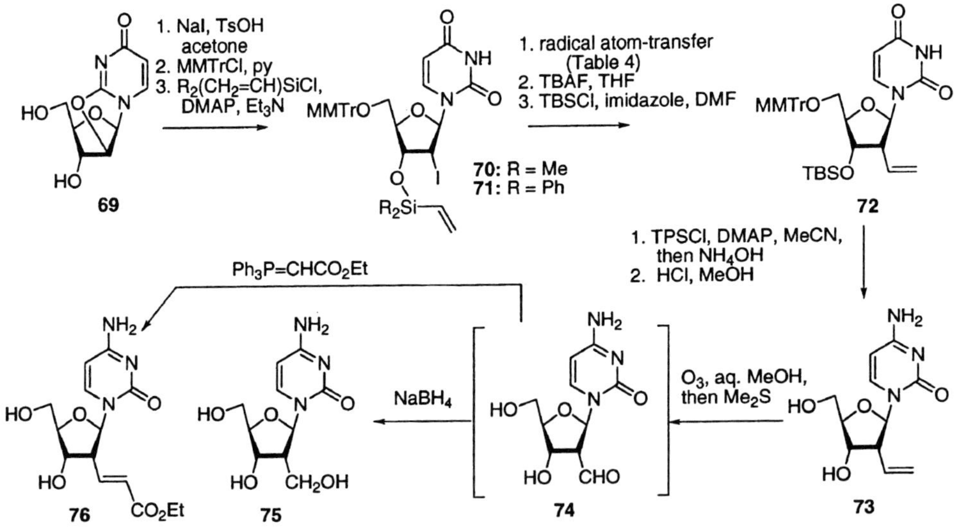

Scheme 15.

Conversion of 72 into the cytosine congener 73 was performed under usual conditions (Matsuda *et al.*, 1992). We next tried to synthesize 74. Thus, 73 was successively treated with O_3 and Me_2S in aqueous MeOH, and the purification of the products by usual silica gel column chromatography and/or reverse-phase HPLC was attempted. However, the desired 74 was not obtained in pure form. Ozonolysis of 73 in aqueous MeOH at -78 °C, followed by reductive treatment with $NaBH_4$, successfully gave the desired 75. Similarly, the crude 74 was treated with $Ph_3P=CHCO_2Et$ to give the corresponding Wittig reaction product 76. These results showed that although 74 could be produced under these conditions, it was too unstable to be isolated.

6.   Synthesis of 1'α-branched-chain sugar uridine derivatives

Although a number of procedures for preparing branched-chain sugar nucleosides have been developed, examples of synthesis of 1'-branched-chain sugar nucleosides are rare (Elliott *et al.*, 1992; Itoh *et al.*, 1995; Goodman and Greenberg, 1996). Furthermore, because of the lack of efficient synthetic methods for their preparation, the biological activities of 1'-branched-chain sugar nucleosides have not been systematically investigated. With this in mind, we decided to prepare 1'α-branched-chain sugar nucleosides using the radical atom-transfer cyclization reaction.

6.1.   Introduction of a phenylseleno group at the 1'-position

To provide the substrate for the radical reaction, a method for introducing a phenylseleno group at the 1'-positon of nucleosides was needed. We assumed that treatment of the 2'-ketouridine derivative 77 with a strong base would produce the corresponding 1'-enolate 78, and that the subsequent reaction with PhSeCl as an electrophile would give the 1'-phenylseleno-2'-ketouridine derivative 81 (Scheme 16). The stereoselective hydride reduction of the 2'-carbonyl from the β-face, followed by introduction of a dimethylvinylsilyl group at the 2'-hydroxyl, would provide the desired radical reaction substrate 84 (Scheme 17).

The 3',5'-*O*-(1,1,3,3-tetraisopropyldisiloxane-1,3-diyl) (TIPDS)-2'-ketouridine 77, first prepared by our group (Ueda *et al.*, 1984), has been widely used for the synthesis of 2'-modified nucleosides via nucleophilic addition reactions at the 2'-carbonyl (Matsuda *et al.*, 1993). We first examined whether enolization at the 1'-position of 77 occurred by deuterium-labeling experiments (Scheme 16). A mixture of 77 and LiHMDS (2.1 equiv) in THF was stirred at < −70 °C for 1 h and quenched with CD_3CO_2D/CD_3OD. The 2'-ketouridine 79 was obtained in about 90% yield, of which 54% of the 1'-proton was replaced by a deuterium atom. Furthermore, when the reaction mixture of 77 and LiHMDS in THF was treated with BzCl at < −70 °C, the enol-*O*-benzoate 80 was obtained in 73% yield. These experiments clearly showed that the 1'-enolate 78 was produced under these conditions, as predicted. As far as we know, this is the first example demonstrating enolization at the 1'-position of a 2'-ketonucleoside (Kodama *et al.* 2000).

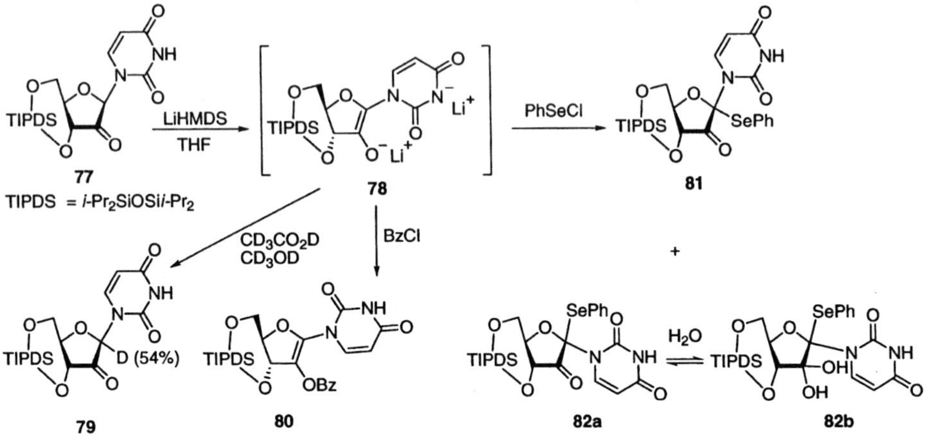

Scheme 16.

Introduction of a phenylseleno group at the 1'-position of 77 via its enolization was next investigated. The reaction of 77 with LiHMDS in THF was stirred at < -70 °C for 1 h. PhSeCl (2 equiv) was added, and the resulting mixture was further stirred at the same temperature for 1 h to give the desired 1'-phenylseleno derivative in 85% yield as an anomeric mixture (81:82 = 2.5:1). It is worth noting that the 1'β-phenylseleno product 82 was in equilibrium between the 2'-keto form 82a and its 2'-hydrate 82b. The 1'α-phenylseleno product 81 is probably produced via a chelation-controlled reaction pathway, since the facial selectivity was almost lost when HMPA was added to the reaction mixture. While the structure of the chelation intermediate has not been elucidated, one might postulate a chelation of Li^+ between the 2'-enol-oxygen and 2-carbonyl-oxygen of the uracil moiety.

The 1'α-phenyseleno product 81 was then subjected to reduction at the 2'-keto moiety with hydride reagents, such as $NaBH_4$, $LiBH_4$, $NaBH_3CN$, DIBAL-H, or $LiAlH(OEt)_3$ (Scheme 17). We investigated various reaction conditions and found that the 2'-carbonyl group of 81 was chemo- and stereoselectively reduced from the β-face when it was treated with $NaBH_4$ in the presence of $CeCl_3$ in MeOH (Luche, 1978) at < -70 °C to give the desired sugar-protected 1'-phenylselenouridine 83 in 90% yield as the sole product.

The vinylsilyl group was introduced at the 2'-hydroxyl group by treating 83 with a dimethylchlorovinylsilane, DMAP, and Et_3N in toluene to give 84.

6.2. Introduction of a vinyl group via radical atom-transfer cyclization

We next investigated the introduction of a vinyl group at the 1'-position of uridine via radical atom-transfer cyclization and subsequent elimination of the phenylseleno group (Sugimoto *et al.*, 1999c). When the radical atom-transfer cyclization reaction was performed by irradiation of a solution of 84 in benzene with a high-pressure mercury lamp through a Pyrex filter in the presence of $(PhSe)_2$ (0.7 equiv) at room temperature, the starting material 84 disappeared within 4 h. The product, without purification, was immediately treated successively with H_2O_2 in THF and with $Ac_2O/DMAP/Et_3N$ in MeCN at room temperature to give the desired 1'-*C*-vinyl derivative 86 in 52% yield (Scheme 17) (Kodama *et al.* 2001).

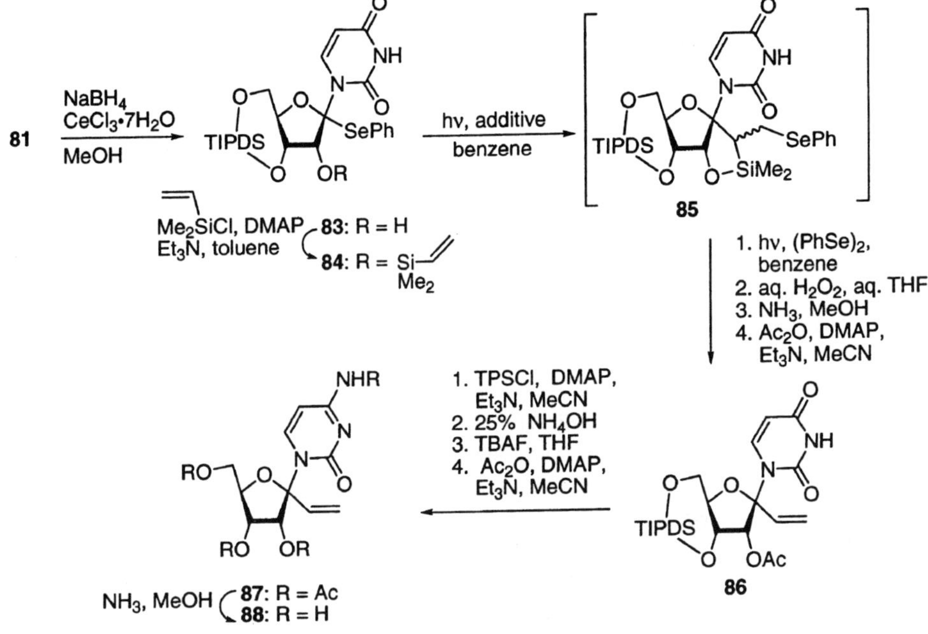

Scheme 17.

The 1'-*C*-vinyluridine derivative 86 was treated with 2,4,6-triisopropylbenzenesulfonyl chloride (TPSCl)/DMAP/Et_3N in MeCN followed by ammonolysis (Matsuda *et al.*, 1992) to give the corresponding cytidine derivative, which was isolated as the tetraacetate 87 in 78% yield. Removal of the acetyl groups of 87 with saturated NH_3 in MeOH afforded the 1'-*C*-vinylcytidine (88).

Conclusion

We have successfully developed a regio- and stereoselective method for introducing a 1-hydroxyethyl or 2-hydroxyethyl group at the β-position of a hydroxyl group in halohydrins or α-phenylselenoalkanols by using an intramolecular radical cyclization reaction with a dimethyl or diphenylvinylsilyl group as a radical acceptor using indane

48   *S. Shuto et al.*

derivatives as models. This radical cyclization reaction involves the ring-enlargement reaction from a (3-oxa-2-silacyclopentyl)methyl radical to a 4-oxa-3-silacyclohexyl radical via a pentavalent silicon-bridging radical transition state. As an extension of this reaction, we have also developed a new simple method for introducing a vinyl group via an atom-transfer radical-cyclization reactions. These reactions have been successfully applied not only to nucleoside chemistry to prepare several types of branched-chain sugar nucleosides but also to sugar chemistry to introduce C2-carbon substituents to hexopyranoses at the anomeric position (Yahiro *et al.*, 1999; Shuto *et al.* 2000b, Shuto *et al.* 2000c). We are working further to introduce other carbon functional groups by radical cyclization reactions including atom-transfer reactions.

References

Agrawal S, Mayrarrd SH, Zamecnik PC, Pederson T. (1990) Site-specific excision from RNA by RNase H and mixed-phosphate-backbone oligodeoxynucleotides. Proc Natl Acad Sci USA; 87: 1401-1405.

Agrawal S. editor. (1996a) Antisense therapeutics. Totowa, NJ: Humana Press.

Agrawal S. (1996b) Antisense oligonucleotides: towards clinical trials. Trends Biotechnol; 14: 376-387.

Azuma A, Nakajima Y, Nishizono N, Minakawa N, Suzuki M, Hanaoka K, Kobayashi T, Tanaka M, Sasaki T, Matsuda A. (1993) 2'-*C*-Cyano-2'-deoxy-1-β-D-arabinofuranosylcytosine and its derivatives: A new class of nucleoside with a broad antitumor spectrum. J Med Chem; 36: 4183-4189.

Azuma A, Hanaoka K, Kurihara A, Kobayashi T, Miyauchi S, Kamo N, Tanaka M, Sasaki T, Matsuda A. (1995) Chemical stability of a new antitumor nucleoside, 2'-*C*-cyano-2'-deoxy-1-β-D-*arabino*-pentofuranosylcytosine (CNDAC) in alkaline medium: formation of 2'-*C*-cyano-2'-deoxy-1-β-D-*ribo*-pentofuranosylcytosine (CNDC) and its antitumor activity. J Med Chem; 38: 3391-3397.

Baldwin JE. (1976) Rules for ring closure. J Chem Soc Chem Commun; 734-736.

Barawkar DA, Kumar VA, Ganesh KN. (1994) Triplex formation at physiological pH by oligonucleotides incorporating 5-Me-dC-(N^4-spermine). Biochem Biophys Res Commun; 205: 1665-1670.

Barawkar DA, Rajeev KG, Kumar VA, Ganesh KN. (1996) Triplex formation at physiological pH by 5-Me-dC-N^4-(spermine) [X] oligodeoxynucleotides: non protonation of N3 in X of X*G:C triad and effect of base mismatch/ionic strength on triplex stabilities. Nucleic Acids Res; 24: 1229-1237.

Barton TJ, Revis A. (1984) Regiochemistry of alkenylsilyl and alkenyldisilanyl radical cyclizations. J Am Chem Soc; 106: 3802-3805.

Beaucage SL, Caruthers MH. (1981) Studies on nucleotide chemistry V. Deoxynucleoside phosphoramidites. A new class of key intermediates for deoxypolynucleotide synthesis. Tetrahedron Lett; 22: 1859-1862.

Beaucage SL, Iyer RP. (1993a) The functionalization of oligonucleotides via phosphoramidite derivatives. Tetrahedron; 49: 1925-1963.

Beaucage SL, Iyer RP. (1993b) The synthesis of modified oligonucleotides by the phosphoramidite approach and their applications. Tetrahedron; 49: 6123-6194.

Beckwith ALJ, Gream GE, Struble DL. (1972) Cyclization and cupric-ion oxidation of 4-(cyclohex-1-enyl)butyl radical. Aust J Chem; 25: 1081-1105.

Beckwith ALJ. (1981) Regio-selectivity and stereo-selectivity in radical reactions. Tetrahedron; 37: 3073-3100.

Beckwith ALJ, Schiesser CH. (1985) Regio- and stereo-selectivity of alkenyl radical ring closure: a theoretical study. Tetrahedron; 41: 3925-3941.

Brook AG, Duff JM. (1969) The optical stability of asymmetric silyl radicals. J Am Chem Soc; 91: 2118-2119.

Cai Y, Roberts BP. (1998) Intramolecular radical-chain hydrosilylation catalysed by thiol: cyclisation of alkenyloxysilanes. J Chem Soc Perkin Trans 1; 467-475.

Chatgilialoglu C, Ingold KU, Scainao JC. (1982) Absolute rate constants for the reaction of triethylsilyl radicals with organic halides. J Am Chem Soc; 104: 5123-5127.

Chatgilialoglu C, Woynar H, Ingold KU, Davies AG. (1983) Intramolecular and intermolecular reactions of alkenylsilyl radicals. J Chem Soc Perkin Trans 2; 555-561.

Chiang MY, Chan H, Zounes MA, Freier SM, Lima WF, Bennett CF. (1991) Antisense oligonucleotides inhibit intercellular adhesion molecule 1 expression by two distinct mechanisms. J Biol Chem; 255: 9434-9443.

Cosstick R, Eckstein F. (1985) Synthesis of d(GC) and d(CG) octamers containing alternating phosphorothioate linkages: effect of the phosphorothioate group on the B-Z transition. Biochemistry; 24: 3630-3638.

Crooke ST, Lebleu B, editors. (1993) Antisense research and applications. Boca Raton, FL: CRC Press.

Cuenoud B, Casset F, Husken D, Natt F, Wolf RM, Altmann K-H, Martin P, Moser HE. (1998) Dual recognition of double-stranded DNA by 2'-aminoethoxy-modified oligonucleotides. Angew Chim Int Ed Engl; 37: 1288-1291.

Curran DP, Chang C-T. (1989) Atom transfer cyclization reactions of α-iodo esters, ketones, and malonates: examples of selective 5-exo, 6-endo, 6-exo, and 7-endo ring closures. J Org Chem; 54: 3140-3157.

Dan A, Yoshimura Y, Ono A, Matsuda A. (1993) Synthesis of oligonucleotides containing a novel 2'-deoxyuridine analogue that carries an aminoalkyl group at 1'-position; stabilization of duplex formation by an intercalating group accommodated in the minor groove. Bioorg Med Chem Lett; 3: 615-618.

Donehower RC, Dees EC, Baker SD, Summerson L, Carducci MA, Izumi T, Kobayashi T. (2000) A phase I study of CS-682, an oral antimetabolite, in patients with refractory solid tumors. Proc Amer Soc Clin Oncol; 19: 196a.

Dowd P, Zhang W. (1993) Free radical-mediated ring expansion and related annulations. Chem Rev; 93: 2091-2115.

Eda H, Ura M, Ouchi K, Tanaka Y, Miwa M, Ishitsuka H. (1998) The antiproliferative activity of DMDC is modulated by inhibition of cytidine deaminase. Cancer Res; 58: 1165-1169.

Elliott RD, Niwas S, Riordan JM, Montgomery JA, Secrist JA. III. (1992) Synthesis of 9-(1-deoxy-1-phosphono-β-D-psicofuranosyl)-1,9-dihydro-6*H*-purin-6-one as a potential transition state analog inhibitor of purine nucleoside phosphorylase. Nucleosides Nucleotides; 11: 97-119.

Etter MC. (1990) Encoding and decoding hydrogen-bond patterns of organic compounds. Acc Chem Res; 23: 120-126.

Furdon PJ, Dominaki Z, Kole R. (1989) RNase H cleavage of RNA hybridized to oligonucleotides containing methylphosphonate, phosphorothioate and phosphodiester bonds. Nucleic Acids Res; 17: 9193-9204.

Gait M, editor. (1984) Oligonucleotides synthesis: a practical approach. Oxford, UK: IRL Press.

Giese B, Erdmann P, Schafer T, Schwitter U. (1994) Synthesis of a 4'-selenated 2'-deoxyadenosine derivative: a novel precursor suitable for the preparation of modified oligonucleotides. Synthesis; 1310-1312.

Goodman BK, Greenberg MM. (1996) Independent generation and reactivity of 2'-deoxyuridin-1'-yl. J Org Chem; 61: 2-3.

Griffey RH, Monia B, Cummins LL, Freier S, Greig MJ, Guinosso CJ, Lesnik E, Manalili SM, Mohan V, Owens S, Ross BR, Sasmor H, Wancewicz E, Weiler K, Wheeler PD, Cook PD. (1996) 2'-*O*-Aminopropyl ribonucleotides: a zwitterionic modification that enhances the exonuclease resistance and biological activity of antisense oligonucleotides. J Med Chem; 39: 5100-5109.

Hacia JG, Wold BJ, Dervan PB. (1994) Phosphorothioate oligonucleotide-directed triple helix formation. Biochemistry; 33: 5367-5369.

Haginoya N, Ono A, Nomura Y, Ueno Y, Matsuda A. (1997) The synthesis of oligodeoxyribonucleotides containing 5-(*N*-aminoalkyl)carbamoyl-2'-deoxyuridines by a new post-modification method, and their thermal stability and nuclease-resistance properties. Bioconjugate Chem; 8: 271-280.

Hanaoka K, Suzuki M, Kobayashi T, Tanzawa F, Tanaka K, Shibayama T, Miura S, Ikeda T, Iwabuchi H, Nakagawa A, Mitsuhashi Y, Hisaoka M, Kaneko M, Tomida M, Wataya Y, Nomura T, Sasaki T, Matsuda A, Tsuruo T, Kurakata S. (1999) Antitumor activity and novel DNA-self-strand-breaking mechanism of CNDAC (1-(2-*C*-cyano-2-deoxy-β-D-*arabino*-pentofuranosyl)cytosine) and its N^4-palmitoyl derivative (CS-682). Int J Cancer; 82: 226-236.

Harris JM, MacInnes I, Walton JC, Maillard B. (1991) 1,2-Migration of the trimethylsilyl group in free radicals. J Organometal Chem; 403: C25-C28.

Harris JM, Walton JC, Maillard B, Grelier S, Picard J-P. (1993) Hydrogen abstraction from silylamines; an investigation of the 1,2-migration of the trimethylsilyl group in aminyl radicals. J Chem Soc Perkin Trans 2; 2119-2123.

Hashimoto H, Nelson MG, Switzer C. (1993a) Formation of chimeric duplexes between zwitterionic and natural DNA. J Org Chem; 58: 4194-4195.

Hashimoto H, Nelson MG, Switzer C. (1993b) Zwitterionic DNA. J Am Chem Soc; 115: 7128-7134.

Hattori H, Tanaka M, Fukushima M, Sasaki T, Matsuda A. (1996) 1-(3-*C*-Ethynyl-β-D-*ribo*-pentofuranosyl)cytosine (ECyd), 1-(3-*C*-ethynyl-β-D-*ribo*-pentofuranosyl)uracil (EUrd), and their nucleobase analogues as new potential multifunctional antitumor nucleosides with a broad spectrum of activity. J Med Chem; 39: 5005-5011.

Hattori H, Nozawa E, Iino T, Yoshimura Y, Shuto S, Shimamoto Y, Nomura M, Fukushima M, Tanaka M, Sasaki T, Matsuda, A. (1998) The structural requirements of the sugar moiety for the antitumor activities of new nucleoside antimetabolites, 1-(3-*C*-ethynyl-β-D-*ribo*-pentofuranosyl)cytosine and -uracil. J Med Chem; 41: 2892-2902.

Hayakawa Y, Kawai R, Otsuki K, Kataoka, M, Matsuda A. (1998) Evidence supporting the activity of 2'-*C*-cyano-2'-deoxy-1-β-D-*arabino*-pentofuranosylcytosine as a terminatior in enzymatic DNA-chain elongation. Bioorg Med Chem Lett; 8: 2559-2562.

Inoue H, Hayase Y, Iwai S, Ohtsuka E. (1987) Sequence-dependent hydrolysis of RNA using modified oligonucleotide splints and RNase H. FEBS Lett; 215: 327-330.

Itoh Y, Haraguchi K, Tanaka H, Gen E, Miyasaka T. (1995) Divergent and stereocontrolled approach to the synthesis of uracil nucleosides branched at the anomeric position. J Org Chem; 60: 656-662.

Johnson GC, Stofko JJ, Jr, Lockhart TP, Brown DW, Bergman RG. (1979) Evidence for the formation of 1,3-dehydrobenzene diradicals by silicon shift in 1,4-dehydrobenzenes at high temperature. J Org Chem; 44: 4215-4218.

Julia M. (1967) Free radical cyclizations. Pure Appl Chem; 15: 167-183.

Kanazaki M, Ueno Y, Shuto S, Matsuda A. (2000) Highly nuclease-resistant phosphodiester-type oligo-deoxynucleotides containing 4'α-*C*-aminoalkylthymidines form thermally stable duplexes with DNA and RNA. A candidate for potent antisense molecules. J Am Chem Soc; 122: 2422-2434.

Kodama T, Shuto S, Momura M, Matsuda A. (2000) Synthesis of a 1'α-phenylselenouridine derivative as a synthetic precursor for various 1'-modified nucleosides, via enolization at the 1'-position of 3',5'-*O*-TIPDS-2'-ketouridine. Tetrahedron Lett; 41: 3643-3646.

Kodama T, Shuto S, Nomura M, Matsuda A. (2001) Synthesis of the 1'-phenylselenouridine derivative via enolization at the 1'-position of 3',5'-TIPDS-2'-ketouridine and its radical atom-transfer cyclization reaction. An efficient method for preparing 1'α-branched-chain sugar pyrimidine nucleosides. Chem Eur J; 7: 2332-2340.

Krieg AM, Stein CA. (1995) Phosphorothioate oligodeoxynucleotides: antisense or anti-protein? Antisense Res Dev; 5: 241 and references therein.

Kulicle KJ, Chatgiliatoglu C, Kopping B, Giese B. (1992) Homolytic substitution reaction at a silicon atom. Helv Chim Acta; 75: 935-939.

LaPlanche LA, James TL, Powell C, Wilson WD, Uznanski B, Stec W, Summers MF, Zon G. (1986) Phosphorothioate-modified oligodeoxyribonucleotides. III. NMR and UV spectroscopic studies of the Rp-Rp, Sp-Sp, and Rp-Sp duplexes, [d(GGSAATTCC)]$_2$, derived from diastereomeric *O*-ethyl phosphorothioates. Nucleic Acids Res; 14: 9081-9093.

Latimer LJP, Hampel K, Lee JS. (1989) Synthetic repeating sequence DNAs containing phosphorothioates: nuclease sensitivity and triples formation. Nucleic Acids Res; 17: 1549-1561.

Luche J-L. (1978) Lanthanides in organic chemistry. 1. Selective 1,2 reductions of conjugated ketones. J Am Chem Soc; 100: 2226-2227.

Matsuda A, Muneyama T, Nishida T, Sato T, Ueda T. (1976) A new synthesis of 5'-deoxy-8,5'-cycloadenosine and -inosine: conformationally fixed purine nucleosides. Nucleic Acids Res; 3: 3349-3357.

Matsuda A, Tezuka M, Niizuma K, Sugiyama E, Ueda T. (1978) Synthesis of carbon-bridged 8,5'-cyclopurine nucleosides. Tetrahedron; 34: 2633-2637.

Matsuda A, Takenuki K, Sasaki T, Ueda T. (1991a) Synthesis of a new broad spectrum of antineoplastic nucleoside, 2'-deoxy-2'-methylidenecytidine (DMDC) and its derivatives. J Med Chem; 34: 812-819.

Matsuda A, Nakajima Y, Azuma A, Tanaka M, Sasaki T. (1991b) 2'-*C*-Cyano-2'-deoxy-1-β-D-arabino-furanosylcytosine (CNDAC): Design of a potential mechanism-based DNA-strand breaking antineoplastic nucleoside. J Med Chem; 34: 2917-2919.

Matsuda A, Okajima H, Masuda A, Kakefuda A, Yoshimura Y, Ueda T. (1992) Radical and palladium-catalyzed deoxygenation of the allylic alcohol systems in the sugar moiety of pyrimidine nucleosides. Nucleosides Nucleotides; 11: 197-226.

Matsuda A, Azuma A, Nakajima Y, Takenuki K, Dan A, Iino T, Yoshimura Y, Minakawa N, Tanaka M, Sasaki T. (1993) Design of new types of antitumor nucleosides: The synthesis and antitumor activity of 2'-deoxy-(2'-*C*-substituted)cytidines. In: Chu CK, Baker DC, editors. Nucleosides and Nucleotides as Antitumor and Antiviral Agents. New York: Plenum Press; p. 1-22.

Matsuda A, Azuma A. (1995) 2'-*C*-Cyano-2'-deoxy-1-β-D-arabinofuranosylcytosine (CNDAC): A mechanism-based DNA-strand-breaking antitumor nucleoside. Nucleosides Nucleotides; 14: 461-471.

Matsuda A, Hattori H, Tanaka M, Sasaki T. (1996) 1-(3-*C*-Ethynyl-β-D-*ribo*-pentofuranosyl)uracil as a potential broad spectrum multifunctional antitumor nucleoside. Bioorg Med Chem Lett; 6: 1887-1892.

Matsuda A, Fukushima M, Wataya Y, Sasaki T. (1999) A new antitumor nucleoside, 1-(3-*C*-ethynyl-β-D-*ribo*-pentofuranosyl)cytosine (ECyd), is a potent inhibitor of RNA polymerase. Nucleosides Nucleotides; 18: 811-814.

Milligan JF, Matteucci MD, Martin JC. (1993) Current concepts in antisense drug design. J Med Chem; 36: 1923-1937.

Miwa M, Eda H, Ura M, Ouchi KF, Keith DD, Foley LH, Ishitsuka H. (1998) High susceptibility of human cancer xenografts with higher levels of cytidine deaminase to a 2'-deoxycytidine antimetabolite, 2'-deoxy-2'-methylidenecytidine. Clin Cancer Res; 4: 493-497.

Monia BP, Johnston JF, Ecker DJ, Zounes MA, Lima WF, Freier SM. (1992) Selective inhibition of mutant Ha-ras mRNA expression by antisense oligonucleotides. J Biol Chem; 267: 19954-19962.

Monia BP, Lesnik EA, Gonzalez C, Lima WF, McGee D, Guinosso CJ, Kawasaki AK, Cook PD, Freier SM. (1993) Evaluation of 2'-modified oligonucleotides containing 2'-deoxy gaps as antisense inhibitors of gene expression. J Biol Chem; 268: 14514-14522.

Neckers L, Whitesell L, Rosolen A, Geselowitz DA. (1991) Antisense inhibition of oncogene expression. Crit Rev Oncol; 3: 175-231.

Nomura Y, Ueno Y, Matsuda A. (1997) Site-specific introduction of functional groups into phosphodiester oligodeoxynucleotides and their thermal stability and nuclease-resistance properties. Nucleic Acids Res; 25: 2784-2791.

Nomura M, Shuto S, Tanaka M, Sasaki T, Mori S, Shigeta S, Matsuda A. (1999a) Synthesis and biological activities of 4'α-*C*-branched-chain sugar pyrimidine nucleosides. J Med Chem; 42: 2901-2908.

Nomura M, Endo K, Shuto S, Matsuda A. (1999b) Ring expansion reaction of 1-[2,3,5-tri-*O*-TBS-4α-formyl-β-D-*ribo*-pentofuranosyl]uracil by treating with (methylene)triphenylphosphorane to give a new nucleoside containing dihydrooxepine ring at the sugar moiety. Tetrahedron; 55: 14847-14854.

Obata T, Endo Y, Tanaka M, Matsuda A, Sasaki T. (1998) Development and biochemical characterization of a 2'-*C*-cyano-2'-deoxy-1-β-D-*arabino*-pentofuranosylcytosine (CNDAC)-resistant variant of the human fibrosarcoma cell line HT-1080. Cancer Lett; 123: 53-61.

Ono A, Dan A, Matsuda A. (1993) Synthesis of oligonucleotides carrying linker groups at the 1'-position of sugar residue. Bioconjugate Chem; 4: 499-508.

Ono A, Haginoya N, Kiyokawa M, Minakawa N, Matsuda A. (1994) A novel and convenient post-synthetic modification method for the synthesis of oligodeoxyribonucleotides carrying amino linkers at the 5-position of 2'-deoxyuridine. Bioorg Med Chem Lett; 4: 361-366.

Ono T, Fujii A, Hosoya M, Okumoto T, Sakata S, Matsuda A, Sasaki T. (1996) Cell kill kinetics of an antineoplastic nucleoside, 1-(2-deoxy-2-methylene-β-D-*erythoro*-pentofuranosyl)cytosine. Biochem Pharmacol; 52: 1279-1285.

Ozaki H, Nakamura A, Arai M, Endo M, Sawai H. (1995) Novel C5-substituted 2'-deoxyuridine derivatives bearing amino-linker arms: synthesis, incorporation into oligodeoxyribonucleotides, and their hybridization properties. Bull Chem Soc Jpn; 68: 1981-1987.

Pitt CG, Fowler MS. (1968) The radical-catalyzed rearrangement of trisilanethiol. J Am Chem Soc; 90: 1928-1930.

Prakash TP, Barawkar DA, Kumar VA, Ganesh KN. (1994) Synthesis of site-specific oligonucleotide-polyamine conjugates. Bioorg Med Chem Lett; 4: 1733-1738.

Roberts BP, Vazquez-Persaud AR (1995) EPR spectroscopic studies of *N*-alkyl-*N*-trialkylsilylmethylaminyl radicals in solution. J Chem Soc Perkin Trans 2; 1087-1095.

Sakurai H, Murakami M, Kumada M. (1969) Stereochemical course of the reaction of an optically active hydrosilane with carbon tetrachloride by a free-radical mechanism. Retention of configuration. J Am Chem Soc; 91: 519-520.

Sakurai H, Murakami M, Kumada M. (1977) Silyl radicals. XV. Stereochemical course of chlorine abstraction by 1-methyl-4-*t*-butyl-1-silacyclohexyl radicals from polychloroalkane. Bull Chem Soc Jpn; 50: 3384-3387.

Sarasa JP, Igual J, Poblet AM. (1986) Electronic effects in the regiochemistry of the alkenylsilyl radical cyclizations. J Chem Soc Parkin Trans 2; 861-865.

Schmid N, Behr J-P. (1995) Recognition of DNA sequences by strand replacement with polyamino-oligonucleotides. Tetrahedron Lett; 36: 1447-1450.

Shinozuka K, Umeda A, Aoki T, Sawai H. (1998) Facile post-synthetic derivatization of oligodeoxynucleotide containing 5-methoxycarbonylmethyl-2'-deoxyuridinemethyl-2'deoxyuridine. Nucleosides Nucleotides;

17: 291-300.

Shuto S, Kanazaki M, Ichikawa S, Matsuda A. (1997) A novel ring-enlargement reaction of (3-oxa-2-silacyclopentyl)methyl radicals. Stereoselective introduction of a hydroxyethyl group via unusual 6-endo-cyclization products derived from 3-oxa-4-silahexenyl radicals and its application to the synthesis of a 4'α-branched nucleoside. J Org Chem; 62: 5676-5677.

Shuto S, Kanazaki M, Ichikawa S, Minakawa N, Matsuda A. (1998) Stereo- and regioselective introduction of 1- or 2-hydroxyethyl group via intramolecular radical cyclization reaction with a novel silicon-containing tether. An efficient synthesis of 4'α-branched 2' deoxyadenosines. J Org Chem; 63: 746-754.

Shuto S, Sugimoto I, Abe H, Matsuda A. (2000a) Mechanistic study of the ring-enlargement reaction of (3-oxa-2-silacyclopentyl)methyl radicals. Evidence for a pentavalent silicon-bridging radical transition state in 1,2-rearrangement reactions of β-silyl radicals. J Am Chem Soc; 122: 1343-1351.

Shuto S, Terauchi M, Yahiro Y, Abe H, Ichikawa S, Matsuda A. (2000b) Stereoselective synthesis of α- and β-*C*-glucosides via radical cyclization with an allylsilyl tether. Control of the stereoselectivity by changing the conformation of the pyranose ring. Tetrahedron Lett; 41: 4151-4155.

Shuto S, Yahiro Y, Ichikawa S, Matsuda A. (2000c) Synthesis of 3,7-anhydro-D-*glycero*-D-*ido*-octitol 1,5,6-trisphosphate as an IP$_3$ receptor ligand using a radical cyclization reaction with a vinylsilyl tether as the key step. Conformational restriction strategy using steric repulsion between adjacent bulky protecting groups on a pyranose ring. J Org Chem; 65: 5547-5557.

Sieburth SM, Fensterbank L. (1992) An intramolecular Diels-Alder reaction of vinylsilanes. J Org Chem; 57: 5279-5281.

Sommer LH, Ulland LA. (1972) Chirality and structure of organosilicon radicals. J Org Chem; 37: 3878-3881.

Spellmeyer SC, Houk KN. (1987) A force-field model for intramolecular radical additions. J Org Chem; 52: 959-974.

Stein CA, Tonkinson JL, Yakubov L. (1991) Phosphorothioate oligodeoxynucleotides-antisense Inhibitors of gene expression?, Pharmac Ther; 52: 365-384.

Stein CA, Narayanan R. (1996) Antisense oligodeoxynucleotides: internalization, compartmentalization and non-sequence specificity. Perspectives Drug Discovery Design; 4: 41-50.

Sugimoto I, Shuto S, Mori S, Shigeta S, Matsuda A. (1999a) Synthesis of 4'α-branched thymidines as a new type of antiviral agent. Bioorg Med Chem Lett; 9: 385-388.

Sugimoto I, Shuto S, Matsuda A. (1999b) Kinetics of a novel 1,2-rearrangement reaction of β-silyl radicals. The ring-expansion of (3-oxa-2-silacyclopentyl)methyl radical into 4-oxa-3-silacyclohexyl radical is irreversible. Synlett; 11: 1766-1768.

Sugimoto I, Shuto S, Matsuda A. (1999c) A one-pot method for the stereoselective introduction of a vinyl group via an atom-transfer radical-cyclization reaction with a diphenylvinylsilyl group as a temporary connecting tether. Synthesis of 4'α-C-vinylthymidine, a potent antiviral nucleoside. J Org Chem; 64: 7153-7157.

Sukeda M, Shuto S, Sugimoto I, Ichikawa S, Matsuda A. (2000) Synthesis of 2'α-branched-chain sugar pyrimidine nucleosides via radical atom-transfer cyclization with a vinylsilyl tether as the key step. J Org Chem; 65: 8988-8996.

Sund C, Puri N, Chattopadhyaya J. (1996) Synthesis of *C*-branched spermine tethered oligo-DNA and the thermal stability of the duplexes and triplexes. Tetrahedron; 52: 12275-12290.

Tabata S, Tanaka M, Matsuda A, Fukushima M, Sasaki T. (1996) Antitumor effect of a novel multifunctional antitumor nucleoside, 3'-ethynylcytidine, on human cancers. Oncology Rep; 3: 1029-1034.

Tabata S, Tanaka M, Endo Y, Obata T, Matsuda A, Sasaki T. (1997) Antitumor mechanisms of

3'-ethynyluridine and 3'-ethynylcytidine as RNA synthesis inhibitors: development and characterization of 3'-ethynyluridine-resistant cells. Cancer Lett; 116: 225-231.

Tabor CW, Tabor H. (1976) 1,4-Diaminobutane (putrescine), spermidine, and spermine. Annu Rev Biochem; 45: 285-306.

Tabor CW, Tabor H. (1984) Polyamines. Annu Rev Biochem; 53: 749-790.

Takatori S, Tsutsumi S, Hidaka M, Kanda H, Matsuda A, Fukushima M, Wataya Y. (1998) The characterization of cell death induced by 1-(3-*C*-ethynyl-β-D-*ribo*-pentofuranosyl)cytosine (ECyd) in FM3A cells. Nucleosides Nucleotides; 17: 1309-1317.

Takatori S, Kanda H, Takenaka K, Wataya Y, Matsuda A, Fukushima M, Shimamoto Y, Tanaka M, Sasaki T. (1999) Antitumor mechanisms and metabolism of novel antitumor nucleoside analogues, 1-(3-*C*-ethynyl-β-D-*ribo*-pentofuranosyl)cytosine and 1-(3-*C*-ethynyl-β-D-*ribo*-pentofuranosyl)uracil. Cancer Chemother Pharmacol; 44: 97-104.

Takenuki K, Matsuda A, Ueda T, Sasaki T, Fujii A, Yamagami K. (1988) Design, synthesis, and antineoplastic activity of 2'-deoxy-2'-methylidenecytidine. J Med Chem; 31: 1063-1064.

Tamao K, Ishida N, Tanaka T, Kumada M. (1983) Hydrogen peroxide oxidation of the silicon-carbon bond in organoalkoxysilanes. Organometallics; 2: 1694-1696.

Tanaka M, Matsuda A, Terao T, Sasaki T. (1992) Antitumor activity of a novel nucleoside, 2'-*C*-cyano-2'-deoxy-1-β-D-arabinofuranosylcytosine (CNDAC) against murine and human tumors. Cancer Lett; 64: 67-74.

Thuong NT, Helene C. (1993) Sequence-specific recognition and modification of double-helical DNA by oligonucleotides. Angew Chem Int Ed Engl; 32: 666-690.

Tidd SM, Hawley P, Warenius HM, Gibson I. (1988) Evaluation of N-ras oncogene antisense and response sequence methylphosphonate oligonucleotide analogues. Anti-Cancer Drug Des; 3: 117-127.

Tsai Y-M, Cherng C-D. (1991) Intramolecular free radical cyclizations using acylsilanes as radicalphiles. Tetrahedron Lett; 32: 3515-3518.

Tung CH, Breslauer KJ, Stein S. (1993) Polyamine-linked oligonucleotides for DNA triple helix formation. Nucleic Acids Res; 21: 5489-5494.

Ueda T, Shuto S, Inoue H. (1984) Synthesis of 2'-*C*-nitromethyl derivatives of uridine and the structure of a carbon-bridged cyclonucleoside derived therefrom. Nucleosides Nucleotides; 3: 173-182.

Ueno Y, Kumagai I, Haginoya N, Matsuda A. (1997a) Effects of 5-(*N*-aminohexyl)carbamoyl-2'-deoxyuridine on endonuclease stability and an ability of oligonucleotide to activate RNase H. Nucleic Acids Res; 25: 37770-3782.

Ueno Y, Mikawa M, Matsuda A. (1997b) Thermal and nuclease stabilities of G-quadruplexes consisting of oligodeoxynucleotides containing 5-[*N*-[2-[*N*,*N*-bis(2-aminoethyl)amino]ethyl]carbamoyl]-2'-deoxyuridine or 5-[*N*-[2-[*N*,*N*-bis(2-aminopropyl)amino]propyl]carbamoyl]-2'-deoxyuridine. Bioorg Med Chem Lett; 7: 2863-2866.

Ueno Y, Mikawa M, Matsuda A. (1998a) Synthesis and properties of oligodeoxynucleotides containing 5-[*N*-[*N*,*N*-bis(2-aminoethyl)amino]ethyl]carbamoyl]-2'-deoxyuridine and 5-[*N*-[*N*,*N*-bis(2-aminopropyl)amino]propyl]carbamoyl]-2'-deoxyuridine. Bioconjugate Chem; 9: 33-39.

Ueno Y, Nagasawa Y, Sugimoto I, Kojima N, Kanazaki M, Shuto S, Matsuda A. (1998b) Synthesis of oligodeoxynucleotides containing 4'-*C*-[2-[[*N*-(2-aminoethyl)carbamoyl]oxy]ethyl]thymidine and their thermal stability and nuclease-resistance properties. J Org Chem; 63: 1660-1667.

Uhlmann E, Peyman A. (1990) Antisense oligonucleotide: a new therapeutic principle. Chem Rev; 90: 544-584.

Usui H, Ueda T. (1986a) Synthesis of 2'-deoxy-8,2'-ethanoadenosine and 3'-deoxy-8,3'-ethanoadenosine.

Chem Pharm Bull; 34: 15-23.

Usui H, Ueda T. (1986b) Synthesis of 8,2'-methano- and 8,2'-ethanoadenosine. Chem Pharm Bull; 34: 1518-1523.

Usui H, Matsuda A, Ueda T. (1986) Synthesis of 8,2'-methanoguanosine and 9-(α-arabinofuranosyl)-8,2'-methanoguanosine. Chem Pharm Bull; 34: 1961-1967.

Walder RY, Walder JA. (1988) Role of RNase H in hybrid-arrested translation by antisense oligonucleotides. Proc Natl Acad Sci USA; 85: 5011-5015.

Wang G, Seifert WF. (1996) Synthesis and evaluation of oligodeoxynucleotides containing 4'-*C*-substituted thymidines. Tetrahedron Lett; 37: 6515-6518.

West R, Boudjouk P. (1973) Organosilyl and organogermyl nitroxides. A new radical rearrangement. J Am Chem Soc; 95: 3938-3987.

Xi Z, Agback P, Plavec J, Sandstrom A, Chattopadhyaya J. (1992) New stereocontrolled synthesis of isomeric *C*-branched-β-D-nucleosides by intramolecular free-radical cyclization-opening reactions based on temporary silicon connection. Tetrahedron; 48: 349-370.

Yahiro Y, Ichikawa S, Shuto S, Matsuda A. (1999) Synthesis of *C*-glycosides via radical cyclization reactions with a vinylsilyl tether. Control of the reaction course by a change in the conformation of the pyranose ring due to steric repulsion between adjacent bulky protecting groups. Tetrahedron Lett; 40: 5527-5531.

Yamagami K, Fujii A, Arita M, Okumoto T, Sakata S, Matsuda A, Ueda T, Sasaki, T. (1991) Antitumor activity of 2'-deoxy-2'-methylidenecytidine. A new 2'-deoxycytidine derivatives. Cancer Res; 51: 2319-2323.

SYNTHESIS OF PURINE ACYCLONUCLEOSIDES *VIA* RIBOFURANOSE-RING CLEAVAGE OF PURINE NUCLEOSIDES BY DIISOBUTYLALUMINUM HYDRIDE

KOSAKU HIROTA,

YASUNARI MONGUCHI and HIRONAO SAJIKI

Laboratory of Medicinal Chemistry, Gifu Pharmaceutical University, Mitahora-higashi, Gifu 502-8585, JAPAN

Introduction

In the search for effective, selective, and nontoxic antiviral agents, a variety of strategies have been devised to design nucleoside analogs.[1] Among them, acyclonucleosides such as acyclovir[2] and ganciclovir[3] have been developed for the treatment of certain herpes virus infections. The discovery of acyclovir has stimulated extensive research in the synthesis of new acyclonucleosides in which the carbohydrate moieties are acyclic chains mimicking the sugar portion of naturally occurring nucleosides. Most synthetic methods for the preparation of such acyclonucleosides involve the condensation of the base moiety with the appropriate side chain moiety.[4] Synthetic methods starting from commercially available nucleosides such as adenosine and guanosine have been unprecedent except for an example of oxidative cleavage of the 2',3'-*cis*-diol portion of ribonuculeosides with $NaIO_4$.[5] In the course of our study on the synthesis of 1,6-dihydropurine nucleosides using various reducing agents, we found that reduction of purine nucleosides with diisobutylaluminum hydride (DIBAL-H) caused a selective cleavage of the C-1'– O-4' bond in the ribose ring to give the corresponding 9-D-ribitylpurines. The scope and limitation of this methodology for the synthesis of 9-D-ribitylpurines and the application to the synthesis of acyclic nucleoside are described here.

Reduction of purine nucleosides with DIBAL-H to 9-D-Ribitylpurines [6,7]

The reaction of 2',3'-*O*-isopropylideneinosine (1a) with 5 equiv of DIBAL-H in anhydrous tetrahydrofuran (THF) under argon atmosphere for 24 h caused the cleavage of the C-1'– O-4' bond in its ribofuranose ring to give 9-(2,3-*O*-isopropylidene-D-ribityl)hypoxanthine (2a) in 67% isolated yield. Although numerous examples of the reductive cleavage of acetals,[8] aminals,[9] and ethers[10] by DIBAL-H have been reported, no applications of DIBAL-H to such a ribofuranose ring-cleavage of the nucleosides have been reported.[4]

Recent Advances in Nucleosides: Chemistry and Chemotherapy, Ed. by C.K. Chu. 57 — 70

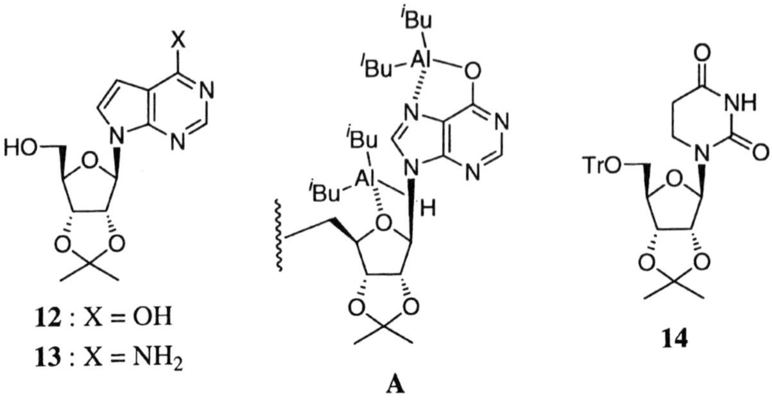

Scheme 1

Table 1.   Solvent Effect on the DIBAL-H Reduction of 2',3'-*O*-Isopropyli-deneinosine (**1a**).[a]

| | Yield (%)[b] | |
Solvent	Product **2a**	Recovery of **1a**
THF	68	24
Et_2O	35	8
CH_2Cl_2	3	14
Toluene	1	13

[a]   These reactions were carried out using 5 equiv of DIBAL-H in the stated medium under argon atmosphere at 25 °C for 24 h.

[b]   Determined by TLC scanner.

The solvent effect of the DIBAL-H reduction of **1a** was investigated (Table 1). In general, solvents possessing no oxygen atom, *e.g.*, dichloromethane and toluene, have been employed for the DIBAL-H reductions,[11] because oxygenophilicity of the aluminum atom facilitates the coordination of DIBAL-H with the oxygen atom of solvents such as diethyl ether and THF and, consequently, decreases the reductive activity of the reagent.[10b] However, employment of the ethereal solvents such as Et_2O and THF favored the present DIBAL-H reduction of purine nucleosides over the use of CH_2Cl_2 and toluene (Table 1). Upon employment of CH_2Cl_2 or toluene as a solvent, TLC analyses of the reaction mixture indicated the weakness of UV absorption spots accompanied by the over-reduction. Of two ethereal solvents in our trials, THF was found to be the most effective solvent, because it led to a smoother and cleaner conversion into 2a than Et_2O. The coordination of DIBAL-H with THF oxygen could prevent the over-reduction of 1a and result in the selective cleavage of the furanose ring. Further, stoichiometric study on the reduction of 1a with DIBAL-H in THF showed that the use of excess DIBAL-H (>5 equiv) is necessary for the conversion.

The best reaction conditions obtained for 1a were utilized to reduce a variety of 6-substituted purine nucleosides and purine 5'-deoxynucleosides; the results are summarized in Table 2. O^6-Alkyl substituted inosine derivatives 1c–1e and 6-thioinosine derivatives 1f and 1g were smoothly reduced to give the corresponding 9-D-ribitylinosine derivatives 2c-2g, respectively.

Table 2.   DIBAL-H Reductions of 6-Substituted Purine Nucleosides and Purine 5'-Deoxynucleosides.[a]

DIBAL-H (5 equiv) in THF, 25 °C, 24 h

1a-n → 2a-n

No.	X	Y	Product 2	Recovery 1
a	OH	OH	67 (68)[c]	23 (24)[c]
b	OH	Cl	82	9
c	OCH_3	OH	64	ND[d]
d	OCH_3	Br	57	ND[d]
e	$OCH(CH_3)_2$	OH	65	ND[d]
f	SH	OH	61	6
g	SCH_3	OH	41	8
h	NH_2	OH	40 (44)[c]	43 (47)[c]
			59[e]	24[e]
i	NH_2	H	29	71
j	NH_2	Cl	26	58
k	$NHCH_3$	OH	14 (14)[c]	75 (81)[c]
l	$N(CH_3)_2$	OH	2 (trace)[c]	54 (55)[c]
m	CH_3	OH	38	ND[d]
n	Ph	OH	28	29

Yield (%)[b]

[a]   These reactions were carried out using 5 equiv of DIBAL-H in THF under argon at 25 °C for 24 h, unless otherwise noted. [b]Isolated yield. [c]The yields in parenthesis were determined by TLC scanner (Shimadzu CS-9000). [d]Not determined. [e]The reaction was performed in the presence of HMPA (3 equiv).

On the other hand, similar treatment of N^6-methylated adenosines 1k and 1l resulted in marked decrease of the yields of the reduction products 2k and 2l, respectively, although the adenosine 1h was reduced in moderate yield. Tsuda *et al.* reported[12] that a remarkable change in the reducing reactivity of DIBAL-H was brought about by addition of hexamethylphosphoric triamide (HMPA). Employment of HMPA in the reduction of the adenosine (1h) led to a sensible improvement of the yield (2h, 40%→59%). The reduction of 6-methylpurine riboside 1m and 6-phenylpurine riboside 1n which have no heteroatom at the 6-position in the base moiety also gave the ribityl products 2m and 2n. The DIBAL-H reduction of 5'-deoxynucleosides 1b, 1d, 1i and 1j eventually afforded 9-D-ribityl derivatives 2b, 2d, 2i and 2j as well as that of 5'-hydroxy derivatives 1a, 1c and 1h.

In the case of the DIBAL-H reduction of non-protected adenosine (3), the significant decrease of the yield of the ribityl derivative 4 was observed (4, 6% and recovery of 3, 93%). On the contrary, treatment of 2'-deoxyadenosine 5 with DIBAL-H gave the ribityl derivative 6 in 39% yield together with recovery of 6 (55%) in analogy with the result of the DIBAL-H reduction using 1h (see Table 2). The reduction of adenosine N_1-oxide (7) proceeded with some complexity to give the corresponding ribityl derivative 8 (28%), its deoxy product 2h (5%), 6-amino-4-(2,3-*O*-isopropylidene-D-ribofuranosylamino)-5-(*N*-methylamino)-pyrimidine N_1-oxide (9, 3%), 1h (trace) and the starting material 3 (33%).

Biopterin and riboflavin are biologically interesting and naturally occurring compounds. Biosyntheses of these compounds are accompanied with two cleavages at the imidazole and ribose rings of guanosine 5'-triphosphate.[13] Therefore, the reductive cleavage of the ribofuranose moiety of 2',3'-*O*-isopropylideneguanosine (10) using DIBAL-H can be considered as a kind of bio-mimetic chemical reaction. Treatment of 10 with 6 equiv of DIBAL-H gave a ribityl derivative 11 in 38% yield. This compound 11 is also expected to be an excellent intermediate for the preparation of antiviral acycloguanosine.

7-Deazainosine and 7-deazaadenosine derivatives 12 and 13 were, however, hardly reduced with DIBAL-H to result in the recovery of the starting materials. These results indicate that electron density or/and structure of the purine base exerts a great influence on the reductive cleavage of the ribose moiety. The coordination of DIBAL-H with a substrate as depicted in Figure 1 seems to be important for the selective cleavage.

The present reduction was examined for the synthesis of pyrimidine acyclonucleosides. When 2',3'-*O*-isopropylidene-5'-*O*-trityluridine was allowed to react with DIBAL-H, the base moiety was reduced in preference to the sugar moiety to afford the corresponding 5,6-dihydrouridine derivative 14 in 48% yield instead of the expected 9-D-ribityl derivative. So far, the synthesis of 5,6-dihydrouridines has been performed by a catalytic hydrogenation of uridine derivatives over rhodium on alumina.[14] The reduction of 2',3'-*O*-isopropylidenecytidine gave a complex mixture and the 1-D-ribitylcytosine derivative could not be isolated. Therefore, the present reduction seems to be inapplicable to the synthesis of pyrimidine acyclonucleosides.

In conclusion, the treatment of a variety of purine nucleosides with DIBAL-H in THF caused the selective cleavage of the C-1'– O-4' bond in the ribose moiety to give 9-D-ribitylpurine derivatives. The reactivity in the reduction was fairly affected by the purine

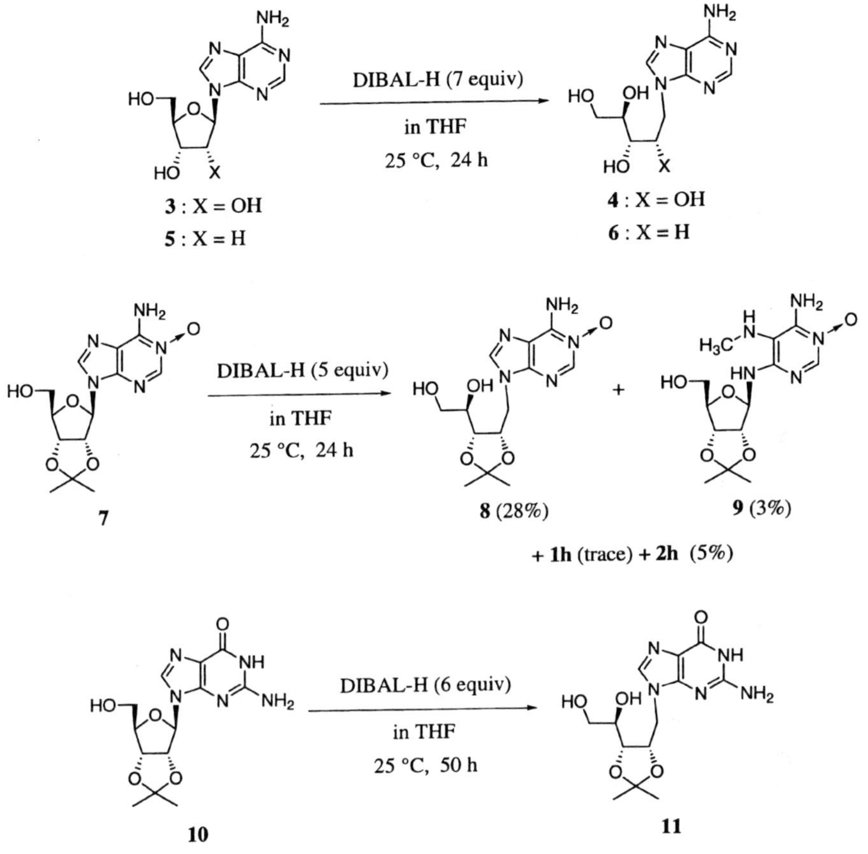

DIBAL-H (7 equiv)
in THF
25 °C, 24 h

3 : X = OH
5 : X = H

4 : X = OH
6 : X = H

DIBAL-H (5 equiv)
in THF
25 °C, 24 h

7

8 (28%)

9 (3%)

+ **1h** (trace) + **2h** (5%)

DIBAL-H (6 equiv)
in THF
25 °C, 50 h

10

11

Scheme 2.

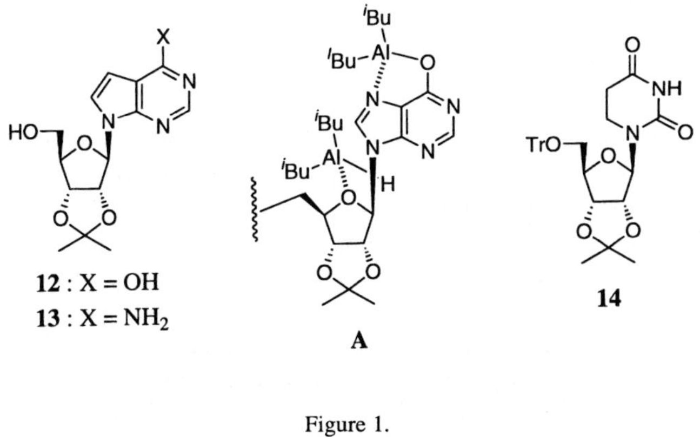

12 : X = OH
13 : X = NH$_2$

A

14

Figure 1.

base moiety. This methodology using the reductive cleavage of purine nucleosides with DIBAL-H was shown to be useful for the synthesis of 9-D-ribitylpurines.[15]

Synthesis of purine acyclonucleosides possessing a chiral 9-Hydroxyalkyl group by sugar modification of 9-D-Ribitylpurines [16,17]

Several purine acyclonucleosides having chiral carbons in the N_9-hydroxyalkyl chain such as D-eritadenine[18,19] and buciclovir[20] have been shown to possess antiviral activity. The 9-D-ribitylpurines 2h and 11, easily prepared as described above, were utilized to synthesize asymmetric 9-(2,3,4-trihydroxybut-1-yl)purines 17a,b and 19a,b and eritadenine analog 21 as potential antiviral agents by taking advantage of two chiral carbons of the ribityl moiety. Our strategy involves two disconnections of the C-1'–O-4' and C-4'–C-5' bonds of purine nucleosides as depicted in Figure 2.

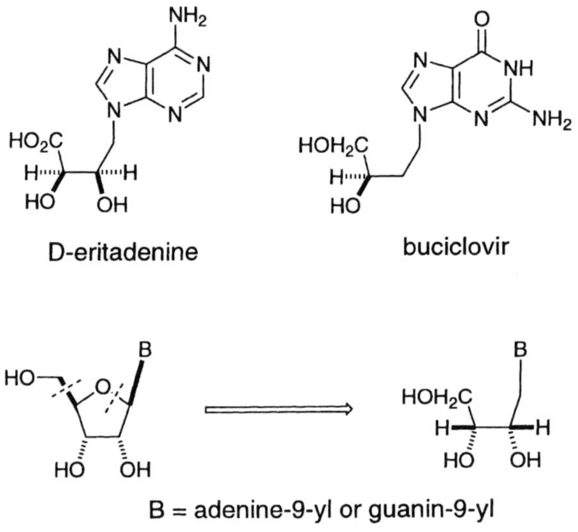

D-eritadenine

buciclovir

B = adenine-9-yl or guanin-9-yl

Figure 2.

First, (2S,3S)-4-(adenin-9-yl)-2,3-dihydroxy-2,3-O-isopropylidenebutanal (15a) was synthesized as a chiral key intermediate for the preparation of acycloadenosines. Oxidation of 2h with $NaIO_4$ afforded 15a in 92% yield. Reduction of 15a with $NaBH_4$ afforded the primary alcohol 16a in 78% yield. The stereochemistry of 16a was confirmed by the conversion into the corresponding (R)-(+)-α-methoxy-α-(trifluoromethyl)phenylacetate (MTPA ester).[21] ^{19}F NMR analyses of the ester showed no formation of any detectable epimeric isomer. This fact evidently indicates that the formation of 15a and 16a proceeds with complete retention of steric configuration. The deprotection of 16a afforded 9-[(2S,3R)-2,3,4-trihydroxybut-1-yl]adenine (17a), quantitatively. On the other hand, an epimer 18a was synthesized by the inversion at the 2-position of the aldehyde 15a and the subsequent reduction according to the known method.[22] Thus, treatment of 15a with NaOMe in MeOH followed by the reduction with $NaBH_4$ gave a mixture of 18a and 16a (18a : 16a = 94 : 6). Both epimers could be separated by silica gel column. Enantiomeric purity of 18a was confirmed by the conversion to MTPA ester and its ^{19}F NMR analysis. Deprotection of 18a afforded 9-[(2*S*,3*S*)-2,3,4-trihydroxybut-1-yl]adenine (19a) in 60% yield.

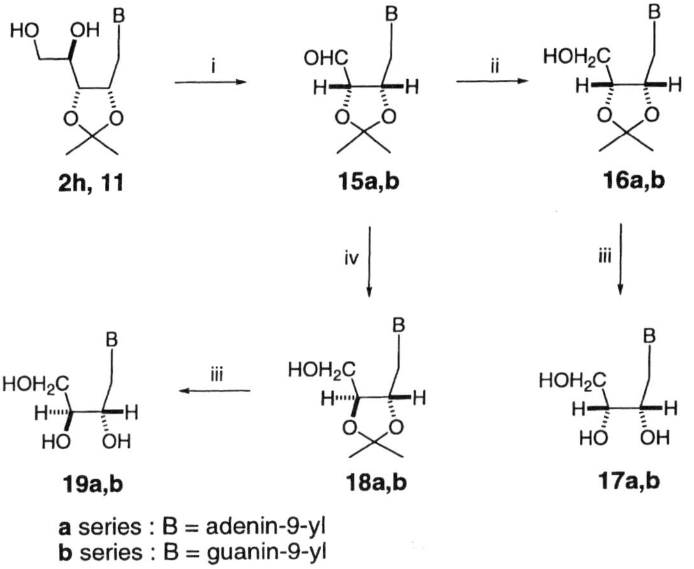

2h, 11   →(i)   15a,b   →(ii)   16a,b

15a,b →(iv) 18a,b

16a,b →(iii) 17a,b

19a,b ←(iii) 18a,b

a series : B = adenin-9-yl
b series : B = guanin-9-yl

Reagents and conditions: i, for **15a**, aq. $NaIO_4$; for **15b**, $NaIO_4$, AcOH-AcONa buffer (pH 4); ii, aq. $NaBH_4$, pH 7-8; iii, 80 % AcOH; iv, for **18a**, NaOMe, MeOH, then aq. $NaBH_4$; for compound **18b**, K_2CO_3, MeOH; then aq. $NaBH_4$

Scheme 3.

This methodology was applied to the synthesis of acycloguanosines. When oxidation of 9-D-ribitylguanine 11 and subsequent reduction were conducted under analogous reaction conditions, a diastereomeric mixture of 16b (*erythro*) and its (3'*S*)-isomer 18b (*threo*) was formed in the ratio of 83 : 17. Therefore, the $NaIO_4$ oxidation of 7 in a sodium acetate buffer (pH 4) gave the aldehyde 15b as a hydrate in 87% yield and subsequent reduction of 15b led to the quantitative formation of a chiral alcohol 16b without epimerization. Deprotection of 16b afforded 9-[(2*S*,3*R*)-2,3,4-trihydroxybut-1-yl]guanine (17b) in 93% yield (Scheme 3). The preparation of a (3'*S*)-epimer 19b was conducted in the modified method described for that of 19a. Thus, treatment of 15b with K_2CO_3 in MeOH gave (2*R*)-epimer of 15b as a diastereomeric mixture with 15b (*threo* : *erythro* = >95 : <5) in 75% yield and subsequent $NaBH_4$ reduction gave the corresponding alcohol 18b. Deprotection of 18b afforded the desired 9-[(2*S*,3*S*)-2,3,4-trihydroxybut-1-yl]guanine (19b) as the sole isomer in 88% yield. It is noteworthy that the stereochemistry at the 3'-position of acyclonucleosides 17 and 19 could be easily controlled by the use of the aldehyde 15 as a chiral pool.

The aldehyde 15a was utilized as a novel approach for the synthesis of L-eritadenine (21), which is an enantiomer of naturally occurring D-eritadenine.[18,19] Votruba and Holy have reported that L-eritadenine, which is the most effective, next to D-eritadenine, of the four stereoisomeric eritadenines, inhibits *S*-adenosyl-L-homocysteine hydrolase.[19a] The epimerization of D-eritadenine methyl ester under basic conditions has been described in the literature.[23] Therefore, the Pt/C catalyzed oxidation of 15a with O_2 was carried out under neutral conditions to afford acid 20 in 46% yield without any detectable epimer. Deprotection of 20 with 10% AcOH gave L-eritadenine (21), quantitatively, which was identical with that reported by Holy and coworkers.[19b]

Another synthetic application of 15a as a chiral intermediate was examined for the synthesis of tetraol 23 which has the sole chiral center in the alkyl chain. Crossed aldol condensation of 15a and formaldehyde in basic media followed by the $NaBH_4$ reduction afforded isopropylidene protected tetraol 22 in 89% yield by one-pot procedure. The desired tetraol 23 was obtained by treatment of 22 with CF_3CO_2H in 83% yield.

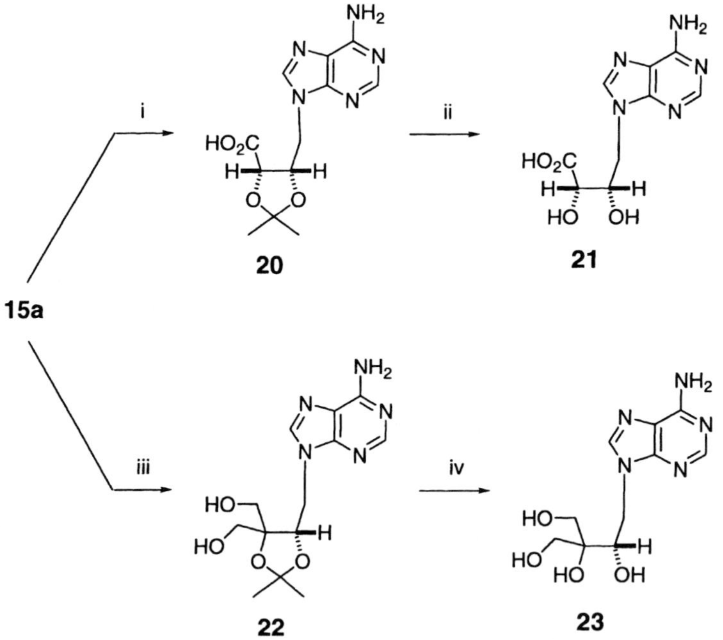

Reagents and conditions: i, Pt/C, O_2, water; ii, 10 % AcOH; iii, HCHO, 2 M NaOH, water; then $NaBH_4$; iv, CF_3CO_2H

Scheme 4.

Among acyclic analogues of adenosine, 17a, 19a, 21, and 23 were virtually inactive against influenza A, respiratory syncytial virus, human immunodeficiency virus, herpes simplex virus type 1, and human cytomegarovirus with EC_{50} values of >40 μg/ml

This methodology using 2',3'-O-isopropylidene protected 9-D-ribitylpurines 2h and 11 as chiral starting materials was shown to be widely applicable to the synthesis of biologically interesting acyclonucleosides. Especially, the aldehydes 15a and 15b are useful intermediates for the preparation of purine acyclonucleosides mimicking ribonucleosides

Synthesis of acyclic adenosines with an unsaturated side chain by modification of 9-D-Ribityladenine [7,24,25]

S-Adenosyl-L-homocysteine (AdoHcy) hydrolase, which catalyses the hydrolysis of AdoHcy to adenosine and L-homocysteine, has been recognized as an attractive target for the development of antiviral agents.[26] This enzyme plays an important role in regulating the S-adenosyl-L-methionine-dependent transmethylation reaction which is involved in the maturation of viral mRNA. Naturally occurring adenosine analogs, D-eritadenine[19] and neplanocin A (NPA),[27] have exhibited antiviral activities through the strong and irreversible inhibition of this enzyme. Here, the 9-D-ribityladenine 2h was also applied to the synthesis of an acyclic NPA analog B (n = 1, R = H). On the other hand, 4'-dehydroxymethyl-NPA (DHCA) has been shown to be a more selective inhibitor of AdoHcy hydrolase than NPA, because lack of the 4'-hydroxymethyl group causes the substrate inactivity for adenosine kinase.[28] Therefore, we also synthesized the acyclic DHCA analogs B (n = 0, R = H, COOR'), though the carboxylic acid B (R = CO_2H) can be also regarded as a vinylog of L-eritadenine.

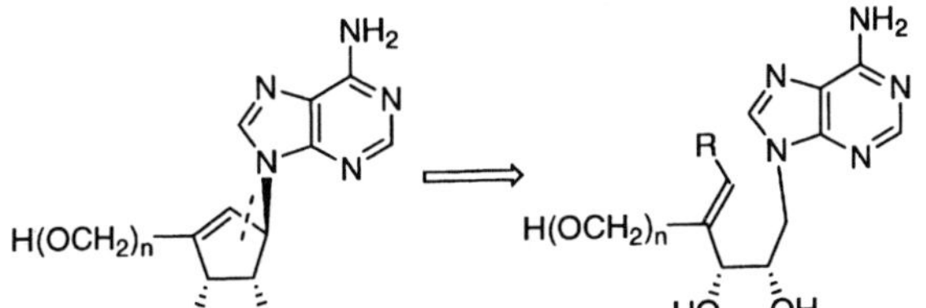

Figure 3.

Thus, *tert*-butyldimethylsilyl (TBDMS) protection at the 5'-position of 9-D-ribityladenine 2h gave 9-(5-O-*tert*-butyldimetylsilyl-2,3-O-isopropylidene-D-ribityl)adenine (24) in 78% yield. Oxidation of 24 with chromic acid afforded the 4'-keto derivative 25 in 52% yield. Wittig reaction of 25 with Ph_3PCH_3Br/ BuLi and subsequent deprotection of the resulting 4'-methylene derivative 26 resulted in the formation of the desired acycloneplanocin A (27).

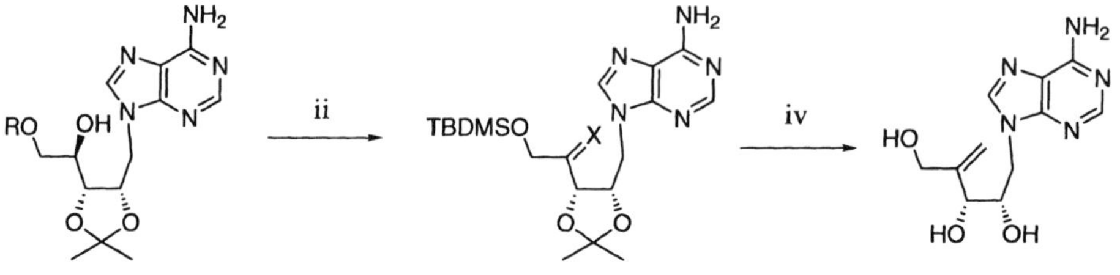

Reagents and conditions: i) *tert*-butyldimetylsilyl chloride (5 equiv), imidazole (10 equiv), in DMF, r.t., 5 min, 78%; ii) CrO₃ (4 equiv), pyridine (8 equiv), Ac₂O (4 equiv), in CH₂Cl₂, r.t., 6 h, 52%; iii) Ph₃PCH₃Br (10 equiv), BuLi (8.3 equiv), in THF, 0 °C, overnight, 69%; iv) 80% AcOH, 60 °C, 6 h, 79%.

Scheme 5.

Although acycloneplanocin A (27) indicated faint inhibitory activity toward AdoHcy hydrolase (rabbit erythrocyte) with IC_{50} values of 350 μM, 27 was virtually inactive against herpes simplex virus type 1 (HSV-1), influenza virus, and human cytomegarovirus (HCMV) with EC_{50} values of >50 μg/mL.

The 1',5'-*seco*-type of DHCA 31 had been synthesized from D-ribonolactone as a chiral pool by Jäger and coworkers.[29] However, their method included non-regioselective condensation of adenine with a side chain after a multi-step procedure, and biological evaluation of 31 was not described. In our first attempt to synthesize 31, the Wittig methylenation of the aldehyde 15a with Ph₃PCH₃Br/BuLi gave a diastereomeric mixture of (2'S,3'R)-*erythro*-isomer 28 and (2'S,3'S)-*threo*-isomer 29 in 16% yield (28 : 29 = 77 : 23).[30] However, the two products could not be separated by column chromatography. In order to obtain 28 as a single diastereomer in high yield, the dideoxygenation at the 4', 5'-position of 2h was investigated alternatively. Among numerous studies on the conversion of 1,2-diols into olefins, we adopted Lerner's method[31] for the preparation of 31. *O*-Mesylation of 2h followed by the treatment with sodium iodide afforded 28 *via* the 4',5'-*O*-dimesylate 30 in good yield. Deprotection of 28 by heating in 80% aqueous AcOH gave the target product 31 in 82% yield.

The Wittig reaction of the aldehyde 15a with Ph₃P=CHCO₂Et at room temperature afforded a mixture of the α,β-unsaturated esters [32 (*E*) and 33 (*Z*)] with the ratio of 67 : 33 in 72% yield, whereas the reaction under reflux resulted in the predominant formation of the (*E*)-isomer 32 in the ratio of 86 : 14. The mixture itself was employed for the preparation of the desired eritadenine vinylog 36 because the respective products could not be isolated. When the mixture was treated under basic conditions and subsequently

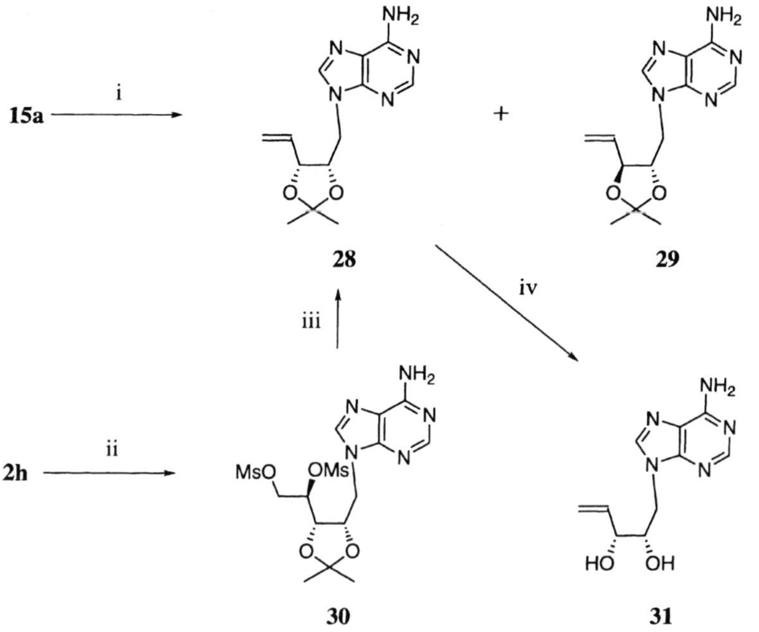

15a —i→ **28** + **29**

2h —ii→ **30**   **31**

Reagents and conditions: i) Ph₃PCH₃Br, BuLi, THF, 0 °C–r. t.; ii) MsCl, pyridine, 0 °C–r. t.; iii) NaI, acetone, reflux; iv) 80% AcOH, 70 °C.

Scheme 6.

acidified with 1N HCl to pH 3–4, the obtained product was not the expected carboxylic acid **36**, but a γ-keto-acid **34** with retention of the 2'-configuration in 79% yield. The keto-acid **34**, which could form *via* an isomerization of the olefins into enolate intermediates under the basic conditions, is of interest in relation to a structural analogy to the 3'-keto-intermediate C proposed for the AdoHcy hydrolase-catalyzed reaction mechanism.[32] On the other hand, the mixture (**32** : **33** = 86 : 14) was treated with trifluoroacetic acid prior to base–treatment to afford the 2',3'-deprotected (*E*)-isomer **35** derived from **32** in 83% yield. Another (*Z*)-isomer derived from **33** was not isolated. Saponification of **35** with LiOH gave an α,β-unsaturated carboxylic acid **36** in 89% yield. Furthermore, Pd/C-catalyzed hydrogenation of **36** furnished the two-carbon elongated L-eritadenine (**37**) in 82% yield.

The synthesized compounds **31**, **34**, and **37** showed no significant activities against influenza A, respiratory syncytial virus, human immunodeficiency virus, herpes simplex virus type 1, and human cytomegarovirus.

In conclusion, 9-(2,3-*O*-isopropylidene-D-ribityl)adenine (**2h**)[33] and (2*S*,3*S*)-4-(adenin-9-yl)- 2,3-dihydroxy-2,3-*O*-isopropylidenebutanal (**15a**) are versatile chiral precursors for the synthesis of biologically interesting acyclic adenosine analogs.

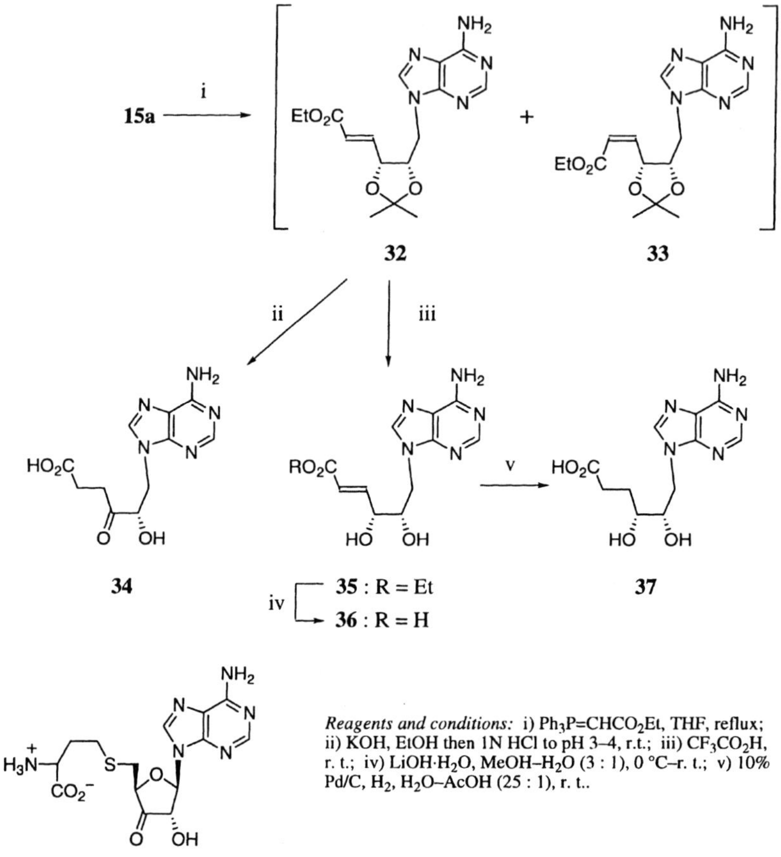

Scheme 7.

Reagents and conditions: i) $Ph_3P=CHCO_2Et$, THF, reflux; ii) KOH, EtOH then 1N HCl to pH 3–4, r.t.; iii) CF_3CO_2H, r. t.; iv) $LiOH \cdot H_2O$, MeOH–H_2O (3 : 1), 0 °C–r. t.; v) 10% Pd/C, H_2, H_2O–AcOH (25 : 1), r. t..

References and notes

1.   (a) *Antiviral Drug Development: A Multidisciplinary Approach*; De Clercq, E.; Walker, R. T., Eds.; Plenum Press: New York, 1988. (b) *Advances in Antiviral Drug Design*; De Clercq, E., Ed.; JAI Press Inc.: Greenwich, Vol.1 (1993), Vol.2 (1996), Vol.3 (1999).

2.   (a) Elion, G. B.; Furman, P. A.; Fyfe, J. A.; De Miranda, P.; Beauchamp, L.; Schaeffer, H. J. *Proc. Natl. Acad. Sci. U.S.A.* 1977, *74*, 5716–5720. (b) Schaeffer, H. J.; Beauchamp, L.; De Miranda, P.; Elion, G. B.; Bauer, D. J.; Collins, P. *Nature* (London) 1978, *272*, 583–585.

3.   (a) Martin, J. C.; Dvorak, C. A.; Smee, D. F.; Matthews, T. R.; Verheyden, J. P. H. *J. Med. Chem.* 1983, *26*, 759–761. (b) Ogilvie, K. K.; Cheriyan, U. O.; Radatus, B. K.; Smith, K. O.; Galloway, K. S.; Kennell, W. L. *Can. J. Chem.* 1982, *60*, 3005–3010. (c) Ashton, W. T.; Karkas, J. D.; Field, A. K.; Tolman, R. L. *Biochem.*

Biophys. Res. Commun. 1982, *108*, 1716–1721. (d) Smith, K. O.; Galloway, K. S.; Kennell, W. L.; Ogilvie, K. K.; Radatus, B. K. *Antimicrob. Ag. Chemoth.* 1982, *22*, 55–61.

4. (a) Chu, C. K.; Cutler, S. J. *J. Heterocycl. Chem.* 1986, *23*, 289–319. (b) El Ashry, E. S. H.; El Kilany, Y. In *Advances in Heterocyclic chemistry*; Katritzky, A. R., Ed.; Academic Press: San Diego, Vol. 67(1997), pp 391–438 and Vol. 68 (1997), pp 1-88.

5. (a) Nemec, J; Rhoades, J. M. *Nucleosides Nucleotides* 1983, *2*, 99–112. (b) Lerner, L. M. *Carbohydr. Res.* 1984, *127*, 141–145. (c) Birnbaum, G. I.; Stolarski, R.; Kazimierczuk, Z; Shugar, D. *Can. J. Chem.* 1985, *63*, 1215–1221. (d) Bessodes, M.; Antonakis, K. *Tetrahedron Lett.* 1985, *26*, 1305–1306. (e) Mikhailov, S. N.; Florentiev, V. L.; Pfleiderer, W. *Synthesis*, 1985, 399–400. (f) McGee, D. P. C.; Martin, J. C. *Can. J. Chem.* 1986, *64*, 1885–1889. (g) Beaton, G.; Jones, S.; Walker, R. T. *Tetrahedron* 1988, *44*, 6419–6428.

6. Kitade, Y.; Hirota, K.; Maki, Y. *Tetrahedron Lett.* 1993, *34*, 4835–4836.

7. Hirota, K.; Monguchi, Y.; Kitade, Y.; Sajiki, H. *Tetrahedron* 1997, *53*, 16683–16698.

8. (a) Mori, A.; Fujiwara, J.; Maruoka, K.; Yamamoto, H. *Tetrahedron Lett.* 1983, *24*, 4581–4584. (b) Ishihara, K.; Mori, A.; Arai, I.; Yamamoto, H. *ibid.* 1986, *27*, 983–986. (c) Mori, A.; Ishihara, K; Arai, I.; Yamamoto, H. *Tetrahedron* 1987, *43*, 755–764. (d) Takano, S.; Akiyama, M.; Sato, S.; Ogasawara, K. *Chem. Lett.* 1983, 1593–1596. (e) Mikami, T.; Asano, H.; Mitsunobu, O. *ibid.* 1987, 2033–2036. (f) Kotsuki, H.; Ushio, Y.; Kadota, I.; Ochi, M. *ibid.* 1988, 927–930. (g) Ishihara, K.; Mori, A.; Yamamoto, H. *Tetrahedron Lett.* 1987, *28*, 6613–6616. (h) Ishihara, K.; Mori, A.; Yamamoto, H. *Tetrahedron* 1990, *46*, 4595–4612.

9. Yamamoto, H.; Maruoka, K. *J. Am. Chem. Soc.* 1981, *103*, 4186–4194.

10. (a) Pino, P.; Lorenzi, G. P. *J. Org. Chem.* 1966, *31*, 329–331. (b) Winterfeldt, E. *Synthesis* 1975, 617–630. (c) Hilscher, J. C. (Schering AG) German Patent 2,409,991, 1975; *Chem. Abstr.* 1976, *84*, 567.

11. Suzuki, T.; Saimoto, H.; Tomioka, H.; Oshima, K.; Nozaki, H. *Tetrahedron Lett.* 1982, *23*, 3597–3600.

12. Tsuda, T.; Hayashi, T.; Satomi, H.; Kawamoto, T.; Saegusa, T. *J. Org. Chem.* 1986, *51*, 537–540.

13. (a) Burg, A. W.; Brown, G. M. *J. Biol. Chem.* 1968, *243*, 2349–2358. (b) Plowman, J.; Cone, J. E.; Guroff, G. *ibid.* 1974, *249*, 5559–5564. (c) Mitsuda, H.; Nakajima, K.; Yamada, Y. *ibid.* 1978, *253*, 2238–2243.

14. (a) Skaric, V.; Gaspert, B.; Hohnjec, M., *J. Chem. Soc. (C)* 1970, 2444–2447. (b) Hanze, A. R. *J. Am. Chem. Soc.* 1967, *89*, 6720–6725.

15. In general, 9-D-ribitylpurines have been synthesized by the multistep method; (a) Davoll, J.; Evans, D. D. *J. Chem. Soc.* 1960, 5041–5049. (b) Ross, D. L.; Skinner, C. G.; Shive, W. *J. Org. Chem.* 1961, *26*, 3582–3583.

16. Hirota, K.; Monguchi, Y.; Sajiki, H.; Kitade, Y., *Synlett* 1997, 697–698.

17. Hirota, K.; Monguchi, Y.; Sajiki, H.; Sako, M.; Kitade, Y., *J. Chem. Soc., Perkin Trams. 1* 1998, 941–946.

18. (a) Rokujo, T.; Kikuchi, H.; Tensho, A.; Tsukitani, Y.; Takenawa, T.; Yoshida, K.; Kamiya, T. *Life Sciences*, 1970, *9*, 379–385. (b) Chibata, I.; Okamura, K.; Takeyama, S.; Kotera, K. *Experientia*, 1969, *25*, 1237-1238. (c) Kamiya, T.; Saito, Y.; Hashimoto, M.; Seki, H. *Tetrahedron Lett.*, 1969, 4729-4732.

19. (a) Votruba, I.; Holy, A. *Collect. Czech. Chem. Commun.* 1982, *47*, 167–172. (b) Holy, A.; Votruba, I.; De Clercq, E. *ibid.* 1982, *47*, 1392–1407. (c) Merta, A.; Votruba, I.; Vesely, J.; Holy, A. *ibid.* 1983, *48*, 2701–2708..

20. Larsson, A.; Öberg, B.; Alenius, S.; Hagberg, C.-E.; Johansson, N.-G.; Lindborg, B.; Stening, G.

Antimicrob. Ag. Chemoth., 1983, *23*, 664-670.

21. (a) Dale, J. A.; Dull, D. L.; Mosher, H. S.; *J. Org. Chem.*, 1969, *34*, 2543-2549. (b) Ward, D. E.; Rhee, C. K. *Tetrahedron Lett.*, 1991, *32*, 7165-7166.

22. (a) Ko, S. Y.; Lee, A. W. M.; Masamune, S.; Reed III, L. A.; Sharpless, K. B.; Walker, F. J. *Tetrahedron*, 1990, *46*, 245-264. (b) Lee, A. W. M. *Magn. Reson. Chem.*, 1985, *23*, 468-469.

23. Hashimoto, M.; Saito, Y.; Seki, H.; Kamiya, T. *Tetrahedron Lett.*, 1970, *16*, 1359-1362.

24. Kitade, Y.; Monguchi, Y.; Hirota, K.; Maki, Y. *Tetrahedron Lett.* 1993, *34*, 6579–6580.

25. Hirota, K.; Monguchi, Y.; Sajiki, H.; Yatome, C; Hiraoka, A.; Kitade, Y., *Nucleosides Nucleotides* 1998, 1333–1345.

26. (a) De Clercq, E. *Biochem. Pharmacol.* 1987, *36*, 2567–2575. (b) Wolfe, M. S.; Borchardt, R. T. *J. Med. Chem.* 1991, *34*, 1521–1530. (c) Yuan, C. S.; Liu, S.; Wnuk, S. F.; Robins, M. J.; Borchardt, R. T. In *Advances in Antiviral Drug Design*; De Clercq, E., Ed.; JAI Press Inc.: Greenwich, 1996; Vol. 2, pp. 41–88.

27. (a) Borchardt, R. T.; Keller, B. T.; Patel-Thombre, U. *J. Biol. Chem.* 1984, *259*, 4353–4358. (b) De Clercq, E. *Antimicrob. Agents Chemother.* 1985, *28*, 84–89.

28. (a) Borcherding, D. R.; Scholtz, S. A.; Borchardt, R. T. *J. Org. Chem.* 1987, *52*, 5457–5461. (b) Hasobe, M.; Mckee, J. G.; Borcherding, D. R.; Borchardt, R. T. *Antimicrob. Agents Chemother.* 1987, *31*, 1849–1851. (c) Narayanan, S. R.; Keller, B. T.; Borcherding, D. R.; Scholtz, S. A.; Borchardt, R. T. *J. Med. Chem.* 1988, *31*, 500–503.

29. (a) Hümmer, W.; Gracza, T.; Jäger, V. *Tetrahedron Lett.* 1989, *30*, 1517–1520. (b) Jäger, V.; Hümmer, W.; Stahl, U.; Gracza, T. *Synthesis*, 1991, 769–776.

30. Lee *et al.* have reported that the easy epimerization of 2,3-*erythro*-aldose acetonide to 2,3-*threo*-aldose acetonide was observed under the basic conditions:[22b]

31. Lerner, L. M. *J. Org. Chem.* 1972, *37*, 470–473.

32. (a) Palmer, J. L.; Abeles, R. H. *J. Biol. Chem.* 1976, *251*, 5817–5819. (b) Palmer, J. L.; Abeles, R. H. *ibid.* 1979, *254*, 1217–1226.

33. Recently, Matsuda *et al.* have utilized the 9-ribityladenine 2 for the synthesis of neplanocin A: Niizuma, S.; Shuto, S.; Matsuda, A. *Tetrahedron* 1997, *53*, 13621–13632.

THE CHEMISTRY OF NUCLEOSIDE AND DINUCLEOTIDE INHIBITORS OF INOSINE MONOPHOSPHATE DEHYDROGENASE (IMPDH)[1]

KRZYSZTOF W. PANKIEWICZ[1] and BARRY M. GOLDSTEIN[2]

[1] *Pharmasset, Inc., 1860 Montreal Road, Tucker, Atlanta, GA 30084, USA.*
[2] *University of Rochester Medical Center,*
601 Elmwood Avenue, Box 712, Rochester, NY 14642

Introduction

At the beginning of the 1980s a vigorous program of synthesis of C-nucleoside analogues of natural N-nucleosides was well advanced at the Memorial Sloan-Kettering Cancer Center. Fox's group had synthesized (Chu, C.K. *et al.*, 1977 and Matsuda, A. *et al.*, 1981) a number of pyrimidine- and purine-C-nucleosides that showed interesting biological activity. For example, ψ-isocytidine (Matsuda, A. *et al.*, 1981) exhibited potent activity against several mouse leukemias resistant to arabinofuranosylcytidine (araC), and 9-deazaadenosine was found to be extremely cytotoxic (Lim M-L. and Klein, R.S., 1981). A convenient method for conversion of ribo-C-nucleosides into the corresponding 2'-deoxy-C-nucleosides was developed (Pankiewicz, K.W., 1982) and the preparation of the last representative of this series, 2'-deoxy-9-deazaguanosine, has been recently published (Gibson, E.S., 1999).

At the same time a novel thiazole-C-nucleoside, tiazofurin (2-β-D-ribofuranosyl-thiazole-4-carboxamide, TR) was synthesized as a potential antiviral agent (Fuertes, M. *et al.*, 1976, Srivastava, P.C. *et al.*, 1977). It was found later that TR showed potent anticancer activity (Robins, R.K., 1982). Tiazofurin is not active as the nucleoside but is uniquely metabolized into an active metabolite, thiazole-4-carboxamide adenine dinucleotide (TAD), an analogue of the cofactor nicotinamide adenine dinucleotide (NAD) (Gebeyehu, G. *et al.*, 1985).

The first step in the biosynthesis of TAD is the phosphorylation of TR by adenosine kinase to give tiazofurin monophosphate (TRMP). TRMP is then coupled with AMP by NMN-adenylyl transferase to give TAD. The replacement of the nicotinamide nucleoside moiety of NAD by tiazofurin afforded a cofactor analogue that cannot participate in hydride transfer. TAD is therefore a potent and specific inhibitor of NAD-dependent inosine monophosphate dehydrogenase (IMPDH), an important target in cancer chemotherapy.

[1.] NAD analogues 18. For part 17 see: Pankiewicz, K. W.; Malinowski, K.; Jayaram, H. N.; Lesiak-Watanabe. K.; Watanabe, K. A. Novel mycophenolic adenine bis(phosphonate)s as potential immunosuppressants. *Curr. Med. Chem.*1999, 6, 629-634.

Recent Advances in Nucleosides: Chemistry and Chemotherapy, Ed. by C.K. Chu. 71 — 90

TAD is not stable as a pyrophosphate. It is cleaved to AMP and TRMP by cellular phosphodiesterases, including specific "TADase". These nucleotides are further degraded to the corresponding nucleosides. Resistance to TR is associated both with decreased anabolic activity by NMN-adenylyl transferase and with increased degradation of TAD by TADase (Jayaram, H.N., 1985).

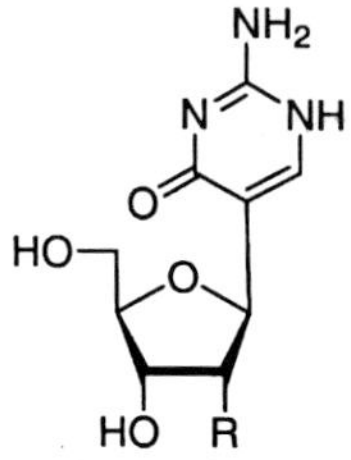

ψ-Isocytidine, R = OH
2'-Deoxy-ψ-isocytidine, R = H

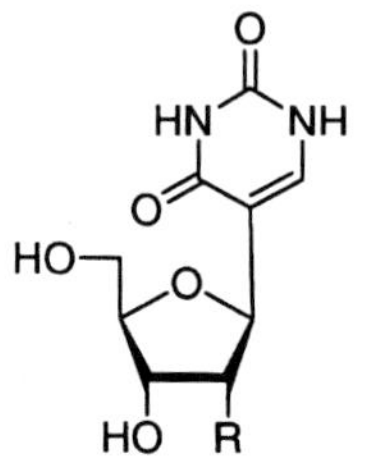

ψ-Uridine, R = OH
2'-Deoxy-ψ-uridine, R = H

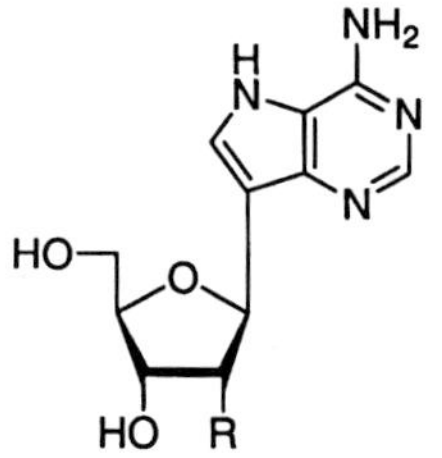

9-Deazaadenosine, R = OH
2'-Deoxy-9-deazaadenosine, R = H

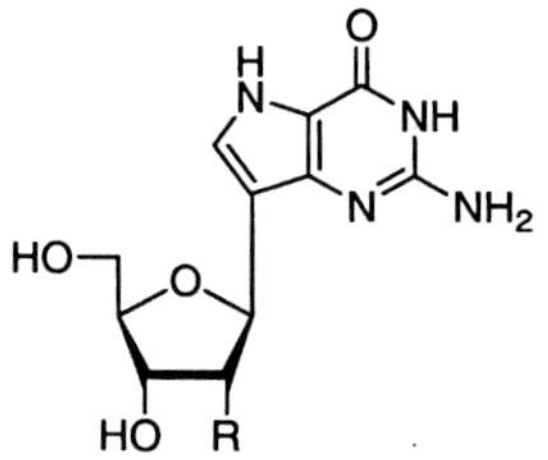

9-Deazaguanosine, R = OH
2'-Deoxy-9-deazaguanosine, R = H

Figure 1.

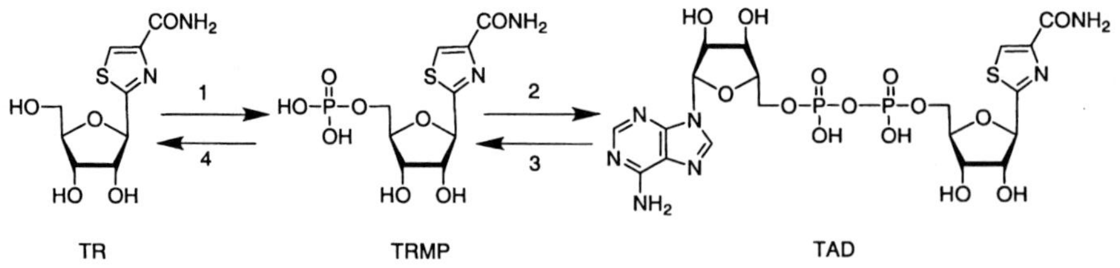

1. Adenosine kinase, 2. NMN - adenylyl transferase, 3. TADase, phosphodiesterase, 4. Phosphatases

Scheme 1.

The chemistry of nucleoside inhibitors

The development of TR prompted us to synthesize pyridine C-nucleosides, such as C-nicotinamide riboside (C-NR), C-picolinamide riboside (C-PR), and C-isonicotinamide riboside (C-IR). These synthetic compounds are isosteric to nicotinamide riboside (NR). We expected that like tiazofurin, these C-nucleosides might be converted into the corresponding NAD analogues, offering an alternative means of IMPDH inhibition.

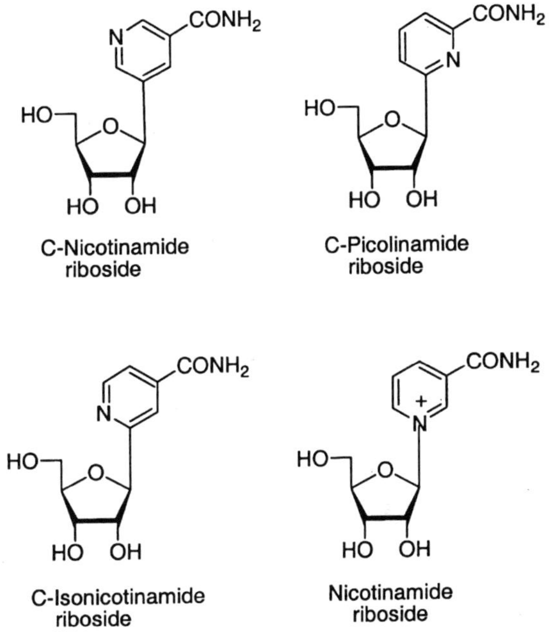

C-Nicotinamide
riboside

C-Picolinamide
riboside

C-Isonicotinamide
riboside

Nicotinamide
riboside

Figure 2.

In our first synthetic approach we used 2,4:3,5-di-O-benzylidene-D-aldehydoribose (1, Scheme 2) as the starting material for condensation with lithiated pyridines (Kabat, M.M. *et al.*, 1987; Kabat, M.M. *et al.*, 1988). Thus, reaction of 1 with 3-bromo-5-lithiopyridine afforded a mixture of allo/altro (45:55) isomers (2a/2b) of the bromopyridine derivative in 57% yield. This mixture was further converted into a mixture of the corresponding nicotinamides 4a and 4b by lithiation, then carboxylation, followed by esterification with CH_2N_2 to the methyl esters 3a and 3b, which were treated with methanolic ammonia. All attempts at debenzylidenation of 4 resulted in the formation of the pentahydroxy derivative instead of the ribofuranosyl structures.

K. W. Pankiewicz and B. M. Goldstein

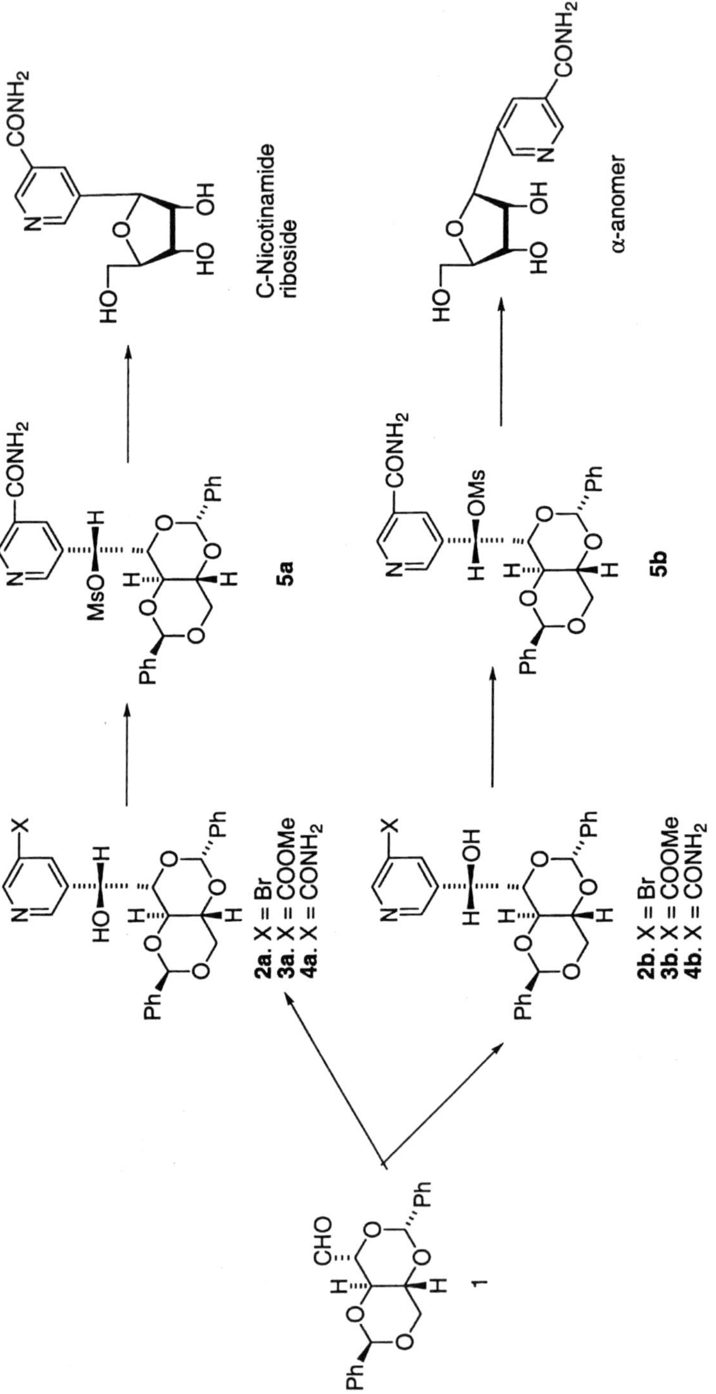

Scheme 2..

However, mesylation of 4 to gave a mixture of 5a and 5b, which upon acid hydrolysis afforded the desired mixture of C-nicotinamide riboside and its α-anomer. Separation of these α,β-anomers was readily achieved on a silica gel column.

In the similar manner, reaction of 1 with 2-bromo-6-lithiopyridine gave a mixture of allo/altro isomers 6a and 6b (Scheme 3), which in this case were separated on a silica gel column (19% and 23% yield, respectively). Altro isomer 6a was then converted into the desired C-picolinamide riboside as described above.

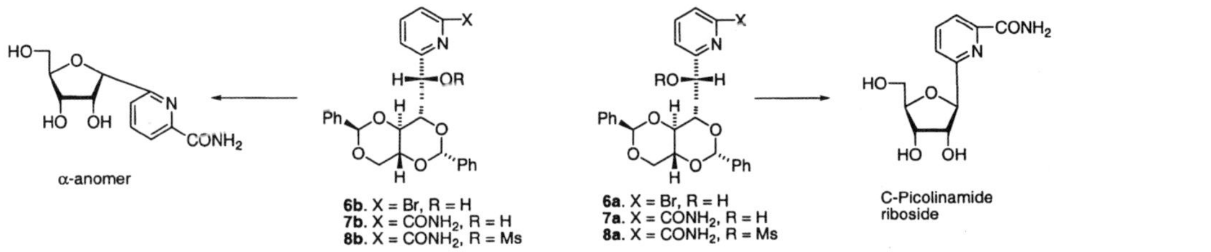

α-anomer

6b. X = Br, R = H
7b. X = $CONH_2$, R = H
8b. X = $CONH_2$, R = Ms

6a. X = Br, R = H
7a. X = $CONH_2$, R = H
8a. X = $CONH_2$, R = Ms

C-Picolinamide
riboside

Scheme 3.

The low yield of the above procedures, and difficulties in separation of the anomeric mixture of C-nicotinamide nucleosides, prompted us to develop an alternative synthesis (Pankiewicz, K.W. *et al.*, 1988). We used commercially available D-ribonolactone for condensation with lithiated pyridines. Thus, D-ribonolactone was protected with the tetrahydropyranyl- and isopropylidene- groups to give derivative 9 (Scheme 4, R = THP), which, under reaction with 3-cyano-5-lithiopyridine afforded nucleoside 10 containing an anomeric hydroxyl group. All our attempts at removal of this group from 10 failed. We found, however, that reduction with $NaBH_4$ gave a mixture of allo/altro isomers. These were separated on a silica gel column to give 11a and 11b in 38% and 35% yield, respectively. After selective removal of the THP group from the altro isomer 11a the product 12a was treated with acetone and pTsOH to give compound 13a in which only the anomeric hydroxyl group is free. Mesylation of 13a followed by treatment with $CF_3COOH/CHCl_3$ afforded the desired C-nicotinamide riboside. In a similar manner the α-anomer was prepared from allo isomer 11b. Later we modified this procedure using, as a starting material, 5-O-(tert-butyldimethylsilyl) -2,3-O-isopropylidene-D-ribonolactone (9, R = tBDMSi) or 2,3,5-tri-O-benzyl-D-ribono-lactone. However, in both cases we were not able to remove the anomeric hydroxyl group from the corresponding condensation products by chlorination, sulfonylation or direct reduction with Et_3SiH.

In the 1990s the synthesis of both ribo- and deoxyribo-C-nucleosides containing aromatic bases received considerable attention. Picicrilli *et al.* (1991) synthesized 3-(β-D-ribofuranosyl)-2,6-diaminopyridine (17) by condensation of 9 (R = tBDMSi) with 2,6-dichloro-3-lithiopyridine, removal of the anomeric hydroxyl group with $Et_3SiH/BF_3(OEt_2)$ and conversion of the 2,6-chloro derivative 16 into the di-amino nucleoside 17.

CONH2
C-Nicotinamide
riboside
α-Isomer
OMs
14a
MsO
14b
CN
OH
H
RO
R1O
11a. R = THP, R1 = H
12a. R = R1 = H
13a. R - R1 = C(CH3)2
CN
H
HO
RO
R1O
11b. R = THP, R1 = H
12b. R = R1 = H
13b. R - R1 = C(CH3)2
OH
10
CONH2
O
9
RO

Scheme 4.

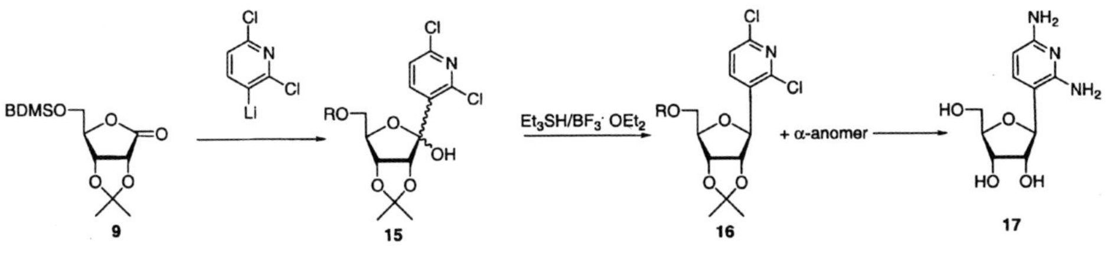
BDMSO
Li
RO
Et3SH/BF3· OEt2
+ α-anomer
HO
9
15
16
17

Scheme 5.

In a similar manner 3-(β-D-ribofuranosyl)-2,6-diaminopyrimidine (Figure 3, 18) was prepared (Piccirilli, J.A. *et al.*, 1990) and converted into the 2'-deoxy-C-nucleoside 19 by the method developed by us (Pankiewicz, K.W. *et al.*, 1982). Compounds 18 and 19 were later incorporated into the synthetic RNA or DNA template and incubated with RNA- or DNA-polymerases to demonstrate the ability of the polymerases to incorporate xantosine triphosphate, an "artificial complementary nucleotide", into the growing chain. This indicated the possibility of an extension of the genetic code (Piccirilli, J.A. *et al.*, 1990).

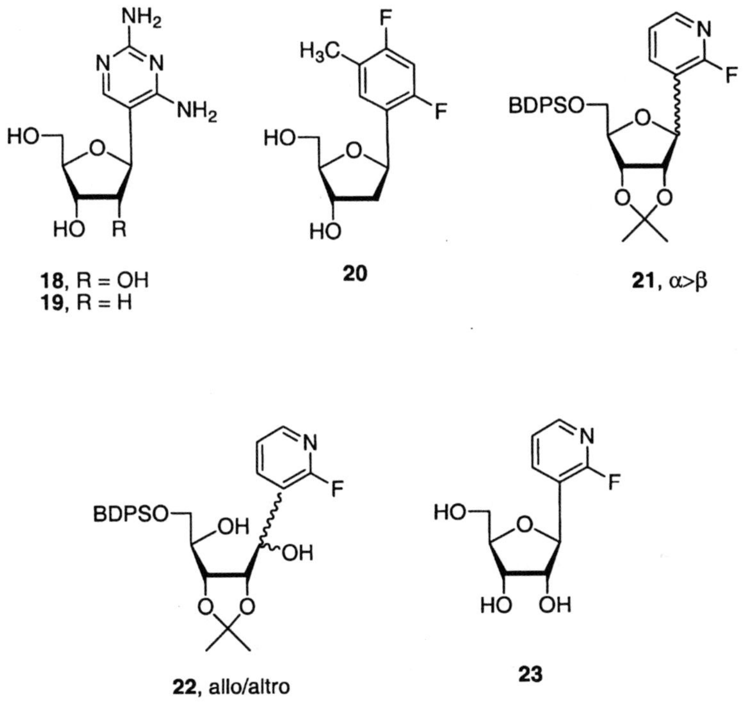
18, R = OH
19, R = H
20
21, α>β
22, allo/altro
23

Figure 3.

Moran, S. *et al.* (1997) reported the synthesis of aromatic 2'-deoxy-C-nucleosides, such as 5-(β-D-ribofuranosyl)-2,4-difluorotoluene (20), a isostere for thymidine, and showed that 20 codes specifically and efficiently for adenine in DNA replication. Using other aromatic nucleosides they demonstrated that shape complementarity is as important in replication as base-base hydrogen bonds (Morales, J.C. and Kool, E.T., 1998). Adamic and Beilgelman (1997a, 1997b) prepared a number of aromatic C-nucleosides and incorporated some of them into the catalytic domain of hammerhead ribozymes (Matulic-Adamic, J. *et al.*, 1996 and Matulic-Adamic, J. and Beigelman, L., 1996). Such ribozymes retained the steric relationship of natural bases without ability or with reduced ability to form hydrogen bonds. They reported use of 5-O-(*tert*-butyldiphenylsilyl)-2,3-O-isopropylidene-D-ribonolactone (9, R = tBDPSi) as an alternative starting material that is more stable to acid. Removal of the anomeric hydroxyl group was usually accomplished by treatment with Et_3SiH/Lewis acid or by the 1'-OH acetylation followed by reductive deacetoxylation with Et_3SiH/TMSOTf. When, however, the ratio of α/β anomers of the product was shifted markedly towards the undesired α-anomer, as in case of 2-fluoropyridine derivative 21, this anomeric mixture was reduced with $NaBH_4$ to generate a 1:1 mixture of allo and altro isomers 22, according to our procedure (Pankiewicz, K.W. *et al.*, 1988). These isomers were efficiently cyclized under Mitsunobu conditions (DEAD/Ph_3P/THF) to give, after separation and deprotection, the desired nucleoside 23 (Mtulic-Adamic, J. and Beilgelman, L., 1997a). Such intramolecular Mitsunobu cyclization was earlier reported by Yokoyama (Yokoyama, M. *et al.*, 1995) and Hurusawa *et al.* (1996) and later "rediscovered" by Benhinda *et al.* (1999).

The synthesis of C-isonicotinamide riboside was accomplished by Joos *et al.* (1991) by coupling 2-bromo-4-(4,4-dimethyloxazolin-2-yl)pyridine with 2,4:3,5-di-O-benzylidene-D-aldehydoribose (Scheme 6, 1) according to our original procedure (Kabat, M.M. *et al.*, 1987). The authors used oxazolidine group (Meyers, A.I. *et al.*, 1974) as protection for 4-carboxyl group of the 2-bromopyridine for reaction with 1 and obtained a mixture of allo/altro isomers 24a and 24b in high yield. Mesylation followed by acidic hydrolysis (CF_3COOH/H_2O) resulted in closure of the ribofuranose ring and opening of the oxazoline ring to give 26a and 26b, which upon treatment with saturated NH_3/CH_3OH in a sealed vessel afforded a mixture of α,β-anomers of C-isonicotinamide riboside. The desired β-anomer was isolated by HPLC.

In contrast to our expectations, the three C-nucleoside isosteres of nicotinamide riboside, C-NR, C-PR, and C-IR, showed only weak inhibitory activity against L1210, P-815, HL-60, CCRF-CEM, MOLT/4F, and MT-4 (Kabat, M.M. *et al.*, 1987; Kabat, M.M. *et al.*, 1988; Joos, P.E. *et al.*, 1991). Thus, these studies indicate that either isosteric compounds are not efficiently converted into their corresponding NAD analogues by cellular enzymes or NAD analogues containing C-NR, C-PR, or C-IR serve as a weak inhibitors of IMPDH.

The chemistry of dinucleotide inhibitors

In order to address this question we synthesized two NAD analogues, 5-(β-D-ribofuranosyl)nicotinamide(5',5")adenosine pyrophosphate and 6-(β-D-ribofuranosyl)-

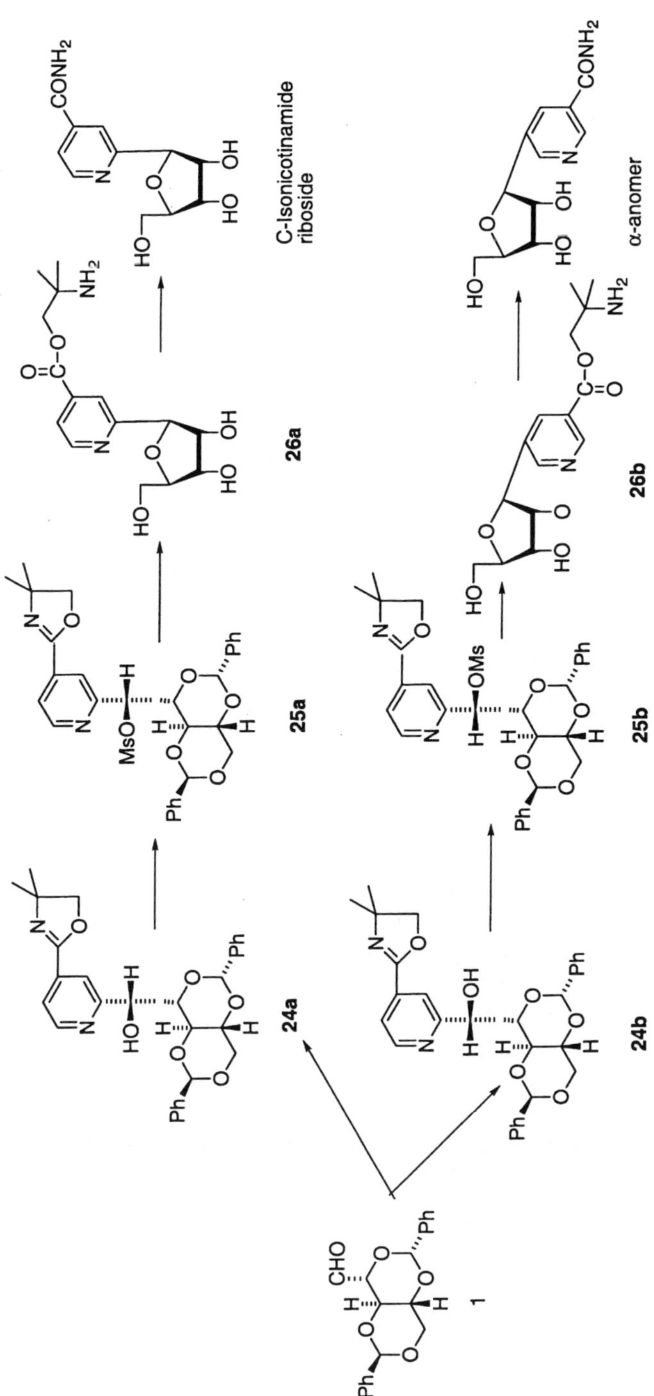

Scheme 6.

picolinamide(5',5")adenosine pyrophosphate (C-NAD and C-PAD) (Pankiewicz, K.W. *et al.*, 1993) C-NR and C-PR were phosphorylated according to Yashikawa's procedure to give mononucleotides that were activated with carbonyldiimidazole (CDI) and coupled with AMP. The desired C-NAD and C-PAD were obtained in 60% and 50% yields, respectively.

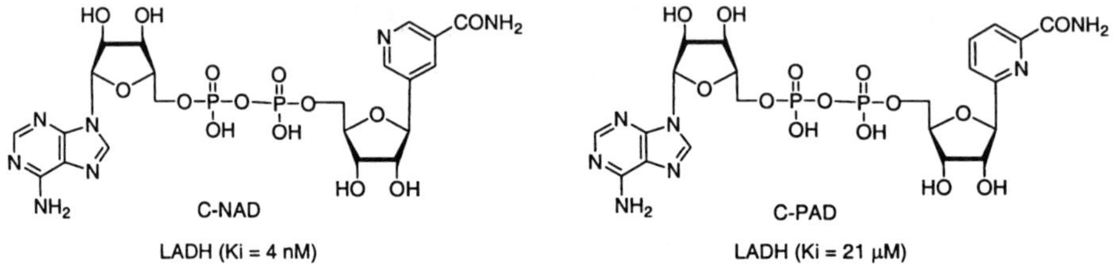

Figure 4.

C-NAD was found to be a general dehydrogenase inhibitor. It showed reasonable inhibition of IMPDH, with an IC_{50} of 20 µM against partially purified IMPDH from L1210 cell extract (Goldstein, B.M. *et al.*, 1994). C-NAD also showed a comparable degree of competitive inhibition with respect to NAD against purified bovine glutamate dehydrogenase (K_i = 15 µM), and a similar pattern of somewhat weaker inhibition against pig heart lactate dehydrogenase (K_i = 188 µM) and cytoplasmic malate dehydrogenase (K_i = 410 µM) (Goldstein, B.M. *et al.*, 1994).

Surprisingly, C-NAD demonstrated a very high degree of specificity against horse liver alcohol dehydrogenase (LADH), showing competitive inhibition with respect to NAD with K_i of 4 nM (Pankiewicz, K.W. *et al.*, 1993; Goldstein, B.M. *et al.*, 1994). In contrast, C-PAD bound LADH five orders of magnitude less tightly, with K_i of 21 µM with respect to NAD. Thus, LADH binding affinity was shown to be exquisitely sensitive to the position of the dinucleotide pyridine nitrogen. The origin of this dependence became apparent upon examination of the crystal structures of ternary complexes of LADH with ethanol and C-PAD and C-NAD.

The crystal structure of C-PAD-bound LADH showed that C-PAD binds to the active site of the enzyme in a conformation virtually identical to that of the native cofactor, NAD (Li, H. *et al.*, 1994). In this position, the C-PAD pyridine ring sits adjacent to substrate ethanol, which in turn forms the fourth coordination ligand to the LADH active-site tetracoordinate Zn cation (Figure 5, left). In LADH bound C-NAD, placement of the nitrogen on the opposite side of the C-NAD pyridine ring allows this pyridine nitrogen to displace ethanol as the fourth Zn ligand (Figure 5, right). Formation of this Zn-C-NAD complex is sufficient to account for the large increase in binding affinity of C-NAD over C-PAD (Li, H. *et al.*, 1994). However, even in this case, the conformation of LADH-bound C-NAD is very similar to that adopted by the native cofactor. Thus, both agents are clearly able to function as isosteric NAD analogues.

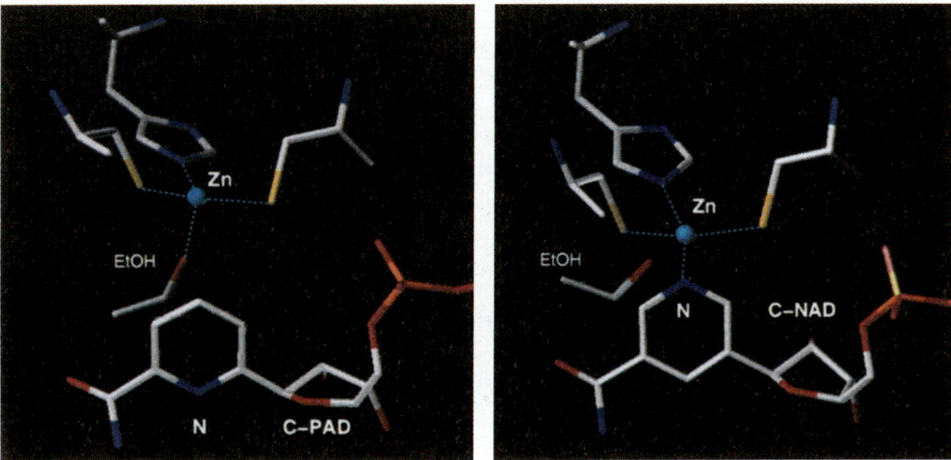

Figure 5. Left. C-PAD binding to LADH. C-PAD binds in the same position as the natural cofactor NAD. The Zn cation (cyan) forms three coordination ligands with two cysteine sulfurs and an imidazole nitrogen (dotted bonds). The fourth ligand is occupied by substrate ethanol (EtOH). Right. C- NAD binding to LADH. Placement of the pyridine nitrogen N to the opposite side of the heterocycle allows this nitrogen to displace EtOH as the fourth Zn coordination ligand. White: carbon, blue: nitrogen, red: oxygen, yellow: sulfur, orange: phosphorus.

The next member of the class of NAD isosteres utilized benzamide riboside instead of nicotinamide riboside. Benzamide is a potent inhibitor of poly(ADP-ribosylation), however, it cannot be used in cancer treatment due to neurological side effects. Benzamide riboside (BR) was synthesized by Krohn et. al (1992) as a potential PARP inhibitor with reduced toxicity. It was found, instead, that in sharp contrast to the C-pyridine ribosides, BR was effectively converted in the cell into the NAD analogue, benzamide adenine dinucleotide (BAD), a potent inhibitor of IMPDH (Jayaram, H.N. *et al.*, 1992; Gharahbengi, K. *et al.*, 1994). Later, a series of studies by Jayaram *et al.* (1999) showed that metabolism of BR is similar to that of tiazofurin. However, BR showed an even more attractive spectrum of anticancer activity than TR indicating, that BR might be a better alternative to tiazofurin in cancer treatment. It was demonstrated that in contrast to tiazofurin, BR induces apoptosis in human ovarian carcinoma N.1 cells (Grush, M. *et al.*, 1999). The synthesis of BR was accomplished by condensation of 3-lithio benzamide protected as oxazoline derivative (27) with 2,3,5-O-benzyl-D-ribonolactone (28) followed be removal of the anomeric hydroxyl group from the condensation product 29 with Et_3SiH and deprotection (Scheme 7). Using BR as a starting material we synthesized BAD by coupling of the 2',3'-O-isopropylidene-benzamide riboside-5'-phosphoimidazolide with 2',3'-O-isopropylideneadenosine-5'-monophosphate followed by deisopropylidenation (Zatorski, A. et al., 1996) We found that indeed BAD was a potent inhibitor of IMPDH with $IC_{50} = 0.8$ μM.

The above studies indicate that metabolic activation of tiazofurin and benzamide riboside into TAD and BAD, respectively, is unique. It is possible that a number of

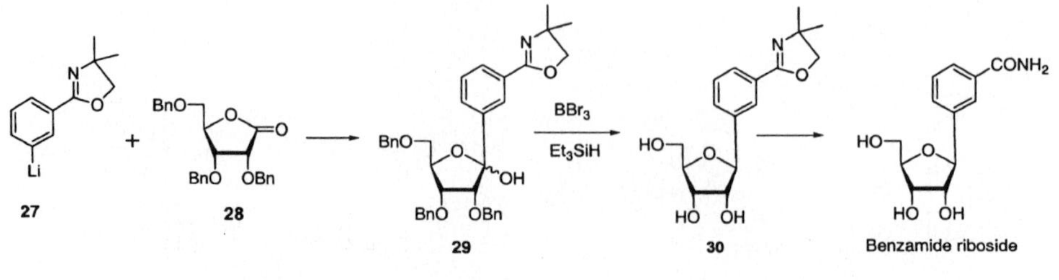

Scheme 7.

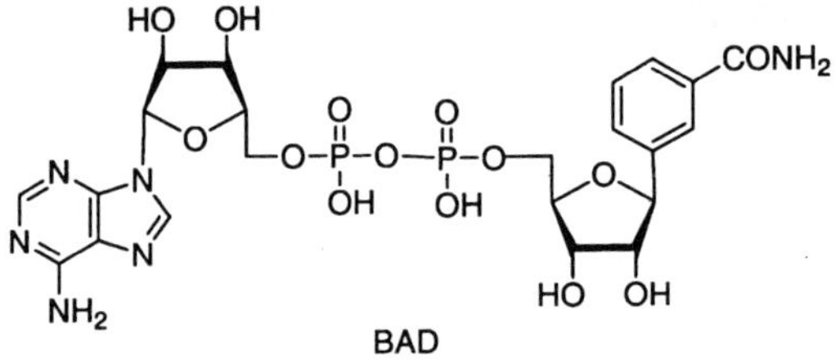

Figure 6.

nucleoside mimics of nicotinamide riboside are not converted in the cell into their corresponding NAD analogues. However, these NAD analogues may show potent inhibitory activity against IMPDH. Therefore, synthetic NAD analogues may be of therapeutic interest. Unfortunately, dinucleotides like TAD and BAD are limited as potential drugs. They are metabolically unstable and do not enter the cell intact. They are quickly degraded by cellular phosphodiesterases, including the TADase expressed in tiazofurin-resistant cell lines. Simple replacement of the pyrophosphate oxygen of TAD and BAD by a -CH$_2$-, or -CF$_2$- group to give the bis(phosphonate) analogues changes this relation dramatically. It was found that the bis(phosphonate)s of TAD (Marquez, V.E. *et al.*, 1986) and BAD (Pankiewicz, K.W. *et al.*, 1997b) are as active inhibitors of IMPDH as the parent dinucleotides and, in contrast to the pyrophosphates, can penetrate cell membranes. In addition, bis(phosphonate) analogues have a built-in resistance to degradation enzymes, and consequently are active in tiazofurin resistant cell lines. For these reasons, we decided to explore the properties of bis(phosphonate) NAD analogues as potential anti-cancer agents or immunosuppressants.

Potential drug development of such compounds requires an efficient synthesis in order to supply enough material for biological testing and animal studies. P^1, P^2-disubstituted bis(phosphonate)s were usually prepared in milligram amounts by

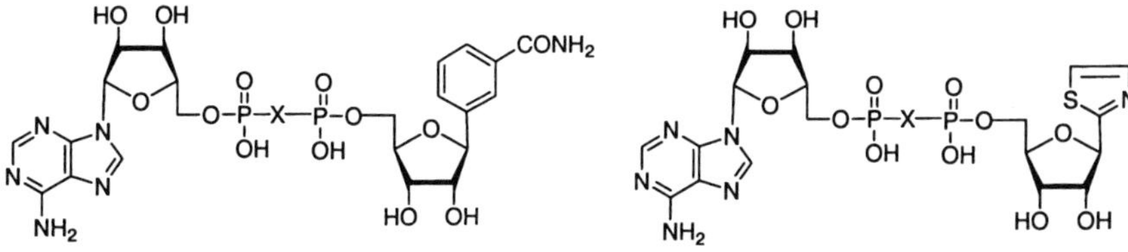

Bis(phosphonate) analogues of BAD

X = CH_2 or CF_2

Bis(phosphonate) analogues of TAD

X = CH_2 or CF_2

Figure 7.

dicyclohexylcarbodiimide (DCC) or diisopropylcarbodiimide (DIC) coupling of mono-substituted bis(phosphonate)s, such as adenosine 5'-methylenebis(phosphonate) with nucleosides such as 2',3'-O-isopropylidene-benzamide riboside. It had been generally accepted that the reaction went through the formation of an amidine intermediate, which was subsequently displaced by the nucleophilic 5'-hydroxyl group of the incoming nucleoside to give the desired product after deprotection (scheme 8). We studied this reaction and found that the first product to be formed is P^1, P^4-diadenosine tetraphosphonate (32, Scheme 9). Further dehydration results in the formation of a cyclic derivative 33 and then a bicyclic intermediate 34. (Pankiewicz, K.W. *et al.*, 1997a). The reaction mixture at this stage shows multisignal resonances in ^{31}P NMR. Since all phosphorus atoms in the structure of 34 are chiral, up to 16 diastereomers can exist. However, the bridgehead phosphorus atoms P^2, and P^3 can adopt only *RR* or *SS* configuration in the rigid structure of 31. They do not fit in this position having *RS* or *SR* geometries. Therefore only 8 of 16 stereoisomers would be expected. Addition of 2',3'-O-isopropylidenebenzamide riboside at this stage causes gradual simplification of ^{31}P. NMR signals, which collapse into two narrow multiplets at δ 8-9 and δ 18-21 ppm when formation of intermediate 35 is completed. Finally, addition of water to the reaction mixture results in the ^{31}P NMR showing almost exclusively an AB system of the desired product 36. De-isopropylidenetion with Dowex 50/H^+ affords the methylene-bis(phosphonate) analogue of BAD in excellent yield.

Using a similar principle, we prepared a new reagent for the synthesis of nucleoside 5'-methylenebis(phosphonate)s (Lesiak, K. *et al.*, 1988a). 2-(4-Nitrophenyl)ethyl alcohol (37, Scheme 10) was treated with an equimolar amount of a commercially available methylenebis(phosphonyl) tetrachloride and tetrazole to give 2-(4-nitrophenyl)ethyl methylenebis(phosphonate) (38) as a major product, which was separated by preparative HPLC. Compound 38 was further converted into the corresponding intermediate 39 by dehydration with DIC. Reaction of 39 with 2',3'-O-isopropylideneadenosine afforded the desired protected derivative 40 from which the nitrophenylethyl group was removed by β-elimination with 1,8-diazabicyclo[5.4.0]undec-7-ene (DBU). Several *ribo-* and *deoxyribo*-nucleoside methylenebis-(phosphonate)s have also been prepared in a similar manner.

Scheme 8..

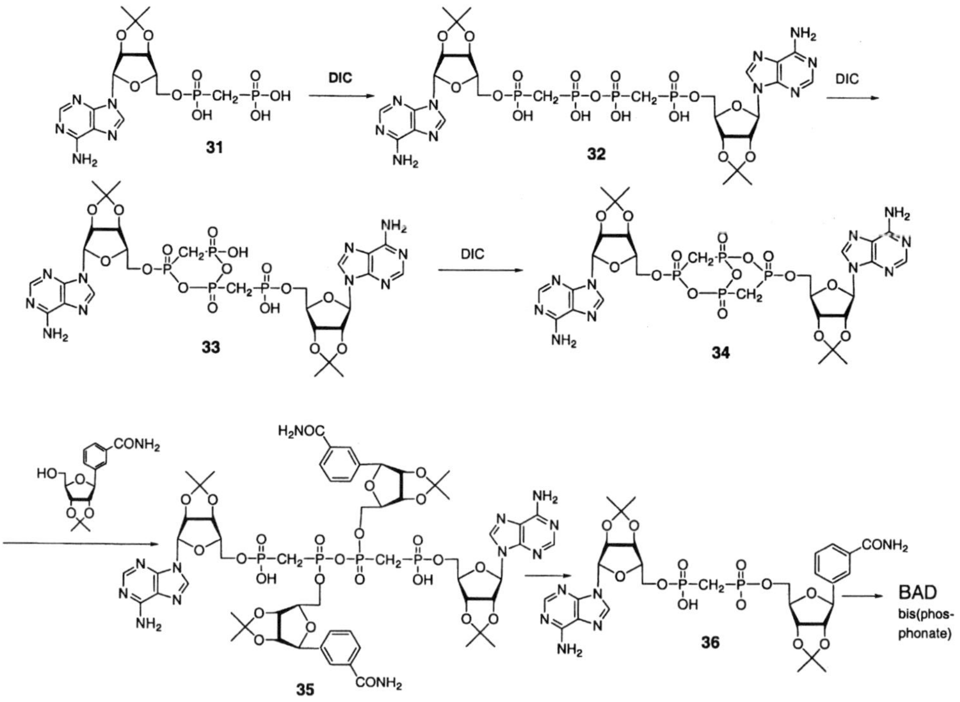

Scheme 9.

Like C-NAD and C-PAD (*vide supra*), BAD is a very close isostere of NAD. The conformation of LADH-bound BAD is very close to that of LADH-bound NAD. Inhibition of LADH by BAD with respect to NAD is also competitive with $K_i = 6$ µM, (Pankiewicz, K.W. *et al.*, 1997b) comparable to the value of 21 µM observed for C-PAD. In IMPDH, BAD shows improved binding over the pyridine dinucleotides, with an $IC_{50} = 0.8$ and 0.9 µM for the type I and type II isoforms, respectively (Zatorski, A. *et al.*, 1996). Of perhaps greater significance is the observation that methylenebis(phosphonate)-BAD is also a potent inhibitor of IMPDH ($IC_{50} = 0.7$ and 0.9 µM for type I and type II, respectively) (Pankiewicz, K.W. *et al.*, 1997b). Thus, methylenebis(phosphonate) analogue of BAD can be added to the list of attractive phosphodiesterase-resistant drug leads, such as bis(phosphonate) analogues of TAD (Lesiak, K. et. al., 1997) and mycophenolic adenine dinucleotide, MAD (Lesiak, K. *et al.*, 1998b).

Scheme 10.

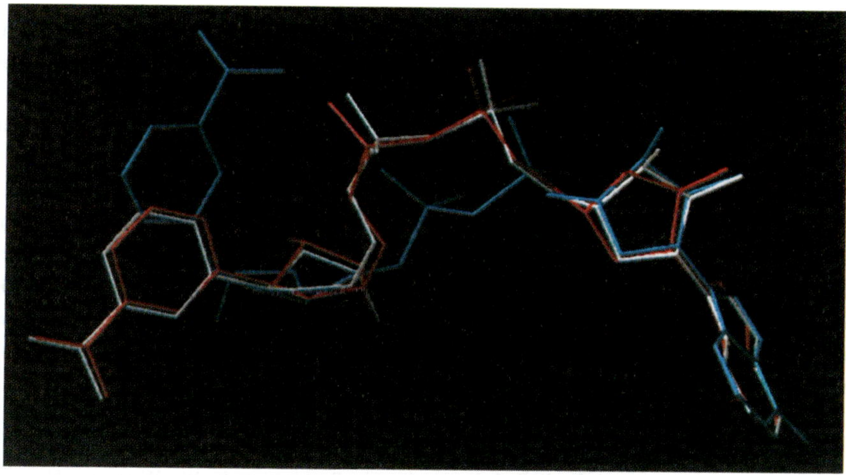

Figure 8.   Overlap of LADH-bound NAD (white), BAD (red), and methylenebis(phosphonate) analogue of BAD (cyan).

The fact that bis(phosphonate) analogues bind IMPDH with affinities comparable to their parent pyrophospates is fortuitous. The P-CH$_2$-P bridge is not a strict isoster of the pyrophosphate (P-O-P) linkage. While the crystal structure of LADH-bound BAD closely mimics that of LADH-bound NAD, the structure of LADH-bound methylenebis(phosphonate) analogue of BAD is distorted relative to that of the native cofactor. This is reflected by a ~ 50-fold weaker binding of methylenebis(phosphonate) BAD to LADH (K = 333 µM) compared to BAD (Pankiewicz, K. W. *et al.*, 1997b) The C-P bonds in the P-C-P linkage are 0.18 A longer then the analogous O-P bonds in the pyrophosphate group, and the P-C-P angle is more acute by 20°. Nevertheless, these differences do not appear to compromise binding to IMPDH. The structure of IMPDH-bound SAD (an analogue of TAD, in which the sulfur atom of the thiazole moiety is replaced by selenium atom) suggests that the phosphate binding region in IMPDH is more tolerant of distortion in this part of the dinucleotide ligand (Colby, T.D. *et al.*, 1999). Thus, P-CH$_2$-P linkage may be exploited to enhance not only phosphodiesterase resistance, but also IMPDH binding specificity among these novel inhibitors.

References

Benhida, R.; Guinvarch, D.; Fourrey, J-L.; Sun, J.S. (1999) Efficient stereoselective synthesis of new C-nucleosides via intramolecular Mitsunobu cyclization, Nucleosides & Nucleotides, 18, 603-604.

Chu, C. K.; Reichman, U.; Watanabe, K. A.; Fox, J. J. (1977) 2'-Deoxy-ψ-isocytidine, 2'-deoxy-ψ-uridine, and 2'-deoxy-1-methyl-ψ-uridine. Isosteres of deoxycytidine, deoxyuridine and thymidine, J. Hetercycl. Chem, 14, 1119.

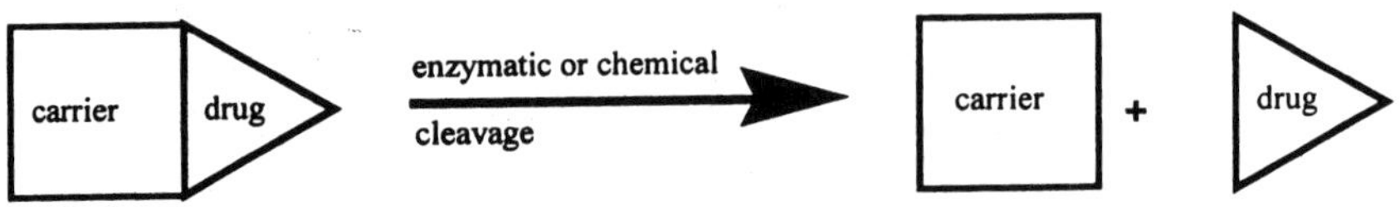

Figure 1. Bipartate Prodrug

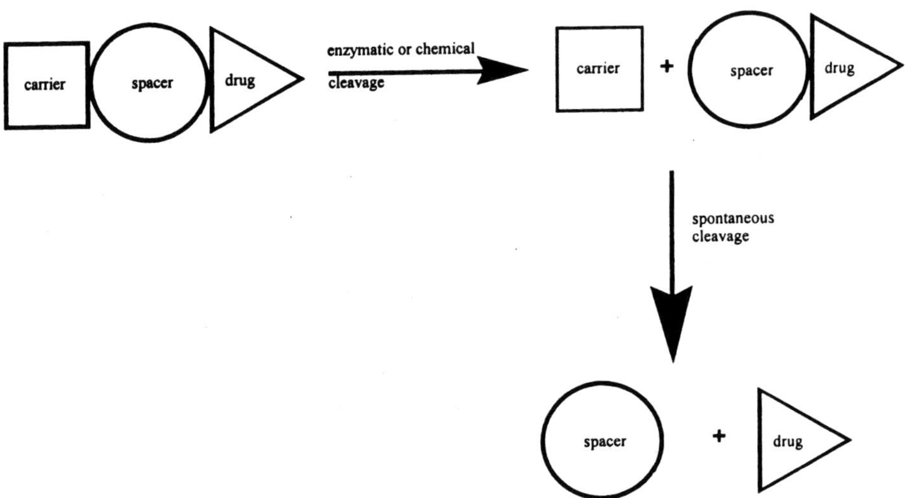

Figure 2. Tripartate Prodrug.

2.   Bipartate approach applied to nucleosides

Many investigators have synthesized derivatives of antiviral and anticancer agents in order to improve their pharmacokinetic properties (Tan *et al.*, 1999). These prodrugs are substrates of esterases or amidases that cleave the bond between carrier and nucleoside analogues. Therefore, it is critical that the prodrug be a suitable substrate for the target enzyme. Besides, when the prodrug is intended for clinical use, the rate of biotransformation to form the drug is a key factor, because it will regulate the rate of release of the parent compound.

2.1. Bipartate approach involving ester, ether and amide linkages between nucleosides and carriers

Lipophilic prodrugs of 2',3'-dideoxycytidine (ddC) such as N^4,5'-diacetyl-ddC (DAC), N^4,5'-ditrimethylacetyl-ddC (DTMAC), N^4,5'-dicyclopentylpropionyl-ddC (DCYPP) and 5'-cholesteryl-ddC (CHOL) were designed to target the brain (Figure 3) (Ibrahim *et al.*, 1996). The partition coefficient values for the compounds increased from 0.03 for ddC to 0.37, 28, 63 and 483 for DAC, DTMAC, DCYPP and CHOL, respectively (Ibrahim *et al.*, 1996). *In vitro* stability studies in phosphate buffered saline solution, pH 7.4 (PBS), human serum, mouse serum, mouse brain homogenate and liver homogenate demonstrated that CHOL was the most stable in all media while DAC, DTMAC and DCYPP were stable only in PBS. Therefore, the latter three prodrugs were suitable substrates for enzymatic hydrolysis (Ibrahim *et al.*, 1996). The half-lives for DCYPP in mouse serum, liver and brain homogenates were 0.04, 0.35 and 0.34 h, respectively. The half-life for DAC in PBS was longer than that of DTMAC (0.82 *vs.* 0.38 h). In mouse brain homogenate, the half-life for DTMAC was 3.9 h while DAC had a half-life of 1.6 h. Nevertheless, both of these prodrugs were rapidly metabolized in mouse liver homogenate with half-lives of 0.36 h and 0.23 h for DAC and DTMAC, respectively (Ibrahim *et al.*, 1996). *In vivo* studies in mice, however, showed that the relative brain exposure (brain/serum concentrations ratio) was not improved by administering DAC and DTMAC prodrugs. DTMAC yielded a relative brain exposure value of 0.023, which was similar to that of ddC (0.028). No ddC was detected in the brain after DAC administration. Thus, although highly lipophilic, these prodrugs were not able to increase ddC brain delivery (Ibrahim *et al.*, 1996).

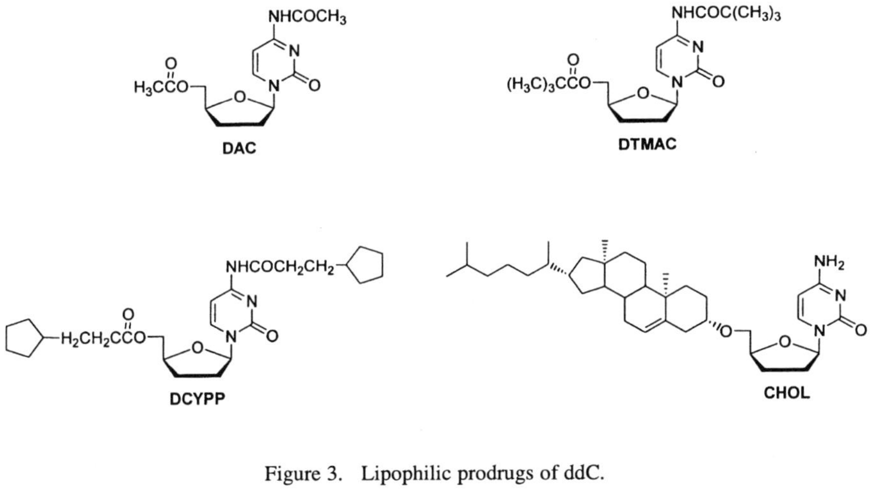

Figure 3. Lipophilic prodrugs of ddC.

Baker *et al.* (1978) synthesized 5'-*O*-acyl derivatives 9-β-D-arabinofuranosyladenine (ara-A, vidarabine) in an effort to improve its pharmacokinetic properties (Figure 4). Vidarabine has been useful in the treatment of fatal herpes encephalitis (Pavan-Langston and Buchanan, 1975) and has demonstrated possible use as a topical agent for ocular herpes keratitis (Whitey *et al.*, 1977). Unfortunately, its therapeutic use is limited by low aqueous solubility, short half-life due to deamination by adenosine deaminase and low lipophilicity, which hinders its use as a topical antiviral agent (Baker *et al.*, 1978). Therefore, target of the prodrug was to increase resistance to deamination and lipophilicity. Among the prodrugs synthesized, the 5'-*O*-valeryl (R = isobutyl in Figure 4) derivative was found to be the most promising because of its aqueous solubility (15-fold more soluble than ara-A), lipophilicity and and anti-herpes simplex virus (HSV) activity.

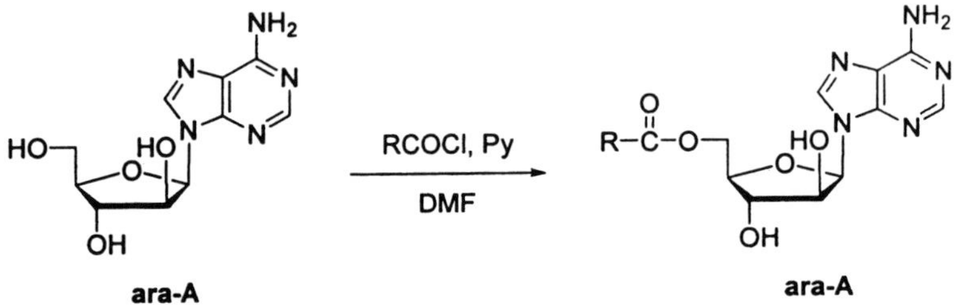

Figure 4.   Synthesis of 5'-*O*-esters of Ara-A.

Kawaguchi *et al.* (1992) synthesized ester prodrugs (Figure 5) to improve the bioavailability of 2',3'-dideoxyinosine (ddI), a drug approved for the treatment of human immunodeficiency virus (HIV) infection (Faulds and Brodden, 1992). Like other dideoxynucleoside analogues, the active form of ddI is the triphosphate metabolite, responsible for eliciting viral suppression by chain termination or competitive inhibition of reverse transcriptase (RT) (Mitsuya *et al.*, 1987). In comparison with AZT, ddI is less toxic towards human hematopoietic progenitor cells (Mitsuya and Broder, 1986). Nevertheless, ddI has a major drawback in that it is very labile to hydrolysis of the C'-N bond under acidic conditions (Anderson *et al.*, 1988). For this reason, the oral bioavailability of ddI in rats is 15.2%, whereas bioavailability following intraduodenal administration (*i.e.* bypassing the acidic gastric environment) is 70.0% (Kawaguchi *et al.*, 1992). Lipophilic esters of ddI were therefore designed and synthesized to slow acidic hydrolysis down, thus increasing bioavailability. Among these prodrugs (Figure 5), the succinate (Suc-ddI) was the only one that possessed a low partition coefficient (-1.5) and ample aqueous solubility (0.1g/ml) making it a promising prodrug candidate (Kawaguchi *et al.*, 1992). Surprisingly, all the prodrugs failed to increase chemical stability of ddI under acidic conditions. Susceptibility studies of prodrugs in rat plasma, liver and duodenum homogenates resulted in quantitative release of the parent drug, ddI. The octanoyl derivative C8-ddI was the most susceptible to enzymatic hydrolysis while Suc-ddI was the least (Kawaguchi *et al.*, 1992). These results were consistent with those reported for other nucleoside esters of 2', 3'-didehydro-3'-deoxythymidine (d4T) (Yajima *et al.*, 1996 and 1998) and AZT (Seki *et al.*, 1990 and Aggarwal *et al.*, 1990). C8-ddI, Bz-ddI (the benzoate) and Suc-ddI prodrugs had relative bioavailabilities of 32%, 31% and 11.5%, respectively after oral administration. The increased bioavailability of C8-ddI and Bz-ddI may be attributed to low water solubility, which slows down chemical decomposition during gastric passage (Kawaguchi *et al.*, 1992). As for the stearic ester C18-ddI, the very low aqueous solubility may prevent efficient absorption in the gastro-intestinal tract as well as slow down chemical and enzymatic hydrolysis. As a result, no parent drug was observed following oral administration of C18-ddI, even though some susceptibility to enzymatic hydrolysis at a very low concentration could be demonstrated (Kawaguchi *et al.*, 1992). At the other extreme, hydrophilic esters Suc-ddI and C2-ddI showed poor oral bioavailability (11.5 and 4.5%, respectively), probably due to extensive hydrolysis in the stomach (Kawaguchi *et al.*, 1992).

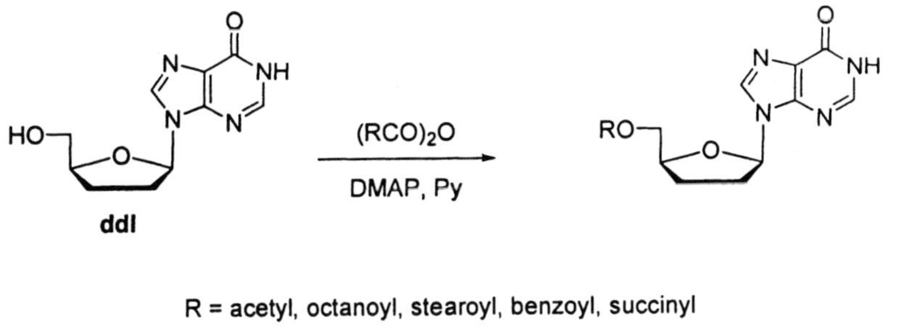

HO

ddI

(RCO)₂O

DMAP, Py

RO

R = acetyl, octanoyl, stearoyl, benzoyl, succinyl

Figure 5. Synthesis of esters of ddI.

In order to improve the solubility of lipophilic prodrugs, while still maintaining a suitable hydrophilic/lipophilic balance to ensure an easy passage through biological membranes, phospholipid-like compounds have been examined. The most recent progresses in this field have been reported by Freeman and co-workers (Cheng *et al.*, 1999a, 1999b, 1999c and 2000) and Hostetler and co-workers (Hostetler *et al.*, 1996, 2000a and 2000b; Beadle *et al.*, 2000). In a rabbit model, the foscarnet (PFA) prodrug 1-*O*-octadecyl-*sn*-glycerol-3-phosphonoformate (ODG-PFA, Figure 6) released and maintained sustained levels of the parent drug upon intravitreal injection for several weeks. The concentration at week 10 was still over 10-fold higher than the IC_{90} of PFA against human cytomegalovirus (HCMV). Although the conversion of ODG-PFA to PFA was efficient in the *in vivo* system considered, it was found negligible *in vitro* (Cheng *et al.*, 1999a). This means that the vitreous only acts as a reservoir of the prodrug, which is metabolized by intracellular enzymes in the retina (Beadle *et al.*, 2000). ODG-PFA-liposome formulations have been suggested as possible long-acting delivery system for the therapy of CMV retinitis, one of the most common opportunistic infections occurring in AIDS patients (Cheng *et al.*, 1999a). Alkylglycerol foscarnet analogues have good bioavailability in mice, with plasma concentrations of the drug well above the EC50 values for foscarnet and several drug-resistant HIV strains (Hostetler *et al.*, 2000b)

Another potential prodrug for the therapy of HSV-1 or HCMV retinitis is 1-*O*-hexadecylpropanediol-3-phosphogancyclovir (HDP-P-GCV, Figure 6), synthesized by coupling of 1-*O*-hexadecylpropanediol-3-phosphate to gancyclovir in pyridine and DCC catalysis. In the rabbit model, intravitreal injections with resultant 0.2 mM intravitreal concentration of prodrugs allowed a 4 to 6 weeks complete protection of the retina against HSV-1 with an IC_{50} of 0.6 µM (Cheng *et al.*, 1999c and 2000). HDP-P-GCV has also been evaluated in HCMV-infected human lung fibroblasts, showing an IC_{50} of 0.6 µM

While acyclovir (ACV) is inactive against HBV, its triphosphate is a potent inhibitor of the same virus. This is due to the fact that HBV does not encode for a thymidine kinase (TK). Thus, it is not surprising that 1-*O*-hexadecylpropanediol-3-phosphoacyclovir (HDP-P-ACV), which delivers ACV monophosphate, is highly active against HBV in 2.2.15 cells. Moreover, in a woodchuck model, a four week treatment with a dosage of 10mg/Kg of body weight of the prodrug twice a day decreased by 95% serum WHV DNA levels and by 54% WHV replicative intermediates, whereas a 5.3-fold molar dosage of the parent drug had no significant activity (Hostetler *et al.*, 2000a).

The favorable antiviral activities of PFA and GCV have been combined in interesting "double" prodrugs (Figure 6), in which GCV is bound to the triglyceride mimetic through PFA. The resulting compounds have the potential to deliver both drugs to infected cells and one of them (n=0) shows anti-HCMV (IC_{50} 0.21 µM) and -HSV-1 (IC_{50} 0.004 µM) activity in infected MRC-5 cells (Beadle *et al.*, 2000).

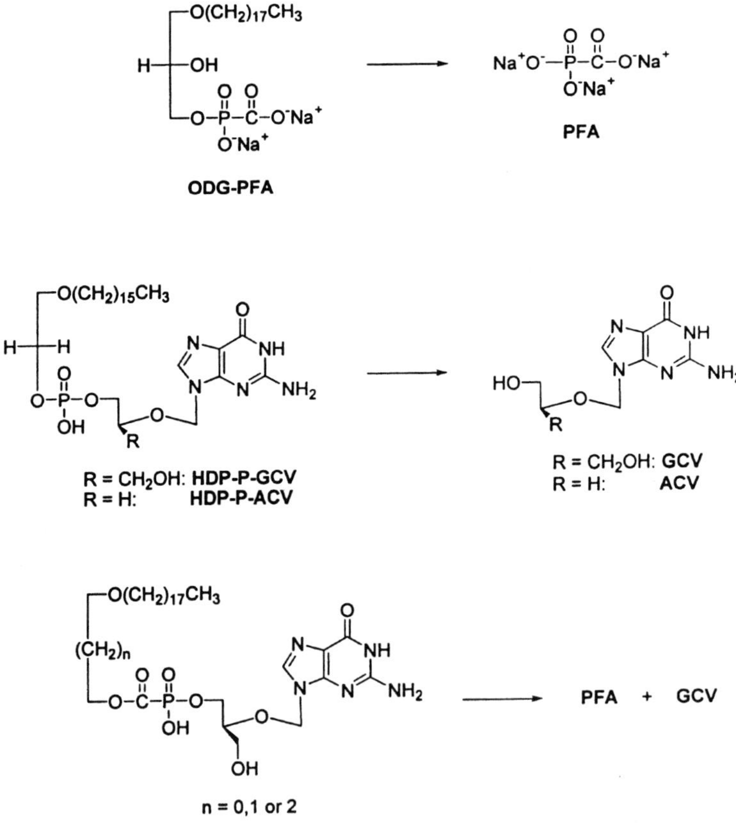

Figure 6.   Phospholipid-mimetic prodrugs and their parent drugs.

Torrence *et al.* (1998) synthesized a dihydropyridine ester of 3'-azido-3'-deoxythymidine (AZT-DHP) in an attempt to increase the brain concentration of AZT by means of a redox delivery system (Figure 7). The 1,4-dihydropyridine-pyridinium approach is based on the capability of the neutral lipophilic 1,4-dihydropyridine adduct to penetrate the blood-brain barrier. Once in the brain, dehydrogenases convert the dihydropyridine to pyridinium-form, which is potentially trapped in the brain due to its positive charge. The pyridinium adduct (AZT-P) is then hydrolyzed, releasing the parent compound (Figure 8). AZT-P was found to be more potent than AZT against murine sarcoma virus (MSV)-induced transformation of murine embryo C3H fibroblasts.

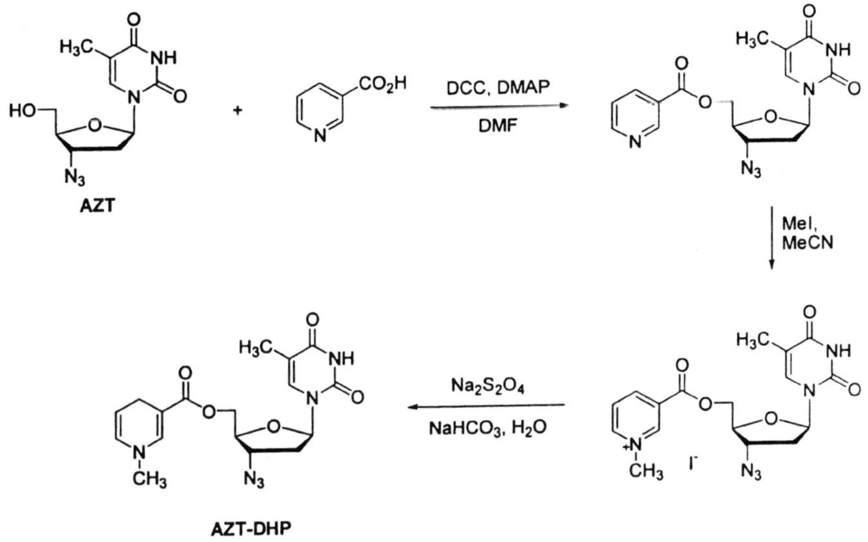

Figure 7. Synthesis of AZT-DHP.

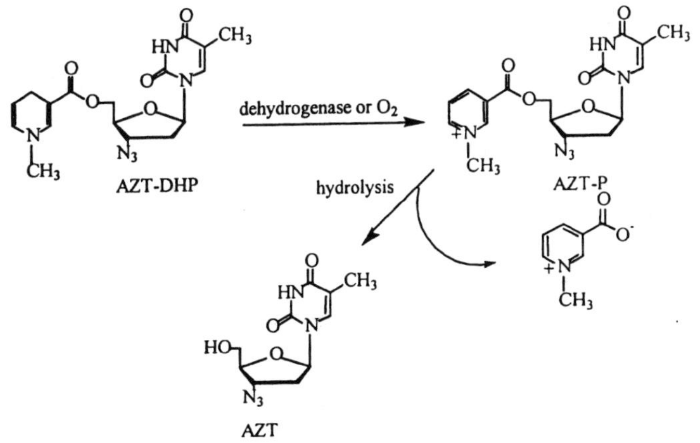

Figure 8. Proposed metabolism of AZT-DHP.

Chu *et al.* (1990) also synthesized a dihydropyridine (DHP) derivative of 3'-azido-2',3'-dideoxyuridine (AZDU) as a prodrug to improve brain delivery of the parent compound. *In vitro*, AZDU-DHP exhibited half-lives of 4.33, 0.56, 0.17 h in human serum, mouse serum, and mouse brain homogenate respectively, whereas AZT-DHP had half-lives of 7.70, 1.40 and 0.18 h. *In vivo,* these prodrugs displayed areas under the serum concentration-time curves (AUCs) in mice similar to those of the parent compounds. Nevertheless, the brain AUCs for both AZDU (11.43 μg h/mL) and AZT (11.28 μg h/mL) after the administration of prodrugs were greater than that of parent compounds AZDU (2.09 μg h/mL) and AZT (1.21 μg h/mL). This indicates a substantial increase in exposure to the anti-HIV agents with relative exposures (re) of 5.47 and 9.32 for AZDU and AZT, respectively. Other dihydropyridine derivatives of nucleosides, such as the bis-DHP derivative of 2',3'-dideoxycytidine (ddC), failed to produce enhanced brain delivery *in vitro* and *in vivo* studies, most likely due to their instabilities (Torrence *et al.*, 1993). 2',3'-Dideoxy-2',3'-didehydro-5'-*O*-[(1,4-dihydro-1-methyl-3-pyridinyl)carbonyl]thymidine (d4T-DHP) showed an increased concentration of the parent compound in the brain of mice (Hamamoto *et al.*, 1987; Palomino *et al.*, 1989).

Ara-C is a well-known drug used in the treatment of human acute myeloblastic and lymphoblastic leukemia (Bodey *et al.*, 1969; Holland and Glidewell, 1970). It has also been reported to be effective in combination against solid tumors (Kodama *et al.*, 1989). However, the clinical activity of ara-C has been limited by cytidine deaminase-catalyzed deaminase catalyzed deamination resulting in a short half-life of the compound (Rivera *et al.*, 1980). N^4-[*N*-Cholesteryloxycarbonyl)glycyl]-ara-C (COCG-ara-C) was designed as a prodrug to provide sustained release delivery following intravenous administration, therefore allowing an increased half-life for ara-C (Tokunaga *et al.*, 1988a). Unfortunately, the N^4-acylpeptidyl conjugates of ara-C proved to be too stable in the presence of α-chymotrypsin.

In the synthesis of COCG-ara-C, ara-C was condensed with *N*-(cholesteryloxycarbonyl)glycine (Tokunaga *et al.*, 1988b) in the presence of ethylchlorocarbonate and triethylamine (Figure 9). COCG-ara-C was found labile to chemical as well as enzymatic hydrolysis (rat, mouse and human plasma). *In vitro* studies showed that this prodrug possessed only one-fifth of the antitumor activity of ara-C against P 388 leukemia cells. In contrast, COCG-ara-C bearing liposomes were found to be superior to ara-C against L 1210 leukemia in mice. Prodrug bearing liposomes displayed antitumor activity against human lung adenocarcinoma A549 xenograft implanted in kidney capsules of mice Sharma *et al.* (1993) also designed and synthesized steroidal esters of AZT to improve its pharmacokinetic profile and to reduce dose-related bone marrow toxicities such as severe anemia and leukemia by influencing its half-life (Figure 10). 5'-*O*-gluronidation is the major metabolic process that results in a rapid elimination of AZT. Therefore, 5'-hydroxyl protection can inhibit this metabolic pathway, which may improve uptake by target tissues. The coupling of AZT with a steroidal moiety was conceptualized as being the ideal approach for increasing the half-life and intracellular delivery of AZT. Preliminary *in vitro* studies showed that the activity of the prodrug was comparable to that of AZT against HIV infected CEM cells (Sharma *et al.*, 1993).

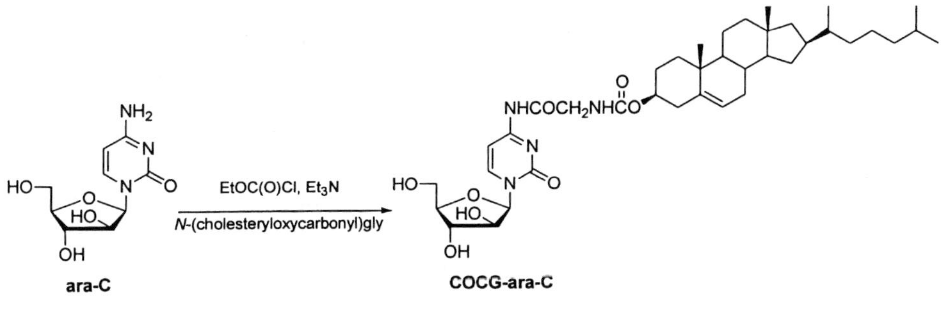

Figure 9. Synthesis of COCG-ara-C.

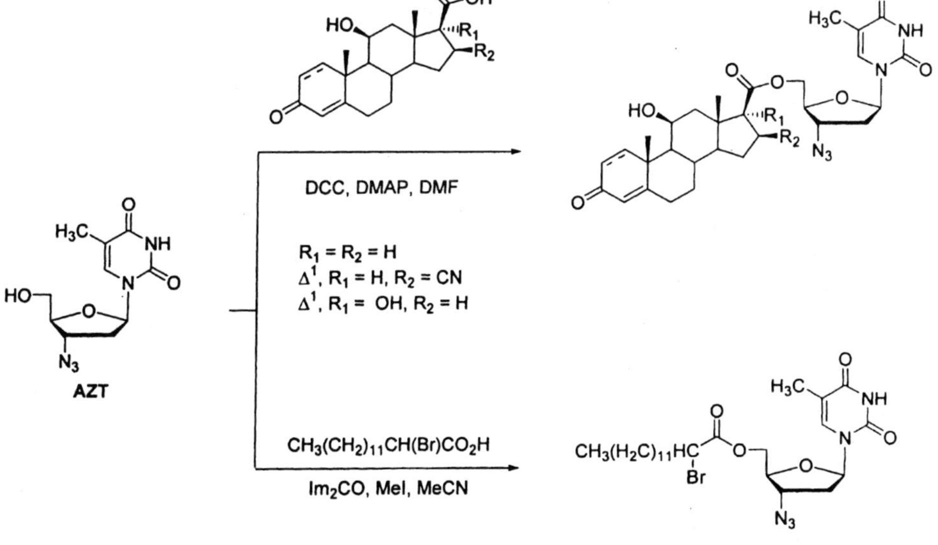

Figure 10. Synthesis of steroidal and bromomyristoyl derivatives of AZT.

Parang *et al.* (1998) synthesized a dually active prodrug, 3'-azido-2',3'-dideoxy-5'-*O*-(2-bromomyristoyl)thymidine by using 1,1'-carbonyldiimidazole in the presence of methyl iodide to couple AZT with 2-bromomyristic acid to yield the ester prodrug (Figure 10). Bromomyristic acid, along with other 2-halotetradecanoic acids, is an inhibitor of *Cryptococcus* neoformans, which attacks the CNS of late-stage HIV patients (Parang *et al.*, 1996). Therefore, the 2-bromomyristic acid portion of the prodrug could serve as a lipophilic carrier and an anti-fungal agent.

In mice, this dual acting prodrug was found to have a distribution half-life of 4.2 min, in comparison with 4.4 min for AZT. The short distribution of the lipophilic prodrug may be associated with sequestration by lipoidal tissues. The elimination half-life (428.5 min) was substantially greater than that of AZT (112.5 min) and, surprisingly, the prodrug was detected in mouse blood throughout the experiment, which indicates that metabolic stability of the prodrug may play a major part in the persistence of this nucleoside ester in the blood along with possible redistribution from peripheral tissues to blood (Parang *et al.*, 1998). High concentrations of prodrug are also distributed in the liver. The 2-bromomyristic acid portion of the prodrug was not detectable in plasma after administration, most probably due to absorption into blood cells and fatty tissues. AZT concentrations within the brain were not significantly changed, but the concentration of AZT increased from 10 to 25 nmol/g one minute after administration of the prodrug. The relative brain exposure (AUC for AZT in the brain after prodrug administration divided by AUC for AZT after parent drug administration) increased by over 100%, which indicates favorable delivery (Parang *et al.*, 1998).

Valaciclovir (2-[(2-amino-1,6-dihydro-6-oxo-9*H*-purin-9-yl)methoxy]ethyl-L-valinate) was synthesized as a prodrug of ACV (Figure 11) (Beaucamp *et al.*, 1992). ACV was the first antiviral agent that possessed potent and selective viral inhibition (Whitley *et al.*, 1992). Nevertheless, its efficacy is limited due to poor oral bioavailability (Lewis *et al.*, 1986). In patients, the approximate bioavailability was 20 to 25% after oral administration. Low bioavailability of ACV may be linked to its lack of sufficient aqueous solubility and possibly to the mechanism of absorption (Lewis *et al.*, 1986). Once in the circulatory system, ACV has a half-life of 2.5 h, with 67% eliminated in the urine *via* renal excretion (De Miranda and Blum, 1983). However, valaciclovir at a 1000 mg dose had an absolute bioavailability of 54% after oral administration to healthy patients (Weller *et al.*, 1993). The absorption of valaciclovir appeared to involve the saturable dipeptide transporter system of the intestinal brush border. In a single-dose crossover escalation study over the dose range 100 to 1000 mg, its absorption was found to be saturable as indicated by a non-proportional increase in AUC. Furthermore, there was a delay in C_{max} (0.88 h at 100 mg to 1.75 h at 1000 mg), which supports the notion of a saturable absorption process.

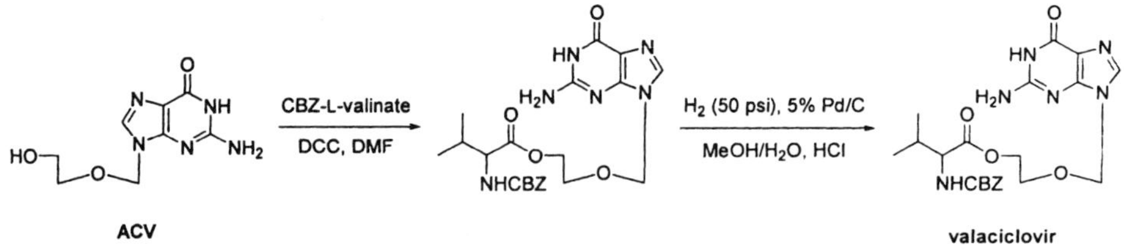

Figure 11.   Synthesis of valaciclovir.

Penciclovir [9-(4-hydroxy-3-hydroxymethyl-but-1-yl)guanine], an acyclic nucleoside, is a broad inhibitor of herpes viruses, which include herpes simplex virus types 1 and 2

(HSV-1 and HSV-2), varicella-zoster virus (VZV), Epstein-Barr virus (EBV) and hepatitis B virus (HBV) (Harnden *et al.*, 1989; Boyd *et al.*, 1988 and 1993). As with ACV, its efficacy is limited by low oral bioavailability in rats and mice (Vere Hodge *et al.*, 1989). As a prodrug, famciclovir [2-amino-9-(4-acetoxy-3-acetoxymethylbut-1-yl)purine] was designed to improve oral absorption (Figure 12), increasing the bioavailability of penciclovir to 77% (Harnden *et al.*, 1987 and 1989; Pue and Benet, 1993).

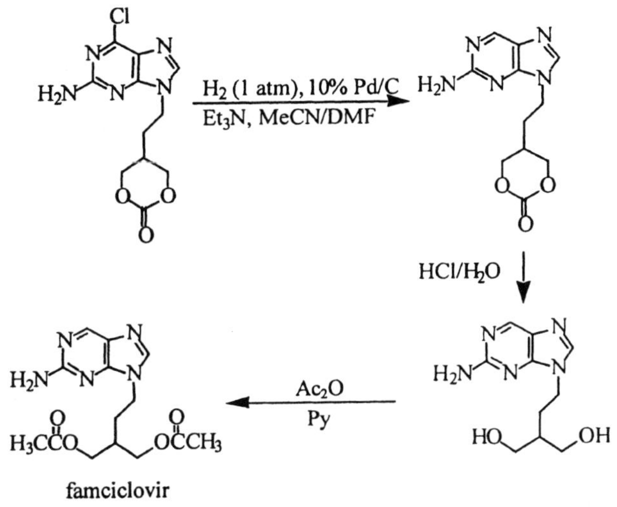

Figure 12.   Synthesis of famciclovir.

Bioconversion of famciclovir to penciclovir requires three metabolic steps: two deacetylations and oxidation of the heterocyclic ring (Figure 13) (Rolan, 1995). Cytosolic enzyme aldehyde oxidase of the liver is responsible for the oxidation of the guanine ring. In patients, between 50 and 67% of famciclovir is converted to penciclovir between 30 min and 1 h after oral administration (Filer *et al.*, 1994). The remaining amount consists mostly of the 6-deoxy monoacetate. Approximately 21% of the oral dose is observed in feces, with 17% and 4% consisting of 6-deoxy-penciclovir and penciclovir, respectively. This observation suggests incomplete absorption of famciclovir and ultimately degradation to 6-deoxy-penciclovir and penciclovir in the digestive tract (Filer *et al.*, 1994).

Kim *et al.* (1998) synthesized racemic mixtures of alkylcarbonate derivatives of 2-amino-9-(3-hydroxymethyl-4-alkoxycarbonyloxybut-1-yl)purines as dual prodrugs of penciclovir. The prodrugs require enzyme-mediated oxidation and hydrolysis in order to generate the parent drug penciclovir as with famciclovir. These prodrugs showed no significant activity at concentrations up to 100 μM against HCMV in human embryonic lung fibroblast (HEL) 299 cells. Nevertheless, no cytotoxicity was exhibited at a maximum concentration of 100 μM in HEL 299 cells (Kim *et al.*, 1998).

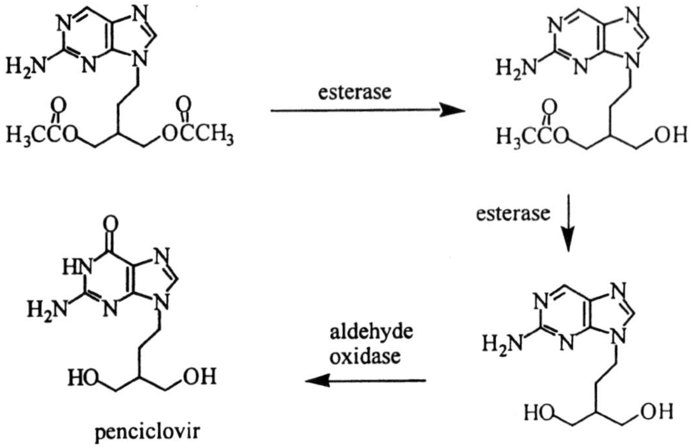

Figure 13.   Major route of metabolism of famciclovir.

Following administration to mice, the urinary recoveries of the compounds were similar: 53% for the isopropyl monocarbonate of penciclovir, 51% for the propyl and isopentyl monocarbonates, 50% for the ethyl monocarbonate, and 48% for famciclovir (Kim *et al.*, 1998). In rats the methyl monocarbonate, ethyl monocarbonate, isopropyl monocarbonate, *n*-butyl monocarbonate and isopentyl monocarbonate showed levels of urinary recovery of penciclovir ranging from 39 to 41%, whereas in the case of famciclovir it was 40% (Kim *et al.*, 1998). The isopropyl monocarbonate was the most stable in aqueous buffer solutions among the compounds synthesized with half-lives of 88, >200, 61 and 26 days at pH 1.2, 6.0, 7.4 and 8.0, respectively (Kim *et al.*, 1998). In *in vitro* antiviral activity studies all compounds had EC_{50} > 125 μM against HSV-1 (KOS strain) in Vero cells with no significant cytotoxicity (Kim *et al.*, 1998).

In the synthesis of monocarbonates of 2-amino-9-(3-hydroxymethyl-4-alkoxycarbonyl-oxybut-1-yl)purines, a cyclic carbonate intermediate was opened by reaction with the appropriate alcohol in the presence of activated silica gel (Figure 14).

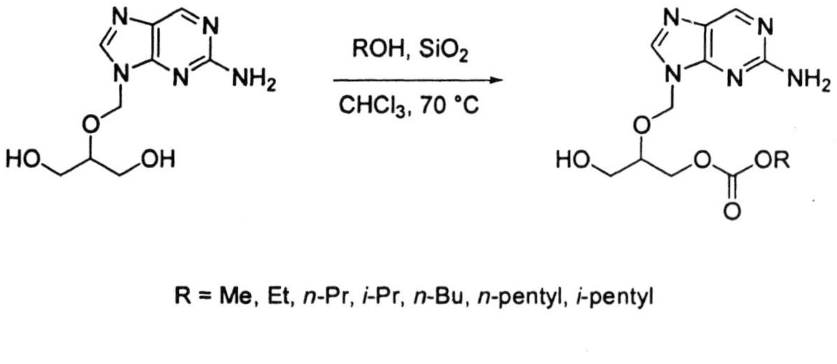

ROH, SiO2; CHCl3, 70 °C

R = Me, Et, *n*-Pr, *i*-Pr, *n*-Bu, *n*-pentyl, *i*-pentyl

Figure 14.   Synthesis of carbonates of famciclovir

Having reported that 2-amino-9-(1,3-dihydroxy-2-propoxymethyl-6-fluoropurine undergoes bioconversion to ganciclovir (GCV) in the presence of calf intestinal mucosal adenosine deaminase (Kim *et al.*, 1994), Kim *et al.* (1999) prepared its mono and diesters as prodrugs of ganciclovir (Figure 15). Ganciclovir [9-(1,3-dihydroxy-2-propoxymethyl)guanine] is the drug of choice for the treatment of Cytomegalovirus (CMV) retinitis (Laskin *et al.*, 1987; Buhles *et al.*, 1988). Recently, oral administration of ganciclovir has been approved by the Food and Drug Administration (FDA) as an alternative to intravenous infusion for the maintenance therapy of CMV retinitis (Drew *et al.*, 1995). However, the bioavailability of orally administered ganciclovir in humans ranges from 2 to 7% (Spector *et al.*, 1995).

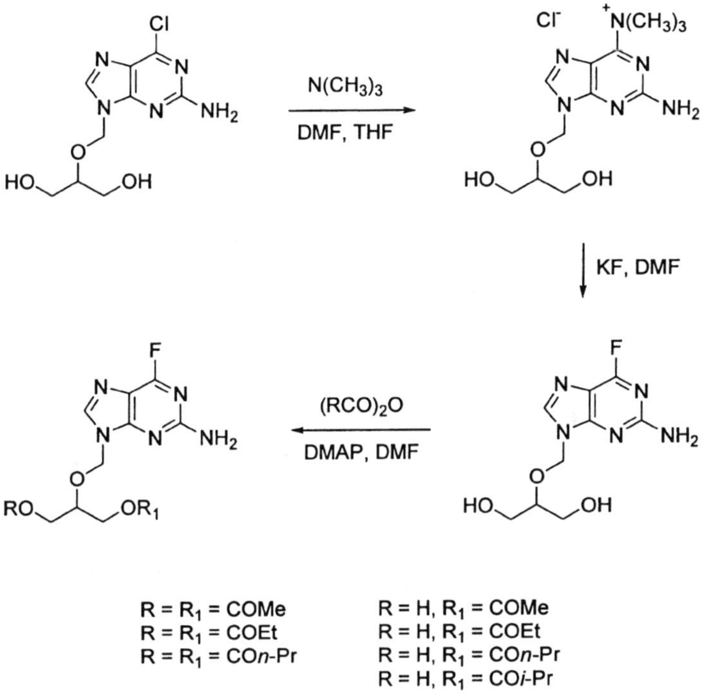

Figure 15.   Synthesis of mono and diesters of the 6-fluoro analogue of ganciclovir.

Among the compounds synthesized and evaluated in rats as prodrugs of ganciclovir, the monoisobutyrate appeared to provide the highest ganciclovir bioavailability (45%) followed by the diisobutyrate (42%), the diacetate (41%), the monobutyrate (41%), the monopropionate (39%), the dipropionate (35%), the dibutyrate (35%) and the monoacetate (29%) (Kim *et al.*, 1999). The monoacetate, monopropionate, monobutyrate and monoisobutyrate prodrugs were very stable at pH 7.4 ($t_{1/2} \geq$ 7days) but had relatively short half-lives at pH 1.2 ($t_{1/2}$ = 60-83 min). The prodrugs had EC_{50} values greater than 100 μM against HCMV in human embryonic lung fibroblast (HEL) 299 cells in comparison with 0.63 μM for ganciclovir. The prodrugs showed no cytoxicity in HEL 299 cells (>100 μM) (Kim *et al.*, 1999).

In summary, the bipartate approach involving hydrolysable linkages between nucleosides and carriers can be an effective way to modify the pharmacokinetic characteristics of a drug improving its stability, increasing its half-life and prolonging its action.

2.2. Bipartate approach involving polymer-linked nucleosides

A potentially improved bipartate approach involves the use of polymeric carriers. This approach is analogous to that discussed in the previous paragraph. Here, however, the physico-chemical properties of the prodrug resemble more those of the carrier, which can "mask" unfavorable characteristics of the parent drug more efficiently.

5-Iodo-2'-deoxyuridine (IDU) is a well-known antiviral drug used in the topical treatment of herpes simplex, and it has also been reported as having antitumoral activity. Rimoli *et al.* (1999) synthesized a polymeric derivative of IDU consisting of poly D,L-lactic acid (PLA) linked by a succinic acid spacer (Figure 16) as a prodrug in an attempt to provide enhanced antitumor activity by preventing rapid metabolism. PLA is a widely applied polymeric carrier with good biocompatibility and rapid clearance (Wade *et al.*, 1977; Kobayashi *et al.*, 1992).

PLA-IDU solubility allows it to form microspheres and nanospheres that can be targeted to specific organs on the basis of their size. PLA-IDU was stable in pH 7.4 phosphate buffer after 30 days with no measurable IDU release (Rimoli *et al.*, 1999). When incubated with porcine esterase in pH 7.4 phosphate buffer for a 30 day period, 20% of IDU was released after a lag-time of about 10 days. *In vivo* studies have yet to be reported.

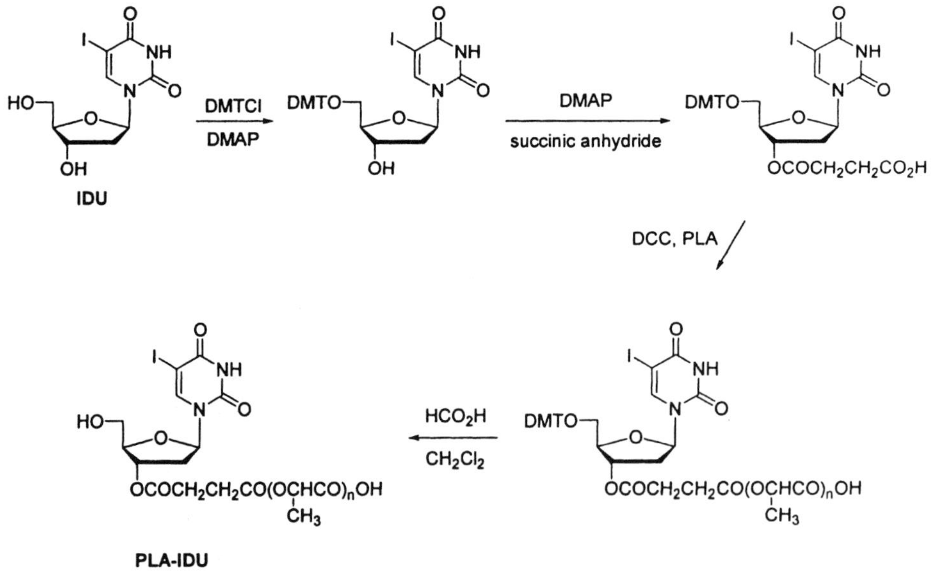

Figure 16.   Synthesis of PLA-IDU.

Having discovered that *N*-Boc-protected polyamine-3TC conjugates elicited potent anti-HIV activity (Kraus *et al.*, 1997), Kraus and co-workers synthesized and evaluated lipophilic *N*-Boc-polyamine conjugates of various dideoxynucleosides such as 3'-azido-2',3'-dideoxythymidine (AZT), 2',3'-dideoxycytidine (ddC), 2',3'-didehydro-2',3'-dideoxythymidine (d4T), 2', 3'-dideoxyinosine (ddI) and 2',3'-dideoxy-3'-thiacytidine (3TC) (Dessolin *et al.*, 1998). These prodrugs, compared to the linear analogues, are more lipophilic and could be better delivered intracellularly. Besides, polyaminic compounds were reported to inhibit HIV-induced membrane fusion (Mayaux *et al.*, 1994), therefore these conjugate could display a dual anti-HIV mode of action.

In the synthesis of the *N*-Boc-tetraazamacrocycle of ddC, 5-[4,8,11-tris(*tert*-butyl-oxycarbonyl)-1,4,8,11-tetraazacyclotetradec-1-yl]-oxo-1-pentanoic acid was synthesized by protecting the cyclam with di-*tert*-butyl dicarbonate followed by alkylation with ethyl-5-bromovalerate in the presence of potassium carbonate and saponification using sodium hydroxide in tetrahydrofuran (Figure 17). This pentanoic acid derivative of *N*-Boc-tetraazamacrocycle was then coupled with ddC using [benzotriazol-1-yloxy-tris(dimethylamino)phosphonium hexafluoro-phosphate] in the presence of a triethylamine/4-dimethylaminopyridine mixture in dichloromethane (Dessolin *et al.*, 1998).

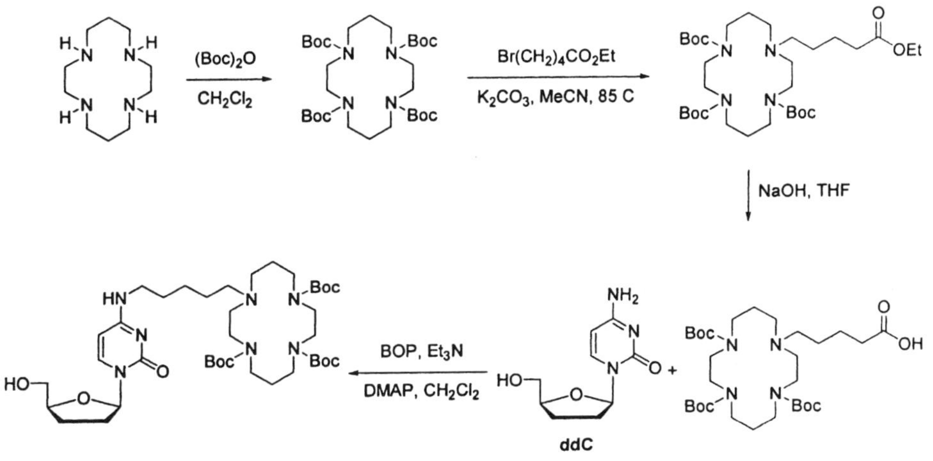

Figure 17. Synthesis of tri-*N*-tetraazamacrocyclic derivative of ddC.

Among the evaluated derivatives, the ddC derivative was the only *N*-Boc-polyamine conjugate found to be more potent than the parent compound against HIV-1 infected MT4 cells with an EC_{50} of 0.005 μM while ddC had an EC_{50} of 0.5 μM. This enhanced potency, however, was accompanied by significant cytotoxicity resulting in a selective index of 20 compared to a selective index of greater than 200 in MT4 cells for ddC (Dessolin *et al.*, 1998). No explanation was given for the increased toxicity of the compounds compared to the parent nucleosides, and no synergistic effect between the hypothetic inhibition of HIV-induced membrane fusion and inhibition of RT was observed (Dessolin *et al.*, 1998).

The macromolecular prodrug approach has been used also to improve the chemotherapeutic properties of the antitumoral agent cytarabine (ara-C) (Shimoyama and Kimura, 1973). The efficacy of this agent is reduced by rapid inactivation *in vivo* by cytidine deaminase and, due to its solubility in water, rapid excretion (Aoshima *et al.*, 1976). Ichikawa and co-workers (1993) coupled N^4-(4-carboxybutyl)-1-β-D-arabino-furanosyl-cytosine with chitosan to obtain the macromolecular prodrug chi-glu-araC (Figure 18). Chitosan is a well-known polymer endowed with good biodegradability and biocompatibility (Ichikawa *et al.*, 1993).

In vitro chemical stability studies indicated that chi-glu-ara-C released 21%, 56% and 76% of ara-C at pH 6, 7.4 and 8, respectively. A small percent of ara-U was generated as a decomposition product of ara-C. The release rate of ara-C was the same for the prodrug in the presence and absence of plasma in pH 7.4 buffer solution, but the amount of ara-U was higher in the presence of plasma, which is most probably due to deamination of the released ara-C by plasma enzymes (Ichikawa *et al.*, 1993). The effectiveness of the prodrug was measured as the increase in the lifespan of P388 leukemia-bearing mice. This increase was equal to 60.7% for the macromolecular drug-treated mice, compared to 3.4% for the ara-C-treated mice (Ichikawa *et al.*, 1993). Unlike ara-C, the prodrug induced an increase of the survival time of the mice, with respect to control. On the other hand, a considerable weight loss in mice treated with the prodrug was reported as a side effect (Ichikawa *et al.*, 1993).

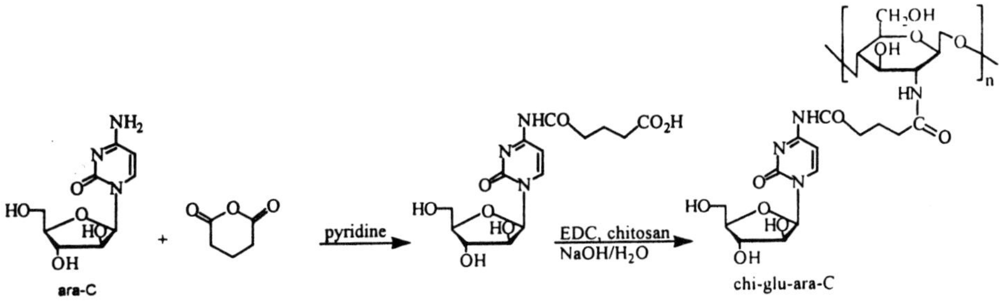

Figure 18.   Synthesis of chitosan derivative of ara-C.

A D-valyl-leucyl-lysyl derivative of ara-C was also synthesized to be a selective substrate of plasmin (Figure 19) (Balajthy *et al.*, 1992). It was also proposed that the peptidic ligand attached at N^4 position would serve to prevent deamination and to selectively deliver ara-C to leukemic cells. It was also conceptualized that the peptidyl ligand would target leukemic cells based upon the fact that plasminogen activators (urokinase and tissue type) are expressed in various kinds of tumor cell lines (Tucker *et al.*, 1978; Wilson *et al.*, 1980). After 48 h of incubation, the prodrug (IC_{50} 0.005 μM) exhibited more potency than ara-C (IC_{50} 0.1 μM) against L1210 lymphoid leukemia cells. But the prodrug was readily hydrolyzed *in vitro* by plasmin as indicated by the decreased potency (IC_{50} 0.01 μM) at 72 h.

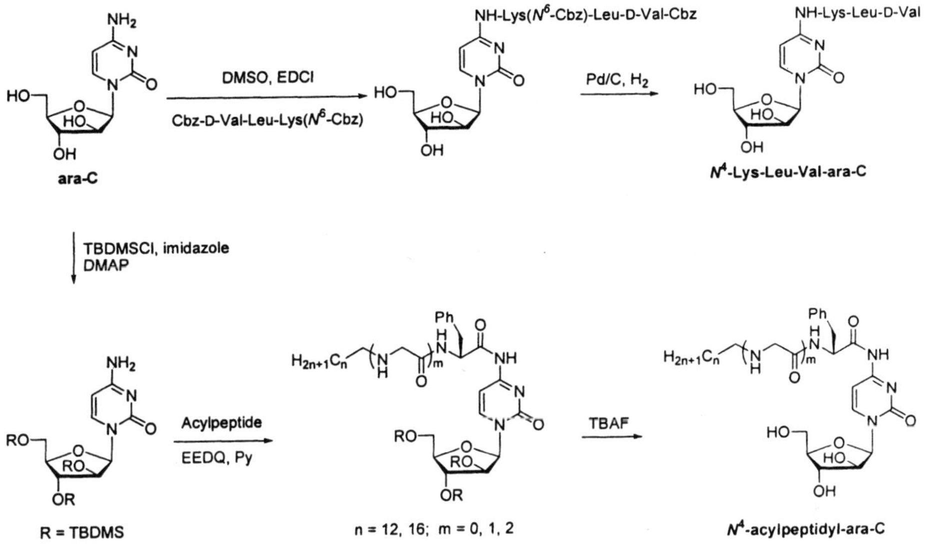

Figure 19.   Synthesis of N^4-tripeptidyl and N^4-acylpeptidyl derivatives of ara-C.

Menger *et al.* (1994) synthesized the N^4-acylpeptidyl conjugate of ara-C (Figure 19) as a means of circumventing deamination, thus increasing its half-life. This prodrug was designed to bind to phospholipid membranes *via* their hydrocarbons and undergo chymotrypsin-induced drug release. In the synthesis of N^4-(acylpeptidyl)conjugates, ara-C was silylated using *tert*-butyldimethylsilyl chloride, imidazole and 4-dimethylami-nopyridine. Silylated ara-C was reacted with the corresponding acylpeptide in the presence of 2-ethoxy-1-(ethoxycarbonyl)1,2-dihydroquinoline and pyridine. Desilylation of N^4-acylated ara-C was carried out using tetrabutylammonium fluoride to give N^4-acylpeptidyl-ara-C (Menger *et al.*, 1994). Unfortunately, all the synthesized compounds were too insoluble in water to show appreciable hydrolysis rates (Menger *et al.*, 1994).

In general, the success of a bipartate approach involving polymer-linked drugs depends on the characteristics of the polymer used. It is important to notice that a polymer may present additional solubility and/or absorption issues, compared to a small molecule. On the other side, biodegradable polymers can constitute effective reservoirs of drugs, and allow a very good control of the release of the therapeutic agent.

3.   Bipartate approach applied to phosphodiester prodrugs

The first phosphorylation of a nucleoside analogue to its monophosphate form is probably the most important step in its bioactivation, because the enzymes catalyzing this transformation have more strict steric requirements than those involved in the syntheses of di- and triphosphates. In fact, there are several examples of nucleoside analogues that are inactive because of inefficient phosporylation *in vivo*, whereas their monophosphates are active. Monophosphates, however, cannot be used as drugs, mainly because of the lability of the phosphate bond in biological media and because their ionic nature prevents their absorption in the gastro-intestinal tract as well as penetration of biological membranes. These drawbacks can be overcome by synthesizing phosphonates and/or masking the charges by conversion to ester prodrugs.

Among phosphate prodrugs, phospholipid derivatives have been extensively explored, because the use of a natural compound as a carrier may improve absorption (*e.g.* when this occurs by a specific transport mechanism) and reduce toxicity (the released carrier is a molecule already present in the organism). Using highly sensitive and quantitative polymerase chain reaction (PCR) techniques, it has been determined that the lymphatic system acts as a reservoir for human immunodeficiency virus (HIV) in asymptomatic infected patients (Pantaleo *et al.*, 1993). Many antiviral agents have been designed and synthesized as means of targeting the lymphatic system by utilizing phospholipids (Hostetler *et al.*, 1994).

Chu and co-workers (Manouilov *et al.*, 1997) synthesized and evaluated the targeting effect of dipalmitoylphosphatidyl-ddI (DPP-ddI) as a prodrug of ddI (Figure 20).

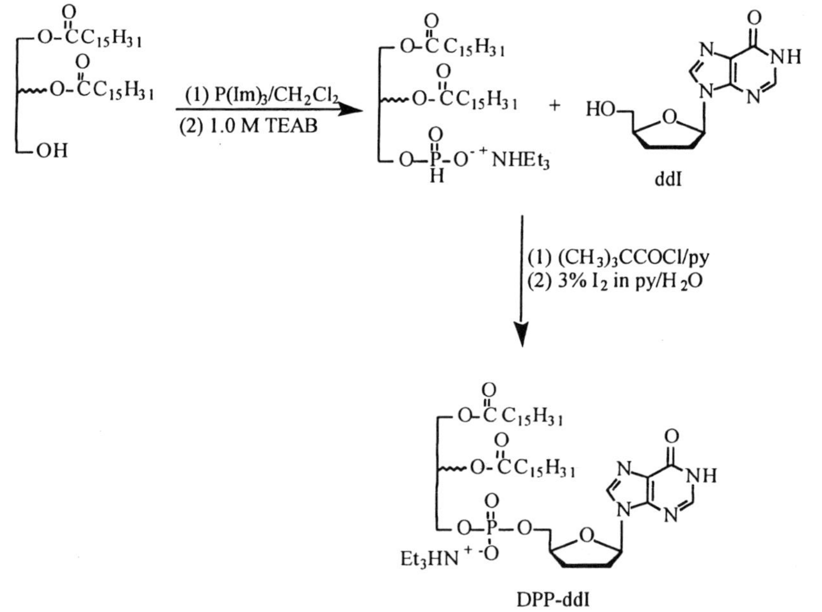

Figure 20.   Synthesis of phosphatidyl derivative of ddI.

Intravenously administered DPP-ddI showed maximum plasma and lymph node concentrations (C_{max}) between 1 and 2 h after injection of the prodrug while C_{max} occurred 5 min after administration of the parent drug (Manouilov *et al.*, 1997). The concentration of ddI declined slowly after C_{max}, yielding a 5 to 9 fold greater terminal half-life in plasma and lymph nodes after administration of the prodrug. Nevertheless, the AUC after intravenous administration of ddI in the plasma was 2-fold greater than that of the prodrug after intravenous administration. In contrast, both ddI and the prodrug had similar AUC values in lymph nodes after intravenous administrations (Manouilov *et al.*, 1997). The oral bioavailability of ddI following administration of DPP-ddI and ddI was 8% and 15%, respectively, which may be related to poor water solubility of the prodrug (Manouilov *et al.*, 1997). DPP-ddI was able to sustain levels of ddI in serum and lymph nodes longer than the parent drug after oral administration, even though C_{max} was lower than that after ddI administration. The lymph nodes relative exposure values of DPP-ddI (ratio of the AUC for ddI in lymph nodes relative to AUC of ddI in blood) were 2-fold higher than ddI administration (Manouilov *et al.*, 1997).

Chu and co-workers (Manouilov *et al.*, 1995) also synthesized dipalmitoylphosphatidyl (DPP) derivatives of 3'-azido-2',3'-dideoxythymidine (AZT) and 3'-azido-2',3'-dideoxyuridine (AZDU) to target the lymphatic system (Figure 21). DPP-AZDU and DPP-AZT exhibited enhanced lymphatic delivery of the parent nucleoside analogues in mice. Furthermore, these prodrugs significantly increased the half-life values of the parent compound. Similarly, Sakai *et al.* (1993) reported lymph nodal disposition of dipalmitoylphosphatidylfluorouridine (DPP-FUrd) and its metabolites in rats. Shuto *et al.* (1988 and 1992) carried out enzymatic one-step synthesis of dipalmitoylphosphatidyl neplanocin A (DPP-NPA) (Figure 21). The transphosphatidylation of neplanocin A was facilitated by phospholipase D-P isolated from *Streptomyces*. Antileukemic potency of DPP-NPA was found to be superior to neplanocin A, an antitumor agent, in mice inoculated with P388 leukemic cells. Dipalmitoylphosphatidyl -(-)-2',3'-dideoxy-3'-thiacytidine (DPP-3TC) was less active against HIV-1 infected human peripheral blood mononuclear cells (PBMC), but it showed anti-HBV activity comparable to that of 3TC in 2.2.15 cells, plus an ability to target the drug to the liver (Xie *et al.*, 1995). DPP-3TC was similar to 3TC in toxicity with no significant cytotoxicity up to 1000 µM in 2.2.15 cells. Currently, it is the most potent and selective anti-HBV lipid-based prodrug (Xie *et al.*, 1995).

Ara-C conjugates of alkylether and thioether phospholipids were designed to couple two cytotoxic groups with different targets to yield synergistic therapeutic effects (Figure 22) (Hong *et al.*, 1990a, 1990b and 1991). Among the compounds screened in L1210 leukemic mice, ara-CDP-DL-1-*S*-octadecyl-2-*O*-palmitoyl-1-thioglycerol (ara-CDP-DL-PTBA, Cytoros) displayed the highest antitumor activity with an increase in life span of 220% at 389 µmol/kg/day (Hong *et al.*, 1990a). These prodrugs displayed more potent anti-tumor activity than the parent drug, ara-C. Their mechanism of action may involve sustained release of ara-C, and their amphiphilic character may facilitate transportation across cell membranes (Hong *et al.*, 1990a). Studies have shown that ara-C triphosphate concentrations were higher after administration of ara-CDP-DL-PTBA than those resulting from ara-C. Other factors that may be involved include resistance to hydrolysis by cytidine deaminase and cytotoxicity of 1-*O*-(or- *S*)-alkyl lysophospholipid (Hong *et al.*, 1990a). In the synthesis of ara-CDP-DL-PTBA, alkylthioglycerol was coupled with ara-CMP morpholidate, which was synthesized by condensing ara-CMP with morpholine using dicyclohexylcarbodiimide (Figure 22).

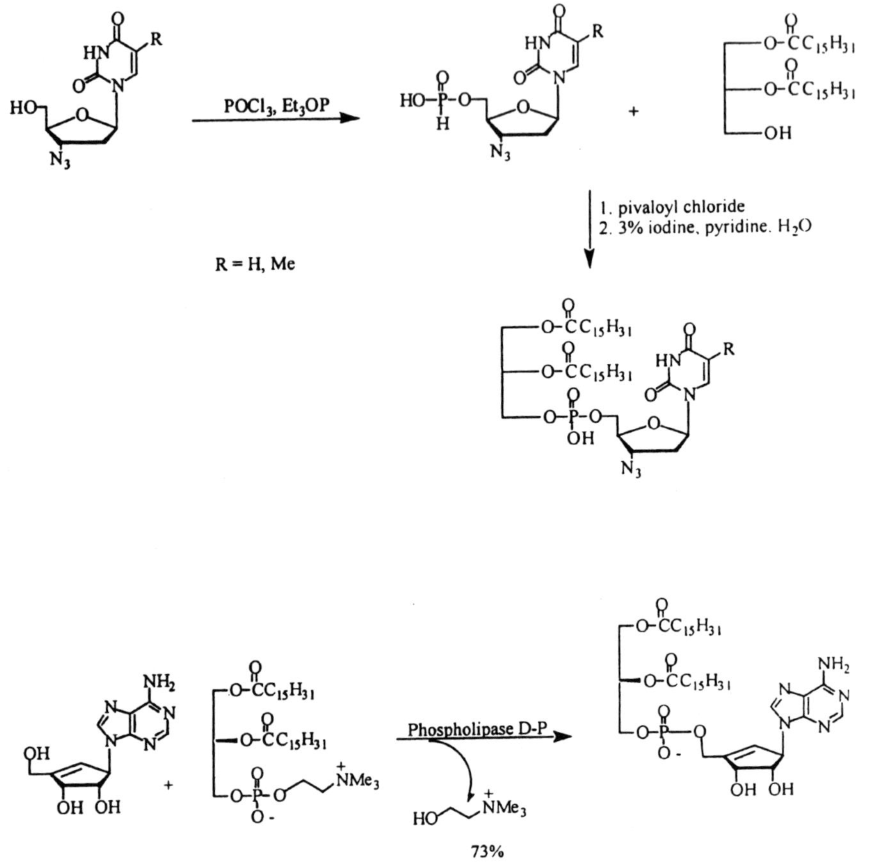

Figure 21.   Synthesis of dipalmitoylphosphatidyl derivatives of AZDU, AZT, and neplanocin A.

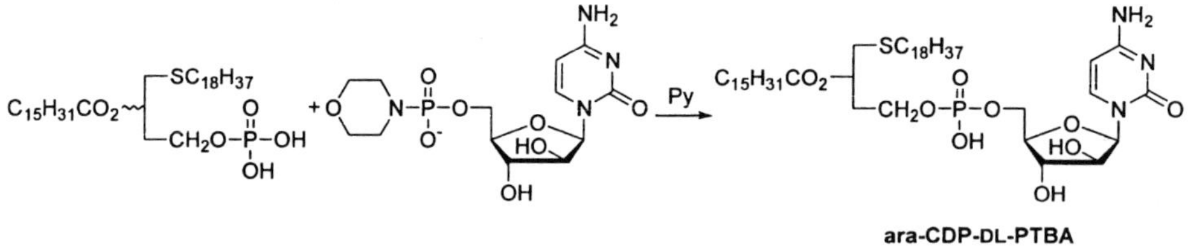

Figure 22.   Synthesis of ara-CDP-PTBA.

Among other synthesized prodrugs, the AZT conjugate protected 80% of HIV infected CEM cells at concentrations as low as 0.58 μM, and showed cytotoxicity at 100 μM in CEM cells (Hong *et al.*, 1996). In pharmacokinetics studies in mice, AZT half-life increased from 0.28 to 5.69 h following administration of AZT and its conjugate, respectively.

Other phosphodiesters including steroidal moieties have been synthesized in attempts to enhance the therapeutic efficacy of anticancer or antiviral agents by improving their pharmacokinetic properties. For example, Hong *et al.* (1979) synthesized prednisolone (X = OH) and prednisone (X = O) phosphodiester conjugates of ara-C (Figure 23).

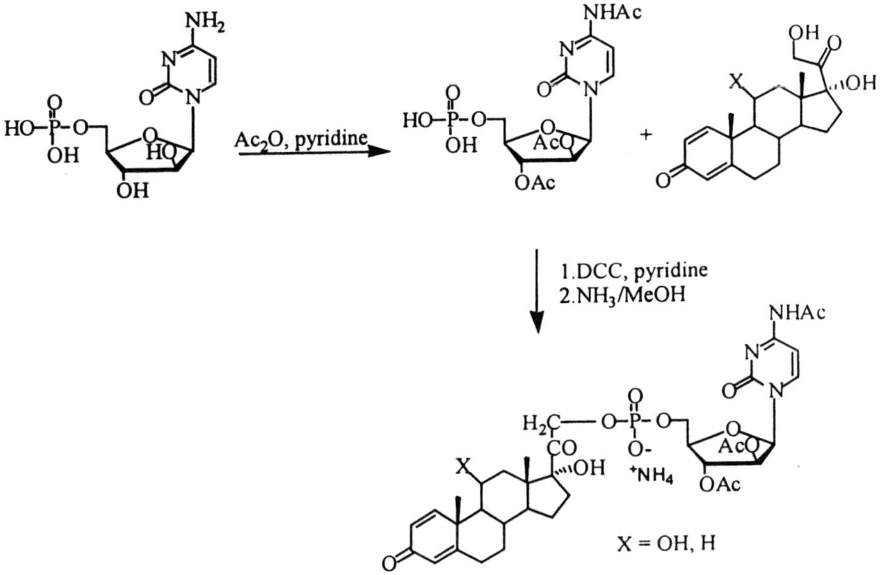

Figure 23.   Synthesis of prednisolone and prednisone phosphodiesters of ara-C.

These phosphodiesters were found to be susceptible to enzymatic hydrolysis in the presence of phosphodiesterase I, snake venom, 5'-nucleotidase and acid phosphatase. The initially formed ara-CMP was converted ultimately to ara-C. The activity of conjugates against L1210 lymphoid leukemia in mice was greater than that of the parent compound alone or in combination with the steroid. The increase in life span values for prednisolone, prednisone and ara-C were 89, 100 and 45%, respectively Hong *et al.* (1979). Luu and co-workers (Ji *et al.*, 1990) synthesized 7β-hydroxy-cholestrol phosphodiester conjugates of 5-fluoro-2'-deoxyuridine (5-FdUrd) (Figure 24). The objective of this phosphodiester was to produce a synergistic effect in view of the fact that both 7-β-hydroxycholesterol (Kumar *et al.*, 1987) and 5-FdUrd (Uchida and Kreis, 1969) have antitumor activity. The conjugate was found to have less antitumor activity than 5-FdUrd against EL-4 murine leukemia cells but exhibited activity similar to that of 7β-hydroxycholesterol. In mice, the prodrug at a dose of 80 mmol/kg per day produced a rate of recovery (defined as a survival of more than 80 days) greater than 90% against carcinoma Krebs II cells-inoculated mice after 2 days. Similarly, Luu and co-workers (Pannecoucke *et al.*, 1994) synthesized the 7-β-hydroxycholestrol conjugate of AZT in an attempt to target lymphocytes.

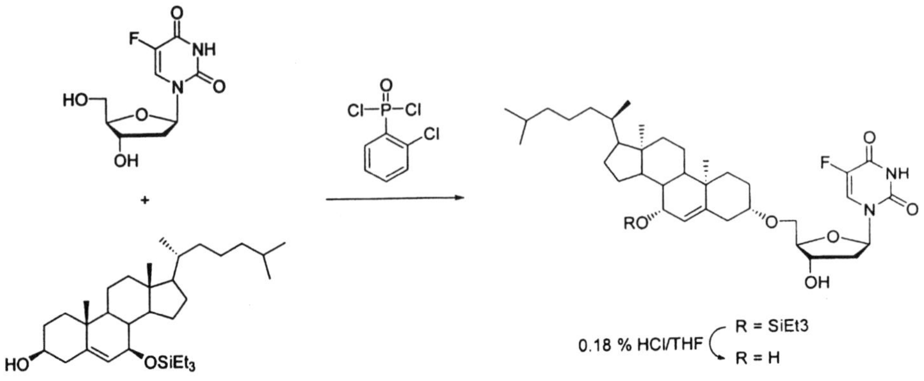

Figure 24.   Synthesis of the 7-β-hydroxycholestrol conjugate of 5Furd.

4.   Tripartate prodrug approach applied to phosphotriester

4.1. Bis-POM and -POC pronucleotides

The approach of using a double ester as a prodrug was first used to improve the bioavailability of marketed β-lactam antibiotics (Higuchi and Stella, 1975) and non-steroidal anti-inflammatory agents (Roche, 1977). In this approach, the diester undergoes enzymatic cleavage, releasing the unstable hydroxyalkyl ester, which spontaneously disengages releasing the parent compound (Figure 25). Recently this approach has been applied to facilitate the intracellular delivery of monophosphates such as 2',3'-dideoxy-2',3'-didehydrouridine monophosphate (ddUMP) (Sastry *et al.*, 1992), 3'-azido-2',3'-dideoxythymidine monophosphate (AZTMP) (Pompon *et al.*, 1994), 5-fluoro-2'-deoxyuridine monophosphate (5-FdUMP) (Farquhar *et al.*, 1994) (*R*)-9-(2-phosphono-methoxypropyl)adenine (PMPA) (Srinivas *et al.*, 1993) and its 2,6-diamino-purine analogue (PMPDAP) as well as [9-(2-phosphonylmethoxyethyl)adenine] PMEA (Naesens *et al.*, 1994).

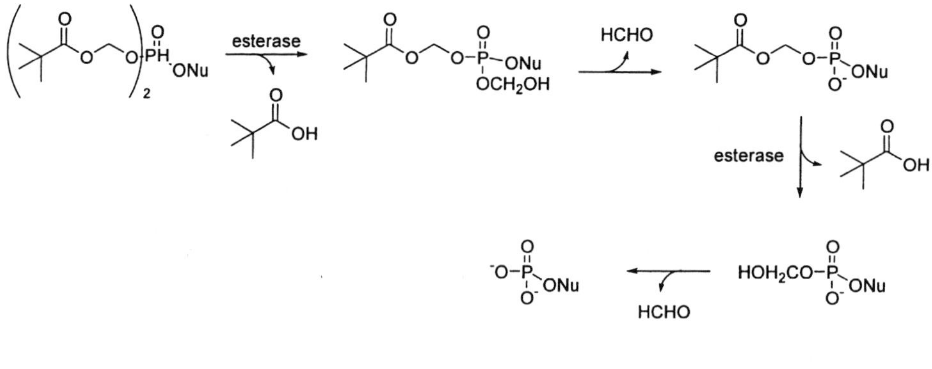

Figure 25. Decomposition of bis-POM pro-nucleotides.

PMEA has a broad spectrum of antiviral activity, which includes retroviruses, hepadnaviruses, and herpesviruses (De Clercq, 1991; Naesens *et al.*, 1994). In phase I/II clinical trials, it appears to be a promising anti-HIV candidate (Walker *et al.*, 1993). Nevertheless, the possibility of PMEA becoming an orally administered drug is limited by its poor bioavailability as shown in monkeys (<1%) (Balzarini *et al.*, 1991) and rats (7-11%) (Bronson *et al.*, 1989; Starrett *et al.*, 1994). Its limited bioavailability is due to the negative charge of the phosphonate functionality at physiological pH, which prevents its penetration through biological membranes.

Srinivas *et al.* (1993) explored the approach of acyloxyalkyl ester pro-nucleotides as a means of masking the phosphonate negative charges of PMEA, thus forming a more lipophilic derivative with the capacity of crossing the gastrointestinal wall and releasing the parent compound in the plasma (Figure 26). Preliminary *in vitro* studies demonstrated that bis-POM-PMEA provided a 100-fold intracellular increase of PMEA concentration (Naesens *et al.*, 1994). *In vitro* studies also showed that bis-pivaloyloxymethyl-PMEA (bis-POM-PMEA) had comparable activity to that of PMEA against human immunodeficiency virus type 1 (HIV-1) infected CEM cells and HCMV-infected MRC-5 cells. Bis-POM-PMEA was substantially more potent than PMEA against HSV-1 and -2- infected Vero cells. Its persistence of HSV-2 inhibition, 20 times longer than that of the parent compound, correlates well with the reported enhanced cellular uptake (Starrett *et al.*, 1992). Bis-POM-PMEA also affected the growth of CEM cells. The growth of CEM cells was completely suppressed at a 2 μM concentration of bis-POM-PMEA, which may result from the liberation of two equivalents of formaldehyde and pivalic acid (Figure 25) or, more likely, from an increase in the cellular uptake of the prodrug (Starrett *et al.*, 1992; Srinivas *et al.*, 1993). Furthermore, *in vivo* bis-POM-PMEA demonstrated a 2-fold and 5-fold enhancement in bioavailability in rats (Jones *et al.*, 1996) and monkeys (Naesens *et al.*, 1994), respectively. At a single 500 mg dose of bis-(POM)-PMEA, oral bioavailability was greater than 40% in clinical trials involving well fed subjects (Starrett *et al.*, 1992).

The synthesis of bis-POM-PMEA was carried out by reacting PMEA with chloromethylpivalate in the presence of the bulky base *N,N'*-dicyclohexyl-morpholine carboxamidine in 32% yield (Figure 26). Bis-POM-PMEA has also been synthesized in lower yields by condensing various salts of PMEA with either chloromethyl pivalate or iodomethyl pivalate (Srivastva and Farquhar, 1984).

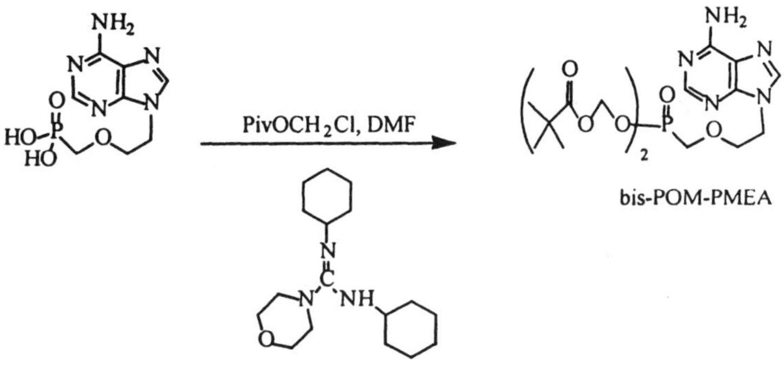

Figure 26.   Synthesis of bis-POM-PMEA.

A recent study compares the disposition of different PMEA ester prodrugs in Caco-2 monolayers (Annaert *et al.*, 1998). All the derivatives were able to deliver PMEA through the monolayer, with better efficiency in the case of more lipophilic PMEA derivatives such as the bis[S(phenyl)ATE]- ester, and a relationship between transport efficiency and octanol/water partition coefficient could be determined.

Bis-(isopropyloxycarbonyloxymethyl) (bis-POC) nucleotide analogue prodrugs are a modification of the bis-POM pronucleotides designed to reduce the cytostatic effect which may be caused by the release of pivalic acid (Figure 28). Bis-POC prodrugs are composed of a carbonate diester that undergoes esterase-catalyzed cleavage of the isopropyl ester to yield two equivalents of 2-propanol and formaldehyde (Figure 27). The bis-POC approach was applied to the anti-HIV agent PMPA (Balzarini *et al.*, 1993 and 1996a). PMPA was reported to completely prevent simian immunodeficiency virus (SIV) infection in monkeys even as late as 24 h after inoculation occurred (Tsai *et al.*, 1995). PMPA showed efficacy without significant toxicity in long-term treatment (13 months) of SIV-infected newborn monkeys (Van Rompay *et al.*, 1996). Furthermore, PMPA exhibited an effect against chronic SIV infection in monkeys (Tsai *et al.*, 1997). PMPA was also found to be active against acute and chronic feline immunodeficiency virus (FIV) infections in cats (Myles *et al.*, 1996). To "Recently approved by the FDA for the treatment of AIDS, PMPA exhibited a 1.1 log reduction in HIV RNA levels after administration of only eight doses (Arimilli *et al.*, 1997). Nevertheless, PMPA displayed low bioavailability in animals.

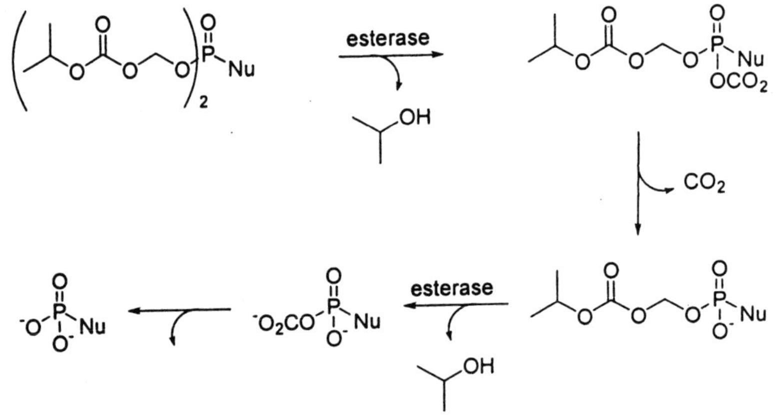

Figure 27.   Decomposition of bis-POC phosphonates.

In an effort to improve the oral absorption of PMPA, Arimilli *et al.* (1997) synthesized various acyloxyalkyl esters of PMPA. The most promising was the bis-isopropyloxycarbonyloxymethyl ester derivative (bis-POC), based upon its chemical stability ($t_{1/2}$ = 9.2 h at pH 7.4), partition coefficient (log P = 1.3), bioavailability (30% in dogs), efficacy, and low toxicity (Arimilli *et al.*, 1997; Shaw *et al.*, 1997). Unfortunately, all the prodrugs were unstable in dog tissues, with bis-POC-PMPA still showing the best stability in intestinal homogenate ($t_{1/2}$ = 52.6 min) and the second best in plasma ($t_{1/2}$ = 20.5 min). *In vitro* studies of the metabolism of radiolabeled PMPA showed that it was hydrolyzed to PMPA and subsequently underwent phosphorylation to the mono and diphosphate derivatives (Fridland *et al.*, 1997). In the synthesis of bis-POC-PMPA, isopropylchloromethyl carbonate (Bohme and Budde, 1971) was prepared by adding pyridine to a cold ethereal solution of chloromethyl chloroformate and 2-propanol. The carbonate was then reacted with PMPA in dimethylformamide (DMF) in the presence of triethylamine or diisopropylethylamine at 50 °C for 20 h (Figure 28).

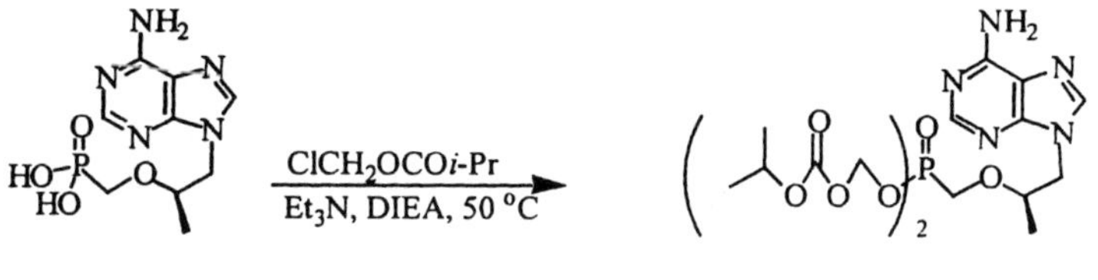

Figure 28.   Synthesis of bis-POC-PMPA.

4.2. Bis-SDTE and -SATE pronucleotides

In an effort to improve the pharmacokinetics of some nucleotides, Imbach and co-workers (Peuch *et al.*, 1993; Périgaud *et al.*, 1997) designed enzymatically activated pronucleotides, bis-[*S*-(2-hydroxyethylsulfidyl)-2-thioethyl]- (bis-SDTE) and bis-[*S*-acyl-2-thioethyl] (bis-SATE). The Bis-SDTE concept was designed to take advantage of the concentration of reductase within the cytosol to liberate the nucleotide (Figure 29). The thioethyl phosphotriester, formed after reductive cleavage of the disulfide bond, spontaneously decomposes to the phosphodiester releasing episulfide. The phosphodiester then undergoes an identical sequence of enzymatic activation steps to yield the

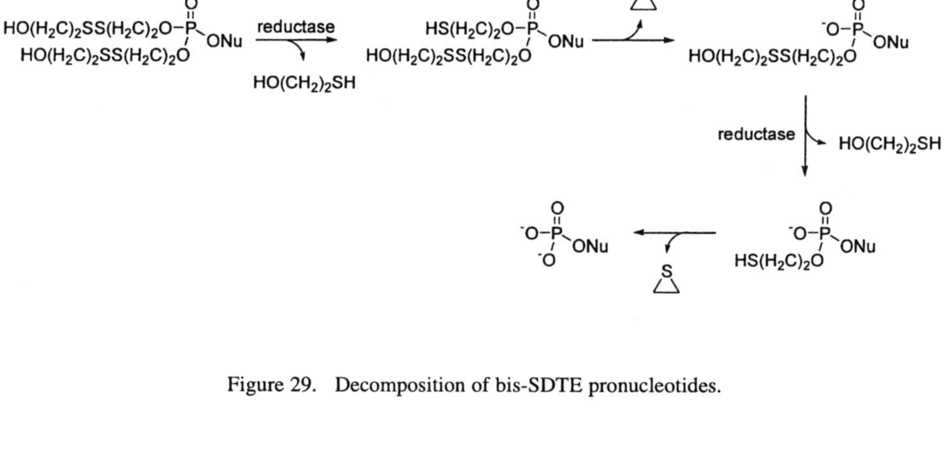

Figure 29.   Decomposition of bis-SDTE pronucleotides.

nucleotide. Both bis-SDTE triesters of 2',3'-dideoxy-2',3'-didehydrouridine monophosphate (d4UMP) and 3'-azido-2',3'-dideoxythymidine monophosphate (AZTMP) were found to be cleaved in cell extracts 30-fold faster than in culture medium (Peuch *et al.*, 1993). The bis-SDTE concept failed to improve the antitumor efficacy of 5-fluoro-2',3'-dideoxyuridine (5-FdU) (Girardet *et al.*, 1995a), whereas bis-SDTE-PMEA exhibited higher anti-HIV activity than the parent drug in MT-4 cell lines (Benzaria *et al.*, 1996). Limited success of the bis-SDTE approach is due to chemical instability as well as metabolism in serum.

The bis-SATE concept, similar to the bis-POM approach, requires esterase-mediated activation to aid in nucleotide delivery (Figure 30) (Lefebvre *et al.*, 1995). The esterase cleaves the thioester to form the thioethyl phosphonate triester that spontaneously decomposes to episulfide and the phosphonate diester, which undergoes the same sequence of enzymatic activation ultimately releasing two equivalents of episulfide. It has been shown that the addition of SDTE and SATE moieties resulted in cytotoxicities comparable to that of the parent nucleosides in human bone marrow cells (Périgaud *et al.*, 1996). This toxicity, as discussed in the case of bis-POM derivatives, may depend on the liberation of toxic metabolites (in this case, episulfide), but it is more likely due to an increase in the intracellular concentration of the parent drug, which amplifies its activity as well as its intrinsic toxicity (Benzaria *et al.*, 1996).

Among the various bis-SATE side chains synthesized, bis-*t*-buSATE-AZTMP was the most stable in culture medium and cell extract (Girardet *et al.*, 1995a). Its stability was mostly due to the bulkiness of the *t*-butyl residue, which prevents rapid cleavage. The stability of *t*-buSATE moiety was also demonstrated in the case of the *t*-buSATE derivative of 2',3'-dideoxy-3'-oxoadenosine (isoddA) (Valette *et al.*, 1996). The bis-SATE concept has been proven to be successful in the cases of 2',3'-didehydro-2',3'-dideoxythymidine monophosphate (d4TMP) (Girardet *et al.*, 1995b), AZTMP (Lefebvre *et al.*, 1995), and 2', 3'-didehydro-2',3'-dideoxyadenosine monophosphate (d4AMP) (Périgaud *et al.*, 1995), which where active against HIV-infected TK-deficient CEM (CEM/TK⁻) cells. In addition to enhanced activity, the prodrug of d4A was more stable than d4A itself against acid catalyzed depurination (Périgaud *et al.*, 1994). For many bis-SATE pronucleotides, there was a decrease in activity against HIV infected CEM/TK⁻ cells.

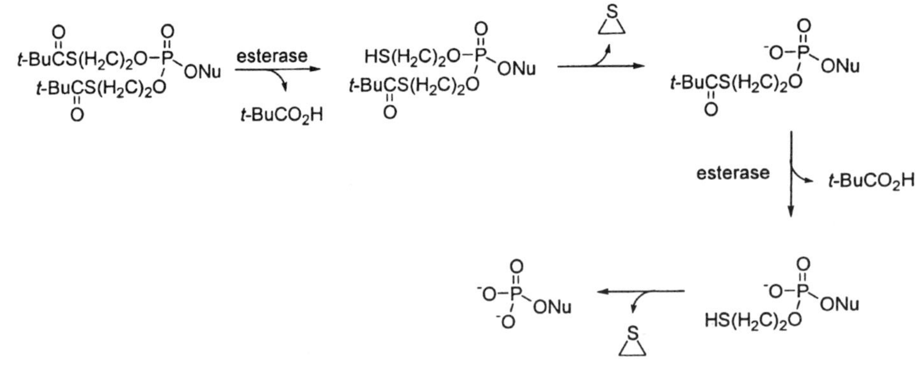

Figure 30. Decomposition of bis-SATE pronucleotides.

Thioesters of bis-SATE pronucleotides of d4T were synthesized by reaction of thiocarboxylic acids with 2-iodoethanol, followed by condensation with *N, N*-diisopropylphosphorodichloridite in tetrahydrofuran in the presence of triethylamine to yield the corresponding phosphoramidites. These were coupled with d4T in the presence of 1*H*-tetrazole and oxidized *in situ* with *tert*-butyl hydroperoxide (TBHP) to obtain the bis-SATE phosphotriester (Figure 31).

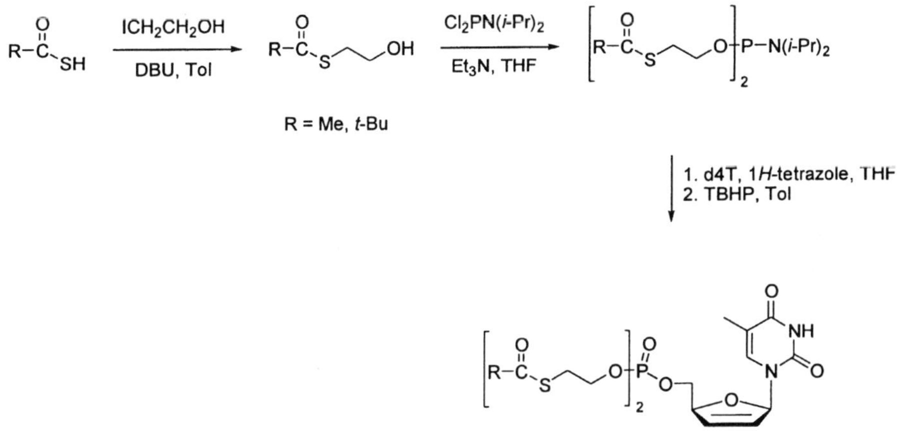

Figure 31.   Synthesis of SATE derivatives of d4TMP.

For the synthesis of the bis-DTE phosphotriester of 2',3'-dideoxyuridine (ddU) (Périgaud *et al.*, 1993), dithiodiethanol phosphodiester was synthesized by protecting dithiodiethanol with monomethoxytrityl chloride in the presence of diisopropylethanol-amine followed by phosphorylation using phosphoryloxy chloride, imidazole and triethylamine. The dithiodiethanol phosphodiester was condensed with ddU in triethylamine in the presence of 1-mesitylene-2-sulfonyl-3-nitro-1,2,4-triazole. Following treatment with acetic acid and aqueous methanol gave bis-DTE ddU (Figure 32).

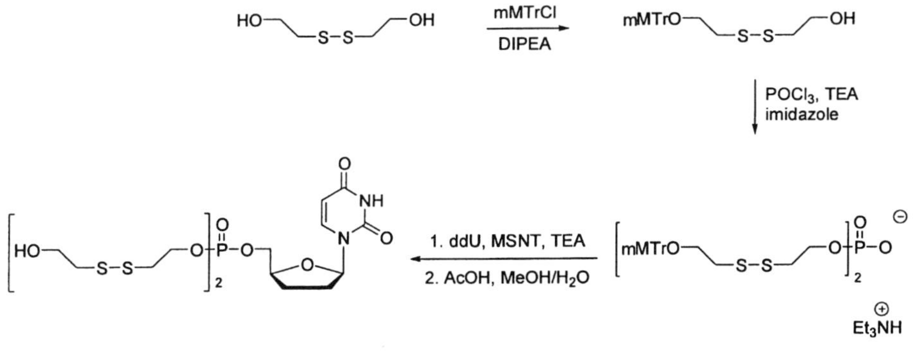

Figure 32.   Synthesis of DTE pronucleotides of ddUMP.

A notable example of the potential of SATE derivatives has been reported by Imbach and co-workers, who showed that SATE pronucleotides of ACV are more potent than the parent drug in inhibiting the replication of HBV in human HepG2.2.15 (whereas ACV itself is inactive) (Périgaud *et al.*, 1996) and of DHBV in duckling hepatocytes (Hantz *et al.*, 1999)

4.3. Bis-Acyloxybenzyl pronucleotides

Bis(4-acyloxybenzyl) pronucleotides were designed by Routledge *et al.* (1995) and Glazier *et al.* (1992) to avoid the close proximity of the negative charge of the intermediate mono-protected phosphodiester and the cleaving site of the carboxyesterase, so as to ease the cleavage of the remaining masking group. Furthermore, Freeman and co-workers (Mitchell *et al.*, 1992; Thomson *et al.*, 1993) calculated the necessary distance for avoidance of this intramolecular electronic repulsion between negative charge of the phosphodiester and carboxyester group to be approximately 4 Å in distance (Figure 33). The process of nucleotide delivery involves first the cleavage at the 4-position of the aromatic ring to yield 4-hydroxybenzyl phosphotriester that spontaneously decomposes to form the phosphodiester. The phosphodiester undergoes the same process again to yield the nucleotide.

Routledge *et al.* (1995) applied this approach for the delivery of AZTMP and found that the prodrug activity against HIV-1 and SIV was comparable to that of AZT *in vitro*. The prodrug ability to deliver AZTMP intracellularly was not determined. Glazier *et al.* (1996) demonstrated that bis(4-acyloxybenzyl) was susceptible to enzymatic cleavage in the case of ACV monophosphate in the presence of porcine liver esterase. Like with DTE pronucleotides, the bis-acyloxybenzyl derivatives were limited by their short half-lives in serum. In addition, these prodrugs are too lipophilic (log P values range from 1 to 4) for systemic administration (Glazier *et al.*, 1996). *In vivo*, the bis(4-acyloxybenzyl) derivatives of ACVMP exhibited no significant toxic side effects at concentrations up to 100 mg/kg of body weight (Glazier *et al.*, 1996) whereas, as already discussed (Hostetler *et al.*, 2000a), the triphosphate of ACV is a potent anti-HBV agent.

In the synthesis of bis-acyloxybenzyl phosphotriester of AZT, the appropriate 4-acyloxybenzyl alcohol is reacted with *N*, *N*-diisopropylphosphorochloridate in the presence of triethylamine to yield the corresponding phosphodiester. The phosphodiester is coupled with AZT in the presence of [1*H*]-tetrazole and oxidized *in situ* with *m*CPBA to obtain the phosphotriester (Figure 34) (Thomson *et al.*, 1993).

4.4. *Cyclo*Sal-pronucleotides

The *cyclo*Sal-pronucleotide concept involves nucleotide delivery based upon pH-driven selective chemical hydrolysis (Figure 35) (Meier, 1996 and 1997a; Meier *et al.*, 1997b). The tandem cleavage originates with the hydrolysis of a phenyl ester followed by hydrolysis of a benzyl ester of the phosphotriester. This concept is based upon the principle that selection of phenyl, benzyl and alkyl phosphate esters can influence the hydrolysis steps of the tripartate approach (Meier *et al.*, 1997c). The phenyl ester is cleaved initially because of stabilization caused by delocalization of the negative charge in the aromatic

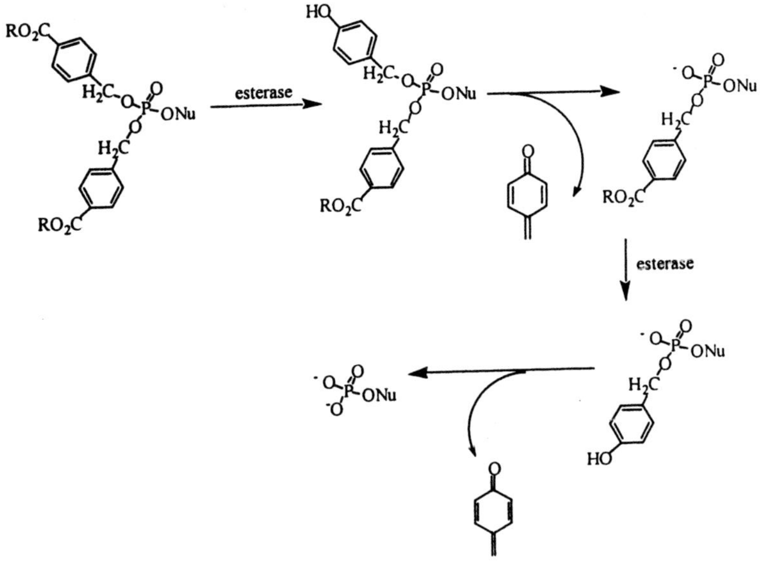

Figure 33. Proposed decomposition of acyloxybenzyl phosphotriesters.

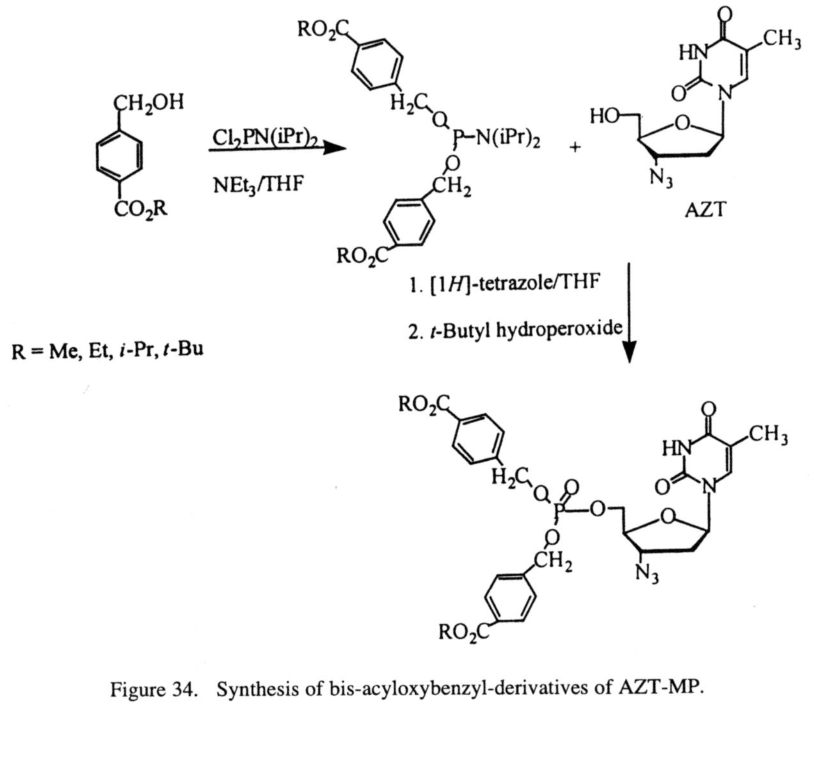

Figure 34. Synthesis of bis-acyloxybenzyl-derivatives of AZT-MP.

ring yielding the 2-hydroxybenzylphosphodiester. The sequence of hydrolytic steps has been verified by multinuclear NMR spectroscopy and mass spectrometry (Lorey *et al.*, 1997; Meier *et al.*, 1998a). This concept has been applied to anti-HIV agents such as d4T (Meier *et al.*, 1997c and 1998a), 5-FUrd (Lorey *et al.*, 1997), AZT (Maier *et al.*, 1997d and 1998b; Balzarini *et al.*, 1999), 2',3'-dideoxyadenosine (ddA) (Maier *et al.*, 1997e and 1999a), d4A (Maier *et al.*, 1999a), and 2'-fluoro-2',3'-dideoxyadenosines (F-ara-ddA and F-ribo-ddA) (Maier *et al.*, 1999b).

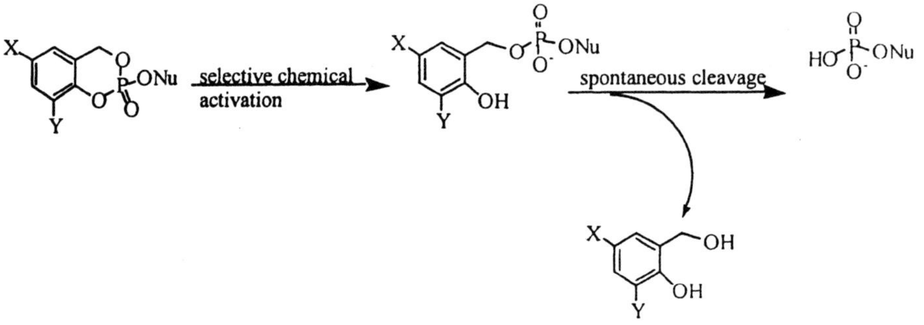

Figure 35. Proposed decomposition of *cyclo*Sal-pronucleotides.

*Cyclo*Sal-Pronucleotides have been successfully used to deliver mono-phosphates of d4T, ddA, F-ara-ddA and F-ribo-ddA. *In vitro* studies of *cyclo*Sal-nucleotides of d4T show that 3-, 5-methyl and 3,5-dimethyl-*cyclo*Sal-d4TMPs have more potent anti-HIV activity than the parent compound (Meier *et al.*, 1997c). Besides, the electron donating capacity of the ring substituent influences the degree of biological activity with the stronger electron-donating group having the greatest potency. More importantly, the potency of *cyclo*Sal-d4TMPs is maintained in CEM/TK⁻ cells and, when drug resistance is based on this kind of deficiency, the use of the *cyclo*Sal-pronucleotide can bypass it (Gröschel *et al.*, 1999). In this regard, it is notable how, although both *cyclo*Sal-d4TMP and *cyclo*Sal-AZTMP pronucleotides can bypass the TK phosphorylation step, only the former shows strong antiviral activity in TK⁻ cells. This fact has been recently explained by the relatively high hydrolysis rate of AZTMP to AZT, combined with the inability of TK⁻cells to re-phosphorylate the nucleoside (Balzarini *et al.*, 2000).

*Cyclo*Sal-ddA has been synthesized to circumvent deamination by adenosine deaminase (ADA) and adenosine monophosphate deaminase (AMPDA). Studies with ADA and AMPDA have demonstrated that the *cyclo*Sal-triesters are not susceptible to enzymatic deamination (Meier *et al.*, 1997e) as reported earlier for 5'-*O*-protected adenosine (Bloch *et al.*, 1967). *Cyclo*Sal-ddAMP and *cyclo*Sal-d4AMP show more potency than the respective parent compounds (Meier *et al.*, 1997e). In addition, this increase in potency is accompanied with a higher selectivity index than the parent compounds. *Cyclo*Sal-derivatives of F-ara-ddA and F-ribo-ddA are stable in the presence of ADA and AMPDA and more potent than the parent compounds (Meier *et al.*, 1999b).

Recently, the *cyclo*Sal approach has been applied to the acyclic anti-HSV and -EBV agents aciclovir and penciclovir (Meerbach *et al.*, 2000a). Among the evaluated derivatives, 3-methyl-*cyclo*Sal-aciclovir monophosphate retained the same activity of ACV in TK-deficient HSV-1 (HSV-1/TK⁻) strains, whereas the penciclovir prodrugs lost most of the antiviral activity of the parent drug. Also lipophilic *cyclo*Sal-BVDUMP derivatives seem to maintain anti-EBV activity compared to the parent agent BVDU (Meerbach *et al.*, 2000b).

In the synthesis of *cyclo*Sal-d4T monophosphates (Meier *et al.*, 1997c), salicylic alcohols were obtained by reduction of salicylaldehydes or salicylic acids using sodium borohydride or lithium aluminum hydride. The salicyl alcohols were reacted with phosphorus trichloride to obtain the cyclic saligenylchlorophosphanes. These were condensed with d4T in the presence of diisopropylethylamine and subsequently oxidized *in situ* by *t*-butylhydroperoxide (TBHP) to obtain the corresponding phosphotriesters (Figure 36).

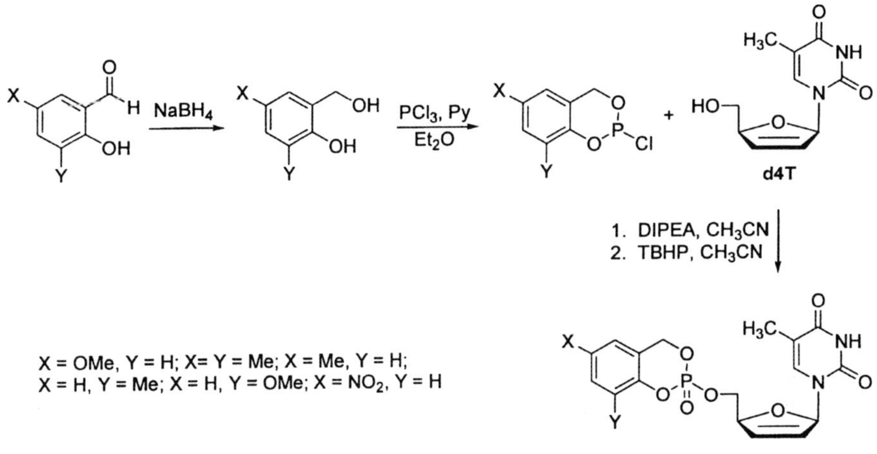

Figure 36. Synthesis of *cyclo*Sal-d4TMP.

4.5. Phosphoramidate and cyclic phosphoroamidate pronucleotides

McGuigan *et al.* (1997a) designed and synthesized phosphoramidate pronucleotides as a means of circumventing the membrane impermeability of negatively charged nucleotides (Figure 37). Phosphoramidate derivatives of d4TMP (Figure 37) (Balzarini *et al.*, 1996b and 1996c; McGuigan *et al.*, 1996a; Valette *et al.*, 1996; Beltran *et al.*, 1999; Siddiqui *et al.*, 1999), 2',3'-dideoxy-3'-thiacytidine monophosphate (3TCMP) (Balzarini *et al.*, 1996d), AZTMP (McGuigan *et al.*, 1993), (ddAMP) (Balzarini *et al.*, 1997) and d4AMP (Balzarini *et al.*, 1997; McGuigan *et al.*, 1996b) have been synthesized in an effort to enhance the delivery of their corresponding monophosphates. Delivery of the nucleotide analogue involves degradation of the prodrug by the liberation of the phenyl group or cleavage of the methyl ester, which ultimately leads to complete unmasking mediated by enzyme or chemical catalysis (Figure 37).

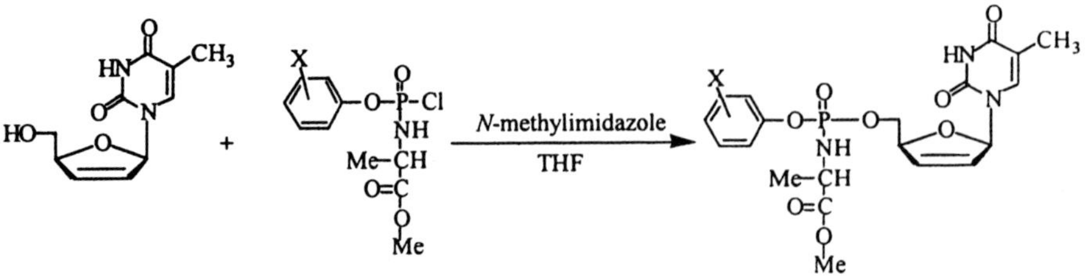

X = p-NO2, p-CN, p-Cl, p-F, p-Me, p-OMe, m-COMe, H p-CO2Me, p-COMe,

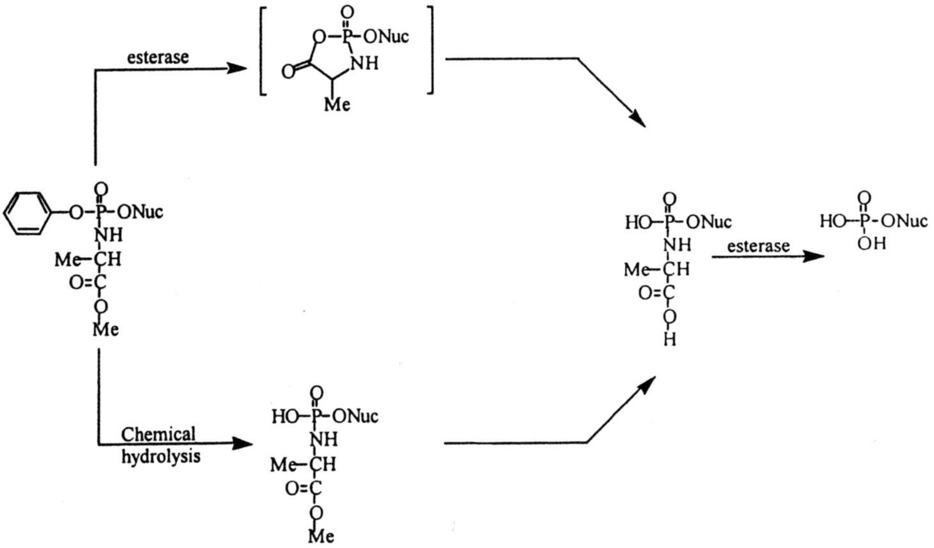

Figure 37. Synthesis of phosphoramidate derivatives of d4T and their decomposition.

In vitro studies of d4T phosphoramidate have shown that the prodrug is potent against HIV-2 infected CEM and CEM/TK$^-$ cells, which supports evidence of efficient intracellular delivery of nucleotide (Balzarini *et al.*, 1996e). The prodrug of d4T is also active against other retroviruses such as SIV, FIV, visna virus (Balzarini *et al.*, 1998) and Moloney murine sarcoma virus *in vitro* (Balzarini *et al.*, 1996e). Among various α-amino acid derivatives synthesized, the L-alanine proved the most efficacy as an antiviral (McGuigan *et al.*, 1997b), while its enantiomer was 30 times less potent. Since for d4T-based phosphoramidate prodrugs a variety of aryl substituents (Figure 37) can be tolerated, McGuigan and co-workers obtained polyether derivatives with unchanged antiviral activity and 30-fold superior water solubility (Siddiqui *et al.*, 2000).

In vitro studies indicated that phosphoramidates of 3TC and AZT provide no benefit as inhibitors of HIV replication (McGuigan *et al.*, 1993; Balzarini *et al.*, 1996d). On the other hand, in the case of ddA and d4A, the potency of the phosphoramidate prodrugs increases significantly (100-1000-fold) (Balzarini *et al.*, 1997). In addition, these phosphoramidates display anti-HBV activity equal in potency to that of 3TC. Both aryloxyphosphoramidate prodrugs show a higher selectivity index (defined as ratio of cytotoxic/antiviral concentration) in comparison with the parent compounds.

A similar effect has been observed in recently synthesized methylenecyclopropane nucleosides (Qiu *et al.*, 1999), whose weak anti-HIV-1 activity is potentiated by conversion to phosphoro-L-alaninates. These prodrugs are active against AZT, ddI, and multi-dideoxynucleosides-resistant clones (Uchida *et al.*, 1999), although they may induce resistant variants (Yoshimura *et al.*, 1999).

A phosphoramidate of ACV-5'-monophosphate has been reported, but its biological activity indicates poor intracellular phosphate delivery (McGuigan *et al.*, 2000).

According to preliminary studies on araA derivatives, the phosphoramidate prodrug may be able to release araAMP, and yet it is approximately 10 times less cytotoxic than the parent compound (Ballatore *et al.*, 2000).

Recently, Wagner and co-workers (Iyer *et al.*, 2000; Wagner *et al.*, 1999; Chang *et al.*, 2001) reported synthesis and antiviral and antitumoral activities of phosphoramidates monoesters of AZT containing amino acid methyl ester and *N*-alkyl amide moieties (Figure 38). The synthesis (Wagner *et al.*, 1999) was carried out using *H*-phosphonate chemistry. Phosphorous (III) is much more reactive than phosphorous (V) and phosphites can be easily oxidized to phosphates. Thus, transesterification of AZT with diphenyl phosphite in pyridine followed by treatment with iodine gave the phosphate. This was condensed with several amino acids methyl esters to get a series of derivatives. Then, in order to improve the stability of the prodrugs towards esterases the esters were converted to methyl amides by treatment with concentrated methylamine in methanol. Triptophan derivatives bearing different alkyl amides were also synthesized by direct condensation with triptophan alkyl amides (Wagner *et al.*, 1999).

The synthesized compounds were stable and water-soluble and displayed potent-HIV activity. Particularly, the activities of the triptophan and phenylalanine methyl esters in CEM cells (EC_{50} = 4 and <1 nm, respectively) were comparable to that of AZT (EC_{50} = <1 nm), but without the toxicity of the latter (IC_{50} > 100 μM *vs.* 14.2 μM for AZT) (Iyer *et al.*, 2000). Measurements of intracellular levels of prodrug, AZT and its phosphorylated metabolites showed that the major metabolites of the prodrugs are AZT and AZTMP (Iyer *et al.*, 2000). Both are then converted to the active triphosphate (AZTTP) (Iyer *et al.*, 2000). The reduced toxicity of the prodrugs compared to AZT correlated with reduced levels of total phosphorylated AZT, and not AZTTP (Iyer *et al.*, 2000). Substitution of the methyl amide for the methyl ester had little impact for the tryptophan derivative, but improved the anti-HIV potency of the phenylalanine derivative 60-166-fold (Iyer *et al.*, 2000). Synthesis of the equally active D-phenylalanine or D-tryptophan-substituted prodrugs proved that activity is not dependent on the stereochemistry of the amino acids (Iyer *et al.*, 2000). Pharmacokinetic studies showed improved half-life and tissue distribution of the prodrugs compared to AZT (5- and 10-fold greater for the phenylalanine methyl ester) (Wagner *et al.*, 1999).

The phosphoramidate derivatives of AZT also showed interesting antitumoral activity, particularly against the breast tumor MCF-7 cell line (Iyer *et al.*, 2000). Unlike antiviral activity, antitumoral effect was higher for L-amino acid-substituted prodrugs. Aromatic amino acid methyl esters were more cytotoxic than the aliphatic analogues (Iyer *et al.*, 2000). The selective cytotoxicity towards MCF-7 compared to CEM cells may be associated with greater intracellular levels of prodrug and/or phosphorylated metabolites of AZT.

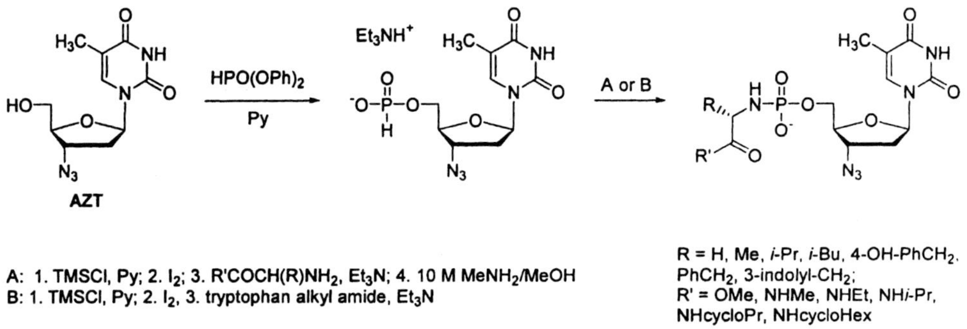

A: 1. TMSCl, Py; 2. I$_2$; 3. R'COCH(R)NH$_2$, Et$_3$N; 4. 10 M MeNH$_2$/MeOH
B: 1. TMSCl, Py; 2. I$_2$, 3. tryptophan alkyl amide, Et$_3$N

R = H, Me, *i*-Pr, *i*-Bu, 4-OH-PhCH$_2$, PhCH$_2$, 3-indolyl-CH$_2$;
R' = OMe, NHMe, NHEt, NH*i*-Pr, NHcycloPr, NHcycloHex

Figure 38.   Synthesis of phosphoramidate derivatives of AZT.

Five and six membered cyclic phosphoramidate derivatives were also used to aid in the delivery of nucleotide analogs (Figure 39). Hunston *et al.* (1984) found that 5'-*O*-3"-methyl-1",3",2"-oxazaphosphacyclopentan-2"-ylthymidine 2"-oxide and other five-membered cyclic phosphotriesters were very unstable at physiological pH. Kumar *et al.* (1990) synthesized 5'-*O*-3"-methyl-1",3",2"-oxazaphosphacyclopentan-2"-yl derivatives of ACV and (*E*)-5-(2-bromovinyl)-2'-deoxyuridine (BVDU) after finding that their hydrolysis rate may be dependent upon the buffer used (Figure 40). The phosphoramidates of BVDU were found to be inactive against HSV-1/TK$^-$ replication in rabbit kidney cells indicating no release of BVDU-monophosphate. Farquhar *et al.* (1983) synthesized the more stable six membered cyclic phosphoramidate as means to delivery 5-FdUMP intracellularly (Figure 41). Both prodrugs were resistant to enzymatic degradation by 5'-nucleotidase, alkaline phosphatase, venom phosphodiesterase and crude snake venom. When administered intraperitoneally for 5 consecutive days, the phosphoramidate prodrug was as effective as 5-fluorouracil (5-FU) at increasing the life spans of mice inoculated intraperitoneally with leukemia P-388 cells. The dioxaphosphorinanyl derivative of 5-FdUrd (X = O) showed virtually no activity in the mice model study. Both prodrugs were inactive against a P-388 mutant resistant to 5-FU in mice.

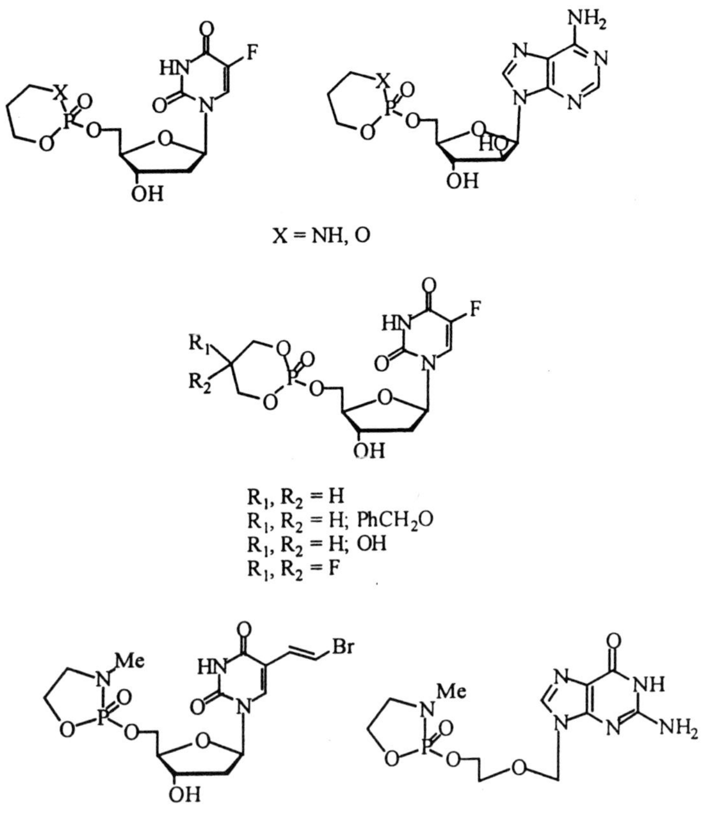

X = NH, O

R₁, R₂ = H
R₁, R₂ = H; PhCH₂O
R₁, R₂ = H; OH
R₁, R₂ = F

Figure 39. Oxazaphosphacyclic and dioxophosphacyclic nucleoside derivatives.

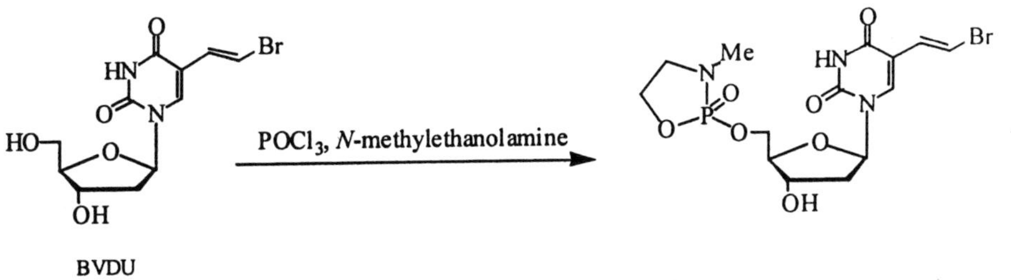

BVDU

POCl₃, *N*-methylethanolamine

Figure 40. Synthesis of oxaphosphacyclopentanyl derivative of BVDU.

Similarly, Farquhar *et al.* (1985) synthesized oxazaphosphocyclohexyl and dioxa-phosphocyclohexyl derivatives of ara-A. The phosphoramidate showed marginal effectiveness against P-388 leukemia-bearing mice. The bioconversion of six membered cyclic phosphoramidates was postulated to be mediated by a cytochrome P-450 mixed function oxidase that oxidizes the ring system forming 4''-hydroxy analogs.

Jones *et al.* (1984) synthesized 5'-substituted dioxaphosphacyclohexyl derivatives of 5-FUrd and found that the 5'',5''-difluoro compound was as potent as the parent compound towards murine leukemia L1210 cells (Figure 39). It was conceptualized that 4''-hydroxy analogs would penetrate into cells by passive diffusion and undergo spontaneous ring opening to yield the acyclic tautomers. The acyclic tautomers would dissociate releasing acrolein and the corresponding phosphoramidate (Jones *et al.*1984; Abraham and Wagner, 1994). However, its activity was minimal against thymidylate synthetase absent cells.

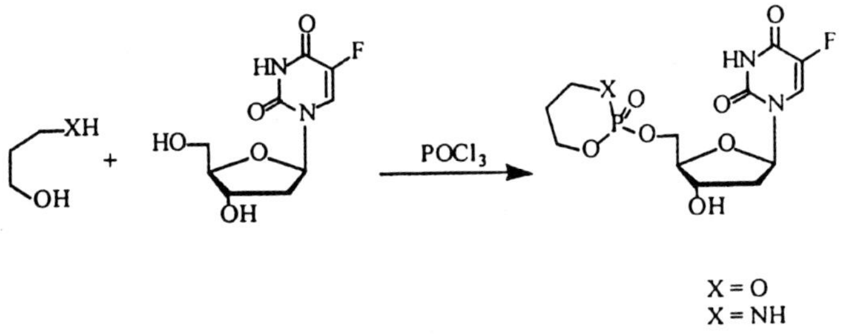

Figure 41.   Synthesis of oxaza- and dioxaphosphacyclohexyl derivatives of 5-Furd.

Lorey *et al.* (1997) have reported the phosphoramidate analogues of *cyclo*Sal-pronucleotides, namely *cyclo*Amb (i.e. *cyclo*aminobenzyl)-d4T-phosphoramidates. The proposed hydrolysis pathway of this prodrug is shown in Figure 42.

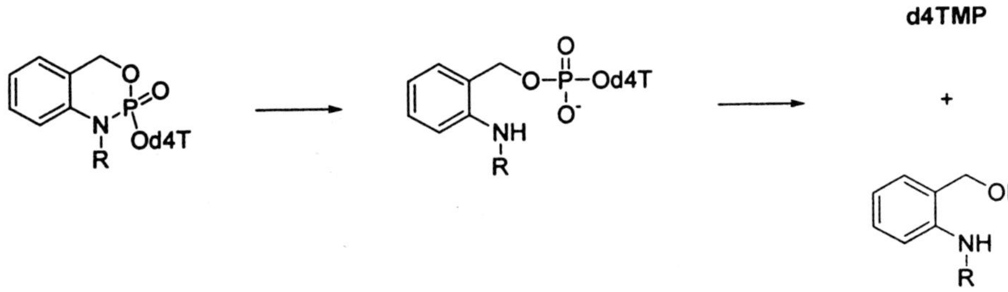

Figure 42.   Proposed mechanism for the hydrolysis of *cyclo*Amb-d4T-phosphoramidates.

Amino acid phosphoramidate diesters were also synthesized to enhance the delivery of nucleotide analogs (Figure 43). Degradation of phosphoramidate diesters is proposed to be mediated by an unknown phosphoramidase which is able to cleave the P-N bond (Abraham and Wagner, 1994; Wagner *et al.*, 1995; Abraham *et al.*, 1996). L-Tryptophan phosphoramidates appeared to be promising by providing stability (blood serum), susceptibility to enzymatic degradation, and potency. The L-Tryptophan phosphoramidate of AZT was reported to be 8 times more potent than AZT against HIV-1 infected PBM cells with no significant toxicity. In addition, the prodrug produced a 4-fold increase of phosphorylated AZT levels in comparison with AZT. In contrast, the L-tryptophan phosphoramidate of 5-FdUrd did not improve the biological activity, most probably due to latency in generation of 5-fluoro-2'-deoxyuridine (5-FUrd) (Abraham *et al.*, 1996).

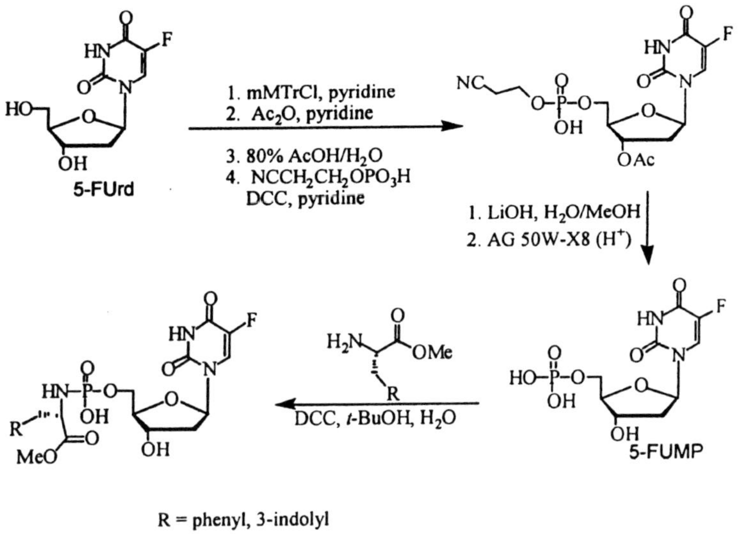

Figure 43. Synthesis of aromatic amino acid phosphoramidates of FUrd.

5. Base-modified nucleoside prodrugs

While in the classic bipartate or tripartate approaches the prodrug is made of an active moiety linked to one or more carriers, there can be another kind of approach, in which an inactive molecule is converted to an active species by transformation of a chemical functionality. The reaction involved can be an enzymatic hydrolysis, oxidation or reduction. One example of this class of prodrugs, famciclovir, has been considered above (Figure 13). Some other important examples of such prodrugs will be considered.

5.1. Prodrugs of carbovir

Carbovir (CBV, Figure 44) is a very active anti-HIV agent (Yeom *et al.*, 1989; Carter *et al.*, 1990). Although very potent *in vitro*, CBV has a low aqueous solubility, which limits its oral absorption, with oral bioavailability of 26 and 23% in rat and monkey, respectively, and may be the cause of *in vivo* toxicity seen in animal models (Daluge *et al.*, 1997). Its prodrug abacavir (1592U89) (Daluge *et al.*, 1997; Dobkin, 1999), much more soluble, is equipotent to CBV *in vitro*, but extremely more effective *in vivo*, with much higher bioavailability (92% in mice and 77% in monkeys, as the succinate salt) and higher brain delivery (Daluge *et al.*, 1997). The cellular metabolism of abacavir is very peculiar (Figure 44) (Faletto *et al.*, 1997). In fact, the concentration of CBV following administration of abacavir is very low, meaning that a direct conversion of the latter to the former is a minor process. Instead the major pathway leading to the active form CBV-TP seems to proceed *via* the monophosphorylation of abacavir by adenosine phosphotransferase, followed by deamination to CBVMP by cytosolic deaminase, and two further phosphorylation steps to CBVTP.

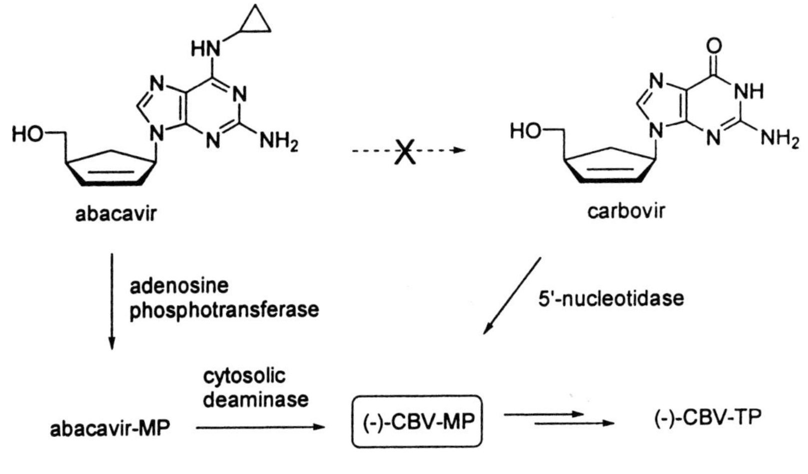

Figure 44.   Metabolism of abacavir.

The favorable pharmacokinetic properties of abacavir have allowed it to become one of the seven approved nucleoside analogue for the therapy of HIV infection. Nevertheless, a number of other prodrugs of carbovir has been synthesized and evaluated (Vince *et al.*, 1995a). Among them, particularly potent *in vitro* is 6-propoxycarbovir (Figure 45), the monophosphate of which seems to be metabolized to CBV-MP by the enzyme adenylic acid deaminase. Also 6-aminocarbovir (6-AC) has been the object of a number of studies (Yeager *et al.*, 1991; Zimmerman *et al.*, 1992 and 2000; Wen *et al.*, 1999). This prodrug is metabolized *in vitro* to CBV by ADA, and *in vivo* studies on rat show that brain delivery of CBV following administration of 6-AC is 2.5-fold higher (Wen *et al.*, 1995). The analogue 6-deoxycarbovir is rapidly converted to CBV by the enzyme xanthine oxidase (Vince *et al.*, 1995b).

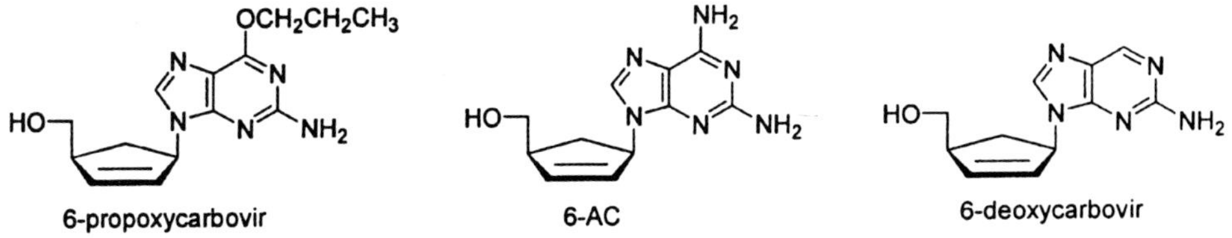

Figure 45. Other prodrugs of carbovir.

5.2. 2,6-Diaminopurine prodrugs

2,6-Diaminopurine nucleosides can be considered as prodrugs of their corresponding guaninc derivatives, to which they are converted by enzymatic oxidation.

The most promising analogue in this class is (DAPD, Figure 46), currently under clinical development as anti-HIV (Kim *et al.*, 1993a and 1993b) and anti-HBV (Schinazi *et al.*, 1994; Chin *et al.*, 1999) agent.

Pharmacokinetic studies suggest that DAPD is converted to the active nucleoside β-D-2,6-diaminopurine dioxolane (DXG) by ADA (Rajagopalan *et al.*, 1996; Gu *et al.*, 1999). The 6-chloro analogue β-D-2-amino-6-chloropurine dioxolane (ACPD) is analogously converted to DXG, whereas β-D-2-aminopurine dioxolane (APD) is metabolized to DXG by xanthine oxidase (Chen *et al.*, 1996).

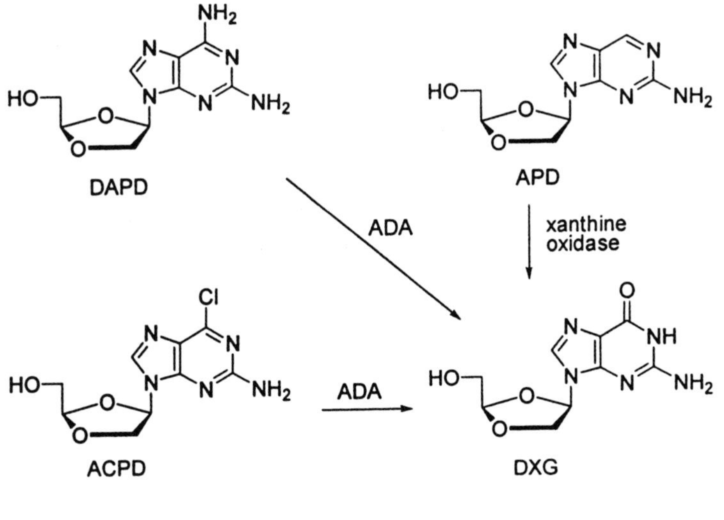

Figure 46. Biotransformation of DAPD, APD, and ACPD.

5.3. 6-Dimethylaminopurine prodrugs

The potent anti-VZV agent 6-Dimethylamino-9-(β-arabinofuranosyl)-9*H*-purine (Koszalka *et al.*, 1991) (ara-DMAP, Figure 47) is demethylated to the 6-aminopurine nucleoside ara-MAP and then to ara-A by liver microsomal enzymes. Ara-MAP and ara-A are further metabolized to the hypoxanthine analogue ara-H (Soike *et al.*, 1993).

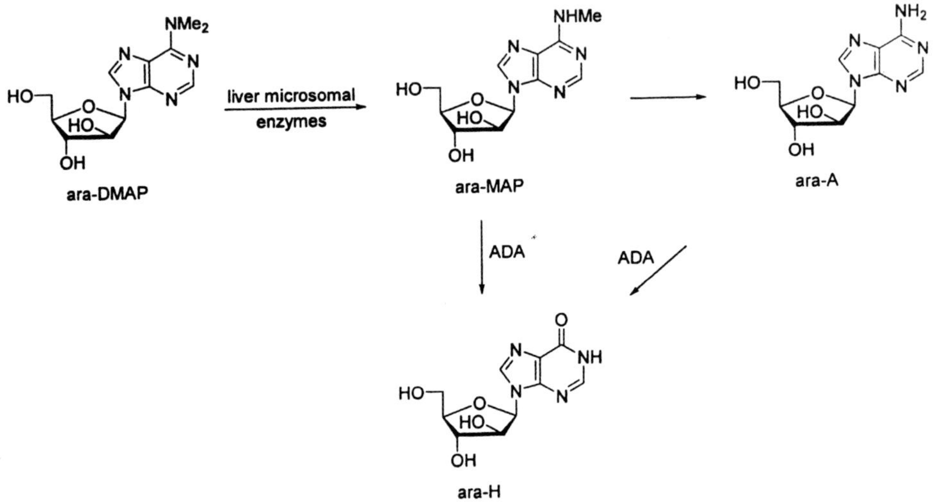

Figure 47.   Metabolism of ara-DMAP.

5.4. 6-Methoxypurine prodrugs

A number of 6-alkoxypurine arabinosides are endowed with anti-VZV activity. Among them, the 6-methoxypurine analogue (ara-M, Figure 48) displays the highest potency (Averett *et al.*, 1991). The active form of ara-M is adenine arabinoside triphosphate ara-ATP, formed with high selectivity in virus-infected cells probably thanks to a greater affinity for the VZV-encoded TK. This enzyme phosphorylates ara-M to the monophosphate, ara-MMP, which is then demethoxylated to ara-AMP by the action of AMP deaminase (De Miranda *et al.*, 1991). This is eventually di- and triphosphorylated to the DNA polymerase inhibitor ara-ATP (Biron *et al.*, 1991). Pharmacokinetic studies on rat and monkeys show poor oral bioavailability of ara-M. This seems to be due to the presystemic action of cellular ADA, responsible for the conversion of ara-M to ara-H, which leads to further metabolic deactivation. This metabolism prevents the use of ara-M orally (Burnette *et al.*, 1991).

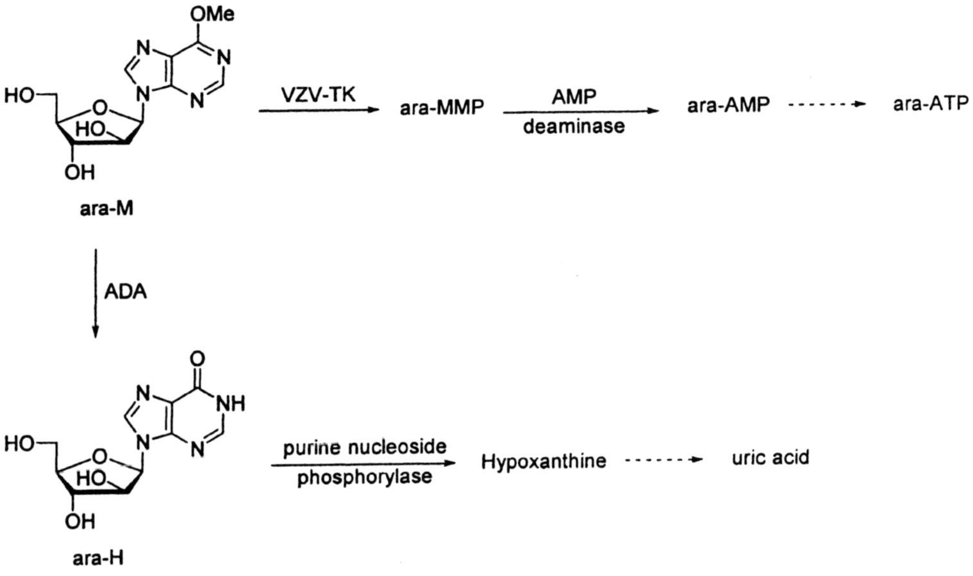

Figure 48. Farmacokinetics of ara-M.

In order to improve water solubility, and thus the bioavailability of ara-M, a series of di- and triesters have been synthesized. Of these, the 2',3'-diacetate showed the best combination of high systemic availability and water solubility (Jones *et al.*, 1992).

2-Amino-9-β-D-arabinosyl-6-methoxy-9*H*-purine (506U78, Figure 49) is a water-soluble prodrug of 9-β-D-arabinosylguanine (ara-G), endowed with antitumoral effect against refractory hematologic malignancies (Kurtzberg *et al.*, 1999).

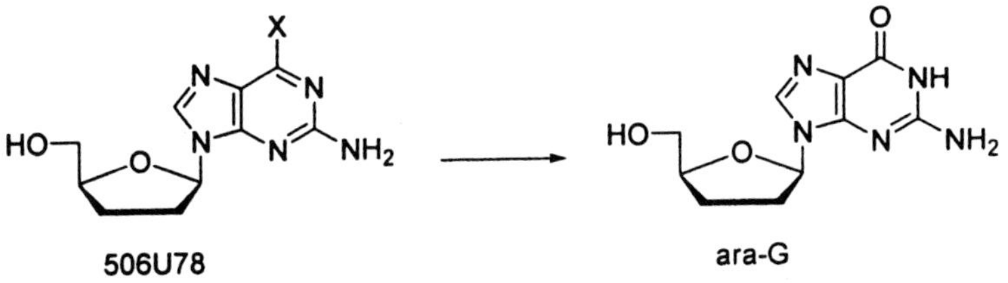

Figure 49. Prodrug of ara-G.

5.5. 6-Azidopurine prodrugs

6-Azidopurine nucleosides have been designed to exploit the azide reduction biotranformation pathway. Thus, 9-(β-D-arabinofuranosyl)-6-azidopurine (6-AAP, Figure 50) is converted to the antiherpes agent vidarabine (ara-A) by liver and brain enzymes. Although the main metabolism is observed in the liver, significant levels of ara-A were found in the brain following oral or intravenous administration of 6-AAP, whereas ara-A itself is not capable to penetrate the blood-brain barrier (BBB) (Kotra *et al.*, 1996).

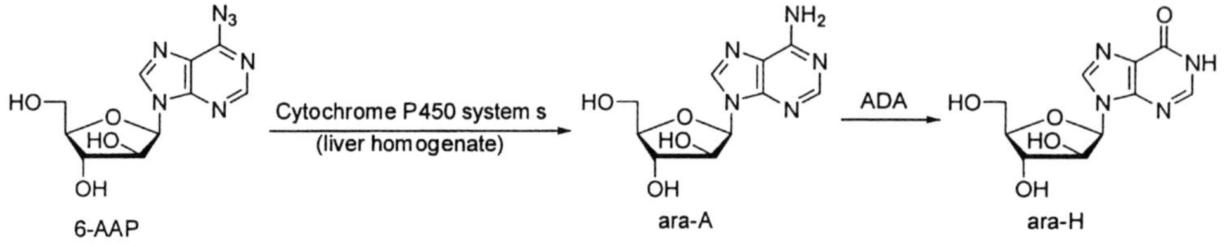

Figure 50.   Metabolism of 6-AAP.

5.6.   6-Halopurine prodrugs

Besides ACPD (Figure 46) other 6-halopurines have been synthesized and tested with the objective of finding compounds able to penetrate the BBB and be active on viral reservoirs in the central nervous system (CNS). The ability of these compounds to act as prodrugs depends on their affinity for ADA, which catalyzes their conversion to hypoxanthine analogues. Moreover, their ability to reach the CNS depends to some extent on their lipofilicity, which increases in the order I>Br>Cl>F. Thus, a series of 6-halopurine and 2-amino-6-halopurine 2',3'-dideoxynucleosides showed anti-HIV activity comparable to that of ddI and ddG (Murakami *et al.*, 1991), and in a series of 6-halo and 6-alkoxy prodrugs of 2'-β-fluoro-2',3'-dideoxyinosine, the anti-HIV activity has been correlated with ADA hydrolysis rates (Figure 51) (Ford *et al.*, 1995).

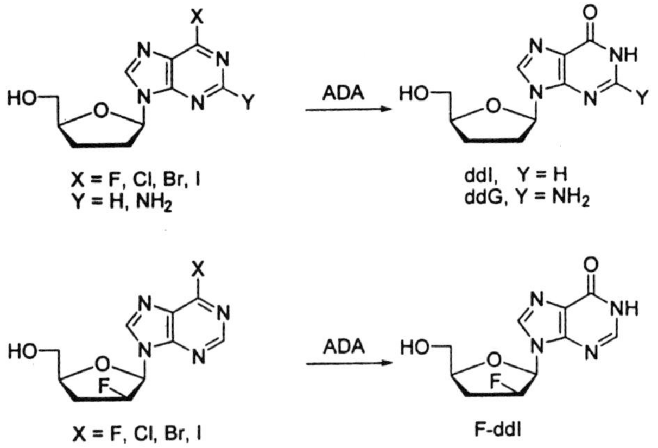

Figure 51.   6-Halo prodrugs.

5.7. Pyrimidine prodrugs

5-Halo-6-alkoxy-5,6-dihydropyrimidine nucleosides have been synthesized as prodrugs of anti-HIV agents such as AZT and ethyldeoxyuridine (EDU) (Figure 52) (Wiebe *et al.*, 1995; Wang *et al.*, 1996). Masking the 5-6 double bond with the contemporary introduction of a halogen allows more lipophilic derivatives, able to penetrate the BBB. In this series of prodrugs, the antiviral activity depends on the possibility to regenerate the parent drug *in vivo*. Thus, while the activities of the bromo and iodo analogues arc similar to those of the parent drugs, regencration rates for the chloro derivatives are too low, and these latter compounds are inactive.

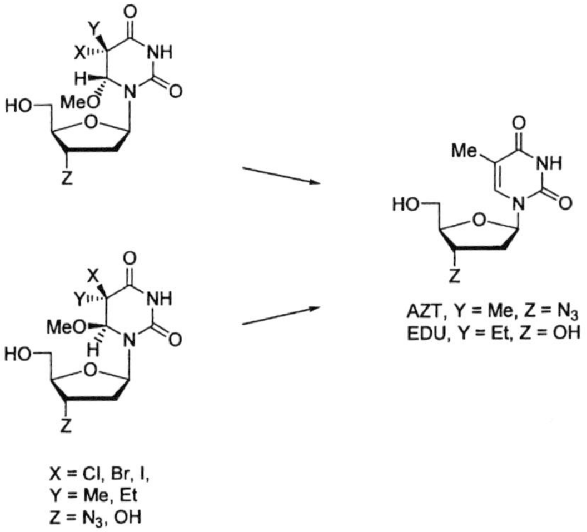

Figure 52. Pyrimidine prodrugs.

6. Conclusions

The prodrug concept has been utilized to overcome specific problems associated with certain drugs. Investigators, through the knowledge of factors influencing drug absorption, distribution, metabolism and excretion, have designed and synthesized prodrugs to resolve some problems associated with the parent drugs. The primary goal of a prodrug is to provide an effective and desired concentration of parent drug at the target organ. This goal is usually achieved by improved absorption, but ideally this must occur without accompanying toxicity. Chemical modification to increase lipid solubility to facilitate oral absorption, however, can be limited by the dissolution rate. Furthermore, a prodrug may apparently be poorly absorbed into systemic circulation as a result of first-pass gastrointestinal and liver metabolism during its initial passage through these organs.

Various alkyl and amino acid esters of nucleosides have been synthesized as a means to inhibit metabolism and improve oral bioavailability. Recent examples of prodrugs such as valaciclovir and famciclovir have shown that chemical manipulation of nucleosides can overcome the shortcomings of parent compounds. Polymer-linked nucleosides appear to be promising *in vitro*, but have not been substantiated as being beneficial *in vivo*. Phospholipids of nucleosides have been successful in targeting the lymphatic system, the brain and the liver, but oral usage of such prodrugs is unlikely due to lack of sufficient water solubility. The tripartate approach applied to pronucleotides appears to be very effective *in vitro*, but toxicity and solubility, as well as synthetic methodology issues should be taken into consideration.

Acknowledgement

This review article was supported by U.S. Public Health Service Research Grants (AI 25899, AI 32351) from the National Institute of Allergy and Infectious Diseases.

References

Abraham, T.W. and Wagner, C.R. (1994) A phosphoramidite-based synthesis of phosphoramidate amino acid diesters of antiviral nucleosides, Nucleosides & Nucleotides 13, 1891-1903.

Abraham, T.W., Kalman, T.I., McIntee, E. J. and Wagner, C.R. (1996) Synthesis and biological activity of aromatic amino acid phosphoramidates of 5-fluoro-2'-deoxyuridine and 1-β-arabinofuranosylcytosine: evidence of phosphoramidase activity, J. Med. Chem. 39, 4569-4575.

Aggarwal, S.K., Gogu, S.R., Rangan, S.R.S. and Agrawal, K.C. (1990) Synthesis and biological evaluation of prodrugs of zidovudine, J. Med. Chem. 33, 1505-1510.

Albert, A. (1958) Chemical aspects of selective toxicity, Nature 182, 421-423.

Anderson, B.D., Wygant, M.B., Xiang, T.X., Waugh, W.A. and Stella, V.J. (1988) Preformulation solubility and kinetic studies of 2',3'-dideoxypurine nucleosides - potential anti-AIDS agents, Int. J. Pharmaceut. 45, 27-37.

Annaert, P., Gosselin, G., Pompon, A., Benzaria, S., Valette, G., Imbach, J.-L., Naesens, L., Hatse, S., De Clercq, E., Van den Mooter, G., Kinget, R. and Augustijns, P. (1998) Comparison of the disposition of ester prodrugs of the antiviral agent 9-(2-phosphonylmethoxyethyl)adenine [PMEA] in Caco-2 monolayers, Pharm. Res. 15, 239-245.

Aoshima, M., Tsukagoshi, S., Sakurai, Y., Oh-ishi, J., Ishida, T. and Kobayashi, H. (1976) Antitumor activities of newly synthesized N^4-acyl-1-β-D-arabinofuranosylcytosine, Cancer Res. 36, 2726-2732.

Arimilli, M.N., Kim, C.U., Dougherty, J., Mulato, A., Oliyai, R., Shaw, J.P., Cundy, K.C. and Bischofberger, N. (1997) Synthesis, in vitro biological evaluation and oral bioavailability of 9-[2-(phosphonomethoxy)propyl]adenine (PMPA) prodrugs, Antiviral Chem. Chemother. 8, 557-564.

Averett, D.R., Doszalka, G.W., Fyfe, J.A., Roberts, G.B., Purifoy, D.J.M. and Krenitsky, T.A. (1991) 6-Methoxypurine arabinoside as a selective and potent inhibitorof varicella-zoster virus, Antimicrob. Agents Chemother. 35, 851-857.

Balajthy, Z., Aradi, J., Kiss, I.T. and Elödi, P. (1992) Synthesis and functional evaluation of a peptide derivative of 1-β-D-arabinofuranosylcytosine, J. Med. Chem. 35, 3344-3349.

Ballatore, C., McGuigan, C., De Clercq, E. and Balzarini, J. (2000) Ara-A-5'-phenyl methoxy alaninyl phosphate as a prodrug of the adenine arabinoside-monophosphate: synthesis and antiviral evaluation, Antiviral Res. 46, A63.

Balzarini, J., Naesens, L., Slachmuylders, J., Niphuis, H., Rosenburg, I., Holy, A., Schellekens, H. and De Clercq, E. (1991) 9-(2-Phosphonylmehtoxyethyl)adenine (PMEA) effectively inhibits retrovirus replication *in vitro* and simian immunodeficiency virus infection in rhesus monkeys, AIDS 5, 21-28.

Balzarini, J., Holy, A., Jindrich, J., Naesens, L., Snoeck, R., Schols, D. and De Clercq, E. (1993) Differential antiherpesvirus and antiretrovirus effects of the (*S*) and (*R*) enantiomers of acyclic nucleoside phosphonates: potent and selective *in vitro* and *in vivo* antiretrovirus activities of (R)-9-(2-Phosphonylmethoxypropyl)-2,6-Diaminopurine, Antimicrob. Agents and Chemother. 37, 332-338.

Balzarini, J., Aquaro, S., Perno, C.-F., Witvrouw, M., Holy, A. and De Clercq, E. (1996a) Activity of the (*R*)-enantiomers of 9-(2-phosphonylmethoxypropyl)adenine and 9-(2-phosphonylmethoxypropyl)-2,6-diaminopurine against human immunodeficiency virus in different human cell systems, Biochem. Biophys. Res. Comm. 219, 337-341.

Balzarini, J., Karlsson, A., Aquaro, S., Perno, C.-F. Cahard, D., Naesens, L., De Clercq, E. and McGuigan, C. (1996b) Mechanism of anti-HIV action of masked alaninyl d4T-MP derivatives, Proc. Natl. Acad. Sci. USA 93, 7295-7299.

Balzarini, J., Egberink, H., Hartmann, K., Cahard, D., Vahlenkamp, T. Thormar, H., De Clercq, E. and McGuigan, C. (1996c) Antiretrovirus specificity and intracellular metabolism of 2',3'-didehydro-2',3'-dideoxythymidine (stavudine) and its 5'-monophosphate triester prodrug So324, Mol. Pharmacol., 50, 1207-1213.

Balzarini, J., Wedgwood, O., Kruining, J., Pelemans, H., Heijtink, R., De Clercq, E. and McGuigan, C. (1996d) Anti-HIV and anti-HBV activity and resistance profile of 2',3'-dideoxy-3'-thiacytidine (3TC) and its arylphosphoramidate derivative CF 1109, Biochem. Biophys. Res. Comm. 225, 363-369.

Balzarini, J., Egberink, H., Hartmann, K., Karlsson, A., Perno, C.-F., Cahard, D., Naesens, L., Thormar, E., De Clercq, E. and McGuigan, C. (1996e) Mechanism of anti-HIV action of masked alaninyl 2',3'-dideoxy-2',3'-didehydrothymidine 5'-monophosphate derivatives, Antiviral Res. 30, A18.

Balzarini, J., Kruining, J., Wedgwood, O., Pannecouque, C., Aquaro, S., Perno, C.-F., Naessens, L., Witvrouw, M., Heijtink, R., De Clercq, E. and McGuigan, C. (1997) Conversion of 2', 3'-dideoxyadenosine (ddA) and 2', 3'-didehydro-2', 3'-dideoxyadenosine (d4A) to their corresponding aryloxyphosphoramidate derivatives markedly potentiates their activity against human immunodeficiency virus and hepatitis B virus. FEBS Lett. 410, 324-328.

Balzarini, J., Cahard, D., Wedgwood, O., Salgado, A., Velázquez, S., Yarnold, C.J., De Clercq, E., McGuigan, C. and Thormar, H. (1998) Marked inhibitory activity of masked aryloxy aminoacyl phosphoramidate derivatives of dideoxynucleoside analogues against visna virus infection, J. Acq. Imm. Def. Syndr. Human Retr. 17, 296-302.

Balzarini, J., Naesens, L., Aquaro, S., Knispel, T., Perno, C.-F., De Clercq, E. and Meier, C. (1999) Intracellular metabolism of cyclosaligenyl 3'-azido-2',3'-dideoxythimidine monophosphate, a prodrug of 3'-azido-2',3'-dideoxythymidine (zidovudine), Mol. Pharmacol. 56, 1354-1361.

Balzarini, J., Knispel, T., Naesens, L., De Clercq and Meier, C. (2000) Intracellular metabolism of 3-methyl-cycloSal-D4TMP and 3-methyl-cycloSal-AZTMP: same concept but different antiviral outcome, Antiviral Res. 46, A46-A46.

Baker, D.C., Haskell, T.H. and Putt, S.R. (1978) Prodrugs of 9-β-D-arabinofuranosyladenine. 1. Synthesis and evaluation of some 5'-(*O*-acyl) derivatives, J. Med. Chem. 21, 1218-1221.

Beadle, J.R., Rodriguez, N., Aldern, K.A. and Hostetler, K.Y. (2000) Phospholipid prodrugs of antiviral

136   *J. S. Cooperwood et al.*

nucleosides: synthesis and activity of octadecyloxyethyl, -propyl and -butyl esters of ganciclovir phosphonoformate, Antiviral Res. 46, A79-A79.

Beaucamp, L.M., Orr, G.F., De Miranda, P., Burnette, T. and Krenitsky, T.A. (1992) Amino acid ester prodrugs of acyclovir, Antiviral Chem. Chemother. 3, 157-164.

Beltran, T., Egron, D., Lefebvre, I., Périgaud, C., Pompon, A., Gosselin, G., Aubertin, A.-M. and Imbach, J.-L. (1999) Anti-HIV pro-nucleotides: SATE versus phenyl as a protecting group of AZT phosphoramidate derivatives, Nucleosides Nucleotides 4-5, 973-975.

Benzaria, S., Pélicano, H., Johnson, R., Maury, G., Imbach, J.-L., Aubertin, A.-M., Obert, G. and Gosselin, G. (1996) Synthesis, *in vitro* antiviral evaluation, and stability studies of bis(*S*-acyl-2-thioethyl) ester derivatives of 9-[2-(phosphonomethoxy)ethyl]adenine (PMEA) as potential PMEA prodrugs with improved oral bioavailability, J. Med. Chem. 39, 4958-4965.

Biron, K.K., De Miranda, P., Burnette, T.C. and Krenitsky, T.A. (1991) Selective anabolism of 6-methoxypurine arabinoside in varicella-zoster virus-infected cells, Antimicrob. Agents Chemother. 35, 2116-2120.

Bloch, A., Robins, M.J. and McCarthy, J.R., Jr. (1967) The role of the 5'-hydroxyl group of adenosine in determining substrate specificity for adenosine deaminase, J. Med. Chem. 10, 908-912.

Bodey, G.P., Freireich, E.J., Monto, R.W. and Hewlett, J.S. (1969) Cytosine arabinoside (NSC-63878) therapy for acute leukemia in adults, Cancer Chemother. Rep. 53, 59.

Bohme, H. and Budde, J. (1971) Chloromethyl-, mercaptomethyl-and imidomethyl-carbonate, Synthesis 588-590.

Boyd, M.R., Bacon, T.H. and Sutton, D. (1988) Antiherpesvirus activity of 9-(4-hydroxy-3-hydroxymethylbut-1-yl)guanine (BRL 39123) in animals, Antimicrob. Agents Chemother. 32, 358-363.

Boyd, M.R., Bacon, T.H. and Sutton, D. (1993) Comparative activity of penciclovir and acyclovir in mice infected intraperitoneally with herpes simplex virus type 1 SC 16, Antimicrob. Agents Chemother. 37, 642-645.

Bronson, J.J., Ghazzouli, I., Hitchcock, M.J.M., Russel, J.W., Klunk, L.J., Kern E.R. and Martin, J.C. (1989) *In vivo* anti-retrovirus and anti-cytomegalovirus activity of 9-((2-Phosphonylmethoxy)-ethyl)adenine (PMEA), Abstracts of Papers, 5th International Conference on AIDS, Montreal, Canada Abstract M.C.P. 74.

Buhles, W.C., Mastre, B.J. Jr. and Tinker, A.J. (1988) Ganciclovir treatment of life or sight-threatening cytomegalovirus infection: experience in 314 immunocompromised patients, Rev. Infect. Dis. 10 (Suppl. 3), S495-506.

Burnette, T.C., Koszalka, G.W., Krenitsky, T.A. and De Miranda, P. (1991) Metabolic disposition and pharmacokinetics of the antiviral agent 6-methoxypurine arabinoside in ratsand monkeys, Antimicrob. Agents Chemother. 35, 1165-1173.

Carl, P.L., Chakravarty, P.K. and Katzenellenbogen, J.A. (1981) A novel connector linkage applicable in prodrug design, J. Med. Chem. 24, 480-481.

Carter, S.G., Kessler, J.A. and Rankin, C.D. (1990) Activities of (-)-carbovir and 3'-azido-3'-deoxythymidine against human immunodeficiency virus *in vitro*, Antimicrob. Agents Chemother. 34, 1297-1300.

Chang, S.-L., Griesgraber, G.W., Southern, P.J. and Wagner, C.R. (2001) Amino acid phosphoramidate monoesters of 3'-azido-3'-deoxythymidine: relationship between antiviral potency and intracellular metabolism, J. Med. Chem. 44, 223-231.

Chen, H., Boudinot, F.D., Chu, C.K., McClure, H.M. and Schinazi, R.F. (1996) Pharmacokinetics of (-)-β-D-2-aminopurine dioxolane and (-)-β-D-2-amino-6-chloropurine dioxolane and their antiviral metabolite (-)-β-D-dioxolane guanine in rhesus monkeys, Antimicrob. Agents Chemother. 40, 2332-2336.

Cheng, L., Hostetler, K.Y., Gardner, M.F., Avila, Jr., C.P., Bergeron-Lynn, G., Keefe, K.S., Wiley, C.A.

and Freeman, W.R. (1999a) Intravitreal toxicology in rabbits of two preparations of 1-*O*-octadecyl-sn-glycerol-3-phosphonoformate, a sustained-delivery anti-CMV drug. Invest. Ophtalmol. Vis. Sci. 40, 1487-1495.

Cheng, L., Hostetler, K. Y., Gardner, M. F., Avila, Jr., C. P., Bergeron-Lynn, G., Severson, G. M., Freeman, W. R. (1999b) Intravitreal pharmacokinetics in rabbits of the foscarnet lipid prodrug: 1-*O*-octadecyl-sn-glycerol-3-phosphonoformate (ODG-PFA), Curr. Eye Res. 18, 161-167.

Cheng, L.Y., Hostetler, K.Y., Ozerdem, U., Gardner, M.F., Mach-Hofacre, B., Freeman, W.R. (1999c) Intravitreal toxicology and treatment efficacy of a long acting anti-viral lipid prodrug of ganciclovir (HDP-GCV) in liposome formulation, IOVS 40, S872-S872.

Cheng, L., Hostetler, K.Y., Chaidhawangul, S., Gardner, M.F., Beadle, J.R., Keefe, K.S., Bergeron-Lynn, G., Severson, G.M., Soules, K.A., Mueller, A.J. and Freeman, W. (2000) Intravitreal toxicology and duration of efficacy of a novel antiviral lipid prodrug of ganciclovir in liposome formulation, Invest. Ophtalmol. Vis. Sci. 41, 1523-1532.

Chin, R., Torresi, J., Earnest-Silveira, L., Trautwein, C., Bock, T., Manns, M.P., Furman, P.A., Cleary, D. and Locarnini, S.A. (1999) Two novel antiviral agents, DAPD and L-FMAU, which are active against wild-type and lamivudine-resistant HBV, Hepatology 30, 421A-421A.

Chu, C.K., Bhadti, V.S., Doshi, K.J., Etse, J.T., Gallo, J.M., Boudinot, F.D. and Schinazi, R.F. (1990) Brain targeting of anti-HIV nucleosides: synthesis and *in vitro* and *in vivo* studies of dihydopyridine derivatives of 3'-azido-2',3'-dideoxyuridine and 3'-azido-3'-deoxythymidine, J. Med. Chem. 33, 2188-2192.

Daluge, S.M., Good, S.S., Faletto, M.B., Miller, W.H., St. Clair, M.H., Boone, L.R., Tisdale, M., Parry, N.R., Reardon, J.E., Dornsife, R.E., Averett, D.R. and Krenitsky, T.A. (1997) 1592U89, A novel carbocyclic nucleoside analog with potent, selective anti-human immunodeficiency virus activity, Antimicrob. Agents Chemother. 41, 1082-1093.

De Clercq, E. (1983) Broad-spectrum anti-DNA virus and anti-retrovirus activity of phosphonylmethoxyalkylpurine and pyrimidines, Biochem. Pharmacol. 1991, 42, 963-972.

De Miranda, P. and Blum, M.R. (1983) Pharmacokinetics of acyclovir after intravenous and oral administration, J. Antimicrob. Chemother. 12 (Suppl. B) 29-37.

De Miranda, P., Burnette, T.C., Biron, K.K., Miller, R.L., Averett, D.R. and Krenitsky, T.A. (1991) Anabolic pathway of 6-methoxypurine arabinoside in cells infected with varicella-zoster virus, Antimicrob. Agents Chemother. 35, 2121-2124.

Dessolin, J., Vlieghe, P., Bouygues, M., Medou, M., Quéléver, G., Camplo, M., Chermann, J.C. and Kraus, J.L. (1998) Tri-*N*-Boc-tetraazamacrocycle-nucleoside conjugates: synthesis and anti-HIV activities, Nucleosides & Nucleotides 17, 957-968.

Dobkin, J.F. (1999) Abacavir enters the clinic, Infect. Med. 16, 7-7.

Drew, W.L, Ives, D., Lalezari, J.P., Crumpacker, C., Follansbee, S.E., Spector, S.A., Benson, C. A., Friedberg, D.N. and Hubbard, L. (1995) Oral ganciclovir as maintenance treatment for cytomegalovirus retinitis in patients with AIDS, N. Engl. J. Med. 333, 615-620.

Faletto, M.B., Miller, W.H., Garvey, E.P., St. Clair, M.H., Daluge, S.M. and Good, S.S. (1997) Unique intracellular activation of the potent anti-human immunodeficiency virus agent 1592U89, Antimicrob. Agents Chemother. 41, 1099-1107.

Farquhar, D., Kuttesch, N.J., Wilkerson, M.G. and Winkler, T. (1983) Synthesis and biological evaluation of neutral derivatives of 5'-fluoro-2'-deoxyuridine 5'-phosphate, J. Med. Chem. 26, 1153-1158.

Farquhar, D. and Smith, R. (1985) Synthesis and biological evaluation of 9-[5'-(2-oxo-1,3,2-oxazaphosphorinan-2-yl)-β-D-arabinosyl]adenine and 9-[5'-(2-Oxo-1,3,2,-dioxaphosphorinan-2-yl)-β-D-arabinosyl]adenine: potential neutral precursors of 9-[β-D-arabinofuranosyl]adenine 5'-monophosphate, J. Med. Chem.

28, 1358-1361.

Farquhar, D., Khan, S., Srivastva, D.N. and Saunders, P.P. (1994) Synthesis and antitumor evaluation of bis[(pivaloyloxy)methyl]-2'-deoxy-5-fluorouridine 5'-monophosphate (FdUMP): A strategy to introduce nucleotides into cells, J. Med. Chem. 37, 3902-3909.

Faulds, D. and Brodden, R.N. (1992) Didanosine. A review of its antiviral activity, pharmacokinetic properties and therapeutic potential in human immunodeficiency virus infection, Drugs 44, 94-116.

Filer, C.W., Allen, G.D. and Brown, T.A. (1994) Metabolic and pharmacokinetic studies following oral administration of ^{14}C-famciclovir to healthy subjects, Xenobiotica 24, 357-368.

Ford, Jr., H., Siddiqui, M.A., Driscoll, J.S., Marquez, V.E., Kelley, J.A., Mitsuya, H. and Shirasaka, T. (1995) Lipophilic, acid-stable, adenosine deaminase-activated anti-HIV prodrugs for central nervous system delivery. 2. 6-Halo and 6-alkoxy prodrugs of 2'-β-fluoro-2',3'-dideoxyinosine, J. Med. Chem. 38, 1189-1195.

Fridland, A., Robbins, B.L., Srinivas, R.V., Arimilli, M., Kim, C. and Bischofberger, N. (1997) Antiretroviral activity and metabolism of bis(POC)PMPA, an oral bioavailable prodrug of PMPA, Antiviral Res. 34, A49, 27.

Girardet, J.-L., Gosselin, G., Périgaud, C., Balzarini, J., De Clercq, E. and Imbach, J.-L. (1995a) Synthesis and antitumor properties of some neutral triesters of 5-fluoro-2'-deoxyuridine-5'-monophosphate and 3',5'-cyclic monophosphate, Nucleosides Nucleotides 14, 645-647.

Girardet, J.-L., Périgaud, C., Aubertin, A.-M., Gosselin, G., Kirn, A. and Imbach, J.-L. (1995b) Increase of the anti-HIV activity of D4T in human T-cell culture by the use of the SATE pro-nucleotide approach, Bioorg. Med. Chem. Lett. 5, 2981-2984.

Glazier, A., Kwong, C., Rose, J. and Buckheit, R. (1992) Antiviral Res. 17(Suppl. 1), 66.

Glazier, A., Yanachkova, M., Yanachkov, I., Wright, G.E., Kern, E.R., Sidwell, R., Smee, D., McKeough, M. and Sprunance, S.L. (1996) 9th International Conference on Antiviral Research, Japan, May 19-24.

Gröschel, B., Meier, C., Zehner, R., Cinatl, J., Doerr, H.W. and Cinatl Jr., J. (1999) Effects of cycloSal-D4TMP derivatives in H9 cells with induced AZT resistance phenotype, Nucleosides Nucleotides 18, 933-936.

Gu, Z., Wainberg, M.A., Nguyen-Ba, N., L'Heureux, L., De Muys, J.-M., Bowlin, T.L. and Rando, R.F. (1999) Mechanism of action and in vitro activity of 1',3'-dioxolanylpurine nucleoside analogues against sensitive and drug-resistant human immunodeficiency virus type 1 variants, Antimicrob. Agents Chemother. 43, 2376-2382.

Hamamoto, U., Nakashima, H., Matsui, A., Matsuda, A., Ueda, T. and Yamamoto, N. (1987) Inhibitory effects of 2',3'-didehydro-2',3'-dideoxynucleosides on infectivity, cytopathic effects, and replication of human immunodeficiency virus, Antimicrob. Agents Chemother. 31, 907-910.

Hantz, O., Périgaud, C., Borel, C., Jamard, C., Zoulim, F., Trépo, C., Imbach, J.-L. and Gosselin, G. (1999) The SATE pro-nucleotide approach applied to acyclovir. Part II. Effects of bis (SATE)phosphotriester derivatives of acyclovir on duck hepatitis B virus replication in vitro and in vivo. Antiviral Res. 40, 179-187.

Harnden, M.R., Jarvest, R.L., Bacon, T.H. and Boyd, M.R. (1987) Synthesis and antiviral activity of 9-[4-Hydroxy-3-(hydroxymethyl)but-1-yl]purines, J. Med. Chem. 30, 1636-1642.

Harnden, M.R., Jarvest, R.L., Boyd, M.R., Sutton, D. and Vere Hodge, R.A. (1989) Prodrugs of the selective antiherpes virus agent 9-[4-hydroxy-3-(hydroxymethyl)but-1-yl]guanine (BRL 39123) with improved gastrointestinal absorption properties, J. Med. Chem. 32, 1738-1743.

Holland, J.F. and Glidewell, O. (1970) Complementary chemotherapy in acute leukemia. Recent results, Cancer Res. 30, 95.

Hong, C.I., Nechaev, A. and West, C.R. (1979) Nucleoside conjugates as potential antitumor agents. 2. Synthesis and biological activity of 1-(β-D-arabinofuranosyl)cytosine conjugates of prednisolone and prednisone, J. Med. Chem. 22, 1428-1432.

Hong, C.I., Kirisits, A.J., Nechaev, A., Buchheit, D.J. and West, C.R. (1990a) Nucleoside conjugates. 11. Synthesis and antitumor activity of 1-β-D-arabinofuranosylcytosine and cytidine conjugates of thioether lipids, J. Med. Chem. 33, 1380-1386.

Hong, C.I., Bernacki, R.J., Hui, S.-W., Rustum, Y. and West, C.R. (1990b) Formulation, stability, and antitumor activity of 1-β-D-arabinofuranosylcytosine conjugate of thioether phospholipid, Cancer Res. 50, 4401-4406.

Hong, C.I., West, C.R, Bernacki, R.J., Tebbi, C. and Berdel, W.E. (1991) 1-β-D-Arabinofuranosylcytosine conjugates of ether and thioether phospholipids. A new class of ara-C prodrug with improved antitumor activity, Lipids 26, 1437-1444.

Hong, C.I., Nechaev, A., Kirisits, A.J., Vig, R., West, C.R., Manouilov, K.K. and Chu, C.K. (1996) Nucleoside conjugates. 15. Synthesis and biological activity of anti-HIV nucleoside conjugates of ether and thioether phospholipids, J. Med. Chem. 39, 1771-1777.

Hostetler, K.Y., Richman, D.D., Sridhar, C.N. and Gardner, M.F. (1994) Antiviral activity of phosphatidyl-dideoxycytidine in hepatitis B infected cells and enhanced hepatitic uptake in mice, Antiviral Res. 24, 59-67.

Hostetler, K.Y., Kini, G.D., Beadle, J.R., Aldern, K.A., Gardner, M.F., Border, R., Kumar, R., Barshak, L., Sridhar, C.N. and Wheeler, C.J. (1996) Lipid prodrugs of phosphonoacids: greatly enhanced antiviral activity of 1-*O*-octadecyl-*sn*-glycero-3-phosphonoformate in HIV-1, HSV-1 and HCMV-infected cells, *in vitro*, Antiviral Res. 31, 59-67.

Hostetler, K.Y., Beadle, J.R., Hornbuckle, W.E., Bellezza, C.A., Tochkov, I.A., Cote, P.J., Gerin, J.L., Korba, B.E. and Tennant, B.C. (2000a) Antiviral activities of oral 1-*O*-hexadecylpropanediol-3-phosphoacyclovir and acyclovir in woodchucks with chronic woodchuck hepatitis virus infection, Antimicrob. Agents Chemother. 44, 1964-1969.

Hostetler, K.Y., Wright, K.N., Gardner, M.F. and Beadle, J.R. (2000b) Alkylglycerol foscarnet analogs active against drug-resistant HIV-1 are orally bioavailable in mice, Antiviral Res. 46, A70-A70.

Hunston, R.N., Jones, A.S., McGuigan, C., Walker, R.T., Balzarini, J. and De Clercq, E. (1984) Synthesis and biological properties of some cyclic phosphotriesters derived from 2'-deoxy-5-fluorouridine, J. Med. Chem. 27, 440-444.

Ibrahim, S.S., Boudinot, F.D., Schinazi, R.F. and Chu, C.K. (1996) Physicochemical properties, bioconversion and disposition of lipophilic prodrugs of 2',3'-dideoxycytidine, Antiviral Chem. Chemother. 7, 167-172.

Ichikawa, H., Onishi, H., Takahata, T., Machida, Y. and Nagai, T. (1993) Evaluation of the conjugate between N^4-(4-carboxybutyryl)-1-β-D-arabinofuranosylcytosine and chitosan as a macromolecular prodrug of 1-β-D-arabinofuranosylcytosine, Drug Des. Disc. 10, 343-353.

Iyer, V.V., Griesgraber, G.W., Radmer, M.R., McIntee, E.J. and Wagner, C.R. (2000) Synthesis, in vitro anti-breast cancer activity, and intracellular decomposition of amino acid methyl ester and alkyl amide phosphoramidate monoesters of 3'-azido-3'-deoxythymidine (AZT), J. Med. Chem. 43, 2266-2274.

Ji Y., Moog, C. Schmitt, G., Bischoff, P. and Luu, B. (1990) Monophosphoric acid diesters of 7β-hydroxycholestrol and of pyrimidine nucleosides as potential antitumor agents: synthesis and preliminary evaluation of antitumor activity, J. Med. Chem. 33, 2264-2270.

Jones, A.S., McGuigan, C., Walker, R.T., Balzarini, J. and De Clercq, E. (1984) Synthesis, properties and biological activity of some cyclic phosphoramidates nucleoside, J. Chem. Soc. Perkin Trans 1 1471-1474.

Jones, L.A., Moorman, A.R., Chamberlain, S.D., De Miranda, P., Reynolds, D.J., Burns, C.L. and Krenitsky, T.A. Di- and triester prodrugs of the varicellzoster antiviral agent 6-methoxypurine arabinoside, J. Med. Chem. 1992, 35, 56-63.

Jones, R.J., Arimilli, M.N., Lin, K.-Y., Louie, M.S., McGee, L.R., Shaw, J.-P., Burman, D., Lee, M., Kennedy, J.A., Prisbe, E.J., Bischofberger, N., Lee, W.A. and Cundy, K.C. (1996) Abstract Op11, XII International Roundtable Nucleosides, Nucleotides and their biological Application, September 15-19, La Jolla, USA.

Kawaguchi, T., Hasegawa, T., Seki, T. Juni, K., Morimoto, Y., Miyakawa, A, Saneyoshi, M. Prodrugs of 2',3'-Dideoxyinosine (DDI): Improved oral bioavailability *via* hydrophobic esters, Chem. Pharm. Bull. 1992, 40, 1338-1340.

Kim, H.O., Schinazi, R.F., Nampalli, S., Shanmuganathan, K., Cannon, D.L., Alves, A.J., Jeong, L.S., Beach, J.W. and Chu, C.K. (1993a) 1,3-Dioxolanylpurine nucleosides (2R,4R) and (2R,4S) with selective anti-HIV-1 activity in human lymphocytes, J. Med. Chem. 36, 30-37.

Kim, H.O., Schinazi, R.F., Shanmuganathan, K., Jeong, L.S., Beach, J.W., Nampalli, S., Cannon, D.L., Chu, C.K. (1993b) L-β-(2S,4S)- and L-α-(2S,4R)-Dioxolanyl nucleosides as potential anti-HIV agents: asymmetric synthesis and structure-activity relationships, J. Med. Chem. 36, 519-528.

Kim, D.-K., Kim, H.-K. and Chae, Y.-B. (1994) Design and synthesis of 6-fluoropurine acyclonucleosides: potential prodrugs of acyclovir and ganciclovir, Bioorg. Med. Chem. Lett. 4, 1309-1312.

Kim, D,-K., Lee, N., Kim, Y.-K., Chang, K., Kim, J.-S., Im, G.-J., Choi, W.-S., Jung, I., Kim, T.-S., Hwang, Y.-Y., Min, D.-S., Um, K.A., Cho, Y.-B. and Kim, K.H. (1998) Synthesis and evaluation of 2-amino-9-(3-hydroxymethyl-4-alkoxycarbonyloxybut-1-yl)purines as potential prodrugs of penciclovir, J. Med. Chem. 41, 3435-3441.

Kim, D.-K., Chang, K., Im, G.-J., Kim, H.-T., Lee, N. and Kim, K.H. (1999) Synthesis and evaluation of 2-amino-9-(1,3-dihydroxy-2-propoxymethyl)-6-fluoropurine mono- and diesters as potential prodrugs of ganciclovir, J. Med. Chem. 42, 324-328.

Kobayashi, H., Shiraki, K. and Ikada, Y. (1992) Toxicity test of biodegradable polymers by implantation in rabbit cornea, J. Biomed. Mater. Res. 26, 1463-1476.

Kodama, K., Morozumi, M., Saitoh, K., Kuninaka, A., Yoshino, H. and Saneyoshi, M. (1989) Antitumor activity and pharmacology of 1-β-D-arabinofuranosylcytosine-5'-stearylphosphate: an orally active derivative of 1-β-D-arabinofuranosylcytosine, Jpn. J. Cancer Res. 80, 679-685.

Koszalka, G.W., Averett, D.R., Fyfe, J.A., Roberts, G.B., Spector, T., Biron, K. and Krenitsky, T.A. (1991) 6-*N*-Substituted derivatives of adenine arabinoside as selective inhibitors of varicella-zoster virus, Antimicrob. Agents Chemother. 35, 1437-1443.

Kotra, L.P., Manouilov, K.K., Cretton-Scott, E., Sommadossi, J.-P., Boudinot, F.D., Schinazi, R.F. and Chu, C.K. (1996) Synthesis, biotransformation, and pharmacokinetic studies of 9-(beta-D-arabinofuranosyl)-6-azidopurine: a prodrug for ara-A designed to utilize the azide reduction pathway, J. Med. Chem. 39, 5202-5207.

Kraus, J.L., Camplo, M. and Mourier, N. (1997) "Derivati 1,3-ossatiolanici ad attività antivirale", Patent Inserm-Zambon Italie, pending.

Kumar, V., Amann, A., Ourisson, G. and Luu, B. (1987) Stereospecific syntheses of 7-β-hydroxycholesterol and 7-α-hydroxycholesterol, Synth. Commun. 17, 1279-1286.

Kumar, A., Coe, P.L., Jones, A.S., Walker, R.T., Balzarini, J., De Clercq, E. (1990) Synthesis and biological evaluation of some cyclic phosphoramidate nucleoside derivatives, J. Med. Chem. 33, 2368-2375.

Kurtzberg, J., Ernst, T.J., Keating, M.J., Gandhi, V., Hodge, J.P., Kisor, D.F., Therriault, R., Stephens, C., Levin, J., Krenitsky, T., Elion, G.B. and Mitchell, B.S. (1999) A phase I study of 2-amino-9-β-D-arabinosyl-6-methoxy-9H-purine (506U78) administered on a consecutive five day schedule in children and adults with refractory hematologic malignancies, Blood 94, 629A-629A.

Laskin, O.L., Cederberg, D.M., Mills, J., Eron, L.J., Mildvan, D., Spector, S.S. (1987) Ganciclovir for the treatment and suppression of serious infections caused by cytomegaloviruses, Am. J. Med. 83, 201-207.

Lefebvre, I., Périgaud, C., Pompon, A., Aubertin, A.-M., Girardet, J.-L., Kirn, A., Gosselin, G. and Imbach, J.-L. (1995) Mononucleoside phosphotriester deravitives with *S*-acyl-2-thioethyl bioreversible phosphate protecting groups: intracellular delivery of 3'-azido-2',3'-dideoxythymidine 5'-monophosphate, J. Med. Chem. 38, 3941-3950.

Lewis, L.D., Fowle, A.S.E. and Bittiner, S.B. (1986) Human gastrointestinal absorption of acyclovir from tablet duodenal infusion and sipped solution, Br. J. Clin. Pharmacol. 21, 459-462.

Lorey, M., Meier, C., De Clercq, E. and Balzarini, J. (1997) New synthesis and antitumor activity of cycloSal-derivatives of 5-fluoro-2'-deoxyuridinemonophosphate, Nucleosides Nucleotides 16, 789-792.

Manouilov, K.K., Fedorov, I.I., Boudinot, F.D., White, C.A., Kotra, K.P., Schinazi, R.F., Hong, C.I. and Chu, C.K. (1995) Lymphatic targeting of anti-HIV nucleosides: distribution of 3'-azido-3'-deoxythymidine (AZT) and 3'-azido-2',3'-deoxyuridine (AzdU) after administration of dipalmitoylphosphatidyl prodrugs to mice, Antiviral Chem. Chemother. 6, 230-238.

Manouilov, K.K., Xu, Z.-S., Boudinot, F.D., Schinazi, R.F. and Chu, C.K. (1997) Lymphatic targeting of anti-HIV nucleosides: distribution of 2',3'-dideoxyinosine after intravenous and oral administration of dipalmitoylphosphatidyl prodrug in mice, Antiviral Res. 34, 91-99.

Mayaux, J.F., Bousseau, A., Pauwels, R., Huet, T., Henin, Y., Dereu, N., Evers, M., Soler, F., Poujade, C., De Clercq, E., Le Pecq, J.B. (1994) Triterpene derivatives that block entry of human immunodeficiency virus type 1 into cells, Proc. Natl. Acad. Sci. U.S.A. 91, 3564-3568.

McGuigan, C., Pathirana, R.N., Balzarini, J. and De Clercq, E. (1993) Intracellular delivery of bioactive AZT nucleotides by aryl phosphate derivatives of AZT, J. Med. Chem. 36, 1048-1052.

McGuigan, C., Cahard, D., Sheeka, H.M., De Clercq, E. and Balzarini, J. (1996a) Aryl phosphoramidate derivatives of d4T have improved anti-HIV efficacy in tissue culture and may act by the generation of a novel intracellular metabolite, J. Med. Chem. 39, 1748-1753.

McGuigan, C., Wedgwood, O.M., De Clercq, E. and Balzarini, J. (1996b) Phosphoramidate derivatives of 2',3'-didehydro-2',3'-dideoxyadenosine [d4a] have markedly improved anti-HIV potency and selectivity, Bioorg. Med. Chem. Lett. 6, 2359-2362.

McGuigan, C., Camarasa, M.-J., Egberink, H., Hartmann, K., Karlsson, A., Perno, C. F. and Balzarini, J. (1997a) Design, synthesis and biological evaluation of novel nucleotide prodrugs as inhibitors of HIV replication, Int. Antiviral News 5, 19-21.

McGuigan, C., Tsang, H.-W., Cahard, D., Turner, K., Velazquez, S., Salgado, A., Bidois, L., Naesens, L., De Clercq, E. and Balzarini, J. (1997b) Phosphoramidates derivatives of d4T as inhibitors of HIV: the effect of amino acid variation, Antiviral Res. 35, 195-204.

McGuigan, C., Slater, M.J., Parry, N.R., Perry, A. and Harris, S. (2000) Synthesis and antiviral activity of acyclovir-5'-(phenyl methoxy alaninyl) phosphate as a possible membrane-soluble nucleotide prodrug, Bioorg. Med. Chem. Lett. 10, 645-647.

Meerbach, A., Klöcking, R., Meier, C., Lomp, A., Helbig, B. and Wutzler, P. (2000a) Inhibitory effect of cycloSaligenyl-nucleoside monophosphates (cycloSal-NMP) of acyclic nucleoside analogues on HSV-1 and EBV, Antiviral Res. 45, 69-77.

Meerbach, A., Wutzler, P., Lomp, A. and Meier, C. (2000b) Synthesis, properties and anti-EBV activity of a new series of 3'-modified cycloSal-BVDUMP pro-nucleotides, Antiviral Res. 46, A82.

Meier, C. (1996) 4H-1.3.2.-Benzodioxaphosphorin-2-nucleosyl-2-oxide - a new concept for lipophilic, potential prodrugs of biologically active nucleoside monophosphates, Angew. Chem. 108, 77-79.

Meier, C. (1997a) Pro-nucleotides - recent advances in the design of efficient tools for the delivery of biologically active nucleoside monophosphate, Synlett 233-242.

Meier, C., Knispel, T., Lorey, M. and Balzarini, J. (1997b) CycloSal-pro-nucleotides: the design and biological evaluation of a new class of lipophilic nucleotide prodrugs, Int. Antiviral News 5, 183-186.

Meier, C., Lorey, M., De Clercq, E. and Balzarini, J. (1997c) Cyclic saligenyl phosphotriesters of 2',3'-dideoxy-2',3'-didehydrothymidine (d4T) - a new pro-nucleotide approach, Bioorg. Med. Chem. Lett. 7, 99-104.

Meier, C., De Clercq, E. and Balzarini, J. (1997d) Cyclo-saligenyl-3'-azido-2',3'-dideoxythymidine monophosphate (CycloSal-AZTMP) - a new pro-nucleotide approach, Nucleosides Nucleotides 16, 793-796.

Meier, C., Knispel, T., De Clercq, E., Balzarini, J. (1997e) ADA-Bypass by lipophilic cycloSal-ddAMP pro-nucleotides - a second example of the efficiency of the cycloSal concept, Bioorg. Med. Chem. Lett. 1577-1582.

Meier, C., Lorey, M., De Clercq, E. and Balzarini, J. (1998a) CycloSal-2',3'-dideoxy-2',3'-didehydrothymidine monophosphate (cycloSal-d4TMP): synthesis and antiviral evaluation of a new d4TMP delivery system, J. Med. Chem. 41, 1417-1427.

Meier, C., De Clercq, E. and Balzarini, J. (1998b) Nucleotide delivery from cycloSaligenyl-3'-azido-3'-deoxythymidine monophosphates (cycloSal-AZTMP), Eur. J. Org. Chem. 837-846.

Meier, C., Knispel, T., De Clercq, E. and Balzarini, J. (1999a) CycloSal-pro-nucleotides of 2',3'-dideoxyadenosine and 2', 3'-dideoxy-2',3'-didehydroadenosine: synthesis and Antiviral evaluation of a highly efficient nucleotide delivery system, J. Med. Chem. 42, 1604-1614.

Meier, C., Knispel, T., Marquez, V. E., Siddiqui, M.A., De Clercq, E. and Balzarini, J. (1999b) CycloSal-pro-nucleotides of 2'-fluoro-*ara*- and 2'-fluoro-*ribo*-2',3'-dideoxyadenosine as a strategy to bypass a metabolic blockade, J. Med. Chem. 42, 1615-1624.

Menger, F.M., Guo, Y. and Lee, A.S. (1994) Synthesis of a lipid/peptide/drug conjugate: N^4-(acylpeptidyl)-Ara-C, Bioconj. Chem. 5, 162-166.

Mitchell, A.G., Thomson, W., Nicholls, D., Irwin, W.J. and Freeman, S. (1992) Prodrugs of phosphonoformate: the effect of *para*-substituents on the products, kinetics and mechanism of hydrolysis of dibenzylmethoxycarbonyl phosphonate, J. Chem. Soc. Perkin Trans. 2 1145-1150.

Mitsuya, H. and Broder, S. (1986) Inhibition of the *in vitro* infectivity and cytopathic effect of human T-lymphotrophic virus type III/lymphadenopathy-associated virus (HTLV-III/LAV) by 2',3'-dideoxynucleosides, Proc. Natl. Acad. Sci. USA. 83, 1911-1915.

Mitsuya, H., Jarrett, R.F. and Mastsukura, M. (1987) Long-term inhibition of human T-lymphotropic virus type III/lymphoadenopathy-associated virus (human immunodeficiency virus) DNA synthesis and RNA expression in T cells protected by 2',3'-dideoxynucleosides *in vitro*, Proc. Natl. Acad. Sci. U.S.A. 84, 2033.

Murakami, K., Shirasaka, T., Yoshioka, H., Kojima, E., Aoki, S., Ford, Jr., H., Driscoll, J.S., Kelley, J.A. and Mitsuya, H. (1991) Escherichia coli mediated biosynthesis and *in vitro* anti-HIV activity of lipophilic 6-halo-2',3'-dideoxypurine nucleosides, J. Med. Chem. 34, 1606-1612.

Myles, M.H., Bischofberger, N. and Hoover, E.A. (1996) Evaluation of 9-(2-phosphonylmethoxypropyl)adenine antiviral therapy for acute feline immunodeficiency virus infection in cats, 3rd International Feline Retrovirus Research Symposium Fort Collins, Col. USA.

Naesens, L., Balzarini, J. and De Clercq, E. (1994) Therapeutic potential of PMEA as an antiviral drug, Med. Virol. 4, 147-159.

Palomino, E., Kessel, D. and Horwitz, J. P. (1989) A dihydropyridine carriers system for sustained delivery of 2',3'-dideoxynucleosides to the brain. J. Med. Chem. 32, 622-625.

Pannecoucke, X., Parmentier, G., Schmitt, G., Dolle, F. and Luu, B. (1994) Synthesis of the phosphodiester of 3β-(7β-hydroxycholesterol) and of 5'-(3'deoxy, 3'azido-thymidine), Tetrahedron 50, 1173-1178.

Pantaleo, G., Graziosi, C., Demarest, J.F., Butini, L., Montroni, M., Fox, C.H., Orenstein, J.M., Kotler, D.P. and Fauci, A.S. (1993) HIV Infection is active and progressive in lymphoid tissue during the clinical latent stage of disease, Nature 392, 355-358.

Parang, K., Knaus, E.E., Wiebe, L.I., Sardari, S., Daneshtalab, M. and Csizmadia, F. (1996) Synthesis and antifungal activities of myristic acid analogs, Arch. Pharm.-Pharm. Med. Chem. 329, 475-482.

Parang, K., Wiebe, L.I. and Knaus, E.E. (1998) Pharmacokinetics and tissue distribution of (±)-3'-azido-2',3'-dideoxy-5'-*O*-(2-bromomyristoyl)thymidine, a prodrug of 3'-azido-2',3'-dideoxythymidine (AZT) in mice, J. Pharm. Pharmacol. 50, 989-996.

Pavan-Langston, D. and Buchanan, R.A. (1975) Adenine arabinoside, an antiviral agent. Alford, C.A. Eds. Raven Press, New York, N.Y. 381-392.

Périgaud, C., Gosselin, G., Lefebvre, I., Girardet, J.-L., Benzaria, S., Barber, I. and Imbach, J.-L. (1993) Rational design for cytosolic delivery of nucleoside monophospates: "SATE" and "DTE" as enzyme-labile transient phosphate protecting groups, Bioorg. Med. Chem. Lett. 3, 2521-2526.

Périgaud, C., Aubertin, A.-M., Benzaria, S., Pélicano, H., Girardet, J.-L., Maury, G., Gosselin, G., Kirn, A., Imbach, J.-L. (1994) Equal inhibition of the replication of human immunodeficiency virus in human T-cell culture by ddA Bis (SATE) phosphotriester and 3'-azido-2',3'-dideoxythymidine, Biochem. Pharmacol. 48, 11-14.

Périgaud, C., Gosselin, G., Benzaria, S., Girardet, J.-L., Maury, G., Pélicano, H., Aubertin, A.-M., Kirn, A., Imbach, J.-L. (1995) Nucleosides Nucleotides. Bis(*S*-acyl-2'-thioethyl)esters of 2',3'-dideoxyadenosine 5'-monophosphate are potent anti-HIV agents in cell culture, 14, 789-791.

Périgaud, C., Girardet, J.-L., Lefebvre, I., Xie, M.-Y., Aubertin, A.-M., Kirn, A., Gosselin, G., Imbach, J.-L. and Sommadossi, J.-P. (1996) Comparison of cytotoxicity of mononucleoside phosphotriester derivatives bearing biolabile phosphate protecting groups in normal human bone marrow progenitor cells, Antiviral Chem. Chemother. 7, 338-345.

Périgaud, C., Gosselin, G. and Imbach, J.-L. (1997) Minireview: from the pro-nucleotide concept to the SATE phosphate protecting groups, Curr. Topics in Med. Chem. 2, 15-29.

Périgaud, C., Gosselin, G., Girardet, J.-L., Korba, B. E. and Imbach, J.-L. (1999) The SATE pro-nucleotide approach applied to acyclovir. Part I. Synthesis and in vitro anti-HBV activity of bis(SATE)phosphotriester derivatives of acyclovir, Antiviral Res. 40, 167-178.

Peuch, F., Gosselin, G., Lefebvre, I., Pompon, A., Aubertin, A.-M., Kirn, A. and Imbach, J.-L. (1993) Intracellular delivery of nucleoside monophosphates through a reductase-mediated activation process, Antiviral Res. 22, 155-174.

Pompon, A., Lefebvre, I., Imbach, J.-L., Khan, S. and Farquhar, D. (1994) Decomposition pathways of the mono-(pivaloyl-oxymethyl) and bis(pivaloyl-oxymethyl) esters of azidothymidine 5'-monophosphate in cell extract and in tissue-culture medium - an application of the online ISRP-cleaning HPLC technique, Antiviral Chem. Chemother. 5, 91-98.

Pue, M.A. and Benet, L.Z. (1993) Pharmacokinetics of famciclovir in man, Antiviral Chem. Chemother. 4 (Suppl. 1), 47-55.

Qiu, Y.-L., Ptak, R.G., Breitenbach, J.M., Lin, J.-S., Cheng, Y.-C., Drach, J.C., Kern, E.R. and Zemlicka, J. (1999) Synthesis and antiviral activity of phosphoralaninate derivatives of methylenecyclopropane analogues of nucleosides, Antiviral Res. 43, 37-53.

Rajagopalan, P., Boudinot, F.D., Chu, C.K., Tennant, B.C., Baldwin, B.H. and Schinazi, R.F. (1996) Pharmacokinetics of (-)-β-D-2,6-diaminopurine dioxolane and its metabolite, dioxolane guanosine, in woodchucks (Marmota monax), Antiviral Chem. Chemother. 7, 65-70.

Rimoli, M.G., Avallone, L., De Carpariis, P., Galeone, A., Forni, F. and Vandelli, M.A. (1999) Synthesis and characterization of poly(D,L-lactic acid)-idoxuridine conjugate, J. Controlled Release 58, 61-68.

Rivera, G., Rhomes, J.A., Dahl, G.V., Pratt, C.B., Wood, A. and Avery, T.L. (1980) Combined VM-26 and cytosine arabinoside in treatment of refractory childhood lymphocytic leukemia, Cancer 45, 1284.

Roche, E.B. (1977) Design of biopharmaceutical properties through prodrugs and analogs, Academy of Pharmaceutical Sciences Symposium, Washington DC.

Rolan, P. (1995) Pharmacokinetics of new antiherpetic agents, Clin. Pharmacokinet. 29, 333-340.

Routledge, A., Walker, I., Freeman, S., Hay, A. and Mahmood, N. (1995) Synthesis, bioactivation and anti-HIV activity of 4-acyloxybenzyl, bis(nucleosidyl-5'-yl) phosphates, Nucleosides Nucleotides 14, 1545-1558.

Sakai, A., Mori, N., Shuto, S. and Suzuki, T. (1993) Deacylation-reacylation cycle: a possible absorption mechanism for the novel lymphotropic antitumor agent dipalmitoylphosphatidylfluorouridine in rats, J. Pharm. Sci. 83, 575-578.

Sastry, J.K., Nehete, P.N., Khan, S., Nowak, B.J., Plunkett, W., Arlinghaus, R.B. and Farquhar, D. (1992) Membrane-permeable dideoxyuridine 5'-monophosphate analogue inhibits human immunodeficiency virus infection, Mol. Pharmacol. 41, 441-445.

Schinazi, R.F., McClure, H.M., Boudinot, F.D., Xiang, Y.-J. and Chu, C.K. (1994) Development of (-)-β-D-2,6-diaminopurine dioxolane as a potential antiviral agent, Antiviral Research 23, (Suppl.) 81-81.

Seki, T., Kawaguchi, T. and Juni, K. (1990) Enhanced delivery of zidovudine through rat and human skin *via* ester prodrugs, Pharm. Res. 9, 948-952.

Sharma, A.P., Ollapally, A.P. and Lee, H.J. (1993) Synthesis and anti-HIV activity of prodrugs of azidothymidine, Antiviral Chem. Chemother. 4, 93-96.

Shaw, J.P., Sueoka, C.M., Oliyai, R., Lee, W.A., Arimilli, M.N., Kim, C.U. and Cundy, K.C. (1997) Metabolism and pharmacokinetics of novel oral prodrugs of 9-[(R)-2-(phosphonomethoxy)propyl]adenine (PMPA) in dogs, Pharm. Res. 14, 1824-1829.

Shimoyama, M. and Kimura, K. (1973) Quantitative study on cytocidal action of anticancer agents, Saishin Igaku 28, 1024-1040.

Shuto, S., Itoh, H., Ueda, S., Imamura, S., Fukukawa, K., Tsujino, M., Matsuda, A. and Ueda, T. (1988) A facile enzymatic synthesis of 5'-(3-*sn*-phosphatidyl)nucleosides and their antileukemic activities, Chem. Pharm. Bull. 36, 209-217.

Shuto, S., Itoh, H., Obara, T., Nakagami, K., Yaso, M., Yaginuma, S., Tsujino, M. and Saito, T. (1992) New neplanocin analogues II. Enzymatic one-step synthesis and antitumor activity of 6'-(3-*sn*-phosphatidyl)neplanocin A derivatives, Nucleosides Nucleotides 11, 437-446.

Siddiqui, A.Q., Ballatore, C., McGuigan, C., De Clercq, E. and Balzarini, J. (1999) The presence of substituents on the aryl moiety of the aryl phosphoramidate derivative of d4T enhances anti-HIV efficacy in cell culture: a structure-activity relationship, J. Med. Chem. 42, 393-399.

Siddiqui, A., McGuigan, C., Ballatore, C., Srinivasan, S., De Clercq, E. and Balzarini, J. (2000) Enhancing the aqueous solubility of d4T-based phosphoramidate prodrugs, Bioorg. Med. Chem. Lett. 10, 381-384.

Sinkula, A.A. and Yalkowsky, S.H. (1975) Rationale for design of biologically reversible drug derivatives – prodrugs, J. Pharm. Sci. 64, 181-210.

Soike, K.F., Huang, J.-L., Lambe, C.U., Nelson, D.J., Ellis, M.N., Krenitsky, T.A. and Koszalka, G. W. (1993) 6-Dimethylamino-9-(β-D-arabinofuranosyl)-9H-purine: pharmacokinetics and antiviral activity in simian varicella virus-infected monkeys, Antiviral Res. 20, 13-20.

Spector, S.A., Busch, D.F., Follansbee, S., Squires, K., Lalezari, J.P., Jacobson, M.A., Connor, J. D., Jung, D., Shadman, A., Mastre, B., Buhles, W. and Drew, W.L. (1995) Pharmacokinetics, safety, and antiviral profiles of oral ganciclovir in persons infected with human immunodeficiency virus: a phase II study, J. Infect. Dis. 171, 1431-1437.

Srinivas, R.V., Robbins, B.L., Connelly, M.C., Gong, Y.-F., Bischofberger, N. and Fridland, A. (1993) Metabolism and *in vitro* antiretroviral activities of bis(pivaloyloxymethyl) prodrugs of acyclic nucleoside phosphonates, Antimicrob. Agents Chemother. 37, 2247-2250.

Srivastva, D. and Farquhar, D. (1984) Bioreversible phosphate protective groups: synthesis and stability of model acyloxymethyl phosphates, Bioorg. Chem. 12, 118-129.

Starrett, J.E. Jr., Tortolani, D.R., Hitchcock, M.J.M., Martin, J.C. and Mansuri, M.M. (1992) Synthesis and *in vitro* evaluation of a phosphonate prodrug: bis(pivaloyloxymethyl) 9-(2-phosphonylmethoxyethyl)adenine, Antiviral Res. 19, 267-273.

Starrett, J.E., Jr., Tortolani, D.R., Russel, J., Hitchcock, M.J.M., Whiterock, V., Martin, J.C. and Mansuri, M.M. (1994) Synthesis, oral bioavailability determination, and *in vitro* evaluation of prodrugs of the antiviral agent 9-[2-(phosphonomehtoxy)-ethyl]adenine (PMEA), J. Med. Chem. 37, 1857-1864.

Stella, V. and Himmelstein, K.J. (1980) Prodrugs and site-specific drug delivery, J. Med. Chem. 23, 1276-1282.

Tan, X.: Chu, C.K. and Boudinot, F.D. (1999) Development and optimization of anti-HIV nucleoside analogs and prodrugs: a review of their cellular pharmacology, structure-activity relationships and pharmacokinetics, Adv. Drug Deliv. Rev. 39, 117-151.

Thomson, W., Nicholls, D., Irwin, W.J., Al-Mushadani, J.S., Freeman, S., Karpas, A., Petrik, J., Mahmood, N. and Hay, A.J. (1993) Synthesis, bioactivation and anti-HIV activity of the bis(4-acyloxybenzyl) and mono(4-acyloxybenzyl) esters of the 5'-monophosphate of AZT, J. Chem. Soc. Perkin Trans. 1 1239-1245.

Thomson, W., Nicholls, D., Mitchell, A.G., Corner, J.A., Irwin, W.J. and Freeman, S. (1993) Synthesis and bioactivation of bis(aroyloxymethyl) and mono(aroyloxymethyl) esters of benzylphosphonate and phosphonoacetate, J. Chem. Soc. Perkin Trans. 1 2303-2308.

Tokunaga, Y., Iwasa, T., Fujisaki, J., Sawai, S. and Kagayama, A. (1988a) Liposomal sustained-release delivery systems for intravenous injection. IV. Antitumor activity of newly synthesized lipophilic 1-β-D-arabinofuranosylcytosine prodrug-bearing liposomes, Chem. Pharm. Bull. 36, 3574-3583.

Tokunaga, Y., Iwasa, T., Fujisaki, J., Sawai, S. and Kagayama, A. (1988b) Liposomal sustained-release delivery systems for intravenous injection. III. Antitumor activity of lipophilic mitomycin C prodrug bearing liposomes, Chem. Pharm. Bull. 36, 3565-3574.

Torrence, P.F., Kinjo, J.-E., Lesiak, K., Balazarine, J. and De Clercq, E. (1988) AIDS Dementia: synthesis and properties of a derivative of 3'-azido-3'-deoxythymidine (AZT) that may become 'locked' in the central nervous system, FEBS Lett. 234, 135-140.

Torrence, P.F., Kinjo, J., Khamnei, S. and Greig, N.H. (1993) Synthesis and pharmacokinetics of a dihydropyridine chemical delivery system for the antiimmunodeficiency virus agent dideoxycytidine, J. Med. Chem. 36, 529-537.

Tsai, C.C., Follis, K.E., Sabo, A., Beck, T.W., Grant, R.F., Bischofberger, N., Benveniste, R.E. and Black, R. (1995) Prevention of SIV infection in macaques by (R)-9-(2-phosphonylmethoxypropyl)adenine, Science, 270, 1197-1199.

Tsai, C.C., Follis, K.E., Beck, T.W., Sabo, A., Bischofberger, N. and Dailey, P.J. (1997) Effects of (R)-9-(2-phosphonylmethoxypropyl)adenine monotherapy on chronic SIV infection in macaques, AIDS Research and Human Retroviruses 13, 707-712.

Tucker, W.S., Kirsch, W.M., Martinez-Hernandez, A. and Fink, L.M. (1978) *In vitro* plasminogen activator activity in human brain tumors. Cancer Res. 38, 297-302.

Uchida, H., Kodama, E.N., Yoshimura, K., Maeda, Y., Kosalaraksa, P., Maroun, V., Qiu, Y.-L., Zemlicka, J. and Mitsuya, H. (1999) *In vitro* anti-human immunodeficiency virus activities of *Z*- and *E*-methylenecyclopropane nucleoside analogues and their phosphoro-L-alaninate diesters, Antimicrob. Agents Chemother. 43, 1487-1490.

Uchida, K. and Kreis, W. (1969) Distribution of 1-β-D-arabinofuranosylcytosine, cytidine and deoxycytidine in mice bearing ara-C sensitive and resistant P815 neoplasms, Biochem. Pharmacol. 18, 1115.

Valette, G., Pompon, A., Girardet, J.-L., Gosselin, G., Périgaud, C., Gosselin, G., Imbach, J.-L. (1996) Decomposition pathways and *in vitro* HIV inhibitory effects of IsoddA pro-nucleotides: toward a rational approach for intracellular delivery of nucleoside 5'-monophosphates, J. Med. Chem. 39, 1981-1990.

Van Rompay, K.K.A., Cherrington, J.M., Marthas, M.L., Berardi, C.J., Mulato, A.S., Spinner, A., Tarara, R.P., Canfield, D.R., Telm, S., Bischofberger, N. and Pederson, N.C. (1996) 9-[2-(Phosphomethoxy)propyl]-adenine therapy of established simian immunodeficiency virus infection in infant rhesus macaques, Antimicrob. Agents Chemother. 40, 2586-2591.

Vere Hodge, R.A., Sutton, D., Boyd, M.R., Harnden, M.R. and Jarvest, R.L. (1989) Selection of an oral prodrug (BRL 42810, famciclovir) for the antiherpesvirus agent BRL 39123 [9-(4-hydroxy-3-hydroxymethylbut-1-yl)guanine, penciclovir], Antimicrob. Agents Chemother. 33, 1765-1773.

Vince, R., Kilama, J., Pham, P.T. and Beers, S.A. (1995a) 6-Substituted derivatives of carbovir: anti-HIV activity. Nucleosides Nucleotides 14, 1703-1708.

Vince, R., Brownell, J. and Beers, S.A. (1995b) 6-Deoxycarbovir: a xanthine oxidase activated prodrug of carbovir, Nucleosides Nucleotides 14, 39-44.

Wade, C.W.R., Hegyeli, A.F. and Kulkarni, R.K. (1977) Standards for *in vitro* and *in vivo* comparison and qualification of bioabsorbable polymers. J. Test. Eval. 5, 397-400.

Wagner, C.R., McIntee, E.J., Schinazi, R.F. and Abraham, T.W. (1995) Aromatic aminoacid phosphoramidate diesters and triesters of 3'-azido-3'-deoxythymidine (AZT) are nontoxic inhibitors of HIV-1 replication, Bioorg. Med. Chem. Lett. 5, 1819-1824.

Wagner, C.R., Chang, S.-L., Griesgraber, G.W., Song, H., McIntee, E. and Zimmerman, C.L. (1999) Antiviral nucleoside drug delivery via amino acid phosphoramidates, Nucleosides Nucleotides 18, 913-919.

Walker, R.E., Vogel, S.E., Jaffe, H.S., Polis, M.A., Kovacs, J., Markowitz, N., Masur, H. and Lane, H.C. (1993) A Phase I/II Study of PMEA in HIV Infected Patients, Abstracts of Papers, 1st National Conference on Human Retroviruses and Related Infections, Washington DC.

Wang, L., Morin, K.W., Kumar, R., Cheraghali, M., Todd, K.G., Baker, G.B., Knaus, E.E. and Wiebe, L.I. (1996) *In vivo* biodistribution, pharmacokinetic parameters, and brain uptake of 5-halo-6-methoxy(or ethoxy)-5,6-dihydro-3'-azido-3'-deoxythymidine diastereomers as potential prodrugs of 3'-azido-3'-deoxythymidine, J. Med. Chem. 39, 826-833.

Weller, S., Blum, M.R. and Doucette, M. (1993) Pharmacokinetics of the acyclovir pro-drug valaciclovir after escalating single and mutiple-dose administration to normal volunteers, Clin. Pharmacol. Ther. 54, 595-605.

Wen, Y.-D., Remmel, R.P., Pham. P.T., Vince, R. and Zimmerman, C.L. (1995) Comparative brain exposure to (-)-carbovir after (-)-carbovir or (-)-6-aminocarbovir intravenous infusion in rats, Pharm. Res. 12, 911-915.

Wen, Y.-D., Remmel, R.P. and Zimmerman, C.L. (1999) First-pass disposition of (-)-6-aminocarbovir in rats. I. Prodrug activation may be limited by access to enzyme, Drug Metab. Dispos. 27, 113-121.

Whitley, R.J., Soong, S., Dolin, R., Galasso, G.J., Chuien and Alford, C.A. (1977) Adenine arabinoside therapy of biopsy-proven herpes simplex encephalitis, N. Engl. J. Med. 297, 289.

Whitley, R.J. and Gnann, J.W. (1992) Acyclovir: A Decade Later, N. Engl. J. Med. 327, 782-788.

Wiebe, L.I., Knaus, E.E., Cheraghali, A.M., Kumar, R., Morin, K.W. and Wang, L. (1995) 5-Halo-6-alkoxy-5,6-dihydro-pyrimidine nucleosides: antiviral nucleosides or nucleoside prodrugs? Nucleosides Nucleotides 14, 501-505.

Wilson, E.L., Becker, M.L., Hoal, E.G. and Dowdle, E.B. (1980) Molecular species of plasminogen activators secreted by normal and neoplastic human cells, Cancer Res. 40, 933-938.

Xie, H., Voronkov, M., Liotta, D.C., Korba, B.A., Schinazi, R.F., Richman, D.D. and Hostetler, K.Y. (1995) Phosphatidyl-2', 3'-dideoxy-3'-thiacytidine: synthesis and antiviral activity in hepatitis B- and HIV-1 infected cells, Antiviral Res. 28, 113-120.

Yajima, T., Hasegawa, T., Juni, K., Saneyoshi, M. and Kawaguchi, T. (1996) Nasal absorption of 2',3'-didehydro-3'-deoxythymidine (D4T) and its esters in rats, Biol. Pharm. Bull. 19, 1234-1237.

Yajima, T., Juni, K., Saneyoshi, M., Hasegawa, T. and Kawaguchi, T. (1998) Direct transport of 2',3'-didehydro-3'-deoxythymidine (D4T) and its ester derivatives to the cerebrospinal fluid *via* the nasal mucous membrane in rats, Biol. Pharm. Bull. 21, 272-277.

Yeager, R.L., Brouwer, K.R. and Miwa, G.T. (1991) 6-(-)-Aminocarbovir pharmacokinetics and relative carbovir exposure in rats, Drug Metab. Dispos. 19, 462-466.

Yeom, Y.-H., Remmel, R.P., Huang, S.-H., Hua, M., Vince, R. and Zimmerman, C.L. (1989) Pharmacokinetics and bioavailability of carbovir, a carbocyclic nucleoside active against human immunodeficiency virus, in rats. Antimicrob. Agents Chemother. 33, 171-175.

Yoshimura, K., Feldman, R., Kodama, E., Kavlick, M.F., Qiu, Y.-L., Zemlicka, J. and Mitsuya, H. (1999) *In vitro* induction of human immunodeficiency virus type 1 variants resistant to phosphoralaninate prodrugs of Z-methylenecyclopropane nucleoside analogues, Antimicrob. Agents Chemother. 43, 2479-2483.

Zimmerman, C.L., Remmel, R.P., Ibrahim, S.S., Beers, S.A. and Vince, R. (1992) Pharmacokinetic evaluation of (-)-6-aminocarbovir as a prodrug for (-)-carbovir in rats, Drug Metab. Dispos. 20, 47-51.

Zimmerman, C.L., Wen, Y.-D. and Remmel, R.P. (2000) First-pass disposition of (-)-6-aminocarbovir in rats: II. Inhibition of intestinal first-pass metabolism, Drug Metab. Dispos. 28, 672-679.

ANTIVIRAL ISONUCLEOSIDES: DISCOVERY, CHEMISTRY AND CHEMICAL BIOLOGY

VASU NAIR

*Department of Chemistry, The University of Iowa,
Iowa City, Iowa 52242, USA*

1.   Introduction

The *pol* gene of the human immunodeficiency virus (HIV) encodes the viral enzyme, HIV reverse transcriptase (HIV RT), which is essential for viral replication (Katz and Skalka 1994; Frankel and Young 1998). This is a multifunctional enzyme which has RNA-dependent DNA polymerase, DNA-dependent DNA polymerase and RNaseH activities. Inhibitors of this viral enzyme have been a major source of therapeutic agents for the treatment of acquired immunodeficiency syndrome (AIDS) (De Clercq 1997; Johnson and Gerber 2000). Most of these therapeutic agents belong to the nucleoside/ nucleotide family. The finding that the replication of HIV was inhibited by the nucleoside, 3′-deoxy-3′-azidothymidine (AZT), led to the development of the first chemotherapeutic agent for the treatment of AIDS (Mitsuya *et al.*, 1985). This discovery stimulated the quest for and the finding of other β-D-dideoxynucleosides as clinically useful anti-HIV agents. Among those discovered were 2′,3′-dideoxyinosine (ddI), 2′,3′-dideoxycytidine (ddC) and 2′,3′-didehydro-3′-deoxythymidine (d4T) (Mitsuya and Broder 1986; Baba *et al.*, 1987; Lin *et al.*, 1987). The mechanism of action of all of these dideoxynucleosides apparently involves intracellular phosphorylation (bioactivation) to their 5′-triphosphates which may serve as chain terminators and/or inhibitors/alternate substrates for the viral DNA polymerase, HIV reverse transcriptase (Cooney *et al.*, 1986; Ahluwalia *et al.*, 1987; Cooney *et al.*, 1987; Nanni *et al.*, 1993; De Clercq 1995).

Subsequent discovery and development of (-)-2′,3′-dideoxy-3′-thiacytidine [(-)-3TC] as a clinically useful anti-HIV agent suggested that even major modifications in the carbohydrate moiety, including changes in absolute stereochemistry from D- to L-, could lead to active compounds (Sougdeyns *et al.*, 1991; Coates *et al.*, 1992; Schinazi *et al.*, 1992). Two anti-HIV nucleosides with modifications in the carbohydrate moiety and that are referred to as carbocyclic nucleosides are also of relevance. The first of these is (-)-carbovir, a cyclopentenyl dideoxynucleoside bearing the guanine moiety (Vince and Brownell 1990). The second compound is abacavir, a cyclopropylamino analog of carbovir (Daluge *et al.*, 1997).

This chapter will focus on the chemistry and biology of another family of antiviral dideoxynucleosides, i.e., those in which the endocyclic oxygen has been transposed

Recent Advances in Nucleosides: Chemistry and Chemotherapy, Ed. by C.K. Chu. 149 — 166

150   *V. Nair*

from the natural position to the 2′- or 3′-position. The review will summarize, in large
part, the discoveries from our laboratory in this area in the last several years.

2.   Classification of isonucleosides

Four representative classes of isomeric dideoxynucleosides (isodideoxynucleosides)
and their relationship to normal nucleosides are illustrated in Figure 1. The structural
relationship of the enantiomeric compounds of Classes I and II to the "natural" D- and
and "non-natural" L-nucleosides may be explained as arising from the transposition
of the nucleic acid base from C-1' to C-2'. These compounds can also be viewed
as arising from the transposition of the endocyclic oxygen from the normal position
to the 3′-position. Compounds of Class I are viewed as being L-related and those
of Class II are D-related because of their relationship to L- and D-nucleosides as
shown in Figure 1.

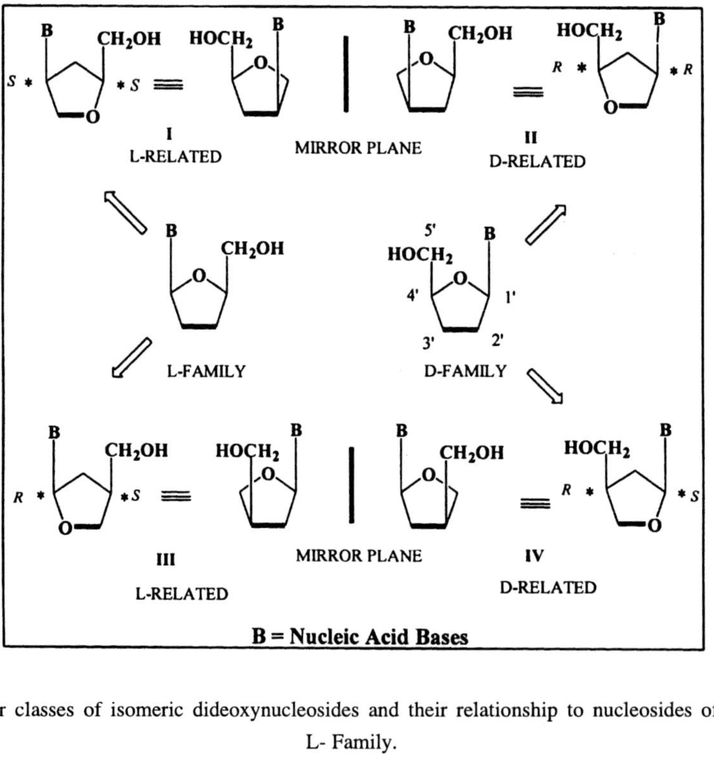

Figure 1.   ur classes of isomeric dideoxynucleosides and their relationship to nucleosides of the D- and
L- Family.

Compounds of Classes III and IV can be viewed as emerging from the transposition
of the CH₂OH from C-4′ to C-3′ or the transposition of the endocyclic oxygen from
the 0-1′ to the 2′-position. Compounds of Classes III and IV may also be referred

to as apiodideoxynucleosides because of the relationship of the carbohydrate moiety to apio sugars. Although the *cis*-structure of the base and the 4'-CH$_2$OH and their 1,3-relationship are maintained as in D- and L-nucleosides, the corresponding *trans*-diastereoisomeric structures are also possible. Finally, it should be mentioned that the earliest reference to isonucleosides involving the sugar moiety can be traced back to 1975 (Montgomery *et al.*, 1975). However, the interest in isodideoxynucleosides is much more recent.

3.　Synthesis

Synthesis of compounds of Class I is illustrated in Schemes 1 and 2 (Nair and Nuesca 1992; Nair 1993; Bolon *et al.*, 1994; Nair and Sharma 2000). For example, the synthesis of (*S,S*)-isoddA required ten steps and commenced with readily available D-xylose (1), which was initially converted in three steps to the benzoate 2 (Nair and Emanuel 1977). Compound 2 was deoxygenated at the 3-position to give 3 by conversion of the 3-hydroxy group to its imidazole thiocarbonyl ester and subsequent treatment with tri-*n*-butyltin hydride and AIBN in refluxing toluene. Acid-catalyzed methanolysis of the remaining acetonide group of 3 was followed by reductive demethoxylation of the resulting α- and β-methyl glycosides. This demethoxylation methodology involved protection of the 2-hydroxy group by silylation (HMDS, TMSCl), followed *in situ* by treatment of this intermediate with triethylsilane and TMS-triflate in dichloroethane, which produced the tetrahydrofuran derivative 4. This is the key intermediate in the synthesis of the entire series of compounds of Class I. Tosylation or mesylation of compound 4 gave 5 which was condensed with adenine in the presence of K$_2$CO$_3$ and 18-crown-6 to give 6 which was deprotected with catalytic NaOMe in MeOH to give (*S,S*)-isoddA 7. Compound 7 could also be produced through the Mitsunobu reaction (Ph$_3$P, DEAD with adenine or 6-chloropurine) of 4 followed by deprotection with NaOMe/MeOH (for 6) or NH$_3$/EtOH (for 8). The absolute stereochemistry of 7 and its solid state conformation were confirmed by single crystal X-ray data. Since this first report of the synthesis of (*S,S*)-isoddA, related preparations have also been reported (see for example, Diaz *et al.*, 1999).

Other isodideoxynucleosides of the (*S,S*)-series can be synthesized as summarized in Scheme 2. While the purine isonucleosides can be prepared by direct coupling procedures (Mitsunobu or K$_2$CO$_3$ and 18-crown-6), the pyrimidine isonucleosides are best prepared by utilizing the base construction methodology starting from the corresponding β-amino compound.

Many analogs of (*S,S*)-isodideoxynucleosides (Figure 2) have also been reported by us (Purdy *et al.*, 1994; Zintek *et al.*, 1994; Nuesca and Nair 1994; Bolon *et al.*, 1995; Jahnke and Nair 1995; Zintek *et al.*, 1996; Jeon and Nair 1996; Zheng and Nair 1999 a,b; Bera *et al.*, 1999; Bera and Nair 2000; Taktakishvili *et al.* 2000; Bera and Nair, unpublished results).

Synthesis of compounds of Class II (the (*R,R*)-enantiomers of Class I) (Huryn *et al.*, 1992) relied on a rearrangement reaction of the tosylate 9 to the acetal 10, the key intermediate for entry into this series (Scheme 3). However, this type of

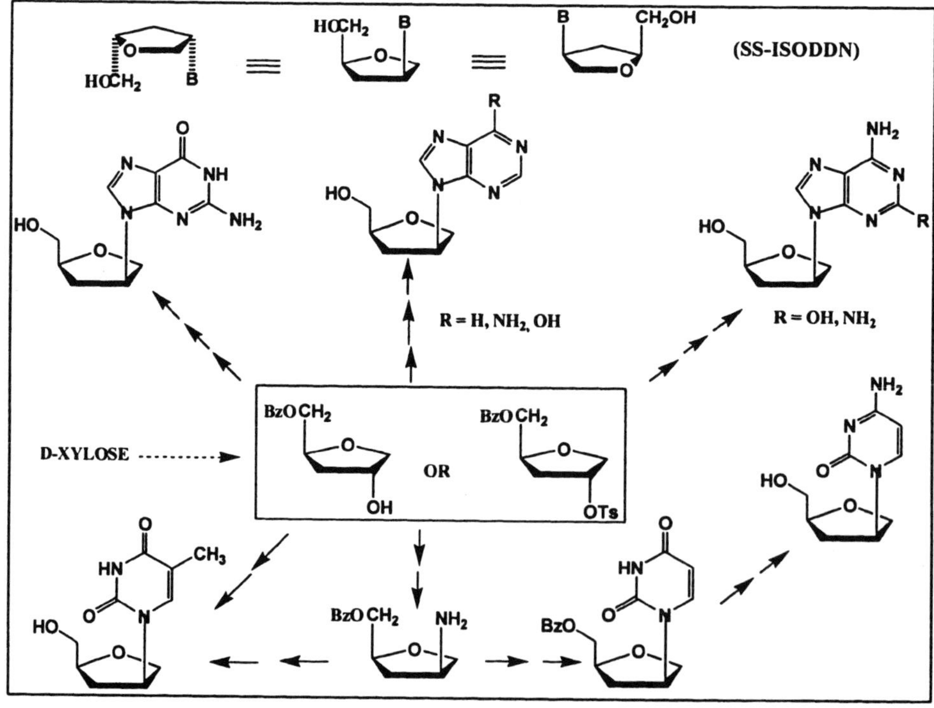

Scheme 1.   Summary of key steps in the synthesis of (*S,S*)-IsoddA.

Scheme 2.   Summary of the synthesis of (*S,S*)-Isodideoxynucleosides of Class I.

rearrangement had been reported as early as 1971 (Defaye *et al.*, 1971). Intermediate 10 was transformed in several well-known steps to 11. Conversion of 11 to the D-series of isodideoxynucleosides involved steps similar to those described above for the L-series.

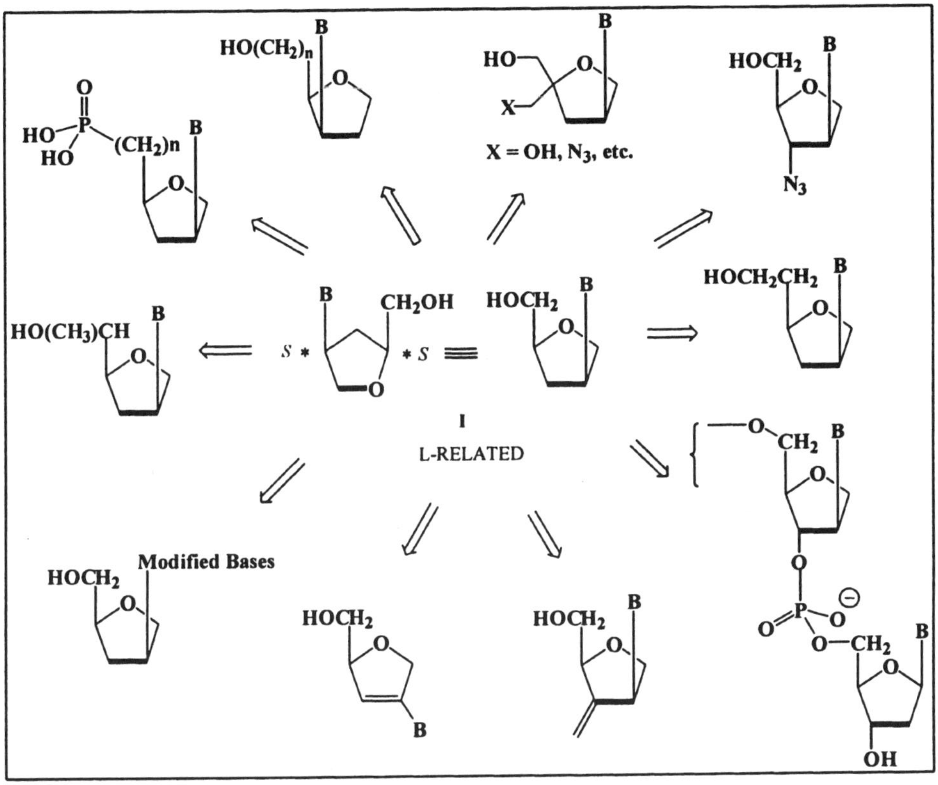

Figure 2. Selected examples of new analogs of (*S,S*)-isodideoxynucleosides synthesized in our laboratory.

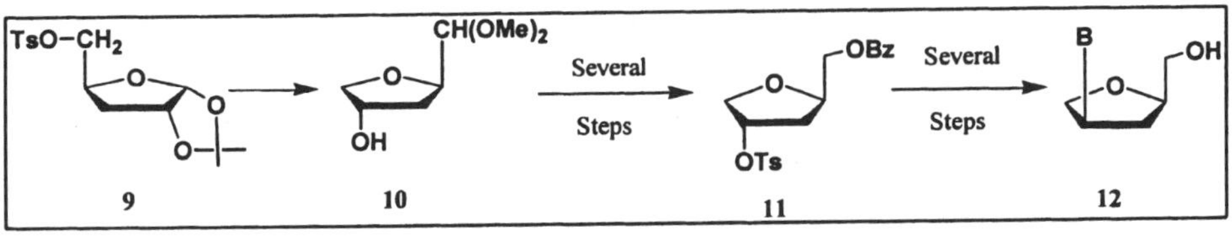

Scheme 3. Summary of steps in the synthesis of (*R,R*)-isodideoxynucleosides.

Another route to (***R,R***)-isodideoxynucleosides may be realized through the epoxide 15 (Scheme 4). Coupling of epoxide 15 with nucleobases produces 16 as the major product and 17 as the minor product (Yu *et al.*, 1996). Compound 16 can be converted in several steps to compounds of Class II (Chun *et al.* unpublished results)

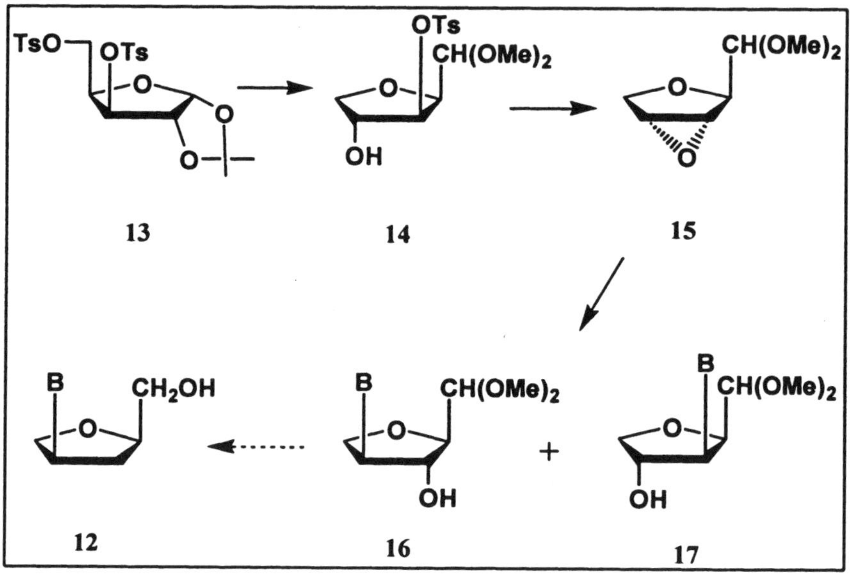

Scheme 4.   Synthesis of (***R,R***)-isodeoxynucleosides *via* epoxide

In the synthesis of isodideoxynucleosides where the sugar moiety is an apio sugar (Classes III and IV), Sells and Nair (1994) used a divergent synthetic plan that employed the cyclization of a chiral derivative of a prochiral aldodiol. The synthesis is summarized in Scheme 5. Lactol formation from the pro-*S*- or pro-*R*- hydroxymethyl group allowed entry into both series of apiodideoxynucleosides. The starting material was commercially available allyl diethylmalonate which was converted by reduction and acetylation to the ***meso***-diacetate 18. Enantioselective ester hydrolysis with the lipase from ***Candida cylindracea*** gave the (*S*) *or* (-)-monoacetate 19 which served as the key intermediate for the enantiomeric series of compounds of Class III and IV. Silylation of 19 and chemical hydrolysis of the acetate gave the chiral precursor 20. Treatment of 20 with sodium periodate and catalytic osmium tetroxide gave, after spontaneous cyclization, compound 21 which was acetylated to 22. Vorbruggen type coupling of 22 with nucleobases and deprotection afforded 23 (L-related, Class III) and the corresponding ***trans***-isomers which were separable. Periodate/OsO$_4$ (catalytic) oxidation of 19 provided entry into the D-related series 26 (Class IV) *via* compound 24.

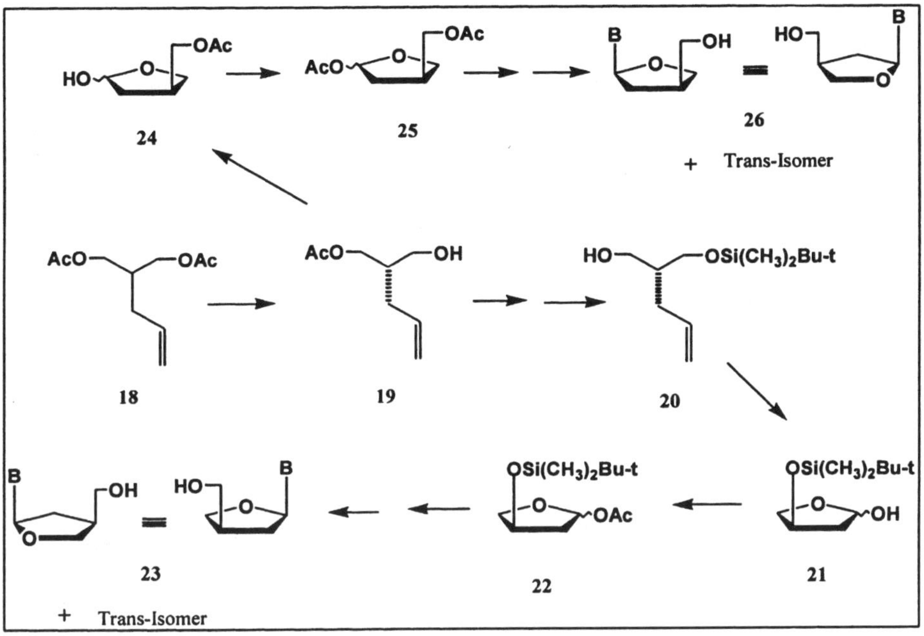

Scheme 5. Synthetic routes to isodideoxynucleosides of the apio family

4. Enzymology

As the most active isodideoxynucleosides contain adenine as the nucleobase, this enzymology section will focus on isodideoxyadenosine analogs, with particular emphasis on (S,S)-isoddA. The consequences of substrate activity with adenosine deaminase is shown in Figure 3 (Nair and Sells 1992).

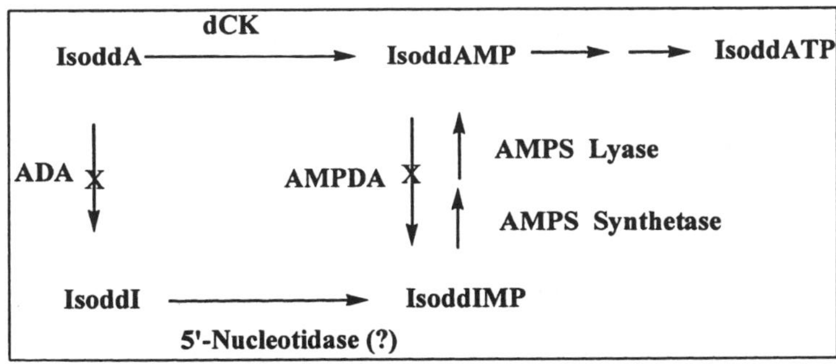

Figure 3. Consequences of deamination of isodideoxyadenosines

Determination of the substrate kinetic constants for (*S,S*)-isoddA were made with calf intestinal adenosine deaminase (ADA) and compared in the same study with the data for adenosine and dideoxyadenosine (ddA) (Nair *et al.*, 1992; Nair *et al.*, 1995). The K_m for isoddA was 250 μM and the V_{max} was 7.8 × 10^{-5} μmol/min/unit enzyme. The corresponding values for adenosine and dideoxyadenosine were 30 μM and 1.3 μmol/min/unit enzyme and 160 μM and 0.8 μmol/min/unit enzyme, respectively. Thus, the relative substrate efficiency (V_{max}/ K_m) of isoddA was 0.0008% of that for adenosine. This substrate efficiency is also much lower than that for ddA (V_{max}/ K_m is 11.5% of dideoxyadenosine). These findings were consistent with both *in vitro* and *in vivo* studies (in rats) of metabolites of isoddA. It is thus clear that for (*S,S*)-isoddA, the pathway through the deaminated product, isoddI, is almost non-existent. We have also discovered that (*S,S*)-isoddAMP is not a substrate for AMP deaminase (AMPDA) and that (*S,S*)-isoddA is not a substrate for purine nucleoside phosphorylase (PNP). Thus, it appears that the major pathway for the production of (*S,S*)-isoddATP is through the three-step phosphorylation of the nucleoside. The enantiomer of (*S,S*)-isoddA is also resistant to deamination by ADA (Huryn *et al.*, 1992). However, the apiodideoxyadenosines were somewhat less resistant, being deaminated at 0.12% (L-related) and 0.10% (D-related) of the rate of adenosine. Interestingly, while the apiodideoxynucleosides are unstable in aqueous acidic solutions like their "natural" counterparts, the isomeric compounds of Classes I and II are extremely stable with respect to "glycosidic bond" cleavage in acid. For example, (*S,S*)-isoddA has a $t_{1/2}$ of >16 days at pH 1 and it is also stable in aqueous base. This is not surprising because the "glycosidic bond" in isoddA is similar to that found in carbocyclic nucleosides and thus participation of the endocyclic oxygen in the departure of the base during hydrolysis is not possible

The issue of the enzymatic phosphorylation of (*S,S*)-isoddA was also investigated. Because it is well established that deoxycytidine kinase (dCK, E.C. 2.7.1.74) plays a major role in the initial phosphorylation of several nucleoside analogs of anti-HIV therapeutic interest, we studied the phosphorylation of (*S,S*)-isoddA by human recombinant deoxycytidine kinase (dCK) (Pal and Nair 2000). These results are summarized in Table 1 and indicate that isoddA is a poor substrate for dCK. In addition, (*S,S*)-isoddA is not a substrate for 5′-nucleotidase or adenosine kinase. The isolated enzyme studies are consistent with the results of the *in vitro* anabolism of isoddA in CEM cells. Incubation of cells with 10 μM isoddA for 24 h resulted in an intracellular level of isoddATP of approximately 4.4 nM. In comparison, the intracellular level of ddATP is 56 nM following incubation of MOLT-4 cells with only 5 μM ddI (Ahluwalia *et al.*, 1987; Nair *et al.*, 1995). The low level of phosphorylation was apparently not due to inefficient transport of the nucleoside as the concentration of [^{3}H]isoddA in the media and inside the cells was approximately equal.

5.   Antiviral and related studies

The *in vitro* anti-HIV data for selected isodideoxynucleosides are shown in Table 2. The active isodideoxynucleosides belong to Classes I and II. No active compounds were discovered in the apio series presumably because of very poor phosphorylation

Table 1.   Substrate specificity studies of selected nucleosides with 2′-deoxycytidine kinase

Substrate	K_m (µM)	V_{max} (nmol/min/mg)	V_{max}/K_m (nmol/min/mg/µM)
ddA	237	137	0.58
(*S,S*)-IsoddA	603	200	0.33
dA (Kierdaszuk *et al.*, 1992)	110	800	7.3
dC (Kierdaszuk *et al.*, 1992)	1.5	185	123.3

and/or because of the difficulty of the triphosphates to be recognized by HIV reverse transcriptase. The latter may be the result of the presence of the endocyclic oxygen at the 2-position which may make these triphosphates mimic ribonucleoside triphosphates (e.g., cordycepin triphosphate).

Table 2.   Selected examples of *in vitro* anti-HIV-1 data of isodideoxynucleosides

Compound	IC_{50} µM	CC_{50} µM	Cell Line	References
(*S,S*)-IsoddA	0.67	>100	PBL, MT4	(Nair *et al.*, 1995; Nair and Jahnke 1995)
(*S,S*)-IsoddG	>200	>200	MT4	(Nair and Jahnke 1995)
(*R,R*)-IsoddA	5-15	>100	ATH8	(Huryn *et al.*, 1992)
(*R,R*)-IsoddA	43	-	MT4	(Jones *et al.*, 1992)
(*R,R*)-IsoddG	10-50	>500	ATH8	(Huryn *et al.*, 1992)
(*R,S*)-ApioddA*	>200	>200	MT4	(Nair and Jahnke 1995)
(*S,R*)-ApioddA*	>200	>200	MT4	(Nair and Jahnke 1995)
(*R,R*)-ThiaisoddA	>400	-	MT4	(Jones *et al.*, 1991)

*   The first letter represents the absolute stereochemistry of the carbon bearing the base.

The most active compound in the entire isomeric nucleoside family was (*S,S*)-isoddA discovered in our laboratory. Figure 2 shows various structural representations of this compound. Single crystal X-ray data showed that the base was in the preferred anti-conformation and the carbohydrate moiety was in the C-5′-exo/O-1′-endo conformation. Discussion of the biological data will focus on this compound.

Anti-HIV studies were carried out using an acute-infection assay in phytohemagglu-tinin-stimulated peripheral blood lymphocytes (PBLs) and in human T-cell lymphotropic virus type 1-transformed cell line (MT4) (Nair *et al.*, 1995). In PBLs, the IC_{50} against

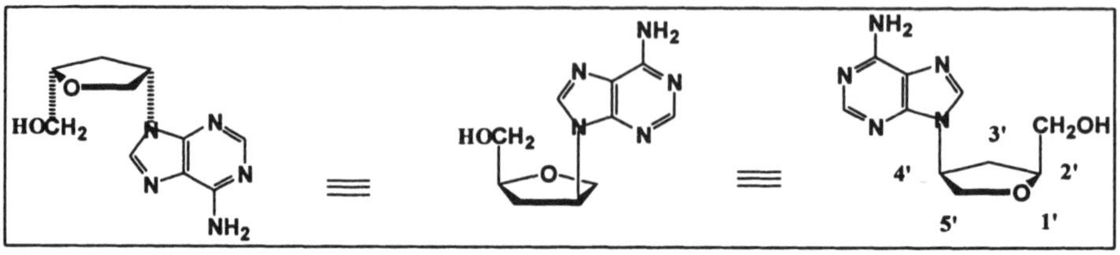

Figure 4.   Structural representation of (*S,S*)-IsoddA illustrating base and endocyclic oxygen transpositions, absolute stereochemistry and relationship to 2′-deoxyadenosine.

HIV-1 was 0.7 ± 0.2 μM. The activity in MT4 cells was comparable to ddI. IsoddA was also active against HIV-2. The IC_{50} against HIV-2 was 3.4 ± 0.9 μM in MT4 cells. The activity against clinical isolates of HIV (three AZT-sensitive and three AZT-resistant isolates) was also measured. The IC_{50} values for isoddA with AZT-sensitive isolates were 13 ± 2, 7 ± 3, and 1.3 ± 0.4 μM (mean IC_{50} of 7.1 μM), whereas the IC_{50} values for AZT-resistant isolates were 40 ± 30, 31 ± 8, and 10 ± 2 μM (mean IC_{50} of 27 μM). IsoddA at 100 μM did not inhibit significantly the growth of the human leukemic cell lines CEM, MOLT-4, and IM9. Isobolograms for combinations of isoddA and AZT, ddI, or β-L-FTC in anti-HIV-1 assays gave clear evidence for synergistic behavior. (*S,S*)-IsoddA also showed *in vitro* anti-HBV activity against HBV-producing 2.2.15 cells. The IC_{50} value for HBV inhibition averaged 3.4 μM and that for cell growth inhibition (cytotoxicity) was 148 μM.

In order to obtain further information on the inhibition of HIV reverse transcriptase by cellularly produced (*S,S*)-isoddA 5′-triphosphate (isoddATP), we synthesized this compound by established chemical methods and purified it by reversed-phase HPLC (Scheme 6). IsoddATP is a potent inhibitor of HIV-1 reverse transcriptase with a K_i of 16 nM! (Nair *et al.*, 1995). Inhibition constants for isoddATP and ddATP with human DNA polymerases α, β, γ, and HIV-1 RT were determined using activated calf thymus DNA as the nucleic acid substrate. Significant differences were observed with the human enzymes. ddATP was a potent inhibitor of polymerases β and γ with K_i values of 1.1 ± 0.2 μM and 0.018 ± 0.002 μM, respectively, whereas it was a relatively weak inhibitor of polymerase α with a K_i value of 64 ± 8 μM (Martin *et al.*, 1994). IsoddATP was a relatively weaker inhibitor of polymerases β and γ with K_i values of 18 ± 2 μM and 0.36 ± 0.06 μM, respectively, whereas it was a more potent inhibitor of polymerase α with a K_i value of 0.63 ± 0.08 μM. Data for the inhibition of mitochondrial DNA polymerase is $1/20^{th}$ that of ddATP. The intracellular half-life of isoddATP in human T-lymphocytic cells was assessed to be 9.4 h which is shorter than the value of 24 h for ddATP. The latter study was carried out under similar conditions in human T cells that were incubated with ddI (Ahluwalia *et al.*, 1993). However, the half-lives of the 5′-triphosphates of ddC, AZT, and d4T measured in MOLT-4 and CEM cells are much shorter compared to that of isoddATP and are of the order of 3 h (Starnes and Cheng 1987; Ho and Hitchcock 1989).

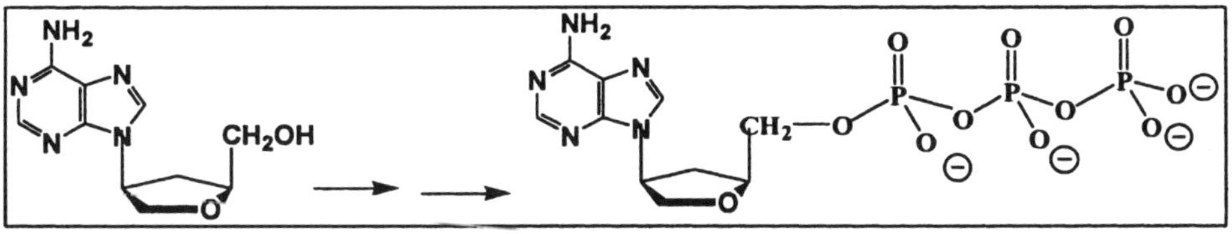

Scheme 6. Two step synthesis of (*S,S*)-isoddATP by chemical phosphorylation methods.

In order to understand more about the mechanism of apparent incorporation of this L-related nucleoside in a D-related viral DNA structure in the chain termination process by (*S,S*)-isoddATP, we have incorporated isodA into a 12 mer [d(G-T-A-G-isoA-isoA-T-T-C-T-A-C)] (Scheme 7). Interestingly, we have discovered that all of the internucleotide phosphate bonds involving the isonucleotides exhibited resistance to cleavage by exonucleases. The CD spectral data suggest a β-DNA-like structure in which two double-helical tetramers are interrupted by a central tetramer, d(IsoA-IsoA-T-T)$_2$, which does not appear to form a duplex with a different secondary structure (T$_m$ 43 °C, cf., unmodified self-complementary oligomer, T$_m$ 46 °C) (Wenzel and Nair, 1997, 1998).

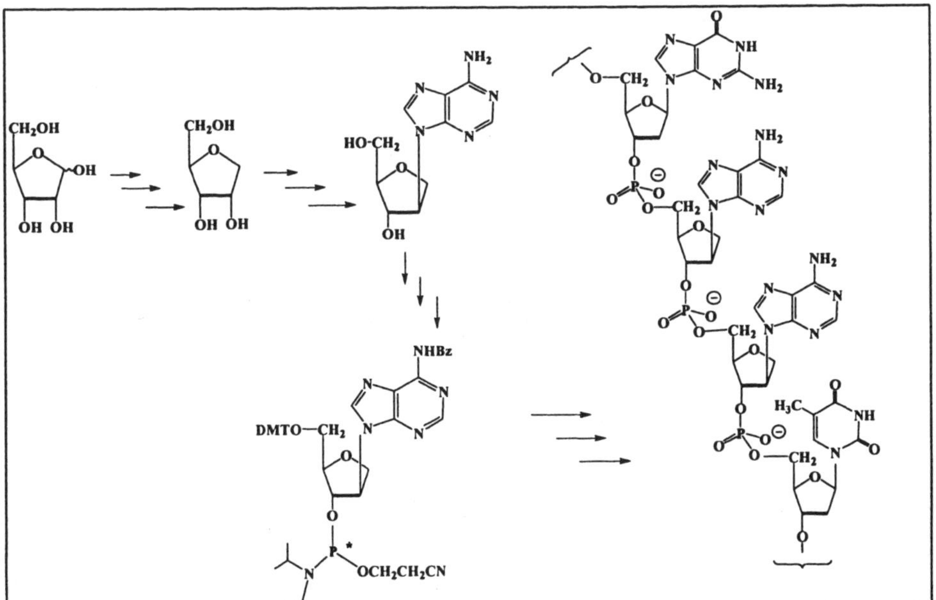

Scheme 7. Incorporation of isoddA into a 12-Mer oligonucleotide.

(*S,S*)-IsoddA was well absorbed orally (81% bioavailability in mice) and rapidly eliminated from the plasma *via* renal excretion (>70% of dose recovered in both mouse and rat urine as isoddA). No metabolites of isoddA were found in the urine in these *in vivo* studies (Nair *et al.*, 1995).

In vitro selection of isoddA-resistant HIV-1 (HXB2) was carried out through multiple passages in the presence of drug. The passaged virus exhibited IC_{50} values that were 4 to 8-fold higher than that for wild-type HXB2. Genetic analysis of the appropriate RT region of the HIV-1 *pol* gene indicated that the change in sensitivity to isoddA was due to a single amino acid change at codon 184 from Met $\rightarrow$ Val [M184V]. The change from methionine to valine introduces a β-methyl side chain with accompanying increase in steric size of this amino acid at the nucleotide binding site of HIV reverse transcriptase. This mutation, [M184V], has been reported previously in a number of cases involving multiple passages of drugs in HIV infected cells and examples include ddI, 3TC and FTC (Gu *et al.*, 1992; Gao *et al.*, 1993; Tisdale *et al.*, 1993). In the case of ddI, this mutation results in a five-fold increase in the IC_{50} value for ddI with accompanying cross-resistance to ddC (Nair *et al.*, 1995).

6.   Correlation of anti-HIV activity with electrostatic potential surfaces

Correlation of the anti-HIV activity of nucleosides with specific structural characteristics is of significance in contributing to the understanding of the mechanism of action of these compounds. Such investigations may also provide predictive information on structural characteristics most likely to elicit activity. Examination of the anti-HIV data of some normal and isomeric dideoxynucleosides (isoddNs) (Figure 5), their 3-D electron density patterns, their electrostatic potential surfaces (EPS), and their conformational maps reveals some interesting correlations (Mickle and Nair, 1999; Mickle and Nair 2000 a,b). For example, the electrostatic potential surface of (*S,S*)-isoddA shows remarkably similar regions of high and low electrostatic potential as AZT, (-) oxetanocin A and (-) carbovir (Figure 6) (Mickle and Nair, 1999).

Correlations involving EPS data and anti-HIV activity were also found with many other active nucleosides. Interestingly, inactive compounds had different EPS to those in the same series that were active (Figure 7). For example, the anti-HIV inactive apio-ddNs exhibit a clear difference in electrostatic potential and 3-D electron density shape compared to the anti-HIV active isoddNs. Additionally, the inactivity of (*S,S*)-isoddC and (*S,S*)-isoddT can be correlated convincingly through a combination of their EPS data and their conformational energy maps (Mickle and Nair, 1999).

The electrostatic potential distributions of anti-HIV active nucleoside triphosphates, the cellularly active anabolic products, also show remarkable correlations (Mickle and Nair 1999). For example, (*S,S*)-isoddATP, AZTTP and oxetanocin A TP have similar 3D-electron density surface patterns and similar high and low regions of electrostatic potential (Figure 8), which may suggest that these compounds proceed through related mechanisms in their interaction and inhibition of HIV reverse transcriptase. Docking of AZTTP, (*S,S*)-isoddATP and other active triphosphates into the active site of HIV RT and calculation of the EPS of both the nucleotide and active site show that there is

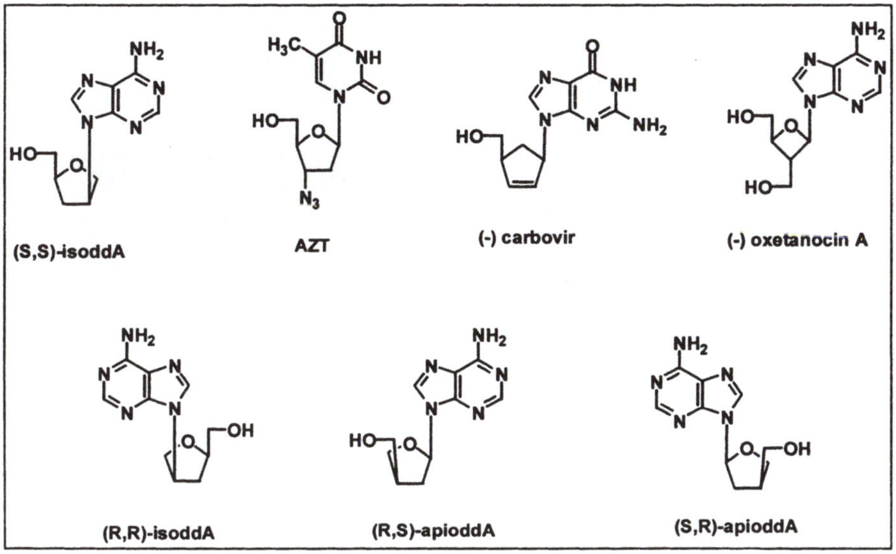

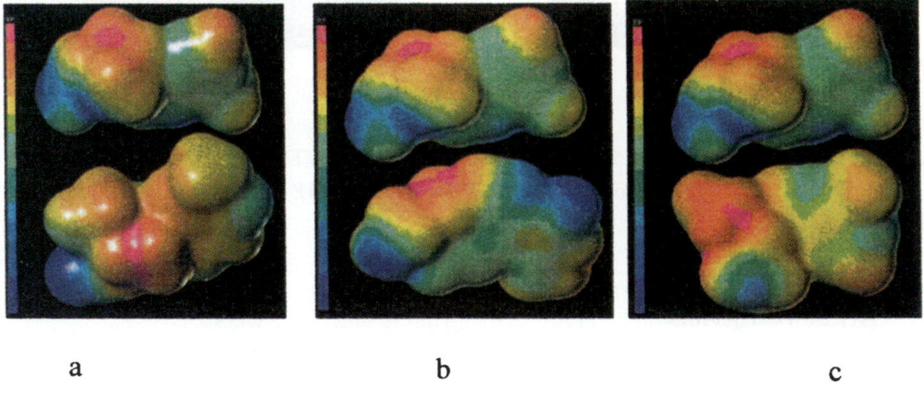

Figure 5.   Some representative anti-HIV active and inactive compounds examined.

excellent matching between inhibitor and enzyme binding site EPS data. The structure-activity profile discovered has contributed to the development of a first predictive QSAR analysis in the area (Mickle and Nair, 2000 a,b).

Figure 6.   Electrostatic Potential Surfaces of (S,S)-isoddA (top) and AZT (bottom). b. Electrostatic Potential Surfaces of (*S,S*)-isoddA (top) and (-) carbovir (bottom). c. Electrostatic Potential Surfaces of (S,S)-isoddA (top) and oxetanocin A (bottom). Note: Relative electrostatic potentials are indicated on the left-hand side. High (positive) regions are designated in red/ orange while low (negative) regions are indicated in purple/ blue.

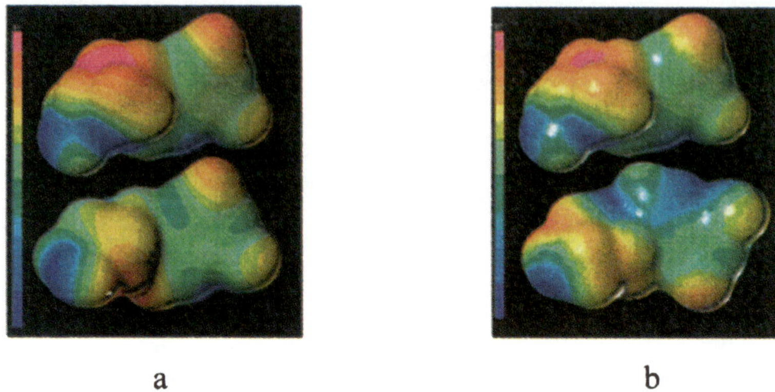

a   b

Figure 7.   a. Electrostatic Potential Surfaces of (S,S)-isoddA and (R,R)-isoddA (bottom). b. Electrostatic
Potential Surfaces of (S,S)-isoddA (top) and (R,S)-apio-isoddA (bottom).

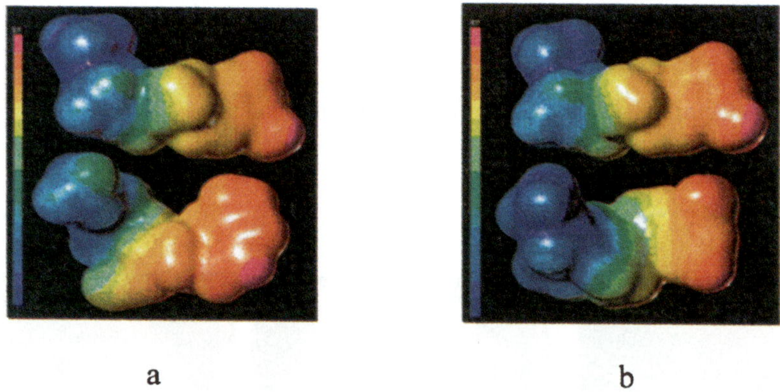

a   b

Figure 8.   a. Electrostatic Potential Surfaces of *(S,*S)-IsoddATP and AZTTP (left). b. Electrostatic Potential
Surfaces of *(S,*S)-IsoddATP and Oxetanocin TP (right).

7.   Acknowledgments

I thank the National Institutes of Health (NIAID) for support of this research work
on the discovery of novel anti-HIV compounds. It is a pleasure to acknowledge the
contributions of my many able coworkers whose names appear in the publications
that are cited. Collaborative anti-HIV studies were carried out at the National Cancer
Institute (Dr. Shoemaker), the Rega Institute for Medical Research, Leuven, Belgium
(Drs. Erik De Clercq and Jan Balzarini), and the Wellcome Laboratories (Dr. Marty
St. Clair).

8. References

Ahluwalia, G.; Cooney, D. A.; Mitsuya, H.; Fridland, A.; Flora, K. P.; Hao, Z.; Dalal, M.; Broder, S. and Johns, D. G. (1987) Initial studies on the cellular pharmacology of 2',3'-dideoxyinosine, an inhibitor of HIV infectivity, Biochem. Pharmacol. 36, 3797-3800.

Ahluwalia, G.; Cooney, D. A.; Hartman, N. R.; Mitsuya, H.; Yarchoan, R.; Fridland, A.; Broder, S. and Johns, D. G. (1993) Anomalous accumulation and decay of 2',3'-dideoxy- adenosine 5'-triphosphate in human T-cell cultures exposed to the anti-HIV drug 2',3'-dideoxyinosine, Drug. Metab. Dispos. 21, 369-376.

Baba, M.; Pauwels, R.; Herdewijn, P.; De Clercq, E.; Desmyter, J. and Vandeputte, M. (1987) Both 2',3'-dideoxythymidine and its 2',3'-unsaturated derivative (2',3'-dideoxythymidinene) are potent and selective inhibitors of human immuno-deficiency virus replication *in vitro*. Biochem. Biophys. Res. Commun. 142, 128-134.

Bera, B. and Nair, V. Synthesis of analogs of isoddA with replacement of $-CH_2OH$ with a secondary hydroxy functionality, Unpublished results.

Bera, S.; Mickle, T. and Nair, V. (1999) Synthesis and antiviral studies of unsaturated analogues of isomeric dideoxynucleosides, Nucleosides & Nucleotides 18, 2379-2395.

Bera, S. and Nair, V. (2000) Isonucleosides with exocyclic methylene groups, Helv. Chim. Acta 83, 1398-1467.

Bolon, P. J.; Sells, T. B.; Nuesca, Z. M.; Purdy, D. F. and Nair, V. (1994) Novel isomeric dideoxynucleosides as potential antiviral agents, Tetrahedron 50, 7747-7764.

Bolon, P.J.; Jahnke, T. S. and Nair, V. (1995) Homologues of anti-HIV active isodideoxy-nucleosides, Tetrahedron 51, 10443-10452.

Chun, B.-K.; Seo, B.-I. and Nair V. Synthesis of (*R,R*)-isodideoxynucleosides, Unpublished results.

Coates, J. A. V.; Cammack, N.; Jenkinson, H. J.; Mutton, I. M.; Pearson, B. A.; Storer, R.; Cameron, J. M. and Penn, C. R. (1992) The separated enantiomers of 2'-deoxy-3'-thiacytidine (BCH 89) both inhibit human immunodeficiency virus replication *in vitro*, Antimicrob. Agents Chemother. 36, 202-205.

Cooney, D. A.; Dalal, M.; Mitsuya, H.; McMahon, J. B.; Nadkarni, M.; Balzarini, J. and Broder, S. (1986) Initial studies on the cellular pharmacology of 2',3'-dideoxycytidine, Biochem. Pharmacol. 35, 2065-2068.

Cooney, D. A.; Ahluwalia, G.; Mitsuya, H.; Fridland, A.; Johnson, M.; Hao, Z.; Dalal, M.; Balzarini, J.; Broder, S. and Johns, D. G. (1987) Initial studies on the cellular pharmacology of 2',3'-dideoxyadenosine, an inhibitor of HTLV-III infectivity, Biochem. Pharmacol. 36, 1765-1768.

Daluge, S. M.; Good, S. S.; Faletto, M. B.; Miller, W. H.; St. Clair, M. H.; Boone, L. R.; Tisdale, M.; Parry, N. R.; Reardon, J. E.; Dornsife, R. E.; Averett, D. R. and Krenitsky, T. A. (1997) 1592U89, A novel carbocyclic analog with potent selective anti-human immunodeficiency virus activity, Antimicrob. Agents Chemother. 41, 1082-1093.

De Clercq, E. (1995) Toward improved anti-HIV chemotherapy: therapeutic strategies for intervention with HIV infections, J. Med. Chem. 38, 2491-2517.

DeClercq, E. (1997) In search of a selective antiviral chemotherapy, Clin. Microbiol. Rev. 10, 674-693.

Defaye, J.; Horton, D. and Musser, M. (1971) Methanolyse des 1,2-*O*-isopropylidene-5-*O*-*p*-tolylsulfonyl-pentofuranoses: Influence des facteurs steriques sur la formation des 2,5-anhydropentoses dimethyl acetals, Carbohydrate Research 20, 305-318.

Diaz, Y.; Bravo, F. and Castillon, S. (1999) Synthesis of purine and pyrimidine isodideoxy-nucleosides from (*S*)-glycidol using iodoetherification as key step. Synthesis of (*S,S*)-isoddA, J. Org. Chem. 64,

6508-6511.

Frankel, A. D. and Young, J. A. T. (1998) HIV-1: Fifteen proteins and an RNA, Annu. Rev. Biochem. 67, 1-25.

Gao, Q.; Gu, Z. X.; Parniak, M. A.; Cameron, J.; Cammack, N.; Boucher, C. and Wainberg, M. A. (1993) The same mutation that encodes low-level human immuno-deficiency virus type I resistance to 2',3'-dideoxyinosine and 2',3'-dideoxy-3'-thiacytidine, Antimicrob. Agents. Chemother. 37, 1390-1392.

Gu, Z.; Gao, Q.; Li, X.; Parniak, M. A. and Wainberg, M. A. (1992) Novel mutation in the human immunodeficiency virus type I reverse transcriptase gene that encodes cross-resistance to 2',3'-dideoxyinosine and 2',3'-dideoxycytidine, J. Virol. 66, 7128-7135.

Ho, H.-T. and Hitchcock, M. J. M. (1989) Cellular pharmacology of 2',3'-dideoxy-2',3'- didehydro-thymidine, a nucleoside analog active against human immunodeficiency virus, Antimicrob. Agents Chemother. 33, 844-849.

Huryn, D. M.; Sluboski, B. C.; Tam, S. Y.; Todaro, L. J.; Weigele, M.; Sim, I.; Anderson, B. D.; Mitsuya, H. and Broder, S. (1992) Synthesis and anti-HIV activity of isonucleosides, J. Med. Chem. 35, 2347-2354.

Jahnke, T. S. and Nair, V. (1995) 2'-Deoxyadenylyl-(3'→5')-isodideoxyadenosine, a unique dinucleotide: synthesis and characterization, Bioorg. Med. Chem. Lett. 5, 2235-2238.

Jeon, G. S. and Nair, V. (1996) New isomeric analogues of anti-HIV active azidonucleosides, Tetrahedron 52, 12643-12650.

Johnson, S. C. and Gerber, J. G. (2000) Advances in HIV/AIDS therapy. In R. W. Schrier, J. D. Baxter, V. J. Dzau and A. S. Fauci (Eds), Advances in Internal Medicine, Volume 44 pp. 1-40, Mosby, St. Louis.

Jones, M. F.; Nobel, S. A.; Robertson, C. A. and Storer, R. (1991) Tetrahydrothiophene nucleosides as potential anti-HIV agents, Tetrahedron Lett. 32, 247-250.

Jones, M. F.; Nobel, S. A.; Robertson, C. A.; Storer, R.; Highcock, R. M. and Lamont, R. B. (1992) Enantiospecific synthesis of 3'-hetero-dideoxy nucleoside analogues as potential anti-HIV agents, J. Chem. Soc. Perkin Trans. 1, 1427-1436.

Katz, R. A. and Skalka, A. M. (1994) The retroviral enzymes, Annu. Rev. Biochem. 63, 133-173.

Kierdaszuk, B.; Bohman, C.; Ullman, B. and Eriksson, S. (1992) Substrate specificity of human deoxycytidine kinase toward antiviral 2',3'-dideoxynucleoside analogs, Biochem. Pharmacol. 43, 197-206.

Lin, T.-S.; Schinazi, R. F. and Prusoff, W. H. (1987) Potent and selective *in vitro* activity of 3'-deoxythymidin-2'-ene (3'-deoxy-2',3'-didehydrothymidine) against human immuno-deficiency virus, Biochem. Pharmacol. 36, 2713-2718.

Martin, J. L.; Brown, C. E.; Matthews-Davis, N. and Reardon, J. E. (1994) Effects of antiviral nucleoside analogs on human DNA polymerases and mitochondrial DNA synthesis, Antimicrob. Agents Chemother. 38, 2743-2749.

Mickle, T. and Nair, V. (1999) Correlation of anti-HIV activity with structure: Use of electrostatic potential and conformational analysis, Bioorg. Med. Chem. Lett. 9, 1963-1968.

Mickle, T. and Nair, V. (2000a) Anti-HIV activity of nucleosides and nucleotides: Correlation with molecular electrostatic potential data, Antimicrob. Agents Chemother. 44, 2939-2947.

Mickle, T. and Nair, V. (2000b) Predictive QSAR analysis of anti-HIV agents, Drugs of the Future 25, 393-400.

Mitsuya, H.; Weinhold, K. J.; Furman, P. A.; St. Clair, M. H.; Lehrman, S. N.; Gallo, R. C.; Bolognesi, D.; Barry, D. W. and Broder, S. (1985) 3-Azido-3'-deoxythymidine (BW A509U): An antiviral agent that inhibits the infectivity and cytopathic effect of human T-lymphotropic virus type III/ lymphadenopathy-associated virus *in vitro*, Proc. Natl. Acad. Sci. U.S.A. 82, 7096-7100.

Mitsuya, H. and Broder, S. (1986) Inhibition of the *in vitro* infectivity and cytopathic effect of HTLVIII/LAV by 2',3'-dideoxynucleosides, Proc. Natl. Acad. Sci. U.S.A. 83, 1911-1915.

Montgomery, J. A.; Clayton, S. D. and Thomas, H. J. (1975) Preparation of methyl 2-deoxy-2-(purin-9-yl)arabinofuranosides and methyl 3-deoxy-3-(purin-9-yl)xylofuranosides, J. Org. Chem. 40, 1923-1927.

Nair, V. and Emanuel, D. J. (1977) Synthetic design, stereochemistry, and enzymatic activity of a reversed aminoacyl nucleoside: An analogue of puromycin, J. Am. Chem. Soc. 99, 1571-1576.

Nair, V. and Nuesca, Z. M. (1992) Isodideoxynucleosides: A conceptually new class of nucleoside antiviral agents, J. Am. Chem. Soc. 114, 7951-7953.

Nair, V.; Nuesca, Z. M.; Purdy, D. F.; Sells, T. B. and Zintek, L. B. (1992) Optically active isodideoxy-nucleosides: a new family of anti-HIV agents, Antiviral Res., Suppl. I 17, 44.

Nair, V. and Sells, T. B. (1992) Interpretation of the roles of adenylosuccinate lyase and of AMP deaminase in the anti-HIV activities of 2',3'-dideoxyadenosine and 2',3'-dideoxyinosine, Biochim. Biophys. Acta 1119, 201-204.

Nair, V. (1993) Approaches to novel isomeric nucleosides as antiviral agents. In C. K. Chu and D. C. Baker (Eds), Nucleosides and Nucleotides as Antitumor and Antiviral Agents, pp 127-140, Plenum Press, New York.

Nair, V.; St. Clair, M.; Reardon, J. E.; Krasny, H. C.; Hazen, R. J.; Paff, M. T.; Boone, L. R.; Tisdale, M.; Najera, I.; Dornsife, R. E.; Everett, D. R.; Borroto-Esoda, K.; Yale, J. L.; Zimmerman, T. P. and Rideout, J. L. (1995) Antiviral, metabolic and pharmacokinetic studies on the isomeric dideoxynucleoside, 4(S)-(6-amino-9H-purin-9-tetrahydro-2(S)-furanmethanol, Antimicrob. Agents Chemother. 39, 1993-1999.

Nair, V. and Jahnke, T. S. (1995) Antiviral activity of isomeric dideoxynucleosides of D-and L-related stereochemistry, Antimicrob. Agents Chemother. 39, 1017-1029.

Nair, V. and Sharma, P. K. (2000) Methodologies in the Synthesis of Anti-HIV Nucleosides. In Recent Developments in Organic Chemistry 4, 53-86.

Nanni, R. G.; Ding, J.; Jacobo-Molina, A.; Hughes, S. H. and Arnold, E. (1993) Review of HIV-1 reverse transcriptase three-dimensional structure: implications for drug design, Perspect. Drug Discovery Design 1, 129-150.

Nuesca, Z. M. and Nair, V. (1994) Synthesis of novel 3'-isomeric dideoxynucleosides, Tetrahedron Lett. 35, 2485-2488.

Pal, S. and Nair, V. (2000) Phosphorylation of the anti-HIV compound (S,S)-isodideoxy-adenosine by human recombinant deoxycytidine kinase, Biochem. Pharmacol. 60, 1505-1508.

Purdy, D. F.; Zintek, L. B. and Nair, V. (1994) Synthesis of isonucleosides related to AZT and AZU, Nucleosides & Nucleotides 13, 109-126.

Schinazi, R. F.; Chu, C. K.; Peck, A.; McMillan, A.; Mathis, R.; Cannon, D.; Jeong, L.-S.; Beach, J. W.; Choi, W.-B.; Yeola, S. and Liotta, D. C. (1992) Activities of the four optical isomers of 2',3'-dideoxy-3'-thiacytidine (BCH-189) against HIV-1 in human lymphocytes, Antimicrob. Agents Chemother. 36, 672-676.

Sells, T. B. and Nair, V. (1994) Synthetic approaches to novel cis and trans dideoxynucleosides of the apiose family, Tetrahedron 50, 117-138.

Sougdeyns, H.; Yao, X.-J.; Gao, Q.; Belleau, B.; Kraus, J.-L.; Nguyen-Ba, N.; Spira, B. and Wainberg, M.A. (1991) Anti-HIV-1 activity and *in vitro* toxicity of BCH-189, a novel heterocyclic nucleoside analog, Antimicrob. Agents Chemother. 35, 1386-1390.

Starnes, M. C. and Cheng, Y.-C. (1987) Cellular metabolism of 2',3'-dideoxycytidine, a compound active against human immunodeficiency virus *in vitro*, J. Biol. Chem. 262, 988-991.

Taktakishvili, M.; Neamati, N.; Pommier, Y.; Pal, S. and Nair, V. (2000) Recognition and inhibition of HIV integrase by novel dinucleotides, J. Am. Chem. Soc. 122, 5671-5677.

Tisdale, M.; Kemp, S. D.; Parry, N. R. and Larder, B. A. (1993) Rapid *in vitro* selection of human immunodeficiency virus type I resistant to 3′-thiacytidine inhibitors due to a mutation in theYMDD region of reverse transcriptase, Proc. Natl. Acad. Sci. U.S.A. 90, 5653-5656.

Vince, R. and Brownell, J. (1990) Resolution of racemic carbovir and selective inhibition of HIV by the (-) enantiomers, Biochem. Biophys. Res. Commun. 168, 912-916.

Wenzel, T. and Nair, V. (1997) Novel oligodeoxyribonucleotides incorporating L-related isodeoxy-nucleosides: Solid phase synthesis, enzymology, and CD studies, Bioorg. Med. Chem. Lett. 7, 3195-3198.

Wenzel, T. and Nair, V. (1998) Self-complementary oligodeoxyribonucleotides incorporating L-related isodideoxynucleosides: Synthesis, physical characterization, enzymology, and CD studies, Bioconjugate Chemistry 9, 683-690.

Yu, H.-W.; Zhang, L.-R.; Zhou, J.-C.; Ma, L.-T. and Zhang, L.-H. (1996) Studies on the synthesis and biological activities of 4′-(R)-hydroxy-5′-(S)-hydroxymethyltetrahydro-furanyl purines and pyrimidines, Bioorg. Med. Chem. 4, 609-614.

Zheng, X. and Nair, V. (1999) Homologues of isomeric dideoxynucleosides as potential antiviral agents: Synthesis of isodideoxynucleosides with a furanethanol sugar moiety, Nucleosides & Nucleotides 18, 1961-1976.

Zheng, X. and Nair, V. (1999) Synthesis of isomeric nucleoside phosphonates: Cyclic analogs of the anti-HIV active compound, PMEA, Tetrahedron 55, 11803-11818.

Zintek, L. B.; Jeon, G. S. and Nair, V. (1994) The synthesis of (S) and (R) enantiomers of novel hydroxymethylated isodideoxynucleosides, Heterocycles 37, 1853-1864.

Zintek, L. B.; Jahnke, T. S. and Nair, V. (1996) Synthesis and conformational studies of new purine isodideoxynucleosides, Nucleosides & Nucleotides 15, 69-84.

SYNTHESIS AND BIOLOGICAL ACTIVITY OF ISOPOLAR ACYCLIC NUCLEOTIDE ANALOGS

ANTONÍN HOLÝ[1]

*Institute of Organic Chemistry and Biochemistry,
Prague, CZECH REPUBLIC*

1. Introduction

Modified nucleosides attracted attention since the early phase of rational drug design that was directed predominantly to cancer and leukemia chemotherapy. It soon became obvious that in the cells these molecules are regularly transformed to their 5'-phosphate esters (5'-nucleotides) - precursors for subsequent transformation to the *active antimetabolites* (usually 5'-triphosphates). Much later, an identical mechanism of activation was demonstrated for the majority of nucleoside antimetabolites active against cellular parasites (viruses, protozoa). This phosphorylation is usually catalyzed by nucleoside kinases (cellular or parasite-specific), but in some cases, reverse (synthetic) reactions of enzymes catalyzing dephosphorylation (e.g., 5'-nucleotidases) or phosphate transfer from 5'-nucleotides (AMP), or other phosphoric acid monoesters (phospho-transferases) can participate in this anabolic process. This enzymatic 5'-phosphorylation is not limited to ribo- or 2-deoxyribonucleosides but occurs also with sugar-modified nucleoside analogues (arabinosides, 3-deoxynucleosides, 2,3-dideoxynucleosides, carbocyclic nucleosides, etc.), and even with opposite enantiomers (L-ribonucleosides) (Jurovčík *et al.,* 1976).

The importance of 5'-nucleotides (and their higher anabolites) as active species for the nucleoside activity *in vitro* and *in vivo* influenced several lines of fundamental research in the medicinal chemistry of nucleosides:

a) the efficient degradation of modified 5'-nucleotides back to nucleosides catalyzed by diverse specific or non-specific phosphomonoesterases ("phosphatases") leveling down the intracellular concentration of the active antimetabolite was therapeutically counterproductive; it enforced potentially hazardous increased dosages of nucleosides. Thus it seemed imperative to search for methods combating this catabolic instability.

b) the direct therapeutic use of 5'-nucleotides was evidently out of the question in general, due to their enzymatic instability *per se* (in the cell pool as well as during

---

[1] Institute of Organic Chemistry and Biochemistry, Academy of Sciences of the Czech Republic, 16610 Praha 6 (Czech Republic). Phone: (4202)-20183384; fax: (4202)-24310090; E-mail: holy@uochb.cas.cz

Recent Advances in Nucleosides: Chemistry and Chemotherapy, Ed. by C.K. Chu. 167 — 238

membrane transport); their dephosphorylation occurs readily in the blood plasma. Thus, it seemed imperative to search for *nucleotide prodrugs* which would liberate the required (base, sugar)-modified nucleotide following its parenteral application.

c) the polar character of nucleotides complicates their resorption through the intestine on enteric application and makes their transport through the cell membrane nearly impossible. Thus, it seemed imperative to develop their *unpolar prodrugs* (preferentially diesters) which would liberate the modified nucleotide at the target. This led eventually to nucleotide diester prodrugs which decompose spontaneously in the cell or under catalysis by cellular enzymes (esterases, etc.).

These three topics influenced fundamental research in the medicinal chemistry of nucleosides and nucleotides for the last 30 years. The importance of chemical modifications on nucleotides is witnessed by the monograph edited in 1980 which comprized the contemporary knowledge of the topics (Scheit, 1980).

We have been engaged in research of nucleotide analogues for a long time, studying originally the substrate specificity of diverse enzymes involved in the catabolic reactions of nucleic acids and their constituents. Studies of numerous compounds modified at the heterocyclic base or sugar moiety as substrates of such enzymes resulted in the finding that these parts of the nucleotide molecule are not decisive for recognition by dephosphorylating enzymes. The enzymatic stability of the nucleotide can be achieved solely by a suitable alteration of the phosphoric acid (or its ester linkage). This conclusion brought to the forefront a search for nucleotide analogues modified at the phosphate group as potential inhibitors of nucleotide dephosphorylation; such compounds could also interfere with nucleotide biosynthetic pathways and achieve a broader applicability. Some of the characteristic structures of this early stage are depicted in Chart 1. We have contributed by investigation of nucleoside-5'-phosphonates (Holý, 1967b) but our main study focused on phosphonic acid esters: synthesis and investigation of nucleoside phosphites (Holý *et al.*, 1966), methylphosphonates (Holý *et al.*, 1967a), hydroxymethylphosphonates (Holý *et al.*, 1971) and aminomethyl phosphonates (Gulyaev *et al.*, 1972). The outcome from these studies was recognition of the *isosteric and isopolar* (*isoelectronic*) character of the nucleotide analogue as a *conditione sine qua non* for the optimum recognition by the enzymes. It is witnessed by the inhibitory activity of nucleoside phosphothioates discovered by Eckstein (Eckstein, 1985) or substituted "homonucleoside phosphonates" (in fact carba analogues of phosphate esters) (Hampton *et al.*, 1976). Our empirical knowledge at that time implied also the importance of the oxygen atom in the vicinity of phosphorus for the enzyme recognition. [At that time the conclusion was empirical; much later it was demonstrated that the ether oxygen in phosphonomethyl ethers participates in complex formation of the phosph(on)ate with bivalent metal cations (Sigel *et al.*, 1999)].

This brought us later to design and synthesize a novel type of nucleotide analogues, nucleoside O-phosphonomethyl ethers, which are isopolar with nucleotides and whose sp3-hybridized CH_2 grouping should warrant the conformational adaptability of nucleotide molecules (Rosenberg *et al.*, 1982, 1983). It is quite evident that the ether linkage should be stable against enzymatic degradation (it is also chemically stable). In pilot studies, we have demonstrated that the 5'-O-phosphonomethyl ribonucleosides inhibit

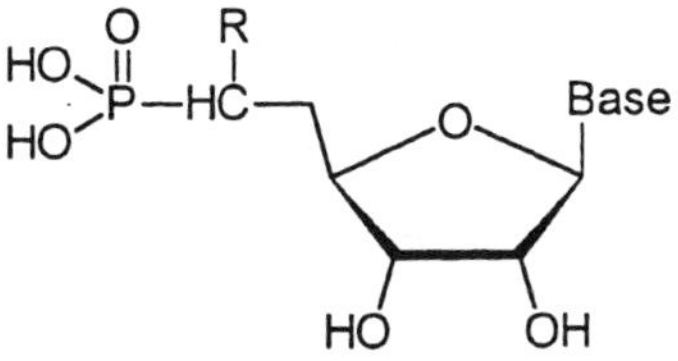

R = H, CH₃, OH, CN,
CH₂NH₂, CH₂NHCOR
Moffatt 1968, Hampton 1973

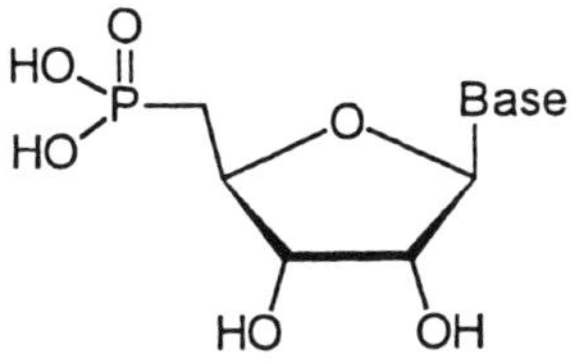

Rammler 1966
Holý 1967

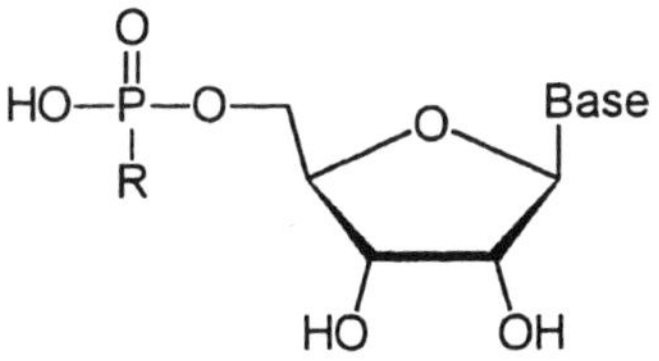

R = H, CH₃, CH₂OH, CH₂NH₂
Holý 1966-1972

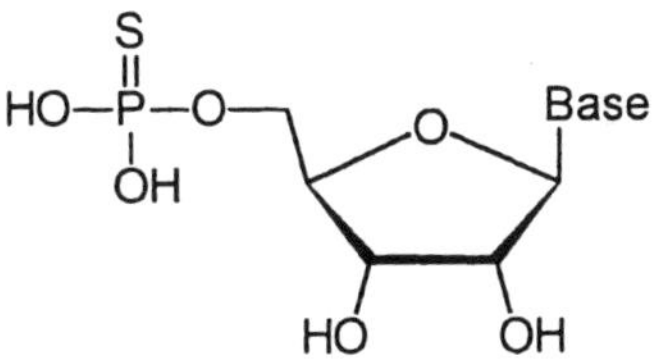

Eckstein 1966

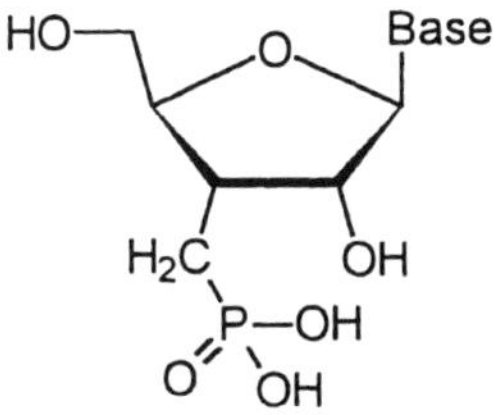

Moffatt 1970

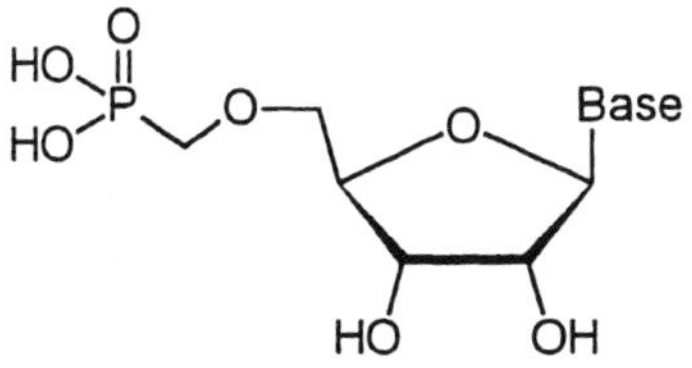

Rosenberg & Holý, 1982

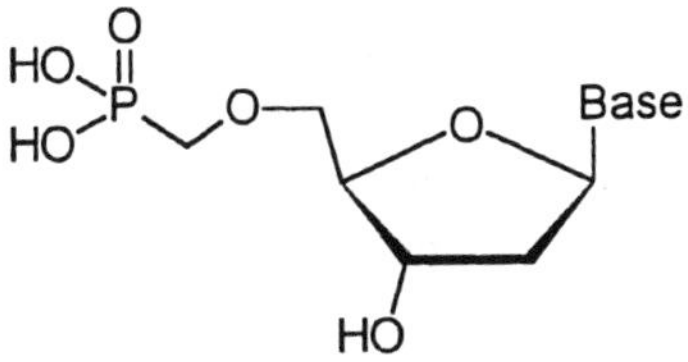

Krečmerová 1990

Chart 1.

some nucleotidases (Holý, 1980) and that the triphosphate analogues of GTP and ATP containing a modified α-phosphorus atom can serve as phosphate donors in uridine kinase catalyzed reactions (Veselý *et al.*, 1982), whereas the analogues of UTP and/or CTP inhibit this process (Veselý *et al.*, 1983). We were able to demonstrate that such ATP or CTP analogues inhibit DNA-dependent RNA polymerase catalyzed RNA synthesis, but can be incorporated into the growing RNA chain (Horská *et al.*, 1983, 1890). The 3'-O-phosphonomethyl and 5'-O-phosphonomethyl analogues of (3'-5')-ApU and UpA are substrates for the *E.coli* DNA-dependent RNA polymerase in the primed abortive synthesis on a poly[d(A-T)] template (Cvekl *et al.*, 1989).

Despite these significant *in vitro* activities, neither the 5'-O-phosphonomethyl ethers of natural ribonucleosides, nor analogous compounds derived from biologically active (e.g., antiviral or cytostatic) nucleosides (Rosenberg *et al.*, 1982) exhibited any significant cytostatic or antiviral properties *in vitro* (Jie *et al.*, 1990*)*. This disappointment was compensated by our discovery of potent antiviral activity of HPMPA (De Clercq, 1986), a structurally related molecule bearing the phosphonomethyl ether group linked to the side-chain of the acyclic adenosine analogue DHPA [9-(2,3-dihydroxypropyl)adenine]. However, while the parent metabolically inert adenosine analogue (Holý *et al.*, 1981) inhibits RNA viruses (De Clercq *et al.*, 1978) by virtue of interfering with cap-methylation of viral mRNA (Votruba *et al.*, 1990) (*via* inhibition of SAH hydrolase (Votruba *et al.*, 1980), the activity of its new phosphonate derivative was directed specifically against DNA viruses (De Clercq, 1986). The regio- and enantiospecific syntheses proved that the antiviral activity is linked to the 2'-*(S)*-isomer only (Holý *et al.*, 1987c).

2.   Early development

During the first developmental stage, much of our effort was in the rapid estimation of the approximate structural limits of biological activity in this class of analogues. The simplest member of this family, 9-[2-(phosphonomethoxy)ethyl]adenine (PMEA) (Holý *et al.*, 1987b), also exhibited antiviral activity against DNA viruses (De Clercq *et al.*, 1986). None of the tested RNA viruses responded to its action; however – contrary to HPMPA - it showed a significant effect against retroviruses (Pauwels *et al.*, 1988; Balzarini *et al.*, 1989). In the adenine series, alterations at the side chain, in particular omission of the ether oxygen atom (e.g., in ω-phosphonoalkyladenines) resulted in antivirally inert compounds (Rosenberg *et al.*, 1988; Holý *et al.*, 1989) (Chart 2).

Abbreviations used in the literature for the ANPs (acyclic nucleoside phosphonates) are composed of two parts: the first letters are derived from the chemical name of the side chain (PME for 2-(phosphonomethoxy)ethyl, HPMP for 3-hydroxy-2-(phosphonomethoxy)propyl, FPMP for 3-fluoro-2-(phosphonomethoxy)propyl and PMP for 2-(phosphonomethoxy)propyl residue; the last letter is the standard one-letter abbreviation of the nucleobase (C,A,G,T), DAP stands for 2,6-diaminopurine.

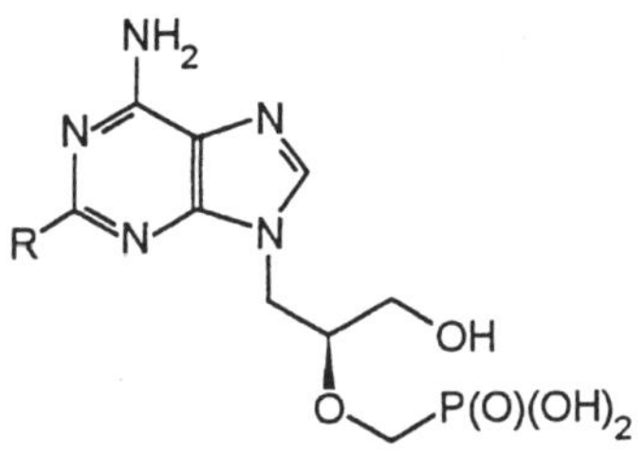

R = H   (S)-HPMPA
R = NH$_2$ (S)-HPMPDAP

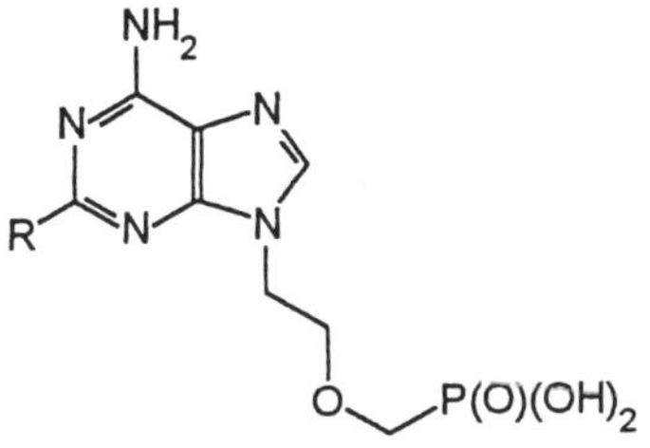

R = H   PMEA
R = NH$_2$ PMEDAP

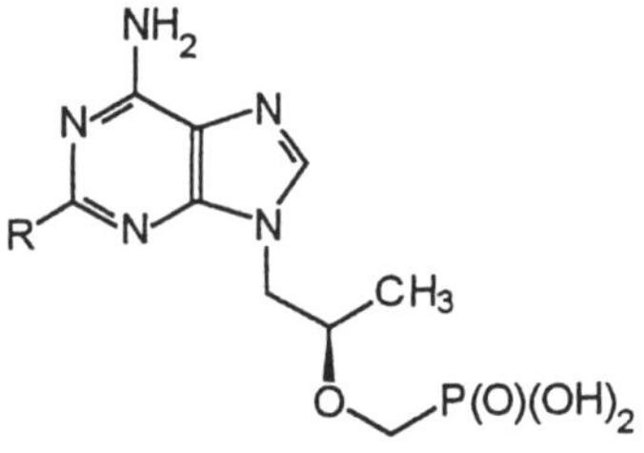

R = H   (R)-PMPA
R = NH$_2$ (R)-PMPDAP

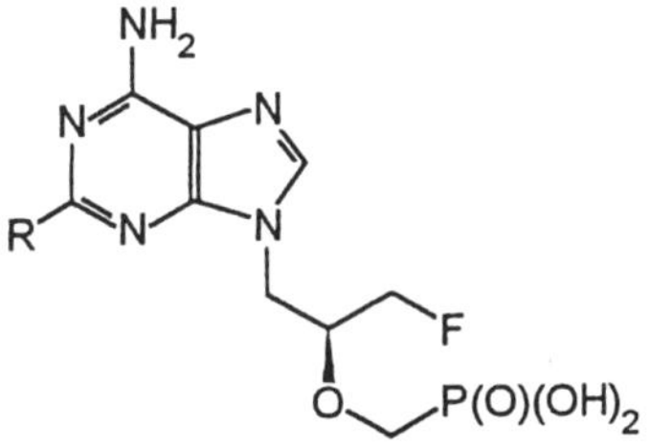

R = H   (S)-FPMPA
R = NH$_2$ (S)-FPMPDAP

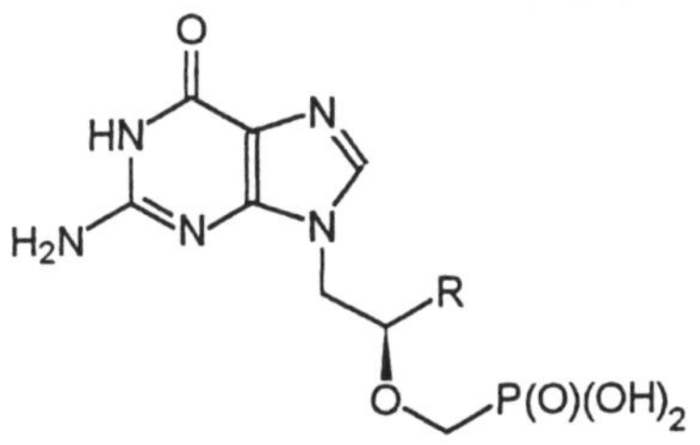

R = H   PMEG
R = CH$_2$OH (S)-HPMPG
R = CH$_3$ (R)-PMPG
R = CH$_2$F   (S)-FPMPG

Chart 2.

The credit for inventing this easily understandable system belongs to Professor E. De Clercq. The absolute configuration at the side-chain is described by the Ingold-Prelog-Cahn system. If not specified otherwise, the purine heterocyclic bases are substituted at position N9, pyrimidine derivatives are N1-regioisomers. It should be noted that the order of substituent preference in the HPMP and PMP series is opposite; therefore, (S)-HPMP, (S)-FPMP but (R)-PMP-compounds have the same absolute configuration at the side chain.

It was soon recognized that biological activity was connected with 9-regioisomers of adenine, guanine and 2,6-diaminopurine, while their N3 and N7-isomers, hypoxanthine and/or xanthine derivatives were devoid of antiviral activity; in the pyrimidine series, uracil and thymine derivatives were inactive in all series examined (Pauwels *et al.,* 1988; De Clercq *et al.,* 1987). Surprizingly, an exceptionally high antiviral effect was noted with the cytosine derivative HPMPC (Snoeck *et al.,* 1988). At the end of this stage, preclinical investigation was initialized with several compounds. It was paralleled by the development of synthetic procedures, biochemical investigations and detailed examination of biological activity of ANPs in general.

3.   Part A.   Chemical synthesis

There are five general approaches to the synthesis of acyclic nucleoside phosphonates:
a)   introduction of the phosphonomethyl ether residue at the hydroxyl group of N-(hydroxyalkyl) derivatives of purine or pyrimidine bases
b)   alkylation of the appropriate heterocyclic base by a synthon bearing all characteristic features of the side chain
c)   ring-closure of the heterocyclic base at the amino group of suitably protected aminoalkylphosphonates
d)   transformation of functional groups in the side-chain of ANPs
e)   transformation of reactive functional groups at the heterocyclic bases in ANPs.

3.1. Etherification of the hydroxyl group in hydroxyalkyl derivatives

Hydroxyalkyl derivatives of purine and pyrimidine bases needed as parent structures for this approach are easily accessible by numerous synthetic procedures, including alkylation with suitably protected hydroxyalkyl halides, sulfonates, carbonates, oxiranes, etc. (e.g., Holý *et al.,* 1974; Holý 1978a, b, c, Holý 1979; Holý *et al.,* 1985 and references therein). The introduction of a phosphonomethyl ether residue onto an isolated hydroxyl function can be achieved by condensation with dialkyl *p*-tolylsulfonyloxymethyl-phosphonates in the presence of excess (2.5-3 equivs.) sodium hydride. The reaction is best performed in DMF or THF solution. The reagents can be easily synthesized by triethylamine-catalyzed reaction of dialkyl phosphites with paraformaldehyde or 1,3,5-trioxane followed by tosylation. Originally, we used the diethyl ester. In some cases, depending on the character of the heterocyclic base, the resulting phosphonate diethyl ester can be hydrolyzed to the free phosphonates by heating with aqueous hydrochloric acid. This variant was applied to the manufacture of PMEA (Adefovir) making use of 9-(2-hydroxyethyl)adenine as starting material. Later on, an improved manufacturing procedure was described which replaces NaH in the coupling reaction with sodium *tert*-butoxide The optimized process was scaled-up to batch sizes of >100 kg (Yu *et al.,* 1999). A similar approach was later applied to the synthesis of another drug candidate, *(R)*-PMPA (Tenofovir) from 9-(2-hydroxypropyl)adenine (Schultze *et al.,* 1998) (Scheme 1).

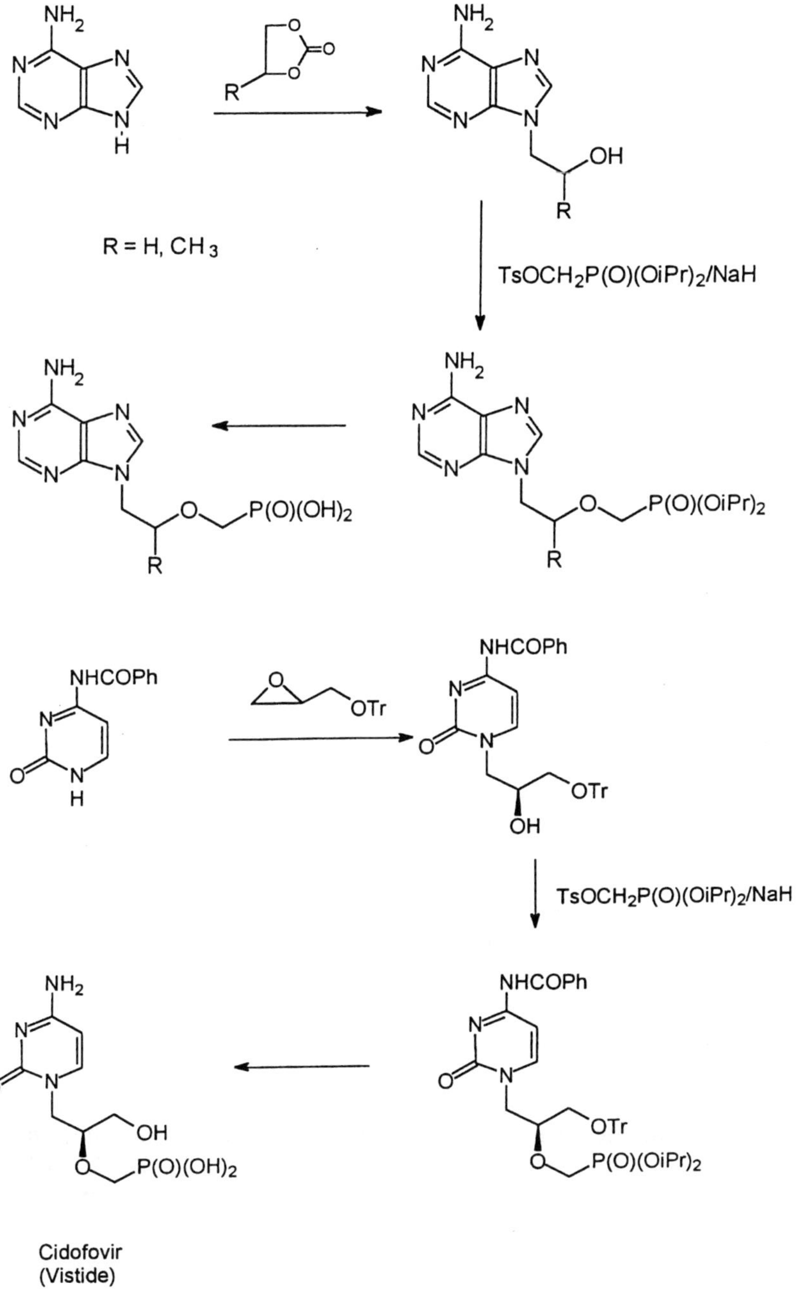
NH2
R
R = H, CH3
OH
TsOCH2P(O)(OiPr)2/NaH
P(O)(OH)2
P(O)(OiPr)2
NHCOPh
OTr
Cidofovir
(Vistide)

Scheme 1.

It is preferable to protect the amino functions at the base by suitable protecting groups (acyl, amidino, etc.). In the HPMP-series, the starting 2,3-dihydroxypropyl derivatives must be protected both at the 3'-hydroxyl group of the side-chain and at the amino group of the heterocyclic base; *(S)*-HPMPA was prepared by this approach from 9-*(S)*-(2-hydroxy-3-triphenylmethoxypropyl)-N6-benzoyladenine (Holý *et al.*, 1987a) or from O3',N6-ditrityl *(S)*-DHPA obtained by tritylation of *(S)*-DHPA with excess trityl chloride (Webb *et al.*, 1987), while the synthesis of *(S)*-HPMPC (Cidofovir) started from 1-*(S)*-(2-hydroxy-3-trityloxypropyl)-N4-benzoylcytosine which was obtained directly by the reaction of N4-benzoylcytosine with trityloxymethyl-*(R)*-glycidol (Webb *et al.*, 1988; Brodfuehrer *et al.*, 1994).

A certain handicap of the application of diethyl *p*-tolylsulfonyloxymethylphosphonate in the above condensation is the oily character of the reagent as well as a limited stability of the resulting intermediary diethyl ester which can partially hydrolyze to the monoethyl ester during the work-up of the reaction mixtures. Therefore, we prefer to use the corresponding diisopropyl ester which is a crystalline material stable on storage; the resulting products are much more stable to hydrolysis. Both phosphonate ester groups can be easily removed by transsilylation with iodo- or bromotrimethylsilane (Rosenberg *et al.*, 1982) followed by hydrolysis. This synthetic alternative was successfully applied to the preparation of several ANP types, e.g., *(R)*- and *(S)*-enantiomers of HPMP-derivatives (HPMPA, HPMPG, HPMPDAP, HPMPC and others) (Holý, 1993) as well as their deaza analogues (1-deazaHPMPA, 3-deazaHPMPA) (Dvořáková *et al.*, 1993) and/or PMP-derivatives (Holý *et al.*, 1995b). Other applications of this general synthetic route comprize racemic pyrimidine and purine N-[3-fluoro-2-(phosphonomethoxy)-propyl]- (FPMP) derivatives (Jindřich *et al.*, 1993), analogous 3-amino-, dialkylamino and trialkylammonium derivatives of HPMPA and its congeners (Dvořáková *et al.*, 1996), as well as the 2'-alkyl, cycloalkyl, aralkyl and aryl compounds of the PME-series (Dvořáková *et al.*, 1994).

As illustrated by the above examples, the "stepwise synthesis" of the target compounds is preferable for the scale-up of the individual compounds, specifically for large-scale manufacturing processes, and in those cases where the starting hydroxyalkyl derivatives are easily available. The latter compounds are usually prepared by alkylation of the heterocyclic bases (sodium or cesium salts, respectively), a reaction largely regiospecific for cytosine, adenine or 2,6-diaminopurine, much less so for uracil or guanine. In these cases, suitable modification of the base (4-methoxy-2-pyrimidone, 6-halogenopurines, 6-alkoxypurines, etc.) is indicated. In addition to the widely used synthons bearing a *p*-tolylsulfonyloxy leaving group, a closely related (diethylphosphonyl)methyl triflate was described in the literature; it was used to introduce the phosphonomethyl ether group both at the hydroxyalkyl group of the acyclic nucleoside (Kim *et al.*, 1991c) or onto the side-chain-type synthon (*vide infra*) (Kim *et al.*, 1990a) (Scheme 2).

Related to this group is an intramolecular etherification reaction which takes place with vicinal diol bearing derivatives: their reaction with chloromethylphosphonyl dichloride in triethyl phosphate gives rise to the ester chlorides which hydrolyze in alkaline solution with simultaneous etherification of the chloromethyl group by the neighboring hydroxyl. This reaction proceeds easily at the 2',3'-cis-diol group of

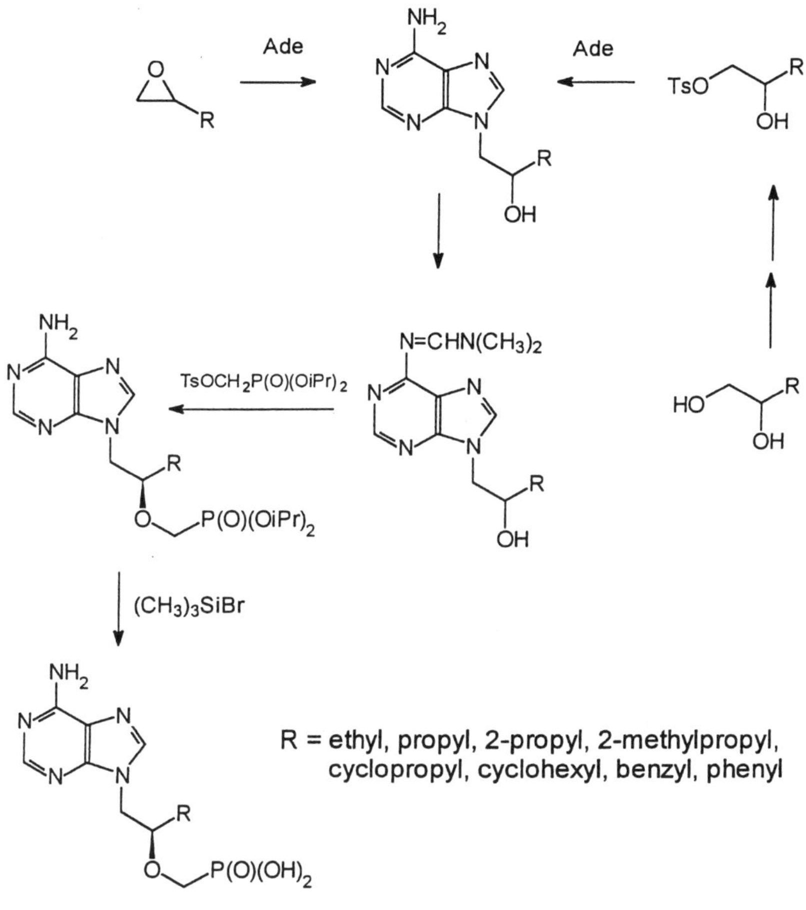

R = ethyl, propyl, 2-propyl, 2-methylpropyl, cyclopropyl, cyclohexyl, benzyl, phenyl

Scheme 2.

ribonucleosides to give a mixture of isomeric 2'- and/or 3'-O-phosphonomethyl ether (Rosenberg *et al.*, 1983). However, its main application is in the preparation of the HPMP-derivatives from N-(2,3-dihydroxypropyl)purines or pyrimidines. The primary esterification takes place predominantly at the 3'-hydroxyl function. Hydrolysis of the intermediary ester chloride by boiling in water causes additional isomerisation of the mixture of chloromethylphosphonate isomers wherein the required 3'-isomer can often be enriched to >90%. Subsequent treatment with warm aqueous alkali produces a mixture of isomeric O-phosphonomethyl derivatives with significantly predominant 2'-isomers (HPMP-derivative) (Holý *et al.*, 1987c). The etherification process involves intermediary formation of a cyclic phosphonate. The individual isomeric chloromethylphosphonates can be separated by ion exchange chromatography or preparative HPLC. Thus, this reaction can be applied to the synthesis of pure HPMP-derivatives (Holý *et al.*, 1989). Alternatively, regiospecific synthesis of HPMP-derivatives can be performed by the following reaction sequence: N-(2,3-dihydroxypropyl) compounds are transformed to their 2'-*O*-benzoyl derivatives by 3'-*O*-dimethoxy-tritylation, benzoylation with benzoyl cyanide (Holý *et al.*, 1971a) and mild acid detritylation; the following reaction with chloromethylphosphonyl dichloride thus proceeds solely at the 3'-hydroxyl group; heating of the ester chloride intermediate with sodium methoxide in MeOH effects removal of the benzoyl group by methanolysis which is followed by etherification. The mixture of cyclic phosphonate and phosphonate methyl esters is easily converted to the phosphonomethyl ether by alkaline hydrolysis (Holý *et al.*, 1989b) (Scheme 3).

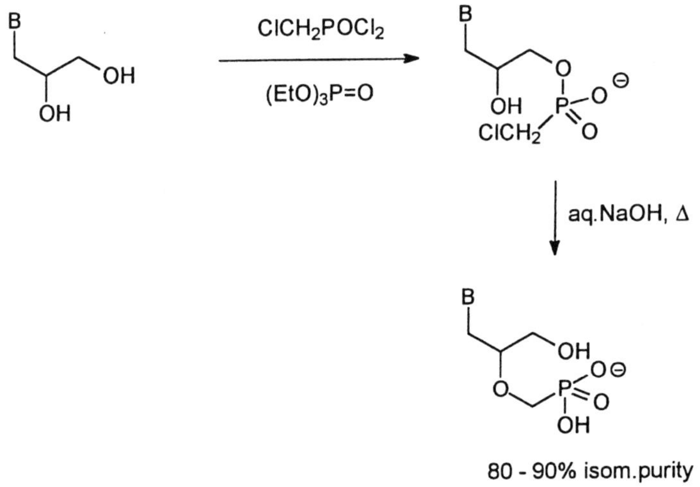

$ClCH_2POCl_2$

$(EtO)_3P=O$

aq.NaOH, Δ

80 - 90% isom.purity

Scheme 3.

3.2. Alkylation of the heterocyclic base by a side-chain type synthon

The general strategy of this approach is based on alkylation of an appropriate heterocyclic base with a reagent which contains a leaving group at a (protected) alkyl chain bearing an esterified phosphonate residue linked through an ether bridge. The indubitable advantage of this process is its shortness which makes it possible to perform the synthesis with large numbers of different bases on a small and/or medium scale and isolate the phosphonates after suitable deprotection steps. The method is especially applicable for rare bases and isotopically labelled compounds. Furthermore, this approach in most cases does not require any protection at the base and can be applied to sensitive heterocyclic systems or systems bearing reactive substituents (e.g., 6-chloropurines). In some cases, regioselectivity of the alkylation reaction is limited, which necessitates separation of regioisomers. The alkylation reaction is best performed at elevated temperatures in DMF with the sodium salt of the heterocyclic base generated *in situ* by the action of NaH, or in the presence of an inorganic base; the best results are achieved by the use of Cs_2CO_3. Recently, we have described an appreciable acceleration of the reaction rate and more favorable regioisomer ratios with 1,8-diazabicyclo-[5,4,0]undec-7-ene (DBU) (Holý *et al.*, 1999). The choice of the phosphonate ester group is determined not only by its stability in the reaction product (*vide supra*): dialkyl phosphonates are able to N-alkylate heterocyclic bases similar to phosphoric acid triesters. Alkylations of adenine or 2,6-diaminopurine with phosphonate dimethyl or diethyl esters gave substantial amounts of N-methyl or N-ethyl derivatives of the corresponding base. This side-reaction can be substantially suppressed by the use of diisopropyl esters.

a)   Purine and pyrimidine N-[2-(phosphonomethoxy)ethyl (PME) derivatives

were synthesized from the heterocyclic bases by alkylation with dialkyl 2-chloro-, 2-bromo- or 2-(*p*-tolylsulfonyloxy)ethoxymethyl phosphonate in the presence of NaH, K_2CO_3 or Cs_2CO_3 (Holý *et al.*, 1987b, 1989c). The most practical reagents are dialkyl 2-chloroethoxymethylphosphonates, easily accessible by the Arbuzov reaction of trialkyl phosphites with 2-chloroethyl chloromethyl ether (Holý *et al.*, 1989c). The optimum yields are obtained with the diisopropyl ester (Holý *et al.*, 1999). This reagent was applied for the preparation of numerous PME derivatives of natural nucleobases, their C-substituted derivatives (Holý *et al.*, 1989c, 1999), 2-azapurine(Hocková *et al.*, 1995), 8-azapurine (Holý *et al.*, 1996b), 1-deazapurine and 3-deazapurine analogues (Dvořáková *et al.*, 1993) as well as of 2-aminomethyl- (Hocek *et al.*, 1995) or 6-aminomethylpurines (Hocek *et al.*, 1996). In all these cases, deprotection was achieved by treatment with bromotrimethylsilane followed by hydrolysis. The synthon containing mesyl as a leaving group (Holý *et al.*, 1987b) was also used for the synthesis of the cytosine derivative PMEC by other authors (Bronson *et al.*, 1989a) (Scheme 4).

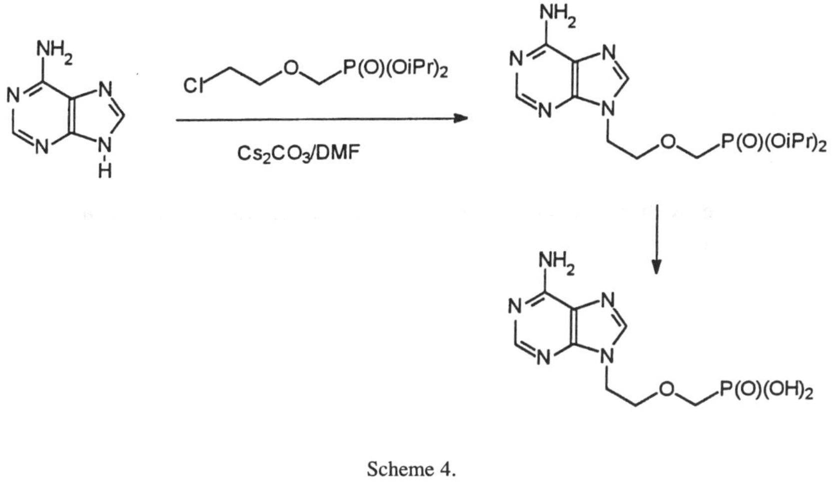

Scheme 4.

An alternative to this alkylation consists of reaction of the heterocyclic bases with dialkyl 2-hydroxyethylphosphonate under Mitsunobu conditions (Chen *et al.*, 1996). The authors stress an increased regiospecificity of the reaction.

Diisopropyl 2-chloroethoxymethylphosphonate was used for quaternization reactions of 2,4-diaminopyrimidine, 4,6-diaminopyrimidine, 2-aminopyrimidine and related compounds. The resulting free phosphonic acids possess pronounced zwitterionic character (Holý *et al.*, 1998b) (Scheme 5).

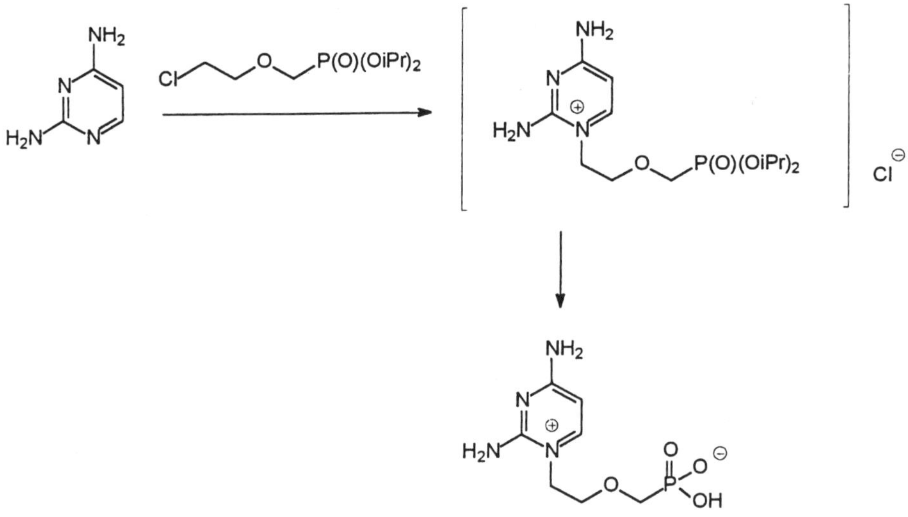

Scheme 5.

b)   *N-[2-(2-phosphonoethoxy)ethyl] derivatives of nucleobases*

are homologues of the PME compounds. Their syntheses were accomplished by alkylation of the bases with synthons obtained from bis(2-chloroethyl) ether and trialkyl phosphites in the presence of inorganic bases (Holý *et al.*, 1990) or DBU (Holý *et al.*, 1999) (Scheme 6).

c)   *Enantiomeric N-[2-(phosphonomethoxy)propyl] derivatives of nucleobases*

This important class of ANP derivatives can be synthesized using a common side-chain synthon which is accessible by multistep synthesis starting from commercially available alkyl *(R)-* or *(S)*-lactates: protection by a tetrahydropyranyl group followed by reduction with Red-Al gives 2-*O*-tetrahydropyranylpropan-1,2-diols which are benzylated and deprotected under acid catalysis to afford 1-*O*-benzylpropan-1,2-diols. Chloromethylation by HCl/1,3,5-trioxane followed by Arbuzov reaction with triisopropyl phosphite, or condensation with diisopropyl *p*-tolylsulfonyloxymethylphosphonate gives

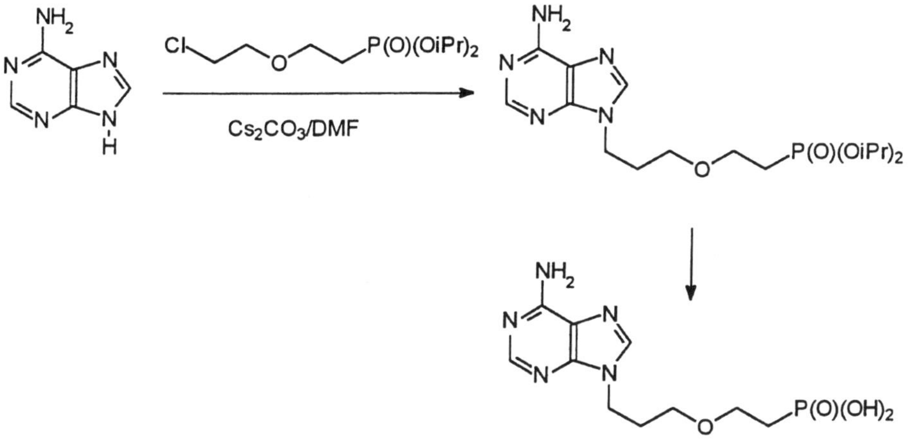

Scheme 6.

diisopropyl 1-*O*-benzyl-2-*O*-(phosphonomethoxy)propan-1,2-diol. The target synthons are then prepared by catalytic hydrogenolysis followed by tosylation. This approach was successfully applied to the synthesis of *(S)*- and *(R)*-PMP-derivatives of nucleobases (Holý *et al.*, 1995b), purine 2-aminomethyl derivatives (Hocek *et al.*, 1995) and/or their 8-aza analogues (Holý *et al.*, 1996b) (Scheme 7).

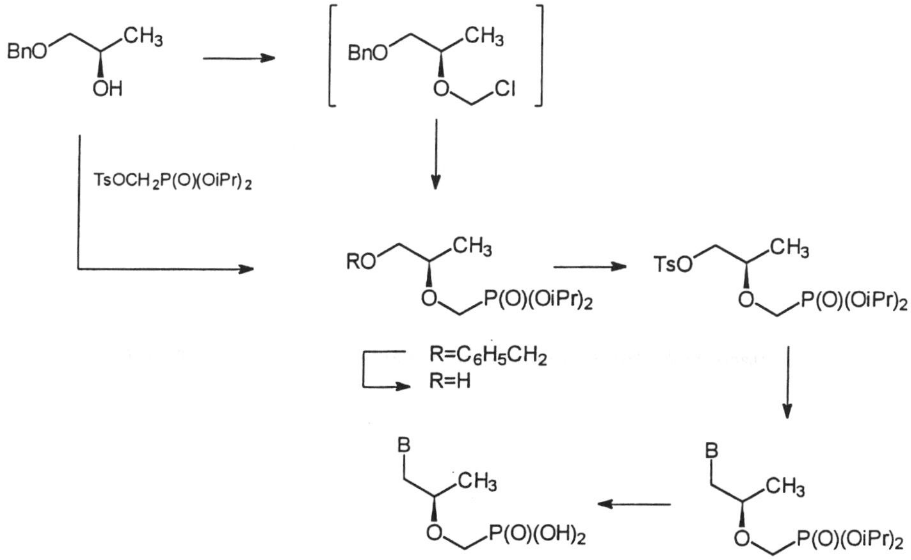

Scheme 7.

d)   N-[3-Fluoro-2-(phosphonomethoxy)propyl] derivatives of nucleobases

The key-compound for preparation of this prominent class of ANPs was prepared from *(S)-* or *(R)*-1-O-p-tolylsulfonylglycerols by treatment with KF and tosylation, followed by reaction of the resulting 1-fluoro-2-hydroxypropyl tosylates with dimethoxymethane. Acetolysis of the MOM-derivatives gave acetoxymethoxymethyl derivatives which were transformed by successive reaction with bromotrimethylsilane and triisopropyl phosphite to the ultimate synthons (Jindřich *et al.*, 1993). Alkylation of the sodium salts of heterocyclic bases proceeded with displacement of the tosyl leaving group (Scheme 8).

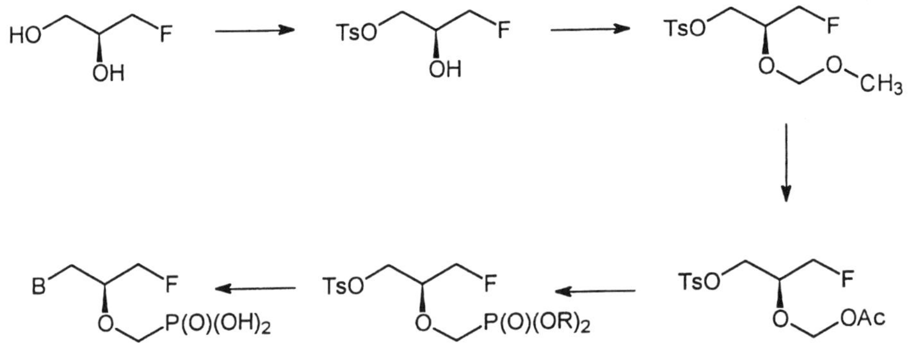

Scheme 8.

e)   N-[3-Hydroxy-2-(phosphonomethoxy)propyl] derivatives of nucleobases

Despite several methods available for the synthesis of this class of ANPs, it was thought desirable to elaborate a side-chain synthon approach in order to easily synthesize various base-modified HPMP-derivatives. Such a method was originally developed for the preparation of HPMPC. Briefly, *(S)*-3-O-benzylglycerol was tritylated and condensed with diethyl *p*-tolylsulfonyloxymethylphosphonate in THF in the presence of NaH: the reaction product was detritylated and transformed to the 1-O-mesyl derivative. Alkylation of cytosine in the presence of Cs_2CO_3 gave the 3-O-benzyl protected N^1-substituted derivative together with the O^2-isomer. The former was deprotected with Pd/cyclohexene followed by transsilylation (Bronson *et al.*, 1989a, 1989b). This approach was applied also to the synthesis of [cytosine-^{14}C]-HPMPC (Haynes *et al.*, 1991) (Scheme 9).

   The obvious disadvantage of this procedure is the debenzylation step, which precludes its application for sensitive bases (including cytosine) (Ho *et al.*, 1992). In order to avoid this complication, we designed synthons bearing instead an alkali labile O-protecting group. Thus, 3-O-p-tolylsulfonyl-*(R)*-glycerol was converted to its 1-O-trimethylacetyl derivative, and the secondary hydroxyl group was protected as its MOM-derivative.

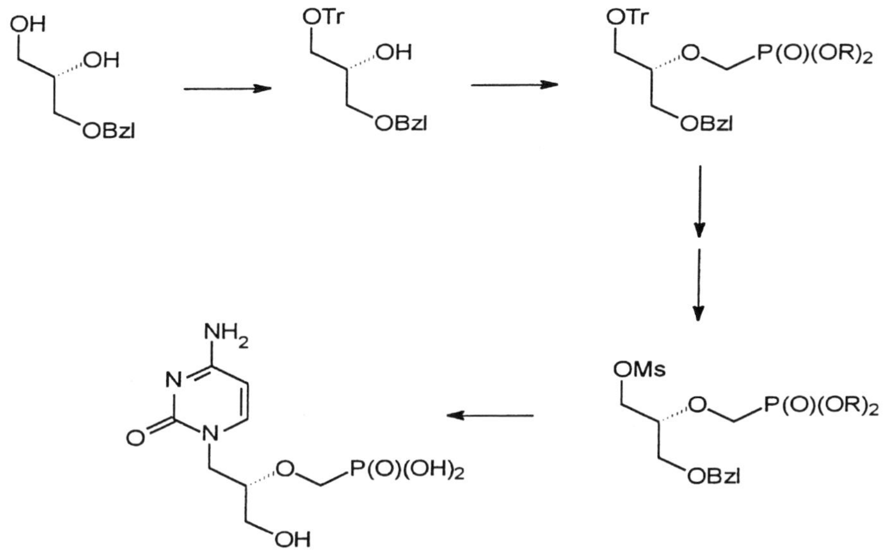

Scheme 9.

Acetolysis followed by bromotrimethylsilane treatment, and Arbuzov reaction with triisopropyl phosphite gave the required side-chain HPMP-synthon. Alkylation of the heterocyclic bases with this reagent in DMF in the presence of NaH or Cs_2CO_3 followed by methanolysis and transsilylation gave *(S)*-HPMP-derivatives of adenine, 2,6-diaminopurine, 2-aminopurine, cytosine, 6-azacytosine (Alexander *et al.*, 1993) and 3-deazaadenine (Dvořáková *et al.*, 1993) (Scheme 10).

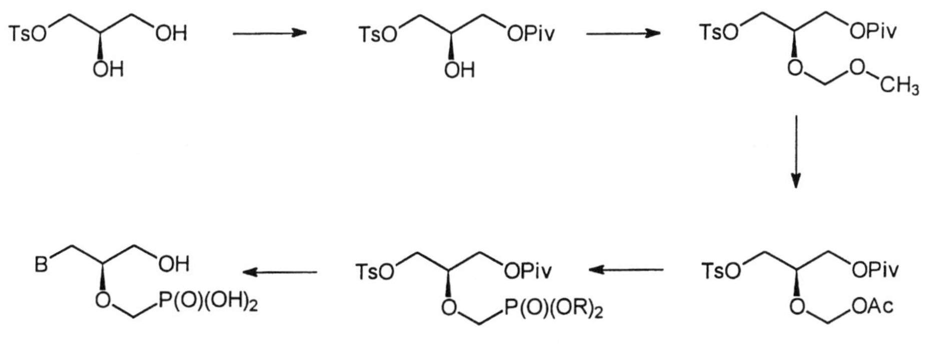

Scheme 10.

A similar synthon protected by benzoyl instead of trimethylacetyl was prepared from 1-*O*-benzyl-3-*O*-trityl-*(R)*-glycerol by reaction with diisopropyl *p*-tolylsulfonyloxymethyl-phosphonate in THF, in the presence of NaH; removal of the trityl function by acid hydrolysis gave an intermediate which can be used for the synthesis of both enantiomeric HPMP-synthons: (a) tosylation followed by hydrogenolysis and benzoylation affords the *(S)*-HPMP synthon while (b) benzoylation followed by hydrogenolysis and tosylation leads to the *(R)*-enantiomer (Holý *et al.*, 1995a) (Scheme 11).

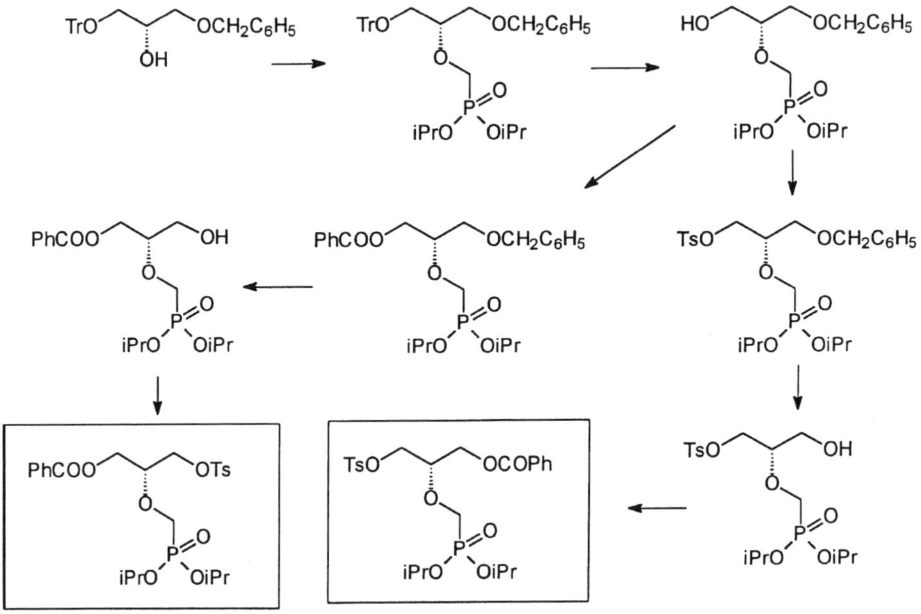

Scheme 11.

f)   Other acyclic nucleoside phosphonate analogues

Large series of 2'-substituted 9-[2-(phosphonomethoxy)ethyl]guanines were synthesized by alkylation of 2-amino-6-chloropurine with suitable side-chain synthons (bearing the mesyloxy function as a leaving group) in the presence of Cs_2CO_3 (Yu *et al.*, 1992, 1993).

The 1-oxa analogue of PMEG (1) was synthesized from 6-alkoxy-2-amino-9-hydroxypurine by reaction with diethyl 2-chloromethoxyethylphosphonate (Kim *et al.*, 1994). The related 1-oxa analogue of HPMPA (2) was prepared by addition of 9-hydroxy-N6-trityladenine to diethyl vinyloxymethylphosphonate and *N*-iodosuccinimide followed by acetolysis and deprotection (Kim *et al.*, 1992b). 2'-Alkoxy-, 2'-aryloxy or 2'-azido derivatives of PMEA (3) were prepared by alkylation of adenine with synthons obtained by other transformations of diisopropyl vinyloxymethylphosphonate (Rejman *et al.*, 1996).

Analogues of PMEA substituted at the carbon atom of the phosphonomethyl grouping [2-(1-phosphonoalkoxy)ethyl derivatives (4)] were obtained from the corresponding side-chain synthons bearing tosyl group (Rosenberg *et al.*, 1996). Isomers of HPMP derivatives with hydroxyethyl group branching at the methylene group of the phosphonomethyl function (5) were prepared by Cs_2CO_3-mediated alkylation of the base with the synthon prepared from 2-benzyloxymethyl-1,3-dioxolane (Kim *et al.*, 1996).

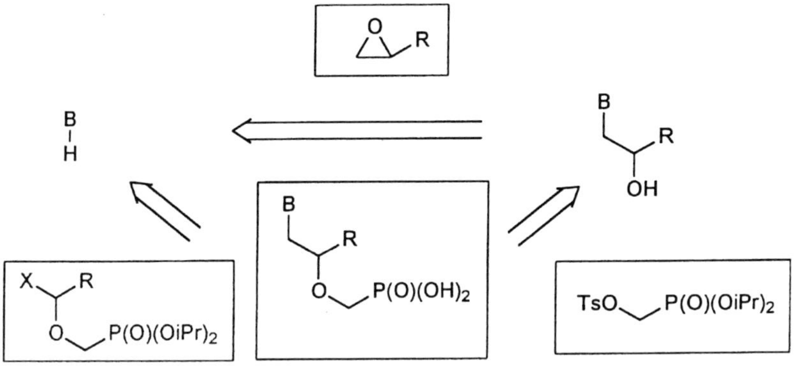

Chart 3.

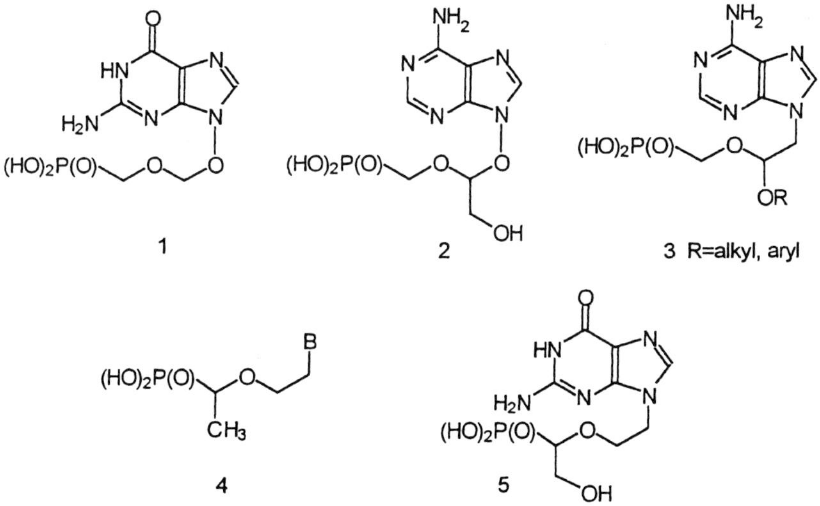

Chart 4.

Mitsunobu coupling of 1-hydroxypyrimidines with diethyl 2-hydroxyethoxymethyl-phosphonate or 3-acetoxy-2-(diethylphosphonylmethoxy)propanol gave 1-oxa analogues of pyrimidine (Harnden *et al.*, 1991) or purine ANPs (6) (Duckworth *et al.*, 1991). Similar reactions using dialkyl 2-hydroxyethylthiomethylphosphonate gave 9-[2-(phosphonomethylthio)ethoxy]guanine and -adenine (7) (Harnden *et al.*, 1993). 9-[2-(Phosphonomethoxy)ethylamino]adenine was prepared by condensation of diethylphosphonylmethoxyacetaldehyde (formed *in situ* from its dimethyl acetal) with 9-aminoadenine followed by $NaBH_4$ reduction and transsilylation (Harnden *et al.*, 1992).

Upon treatment with diethyl chloromethoxymethoxymethylphosphonate followed by alkaline hydrolysis and transsilylation 2-amino-6-chloropurine gave an 1-oxa analogue of 9-[3-(phosphonomethoxy)propyl]guanine (8) (Kim *et al.*, 1991c)

Finally, a thio analogue of PMEA, 9-[2-(phosphonomethylsulfanyl)ethyl]adenine (9), was synthesized by alkylation of adenine with a side-chain synthon which was obtained from dialkyl chloromethylphosphonate by reaction with sodium 2-hydroxyethanethiolate and subsequent mesylation of the dialkyl [(2-hydroxyethyl)sulfanyl]methylphosphonate (Villemin *et al.*, 1993).

There are many more formal analogues of ANPs (e.g., numerous ω-phosphonoalkyl derivatives or "carba analogues", but their discussion is out of the scope of this review. However, two more structural types should be mentioned which are related to the discussed group by virtue of incorporating the O-methylphosphonic acid ether residue: (a) "rigid" acyclonucleotide analogues" (i.e., guanine and adenine 9-phosphonomethoxyalkenyl, -alkinyl and methylidenealkyl derivatives) which were synthesized by alkylation of the bases with diethoxyphosphonylmethoxymethyl)alkenyl (alkinyl) chlorides or mesylates (Casara *et al.*, 1995). Their structural rigidity contradicts one of the above important premises – conformational adaptability. 9-[2-(Phosphinomethoxy)ethyl]adenine (10) is another analogue of PMEA which does not fulfill the requirement of iso-polarity; it was synthesized by alkylation of adenine with isopropyl 2-chloroethoxymethyl[bis(alkoxy)methyl]phosphinate followed by acid hydrolysis (Alexander *et al.*, 1994).

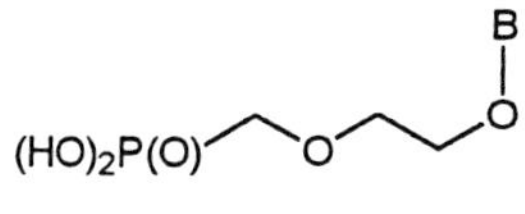

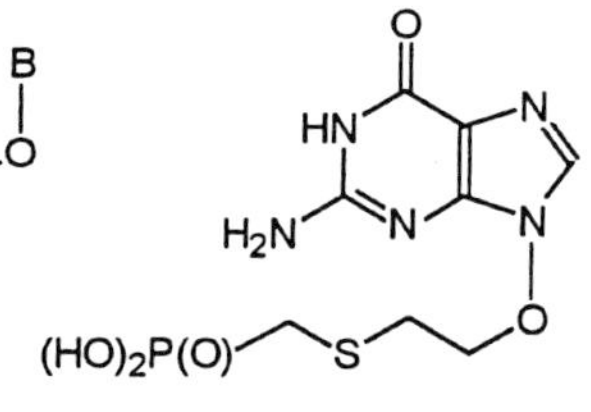

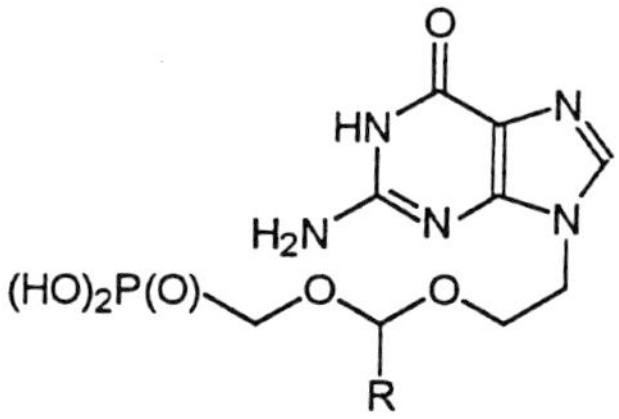

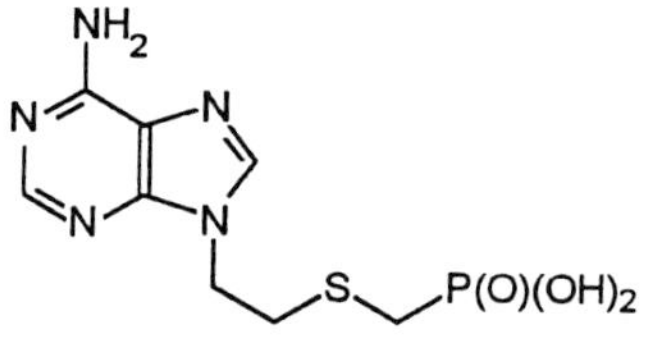

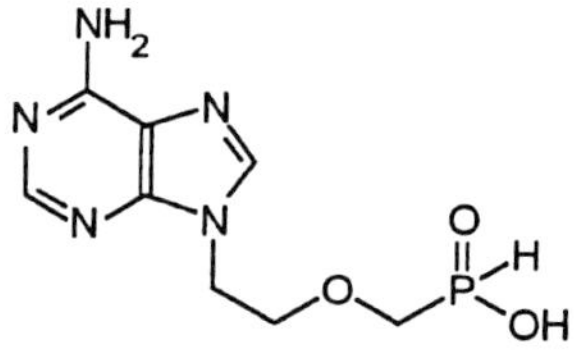

Chart 5.

3.3. Ring-closure of the heterocyclic base at the amino group of suitably protected aminoalkylphosphonates

Due to the comparatively easy accessibility of the above ANPs by methods starting from the heterocyclic base, this synthetic alternative has attracted but limited attention. The synthesis of PMEA described in the literature transforms diethyl 2-chloroethyloxymethylphosphonate by a Gabriel reaction to diethyl 2-aminoethyloxymethylphosphonate and builds the purine ring therefrom by Montgomery procedure (Dang *et al.*, 1998). This reaction has limited importance; contrary to the authors' arguments we have never experienced "low regioselectivity" in adenine alkylations. The ring-closure approach was used also for the synthesis of 8-aza analogs of PMEA (11) and PMEG (12) (Franchetti *et al.*, 1994); in this case, the ring-closure procedure leading to the 9-isomer is more justifiable because direct alkylation of 8-azapurine bases affords mixtures of 9- and 8-substituted products (Holý *et al.*, 1996b). Also 8-aza analogues of chiral PMPG (13) (Franchetti *et al.*, 1995) and additional 8-azapurine ANPs modified at the side chain (Franchetti *et al.*, 1997) have been prepared by these procedures.

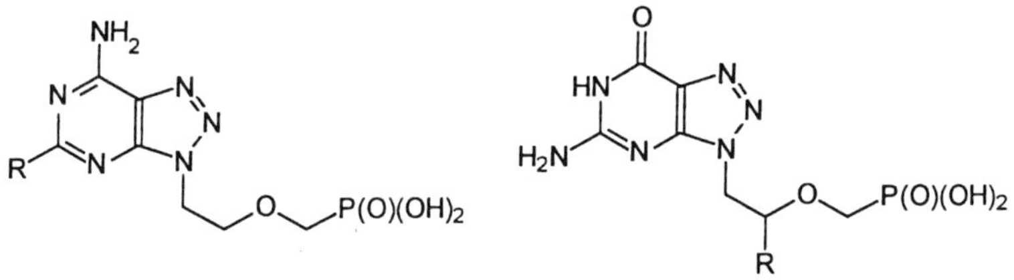

Chart 6.

3.4. Transformation of functional groups of the side-chain in ANPs

The 3'-azido analogue of HPMPA was prepared by condensation of 9-(3-azido-2-hydroxypropyl)adenine with dimethyl *p*-tolylsulfonyloxymethylphosphonate followed by deprotection (Holý, 1989). Its hydrogenation afforded racemic 9-(3-amino-2-(phosphonomethoxy)propyl)adenine, a compound which was later synthesized from 9-[3-(phtalimido)-2-hydroxypropyl]adenine (Dvořáková *et al.*, 1996). Optically active forms of the 3'-azido derivatives were prepared either by alkylation of adenine with diisopropyl *(S)*-[1-azidomethyl-2-(p-tolylsulfonyloxy)ethoxymethyl]phosphonate [*(R)*-enantiomer], while the *(S)*-enantiomer was obtained from diisopropyl *(S)*-HPMPA by tosylation, and sodium azide treatment (Hocková *et al.*, 1996). These compounds were converted by hydrogenation and deprotection in the enantiomeric 3'-amino derivatives. They were then utilized for the synthesis of NAD analogues containing phosphonate residues at the side-chain connecting the adenine and quaternary nicotinamidinium heterocyclic systems. The Zincke reaction was used to construct the nicotinamide ring from the terminal amino group. Due to the much stronger basicity of the amino group at the side-chain, the adenine amino group did not interfere in this reaction (Scheme 12).

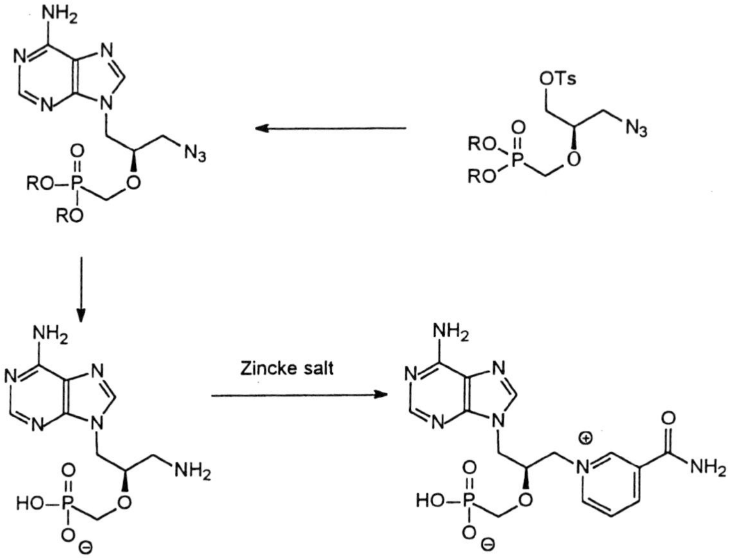

Scheme 12.

Diisopropyl *(S)*-HPMPA can be brominated at C-8 of the purine base with preservation of the ester groups. Treatment of this compound with NaH in inert solvent results in the formation of 3',O8-anhydro derivative which can be either deprotected by transsilylation to 3',O8-anhydro-*(S)*-HPMPA or hydrolyzed under acid catalysis and deprotected to 8-oxo-*(S)*-HPMPA (Janeba *et al.*, 1996).

3.5. Transformation of functional groups at the heterocyclic bases

All transformation reactions which are characteristic for purine or pyrimidine bases can be performed also in the ANP series. These comprise deamination of adenine to hypoxanthine, guanine to xanthine, etc., alkylation of NH-functions, halogenations at C-5 of pyrimidines or C-8 of purines and subsequent transformations of these halogeno derivatives. Special attention was paid to the reactivity of 6-halogeno functions in purine (6-chloropurine, 2-amino-6-chloropurine) ANPs which give rise to numerous biologically attractive compounds. Routinely, the transformations are performed with the corresponding neutral diesters. Thus, diisopropyl 9-[2-(phosphonomethoxy)ethyl]-6-chloropurine and -2-amino-6-chloropurine transformed to the 6-alkylamino, alkenyl-amino, alkinylamino, cycloalkylamino, arylamino, aralkylamino, dialkylamino or

N-hetaryl derivatives. Analogous conversions were made in the *(R)*- and *(S)*-PMP series (Holý *et al.*, 1995c, 1996a). The free phosphonates were obtained from the intermediates by transsilylation and hydrolysis (Scheme 13).

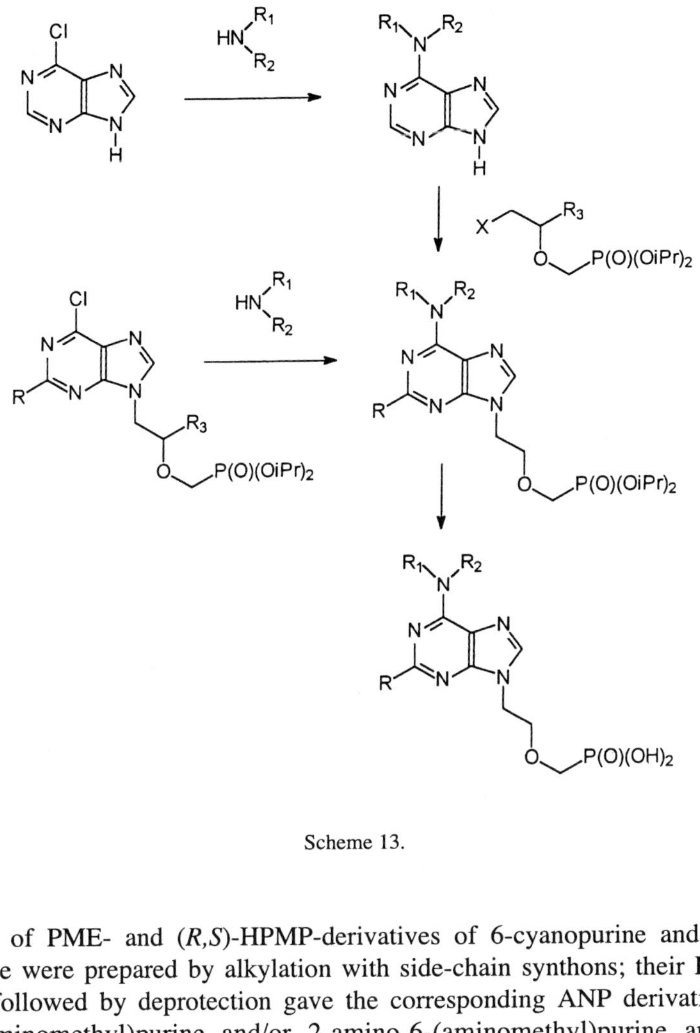

Scheme 13.

Diesters of PME- and *(R,S)*-HPMP-derivatives of 6-cyanopurine and 2-amino-6-cyanopurine were prepared by alkylation with side-chain synthons; their Pd-catalyzed reduction followed by deprotection gave the corresponding ANP derivatives derived from 6-(aminomethyl)purine and/or 2-amino-6-(aminomethyl)purine analogues of PMEA, PMEDAP and HPMPA, HPMPDAP (Hocek *et al.*, 1996). Upon treatment with catalytic amounts of sodium methoxide in methanol followed by reaction with ammonium chloride the 6-cyanopurine derivatives were converted into 6-amidinopurine or 2-amino-6-amidinopurine ANPs, respectively (Hocek *et al.*, 1996) (Scheme 14).

Reactivity of the 6-halogeno atom in purines was exploited for introduction of C-substituents: thus, coupling of 9-[2-(diisopropyloxyphosphonylmethoxy)ethyl]-6-chloropurine and *(R)*-9-[2-(diisopropyloxyphosphonylmethoxy)propyl]-6-chloropurine with organocuprates derived from Grignard reagents afforded 6-(sec- or tert-alkyl)

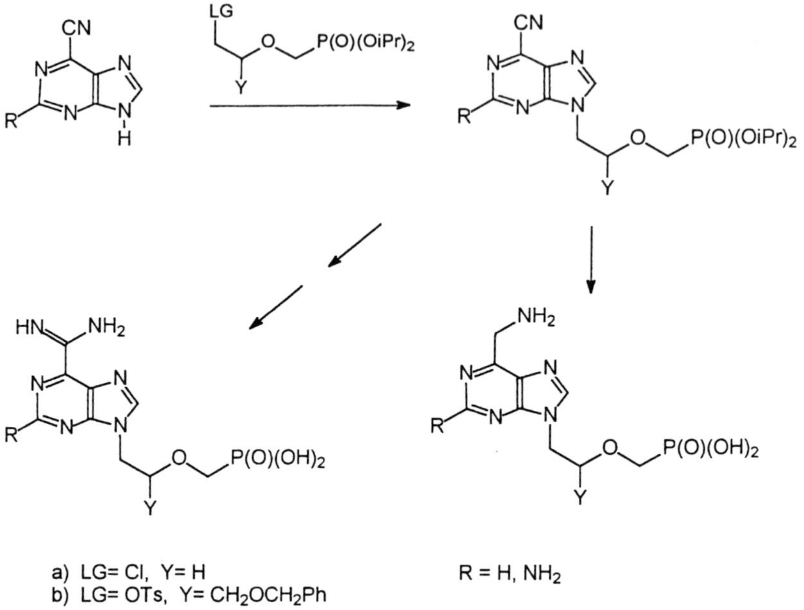

Scheme 14.

substituted phosphonates after deprotection (Dvořáková *et al.*, 1998). In a CuI-KF mediated perfluoroalkylation with trimethyl(trifluoromethyl)silane followed by transsilylation, 9-[2-(diethoxyphosphonylmethoxy)ethyl]-6-iodopurine (prepared by iododeamination of the adenine congener) was transformed to 9-[2-phosphonomethoxy)ethyl]-6-trifluoromethylpurine (Hocek *et al.*, 1999). Pd(0)-catalyzed cross-coupling of the protected phosphonate intermediate with hetarylzinc chlorides or hetarylstannanes gave 9-[2-phosphonomethoxy)ethyl]-6-hetarylpurines (Hocek *et al.*, 1997) (Scheme 15).

Stille coupling of 9-[2-(diisopropyloxyphosphonylmethoxy)ethyl]-6-chloropurine with 1-(ethoxyvinyl)tributyltin followed by acid hydrolysis gave a 6-acetylpurine derivative which was by reductive amination with primary and secondary amine hydrochlorides and sodium cyanoborohydride, followed by removal of the phosphonate diester groups, transformed into N-substituted 9-[2-phosphonomethoxy)ethyl]-6-(1-aminoethyl)purines (Hocek *et al.*, 1997) (Scheme 16).

Also, 9-[2-(diisopropyloxyphosphonylmethoxy)ethyl]-2-amino-6-chloropurine gives the corresponding 6-C substituted products in fair yields in Pd(0)-catalyzed cross coupling reactions with alkynylstannanes, hetarylstannanes, trialkylaluminum reagents and aryl-boronic acids; deprotection yielded the modified ANPs (Česnek *et al.*, 1999, 2000).

It should be mentioned that a method for determination of PMEA and its metabolites in samples of biological fluids was developed which is based on modification of the adenine residue in PMEA to the N^1,N^6-etheno derivative (14) by the Leonard reaction

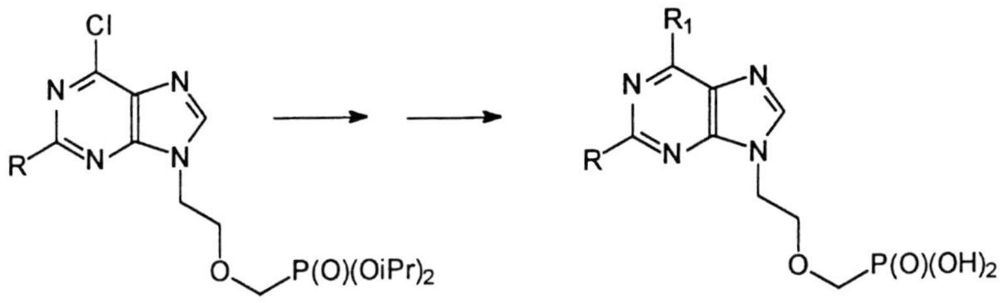

R = H, NH$_2$; R$_1$ = alkyl, aryl, hetaryl, CF$_3$

Scheme 15.

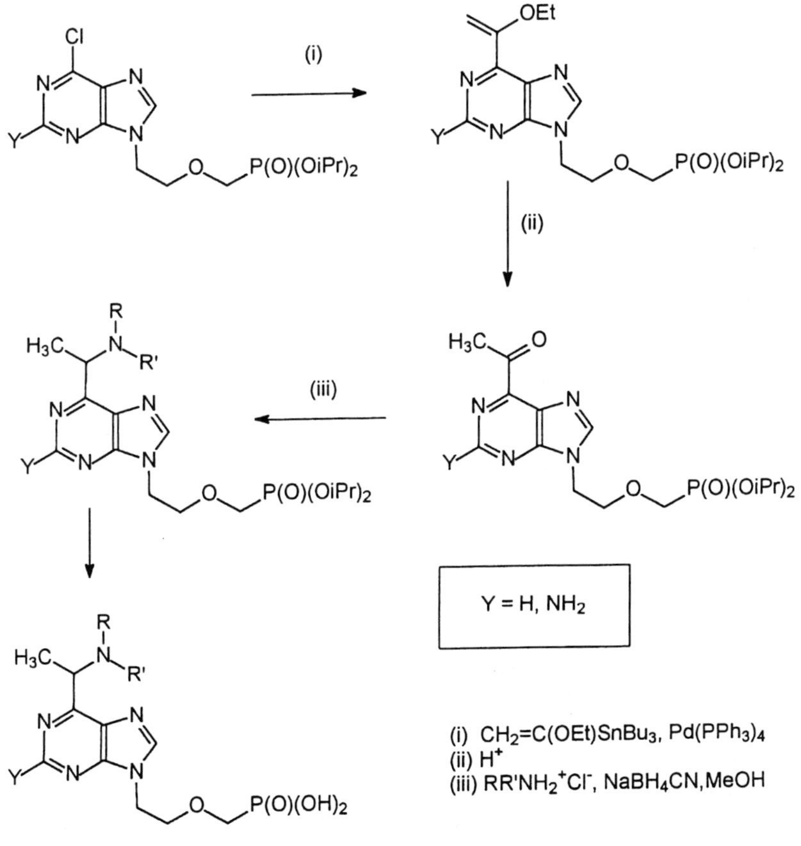

Scheme 16.

with chloroacetaldehyde (Russell *et al.*, 1991); it is possible to apply an HPLC method with fluorescent detection for analysis of urine or blood samples, etc. (Naesens *et al.*, 1992; Sparidans *et al.*, 1999).

4.   Prodrugs and metabolites of acyclic nucleoside phosphonates

It has been mentioned that the polarity of acyclic nucleoside phosphonates is the major obstacle to their wider applications. However efficient is their conversion to active metabolites, whatever their inhibitory activity in the *in vitro* enzymatic assays may be, their overall biological effect depends primarily on their transport through cellular membranes. Moreover, the pharmacological parameters (elimination rate, etc.) is very short. The physico-chemical parameters of the free phosphonate forms practically preclude their resorption from stomach and intestine and limit the application of their oral formulations. The concept of prodrugs based on lowering the polarity by esterification of the phosphonate group and thereby increasing the permeability suggests itself in the first instance. The crucial issue is the need for their suitable degradation to the free phosphonates inside the cell either by enzymes, or, by spontaneous decomposition.

Modification of the phosphonate group can result in compounds bearing one negative charge and/or in neutral prodrugs (diesters, esteramidates, etc.). The former category is represented in the HPMP-series by cyclic phosphonates characterized by an intramolecular ester-linkage between the phosphonate group and the hydroxymethyl moiety. These compounds which bear one negative charge can be prepared by standard methodology known from the chemistry of ribonucleoside 2'(3')-phosphates by treatment with DCC, water-soluble carbodiimides, alkyl chloroformates, etc. (Rosenberg *et al.*, 1987). This possibility was exploited in developing the cyclic phosphonate of the approved antiviral *cidofovir* [(*S*)-HPMPC] which exhibited improved pharmaco-logical and antiviral parameters over the parent drug (Bischofberger, 1994). It is stable against the action of ribonucleases, but a base-nonspecific cyclic (decyclizing) phosphodiesterase was identified which catalyzes the ring opening of ribonucleoside 3',5'-cyclic phosphates; this enzyme is capable of opening the ring of cHPMPC to HPMPC. cHPMPC is also less prone to chemical degradation (deamination to uracil derivative) (Oliyai *et al.*, 1999). Other cyclic HPMP derivatives have been synthesized and examined in several biological models (*vide infra*); their activity always paralleled or exceeded those of the open forms of HPMP derivatives (Chart 7).

The cyclic prodrug concept applies to HPMP derivatives only. In all other cases, it is possible to synthesize non-cyclic ANP monoesters; a method making use of triphosgene (trichloromethyl chloroformate) in DMF for phosphonate group activation was developed for the synthesis of a series of PMEA monoesters derived from aliphatic and alicyclic alcohols and/or carbohydrates (Alexander *et al.*, 1994). The phosphonate diesters of the PME- and PMP-series can be easily and specifically cleaved to monoesters by heating with sodium or lithium azide in DMF (Holý, 1998). As the diesters are easily accessible, this method was used also for the preparation of the hydrophobic octyl ester of PMEA (Holý *et al.*, 1999).

An "abbreviated NAD analogue" (15) containing PMEA instead of AMP was synthesized from PMEA 2-aminoethyl ester prepared by successive reduction of the 2-azido ethyl ester and application of Zincke reaction to the intermediary amino derivative; ring closure led to the 2-(1-nicotinamidinium)ethyl ester of PMEA (Hocková *et al.*, 1996).

Direct access to ANP monoesters is theoretically possible by condensation of the nucleoside hydroxyl group with a monoester synthon, (e.g., ethyl *p*-tolylsulfonyl-oxymethylphosphonate) in the presence of NaH (Jasko *et al.*, 1993), but application of this alternative to acyclic nucleosides was not described.

Simple aliphatic (methyl, ethyl, isopropyl, octyl) diesters of all types of ANPs are obtained as intermediates during their preparation. Various dialkyl and diaryl esters of PMEA were prepared from crude PMEA dichlorophosphonate obtained by treatment of PMEA with thionyl chloride and catalytic DMF. This dichloridate reacted with alcohols to give the diester, while the treatment with amines afforded the corresponding (acid-labile) amidates (Starrett *et al.*, 1994). In another approach, PMEA was activated *in situ* by triphosgene in DMF solution and the intermediate was treated with an alcohol or phenol in the absence or presence of base (Alexander *et al.*, 1994).

Though marked biological activity was demonstrated with some of the dialkyl esters (Holý *et al.*, 1999), the main effort was focused on "cascade-type" prodrugs which are degraded in two steps: the enzymatic equipment of the cell begins the process by degradation of the acyl ester in the complex prodrug and the resulting unstable intermediate decomposes spontaneously to the parent phosphonate. A classical example of such prodrug design is the pivaloyloxymethyl ester used originally in penicillin chemistry. The bis(pivaloyloxymethyl) ester of PMEA (Bis-POM-PMEA, Adefovir dipivoxil) affords pivalic acid and PMEA di- and/or monohydroxymethyl esters on hydrolysis by cellular esterase(s) and these spontaneously decompose to formaldehyde and PMEA. The prodrug diester is easily obtained by alkylation of PMEA with chloromethyl trimethylacetate in the presence of N,N'-dicyclohexylmorpholinocarboxamidine. Additional acyloxymethyl esters of PMEA were prepared by the same procedure (Starrett *et al.*, 1992, 1994).

A similar principle was applied to the design of neutral prodrugs of another antiviral drug candidate, *(R)*-PMPA (Tenofovir): chloromethyl alkylcarbonates were prepared from chloromethyl chloroformate and an alcohol in the presence of pyridine; their reaction with PMPA and diisopropylethylamine in DMF gave bis(alkoxycarbonyloxymethyl) prodrugs (Arimilli *et al.*, 1997). Upon enzymatic hydrolysis, these compounds spontaneously decompose giving the parent phosphonate with simultaneous formation of CO_2 and formaldehyde. Replacement of the alcohol with an amine in this scheme gives rise to the corresponding carbamates. Investigation of the plasma stability and cellular permeability in these series of PMPA prodrugs selected the bis(isopropyloxycarbonyloxymethyl) ester [bis(POC)-PMPA] as the PMPA prodrug tenofovir disoproxil (Arimilli *et al.*, 1999).

Another type of diester prodrugs for transient protection of a nucleotide phospho-monoester residue was developed which is based on the carboxyesterase-labile acylthioethyl ester groups (SATE-esters). Several bis-SATE esters of PMEA together with the corresponding monoesters were prepared by the reaction of N⁶-monomethoxytrityl-PMEA with 2-acylthioethanols in the presence of 1-mesitylene-2-sulfonyl-3-nitro-1,2,4-triazole, followed by mild acid deprotection (Benzaria *et al.*, 1996). All of the compounds tested exhibited enhanced *in vitro* antiviral activity compared to the parent PMEA.

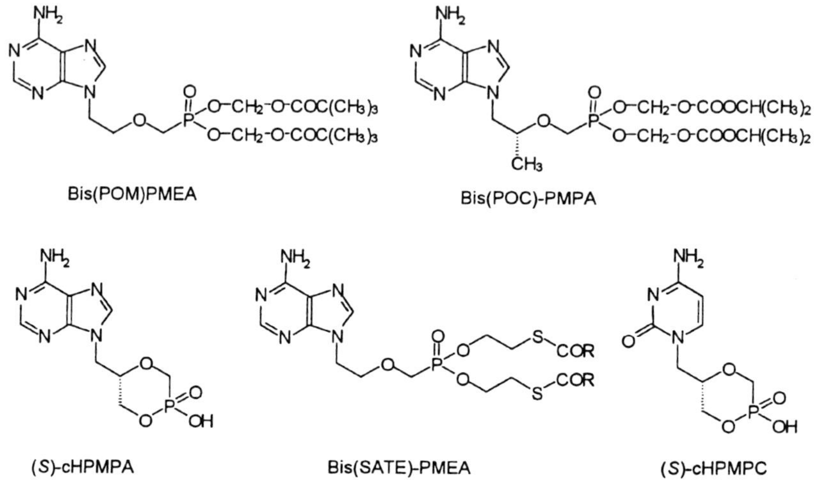

Bis(POM)PMEA

Bis(POC)-PMPA

(S)-cHPMPA

Bis(SATE)-PMEA

(S)-cHPMPC

Chart 7.

Salicylate and aryl ester prodrugs of cyclic HPMPC (16) were recently studied as potential "double prodrugs" of cidofovir (HPMPC) (Oliyai *et al.*, 1999). Another recent paper suggests the use of bis(3-phtalidyl) esters as phosphonate prodrugs which can be synthesized by reaction of the phosphonate with excess 3-bromophtalide in the presence of a tertiary base (Dang *et al.*, 1999).

Another approach to the polar drug transport problem is based on derivatization with a lipophilic residue. Thus, PMEA was converted into its 2-aminoethylamide which gave an acid-labile prodrug in a reaction with an active ester of 3-oleyl-lithocholic acid. It was then incorporated into a lactosylated lipid carrier. This transformation substantially increased preferential uptake of the drug by the liver (de Vrueh *et al.*, 1997, 1999). The antiviral effect of PMEA can be substantially increased by its encapsulation into pH-sensitive liposomes (Düzgünes *et al.*, 1999). Liposome-encapsulated *(S)*-1-[3-hydroxy-2-phosphonomethoxy propyl] cytosine (HPMPC; cidofovir) was evaluated for treatment of experimentally induced HSV-1 retinitis in a rabbit model (Besen *et al.*, 1995); the antiviral effect was remarkably potent and prolonged (Kuppermann *et al.*, 1996). 9-[2-(Phosphonomethoxy)ethyl]adenine (PMEA) was linked to a synthetic glygyl-glycyl-substituted poly-L-lysine polymer bearing mannosyl residues. The PMEA-modified polymer was more efficient *in vitro* compared to PMEA in preventing lysis of human macrophages by HSV-1 (Midoux *et al.*, 1990).

Syntheses of ANP metabolites relate mainly to preparation of the corresponding mono- and diphosphates (analogues of NDP and/or NTP). In the PME- and PMP-series,

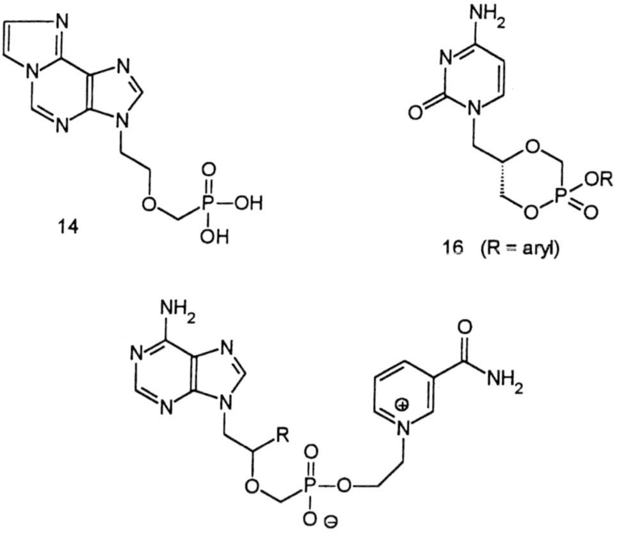

14

16   (R = aryl)

15

Chart 8.

there is no difficulty in applying standard procedures based on phosphonate activation (e.g., by conversion to morfolidates or imidazolidates) and subsequent treatment with inorganic diphosphate (Holý *et al.*, 1987b). However, in the HPMP-series, proximity of the hydroxyl group to the activated phosphonate residue caused formation of the cyclic phosphonate as the main reaction product – a situation parallel to that in ribonucleoside 3'-phosphates. Protection of the hydroxyl group is necessary to prevent this side-reaction. Alkali-labile protecting groups are ruled out due to the sensitivity of the phosphonato-phosphate anhydride linkage to nucleophilic attack. Originally, an O-benzyl group was used for this purpose in the synthesis of HPMPCpp; removal of the protecting group leads to substantial losses due to hydrogenation of the cytosine residue (Bronson *et al.*, 1989a). For the preparation of HPMPApp we have made use of dimethoxytrityl protection, which group can be removed under mild acidic conditions, in combination with diphenyl chlorophosphate activation of the phosphonate (Rosenberg *et al.*, 1987). In the case of HPMPCpp, however, dual protection of the hydroxyl group and the cytosine base was necessary; it was achieved by introduction of the dimethoxytrityl group at both sites (Otmar *et al.*, 1999, 2001).

5.   Related cyclic non-nucleoside phosphonate analogues

The biological activity of acyclic nucleoside phosphonates stimulated syntheses of numerous related compounds based on the same principle. The following information concerns only nucleotide analogues which contain the phosphonomethoxy group linked to a nucleoside analogue bearing a moiety other than aldopentofuranosyl, such as derivatives of dialdopentofuranoses, cyclopentanols, etc. The most important representatives of these analogues are 5-phosphonomethoxy-3,4-dihydrofuran-1-yl derivatives of heterocyclic bases which were synthesized either by the addition of dialkyl hydroxymethylphosphonate to chiral furanoid glycals bearing thymine or adenine (Kim *et al.*, 1991a, 1991d), or 2,6-diaminopurine residue (17) (Kim *et al.*, 1992c). These compounds can be regarded as isosters of antiretroviral d4T (stavudine) or its 5'-monophosphate and its congeners; their hydrogenation resulted in the corresponding tetrahydrofuran derivatives (18), formally related to "dideoxynucleosides". The carba analogue 19 was prepared by reaction of the suitable enantiomer of cis-1-acetoxy-4-hydroxycyclopent-2-ene with 2-methoxyethoxy)methyl chloride, and transformation of the resulting ether with dimethylborondibromide to the bromomethyl ether which gave the allylic acetate bearing a phosphonate function upon Arbuzov reaction with triethylphosphite. Palladium(0) mediated coupling with the sodium salt of the base and deprotection gave the final product (Kim *et al.*, 1992a).

Another group described synthesis of racemic carbocyclic phosphonate nucleosides of adenine, hypoxanthine, guanine, cytosine, uracil, and thymine (20) isosteric (and isoelectronic) with (carbo)cyclic 2,3-dideoxy- and 2,3-didehydro-2,3-dideoxy-nucleoside monophosphates by addition of cyclopentene-1,2-epoxide to the heterocyclic base followed by alkylation of the resulting alcohol with diisopropyl *p*-tolylsulfonyl-oxymethylphosphonate (Jähne *et al.*, 1992). Also the carbocyclic analogues of 2'-deoxy-nucleotides (21) bearing the phosphonomethoxy group at the 4'position were synthesized: racemic (1,2,4)-cyclopentane-1,2,4-triol was protected at the 2,4-diol by a disiloxanyl group and etherified at the remaining function by diethyl phosphonomethyl triflate. The resulting intermediate was resolved by the action of lipase and the appropriate enantiomer was used for alkylation of the bases by the Mitsunobu procedure (Drake *et al.*, 1996). The same group reported on the synthesis of the phosphonomethyl ether isostere of carbocyclic 5-bromovinyl-2'-deoxyuridine monophosphate (Coe *et al.*, 1994). An alternative starts from 3-benzyloxy-4-hydroxycyclopentylamine, builds the guanine or adenine base by ring closure and introduces the phosphonomethyl ether residue by condensation with diethyl *p*-tolylsulfonyloxymethylphosphonate; the final two-step deprotection affords the target compounds 22 as racemates (Elliott *et al.*, 1994).

Another group made use of the phosphonomethyl ether principle by derivatizing another established antiviral nucleoside analogue of the carbocyclic oxetanocin type – Cyclobut-G – by its etherification with diethyl phosphonomethyl triflate followed by deprotection (Norbeck *et al.*, 1992).

The PME system is also contained as a part of the ring in 2-phosphono-4-purin-9-yl)methyl-1,3-dioxolanes (23) and 2-phosphono-5-(purin-9-yl)methyl)tetrahydrofurans (24) (Nguyen-Ba *et al.*, 1998a,b) - the latter compounds are formed by Arbuzov reaction of 5-substituted 2-bromotetrahydrofurans. Homologues of the dioxolane-type (2-phosphonomethyl-1,3-dioxolane derivatives 25, analogues of the 2-phosphonoethoxy ether) were also synthesized (Bednarski *et al.*, 1995).

We had entered this field earlier by synthesizing both [5-(adenin-9-yl)-5-deoxypento-furanosyl]phosphonates with different sugar configuration (26) (Otmar *et al.*, 1993a,b)

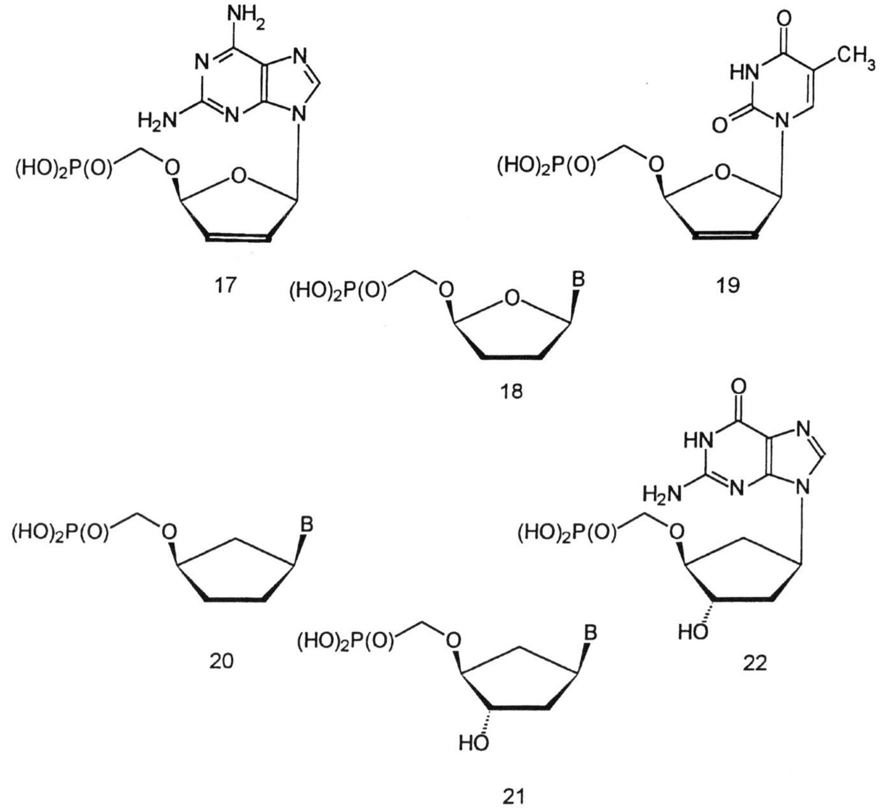

Chart 9.

and later also their carba analogs (27) (Liboska *et al.*, 1996a,b). Isonucleoside-related are (4*R*)-(purin-9-yl)-tetrahydrofuran-(2S)-ylphosphonates 28 synthesized from the appropriate phosphonate-protected (4S)-mesyloxy intermediates by reaction with the base in the presence of K_2CO_3/18-Crown-6 in DMF (Zheng *et al.*, 1999).

Further removed but still formally related are various stereoisomers of 2-hydroxy-methyl-3-(purin-9-yl)-4-cyclohexen-1-phosphonates and their pyrimidine analogues (29); they were made from glycals by addition of triisopropyl phosphite which furnished 1- and -2-enopyranosylphosphonates; they were then substituted with the heterocycle using Mitsunobu chemistry (Alexander *et al.*, 1996).

Phosphonomethyl ethers derived from pyranosyl analogues of 2',3'-didehydro-2',3'-dideoxy- and 2',3'-dideoxyribonucleosides (30) were prepared by condensation of the corresponding 1-(adenin-9-yl) or thymin-1-yl glycal bearing hydroxyl group at the position 4, with dialkyl p-tolylsulfonyloxymethylphosphonate (Pérez-Pérez *et al.*, 1995a, 1995c). Stereocontrolled synthesis of the same compounds was based on introduction of the phosphonomethoxy group into pentopyranosyl glycals through a Ferrier-type rearrangement; the 1,4-*trans* phosphonomethyl glycosides obtained were used for introduction of the heterocyclic base by a Mitsunobu-type condensation (Pérez-Pérez *et al.*, 1995b).

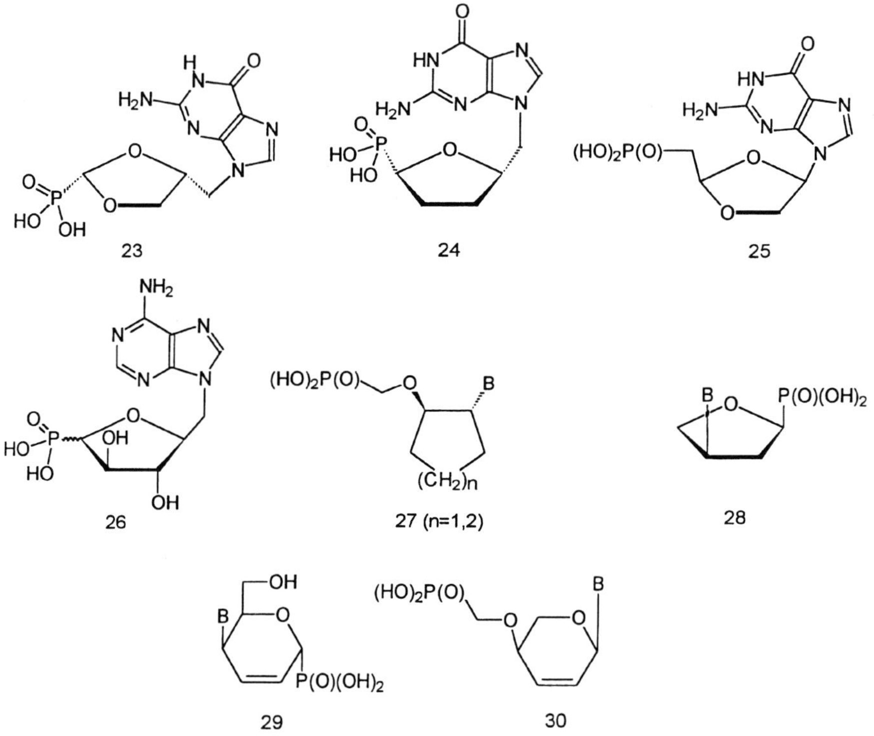

Chart 10.

6.   Part B.   Biological activity

6.1. Antiviral activity of acyclic nucleoside phosphonates

There are numerous reviews regarding antiviral activity of selected ANPs that are in various stages of preclinical or clinical development; they will be mentioned later. However, except for the few cases (Holý *et al.*, 1989a, 1993, 1994), there are no compilative reviews encompassing the present knowledge about the structure – activity relationship in this series. We shall focus mainly on the data which are available on ANPs themselves.

a)   Influence of the side-chain structure

- HPMP derivatives are active against all DNA viruses. The antiviral activity is connected in all cases with *(S)*-enantiomers; however, in some cases - depending on the character of the base - the *(R)*-isomers are also active (Balzarini *et al.,* 1993). Most probably the enantioselectivity is due to the enantiospecificity of nucleotide kinases which catalyze phosphorylation of the analogue. No effect *in vitro* was observed against RNA viruses and retroviruses.
- 3'-isomers of HPMP derivatives [2-hydroxy-3-(phosphonomethoxy)propyl] are inactive against DNA and RNA viruses *in vitro*;
- PME-derivatives are active against DNA viruses although their activity is less pronounced compared to the HPMP-series. The most active compound of the series is the 2,6-diaminopurine derivative PMEDAP. Similarly, as with the *(S)*-HPMP derivatives, the adenine and 2,6-diaminopurine PME-derivatives are active against HHV-6 (Reymen *et al.,* 1995). They suppress also duck hepatitis virus (Yokota *et al.,* 1990), HBV (Yokota *et al.,* 1991; Heijtink *et al.,* 1994) and retroviruses MSV, HIV (Balzarini *et al.,* 1989; Pauwels *et al.,* 1991), FIV (Hartmann *et al.,* 1994), Visna (Thormar *et al.,* 1993, 1995), etc. PMEG is the only ANP which was reported to inhibit measles virus and parainfluenza virus 3 *in vitro* (Barnard *et al.,* 1997).
- FPMP derivatives predominantly suppress the retroviruses (HIV, FIV) (Balzarini *et al.,* 1991b; Hartmann *et al.,* 1994); in the animal model, *(S)*-FPMPA showed better parameters (no hematologic side effects) compared to PMEA (Hartmann *et al.,* 1994). They are inactive *in vitro* against herpesviruses.
- in the PMP-series, the antiviral effect is limited to the *(R)*-enantiomers; by analogy to the FPMP derivatives, their activity is directed solely against retroviruses, DHBV and HBV. PMPA and PMPDAP are the best investigated compounds. (Balzarini *et al.,* 1993, 1996)
- replacement of the hydroxyl group in HPMP-derivatives by other substituents but hydrogen or fluorine (e.g., amino, azido, alkylamino, dialkylamino, trimethylammonium) abolishes the activity;
- replacement of the methyl group in PMP-derivatives by other alkyl, cycloalkyl, aryl or aralkyl substituents results in inactive compounds;
- introduction of a substitutent at the 1' position (methyl, hydroxymethyl) also has a negative effect on the antiviral activity; the guanine derivative exerts certain anti-HIV activity; however, it is much lower compared to PMEG and/or PMPG (Yu *et al.,* 1992; Kim *et al.,* 1990b).
- replacement of the oxygen atom in the PME group by sulfur in the phosphonomethylsulfanylethyl analogue of PMEA results in loss of antiviral activity against HIV (Villemin *et al.,* 1993). The phosphonomethylsulfanyl analogue of PMEG is inactive against both herpesviruses and HIV-1
- insertion of an oxygen atom between the N9-position of the purine base and the PME side-chain preserves the antiviral activity: 9-[2-(phosphonomethoxy)ethoxy]adenine has exceptional activity against HIV-1, HIV-2, SIV and FIV (Perkins *et al.,* 1992); however, activity of the adenine derivative against HBV *in vitro* is substantially

lower than that of PMEA (Balzarini *et al.*, 1994). Also 9-[2-(phosphonomethyl-sulfanyl)ethoxy]guanine has potent activity against herpesviruses, and *(R)*-9-[3-hydroxy-2-(phosphonomethylsulfanyl)propoxy]guanine was reported to be active against Visna virus and HIV-1 (Harnden *et al.*, 1993).

- except for the previous case, extension of the side chain generally leads to loss of activity or its substantial lowering. In some cases, guanine derivatives preserve some activity (Kim *et al.*, 1990a).

b)   Influence of the heterocyclic base on the antiviral activity in HPMP, PME, FPMP and PMP series

- Among the "natural" nucleobases and their nearest congeners, the antiviral activity in all the above series of ANPs is limited to purine (adenine, guanine, 2,6-diaminopurine and 2-aminopurine) derivatives (Table 1). The efficacy follows the order DAP~G>A>>AP. Hypoxanthine and/or xanthine derivatives are inactive.
- Uracil and thymine derivatives exert essentially no antiviral activity; in the cytosine series, however, the HPMP-derivative *(S)*-HPMPC (Cidofovir) exhibits very high effects on herpesviruses, adenoviruses, papillomaviruses as well as on HBV (*vide infra*). Other cytosine ANPs (PMEC, PMPC, FPMPC) are inactive against all viruses tested. Substitution on the base by a methyl group in HPMP-5-methylcytosine suppresses the antiviral effect.
- Replacement of a nitrogen atom by –CH= in the purine ring system results in 1-deaza, 3-deaza and 7-deazapurine derivatives. From the 1-deaza series, 9-[2-(phosphonylmethoxy)ethyl]-1-deazaadenine inhibits HBV DNA synthesis (Yokota *et al.*, 1994); however, it is inactive against HSV-1, HSV-2, CMV or VZV. The HPMP analogue derived from 1-deazaadenine is also essentially inert (Dvořáková *et al.*, 1990). Among the 3-deaza analogues, the anti-DNA-viral activity of the 3-deaza analogue of HPMPA against herpesviruses is comparable to that of the parent compound, HPMPA. HPMPDAP (Gil-Fernández *et al.*, 1987) and its cyclic phosphonate are very active in suppressing African swine fever virus (La Colla *et al.*, 1991). *(S)*-3-deaza-HPMPA and *(S)*-3-deaza-cHPMPA proved highly active against HHV-6 infection (Reymen *et al.*, 1995). Also, 3-deazaPMEA and 3-deazaPMEG are significantly active against herpesviruses; in contrast to the 3-deaza analog, 7-deazaPMEG is not active against DNA viruses, but inhibits MSV and HIV-1 (Holý *et al.*, 1999).
- The reverse modification of the system (i.e., –N= replacing –CH=) at the C2 position of purine ring system is ineffective. None of the 2-aza derivatives tested showed any antiviral activity, but this is not true for 8-azapurines: 9-*(S)*-HPMP-8-azaadenine and PME-8-azaguanine were active against HSV-1, HSV-2, CMV and VZV, as well as against MSV. PME-8-azaguanine and *(R)*-PMP-8-azaguanine protected MT-4 and CEM cells against HIV-1 and HIV-2-induced cytopathicity. None of the 8-isomers exhibited any antiviral activity against herpesviruses, Moloney murine sarcoma virus (MSV), and/or HIV (Holý *et al.*, 1996b)
- The effect of substitution at position 2 of the adenine ring was followed in the HPMP series: except for the 2-amino derivative HPMPDAP, which is highly

antivirally active, neither the introduction of hydroxy, methyl, nor a methylsulfanyl group preserved the activity of the parent (HPMPA) (Holý *et al.*, 1989a). A more detailed study of the influence of 2-substituents on antiviral activity was performed in the PMEA series: except for PMEDAP (2-amino derivative), the introduction of a fluorine or chlorine atom, or a hydroxyl group at the 2 position decreased the anti-DNA-viral activity (Holý *et al.*, 1999).

- In the isomeric 2-aminopurine series, the parent PME derivative is moderately active. Introduction of a methyl group at position 6 leads to an inactive derivative, while the substitution of a hydroxyl function leads to the extremely active antiviral PMEG; also the 6-chloro analogue is very potent (this activity could be due to transformation to PMEG). Also, replacement of the 6-oxo function in PMEG by sulfur in 6-thioPMEG still preserves some of the antiviral activity (Holý *et al.*, 1999).

- Replacement of the amino group at the 6-position of adenine by methyl, sulfanyl, hydroxyl, or methylsulfanyl group reduces the activity. In the HPMPA series, some activity was observed with the 6-hydroxylamino derivative (Holý *et al.*, 1989a).

- Introduction of a bromine atom at the 8 position diminished the antiviral effect of PMEA, PMEDAP, and PMEG. Moderate activity was observed with the 8-bromo-2,6-diaminopurine derivative against VZV, CMV and herpes simplex viruses, and with the 8-bromoguanine derivative and 8-bromoPMEA (Holý *et al.*, 1999).

- A positive effect was observed upon N-substitution of the 6-amino group in purine derivatives of the PME and PMP series; the effect is particularly prominent in the 2,6-diaminopurine derivatives (Holý *et al.*, 1995c, 1996a). Introduction of one or two alkyl, cycloalkyl, aralkyl or hetaryl substituents results in compounds with strong activity against herpesviruses; most sensitive are cytomegaloviruses (Snoeck *et al.*, 1997), VZV and EBV (Meerbach *et al.*, 1994, 1996).

These observations led us to propose the general structure of the heterocyclic part of the pharmacophore (Chart 11). Its characteristic feature is a cumulation of amino groups in or around the pyrimidine part of the purine system. In order to decide whether the role of the amino groups consists in basicity or hydrogen bond formation, we have synthesized 2-aminomethyl analogues of PMEG and PMPG, as well as 6-aminomethyl analogues of PME-, PMP- and HPMP-derivatives of adenine and 2,6-diaminopurine. None of them exerted any antiviral activity, nor was any activity observed with 6-C-hetaryl, or 6-[1-(N-alkylamino)ethyl derivatives. The quaternary PME derivatives of 2,4-diaminopyrimidine and 4,6-diaminopyrimidine which are relevant to the hypothetical pharmacophore, were also devoid of activity.

c)   Influence of the phosphonate linkage

- The length of the linker between phosphorus and oxygen atoms is more critical than the overall length of the side-chain; this is consistent with the expected conformational adaptability of the sp3 system. Therefore, homologs of PME-derivatives containing 2-phosphonoethyl ether residues, PEE-compounds, do not exhibit antiviral activity (Holý *et al.*, 1999), while even the protracted 2'-oxa analogues of 3-(phosphonomethoxy)propyl and 4-hydroxy-3-(phosphonomethoxy)-

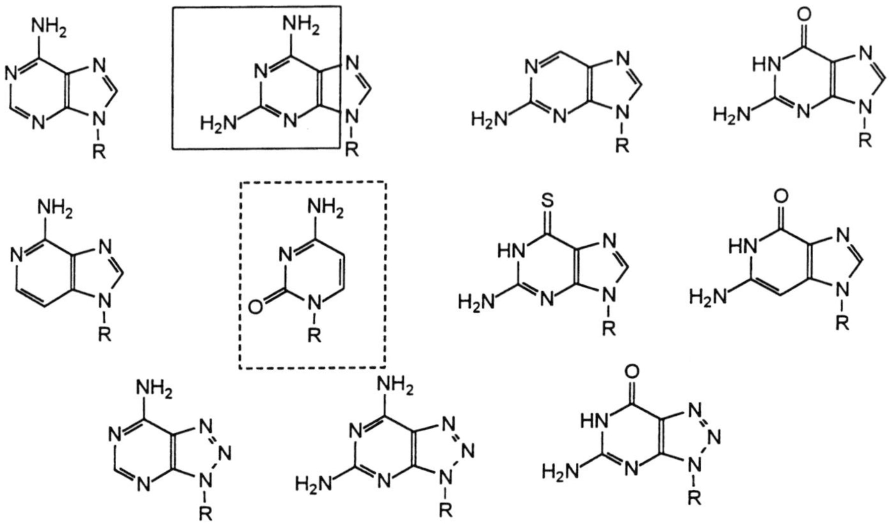

Chart 11.

butyl derivatives are active, at least to some extent (in the guanine series only) (Kim *et al.*, 1991c).

- Introduction of a methyl substituent at the phosphonomethyl function in PMEG had a negative effect on its activity against HCMV, HSV-2 and HIV (Yu *et al.*, 1992).
- Formation of a carbocycle or heterocycle with the participation of the methylene grouping can result in compounds with high antiviral efficacy. Thus, the 5-phosphonomethoxy-3,4-dihydrofuran-1-yl derivatives of thymine, adenine and guanine demostrate high antiviral potency against HIV and Rauscher murine leukemia virus (Kim *et al.*, 1991d). However, their carba analogues were inactive (Bronson *et al.*, 1992).

d)   *Selected compounds* (De Clercq, 1997) (Chart 12)

- **(*S*)-9-(3-Hydroxy-2-phosphonylmethoxypropyl)adenine [(*S*)-HPMPA]** is the parent compound of the ANP group. It was shown to efficiently inhibit herpesviruses (Aduma *et al.*, 1995), not only laboratory strains, but also clinical isolates of varicella zoster virus (Andrei *et al.*, 1995), HHV-6 (Reymen *et al.*, 1995) and HHV-8 (Neyts *et al.*, 1997). HPMPA was effective at inhibiting equine herpesvirus type 1 *in vivo* (de la Fuente *et al.*, 1992). It is one of the few compounds with strong potency against adenoviruses (Baba *et al.*, 1987; Gordon *et al.*, 1991). HPMPA inhibits duck hepatitis B virus (DHBV) core antigen synthesis and DHBV

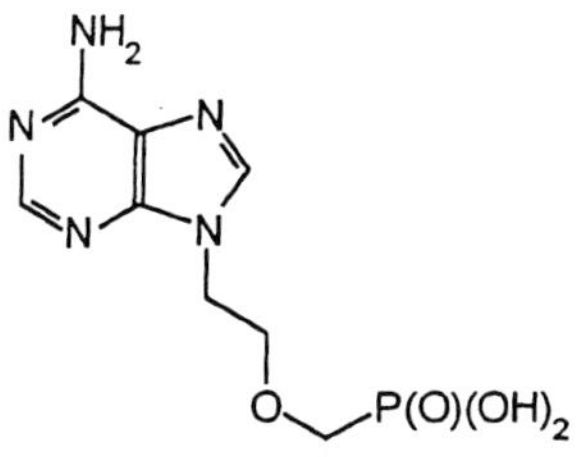

PMEA
Adefovir
active against DNA & retroviruses
oral form Preveon in clinical trials
against hepatitis B)

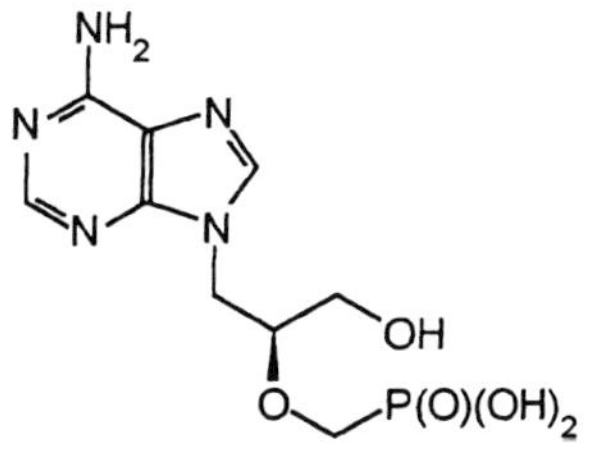

(S)-HPMPA
active against all DNA viruses,
Plasmodium sp., Leishmania sp.
Trypanosoma sp.

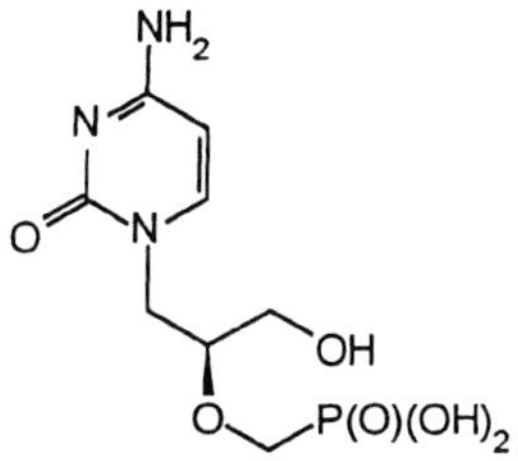

(S)-HPMPC
Vistide, Cidofovir
approved in USA and EU for treatment
of HCMV retinitis in AIDS pacients,

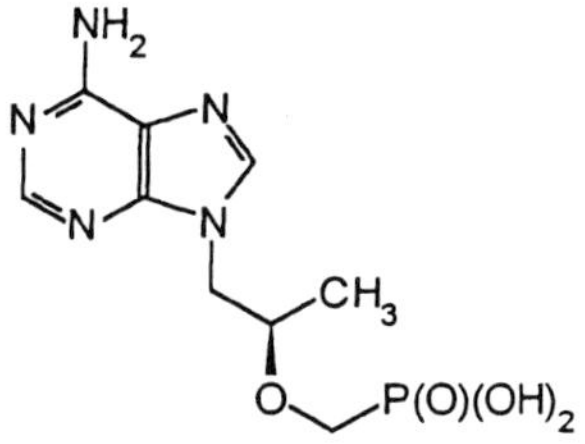

(R)-PMPA
active against HIV, clinical studies
in progress; protects macaques
against SIV infection

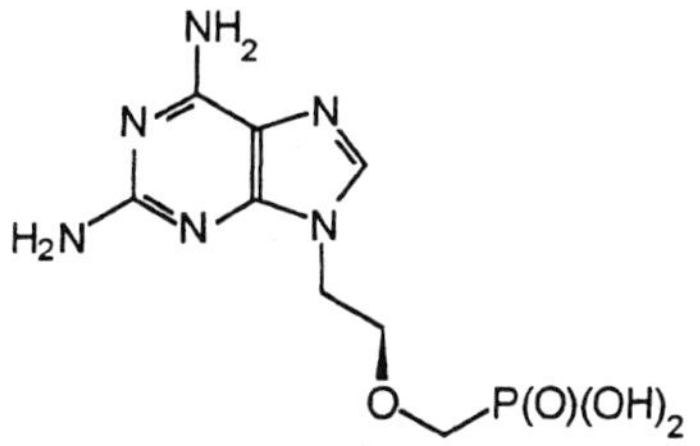

(R)-PMEDAP

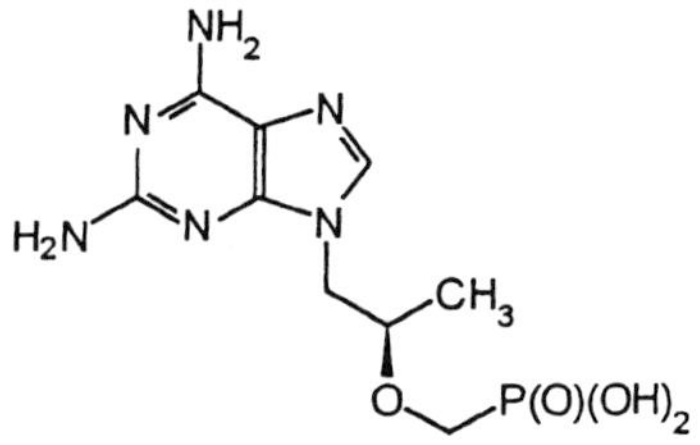

(R)-PMP-DAP

Chart 12.

DNA synthesis in DHBV-infected hepatocytes with a selectivity index > 300,
and hepatitis B virus DNA synthesis in human hepatoblastoma cell line (Yokota
et al., 1990, 1991).

- ***(S)*-9-(3-Hydroxy-2-phosphonylmethoxypropyl)cytosine [*(S)*-HPMPC, Cidofovir]** is the most popular compound of the ANP series. There have been numerous reviews covering its activity (Hitchcock *et al.*, 1996; Naesens *et al.*, 1997a,b, De Clercq 1998), mechanisms of action (Balzarini *et al.*, 1998a), pharmacology (Zabawski *et al.*, 1998; Cundy *et al.*, 1999), etc.; during the last five years there were >300 papers on its various antiviral activities, clinical studies (Wachsman *et al.*, 1996; Lalezari *et al.*, 1997) and therapeutic applications. Cidofovir has been approved for intravenous treatment of cytomegalovirus retinitis in AIDS patients (Garcie *et al.*, 1998; Plosker *et al.*, 1999) both by FDA and EMEA. Its drawback is nefrotoxicity (a common complication with all ANPs) which can be moderated by simultaneous application of probenecid (Lacy *et al.*, 1998; Bagnis *et al.*, 1999). On the other hand, its advantage consists in an unfrequent application regimen which is due to its characteristic metabolic feature (formation of the HPMPC-choline adduct with long intracellular half-time). Cidofovir has been successfully used for various other indications (De Clercq 1996; Safrin *et al.*, 1999) including topical application in the therapy of mucocutaneous herpes simplex virus infections (Snoeck *et al.*, 19994; Javaly *et al.*, 1999), respiratory papillomatosis (Van Cutsem *et al.*, 1995, Snoeck *et al.*, 1998; Pransky *et al.*, 1999), recurrent genital warts (Orlando *et al.*, 1999), progressive multifocal leukoencephalopathy (Happe *et al.*, 1999, 2000), molluscum contagiosum (Davies *et al.*, 1999), other poxvirus infections [e.g., cowpox (Bray *et al.*, 2000)], adenoviral conjunctivitis (Gordon *et al.*, 1996; Romanowski *et al.*, 2000), other adenovirus infections (Ribaud *et al.*, 1999) and Kaposi's sarcoma (Simonart *et al.*, 1998; Fife *et al.*, 1999). Further preclinical studies on HPV (Andrei *et al.*, 1998a; Johnson *et al.*, 1999), and herpesviruses HHV-6, HHV-7 (Yoshida *et al.*, 1998) and HHV-8 (Neyts *et al.*, 1998) are in progress. HPMPC was also examined for its cytostatic activity (*vide infra*).
- **9-[2-(Phosphonomethoxy)ethyl]adenine (PMEA, Adefovir) and its oral prodrug, Adefovir dipivoxil** PMEA was originally developed as an anti-HIV drug (Balzarini *et al.*, 1997), but due to its very low oral availability (Cundy *et al.*, 1994a), its bis(pivaloyloxymethyl) ester prodrug (Cundy *et al.*, 1994b, 1997) was selected as the drug candidate (Tsai *et al.*, 1995; Cundy *et al.*, 1995; Naesens *et al.*, 1995, 1996; Schulman *et al.*, 2000); eventually, the clinical trials for this indication (Deeks *et al.*, 1997) were discontinued. However, a promising response was reported in clinical studies with hepatitis B patients (Colledge *et al.*, 2000; Nicoll *et al.*, 1997; Rizzetto 1999), in particular in the treatment of lamivudine-resistant hepatitis B (Perillo *et al.*, 2000). Several reviews describe the therapeutic potential of this compound (Naesens *et al.*, 1994, 1997a, 1997b; Noble *et al.*, 1999; De Clercq 1999). In addition to antiviral activity, it exhibits other prominent biological effects which will be described later.
- **9-[2-(Phosphonomethoxy)ethyl]-2,6-diaminopurine (PMEDAP)** is a broad-spectrum antiviral agent with potent activity against DNA viruses; it inhibits replication of human immunodeficiency virus (HIV) in human T-lymphocyte MT-4 cells and suppresses tumor formation and mortality in newborn mice inoculated with MSV (Naesens *et al.*, 1993). It is highly efficacious when given orally to mice infected with MSV, Friend leukemia virus (FLV), or murine cytomegalovirus (MCMV).

PMEDAP markedly delayed MSV-induced tumor initiation when administered orally. The therapeutic index of oral PMEDAP is higher than that of intraperitoneal PMEDAP (Naesens *et al.*, 1989). Both PMEA and PMEDAP afford a marked antiviral protection if administered within one day before MSV infection (Naesens *et al.*, 1991). It is effective also against HCMV: although it has no effect on the expression of HCMV-specific immediate early antigens, it inhibited the expression of HCMV late antigens. It also delayed death in severe combined immune deficiency (SCID) mice infected with MCMV (Neyts *et al.*, 1993). PMEDAP inhibits HHV-6 *in vitro* (Reymen *et al.*, 1995). Despite its excellent antiviral parameters, it was never pursued in preclinical phase of drug development. Its main importance now consists in the antitumor activity (*vide infra*).

- **9-*(R)*-[2-(Phosphonomethoxy)propyl]adenine (PMPA, tenofovir) and its bis(isopropoxycarbonyloxymethyl) prodrug (tenofovir disoproxil fumarate)**
  After withdrawal of adefovir from clinical trials in AIDS patients, tenofovir is the most promising ANP candidate for anti-AIDS drug therapy. Although undoubtedly acting as a specific nucleotide-type RT inhibitor (Suo *et al.*, 1998), it is not dependent on activation by nucleoside kinases like the antiretroviral nucleoside analogues (AZT, stavudin, dideoxynucleosides) which act by the same mechanism. Thus, PMPA is expected to be active also against HIV-1 variants resistant to these drugs (Srinivas *et al.*, 1998). The *(R)*-enantiomers of PMPA (and PMPDAP) are much more effective against HIV *in vitro* than their *(S)*-enantiomeric counterparts (Balzarini *et al.*, 1996). PMPA is more efficient than PMEA in preventing murine AIDS (MAIDS) disease progression *in vivo* (Suruga *et al.*, 1998). Interest was enhanced by experimental therapy of simian immunodeficiency virus (SIV) infection in macaques (Tsai *et al.*, 1995a); it was shown that long-term PMPA treatment of four newborn macaques starting 3 weeks after virus inoculation resulted in a rapid, pronounced, and persistent reduction of viremia (Van Rompay *et al.*, 1996). Subcutaneous PMPA therapy in cynomolgus macaques chronically infected with SIV reduced SIV levels by >99% in the plasma or peripheral blood mononuclear cells (Tsai *et al.*, 1997; Silvera *et al.*, 2000). The effectiveness of therapy depends on timing of initiation and on duration of treatment (Tsai *et al.*, 1998; Van Rompay *et al.*, 1999a, 1999b, Hodge *et al.*, 1999). Most importantly, PMPA treatment is efficacious even in those cases where the animals were inoculated with SIV with reduced susceptibility to the drug (Van Rompay *et al.*, 1999b). PMPA crosses the placental barrier (Tarantal *et al.*, 1999); when administered s.c. to gravid rhesus monkeys PMPA reduces viral load in SIV-infected fetuses and infants. However, there is a danger of toxicity at high dosages. A promising feature of PMPA application in the AIDS therapy is the prophylactic activity of the drug (Van Rompay *et al.*, 1998, 2000). However, PMPA acts even when applied several days after infection. Comparative pharmacokinetic studies of PMPA in intravenous, intraperitoneal, and oral administration were performed on animal models (Cundy *et al.*, 1998) and the safety and therapeutic efficacy was followed in HIV-infected adults (Deeks *et al.*, 1998).

Despite the efficacy of PMPA treatment both in animal models and in AIDS-patients, there is a need for an oral formulation which would improve the low oral bioavailability of the free phosphonate (*vide supra*). The solution was found in the lipophilic diester, bis(isopropyloxycarbonyloxymethyl) PMPA [bis(POC)-PMPA] (Robbins *et al.*, 1998; Naesens *et al.*, 1998). Its oral bioavailability is >20% and its antiviral effect *in vitro* is approx. >100-fold greater than that of PMPA. In 2001, bis(POC)-PMPA was approved by FDA for AIDS treatment (Viread).

Adefovir and tenofovir demonstrated strong synergistic anti-HIV activity in combination with AZT. Adefovir showed moderate synergistic inhibition of HIV replication in combination with PMPA, d4T, ddC, and protease inhibitors, while PMPA combined with ddI, nelfinavir and/or adefovir demonstrated minor synergistic inhibition of HIV replication. All other combinations showed additive inhibition of HIV replication *in vitro* (Mulato *et al.*, 1997). No antagonistic interactions were detected for any of the adefovir or PMPA combinations.

6.2. Antiprotozoal activity of acyclic nucleotide analogues

The acyclic adenosine analogue *(S)*-9-[3-hydroxy-2-(phosphonylmethoxy)propyl]adenine [HPMPA] inhibits the growth of cultured *Plasmodium falciparum*. Its 3-deaza analogue is an even stronger inhibitor. A SAR study for *in vitro* antiplasmodial activity with a large series of ANPs confirmed that the activity is limited to *(S)*-9-(3-hydroxy-2-(phosphonomethoxy)propyl]adenine[*(S)*-HPMPA], its 3-deaza analogue *(S)*-3-deaza-HPMPA and their cyclic derivatives. Pyrimidine analogs of the HPMP-series [*(S)*-HPMPT, *(S)*-HPMPU and *(S)*-HPMPC], were inactive. Compounds lacking the hydroxyl group, including PMEA, *(R)*-PMPA and *(S)*-FPMPA, did not show any activity (Smeijsters *et al.*, 1999).

In mice infected with *Plasmodium berghei*, the increase of parasitaemia can be blocked for several days by a single injection of *(S)*-HPMPA. Though *(S)*-HPMPApp efficiently inhibits DNA pol alpha of *P. falciparum* and *P. berghei* and/or DNA pol gamma from *P. falciparum*, it was demonstrated that HPMPApp is not an alternative substrate for plasmodial polymerases; thus, the delayed inhibition of plasmodial schizo-gony cannot result from DNA strand breakage caused by *(S)*-HPMPA incorporation, nor was the *(S)*-HPMPA-induced arrest of DNA replication due to chain termination (Smeijsters *et al.*, 1994). However, the *(S)*-HPMPA-induced nephrotoxicity (even with simultaneous intraperitoneal administration of probenecid) prevented further evaluation of its therapeutic effect in *Plasmodium berghei*-infected mice (Smeijsters *et al.*, 1996).

Some of the phosphonomethoxyalkyl purines showed antitrypanosomal potential at non-toxic dosages. The structural requirements for the antitrypanosomal activity differ from those for *Plasmodium sp.* For *Trypanosoma brucei brucei*, the most active compounds were *(S)*-HPMPA and *(S)*-HPMPDAP. Both compounds were strongly active also against the multidrug-resistant *T. b. brucei* both *in vitro* and in the mice model. PMEA or PMEDAP were less active (and quite inactive against multidrug-resistant T. b. brucei). The most active compound against *Trypanosoma congolense* was PMEDAP (Kaminsky *et al.*, 1994). *(S)*-HPMPA was active *in vitro* against bloodstream forms of *Trypanosoma brucei rhodesiense, T. b. gambiense, T. congolense and T. evansi*, but not against intracellular *T. cruzi* or *Leishmania donovani* (Kaminsky *et al.,* 1996). In the presence of *(S)*-HPMPA, trypanosomes are arrested in the S-phase and cannot enter the G2-phase of the cell cycle. Only nuclear DNA replication is inhibited, while mitochondrial DNA replication and kinetoplast division remains unaffected (Kaminsky *et al.,* 1998).

(S)-HPMPA also inhibits infection of SCID mice by microsporidia *E. cuniculi* (Roučka, 2000).

6.3. Antineoplastic activity of acyclic nucleoside phosphonates

Studies performed in our Laboratory disclosed a significant *in vitro* cytostatic activity with both HPMPA and PMEA: The growth of mouse leukemic cells L1210 *in vitro* was inhibited by *(S)*-HPMPA by 50% at 57.0 µM, while the IC_{50} for PMEA was 15.5 µM. The 2-amino congeners of the above analogs were still more efficient: the corresponding values for PMEDAP and *(S)*-HPMPDAP were 6.0 µM, 19.5 µM, respectively (Veselý *et al.,* 1990). In murine tumor models, *(S)*-HPMPA and PMEA were only modestly active on intraperitoneal P388 leukemia and quite inactive against P388 leukemia implanted intravenously. PMEG which was studied in the same model was more active than the adenine derivatives and increased the life span as well as delayed primary tumor growth of subcutaneously implanted B16 melanoma (Rose *et al.,* 1990).

PMEA was tested in the model of spontaneous acute lymphoblastic leukemia (ALL-type) in the inbred strain of Sprague-Dawley (Prague) rats. The treatment resulted in a significant prolongation of survival time of the treated animals. Histological examination of PMEA-treated and untreated animals indicated that the drug effectively slows down the growth of the lymphoma at the site of inoculation and inhibits the subsequent progression of tumor cells in the lung, liver, spleen and lymph nodes (Otová *et al.,* 1993b). Treatment of inbred LEW rats inoculated with leukemic lymphoblastic cells KPH-Lw-I with PMEA demonstrated a significant cytostatic effect which consisted in a significant prolongation of survival time, suppression of the number of bone marrow leukemia cells with characteristic chromosomal marker of KPH-Lw-I leukemia and a decrease of the number of lymphoblasts in the blood (Otová *et al.,* 1993a). To verify its cytostatic potency, PMEA was evaluated in five rat and mouse experimental tumors: while the growth of spontaneously arisen tumors was significantly inhibited, two chemically induced tumors were not influenced (Otová *et al.,* 1993c).

PMEA has long-lasting and dose-dependent differentiation-inducing properties in human erythroleukemia K562 cell cultures (Hatse *et al.,* 1995). A PMEA-resistant K562 cell line is insensitive to the induction of erythroid differentiation by PMEA (Hatse *et al.,* 1996). PMEA is also a potent inhibitor of growth and differentiation of choriocarcinoma RCHO cells (Hatse *et al.,* 1998b) which cause the highly aggressive choriocarcinoma tumor grafted under the kidney capsule of syngeneic WKA/H rats. Treatment with PMEA afforded a marked reduction in tumor size which lasted for at least 10 days after termination of drug treatment. High-dose treatment with PMEA, started at a time point where choriocarcinoma tumors had already developed, induced regression of the tumors (Hatse *et al.,* 1998a). K562 cells exposed to PMEA displayed retardation of S-phase progression, leading to a severe perturbation of the normal cell cycle distribution pattern and a marked accumulation of cyclin A and cyclins E and B1. A similar picture of cell cycle deregulation was also observed in PMEA-exposed human myeloid THP-1 cells which underwent apoptotic cell death. Evidently, depending on the nature of the tumor cell line, PMEA triggers a process of either differentiation or apoptosis by the uncoupling of normally integrated cell cycle processes through inhibition of DNA replication during the S-phase (Hatse *et al.*, 1999c). Antitumor effects of another ANP, PMEDAP, were examined in an *in vivo* model of transplanted Sprague-Dawley (SD/cub) rat T-cell lymphomas. With three individual SD/cub neoplasias (SD10/96, SD14/97, SD1/90) of different phenotypes PMEDAP exerts positive therapeutic effect in two of the three lymphomas tested. The PMEDAP-insensitive, slowly growing SD1/90 lymphoma differs from the others in a uniform karyotype with trisomy of chromosome 11, CD4-immunophenotype, heterogeneous cellular morphology and constitutive expression of p53 protein in some neoplastic cells. The different anticancer efficacy of PMEDAP treatment among SD/cub lymphomas could be associated with the different phenotypes of individual neoplasias (Bobková *et al.*, 2000).

Comparison of the antitumor activity of PMEDAP with related acyclic nucleotide analogs PMEA, and PMEG on a spontaneous T-cell lymphoma in inbred SD/cub mice revealed significant therapeutic effects of PMEDAP applied to the vicinity of the growing lymphoma. Identical administration of PMEA, or PMEG did not affect the survival of lymphoma-bearing animals. Decrease in the lymphoma weight during drug administration was accompanied by the suppression of mitotic activity in neoplastic cells and increased chromatin condensation. PMEDAP application induces apoptosis in growing lymphomas *in vivo*. The antitumor effect lasts only during the administration of the drug. After its cessation progression of neoplasia was reestablished (Otová *et al.*, 1999). PMEA, PMEDAP and PMEG perturb DNA replication by terminating the growing DNA chain. The drugs suppress the cell growth at low concentrations while inducing apoptotic activity at high concentrations. Activities of the analogues increased in the order PMEA<PMEDAP<PMEG. All compounds reduce the proportion of G1 cell cycle phase. The lymphoid cell line MOLT-4 was more susceptible than the myelogenous cell lines HL-60 and ML-1. In semicontinuous cultures in the presence of low-concentration PMEG the steady-state viable cell concentration was lower and the proportion of G1 phase cells was suppressed. Upon gradual removal of PMEG from the medium, the cell concentration and the DNA profile returned to values characteristic for the control culture (Franěk *et al.*, 1999).

PMEDAP and/or PMEA treatment of SD rat lymphomas significantly prolonged the mean survival time of tumor-bearing animals. Dose-dependent genotoxicity of both PMEDAP and PMEA was not observed in *in vitro* tests on a stabilized diploid MRC-5 cell line. PMEDAP and/or PMEA inhibited proliferation of mitogen-activated T-lymphocytes. Modulation of subpopulations of peripheral blood cells under *in vivo* conditions was found in inbred SD animals. Intraperitoneal administration of PMEDAP and PMEA to young healthy SD animals induced the decrease of the CD4+/CD8+ ratio (Otová *et al.*, 1997b).

Cytostatic activity of another ANP, N6-cyclopropyl-PMEDAP, against a variety of tumor cell lines is more pronounced than that of the parent PMEDAP and similar to that of PMEG. In contrast to PMEDAP, the biological effects of the N6-cyclopropyl derivative were reversed by the adenylate deaminase inhibitor 2'-deoxycoformycin. The drug is probably deaminated to PMEG by a specific adenylate deaminase. Compared to the parent PMEDAP, this PMEG prodrug exhibits superior antiproliferative and differentiation-inducing effects on tumor cells (Compton *et al.*, 1999; Hatse *et al.*, 1999b). Numerous other N6-substituted PMEDAP derivatives exhibit strong cytostatic activity *in vitro* in mouse leukemia L1210 cells, murine L929 cells, human cervix carcinoma HeLa S3 cells and human T-lymphoblastoid CCRF-CEM cell line. The five most active compounds are 9-[2-(phosphonomethoxy)ethyl] derivatives of 2-amino-6-dimethylaminopurine, 2-amino-6-(2,2,2-trifluoroethylamino)purine, 6-allylamino-2-aminopurine, 2-amino-6-cyclopropylaminopurine and 6-amino-2-[2-dimethylamino-ethyl]aminopurine (Holý et al, 2001) (Chart 13).

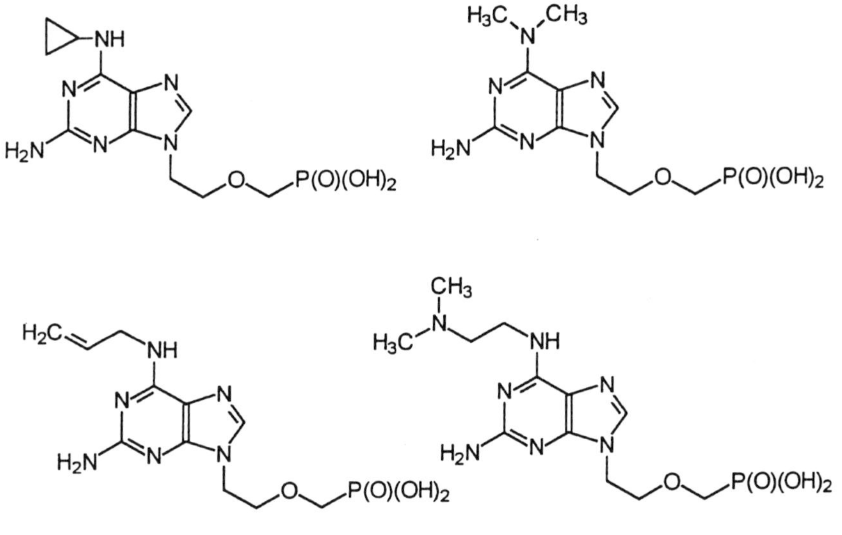

Chart 13.

HPMPC has a strong effect against papillomaviruses. It inhibits proliferation of human cervical keratinocytes immortalized by HPV-33 and the cervical carcinoma cell lines containing HPV-16 or HPV-18. It is followed by HPMPA, PMEG and PMEDAP. The time-dependent antiproliferative effect of ANPs (HPMPC, HPMPA, PMEG) was observed also in HPV-free tumor cell lines (i.e., human melanomas, lung, colon, and breast carcinomas), but to a much lesser extent. Treatment of SV40- and adenovirus-transformed cells with ANPs resulted in the inhibition of cell proliferation as a function of time, similar to that observed with HPV-positive cells, with HPMPC and cHPMPC as the most potent antiproliferative agents (Andrei *et al.*, 1998a,b).

HPMPC also protects newborn rats infected i.p. with murine polyomavirus against hemangioma growth. An antitumor or antiangiogenic effect, rather than inhibition of viral replication, may be the reason for the inhibitory activity of cidofovir in this model (Liekens *et al.*, 1998). Intratumoral injection of HPMPC also arrests tumor growth of the EBV-associated nasopharyngeal carcinoma xenografts in athymic mice. An antitumor effect was also produced by systemic administration; HPMPC probably induces apoptosis in EBV-transformed epithelial cells (Neyts *et al.*, 1998).

The genotoxic and embryotoxic effects of phosphonomethoxyalkylpurines, a new group of antiviral agents, decrease in the following order: PMEG > PME-6-thioG > PMEDAP > PMEA > *(R)*-PMPDAP = *(R)*-PMPA. Results of the present study are fully consistent with the previously found efficacy of their diphosphates to inhibit the replicative DNA polymerases. The marked genotoxicities of PMEG and PME-6-thioG are comparable to that of mitomycin C, whereas the moderate genotoxicity of PMEA is comparable to that of AZT. *(R)*-PMPDAP and *(R)*-PMPA did not induce any structural aberrations of chromosomes under the experimental conditions (Otová *et al.*, 1997a). PMEA and HPMPC differ in their *in vitro* effect on the genetic material of eukaryotic cells. PMEA has a genotoxic activity, while HPMPC exerts a cytostatic effect on eukaryotic cells *in vitro*. These results correspond with the mode of embryotoxic action of these compounds: HPMPC exhibits a general embryolethal effect, whereas PMEA is apparently teratogenic and interacts with the mutant allele producing preaxial polydactyly of the hind limbs (Bílá *et al.*, 1993).

6.4. Immunomodulatory activity of acyclic nucleoside phosphonates

PMEA significantly enhances NK activity and interferon production. Similar enhancement of natural immunity was observed following application of PMEDAP or FPMPA. The immunomodulating effect of PMEA was more pronounced with a single administration compared to repeated administrations. Dose-dependent enhancement of NK activity and IFN production could also be demonstrated during chronic administration of PMEA (more closely resembling the schedule of administration of this drug in patients) (Calio *et al.*, 1994). The levels of natural killer (NK)-cell cytotoxicity from cells isolated from PMEA-treated mice are significantly higher. IFN production was also substantially increased in PMEA-treated animals, while both IL-1 and IL-2 production was decreased (Del Gobbo *et al.*, 1991). Treatment with PMEA significantly decreases the mortality and morbidity of mice challenged with influenza virus (an RNA virus that is non-sensitive to PMEA *in vitro*), evidently through the enhancement of some immune functions (Villani *et al.*, 1994).

Acyclic nucleotide analogues exhibit strong effects on *in vitro* secretion of cytokines and production of nitric oxide by murine peritoneal macrophages. PMEG, *(R)*-PMPA, and *(S)*-PMPA greatly enhanced the secretion of both tumour necrosis factor-alpha (TNF-α) and interleukin-10 (IL-10), *(R)*-PMPDAP stimulated only TNF-α; other tested ANPs, including PMEA, PMEDAP, *(S)*-PMPDAP as well as *(S)*-HPMPC, were ineffective. None of the tested compounds influenced secretion of IL-2 or interferon-gamma (IFN-γ). Both TNF-α and IL-10 are major factors determining the enhancing effects of PMEG, *(R)*-PMPA, and *(S)*-PMPA on production of NO generated by exogenous IFN-γ. There is a possible implication of immunomodulatory properties in the antiviral effects of some acyclic nucleotide analogues (Zídek *et al.*, 1997c, 1999b).

The antiviral drug Tenofovir [*(R)*-PMPA] does not influence IFN-γ and interleukin-2 expression; however, it highly stimulates in a concentration- and time-dependent manner the secretion of TNF-α and interleukin-10. It also substantially enhances the production of NO induced by exogenous IFN-γ. Inhibitory experiments using neutralizing antibodies against TNF-α and/or interleukin-10 demonstrated that these two cytokines are major factors responsible for triggering the underlying mechanisms leading to enhanced NO production (Zídck *et al.*, 1997b).

Severe graft-*versus*-host disease which develops following intravenous administration of parental BN.lx lymphoid cells into (SHR × BN.lx)F1 rats can be completely abrogated by injections of PMEA. Near-to-normal histological findings in PMEA-treated animals contrast with severe damage of bone marrow, lymphoid infiltration of salivary glands and ulceration with haemorrhage of the epidermoid part of the stomach in untreated control animals (Křen *et al.*, 1993). When administered after injection of parental splenocytes, PMEA reduced the development of local graft-*versus*-host reaction (GVHR). PMEA pre-treatment of donors of splenocytes had no influence on GVHR. Modifications in the immune system triggered by PMEA were confirmed in the rat model by evaluation of subsets of white blood cells isolated from peripheral blood by a set of monoclonal antibodies. Enhanced formation of NO was found in both unconditioned and LPS-stimulated macrophage cultures following the drug treatment (Otová *et al.*, 1994).

In a mycobacterial adjuvant-induced rat arthritis model, PMEA delayed the onset, and substantially reduced or nearly completely inhibited the development of arthritic paw swelling. Bis-POM-PMEA expressed much more pronounced beneficial effects after both oral and i.p. administration. HPMPC was inactive (Zídek *et al.*, 1995). PMEA and its prodrug inhibited by > 80% arthritic paw swelling, splenomegaly and fibroadhesive perisplenitis, both in prophylactic and therapeutic dosage regimens. Neither *(R)*-PMPA nor bis(POC)-PMPA suppressed development of arthritic lesions. Substantially reduced nitrite + nitrate levels were detected in serum and urine of PMEA-treated animals compared to those of untreated controls. Also, complete suppression of the disease-associated, greatly enhanced systemic levels of the chemokine, RANTES (regulated upon activation, normal T cell expressed and secreted), was observed in rats injected with PMEA. PMEA does not change, *(R)*-PMPA enhances, and both prodrugs inhibit the immune-activated NO production *in vitro*. Under the same conditions PMEA inhibits, while *(R)*-PMPA slightly stimulates, secretion of RANTES. The *in vivo*-inhibited production of NO seems to be a consequence rather than a mechanism of antiarthritic action of PMEA (Zídek *et al.*, 1999a). Bis-POM-PMEA (Adefovir Dipivoxil) inhibited *in vitro* in a concentration-dependent manner the formation of NO generated by interferon α and lipopolysaccharide. The observed suppressed transcription of mRNA for inducible NO synthase resulted in decreased synthesis of NO synthase protein. The parent PMEA was virtually ineffective in this assay (Zídek *et al.*, 1997a).

7.   Part C.   Biochemical studies

7.1. Transport of acyclic nucleoside phosphonates across cellular membrane

As stated earlier, transport of ANP across cellular membranes is the key-issue for their biological activity. Cellular uptake of the adenine derivatives *(S)*-HPMPA and PMEA has been studied in H9 cells. They exhibited an identical pattern of permeation. Uptake did not occur *via* the nucleoside transport system, but through a different mechanism which, for its slow kinetics and temperature dependence, is compatible with an endocytosis-like process (Palu *et al.,* 1991).

The study of [^{14}C]PMEA uptake in HeLa S3 cells has shown that intracellular levels of the drug plateau after 1 h. Transport across the plasma membrane is saturable and it can operate against a concentration gradient. It is dependent on temperature and on cellular density. Following treatment of cells with proteases, PMEA uptake strongly decreases. The transport process is considerably specific; it is competitively inhibited by certain phosphonate analogs of which PMEDAP is the most efficient. Also, natural nucleotides competitively inhibit PMEA transport to an extent depending on the nature of the nucleobase and on the position of phosphate groups. Nucleosides and nucleobases do not interfere with PMEA uptake. Cellular transport of adenosine and thymidine or uptake of AMP and ATP is not affected by PMEA. By using vectorial labeling of plasma membrane proteins with Na^{125}I and affinity chromatography, a 50-kDa protein which may mediate cellular transport of PMEA has been identified (Cihlář *et al.,* 1995). In contrast to HeLa S3 cells, the uptake of PMEA by human lymphoblastoid cells (CCRF-CEM) proceeds by standard fluid-phase endocytosis: transport is temperature-dependent, takes place at low temperature and depends on the intracellular ATP concentration. It is not competitively inhibited by other PME-derivatives (Olšanská *et al.,* 1997).

A human T lymphoid cell line resistant to the antiproliferative effects of PMEA showed cross-resistance to the related ANPs PMEDAP and PMEG, and to the PMEA-prodrug bis(POM)-PMEA); it was partially resistant to sugar-modified purine nucleosides and adenosine, but not to 2'-deoxyadenosine or araA. Accumulation of major metabolites formed from either PMEA or bis(POM)-PMEA) substantially decreased. Compared with the parental cells, the variant cells showed an increase in the rate of efflux of PMEA and a decrease in the activity of adenylate kinase. Other enzymes of nucleotide metabolism (e.g., adenosine kinase, deoxycytidine kinase, and 5-phosphoribosyl-1-pyrophosphate synthetase) had no significant differences. Thus, the mutation in this PMEA-resistant cell line involves an alteration in the cellular efflux of PMEA (Robbins *et al.,* 1995). A L1210 cell line showing 300-fold resistance to the cytostatic effect of PMEA was also selected. The cytostatic activity was severely impaired only for PMEA and PMEDAP, but not for PMEG or for *(S)*-HPMPA. Our observations point to a compromised and highly specific PMEA/PMEDAP uptake as the molecular basis for the pronounced PMEA resistance of the mutant cells (Balzarini *et al.,* 1998b).

Overexpression and amplification of the MRP4 gene correlates with ATP-dependent efflux of PMEA and AZTMP from cells and, thus, with resistance to these drugs.

It severely impaired the antiviral efficacy of PMEA, azidothymidine and other nucleoside analogs. Increased resistance to PMEA and amplification of the MRP4 gene correlated with enhanced drug efflux; transfer of chromosome 13 containing the amplified MRP4 gene conferred resistance to PMEA (Schuetz *et al.,* 1999). The resistance of cells transfected by human multidrug resistence protein MRP5 against PMEA and/or 6-mercaptopurine is due to an increased extrusion of PMEA and 6-thioinosine 5'-phosphate from the cells that overproduce MRP5. MRP5 might play a role in some cases of unexplained resistance to thiopurines in acute lymphoblastic leukemia and/or to antiretroviral nucleoside analogs in HIV-infected patients (Wijnholds *et al.,* 2000).

The cellular uptake of *(S)*-HPMPC in Vero cells is temperature sensitive: its rate was almost totally inhibited at 4 °C. The time course of the drug uptake was linear and proportional to the concentration in the medium. None of natural nucleosides, nucleotides or ANPs affected the drug uptake at concentrations up to 2000-fold molar excess. From the comparison with transport of [^{14}C]sucrose and the effects of the microtubule antagonist colchicine and the tumor promoting agent phorbol myristate acetate it is concluded that HPMPC enters Vero cells by fluid-phase endocytosis (Connelly *et al.,* 1993).

Intestinal transport, uptake and metabolism of bis(POM)-PMEA in an *in vitro* cell culture system of intestinal mucosa (Caco-2 monolayers) demonstrated considerable improvement of transepithelial transport of total PMEA across Caco-2 monolayers. Uptake studies revealed that only negligible amounts of bis(POM)-PMEA (< 0.2%) were present inside the cells. Very high intracellular concentrations of PMEA were found (approximately 1.2 mM, after a 3 hr incubation with 50 μM bis(POM)-PMEA), which suggests that PMEA was trapped inside the cells probably due to its negative charge. This explains the relatively slow efflux of PMEA. Intracellular trapping of PMEA in the intestinal mucosa may result in slow release of PMEA after oral administration of bis(POM)-PMEA (Annaert *et al.,* 1997). In Caco-2 monolayers, bis(POM)-PMEA is a substrate for a P-glycoprotein-like carrier mechanism while its metabolites mono(POM)-PMEA and PMEA are transported by a non-P-glycoprotein efflux protein (Annaert *et al.,* 1998).

Total transport of the *(R)*-PMPA prodrug bis(POC)-PMPA in the Caco-2 cells was much higher than that of *(R)*-PMPA. The majority of the compound was recovered after transport in the form of the monoester metabolite. Pharmacokinetic studies with mice showed that the oral bioavailability of bis(POC)-PMPA was 20%; it is efficiently converted to the active drug after oral administration. SCID mice infected with Moloney murine sarcoma virus and treated orally with bis(POC)-PMPA showed a significant delay in MSV-induced tumor appearance and tumor-associated death. The antiviral efficacy was not significantly different from that of subcutaneous *(R)*-PMPA given at an equivalent dose (Naesens *et al.,* 1998).

7.2. Intracellular activation of acyclic nucleoside phosphonates

The acyclic nucleotide analogues PMEA and *(S)*-HPMPA are transformed in cells to their mono- and diphosphoryl derivatives. Mouse L1210 cells phosphorylate the compounds in two steps to their diphosphoryl derivatives. Other nucleoside

5'-triphosphates or creatine phosphate could not be substituted for ATP as phosphate donors. At least one other enzyme (creatine kinase) is capable of transforming the monophosphoryl derivatives of the studied compounds to their respective diphosphates (Merta *et al.,* 1992). The donor efficiency decreased in the order CTP > UTP > ATP > GTP. The presence of an ATP regenerating system considerably stimulated the conversion of both compounds. The rate of PMEA phosphorylation was 5-times slower than that of HPMPA both with and without an ATP regenerating system (Merta *et al.,* 1990).

Purification of the nucleotide kinase from L-1210 cells afforded an enzyme identical with AK2 (mitochondrial) adenylate kinase. The efficacy of ANP phosphorylation is low and follows the order *(S)*-HPMPA > *(R)*-PMPA > PMEA > PMEDAP > *(S)*-PMPDAP~*(R)*-PMPDAP (Krejčová *et al.,* 2000a). Also in CEM cells, PMEA is phosphorylated to its mono- and diphosphate in the presence of ATP as the phosphate donor. No other nucleotides or 5-phosphoribosyl pyrophosphate displayed appreciable activity as a phosphate donor. CEM cells contain two nucleotide kinase activities, one in mitochondria and one in the cytosol, which phosphorylated PMEA. The PMEA-resistant CEM mutant has a deficiency in the mitochondrial adenylate kinase activity, indicating the importance of this enzyme in phosphorylation of PMEA. Other effective antiviral purine phosphonate derivatives of PMEA showed a profile of phosphorylating activity similar to that of PMEA. Phosphorylation of *(S)*-HPMPC proceeded by an enzyme present in the cytosol (Robbins *et al.,* 1995a).

9-[2-(Phosphonomethoxy)ethyl]guanine (PMEG), *(R)*- and *(S)*-enantiomers of both 9-[3-hydroxy-2-(phosphonomethoxy)propyl]guanine (HPMPG) and 9-[2-(phosphono-methoxy)propyl]guanine (PMPG) are phosphorylated by GMP kinase isoenzymes from L1210 cells (Krejčová *et al.,* 1999) to the first step. *(R)*-PMPG is a good substrate with a relative phosphorylation efficacy of 12% compared to the natural substrate GMP, whereas PMEG is a poor substrate with a relative phosphorylation efficacy of 1.1%. The structurally related 2,6-diaminopurine analogues PMEDAP and *(R)*- and *(S)*-PMPDAP are not phosphorylated by any of the GMP kinase isoenzymes tested. The inhibitory activity of individual compounds on GMP kinase isoenzymes decreases in the order *(S)*-HPMPG > *(R)*-PMPG > PMEG > *(R)*-HPMPG > *(S)*-PMPG > PMEDAP = *(R)*-PMPDAP = *(S)*-PMPDAP; each compound exerts different type of inhibition (Krejčová *et al.,* 1996, 2000b). PMEG is significantly more cytotoxic than PMEA against human leukemic cells. The diphosphate derivatives are the major metabolites formed in cells with both of these agents, with PMEGpp reaching approximately 4-fold higher cellular concentration than that achieved by PMEApp. There is approx. 30-fold difference in cytotoxicity between the two analogs. PMEGpp is a potent inhibitor of both human polymerases alpha and delta, two key enzymes involved in cellular DNA replication, whereas PMEApp inhibited these enzymes relatively poorly. Factors contributing to the enhanced antileukemic activity of PMEG comprise both its increased anabolic phosphorylation and the increased potency of the diphosphate derivative to target the cellular replicative DNA polymerases (Pisarev *et al.,* 1997).

Phosphorylation of cidofovir [*(S)*-HPMPC] is two-step process catalyzed by several enzymes. (Cihlář *et al.,* 1992). An enzymatic activity phosphorylating the drug to its monophosphate derivative was purified from human liver and identified as pyrimidine nucleoside monophosphate kinase. Pyruvate kinase, creatine kinase and nucleoside diphosphate kinase catalyze HPMPCpp synthesis from HPMPCp, whereas phosphoglycerate kinase and succinyl-CoA synthetase did not. The most efficient phosphorylation is catalyzed by pyruvate kinase. The intracellular levels of HPMPCp and IIPMPCpp increased in cytomegalovirus-infected cells, presumably due to the stimulation of drug uptake and higher activities of phosphorylating enzymes (Cihlář *et al.*, 1996).

In Vero cells, the levels of the PMEA and its metabolites PMEAp and PMEApp reached a plateau during 12 hr and were cleared from the cells with a half-life of 4.9 hr. In contrast, the *(S)*-HPMPC metabolites *(S)*-HPMPC monophosphate (HPMPCp) and *(S)*-HPMPC diphosphate (HPMPCpp) accumulated throughout the 24-hr study period and, reached intracellular levels 2-3-fold greater than those of the PMEA metabolites. *(S)*-HPMPC also differed from other ANPs in its capacity to generate a phosphodiester metabolite (HMPCp-choline) (Ho *et al.*, 1992; Eisenberg *et al.*, 1998) which was a predominant metabolite in *(S)*-HPMPC-treated cells. The decay of HPMPCpp was quite slow and biphasic, while that of HMPCp-choline was monophasic. The HPMPCp-choline adduct may serve as an intracellular store for the long-term maintenance of active HPMPCpp in cells (Aduma *et al.*, 1995). In LEP cells, *(S)*-HPMPC is converted by pyrimidine nucleoside monophosphate kinase to HPMPCp and further to HPMPCpp. In the presence of CTP-phosphorylcholine cytidylyl-transferase, HPMPCpp and choline phosphate are transformed to *(S)*-HPMPCp-choline. Both this metabolite and HPMPCpp persist in the cell long after *(S)*-HPMPC was removed from the medium. Neither *(S)*-HPMPC nor its choline-adduct affect intracellular CDP-choline level (Cihlář *et al.*, 1992).

Radiolabeled *(S)*-[U-^{14}C-adenine]-HPMPA was taken up by HSV-1-infected and mock-infected cells and subsequently converted to its monophosphoryl [*(S)*-HPMPAp] and diphosphoryl [*(S)*- HPMPApp] derivatives by cellular enzymes. It is incorporated to a very low extent into DNA of both mock-infected and HSV-1-infected Vero cells (Votruba *et al.*, 1987).

There is a report on synthesis of HPMPApp and PMEApp from HPMPA and PMEA by direct transfer of diphosphate from 5-phosphoribosyl 1-pyrophosphate (PRPP) in a reversible reaction catalyzed by purified PRPP synthetase. PRPP synthetase was claimed to act non-stereospecifically and to phosphorylate both *(S)*- and *(R)*-HPMPA. Ostensibly, PRPP synthetase should phosphorylate other acyclic adenine and 2,6-diaminopurine phosphonates as substrates (Balzarini *et al.*, 1991a). However, we were unable to confirm any of those findings.

7.3. Principles of antimetabolic activity of acyclic nucleoside phosphonates

Inhibition of HSV-1 DNA polymerase and HeLa DNA polymerases alpha and beta by diphosphoryl derivatives of acyclic phosphonomethoxyalkyl nucleotide analogues was studied and compared with inhibition by ACV-TP, araCTP, ddTTP and AZT-TP. In the series of PME derivatives of heterocyclic bases, the inhibitory effect of their diphosphates on HSV-1 DNA polymerase decreased in the order PMEApp >> PMEGpp

> PMEApp > PMETpp >> PMECpp >> PMEUpp. *(S)*-HPMPApp was a relatively weak inhibitor of HSV-1 DNA polymerase. The inhibitors could be divided into three groups: (a) the diphosphoryl derivatives of acyclic nucleotide analogues (PME-type and HPMPA) and ACV-TP specifically inhibit HSV-1 DNA polymerase and DNA polymerase alpha and do not significantly inhibit DNA polymerase beta; (b) AZT-TP and ddTTP are effective only against DNA polymerase beta, and (c) araCTP inhibits all three enzymes. When dATP was omitted from the reaction mixture, the addition of HPMPApp stimulated DNA synthesis by HSV-1 DNA polymerase indicating that HPMPApp is an alternative substrate for *in vitro* DNA synthesis catalyzed by this enzyme (Merta *et al.,* 1990b).

Diphosphates of PME derivatives of heterocyclic bases inhibit reverse transcriptase from detergent-disrupted AMV(MAV) retrovirions in the endogenous oligo(dT)$_{12-18}$ primed reaction. This inhibition was dependent on the character of the heterocyclic base and decreased in the order: 2,6-diaminopurine > adenine > guanine >> cytosine >> thymine > uracil. PMEDAPpp was more potent than either AZT-TP or ddTTP, while PMEApp had approximately the same potency as the two reference compounds (Votruba *et al.,* 1990b). PMEApp has a relatively long intracellular half-life and much higher affinity for the HIV-specific reverse transcriptase than for cellular DNA polymerase alpha. PMEApp is at least as potent an inhibitor of human immunodeficiency virus reverse transcriptase as 2',3'-ddATP. As an alternative substrate to dATP, PMEApp acts as a potent DNA chain terminator, and this may explain its anti-retrovirus activity (Balzarini *et al.,* 1991c).

The inhibitory potencies of diphosphates of adenine, 2,6-diaminopurine, and guanine ANPs of *(S)*-HPMP and the PME series, [*(S)*-HPMPApp, PMEDAPpp, and PMEGpp] toward cellular DNA polymerases alpha, delta, and epsilon (isolated from tumors of T cell spontaneous acute lymphoblastic leukemia in Sprague-Dawley inbred rats) (Kramata *et al.,* 1995) were estimated by kinetic measurements on synthetic homopolymeric template primers. HPMPApp is a selective and potent inhibitor of DNA pol epsilon, whereas PMEDAPpp strongly inhibits DNA pol delta. Of the nucleotide analogs tested, PMEGpp is the most efficient inhibitor of DNA pol alpha and epsilon, whereas PMEApp inhibits DNA pol alpha and epsilon relatively poorly and exerts only moderate inhibition of DNA pol delta. These data are quite consistent with previously reported cytostatic activity of these nucleotide analogs. All of the enzymes studied catalyze incorporation of PMEA, PMEDAP and *(S)*-HPMPA into DNA chain. PMEApp and PMEDAPpp are DNA chain terminators. HPMPApp formed poly(dT)/oligo(dA$_{(18)}$)-[*(S)*-HPMPA]$_{2-4}$ structures (Kramata *et al.,* 1996).

PMEApp, found to weakly inhibit DNA pol delta/proliferating cell nuclear antigen, is a substrate for pol alpha, delta, epsilon, and epsilon*. A comparison of the V_{max} and K_m for PMEApp and dATP demonstrated that the relative efficiency of incorporation of this analog into the DNA chain was decreasing in the following order: pol delta ~ pol epsilon ~ pol epsilon* > pol alpha. This incorporation amounted to 4.4 to 0.7% of dAMP molecules. Similar K_m values for PMEApp and dATP in pol epsilon and pol epsilon* catalyzed reactions revealed that proteolysis of the enzyme probably does not affect the dNTP binding site. The DNA polymerases tested (Birkuš *et al.,* 1998) were inhibited by the reaction product (PMEA terminated DNA chains) with similar K_i/K_m ratios.

The associated 3'-5'-exonuclease activity of DNA pol delta, epsilon, and epsilon* was able to excise PMEA from the 3'-OH end of DNA with a rate one order of magnitude lower than that of the dAMP residue (Birkuš *et al.*, 1999).

Incubation of CEM cells with PMEG, PMEDAP and PMEA inhibited DNA synthesis with IC_{50} values of 1, 6, and 25 μM, respectively. Reasons for these marked differences might include cellular transport, different efficiencies of phosphorylation, differential effects on 2'-deoxynucleotide (dNTP) pools, and differences in the affinities of the cellular DNA polymerases for the drug diphosphates. CEM cells accumulate higher levels of PMEGpp than PMEDAPpp or PMEApp. Treatment of cells with any of the nucleotide analogs resulted in increased dNTP pools, with PMEG producing the greatest increase. All three analogs had the greatest effect on the dATP pool size, whereas the dGTP pool size was not significantly affected. Comparison of the ratios of nucleotide analog diphosphates to their corresponding dNTPs under conditions where DNA synthesis is inhibited by 50% suggested that cellular DNA polymerases were approximately twice as sensitive to PMEGpp as to PMEDAPpp and 5-fold more sensitive to PMEGpp than to PMEApp. Examination of the efficiencies with which the replicative DNA pol alpha, delta, and epsilon incorporated the analogues showed that DNA pol delta, the most sensitive of the DNA polymerases, incorporated PMEGpp twice as effi ciently as PMEDAPpp and 7-fold more effi ciently than PMEApp (Kramata *et al.*, 1999).

PMEGpp is a competitive inhibitor of cellular DNA polymerases delta and epsilon. Its apparent K_i values for PMEGpp were 3-4 times lower than the K_m values for dGTP. PMEGpp is incorporated into DNA by both enzymes. DNA pol epsilon could elongate PMEG-terminated primers in both matched and mismatched positions with an efficiency equal to 27 and 85% that observed for dGMP-terminated control template-primers. Because PMEG acts as an absolute DNA chain terminator, the elongation of PMEG-terminated primers is possible only by cooperation of the 3'-5'-exonuclease and DNA polymerase activities of the enzyme. In contrast to DNA pol epsilon, DNA pol delta exhibited negligible activity on these template-primers, indicating that DNA pol epsilon, but not DNA pol delta, can repair the incorporated analog (Kramata *et al.*, 1998).

Human DNA polymerases alpha, beta and gamma were able to incorporate PMEApp, PMEGpp, *(R)*-PMPApp and *(R)*-PMPDAPpp as terminators of primer extension into primer/ template DNA of defined sequence. Efficiencies of incorporation (related to the corresponding natural dNTP) by DNA pol alpha reached 51% for PMEGpp. Generally, the lowest incorporation efficiencies with all three DNA polymerases were found for PMPApp (0.06-1.4%) and PMPDAPpp (0.075-2.2%) (Cihlar *et al.*, 1997).

The diphosphate derivative of PMEDAP (PMEDAPpp) selectively inhibited HCMV-induced DNA polymerase (IC_{50} 0.1 μM) (Neyts *et al.*, 1993). *(S)*-HPMPCpp is a competitive inhibitor of dCTP and an alternate substrate for human cytomegalovirus (HCMV) DNA polymerase. HCMV DNA incorporated dCTP approximately 42 times more efficiently than *(S)*-HPMPCpp. HCMV DNA polymerase also utilized a synthetic DNA primer containing a single molecule of HPMPC at the 3'-terminus, with an efficacy of incorporation. approaching 6% of that of dCTP. Incorporation of a single CDV into DNA by HCMV DNA polymerase does not lead to chain termination (Xiong *et al.*, 1996).

216   A. Holý

Diphosphates of *(S)*-HPMP- and PME-derivatives of purine and pyrimidine heterocyclic bases inhibit HSV-1 encoded ribonucleotide reductase; the most efficient inhibitors of CDP reduction (at 5.1 µM) by the HSV-1-encoded enzyme are *(S)*-HPMPApp and PMEApp. PMEApp does not inhibit the enzyme isolated from the mutant HSV-1 KOS strain PMEA[r] which is resistant to PMEA at a concentration of 100 µg/ml (Vonka *et al.*, 1990). The enzyme isolated from the PMEA-resistant virus strain is also insensitive to inhibitory effects of hydroxyurea and *(S)*-HPMPApp. Thus, the inhibitory potency of *(S)*-HPMPApp and PMEApp toward HSV-1 encoded ribonucleotide reductase might be connected with the anti-HSV activity of *(S)*-HPMPA and PMEA (Černý *et al.*, 1990).

Following exposure to PMEA, human erythroleukemia K562, human T-lymphoid CEM and murine leukemia L1210 cells markedly accumulated in the S-phase of the cell cycle. In contrast to DNA replication, RNA synthesis (transcription) and protein synthesis (mRNA translation) were not affected. The NTP pools were slightly elevated, while the intracellular levels of all four dNTPs were 1.5-4-fold increased in PMEA-treated cells. The amount of thymidylate synthase-derived dTTP in the acid soluble pool was 2-4-fold higher in PMEA-treated than in untreated K562 cells, which is in accord with the 3-4- fold expansion of the global dTTP level in the presence of PMEA. Strikingly, the thymidine kinase-derived dTTP accumulated to a much higher extent (i.e., 16-40-fold) in the soluble dTTP pool following PMEA treatment. A markedly increased thymidine kinase activity was estimated in extracts of PMEA-treated K562 cell cultures. Thus, thymidine incorporation may be inappropriate as a cell proliferation marker in the presence of DNA synthesis inhibitors such as PMEA (Hatse *et al.*, 1999a).

8.   Complexes of acyclic nucleoside phosphonates with metal ions

The stability constants for the 1:1 complexes formed between Mg^{2+}, Ca^{2+}, Sr^{2+}, Ba^{2+}, Mn^{2+}, Co^{2+}, Ni^{2+}, Cu^{2+}, Zn^{2+}, Cd^{2+} and (phosphonomethoxy)ethane (PME^{2-}) or 9-[2-(phosphonomethoxy)ethyl]adenine ($PMEA^{2-}$), determined by potentiometric pH titration, show a higher stability than that expected for a single phosphonate coordination with the metal ion in all of the complexes. This increased stability is attributed to the formation of five-membered chelates involving the ether oxygen present in the $-O-CH_2-PO_3^{2-}$ residue of PME^{2-} (and $PMEA^{2-}$). The adenine residue has no influence on the stability of these complexes, with the exception of those with Ni^{2+} and Cu^{2+} (Sigel *et al.*, 1992). The stability study of the 1:1 complexes formed between Ca^{2+}, Cu^{2+}, or Zn^{2+} and *(S)*-9-[3-hydroxy-2-(phosphonomethoxy)propyl]adenine (HPMPA) also revealed (for the $HPMPA^{2-}$ complexes of Ca^{2+} and Zn^{2+}) the presence of five-membered chelates involving the ether oxygen in the $-O-CH_2-PO_3^{2-}$ residue. In the case of Cu(HPMPA) the five-membered chelate reached a degree of formation of 26%. Another 61% of species are present which involve the adenine residue (probably N-3) and/or the oxygen atom of the hydroxymethyl group. The metal ion-binding properties of $HPMPA^{2-}$ (and also of $PMEA^{2-}$) differ considerably from those of adenosine 5'-monophosphate (AMP2-) (Sigel *et al.*, 1994).

In the twofold protonated complexes of the structurally related acyclic nucleotide analogue PMEDAP, the primary binding site of its dianion in complexes with Mg^{2+}, Ca^{2+}, Sr^{2+}, Ba^{2+}, Mn^{2+}, Co^{2+}, Ni^{2+}, Cu^{2+}, Zn^{2+} or Cd^{2+} is the phosphonate group; in all instances 5-membered chelates involving the ether oxygen of the $-CH_2OCH_2PO_3^{2-}$ chain are also formed. In complexes with Co^{2+}, Ni^{2+}, Cu^{2+}, and Zn^{2+}, a third isomer is formed which probably involves participation of the 2,6-diaminopurine residue (Blindauer *et al.*, 1999a).

Stability constants for the 1:1 complexes formed between Mg^{2+} and the anions of the N1, N3, and N7 deaza derivatives of 9-[2-(phosphonomethoxy)ethyl]adenine (PA^{2-}) [i.e., of $Mg(HPA)^+$ and $Mg(PA)$], compared with the results for the corresponding complexes of 9-[2-(phosphonomethoxy)ethyl]adenine ($PMEA^{2-}$) and (phosphonomethoxy)ethane (PME^{2-}) show that in the monoprotonated complexes, Mg^{2+} is coordinated significantly to the nucleobase, H^+ is at the phosphonate group. All these complexes, including those with $PMEA^{2-}$ and PME^{2-}, are more stable than expected from the basicity of the $-PO_3^{2-}$ group. This indicates formation of five-membered chelates involving the ether oxygen of the $-CH_2-O-CH_2-PO_3^{2-}$ chain to about 30-40% in equilibrium with isomers having only phosphonate-Mg^{2+}-coordination (Blindauer *et al.*, 1998). Comparison of the stability constants in the 1:1 complexes formed between Cu^{2+} and the anions of the N1, N3, and N7 deaza derivatives of 9-[2-(phosphonomethoxy)ethyl]adenine with ($PMEA^{2-}$) and (phosphonomethoxy)ethane (PME^{2-}) reveals that Cu^{2+} is coordinated to the nucleobase and H^+ is located at the phosphonate group. All the Cu^{2+} complexes are more stable than expected from the basicity of the $-PO_3^{2-}$ group which indicates a certain amount of formation of five-membered chelates involving the ether oxygen of the $-CH_2-O-CH_2-PO_3^{2-}$ chain; in all complexes an additional nucleobase-Cu^{2+} interaction occurs (Blindauer *et al.*, 1997b). These data are corroborated by NMR determination of the sites of protonation, basicity and conformation, as well as the acidity constants of PMEA and its deaza congeners obtained from the pD dependence of H-1 NMR chemical shifts of the aromatic and aliphatic hydrogens in D_2O. The most basic site in all these compounds is the phosphonate group, PO_3^{2-}, followed by N1 in PMEA, 3- and 7-deaza-PMEA. In 1-deaza-PMEA, protonation occurs at N3. Further protonation in strongly acidic medium is possible with all four PMEA derivatives. About 80% of the H(7-deaza-PMEA)$^-$ species carry the proton at the phosphonate residue and 20% at N1. The H-1 NMR data indicate that PMEA and its deaza analogues occur to some extent in an orientation similar to the anti conformation of 5'-AMP^{2-} (i.e., the phosphonate group is close to H8) (Blindauer *et al.*, 1997a).

Characteristics of metal complexes of the antiviral purine ANPs have been compared with the behavior of antivirally inactive ANPs, with a pyrimidine (cytosine) base replacing adenine. Comparison of PMEC complexes of Mg^{2+}, Ca^{2+}, Sr^{2+}, Ba^{2+}, Mn^{2+}, Co^{2+}, Ni^{2+}, Cu^{2+}, Zn^{2+}, and Cd^{2+} with nucleobase-free compound (phosphonomethoxy)ethane, PME, and the parent nucleotides cytidine 5'-monophosphate and 2'-deoxycytidine 5'-monophosphate shows that the metal ion-binding properties of $PMEC^{2-}$ resemble closely those of PME^{2-}: The primary binding site is the phosphonate group, and with all of the metal ions studied, 5-membered chelates involving the ether oxygen of the $-CH_2-O-CH_2-PO_3^{2-}$ chain are formed. No interaction occurs with the cytosine residue. However, the monoprotonated $M(HPMEC)^+$ as well as the $M(HCMP)^+$ and $M(dCMP)^+$

species carry the metal ion predominantly at the nucleobase, while the proton is at the phosph(on)ate group. The coordinating properties of $PMEC^{2-}$ and CMP^{2-} or $dCMP^{2-}$ differ only with respect to the possible formation of the 5-membered chelates involving the ether oxygen in the M(PMEC) species, a possibility which does not exist in the complexes of the parent nucleotides (Blindauer *et al.*, 1999b).

In the carba analogue of PMEA, 9-(4-phosphonobutyl)adenine (lacking the ether oxygen atom), the primary binding site for metal ion in complexes with Mg^{2+}, Ca^{2+}, Sr^{2+}, Ba^{2+}, Mn^{2+}, Co^{2+}, Ni^{2+}, Cu^{2+}, Zn^{2+} or Cd^{2+} is the phosphonate group; in most instances, the stability is solely determined by the basicity of the phosphonate residue. Additional interaction of the phosphonate-coordinated M^{2+} occurs most probably with N7; $dPMEA^{2-}$ is more similar in its metal ion-binding properties to the parent nucleotide AMP than to the dianion of 9-[2-(phosphonomethoxy)ethyl]adenine ($PMEA^{2-}$) (Gomez-Coca *et al.*, 2000).

PMEA is able to mimic the structure of AMP^{2-} in the reactive metal complex intermediate, $[Cu\text{-}3(ATP)(AMP)(OH)]^-$ formed in promoting the Cu^{2+}-catalyzed dephosphorylation of ATP. In this process, PMEA is about twice as effective as AMP (Sigel *et al.*, 1998b).

9.   Conclusion

This review summarizes the results of the vast effort which has been invested during the past 15 years in the field of acyclic nucleoside phosphonates. Many of the expectations have proven to be true. It is hoped that additional new areas of ANPs in the field of chemotherapy, in particular new therapeutic applications, will add to their utility in the future and warrant further progress in their chemical, biochemical and biological research.

10.   References

Aduma, P, Connelly, MC, Srinivas, RV, Fridland, A (1995) Metabolic diversity and antiviral activities of acyclic nucleoside phosphonates. Mol. Pharmacol.; 47:816-822

Alexander, P and Holý, A (1993) General method for the preparation of N-(3-hydroxy-2-phosphonomethoxypropyl) derivatives of heterocyclic bases. Collect. Czech. Chem. Commun.; 58:1151-1163

Alexander, P, Masojídková, M, Holý, A (1994) Synthesis of 9-(2-Phosphinomethoxyethyl)adenine and related compounds. Collect. Czech. Chem. Commun.; 59:1870-1878

Alexander, P, Masojídková, M, Holý, A (1994) Synthesis of 9-(2-Phosphonomethoxyethyl)adenine Esters as Potential Prodrugs. Collect. Czech. Chem. Commun.; 59:1853-1869

Alexander, P, Krishnamurthy, VV, Prisbe, EJ (1996) Synthesis and antiviral activity of pyranosylphosphonic acid nucleotide analogues. J. Med. Chem.; 39:1321-1330

Andrei, G, Snoeck, R, Reymen, D, Liesnard, C, Goubau, P, Desmyter, J *et al.* (1995) Comparative activity of selected antiviral compounds against clinical isolates of varicella-zoster virus. Eur. J. Clin. Microbiol. Infect. Dis.; 14:318-329

Andrei, G, Snoeck, R, Piette, J, Delvenne, P, De Clercq, E (1998a) Antiproliferative effects of acyclic nucleoside phosphonates on human papillomavirus (HPV)-harboring cell lines compared with HPV-negative cell lines. Oncol. Res.; 10:523-531

Andrei, G, Snoeck, R, Piette, J, Delvenne, P, De Clercq, E (1998b) Inhibiting effects of cidofovir (HPMPC) on the growth of the human cervical carcinoma (SiHa) xenografts in athymic nude mice. Oncol. Res.; 10:533-539

Annaert, P, Kinget, R, Naesens, L, De Clercq, E, Augustijns, P (1997) Transport, uptake, and metabolism of the bis(pivaloyloxymethyl)-ester prodrug of 9-(2-phosphonylmethoxyethyl)adenine in an in vitro cell culture system of the intestinal mucosa (Caco-2). Pharm. Res.; 14:492-496

Annaert, P, Vangelder, J, Naesens, L, De Clercq, E, Vandenmooter, G, Kinget, R *et al.* (1998) Carrier mechanisms involved in the transepithelial transport of bis(POM)-PMEA and its metabolites across Caco-2 monolayers. Pharm. Res.; 15:1168-1173

Arimilli, MN, Kim, CU, Dougherty, J, Mulato, A, Oliyai, R, Shaw, JP *et al.* (1997) Synthesis, in vitro biological evaluation and oral bioavailability of 9-[2-(phosphonomethoxy)propyl]adenine (PMPA) prodrugs. Antivir. Chem. Chemother.; 8:557-564

Arimilli, MN, Dougherty, J, Cundy, KC, Bischofberger, N (1999) Orally bioavailable acyclic nucleoside phosphonate prodrugs: adefovir dipivoxil and bis(POC)PMPA. Adv. Antiviral Drug Res.; 3:69-91

Baba, M, Mori, S, Shigeta, S, De Clercq, E (1987) Selective inhibitory effect of (S)-9-(3-hydroxy-2-phosphonylmethoxypropyl)adenine and 2'-nor-cyclic GMP on adenovirus replication in vitro. Antimicrob. Agents Chemother.; 31:337-339

Bagnis, C, Izzdine, H, Deray, G (1999) Renal tolerance of cidofovir. Therapie; 54:689-691

Balzarini, J, Naesens, L, Herdewijn, P, Rosenberg, I, Holy, A, Pauwels, R *et al.* (1989) Marked in vivo antiretrovirus activity of 9-(2-phosphonylmethoxyethyl)adenine, a selective anti-human immunodeficiency virus agent. Proc. Natl. Acad. Sci. U. S. A.; 86:332-336

Balzarini, J, and De Clercq, E (1991a) 5-Phosphoribosyl 1-pyrophosphate synthetase converts the acyclic nucleoside phosphonates 9-(3-hydroxy-2-phosphonylmethoxypropyl)adenine and 9-(2-phosphonylmethoxyethyl)adenine directly to their antivirally active diphosphate derivatives. J. Biol. Chem.; 266:8686-8689

Balzarini, J, Holý, A, Jindřich, J, Dvořáková, H, Hao, Z, Snoeck, R *et al.* (1991b) 9-[(2RS)-3-fluoro-2-phosphonylmethoxypropyl] derivatives of purines: a class of highly selective antiretroviral agents in vitro and in vivo. Proc. Natl. Acad. Sci. U. S. A.; 88:4961-4965

Balzarini, J, Hao, Z, Herdewijn, P, Johns, DG, De Clercq, E (1991c) Intracellular metabolism and mechanism of anti-retrovirus action of 9-(2-phosphonylmethoxyethyl)adenine, a potent anti-human immunodeficiency virus compound. Proc. Natl. Acad. Sci. U. S. A.; 88 :1499-1503

Balzarini, J, Holý, A, Jindřich, J, Naesens, L, Snoeck, R, Schols, D *et al.* (1993) Differential antiherpesvirus and antiretrovirus effects of the (S) and (R) enantiomers of acyclic nucleoside phosphonates: potent and selective in vitro and in vivo antiretrovirus activities of (R)-9-(2-phosphonomethoxypropyl)-2, 6-diaminopurine. Antimicrob. Agents Chemother.; 37:332-338

Balzarini, J, Kruining, J, Heijtink, R, De Clercq, E (1994) Comparative anti-retrovirus and anti-hepadnavirus activity of three different classes of nucleoside phophonate derivatives. Antivir. Chem. Chemother.; 5:360-365

Balzarini, J, Aquaro, S, Perno, CF, Witwrouw, M, Holý, A, De Clercq, E (1996) Activity of the (R)-enantiomers of 9-(2-phosphonylmethoxypropyl)adenine and 9-(2-phosphonylmethoxypropyl)-2, 6-diaminopurine against human immunodeficiency virus in different human cell systems. Biochem. Biophys. Res. Commun.; 219: 337-341

Balzarini, J, Vahlenkamp, T, Egberink, HF, Hartmann, K, Witwrouw, M, Pannecouque, C *et al.* (1997) Antiretroviral activities of acyclic nucleoside phosphonates [9-(2-phosphonylmethoxyethyl)adenine, 9-(2-phosphonylmethoxyethyl)guanine, (R)-9-(2-phosphonylmethoxypropyl)adenine, and MDL 74, 968] in cell cultures and murine sarcoma virus-infected newborn NMRI mice. Antimicrob. Agents Chemother.; 41:611-616

Balzarini, J, Naesens, L, De Clercq, E (1998a) New antivirals - mechanism of action and resistance development. Curr. Opin. Microbiol.; 1:535-546

Balzarini, J, Hatse, S, Naesens, L, De Clercq, E (1998b) Selection and characterisation of murine leukaemia L1210 cells with high-level resistance to the cytostatic activity of the acyclic nucleoside phosphonate 9-(2-phosphonylmethoxyethyl) adenine (PMEA). Biochim. Biophys. Acta.; 1402:29-38

Barnard, DL, Bischofberger, N, Kim, CU, Huffman, JH, Sidwell, RW, Dougherty, JR *et al.* (1997) Acyclic phosphonomethylether nucleoside inhibitors of respiratory viruses. Antivir. Chem. Chemother.; 8:223-233

Bednarski, K, Dixit, DM, Mansour, TS, Colman, SG, Walcott, SM, Ashman, C (1995) Synthesis of racemic 2-phosphonomethyl-1, 3-dioxolane nucleoside analogues as potential antiviral agents. Bioorg. Medicinal. Chem. Letter.; 5:1741-1744

Benzaria, S, Pelicano, H, Johnson, R, Maury, G, Imbach, JL, Aubertin, AM *et al.* (1996) Synthesis, in vitro antiviral evaluation, and stability studies of bis(S-acyl-2-thioethyl) ester derivatives of 9-[2-(phosphonomethoxy)ethyl]adenine (PMEA) as potential PMEA prodrugs with improved oral bioavailability. J. Med. Chem.; 39:4958-4965

Besen, G, Flores-Aguilar, M, Assil, KK, Kuppermann, BD, Gangan, P, Pursley, M *et al.* (1995) Long-term therapy for herpes retinitis in an animal model with high-concentrated liposome-encapsulated HPMPC. Arch. Ophthalmol.; 113: 661-668

Blindauer, CA, Holý, A, Dvořáková, H, Sigel, H (1997a) Solution properties of antiviral adenine-nucleotide analogues. The acid-base properties of 9-[2-(phosphonomethoxy)ethyl]adenine (PMEA) and of its N1, N3 and N7 deaza derivatives in aqueous solution. J.Chem. Soc. Perkin Trans.2.;2353-2363

Blindauer, CA, Emwas, AH, Holý, A, Dvořáková, H, Sletten, E, Sigel, H (1997b) Complex formation of the antiviral 9-[2-(phosphonomethoxy)ethyl]adenine (PMEA) and of its N1, N3, and N7 deaza derivatives with copper(II) in aqueous solution. Chemistry-Europ. J..; 3:1526-1536

Blindauer, CA, Holý, A, Dvořáková, H, Sigel, H (1998) Magnesium complexes of the antiviral 9-[2-(phosphonomethoxy)ethyl]adenine (PMEA) and of its 1-, 3-, and 7-deaza analogues in aqueous solution. J. Biol. Inorg. Chem.; 3:423-433

Blindauer, CA, Sjastad, TI, Holý, A, Sletten, E, Sigel, H (1999a) Aspects of the co-ordination chemistry of the antiviral nucleotide analogue, 9-[2-(phosphonomethoxy)ethyl]-2, 6-diaminopurine (PMEDAP). J. Chem. Soc.Dalton Trans.;3661-3671

Blindauer, CA, Holý, A, Sigel, H (1999b) Metal ion-binding properties of the nucleotide analogue 1-[2-(phosphonomethoxy)ethyl]cytosine (PMEC) in aqueous solution. Coll. Czech. Chem. Commun.; 64:613-632

Bilá, V, Otová, B, Jelinek, R, Sladká, M, Mejsnarová, B, Holý, A *et al.* (1993) Antimitotic and teratogenic effects of acyclic nucleotide analogues 1-(S)-(3-hydroxy-2-phosphonomethoxypropyl)cytosine (HPMPC) and 9-(2-phosphonomethoxyethyl) adenine (PMEA). Folia Biol. Prague.; 39:150-161

Birkuš, G, Kramata, P, Votruba, I, Otová, B, Otmar, M, Holý, A (1998) Nonproteolyzed form of DNA-polymerase epsilon from T-cell spontaneous lymphoma of Sprague-Dawley inbred rat: Isolation and characterization. Collection of Czechoslovak Chemical Communications.; 63:723-731

Birkuš, G, Votruba, I, Holý, A, Otová, B (1999) 9-[2-(Phosphonomethoxy)ethyl]adenine diphosphate

(PMEApp) as a substrate toward replicative DNA polymerases alpha, delta, epsilon, and epsilon*. Biochem. Pharmacol.; 58:487-492

Bischofberger, N, Hitchcock, MJ, Chen, MS, Barkhimer, DB, Cundy, KC, Kent, KM *et al.* (1994) 1-[((S)-2-hydroxy-2-oxo-1, 4, 2-dioxaphosphorinan-5-yl)methyl]cytosine, an intracellular prodrug for (S)-1-(3-hydroxy-2-phosphonylmethoxypropyl)cytosine with improved therapeutic index in vivo. Antimicrob. Agents Chemother.; 38:2387-2391

Bobková, K, Otová, B, Marinov, I, Mandys, V, Panczak, A, Votruba, I *et al.* (2000) Anticancer effect of PMEDAP - Monitoring of apoptosis. Anticancer Res.; 20:1041-1047

Bray, M, Martinez, M, Smee, DF, Kefauver, D, Thompson, E, Huggins, JW (2000) Cidofovir protects mice against lethal aerosol or intranasal cowpox virus challenge. J. Infect. Dis.; 181:10-19

Brodfuehrer, PR, Howell, HG, Sapino, C, Vemishetti, P (1994) A practical synthesis of (S)-HPMPC. Tetrahedron Lett.; 35:3243-3246

Bronson, JJ, Ghazzouli, I, Hitchcock, MJ, Webb, RR, Martin, JC (1989a) Synthesis and antiviral activity of the nucleotide analogue (S)-1-[3-hydroxy-2-(phosphonylmethoxy)propyl]cytosine. J. Med. Chem.; 32:1457-1463

Bronson, JJ, Kim, CU, Ghazzouli, I, Hitchcock, MJM, Kern, ER, Martin, JC (1989b) Synthesis and antiviral activity of phosphonylmethoxyethyl derivatives of purine and pyrimidine bases. 72 –87. In Martin JC, editor. Nucleotide Analogues as Antiviral Agents. ACS Symposium Series # 401. Washington, D.C., American Chemical Society.

Bronson, JJ, Ferrara, LM, Martin, JC, Mansuri, MM (1992) Synthesis and Biological Activity of Carbocyclic Derivatives of the Potent Antiviral Agent 9-[2-(Phosphonomethoxy)ethyl]guanine (PMEG). Bioorg. Medicinal. Chem. Letter.; 2:685-690

Calio, R, Villani, N, Balestra, E, Sesa, F, Holý, A, Balzarini, J *et al.* (1994) Enhancement of natural killer activity and interferon induction by different acyclic nucleoside phosphonates. Antiviral Res.; 23:77-89

Casara, PJ, Altenburger, J-M, Taylor, DL, Tyms, AS, Kenny, M, Nave, JF (1995) Synthesis and antiviral activity of rigid acyclonucleotide analogs. Bioorg. Medicinal. Chem. Letter.; 5:1275-1280

Chen, W, Flavin, MT, Filler, R, Xu, ZQ (1996) An improved synthesis of 9-[2-(diethoxyphosphono-methoxy)ethyl]adenine and its analogues with other purine bases utilizing the Mitsunobu reaction. Nucleos. Nucleot.; 15:1771-1778

Cihlar, T and Chen, MS (1997) Incorporation of selected nucleoside phosphonates and anti-human immunodeficiency virus nucleotide analogues into DNA by human DNA polymerases alpha, beta and gamma. Antivir. Chem. Chemother.; 8:187-195

Cihlář, T, Votruba, I, Horská, K, Liboska, R, Rosenberg, I, Holý, A (1992) Metabolism of 1-(S)-(3-hydroxy-2-phosphonomethoxypropyl)cytosine (HPMPC) in Human Embryonic Lung Cells. Collect. Czech. Chem. Commun.; 57:661-672

Cihlář, T, Rosenberg, I, Votruba, I, Holý, A (1995) Transport of 9-(2-phosphonomethoxyethyl)adenine across plasma membrane of HeLa S3 cells is protein mediated. Antimicrob. Agents Chemother.; 39: 117-124

Cihlář, T and Chen, MS (1996) Identification of enzymes catalyzing two-step phosphorylation of cidofovir and the effect of cytomegalovirus infection on their activities in host cells. Mol. Pharmacol.; 50:1502-1510

Coe, DM, Garofalo, A, Roberts, SM, Storer, R, Thorpe, AJ (1994) Synthesis of the phosphonate isostere of carbocyclic 5-bromovinyldeoxyuridine monophosphate. J. Chem. Soc. Perkin Trans. 1.; 1994:3061-3063

Colledge, D, Civitico, G, Locarnini, S, Shaw, T (2000) In vitro antihepadnaviral activities of combinations of penciclovir, lamivudine, and adefovir. Antimicrob. Agents Chemother.; 44:551-560

Compton, ML, Toole, JJ, Paborsky, LR (1999) 9-(2-Phosphonylmethoxyethyl)-N6-cyclopropyl-2, 6-diamino-purine (cpr-PMEDAP) as a prodrug of 9-(2-phosphonylmethoxyethyl)guanine (PMEG). Biochem. Pharmacol.; 58:709-714

Connelly, MC, Robbins, BL, Fridland, A (1993) Mechanism of uptake of the phosphonate analog (S)-1-(3-hydroxy-2-phosphonylmethoxypropyl)cytosine (HPMPC) in Vero cells. Biochem. Pharmacol.; 46:1053-1057

Cundy, KC, Fishback, JA, Shaw, JP, Lee, ML, Soike, KF, Visor, GC *et al.* (1994a) Oral bioavailability of the antiretroviral agent 9-(2-phosphonylmethoxyethyl)adenine (PMEA) from three formulations of the prodrug bis(pivaloyloxymethyl)-PMEA in fasted male cynomolgus monkeys. Pharm. Res.; 11:839-843

Cundy, KC, Shaw, JP, Lee, WA (1994b) Oral, subcutaneous, and intramuscular bioavailabilities of the antiviral nucleotide analog 9-(2-phosphonylmethoxyethyl) adenine in cynomolgus monkeys. Antimicrob. Agents Chemother.; 38:365-368.

Cundy, KC, Barditch-Crovo, P, Walker, RE, Collier, AC, Ebeling, D, Toole, J *et al.* (1995) Clinical pharmacokinetics of adefovir in human immunodeficiency virus type 1-infected patients. Antimicrob. Agents Chemother.; 39:2401-2405

Cundy, KC, Sue, IL, Visor, GC, Marshburn, J, Nakamura, C, Lee, WA *et al.* (1997) Oral formulations of adefovir dipivoxil: in vitro dissolution and in vivo bioavailability in dogs. J. Pharm. Sci.; 86:1334-1338

Cundy, KC, Sueoka, C, Lynch, GR, Griffin, L, Lee, WA, Shaw, JP (1998) Pharmacokinetics and bioavailability of the anti-human immunodeficiency virus nucleotide analog 9-[(R)-2-(phosphono-methoxy)propyl]adenine (PMPA) in dogs. Antimicrob. Agents Chemother.; 42:687-690

Cundy, KC (1999) Clinical pharmacokinetics of the antiviral nucleotide analogues cidofovir and adefovir. Clin. Pharmacokinet.; 36:127-143

Cvekl A., Horská, K, Šebesta, K, Rosenberg, I, Holý, A (1989) Phosphonate analogues of dinucleotides as substrates for DNA-dependent RNA polymerase from E.coli in primed abortive initiation reaction. Int. J. Biol. Macromol.; 11:33-38

Černý, J, Votruba, I, Vonka, V, Rosenberg, I, Otmar, M, Holý, A (1990) Phosphonylmethyl ethers of acyclic nucleoside analogues: inhibitors of HSV-1 induced ribonucleotide reductase. Antiviral Res.; 13:253-264

Česnek, M, Hocek, M, Holý, A (1999) Cross-coupling reactions of 2-amino-6-halopurine derivatives with organometallic reagents leading to 6-alkylated purine acyclic nucleotide analogues. Collection Symposium Series.; 2:255-256

Česnek, M, Hocek, M, Holý, A (2000) Synthesis of acyclic nucleotide analogues derived from 2-amino-6-C-substituted purines via cross-coupling reactions of 2-amino-9-[2-(diisopropoxyphosphorylmethoxy)ethyl]-6-halopurines with diverse organometallic reagents. Collect. Czech. Chem. Commun.; in press

Dang, Q, Liu, Y, Erion, MD (1998) A new regio-defined synthesis of PMEA. Nucleos. Nucleot.; 17:1445-1451

Dang, Q, Brown, BS, Vanpoelje, PD, Colby, TJ, Erion, MD (1999) Synthesis of phosphonate 3-phthalidyl esters as prodrugs for potential intracellular delivery of phosphonates. Bioorg. Medicinal. Chem. Letter.; 9:1505-1510

Davies, EG, Thrasher, A, Lacey, K, Harper, J (1999) Topical cidofovir for severe molluscum contagiosum. Lancet.; 353:2042-2042

De Clercq E., Descamps J., De Somer P., Holý A.: (S)-9-(2,3-Dihydroxypropyl)adenine: an aliphatic nucleoside analog with broad-spectrum antiviral activity, Science *200*, 563-564 (1978).

De Clercq E., Holý A., Rosenberg I., Sakuma T., Balzarini J., Maudgal P.C. (1986) A novel selective broad-spectrum anti-DNA virus agent, Nature *323*, 464-467

De Clercq, E, Sakuma, T, Baba, M, Pauwels, R, Balzarini, J, Rosenberg, I *et al.* (1987) Antiviral activity of phosphonylmethoxyalkyl derivatives of purine and pyrimidines. Antiviral Res.; 8:261-272

De Clercq, E (1996) Therapeutic potential of Cidofovir (HPMPC, Vistide) for the treatment of DNA virus (i.e. herpes-, papova-, pox- and adenovirus) infections. Verh. K. Acad. Geneeskd. Belg.; 58:19-47

De Clercq, E (1997) Acyclic nucleoside phosphonates in the chemotherapy of DNA virus and retrovirus infections. Intervirology.; 40:295-303

De Clercq, (1998) Towards an effective chemotherapy of virus infections: Therapeutic potential of cidofovir [(S)-1-[3-hydroxy-2-(phosphonomethoxy)propyl]cytosine, HPMPC] for the treatment of DNA virus infections. Collect. Czech. Chem. Commun.; 63:480-506

De Clercq, E (1999) Perspectives for the treatment of hepatitis B virus infections. Int. J. Antimicrob. Agents; 12:81-95

de la Fuente, R, Awan, AR, Field, HJ (1992) The acyclic nucleoside analogue penciclovir is a potent inhibitor of equine herpesvirus type 1 (EHV-1) in tissue culture and in a murine model. Antiviral Res.; 18:77-89

de Vries, E, Stam, JG, Franssen, FF, Nieuwenhuijs, H, Chavalitshewinkoon, P, De Clercq, E *et al.* (1991) Inhibition of the growth of Plasmodium falciparum and Plasmodium berghei by the DNA polymerase inhibitor HPMPA. Mol. Biochem. Parasitol.; 47:43-50

de Vrueh, RLA, Rump, ET, Biessen, EAL, Balzarini, J, van Berkel, TJC, Bijsterbosch, MK (1997) Lipophilic derivatization of the anti-viral drug 9-(2-phosphonylmethoxyethyl)adenine and its incorporation into a lactosylated lipid carrier to improve its liver uptake. Nucleos. Nucleot.; 16:1279-1282

de Vrueh, RL, Rump, ET, Sliedregt, LA, Biessen, EA, van Berkel, TJ, Bijsterbosch, MK (1999) Synthesis of a lipophilic prodrug of 9-(2-phosphonylmethoxyethyl)adenine (PMEA) and its incorporation into a hepatocyte-specific lipidic carrier. Pharm. Res.; 16:1179-1185

Deeks, SG, Barditch-Crovo, P, Lietman, PS, Hwang, F, Cundy, KC, Rooney, JF *et al.* (1998) Safety, pharmacokinetics, and antiretroviral activity of intravenous 9-[2-(R)-(phosphonomethoxy) propyl]adenine, a novel anti-human immunodeficiency virus (HIV) therapy, in HIV-infected adults. Antimicrob. Agents Chemother.; 42:2380-2384

Deeks, SG, Collier, A, Lalezari, J, Pavia, A, Rodrigue, D, Drew, WL *et al.* (1997) The safety and efficacy of adefovir dipivoxil, a novel anti-human immunodeficiency virus (HIV) therapy, in HIV-infected adults: a randomized, double-blind, placebo-controlled trial. J. Infect. Dis.; 176:1517-1523

Del Gobbo, V, Foli, A, Balzarini, J, De Clercq, E, Balestra, E, Villani, N *et al.* (1991) Immunomodulatory activity of 9-(2-phosphonylmethoxyethyl)adenine (PMEA), a potent anti-HIV nucleotide analogue, on in vivo murine models. Antiviral Res.; 16:65-75

Drake, AF, Garofalo, A, Hillman, JML, Merlo, V, McCague, R, Roberts, SM (1996) Preparation of carbocyclic analogues of 2'-deoxyribonucleotides possessing a phosphonate substituent at the 5'-position. J. Chem. Soc. Perkin Trans. 1.;2739-2745

Duckworth, DM, Harnden, MR, Perkins, RM, Planterose, DN (1991) 9-[2-(phosphonomethoxy)alkoxy]purines, a new series of antiviral acyclonucleotides. Nucleosides & Nucleotides; 10:427-430

Düzgünes, N, Pretzer, E, Simtes, S, Slepushkin, V, Konopka, K, Flasher, D *et al.* (1999) Liposome-mediated delivery of antiviral agents to human immunodeficiency virus-infected cells. Mol. Membr. Biol.; 16:111-118

Dvořáková, H, Holý, A, Snoeck, R, Balzarini, J, De Clercq, E (1990) Acyclic Nucleoside and Nucleotide Analogues Derived from 1-Deaza and 3-Deazaadenine. Collect. Czech. Chem. Commun. Special Issue No. 1.; 55:113-116

Dvořáková, H, Holý, A, Alexander, P (1993) Synthesis and Biological Effects of 9-(3-Hydroxy-2-phosphonomethoxypropyl) Derivatives of Deazapurine Bases. Collect. Czech. Chem. Commun.; 58:1403-1418

Dvořáková, H and Holý, A (1993) Synthesis and Biological Effects of N-(2-Phosphonomethoxyethyl) Derivatives of Deazapurine Bases. Collect. Czech. Chem. Commun.; 58:1419-1429

Dvořáková, H, Holý, A, Rosenberg, I (1994) Synthesis of some 2'-C-alkyl derivatives of 9-(2-phosphono-methoxyethyl)adenine and related compounds. Collect. Czech. Chem. Commun.; 59:2069-2094

Dvořáková, H, Masojídková, M, Holý, A, Balzarini, J, Andrei, G, Snoeck, R *et al.* (1996) Synthesis of 2'-aminomethyl derivatives of N-(2-(phosphonomethoxy)ethyl) nucleotide analogues as potential antiviral agents. J. Med. Chem.; 39:3263-3268

Dvořáková, H, Dvořák, D, Holý, A (1998) Synthesis of acyclic nucleotide analogues derived from 6-(*sec*- or *tert*-alkyl)purines via coupling of 6-chloropurine derivatives with organocuprates. Collect. Czech. Chem. Commun.; 63:2065-2074

Eckstein, F (1985) Nucleoside phosphorothioates. Ann. Rev. Biochem.; 54:367-402

Eisenberg, EJ, Lynch, GR, Bidgood, AM, Krishnamurty, K, Cundy, KC (1998) Isolation and identification of a metabolite of cidofovir from rat kidney. J. Pharm. Biomed. Anal.; 16:1349-1356

Elliott, RD, Rener, GA, Riordan, JM, Secrist, JA, Bennett, LL, Parker, WB *et al.* (1994) Phosphonate Analogs of Carbocyclic Nucleotides. J. Med. Chem.; 37:739-744

Fife, K, Gill, J, Bourboulia, D, Gazzard, B, Nelson, M, Bower, M (1999) Cidofovir for the treatment of Kaposi's sarcoma in an HIV-negative homosexual man. Brit. J. Dermatol.; 141:1148-1150

Franěk, F, Holý, A, Votruba, I, Eckschlager, T (1999) Acyclic nucleotide analogues suppress growth and induce apoptosis in human leukemia cell lines. Int. J. Oncol.; 14:745-752

Franchetti, P, Sheikha, GA, Cappellacci, L, Messina, L, Grifantini, M, Loi, AG *et al.* (1994) 8-Aza-analogues of PMEA and PMEG: Synthesis and in vitro anti-HIV activity. Nucleos. Nucleot.; 13: 1707-171

Franchetti, P, Abu-Sheikha, G, Cappellacci, L, Grifantini, M, Demontis, A, Piras, G *et al.* (1995) Synthesis and antiviral activity of 8-aza analogs of chiral [2-(phosphonomethoxy)propyl]guanines. J. Med. Chem.; 38:4007-4013

Franchetti, P, Abu-Sheikha, G, Perlini, P, Farhat, F, Grifantini, M, Perra, G *et al.* (1997) Synthesis of unsaturated analogues of 9-[2-(phosphonomethoxy)propyl]guanine as antiviral agents. Nucleos. Nucleot.; 16:1921-1932

Garcia, CR, Torriani, FJ, Freeman, WR (1998) Cidofovir in the treatment of cytomegalovirus (CMV) retinitis. Ocul. Immunol. Inflamm.; 6:195-203

Gil-Fernandez, C, Garcia-Villalón, D, De Clercq, E, Rosenberg, I, Holý, A (1987) Phosphonylmethoxy-alkylpurines and -pyrimidines as inhibitors of African swine fever virus replication in vitro. Antiviral Res.; 8:273-281

Gomez-Coca, RB, Kapinos, LE, Holý, A, Vilaplana, RA, Gonzalez-Vilchez, F, Sigel, H (2000) Metal ion-binding properties of 9-(4-phosphonobutyl)adenine (dPMEA), a sister compound of the antiviral nucleotide analogue 9-[2-(phosphonomethoxy)ethyl]adenine (PMEA), and quantification of the equilibria involving four Cu(PMEA) isomers. J.Chem. Soc. Dalton Trans. 2077-20

Gordon, YJ, Romanowski, E, Araullo-Cruz, T, Seaberg, L, Erzurum, S, Tolman, R *et al.* (1991) Inhibitory effect of (S)-HPMPC, (S)-HPMPA, and 2'-nor-cyclic GMP on clinical ocular adenoviral isolates is serotype-dependent in vitro. Antiviral Res.; 16:11-16

Gordon, YJ, Naesens, L, De Clercq, E, Maudgal, PC, Veckeneer, M (1996) Treatment of adenoviral conjunctivitis with topical cidofovir. Cornea; 15:546-546

Gulyaev, NN and Holý, A (1972) Ribonucleoside 5'-aminomethanephosphonates: Synthesis and affinity towards some phosphomonoesterases. FEBS Letters.; 22:294-296

Hampton, A, Sasaki, T, Perini, F, Slotin, LA and Kappler, F (1976) Design of Substrate-Site-directed irreversible inhibitors of adenosine 5'-phosphate aminohydrolase. Effect of substrate substituents on affinity for the substrate site. J. Med. Chem.; 19: 1029-1033

Happe, S, Besselmann, M, Matheja, P, Rickert, CH, Schuierer, G, Reichelt, D *et al.* (1999) Cidofovir (Vistide) in therapy of progressive multifocal leukoencephalopathy in AIDS. Review of the literature and report of 2 cases. Nervenarzt; 70:935-943

Happe, S, Lunenborg, N, Rickert, CH, Heese, C, Reichelt, D, Schuierer, G *et al.* (2000) Progressive multifocal leukoencephalopathy in AIDS. Overview and retrospective analysis of 17 patients. Nervenarzt.; 71:96-104

Harnden, MR, Jennings, LJ, Parkin, A (1991) Synthesis of 1-(Phosphonomethoxyalkoxy)pyrimidines, a Novel Series of Acyclonucleotide Analogues. Synthesis-Stuttgart;947-950

Harnden, MR and Jarvest, RL (1992) Synthesis of 9-[2-(Phosphonomethoxy)ethylamino]adenine and 9-[(Phosphonomethoxy)acetamido]adenine, Analogues of a Potent Anti-HIV Acyclonucleotide. Bioorg. Med. Chem. Lett.; 2:1559-1560

Harnden, MR and Jennings, LJ (1993) Antiviral 9-[2-(phosphonomethylthio)alkoxy]purines. Antivir. Chem. Chemother.; 4:215-227

Hartmann, K, Balzarini, J, Higgins, J, De Clercq, E, Pedersen, NC (1994) In vitro Activity of Acyclic Nucleoside Phosphonate Derivatives Against Feline Immunodeficiency Virus in Crandell Feline Kidney Cells and Feline Peripheral Blood Lymphocytes. Antivir. Chem. Chemother.; 5:13-19

Hartmann, K, Kuffer, M, Balzarini, J, Naesens, L, Goldberg, M, Erfle, V *et al.* (1998) Efficacy of the acyclic nucleoside phosphonates (S)-9-(3-fluoro-2-phosphonylmethoxypropyl)adenine (FPMPA) and 9-(2-phosphonylmethoxyethyl)adenine (PMEA) against feline immunodeficiency virus. J. Acquir. Immune. Defic. Syndr. Hum. Retrovirol.; 17:120-128

Hatse, S, Balzarini, J, De Clercq, E (1995) Induction of erythroid differentiation of human leukemia K562 cells by the acyclic nucleoside phosphonate 9-(2-phosphonylmethoxyethyl)adenine (PMEA). Nucleos. Nucleot.; 14:649-652

Hatse, S, De Clercq, E, Balzarini, J (1996) Evidence for distinction of the differentiation-inducing activities and cytostatic properties of 9-(2-phosphonylmethoxyethyl)adenine and a variety of differentiation-inducing agents in human erythroleukemia K562 cells. Mol. Pharmacol.; 50:1231-1242

Hatse, S, Naesens, L, Degreve, B, Segers, C, Vandeputte, M, Waer, M *et al.* (1998a) Potent antitumor activity of the acyclic nucleoside phosphonate 9-(2-phosphonylmethoxyethyl)adenine in choriocarcinoma-bearing rats. Int. J. Cancer.; 76:595-600

Hatse, S, Naesens, L, Degreve, B, Vandeputte, M, Waer, M, Clercq, ED *et al.* (1998b) In vitro and in vivo inhibitory activity of the differentiation-inducing agent 9-(2-phosphonylmethoxyethyl)adenine (PMEA) against rat choriocarcinoma. Adv. Exp. Med. Biol.; 431:605-609

Hatse, S, De Clercq, E, Balzarini, J (1999a) Impact of 9-(2-phosphonylmethoxyethyl)adenine on (deoxy)ribonucleotide metabolism and nucleic acid synthesis in tumor cells. FEBS Lett.; 445:92-97

Hatse, S, Naesens, L, De Clercq, E, Balzarini, J (1999b) N-6-cyclopropyl-PMEDAP: A novel derivative of 9-(2-phosphonylmethoxyethyl)2-6-diaminopurine (PMEDAP) with distinct metabolic, antiproliferative, and differentiation-inducing properties. Biochem. Pharmacol.; 58:311-323

Hatse, S, Schols, D, De Clercq, E, Balzarini, J (1999c) 9-(2-Phosphonylmethoxyethyl)adenine induces tumor cell differentiation or cell death by blocking cell cycle progression through the S phase. Cell Growth Differ.; 10:435-446

Haynes, UJ, Swigor, JE, Bronson, JJ (1991) Synthesis of (S)-1-[3-hydroxy-2-(phosphonylmethoxy)propyl]-[2-C-14]cytosine ((S)-[C-14] HPMPC). J.Labelled Comp. Radiopharm.; 29:1217-1221

Heijtink, RA, Kruining, J, de Wilde, GA, Balzarini, J, De Clercq, E, Schalm, SW (1994) Inhibitory effects

39:4073-4088

Holý, A (1998a) Simple method for cleavage of phosphonic acid diesters to monoesters. Synthesis-Stuttgart.;381-385

Holý, A, Buděšínský, M, Podlaha, J, Císařová, I (1998b) Synthesis of N1-[2-(phosphonomethoxy)ethyl] derivatives of 2, 4-diaminopyrimidine and related acyclic nucleotide analogs. Collect. Czech. Chem. Commun.; 64:242-256

Holý, A, Günter, J, Dvořáková, H, Masojídková, M, Andrei, G, Snoeck, R *et al.* (1999) Structure-antiviral activity relationship in the series of pyrimidine and purine N-[2-(2-phosphonomethoxy)ethyl] nucleotide analogues. 1. Derivatives substituted at the carbon atoms of the base. J. Med. Chem.; 42:2064-2086

Holý, A, Votruba, I, Tlouaeová, E, Masojídková, M (2001) Synthesis and cytostatic activity of N-[2-(phosphonomethoxy)alkyl] derivatives of N^6-substituted adenines, -2,6-diaminopurines and related compounds. Collect. Czech. Chem. Commun.; 66:1545-1592.

Horská, K, Rosenberg, I, Holý, A, Šebesta, K (1983) 5'-O-diphosphorylphosphonylmethylribonucleosides -a new group of DNA-dependent RNA polymerase inhibitors. Collect. Czech. Chem. Commun.; 48:1352-1357

Horská, K, Cvekl A., Rosenberg, I, Holý, A, Šebesta, K (1990) DNA-dependent RNA polymerase mediated formation of methylphosphonyl internucleotide linkage. Collect. Czech. Chem. Commun.; 55:2100-2109

Jähne, G, Muller, A, Kroha, H, Rosner, M, Holzhauser, O, Meichsner, C *et al.* (1992) Preparation of Carbocyclic Phosphonate Nucleosides. Tetrahedron Lett.; 33:5335-5338

Janeba, Z and Holý, A (1996) Intramolecular cyclization of some (S)-9-(3-hydroxy-2-phosphono-methoxypropyl)adenine derivatives. Collect. Czech. Chem. Commun.; Special Issue 61:S116-S117

Janeba, Z, Holý, A, Masojidková, M (2000) Synthesis of acyclic nucleoside and nucleotide analogs derived from 6-amino-7H-purin-8(9H)-one, Collect. Czech. Chem. Commun.; 65:1126-1144.

Jasko, MV and Tarussova, NB (1993) New possibilities of synthesis of 5'-O-phosphonylmethyl derivatives of nucleosides. Collect. Czech. Chem. Commun. Special Issue No. 1.; 58:105-106

Javaly, K, Wohlfeiler, M, Kalayjian, R, Klein, T, Bryson, Y, Grafford, K *et al.* (1999) Treatment of mucocutaneous herpes simplex virus infections unresponsive to acyclovir with topical foscarnet cream in AIDS patients: A phase I/II study. J. Acquir. Immune. Defic. Syndr. Hum. Retrovirol.; 21:301-306

Jie L., Van Aerschot A., Balzarini J. *et al.* (1990) 5'-O-phosphonomethyl-2', 3'-dideoxynucleosides: synthesis and anti-HIV activity, J Med Chem *33*, 2481-2487

Jurovčík, M and Holý, A (1976) Metabolism of pyrimidine L-nucleosides. Nucleic Acids Res.; 3:2143-2154

Jindřich, J, Holý, A, Dvořáková, H (1993) Synthesis of N-(3-Fluoro-2-phosphonomethoxypropyl) (FPMP) Derivatives of Heterocyclic Bases. Collect. Czech. Chem. Commun.; 58:1645-1667

Johnson, JA and Gangemi, JD (1999) Selective inhibition of human papillomavirus-induced cell proliferation by (S)-1-[3-hydroxy-2-(phosphonylmethoxy)propyl]cytosine. Antimicrob. Agents Chemother.; 43:1198-1205

Kaminsky, R, Zweygarth, E, De Clercq, E (1994) Antitrypanosomal activity of phosphonylmethoxy-alkylpurines. J. Parasitol.; 80:1026-1030

Kaminsky, R, Schmid, C, Grether, Y, Holý, A, De Clercq, E, Naesens, L *et al.* (1996) (S)-9-(3-hydroxy-2-phosphonylmethoxypropyl)adenine [(S)-HPMPA]: a purine analogue with trypanocidal activity in vitro and in vivo. Trop. Med. Int. Health.; 1:255-263

Kaminsky, R, Nickel, B, Holý, A (1998) Arrest of Trypanosoma brucei rhodesiense and T-brucei brucei in the S-phase of the cell cycle by (S)-9-(3-hydroxy-2-phosphonylmethoxypropyl)adenine ((S)-HPMPA). Mol. Biochem. Parasitol.; 93:91-100

Kim, CU, Luh, BY, Martin, JC (1990a) Synthesis and antiviral activity of (S)-9-[4-hydroxy-3-(phosphono-methoxy)butyl]guanine. J. Med. Chem.; 33:1797-1800

Kim, CU, Luh, BY, Misco, PF, Bronson, JJ, Hitchcock, MJ, Ghazzouli, I *et al.* (1990b) Acyclic purine phosphonate analogues as antiviral agents. Synthesis and structure-activity relationships. J. Med. Chem.; 33:1207-1213

Kim, CU, Luh, BY, Misco, PF, Martin, JC (1991a) Phosphonate isosteres of 2',3'-didehydro-2,3-dideoxynucleoside monophosphates - synthesis and anti-HIV activity. Nucleosides & Nucleotides; 10:371-375

Kim, CU, Luh, BY, Misco, PF, Martin, JC (1991b) Phosphonate isosteres of acyclovir and ganciclovir monophosphates - synthesis and anti-herpesvirus activity. Nucleosides & Nucleotides.; 10:483-485

Kim, CU, Misco, PF, Luh, BY, Hitchcock, MJ, Ghazzouli, I, Martin, JC (1991c) A new class of acyclic phosphonate nucleotide analogues: phosphonate isosteres of acyclovir and ganciclovir monophosphates as antiviral agents. J. Med. Chem.; 34 :2286-2294

Kim, CU, Luh, BY, Martin, JC (1991d) Regiospecific and highly stereoselective electrophilic addition to furanoid glycals - synthesis of phosphonate nucleotide analogs with potent activity against HIV. J. Org. Chem.; 56:2642-2647

Kim, CU, Bronson, JJ, Ferrara, LM, Martin, JC (1992a) Synthesis and HIV activity of phosphonate isosteres of D4T monophosphate. Bioorg. Med. Chem. Lett.; 2:367-370

Kim, CU, Luh, BY, Martin, JC (1992b) A Novel Synthesis of 1-Oxa-HPMPA - A Potent Antiviral Agent Against Herpesviruses. Tetrahedron Lett.; 33:25-28

Kim, CU, Luh, BY, Martin, JC (1992c) Synthesis and Anti-HIV Activity of 9-[(2R, 5R)-2, 5-dihydro-5-(phosphonomethoxy)-2-furanyl]-2, 6-diaminopurine. Bioorg. Medicinal. Chem. Letter.; 2:307-310

Kim, DK, Kim, YW, Kim, KH (1994) Synthesis and anti-HCMV activity of 9-[(ethoxyhydroxyphosphinyl)methoxy]methoxy]guanine, an isostere of PMEG monoethyl ester. Bioorg. Medicinal. Chem. Letter.; 4:2241-2244

Kim, DK, Gam, J, Kim, KH (1996) Synthesis of 9-[(2-hydroxy-1-phosphonylethoxy)ethyl]guanine, 1-[(2-hydroxy-1-phosphonylethoxy)ethyl]cytosine and 9-[(2-hydroxy-1-phosphonylethoxy)ethyl] adenine. J. Heterocycl. Chem.; 33:1865-1870

Kramata, P, Černý, J, Birkuš, G, Votruba, I, Otová, B, Holý, A (1995) DNA-polymerases alpha, delta and epsilon from T-cell spontaneous lymphoblastic leukemia of Sprague-Dawley inbred rat: Isolation and characterization. Collect. Czech. Chem. Commun.; 60:1555-1572

Kramata, P, Votruba, I, Otová, B, Holý, A (1996) Different inhibitory potencies of acyclic phosphonomethoxy-alkyl nucleotide analogs toward DNA polymerases alpha, delta and epsilon. Mol. Pharmacol.; 49:1005-1011

Kramata, P, Downey, KM, Paborsky, LR (1998) Incorporation and excision of 9-(2-phosphonyl-methoxyethyl)guanine (PMEG) by DNA polymerase delta and epsilon in vitro. J. Biol. Chem.; 273: 21966-21971

Kramata, P and Downey, KM (1999) 9-(2-phosphonylmethoxyethyl) derivatives of purine nucleotide analogs: A comparison of their metabolism and interaction with cellular DNA synthesis. Mol. Pharmacol.; 56:1262-1270

Krejčová, R, Horská, K, Votruba, I, Holý, A (1996) Phosphorylation of phosphonomethoxyalkyl guanines by GMP-kinase isoenzymes. Collect. Czech. Chem. Commun.; 61:S134-S136

Krejčová, R, Horská, K, Votruba, I, Holý, A (1999) Isoenzymes of GMP kinase from L1210 cells: isolation and characterisation. Collect. Czech. Chem. Commun.; 64:559-570

Krejčová, R, Horská, K, Votruba, I, Holý, A (2000a) Phosphorylation of purine (phosphonomethoxy)alkyl derivatives by mitochondrial AMP kinase (AK2 type) from L1210 cells. Collect. Czech. Chem. Commun.; 65:1653-1668.

Krejčová, R, Horská, K, Votruba, I, Holý, A (2000b) Interaction of Guanine Phosphonomethoxyalkyl Derivatives with GMP Kinase Isoenzymes. Biochem. Pharmacol.; 60:1907-1913.

Křen, V, Otová, B, Elleder, D, Elleder, M, Panczak, A, Holý, A (1993) Prevention of acute graft-versus-host disease in rats using 9-(2-phosphonomethoxyethyl) adenine (PMEA). Folia Biol. Prague.; 39:78-86

Kuppermann, BD, Assil, KK, Vuong, C, Besen, G, Wiley, CA, De Clercq, E *et al.* (1996) Liposome-encapsulated (S)-1-(3-hydroxy-2-phosphonylmethoxypropyl)cytosine for long-acting therapy of viral retinitis. J. Infect. Dis.; 173:18-23

La Colla, P, Holý, A, Jindřich, J, Dvořáková, H, Gelli, L, Marongiu, ME *et al.* (1991) Comparative antiviral activity of HPMPA, PMEA, HPMPG, HPMPC and structurally related compounds against African swine fever virus. Antiviral Res.; 15 Suppl.1:66

Lacy, SA, Hitchcock, MJ, Lee, WA, Tellier, P, Cundy, KC (1998) Effect of oral probenecid coadministration on the chronic toxicity and pharmacokinetics of intravenous cidofovir in cynomolgus monkeys. Toxicol. Sci.; 44:97-106

Lalezari, J, Schacker, T, Feinberg, J, Gathe, J, Lee, S, Cheung, T *et al.* (1997) A randomized, double-blind, placebo-controlled trial of cidofovir gel for the treatment of acyclovir-unresponsive mucocutaneous herpes simplex virus infection in patients with AIDS. J. Infect. Dis.; 176:892-898

Liboska, R, Masojídková, M, Rosenberg, I (1996a) Carbocyclic phosphonate-based nucleotide analogs related to PMEA.1. Racemic trans-configured derivatives. Collect. Czech. Chem. Commun.; 61:313-332

Liboska, R, Masojídková, M, Rosenberg, I (1996b) Carbocyclic phosphonate-based nucleotide analogs related to PMEA.2. Racemic cis-configured derivatives. Collect. Czech. Chem. Commun.; 61:778-790

Liekens, S, Andrei, G, Vandeputte, M, De Clercq, E, Neyts, J (1998) Potent inhibition of hemangioma formation in rats by the acyclic nucleoside phosphonate analogue cidofovir. Cancer Res.; 58:2562-2567

Meerbach, A, Neyts, J, Holý, A, De Clercq, E (1996) Inhibitory effect of 2-phosphonomethoxyalkyl derivatives of N6-substituted 6-aminopurines on Epstein-Barr virus (EBV). Antiviral Res.; 30:A51

Meerbach, A, Holý, A, Wutzler, P, De Clercq, E, Neyts, J (1998) Inhibitory effects of novel nucleoside and nucleotide analogues on Epstein-Barr virus replication. Antivir. Chem. Chemother.; 9:275-282

Mendel, DB, Cihlar, T, Moon, K, Chen, MS (1997) Conversion of 1-[(((S)-2-hydroxy-2-oxo-1,4,2-dioxa-phosphorinan-5-yl)methyl]cytosine to cidofovir by an intracellular cyclic CMP phosphodiesterase. Antimicrob. Agents Chemother.; 41:641-646

Merta, A, Veselý, J, Votruba, I, Rosenberg, I, Holý, A (1990a) Phosphorylation of acyclic nucleotide analogs HPMPA and PMEA in L1210 mouse leukemia cell extracts. Neoplasma; 37:111-120

Merta, A, Votruba, I, Rosenberg, I, Otmar, M, Hřebabecký, H, Bernaerts, R *et al.* (1990b) Inhibition of herpes simplex virus DNA polymerase by diphosphates of acyclic phosphonylmethoxyalkyl nucleotide analogues. Antiviral Res.; 13:209-218

Merta, A, Votruba, I, Jindřich, J, Holý, A, Cihlář, T, Rosenberg, I *et al.* (1992) Phosphorylation of 9-(2-phosphonomethoxyethyl)adenine and 9-(S)-(3-hydroxy-2-phosphonomethoxypropyl)adenine by AMP(dAMP) kinase from L1210 cells. Biochem. Pharmacol.; 44:2067-2077

Midoux, P, Negre, E, Roche, AC, Mayer, R, Monsigny, M, Balzarini, J *et al.* (1990) Drug targeting: anti-HSV-1 activity of mannosylated polymer-bound 9-(2-phosphonylmethoxyethyl)adenine. Biochem. Biophys. Res. Commun.; 167:1044-1049

Mulato, AS and Cherrington, JM (1997) Anti-HIV activity of adefovir (PMEA) and PMPA in combination with antiretroviral compounds: in vitro analyses. Antiviral Res.; 36:91-97

Naesens, L, Balzarini, J, Rosenberg, I, Holý, A, De Clercq, E (1989) 9-(2-Phosphonylmethoxyethyl)-2,6-diaminopurine (PMEDAP): a novel agent with anti-human immunodeficiency virus activity in vitro and potent anti-Moloney murine sarcoma virus activity in vivo. Eur. J. Clin. Microbiol. Infect. Dis.;

8:1043-1047

Naesens, L, Balzarini, J, De Clercq, E (1991) Single-dose administration of 9-(2-phosphonylmethoxy-ethyl)adenine (PMEA) and 9-(2-phosphonylmethoxyethyl)-2,6-diaminopurine (PMEDAP) in the prophylaxis of retrovirus infection in vivo. Antiviral Res.; 16:53-64

Naesens, L, Balzarini, J, De Clercq, E (1992) Acyclic adenine nucleoside phosphonates in plasma determined by high-performance liquid chromatography with fluorescence detection. Clin. Chem.; 38:480-485

Naesens, L, Neyts, J, Balzarini, J, Holý, A, Rosenberg, I, De Clercq, E (1993) Efficacy of oral 9-(2-phosphonylmethoxyethyl)-2, 6-diaminopurine (PMEDAP) in the treatment of retrovirus and cytomegalo-virus infections in mice. J. Med. Virol.; 39:167-172

Naesens, L, Balzarini, J, De Clercq, E (1994) Therapeutic potential of PMEA as an antiviral drug. Rev. Med. Virol.; 4:147-159

Naesens, L, Neyts, J, Balzarini, J, Bischofberger, N, De Clercq, E (1995) In vivo antiretroviral efficacy of oral bis(POM)-PMEA, the bis(pivaloyloxymethyl)prodrug of 9-(2-phosphonylmethoxyethyl)adenine (PMEA). Nucleos. Nucleot.; 14:767-770

Naesens, L, Balzarini, J, Bischofberger, N, De Clercq, E (1996) Antiretroviral activity and pharmacokinetics in mice of oral bis(pivaloyloxymethyl)-9-(2-phosphonylmethoxyethyl)adenine, the bis(pivaloyloxymethyl) ester prodrug of 9-(2-phosphonylmethoxyethyl)adenine. Antimicrob. Agents Chemother.; 40:22-28

Naesens, L, Snoeck, R, Andrei, G, Balzarini, J, Neyts, J, De Clercq, E (1997a) HPMPC (cidofovir), PMEA (adefovir) and related acyclic nucleoside phosphonate analogues: A review of their pharmacology and clinical potential in the treatment of viral infections. Antivir. Chem. Chemother.; 8:1-23

Naesens, L and De Clercq, E (1997b) Therapeutic potential of HPMPC (cidofovir), PMEA (adefovir) and related acyclic nucleoside phosphonate analogues as broad-spectrum antiviral agents. Nucleos. Nucleot.; 16:983-992

Naesens, L, Bischofberger, N, Augustijns, P, Annaert, P, Vandenmooter, G, Arimilli, MN *et al.* (1998) Antiretroviral efficacy and pharmacokinetics of oral bis(isopropyloxycarbonyloxymethyl) 9-(2-phosphonyl-methoxypropyl)adenine in mice. Antimicrob. Agents Chemother.; 42:1568-1573

Neyts, J, Stals, F, Bruggeman, C, De Clercq, E (1993) Activity of the anti-HIV agent 9-(2-phosphonyl-methoxyethyl)-2, 6-diaminopurine against cytomegalovirus in vitro and in vivo. Eur. J. Clin. Microbiol. Infect. Dis.; 12:437-446

Neyts, J and De Clercq, E (1997) Antiviral drug susceptibility of human herpesvirus 8. Antimicrob. Agents Chemother.; 41:2754-2756

Neyts, J, Sadler, R, De Clercq, E, Raab-Traub, N, Pagano, JS (1998) The antiviral agent cidofovir [(S)-1-(3-hydroxy-2-phosphonylmethoxypropyl)cytosine] has pronounced activity against nasopharyngeal carcinoma grown in nude mice. Cancer Res.; 58:384-388

Nguyen-Ba, P, Lee, N, Mitchell, H, Chan, L, Quimpére, M (1998a) Design and synthesis of a novel class of nucleotide analogs with anti-HCMV activity. Bioorg. Med. Chem. Lett.; 8: 3555-3560

Nguyen-Ba, P, Turcotte, N, Yuen, L, Bedard, J, QuimpPre, M, Chan, L (1998b) Identification of novel nucleotide phosphonate analogs with potent anti-HCMV activity. Bioorg. Med. Chem. Lett.; 8:3561-3566

Nicoll, A and Locarnini, S (1997) Review: Present and future directions in the treatment of chronic hepatitis B infection. J. Gastroenterol. Hepatol.; 12:843-854

Noble, S and Goa, KL (1999) Adefovir dipivoxil. Drugs; 58:479-487

Norbeck, DW, Sham, HL, Rosenbrook, W, Herrin, T, Plattner, JJ (1992) Synthesis and Activities of the Phosphonate Isostere of the Monophosphate of (+/-)Cyclobut-G. Nucleos. Nucleot.; 11: 1383-1391

Oliyai, R, Lee, WA, Visor, GC, Yuan, LC (1999a) Enhanced chemical stability of the intracellular prodrug,

1-[((S)-2-hydroxy-2-oxo-1, 4, 2-dioxaphosphorinan-5-yl)methyl]cytosine, relative to its parent compound, cidofovir. Int. J. Pharm.; 179:257-265.

Oliyai, R, Shaw, JP, Sueoka-Lennen, CM, Cundy, KC, Arimilli, MN, Jones, RJ *et al.* (1999b) Aryl ester prodrugs of cyclic HPMPC. I: Physicochemical characterization and in vitro biological stability. Pharm. Res.; 16:1687-1693

Olšanská, L, Cihlář, T, Votruba, I, Holý, A (1997) Transport of adefovir (PMEA) in human T-lymphoblastoid cells. Collect. Czech. Chem. Commun.; 62:821-828.

Orlando, G, Fasolo, MM, Beretta, R, Signori, R, Adriani, B, Zanchetta, N *et al.* (1999) Intralesional or topical cidofovir (HPMPC, Vistide) for the treatment of recurrent genital warts in HIV-1-infected patients [letter]. AIDS.; 13:1978-1980

Otmar, M, Rosenberg, I, Masojídková, M, Holý, A (1993a) [5-(Adenin-9-yl)-5-deoxypentofuranosyl]phosphonates - A Novel Type of Nucleotide Analogs Related to HPMPA.1. Derivatives with L-arabino, D-arabino, 2-Deoxy-L-erythro and 2-Deoxy-L-threo Configuration. Collect. Czech. Chem. Commun.; 58:2159-2179

Otmar, M, Rosenberg, I, Masojídková, M, Holý, A (1993b) [5-(Adenin-9-yl)-5-deoxy-L-pentofuranosyl]phosphonates - A Novel Type of Nucleotide Analogs Related to HPMPA.2. Synthesis of L-ribo and L-xylo Configurated Derivatives by Recyclization of Diethyl (5-RS)-[1,2-O-isopropylidene-5-O-ethanesulfonyl-D-pentofuranos-5-C-yl]phosphonates Under Acidic Conditions. Collect. Czech. Chem. Commun.; 58:2180-2196

Otmar, M, Votruba, I, Holý, A (1999) An alternative synthesis of HPMPC and HPMPA diphosphoryl derivatives. Collection Symposium Series; 2:252-254. XI.Symposium on Chemistry of Nucleic Acid Components, Špindlerùv Mlýn, September 4-9, 1999

Otmar, M, Masojídková, M, Votruba, I, Holý, A (2001) An alternative synthesis of HPMPC and HPMPA diphosphoryl derivatives. Collect. Czech. Chem. Commun.; 66:500-506.

Otová, B, Sladká, M, Votruba, I, Holý, A, Křen, V (1993a) Cytostatic effect of 9-(2-phosphonomethoxyethyl) adenine (PMEA). I. Lymphatic leukemia KHP-Lw-I in Lewis rats. Folia Biol. Prague; 39:136-141

Otová, B, Sladká, M, Blažek, K, Schramlová, J, Votruba, I, Holý, A (1993b) Cytostatic effect of 9-(2-phosphonomethoxyethyl) adenine (PMEA). II. Lymphoblastic leukemia in Sprague-Dawley rats. Folia Biol. Prague; 39:142-149

Otová, B, Křenová, D, Zídek, Z, Holý, A, Votruba, I, Křen, V (1993c) Cytostatic effect of 9-(2-phosphonomethoxyethyl) adenine (PMEA). III. Rat and mouse carcinomas and sarcomas. Folia Biol. Prague; 39:311-314

Otová, B, Zidek, Z, Holý, A, Votruba, I, Franková, D (1994) Immunomodulatory properties of 9-(2-phosphonomethoxyethyl)adenine (PMEA). Folia Biol. Prague; 40:185-192

Otová, B, Holý, A, Votruba, I, Sladká, M, Bilá, V, Mejsnarová, B *et al.* (1997a) Genotoxicity of purine acyclic nucleotide analogs. Folia Biol. Prague; 43:225-229

Otová, B, Zídek, Z, Holý, A, Votruba, I, Sladká, M, Marinov, I *et al.* (1997b) Antitumor activity of novel purine acyclic nucleotide analogs PMEA and PMEDAP. In Vivo; 11:163-167

Otová, B, Francová, K, Franěk, F, Koutník, P, Votruba, I, Holý, A *et al.* (1999) 9-[2-(phosphono-methoxy)ethyl]-2, 6-diaminopurine (PMEDAP) - a potential drug against hematological malignancies - induces apoptosis. Anticancer Res.; 19:3173-3182

Palu, G, Stefanelli, S, Rassu, M, Parolin, C, Balzarini, J, De Clercq, E (1991) Cellular uptake of phosphonylmethoxyalkylpurine derivatives. Antiviral Res.; 16:115-119

Pauwels, R, Balzarini, J, Schols, D, Baba, M, Desmyter, P, Rosenberg, I, Holy, A, De Clercq, E. (1988) Phosphonylmethoxyethyl purine derivatives: A new class of anti-HIV agents. Antimicrob. Agents Chemother.; 32:1025-1030

Pauwels, R and Schols, D (1991) Development of new potent and selective agents against HIV (human immunodeficiency virus). Verh. K. Acad. Geneeskd. Belg.; 53:629-655

Pérez-Pérez, MJ, Doboszewski, B, De Clercq, E, Herdewijn, P (1995a) Phosphonates derivatives of 2', 3'-dideoxy-2', 3'-didehydropentopyranosyl nucleosides. Nucleos. Nucleot.; 14:707-710

Pérez-Pérez, MJ, Doboszewski, B, Rozenski, J, Herdewijn, P (1995b) Stereocontrolled synthesis of phosphonate derivatives of tetrahydro- and dihydro-2H-pyranyl nucleosides: the selectivity of the Ferrier rearrangement. Tetrahedron: Assymetry; 6:973-984

Pérez-Pérez, MJ, Rozenski, J, Busson, R, Herdewijn, P (1995c) Application of the Mitsunobu-type condensation reaction to the synthesis of phosphonate derivatives of cyclohexenyl and cyclohexanyl nucleosides. J. Org.Chem.; 60:1531-1537

Perkins, RM, Duckworth, DM, Elphick, L, Duckworth, DM, Harnden, MR, Kenig, MD *et al.* (1992) Antiviral activity of 9-[2-(phosphonomethoxy)ethoxy]purines against HIV, FIV and Visna virus. Antiviral Res.; 17 Suppl.1:59

Perrillo, R, Schiff, E, Yoshida, E, Statler, A, Hirsch, K, Wright, T *et al.* (2000) Adefovir dipivoxil for the treatment of lamivudine-resistant hepatitis B mutants. Hepatology; 32:129-134

Pisarev, VM, Lee, SH, Connelly, MC, Fridland, A (1997) Intracellular metabolism and action of acyclic nucleoside phosphonates on DNA replication. Mol. Pharmacol.; 52:63-68

Plosker, GL and Noble, S (1999) Cidofovir: a review of its use in cytomegalovirus retinitis in patients with AIDS. Drugs; 58:325-345

Pransky, SM, Magit, AE, Kearns, DB, Kang, DR, Duncan, NO (1999) Intralesional cidofovir for recurrent respiratory papillomatosis in children. Arch. Otolar. Head Neck Surgery; 125:1143-1148

Rejman, D and Rosenberg, I (1996) Synthesis of 2'-alkoxy and 2'-aryloxy derivatives of 9-(2-phosphono-methoxyethyl)adenine (PMEA) and related compounds. Collect. Czech. Chem. Commun.; Special Issue 61:S122-S123

Reymen, D, Naesens, L, Balzarini, J, Holý, A, Dvořáková, H, De Clercq, E (1995) Antiviral activity of selected acyclic nucleoside analogues against human herpesvirus 6. Antiviral Res.; 28:343-357

Ribaud, P, Scieux, C, Freymuth, F, Morinet, F, Gluckman, E (1999) Successful treatment of adenovirus disease with intravenous cidofovir in an unrelated stem-cell transplant recipient. Clin. Infect. Dis.; 28:690-691

Rizzetto, M (1999) Therapy of chronic viral hepatitis: a critical view. Ital. J Gastroenterol. Hepatol.; 31:781-793

Robbins, BL, Connelly, MC, Marshall, DR, Srinivas, RV, Fridland, A (1995a) A human T lymphoid cell variant resistant to the acyclic nucleoside phosphonate 9-(2-phosphonylmethoxyethyl)adenine shows a unique combination of a phosphorylation defect and increased efflux of the agent. Mol. Pharmacol.; 47:391-397

Robbins, BL, Greenhaw, J, Connelly, MC, Fridland, A (1995b) Metabolic pathways for activation of the antiviral agent 9-(2-phosphonylmethoxyethyl)adenine in human lymphoid cells. Antimicrob. Agents Chemother.; 39:2304-2308

Robbins, BL, Srinivas, RV, Kim, C, Bischofberger, N, Fridland, A (1998) Anti-human immunodeficiency virus activity and cellular metabolism of a potential prodrug of the acyclic nucleoside phosphonate 9-R-(2-phosphonomethoxypropyl)adenine (PMPA), Bis(isopropyloxymethylcarbonyl)PMPA. Antimicrob. Agents Chemother.; 42:612 –617

Romanowski, EG and Gordon, YJ (2000) Efficacy of topical cidofovir on multiple adenoviral serotypes in the New Zealand rabbit ocular model. Invest. Ophthalmol. Visual Sci.; 41:460-463

Rose, WC, Crosswell, AR, Bronson, JJ, Martin, JC (1990) In vivo antitumor activity of 9-[(2-

phosphonylmethoxy)ethyl]guanine and related phosphonate nucleotide analogues. J. Natl. Cancer Inst.; 82:510-512

Rosenberg, I and Holý, A (1982) Preparation of 5'-O-phosphonylmethyl analogues of nucleoside 5'-phosphates, 5-diphosphates and 5'-triphosphates. Collect. Czech. Chem. Commun.; 47:3447-3463

Rosenberg, I and Holý, A (1983) Preparation of 2'(3')-O-phosphonylmethyl derivatives of ribonucleosides. Collect. Czech. Chem. Commun.; 48:778-789

Rosenberg, I and Holý, A (1987) Synthesis of potential prodrugs and metabolites of 9-(S)-(3-hydroxy-2-phosphonylmethoxypropyl)adenine. Collect. Czech. Chem. Commun.; 52:2792-2800

Rosenberg, I, Holý, A, Masojídková, M (1988) Phosphonylmethoxyalkyl and phosphonylalkyl derivatives of adenine. Collect. Czech. Chem. Commun.; 53:2753-2777

Rosenberg, and Králíková, S (1996) A simple synthetic route to the preparation of 2-(1-phosphonoalkoxy)ethyl derivatives of heterocyclic bases as novel nucleotide analogs related to PMEA. Collect. Czech. Chem. Commun.; Special Issue 61:S81-S84

Roučka, D, Thesis. Faculty of Biological Sciences, University of South Bohemia Ceské Budějovice

Russell, JW, Marrero, D, Whiterock, VJ, Klunk, LJ, Starrett, JE (1991) Determination of 9-[(2-phosphonylmethoxy)ethyl]adenine in rat urine by high-performance liquid chromatography with fluorescence detection. J. Chromatogr.; 572:321-326

Safrin, S, Cherrington, J, Jaffe, HS (1999) Cidofovir. Review of current and potential clinical uses. Adv. Exp. Med. Biol.; 458:111-120

Scheit, KH (1980) Nucleotide Analogs. Synthesis and Biological Function. New York: Wiley-Interscience.

Shulman, NS, Zolopa, AR, Passaro, DJ, Murlidharan, U, Israelski, DM, Brosgart, CL *et al.* (2000) Efavirenz- and adefovir dipivoxil-based salvage therapy in highly treatment-experienced patients: Clinical and genotypic predictors of virologic response. J. Acq. Immune Defic. Syndr.; 23:221-226

Schuetz, JD, Connelly, MC, Sun, D, Paibir, SG, Flynn, PM, Srinivas, RV *et al.* (1999) MRP4: A previously unidentified factor in resistance to nucleoside-based antiviral drugs. Nat. Med.; 5:1048-1051

Schultze, LM, Chapman, HH, Dubree, NJP, Jones, RJ, Kent, KM, Lee, TT *et al.* (1998) Practical synthesis of the anti-HIV drug, PMPA. Tetrahedron Lett.; 39:1853-1856

Sigel, H, Chen, D, Corfu, NA, Gregáň, F, Holý, A, Strašák, M (1992) Metal-ion-coordinating properties of various phosphonate derivatives, including 9-[2-(phosphonylmethoxy)ethyl]adenine (PMEA) - an adenosine-monophosphate (AMP) analog) with antiviral properties. Helv. Chim. Acta; 75:2634-2656

Sigel, H, Blindauer, CA, Holý, A, Dvořáková, H (1998b) Hydrolysis of nucleoside phosphates - Part 12 - Facilitation of the copper(II)-promoted dephosphorylation of adenosine 5'-triphosphate (ATP(4-)) by the antiviral nucleotide analogue, 9-[2-(phosphonomethoxy)ethyl]adenine (PMEA). Chem. Commun.;1219-1220

Sigel, H, (1999) Metal ion-binding properties of the antiviral nucleotide analogue 9-[2-(phosphono-methoxy)ethyl]adenine (PMEA). Why is its diphosphorylated form, PMEApp(4-), initially a better substrate for nucleic acid polymerases than (2'-deoxy)adenosine 5'-triphosphate (dATP^{4-}/ATP^{4-})? Pure Appl. Chem.; 71:1727-1740

Silvera, P, Racz, P, Racz, K, Bischofberger, N, Crabbs, C, Yalley-Ogunro, J *et al.* (2000) Effect of PMPA and PMEA on the kinetics of viral load in simian immunodeficiency virus-infected macaques. AIDS Res. Hum. Retroviruses; 16:791-800

Simonart, T, Noel, JC, Dedobbeleer, G, Parent, D, Van Vooren, JP, De Clercq, E *et al.* (1998) Treatment of classical Kaposi's sarcoma with intralesional injections of cidofovir: Report of a case. J. Med. Virol.; 55:215-218

Smeijsters, LJ, Zijlstra, NM, de Vries, E, Franssen, FF, Janse, CJ, Overdulve, JP (1994) The effect of

(S)-9-(3-hydroxy-2-phosphonylmethoxypropyl) adenine on nuclear and organellar DNA synthesis in erythrocytic schizogony in malaria. Mol. Biochem. Parasitol.; 67:115-124

Smeijsters, LJ, Nieuwenhuijs, H, Hermsen, RC, Dorrestein, GM, Franssen, FF, Overdulve, JP (1996) Antimalarial and toxic effects of the acyclic nucleoside phosphonate (S)-9-(3-hydroxy-2-phosphonyl-methoxypropyl)adenine in Plasmodium berghei-infected mice. Antimicrob. Agents Chemother.; 40:1584-1588

Smeijsters, LJ, Franssen, FF, Naesens, L, de Vries, E, Holý, A, Balzarini, J *et al.* (1999) Inhibition of the in vitro growth of Plasmodium falciparum by acyclic nucleoside phosphonates. Int. J. Antimicrob. Agents; 12:53-61

Snoeck, R, Sakuma, T, De Clercq, E, Rosenberg, I, Holy, A (1988) (S)-1-(3-hydroxy-2-phosphonylmethoxy-propyl)cytosine, a potent and selective inhibitor of human cytomegalovirus replication. Antimicrob. Agents Chemother.; 32:1839-1844

Snoeck, R, Andrei, G, Gerard, M, Silverman, A, Hedderman, A, Balzarini, J *et al.* (1994) Successful treatment of progressive mucocutaneous infection due to acyclovir- and foscarnet-resistant herpes simplex virus with (S)-1-(3-hydroxy-2-phosphonylmethoxypropyl)cytosine (HPMPC). Clin. Infect. Dis.; 18:570-578

Snoeck, R, Andrei, G, Holý, A, De Clercq, E (1997) Activity of N6-substituted adenine and 2, 6-diaminopurine acyclic nucleoside phosphonates against human cytomegalovirus (CMV) and other herpesviruses. 6th International Cytomegalovirus Workshop, Orange Beach, Alabama, USA. March 5-9, 1997. Proceedings

Snoeck, R, Wellens, W, Desloovere, C, Van Ranst, M, Naesens, L, De Clercq, E *et al.* (1998) Treatment of severe laryngeal papillomatosis with intralesional injections of cidofovir [(S)-1-(3-hydroxy-2-phosphonylmethoxypropyl)cytosine]. J. Med. Virol.; 54:219-225

Song, B, Holý, A, Sigel, H (1994) Metal ion-coordinating properties of the antiviral (S)-9-[3-hydroxy-2-(phosphonomethoxy)propyl]adenine (HPMPA) in aqueous solution. Gazz. Chim. Ital.; 124:387-392

Sparidans, RW, Veldkamp, A, Hoetelmans, RMW, Beijnen, JH (1999) Improved and simplified liquid chromatographic assay for adefovir, a novel antiviral drug, in human plasma using derivatization with chloroacetaldehyde. J. Chromatogr. B.; 736:115-121

Srinivas, V and Fridland, A (1998) Antiviral activities of 9-R-2-phosphonomethoxypropyl adenine (PMPA) and bis(isopropyloxymethylcarbonyl)PMPA against various drug-resistant human immunodeficiency virus strains. Antimicrob. Agents Chemother.; 42:1484-1487

Starrett, JE, Jr., Tortolani, DR, Hitchcock, MJ, Martin, JC, Mansuri, MM (1992) Synthesis and in vitro evaluation of a phosphonate prodrug: bis(pivaloyloxymethyl) 9-(2-phosphonylmethoxyethyl)adenine. Antiviral Res.; 19:267-273

Starrett, JE, Jr., Tortolani, DR, Russell, J, Hitchcock, MJ, Whiterock, V, Martin, JC *et al.* (1994) Synthesis, oral bioavailability determination, and in vitro evaluation of prodrugs of the antiviral agent 9-[2-(phosphonomethoxy)ethyl]adenine (PMEA). J. Med. Chem.; 37:1857-1864

Suo, Z and Johnson, KA (1998) Selective Inhibition of HIV-1 Reverse Transcriptase by an Antiviral Inhibitor, (R)-9-(2-Phosphonylmethoxypropyl)adenine. J. Biol. Chem.; 273:27250-27258

Suruga, Y, Makino, M, Okada, Y, Tanaka, H, De Clercq, E, Baba, M (1998) Prevention of murine AIDS development by (R)-9-(2-phosphonylmethoxypropyl)adenine. J. Acquir. Immune. Defic. Syndr. Hum. Retrovirol.; 18:316-322

Tarantal, AF, Marthas, ML, Shaw, JP, Cundy, K, Bischofberger, N (1999) Administration of 9-[2-(R)-(phosphonomethoxy)propyl]adenine (PMPA) to gravid and infant rhesus macaques (Macaca mulatta): safety and efficacy studies. J. Acquir. Immune Defic. Syndr. Hum. Retrovirol.; 20:323-333

Thormar, H, Balzarini, J, Holý, A, Jindřich, J, Rosenberg, I, Debyser, Z *et al.* (1993) Inhibition of visna virus replication by 2', 3'-dideoxynucleosides and acyclic nucleoside phosphonate analogs. Antimicrob. Agents Chemother.; 37:2540-2544

Thormar, H, Georgsson, G, Palsson, PA, Balzarini, J, Naesens, L, Torsteinsdottir, S *et al.* (1995) Inhibitory effect of 9-(2-phosphonylmethoxyethyl)adenine on visna virus infection in lambs: a model for in vivo testing of candidate anti-human immunodeficiency virus drugs. Proc. Natl. Acad. Sci. U.S.A.; 92:3283-3287

Tsai, CC, Follis, KE, Sabo, A, Beck, TW, Grant, RF, Bischofberger, N *et al.* (1995a) Prevention of SIV infection in macaques by (R)-9-(2-phosphonylmethoxypropyl)adenine. Science; 270:1197-1199

Tsai, CC, Follis, KE, Sabo, A, Grant, R, Bischofberger, N (1995b) Efficacy of 9-(2-phosphonyl-methoxyethyl)adenine treatment against chronic simian immunodeficiency virus infection in macaques. J. Infect. Dis.; 171:1338-1343

Tsai, CC, Follis, KE, Beck, TW, Sabo, A, Bischofberger, N, Dailey, PJ (1997) Effects of (R)-9-(2-phosphonylmethoxypropyl) adenine monotherapy on chronic SIV infection in macaques. AIDS Res. Hum. Retroviruses; 13:707-712

Tsai, CC, Emau, P, Follis, KE, Beck, TW, Benveniste, RE, Bischofberger, N *et al.* (1998) Effectiveness of postinoculation (R)-9-(2-Phosphonylmethoxypropyl) adenine treatment for prevention of persistent simian immunodeficiency virus SIVmne infection depends critically on timing of initiation and duration of treatment. J. Virol.; 72:4265-4273

Van Cutsem, E, Snoeck, R, Van Ranst, M, Fiten, P, Opdenakker, G, Geboes, K *et al.* (1995) Successful treatment of a squamous papilloma of the hypopharynx-esophagus by local injections of (S)-1-(3-hydroxy-2-phosphonylmethoxypropyl)cytosine. J. Med. Virol.; 45:230-235

Van Rompay, KK, Cherrington, JM, Marthas, ML, Berardi, CJ, Mulato, AS, Spinner, A *et al.* (1996) 9-[2-(Phosphonomethoxy)propyl]adenine therapy of established simian immunodeficiency virus infection in infant rhesus macaques. Antimicrob. Agents Chemother.; 40:2586-2591

Van Rompay, KK, Marthas, ML, Lifson, JD, Berardi, CJ, Vasquez, GM, Agatep, E *et al.* (1998) Administration of 9-[2-(phosphonomethoxy)propyl]adenine (PMPA) for prevention of perinatal simian immunodeficiency virus infection in rhesus macaques. AIDS Res. Hum. Retroviruses; 14:761-773

Van Rompay, KK, Dailey, PJ, Tarara, RP, Canfield, DR, Aguirre, NL, Cherrington, JM *et al.* (1999a) Early short-term 9-[2-(R)-(phosphonomethoxy)propyl]adenine treatment favorably alters the subsequent disease course in simian immunodeficiency virus-infected newborn Rhesus macaques. J. Virol.; 73:2947-2955

Van Rompay, KK, Cherrington, JM, Marthas, ML, Lamy, PD, Dailey, PJ, Canfield, DR *et al.* (1999b) 9-[2-(Phosphonomethoxy)propyl]adenine (PMPA) therapy prolongs survival of infant macaques inoculated with simian immunodeficiency virus with reduced susceptibility to PMPA. Antimicrob. Agents Chemother.; 43:802-812

Van Rompay, KK, Miller, MD, Marthas, ML, Margot, NA, Dailey, PJ, Canfield, DR *et al.* (2000) Prophylactic and therapeutic benefits of short-term 9-[2-(R)-(phosphonomethoxy)propyl]adenine (PMPA) administration to newborn macaques following oral inoculation with simian immunodeficiency virus with reduced susceptibility to PMPA. J. Virol.; 74:1767-1774

Veselý, J, Rosenberg, I, Holý, A (1982) Inhibition by CTP and UTP analogues of uridine kinase from mouse leukemic cells. Collect. Czech. Chem. Commun.; 47 :3464-3469

Veselý, J, Rosenberg, I, Holý, A (1983) Donor activity of 5'-O-phosphonylmethyl analogues of ATP and GTP in the phosphorylation of uridine catalyzed by uridine kinase from mouse leukemic cells. Collect. Czech. Chem. Commun.; 48:1783-1787

Veselý, J, Merta, A, Votruba, I, Rosenberg, I, Holý, A (1990) The cytostatic effects and mechanism of action of antiviral acyclic adenine nucleotide analogs in L1210 mouse leukemia cells. Neoplasma; 37:105-110

Villani, N, Calio, R, Balestra, E, Balzarini, J, De Clercq, E, Fabrizi, E *et al.* (1994) 9-(2-Phosphonylmethoxyethyl) adenine increases the survival of influenza virus-infected mice by an enhancement of the immune system. Antiviral Res.; 25:81-89

Villemin, D and Thibault-Starzyk, F (1993) Synthesis of a new sulfur analog of PMEA. Synth. Com.; 23:1053-1059

Vonka, V, Anisimova, E, Černý, J, Holý, A, Rosenberg, I, Votruba, I (1990) Properties of a 9-(2-phosphonylmethoxyethyl)adenine (PMEA)-resistant herpes simplex virus type 1 virus mutant. Antiviral Res.; 14:117-121

Votruba I., Holý A. (1980) Inhibition of S-adenosyl-L-homocysteine hydrolase by the aliphatic nucleoside analogue -9-(S)-(2, 3-dihydroxypropyl)adenine, Collect. Czech. Chem. Commun. *45*, 3039-3044

Votruba, I, Bernaerts, R, Sakuma, T, De Clercq, E, Merta, A, Rosenberg, I *et al.* (1987) Intracellular phosphorylation of broad-spectrum anti-DNA virus agent (S)-9-(3-hydroxy-2-phosphonylmethoxy-propyl)adenine and inhibition of viral DNA synthesis. Mol. Pharmacol.; 32:524-529

Votruba I, Hasobe M., Holý A., Borchardt R.T. (1990a): 2-Methylpropyl ester of 3-(adenin-9-yl)-2-hydroxypropanoic acid. Mechanism of antiviral action in vaccinia virus-infected L929 cells, Biochem. Pharmacol. *39*, 1573-1580

Votruba, I, Trávníček, M, Rosenberg, I, Otmar, M, Merta, A, Hřebabecký, H *et al.* (1990b) Inhibition of AMV(MAV) reverse transcriptase by diphosphates of acyclic nucleotide analogues. Antiviral Res.; 13:287-294

Wachsman, M, Petty, BG, Cundy, KC, Jaffe, HS, Fisher, PE, Pastelak, A *et al.* (1996) Pharmacokinetics, safety and bioavailability of HPMPC (cidofovir) in human immunodeficiency virus-infected subjects. Antiviral Res.; 29:153-161

Webb, RR, II and Martin, JC (1987) A convenient synthesis of S-HPMPA. Tetrahedron Lett.; 28:4963-4964

Webb, RR, II, Wos, JA, Bronson, JJ, Martin, JC (1988) Synthesis of (S)-N-(3-hydroxy-2-phosphonyl-methoxy)propylcytosine, (S)-HPMPC. Tetrahedron Lett.; 29:5475-5478

Wijnholds, J, Mol, CAAM, van Deemter, L, de Haas, M, Scheffer, GL, Baas, F *et al.* (2000) Multidrug-resistance protein 5 is a multispecific organic anion transporter able to transport nucleotide analogs. Proc. Nat. Acad. Sci. USA; 97: 7476-7481

Xiong, X, Smith, JL, Kim, C, Huang, ES, Chen, MS (1996) Kinetic analysis of the interaction of cidofovir diphosphate with human cytomegalovirus DNA polymerase. Biochem. Pharmacol.; 51:1563-1567

Yokota, T, Konno, K, Chonan, E, Mochizuki, S, Kojima, K, Shigeta, S *et al.* (1990) Comparative activities of several nucleoside analogs against duck hepatitis B virus in vitro. Antimicrob. Agents Chemother.; 34:1326-1330

Yokota, T, Mochizuki, S, Konno, K, Mori, S, Shigeta, S, De Clercq, E (1991) Inhibitory effects of selected antiviral compounds on human hepatitis B virus DNA synthesis. Antimicrob. Agents Chemother.; 35:394-397

Yokota, T, Konno, K, Shigeta, S, Holý, A, Balzarini, J, De Clercq, E (1994) Inhibitory Effects of Acyclic Nucleoside Phosphonate Analogues on Hepatitis B Virus DNA Synthesis in Hb611 Cells. Antivir. Chem. Chemother.; 5:57-63

Yoshida, M, Yamada, M, Tsukazaki, T, Chatterjee, S, Lakeman, FD, Nii, S *et al.* (1998) Comparison of antiviral compounds against human herpesvirus 6 and 7. Antiviral Res.; 40:73-84

Yu, KL, Bronson, JJ, Yang, H, Patick, A, Alam, M, Brankovan, V *et al.* (1992) Synthesis and antiviral activity of methyl derivatives of 9-[2-(phosphonomethoxy)ethyl]guanine. J. Med. Chem.; 35:2958-2969

Yu, KL, Bronson, JJ, Yang, H, Patick, A, Alam, M, Brankovan, V *et al.* (1993) Synthesis and antiviral activity of 2'-substituted 9-[2-(phosphonomethoxy)ethyl]guanine analogues. J. Med. Chem.; 36:2726-2738

Yu, RH, Schultze, LM, Rohloff, JC, Dudzinski, PW, Kelly, DE (1999) Process optimization in the synthesis of 9-[2-(Diethylphosphonomethoxy)ethyl]adenine: Replacement of sodium hydride with sodium tert-butoxide as the base for oxygen alkylation. Org. Process. Res. Dev.; 3:53-55

Zabawski, EJ, Jr. and Cockerell, CJ (1998) Topical and intralesional cidofovir: a review of pharmacology and therapeutic effects. J. Am. Acad. Dermatol.; 39:741-745

Zheng, XP and Nair, V (1999) Synthesis of isomeric nucleoside phosphonates: Cyclic analogs of the anti-, HIV active compound PMEA. Tetrahedron; 55:11803-11818

Zidek, Z, Holý, A, Franková, D, Otová, B (1995) Suppression of rat adjuvant arthritis by some acyclic nucleotide analogs. Eur. J. Pharmacol.; 286:307-310

Zídek, Z, Franková, D, Holý, A, Boubelík, M, Dráber, P (1997a) Inhibition of murine macrophage nitric oxide synthase expression by a pivoxil prodrug of antiviral acyclic nucleotide analogue 9-(2-phosphonomethoxyethyl)adenine. Biochem. Pharmacol.; 54:855-861

Zídek, Z, Holý, A, Franková, D (1997b) Antiretroviral agent (R)-9-(2-phosphonomethoxypropyl)adenine stimulates cytokine and nitric oxide production. Eur. J. Pharmacol.; 331:245-252.

Zidek, Z, Holý, A, Franková, D (1997c) Immunomodulatory properties of antiviral acyclic nucleotide analogues: Cytokine stimulatory and nitric oxide costimulatory effects. Int. J. Immunopharmacol.; 19:587-597

Zídek, Z, Franková, D, Holý, A (1999a) Chemokines, nitric oxide and antiarthritic effects of 9-(2-phosphono-methoxyethyl)adenine (Adefovir). Eur. J. Pharmacol.; 376:91-100

Zídek, Z, Franková, D, Holý, A (1999b) Stimulation of cytokine and nitric oxide production by acyclic nucleoside phosphonates. Nucleos. Nucleot.; 18:959-961

SYNTHESIS OF PYRANOSE NUCLEOSIDES AND RELATED NUCLEOSIDES WITH A SIX-MEMBERED CARBOHYDRATE MIMIC

PIET HERDEWIJN

Laboratory of Medical Chemistry, Rega Institute for Medical Research, Minderbroedersstraat 10, B-3000 Leuven, BELGIUM

1.   The early days

The most common natural nucleosides have a pentofuranose sugar moiety and the synthetic nucleoside literature is mostly focussed on the preparation of "five-membered" nucleoside analogues. Pyranose nucleosides, likewise occur in nature and they may exhibit a variety of biological activity. Representative examples are amipurimycin (Czernecki *et al.*, 1997), blasticidin (Onuma *et al.*, 1966), gougerotin (Fox *et al.*, 1966), hikizimycin (Ennifar *et al.*, 1977), mildiomycin (Harada *et al.*, 1978), miharamycin (Seto *et al.*, 1983), pentopyranine (Watanabe *et al.*, 1987) and amicetin (Nakamura and Kondo, 1977). This review, however, does not deal with the synthetic procedures to obtain these antibiotics. It is rather focused on general synthetic schemes to prepare nucleosides with a six-membered carbohydrate (like) moiety.

Fisher and Helferich (1914) described the first chemical synthesis of a nucleoside in 1914, and this was a pyranose nucleoside. They condensed the silver salt of 2,8-dichloroadenine and theophylline with 2,3,4,6-tetra-*O*-acetyl-*gluco*-pyranosyl bromide to obtain 2,8-dichloro-9-(2,3,4,6-tetra-*O*-acetyl-β-D-*glyco*-pyranosyl)adenine and 7-(2,3,4,6-tetra-*O*-acetyl-β-D-*gluco*-pyranosyl)theophylline. In the thirties, Hilbert and Johnson investigated the reaction of 2,4-dialkoxypyrimidines with acetobromoglucose and acetobromogalactose (Hilbert, 1937; Hilbert and Johnson, 1930). Although the analytical instrumentarium at that time was not sophisticated enough to completely characterize the obtained crystalline materials, they tentatively described the compounds as 3-glucosido-uracil and 1-D-galactosido-uracil. 1-(D-*Gluco*-pyranosyl)cytosine **1** was prepared 1936 by Hilbert and Jansen via condensation of acetobromoglucose with 2,4-diethoxypyrimidine followed by reaction with ammonia (Hilbert and Jansen, 1936). The procedure was slightly modified by Fox and Goodman during their synthesis of 1-D-*gluco*-pyranosylcytosine and 1-D-*galacto*-pyranosylcytosine (Fox and Goodman, 1951). In 1949, Davoll and Lythgoe described the synthesis of 2'-deoxy-D-*ribo*-pyranosidotheophylline, 2'-deoxy-D-*gluco*-pyranosidotheophilline, 2'-chloro-2'-deoxy-D-*ribo*-pyranosidotheophylline, 2'-chloro-2'-deoxy-D-*arabino*-pyranosidotheophilline and 3'-deoxy-D-*ribo*-pyranosidotheophilline starting from acetylated 1-halogeno-2-deoxypyranoses and the silver salt of theophylline (Davoll and Lythgoe, 1949). These condensation reactions (between an activated sugar electrophile and a nucleobase,

Recent Advances in Nucleosides: Chemistry and Chemotherapy, Ed. by C.K. Chu. 239 — 290

from which the nucleophilic character can be tuned in different ways) represent the fundamental principle for the synthesis of pyranose nucleosides, which has not changed since then.

2.   Classical methodologies

A first method for the synthesis of pyranose nucleosides starts from pyranosyl halides. 1-(2-Deoxy-β-D-*arabino*-hexopyranosyl)uracil **2a** was prepared by Fox *et al.* from 1-chloro-2-deoxy-3,4,6-triacetyl-D-glucose and 2,4-diethoxypyrimidine (Fox *et al.*, 1953). The same compound was obtained from 1-(2-deoxy-3,4,6-tri-*O-p*-nitrobenzoyl-β-D-*arabino*-hexopyranosyl)-4-ethoxy-2(1H)pyrimidone by Zorbach and Durr (1962). 9-(2-Deoxy-D-*arabino*-hexopyranosyl)adenine **2b** was obtained by reaction of 2-deoxy-3,4,6-tri-*O-p*-nitrobenzoyl-β-D-*arabino*-hexopyranosyl bromide with silver-6-benzamidopurine followed by deprotection with sodium methanolate in methanol. The thymine analogue **2c** was prepared from the same sugar and 2,4-diethoxy-5-methylpyrimidine by the fusion method (Zorbach and Durr 1962). Both anomers of 1-(2-deoxy-D-*arabino*-hexopyranosyl)thymine (**2c** and **3**) could be isolated in pure form from the reaction of the heavy metal (mercury) salt of thymine with 2-deoxy-3,4,6-tri-*O*-acetyl-D-*arabino*-hexopyranosyl bromide in DMF (Etzold and Langen, 1965). Ulbricht and Rogers (1965) demonstrated that the silver salts of cytosine and N-acetylcytosine react with acetobromoglucose to give acetylated *O*-glucosides which rearrange to N-glucosides in the presence of mercuric bromide.

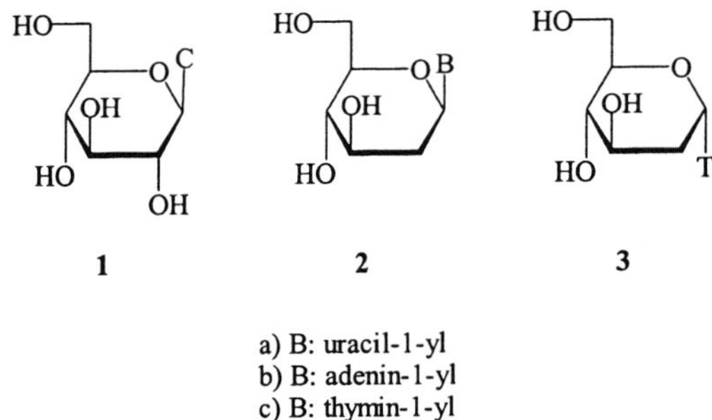

a) B: uracil-1-yl
b) B: adenin-1-yl
c) B: thymin-1-yl

The 2'-deoxy compounds were later resynthesized and used as starting material for the synthesis of 2,3,4-trideoxy-D-*glycero*-hex-3-enopyranosyl nucleosides (**4**) and 2,3,4-trideoxy-D-*glycero*-hexopyranosyl nucleosides (**5**) (Herdewijn and Van Aerschot, 1990) as described in Scheme 1.

Novák and Šorm (1962) described the synthesis of 9-(2-deoxy-1-D-*gluco*-pyranosyl)adenine and 1-(9-adenylo)pseudo-D-glucal from 1-bromo-2-deoxyglucose triacetate and chloromercuribenzamidopurine in DMF. The Hilbert-Johnson procedure was used for the synthesis of chinoline and isochinoline nucleosides with a glucopyranose sugar moiety (Wagner and Schmidt, 1965). Adenine nucleosides derived from α-D-*manno*-pyranose, α-D-*talo*-pyranose, β-D-*gulo*-pyranose, β-D-*allo*-pyranose and α-D-*altro*-

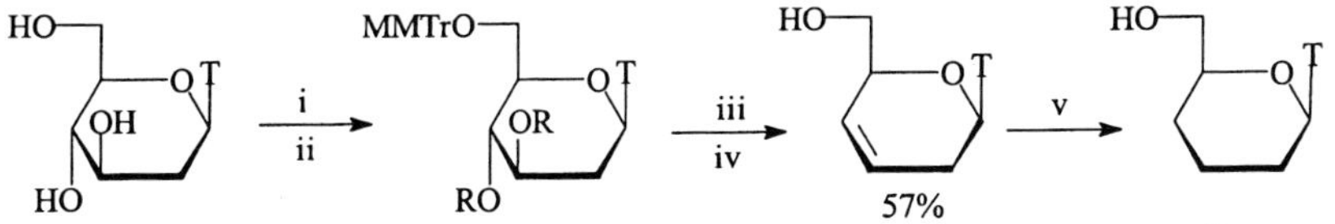

Scheme 1.

i) MMTrCl, pyridine (70%); ii) NaOH, DMSO, CS$_2$; BrCH$_2$CH$_2$CN; iii) Bu$_3$SnH, AIBN, toluene; iv) *p*TsOH, CHCl$_3$; MeOH; v) H$_2$, Raney-Ni, MeOH.

pyranose were synthesized via the respective halides and chloromercuri-6-benzamido-purine (Lerner and Kohn, 1964). The work with D-mannose was preceded by the synthesis of 9-D-*manno*-pyranosidoadenine in 1947 by Lythgoe *et al.* (1947). They synthesized 9-D-*manno*-pyranosido-2-methylthioadenine from 6-amino-4-D-*manno*-pyranosidamino-2-methylthio-yrimidine (Baddiley *et al.* 1943) and converted the compound to the adenine nucleoside by desulphurization with Raney Nickel.

Reaction of α-L-acetobromofucose with chloromercuri-6-benzamidopurine followed by deacylation with sodium methoxide in methanol afforded 9-(β-L-fucopyranosyl)adenine (**6**) (Fisher *et al.*, 1969). Reaction of 1-chloro-2-deoxy-3,4-di-*O*-acetyl-D-*ribo*-pyranose with mercury-thymine in DMF, followed by deacetylation yield the α and β anomer of 1-(2-deoxy-D-*ribo*-pyranosyl)thymine (**7** and **8**) (Etzold and Langen, 1966).

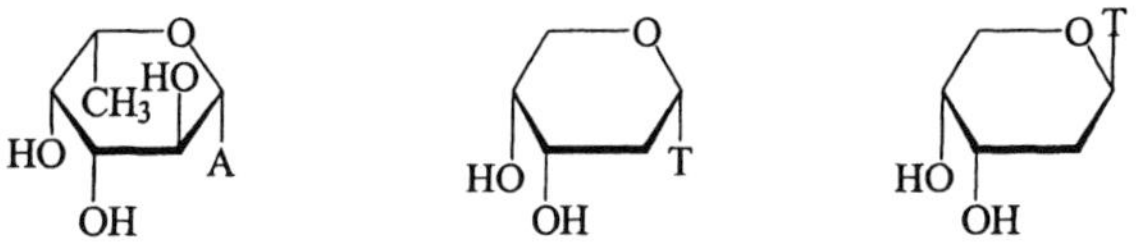

Scheme 2.

i: (PhO)$_2$CO, DMF, NaHCO$_3$, 130°, ii: reflux; iii: NaOH 1N; iv: KOBu^t, DMF, 100 °C; v: H$_2$O, Δ,

Compound **8** was converted into 1-(2-deoxy-β-D-*xylo*pyranosyl)thymin by reaction with diphenylcarbonate in DMF in the presence of NaHCO$_3$ followed by reflux and NaOH treatment (Scheme 2). Reaction of the 3',4'-dimesyl derivative of 1-(2-deoxy-β-D-*ribo*pyranosyl)thymine with potassium tert-butylate in DMF at 100 °C followed by heating in water gives 1-(2-deoxy-α-L-*xylo*pyranosyl)thymine (Etzold *et al.*, 1967) (Scheme 2).

A method to synthesize *gluco*-pyranoside nucleosides without having to prepare the heavy metal salts of the purine bases was described by Yamaoka *et al.* (1965). They described the condensation of a purine with 2,3,4,6-tetra-*O*-acetyl-α-D-*gluco*-pyranosyl halide in a solvent medium (nitromethane) containing a hydrogen halide acceptor (Hg(CN)$_2$). That this classical methodology has stayed useful for a long time is demonstrated by the work of Lerner *et al.* (1987), who used the condensation of acetylated pyranosyl bromide with 6-benzamido-9-(chloromercuri)purine to obtain pyranosyl nucleosides in 1987.

A second approach for the synthesis of pyranose nucleosides starts from glycals. The glycal method has the advantage that the use of toxic mercury salts can be avoided. In 1961, Robins *et al.* (1961) described the reaction of 2-acetoxymethyl-2,3-dihydro-4-H-pyran with 6-chloropurine and *p*-toluenesulphonic acid in ethylacetate to obtain 6-(adenin-9-yl)-tetrahydropyran-2-methanol **9** (after reaction with ammonia).

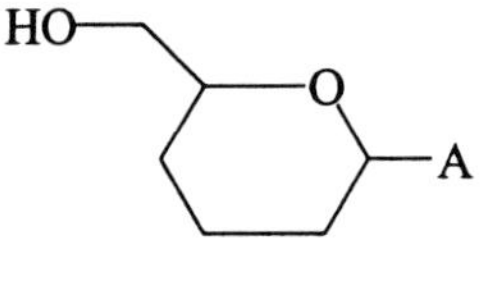

9 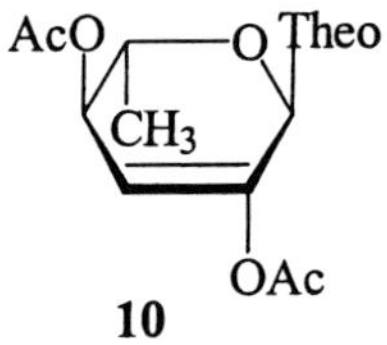   **10**

The use of D-glycal for the synthesis of pyranose nucleosides by an acid catalyzed fusion reaction was first reported by Bowles and Robins (1964). They described the synthesis of 6-chloro-9-(3,4-di-*O*-acetyl-2-deoxy-α- and β-D-*ribo*-pyranosyl)purine from 3,4-di-*O*-acetyl-D-arabinal and 7-(4,6-di-*O*-acetyl-2,3-didehydro-2,3-dideoxy-D-*gluco*-pyranosyl)theophylline from 3,4,6-tri-*O*-acetyl-D-glucal. Fusion of a mixture of 2-acetamido-6-chloropurine and 3,4,6-tri-*O*-acetyl-D-glucal in the presence of a catalytic amount of trichloroacetic acid furnished 2-acetamido-6-chloro-9-(4,6-di-*O*-acetyl-2,3-didehydro-2,3-dideoxy-D-*erythro*-hexopyranosyl)purine which was converted to 2-amino-9-(2,3-dideoxy-2,3-didehydro-D-*erythro*-hexopyranosyl)purin-6-one (Scheme 3) (Leutzinger *et al.*, 1968).

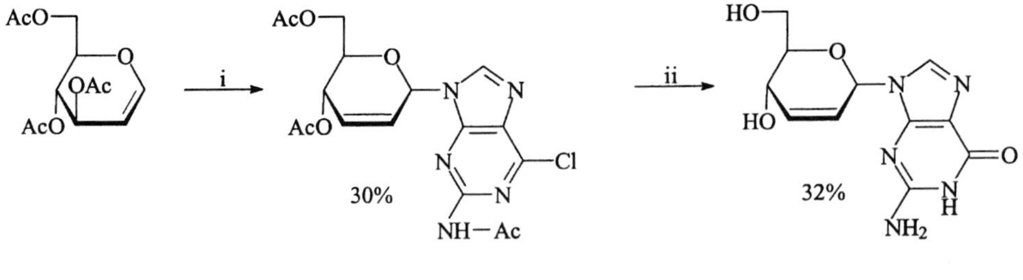

Scheme 3.

i : 2-acetamido-6-chloropurine, 140°, CCl$_3$COOH; ii : NaHS, MeOH; ii : H$_2$O$_2$, NH$_3$, H$_2$O

Leutzinger *et al.*described that the acid catalyzed fusion procedure with acetylated glycals furnishes 2'-deoxypyranosylpurine nucleosides and 2',3'-unsaturated pyranosyl-purine nucleosides (Leutzinger *et al.*, 1968a; Leutzinger *et al.*, 1968b). During these studies it was also observed that fusion of purine bases with D-glucal gave, in addition to the described nucleosides, a 1',2'-unsaturated pyranosyl nucleoside with the purine base attached at position 3 of the pyranose ring (3-deoxy-3-D-*erythro*-hex-1-enopyranosyl purine) (Leutzinger *et al.*, 1970).

The fusion reaction was, likewise, used for the synthesis of several theophylline-nucleosides. Reaction of 1,2,3,4-tetra-*O*-acetyl-L-*rhamno*-pyranose with theophylline in the presence of *p*-toluenesulphonic acid gives 7-(2,4-di-*O*-acetyl-3,6-dideoxy-α-L-*erythro*-hex-2-enopyranosyl)theophylline (**10**) (Onodera *et al.*, 1968). Fusion of 1,2,4,6-tetra-*O*-acetyl-3-deoxy-α-D-*threo*-hex-2-enopyranose with theophylline in the presence of *p*-toluenesulphonic acid yielded the α and β anomer of the protected 3-deoxy-D-*threo*-hex-2-enopyranosyl theophylline (**11** and **12**). The same reaction and starting from tetra-*O*-benzoyl-2-hydroxy-*O*-glucal yielded the 3-deoxy-D-*erythro*-hex-2-enopyranosyl congeners (**13** and **14**) (Ferrier and Ponpipom, 1971).

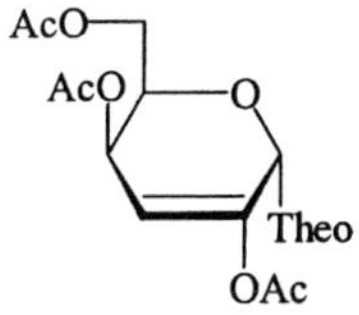

11 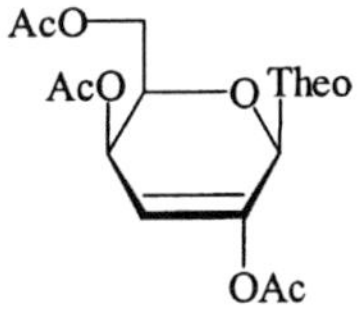   **12** 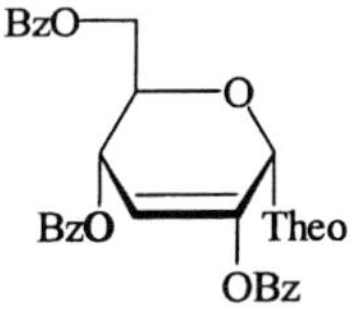   **13** 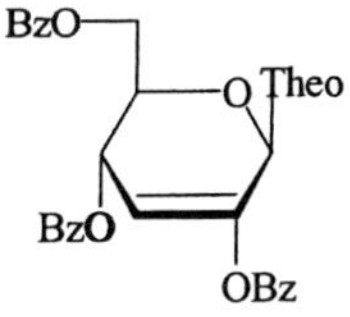   **14**

Also pentopyranoses were used as starting materials in these reactions. The acid-cata-lyzed reaction of 6-chloropurine with 3,4-di-*O*-acetyl-D-xylal yielded mainly 6-chloro-9-(4-*O*-acetyl-2,3-dideoxy-α- and β-D-*glycero*-pent-2-enosyl)purine and 1,2,3-trideoxy-4-*O*-acetyl-3-(6-chloro-9-purinyl)-D-*threo*-pent-1-enopyranose (Scheme 4a) (Fuertes *et al.*, 1970). The same reaction, starting from 3,4-di-*O*-acetyl-L-arabinal yielded, in addition to the enantiomers of the above compounds, 6-chloro-9-(3,4-di-*O*-acetyl-2-deoxy-α-and β-L-*erythro*-pentopyranosyl)purine (Scheme 4b) (Fuertes *et al.*, 1970).

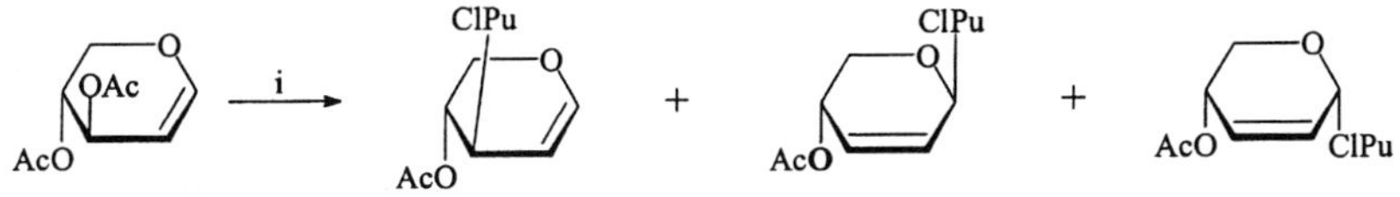

Scheme 4a.

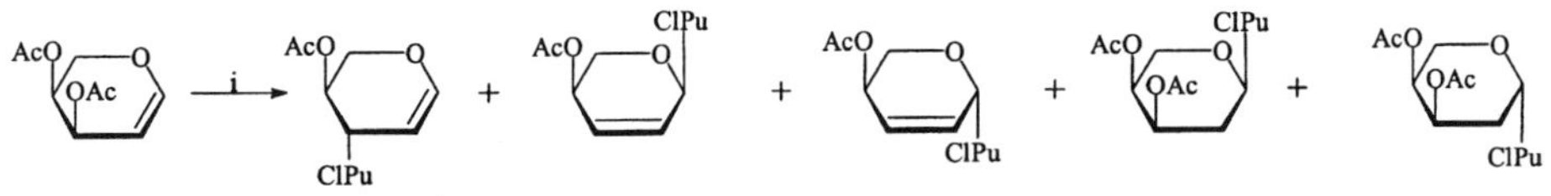

Scheme 4b.
i: 6-chloropurine, CF$_3$COOH, EtOAc, 95°C, 24h

244   *P. Herdewijn*

Mechanistic studies by Ferrier and Ponpipom (1971) revealed that the kinetically controlled reaction products between tri-*O*-acetyl-D-glucal and purine bases are 4,6-di-*O*-acetyl-2,3-dideoxy-D-*erythro*-hex-2-enopyranosyl nucleosides. On heating with acids, those compounds are rearranged to the 3-deoxyglycal structure having the bases attached at C-3.

As a follow up of their previous studies, Leutzinger *et al.* (1972) reported that the fusion reaction of 3,4,6-tri-*O*-acetyl-D-glucal and 6-benzamidopurine in the presence of *p*-toluenesulphonic acid gives a mixture of four isomeric nucleosides in a total yield of 76% (Scheme 5). Hydrogenation of 9-(1,5-anhydro-2,3-dideoxy-D-*arabino*-hex-1-enitol-3-yl)adenine gives 9-(1,5-anhydro-2,3-dideoxy-D-*arabino*-hexitol-3-yl)adenine, which represents an early synthesis of an isonucleoside with a hexitol sugar moiety.

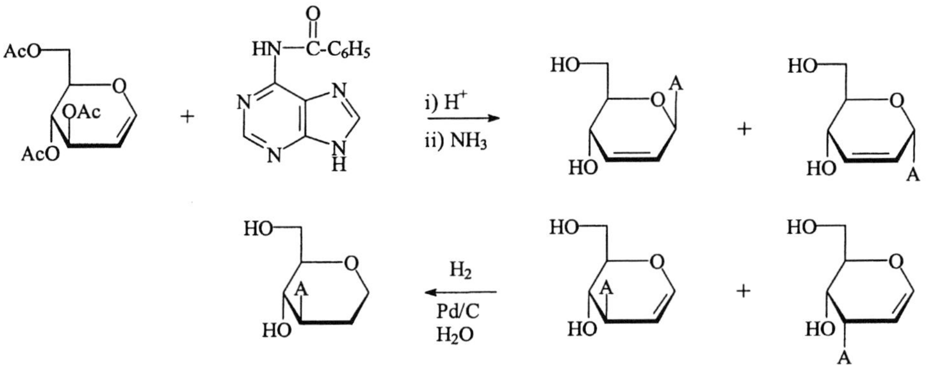

Scheme 5.

As an alternative, several organic solvents can be used and the reaction can be catalyzed by a variety of acids. 9-(2-Deoxy-β-D-*ribo*-pyranosyl)adenine was prepared from 3,4-di-*O*-acetyl-D-arabinal and adenine in DMSO in the presence of HCl (Scheme 6) (Nagasawa *et al.*, 1967).

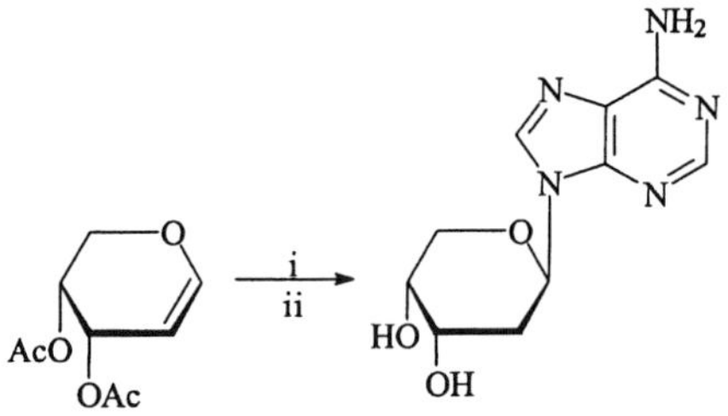

Scheme 6.
i: adenine, DMSO, HCl; ii: NaOMe, MeOH

The results are similar to those obtained by the acid-catalyzed fusion of purine bases and 3,4-di-*O*-acetyl-D-arabinal or 1,3,4-tri-*O*-acetyl-2-deoxy-β-D-*ribo*-pyranose (Leutzinger *et al.*, 1968) i.e. the formation of a mixture of α and β 9-(3,4-di-*O*-acetyl-2-deoxy-D-*ribo*-pyranosyl)nucleosides.

The preparation of 2'-enopyranosylpyrimidine nucleosides directly from triacetylglucal and bis(trimethylsilyl)uracil was described by Kondo *et al.*(1971). The product mixture obtained can be controlled by the Lewis acid [SbCl$_5$ gives 1-(4,6-di-*O*-acetyl-2,3-dideoxy-α- and β-D-*erythro*-hex-2-enopyranosyl)uracil; BF$_3$ gives 3-(4,6-di-*O*-aceyl-2,3-dideoxy-α- and β-D-*erythro*-hex-2-enopyranosyl)uracil and 1,2,3-trideoxy-3-(3-uracilyl)-D-*arabino*-hex-1-enopyranose] (Scheme 7).

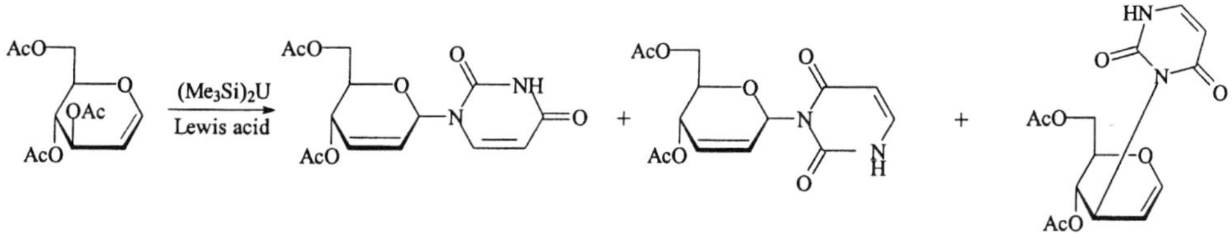

Scheme 7.

2',3'-Unsaturated nucleosides have also been prepared by the condensation of acetylated glycals with silylated purine or pyrimidine bases in the presence of trityl perchlorate (**15-18**) (Herscovici *et al.*, 1988).

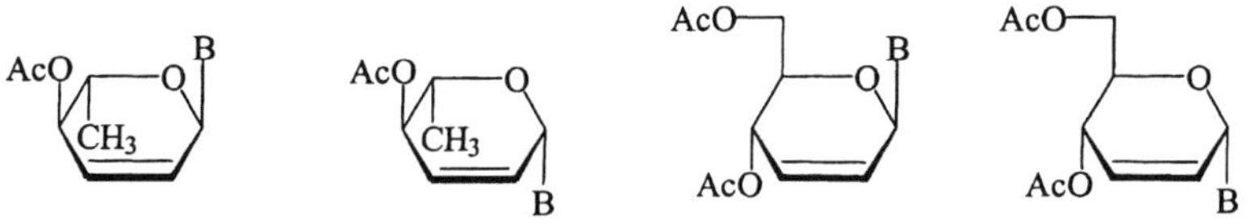

15   **16**   **17**   **18**

Reaction of bis(trimethylsilyl)thymine and 3,4,6-tri-*O*-acetyl-D-glucal in the presence of Lewis acids afforded a mixture of the α and β anomers of 1-(4,6-di-*O*-acetyl-2,3-dideoxy-β-D-*erythro*-hex-2-enopyranosyl)thymine (**19** and **20**) (Ueda and Watanabe, 1985; Augustijns *et al.*, 1992b). Hydrogenation and deacetylation gives 1-(2,3-dideoxy-β-D-*erytho*-hexopyranosyl)thymine (**21**) (Augustijns *et al.*, 1992b).

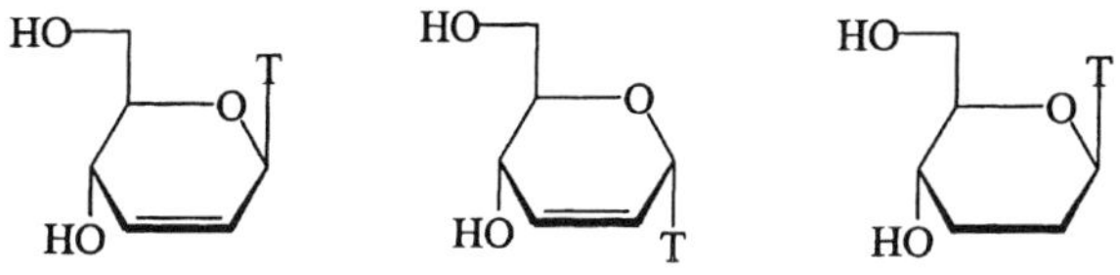

19   **20**   **21**

The use of hex-1-enitol starting material gives more complicated reaction products. Reaction of 2,3,4,6-tetra-O-acetyl-1,5-anhydro-D-*arabino*-hex-1-enitol with bistrimethylsilyl uracil yielded the hex-3-enopyranosid-2-ulose nucleoside (Scheme 8). The yield of the condensation reaction is substantially higher (68%) when starting with the glycal of *lyxo* configuration.

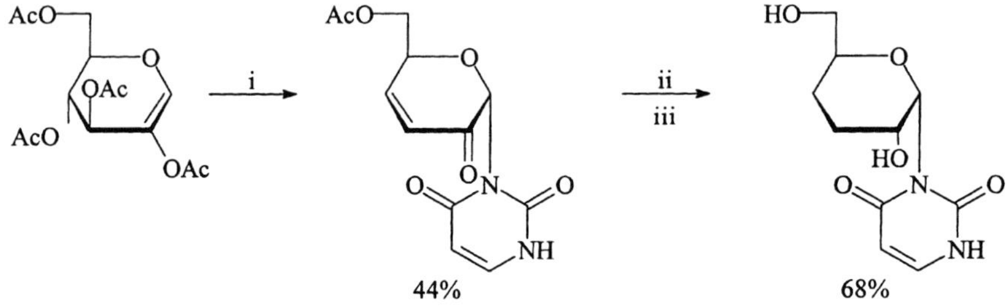

Scheme 8.

i: $(Me_3Si)_2U$, $SnCl_4$; MeOH, Δ; ii: $NaBH_4$, MeOH; iii: NaOMe, MeOH

The high regio- and stereoselectivity of the condensation reaction is explained by the formation of an intermediate acyloxonium in the β-configuration, followed by attack of the pyrimidine base at N^1 and allylic rearrangement, which places the N^3 of the base in a suitable location for attack at 1'-C- of the sugar (De Fina *et al.*, 1994).

A third method to obtain pyranose nucleosides is the condensation reaction between activated nucleobases and acylated sugars in the presence of Lewis acids. Reaction of the trimethylsilyl derivative of uracil with 2-deoxy-1,3,4,6-tetrakis-O-(4-nitrobenzoyl)-β-D-*ribo*-hexopyranose at reflux temperature in the presence of trimethylsilyl trifluoromethanesulfonate gives the 2-deoxy-β-D-*ribo*-hexopyranosyl nucleosides (Scheme 9) (Nord *et al.*, 1987). When the halo sugar 2-deoxy-3,4,6-tris-O-(4-nitrobenzoyl)-α-D-*ribo*-hexopyranosyl bromide was reacted with silylated uracil, at room temperature, the α-D-*ribo*-hexopyranosyl nucleoside was isolated as major compound (Nord *et al.*, 1987).

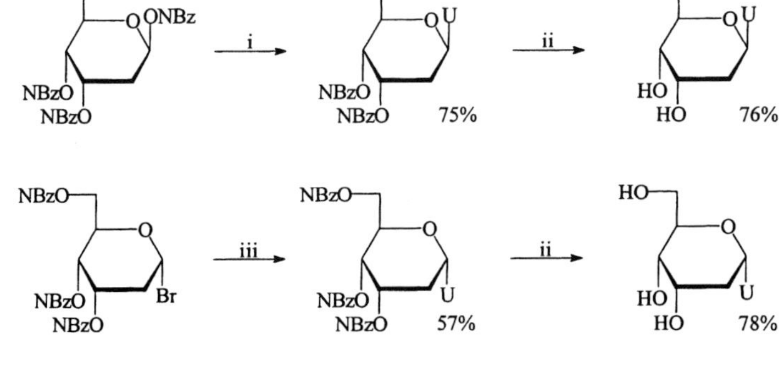

Scheme 9.

i: $(Me_3Si)_2U$, TMSOTf, CH_3CN, reflux, 2 h.; ii: $NaOCH_3$, CH_3OH; iii: $(Me_3Si)_2U$, CH_3CN, R.T., 2 h;

The condensation reaction between silylated nucleobases and acylated sugars or methyl hexopyranosides in the presence of Lewis acids (for example Böhringer *et al.*, 1992 and Al-Masoudi *et al.*, 1995), or, alternatively, reaction of acylated pyranosyl halides and the silylated nucleobases (or the salt of nucleobases) (for example Attia *et al.*, Attia *et al.*, 1995; Buchanan *et al.*, 1994; Mansour *et al.*, 1999; Khodair *et al.*, 1997) are now the most common way to synthesize pyranose nucleosides, and only some selected examples from the literature are given here as reference.

Condensation reaction of appropriate protected carbohydrate precursors with silylated bases in the presence of Lewis acids was used for the synthesis of 2-deoxy-D-*ribo*-pyranosyl nucleosides (**22**, **23**) and 2-deoxy-2-fluoro-D-*arabino*-pyranosyl nucleosides (**24**, **25**) (Herdewijn *et al.*, 1991).

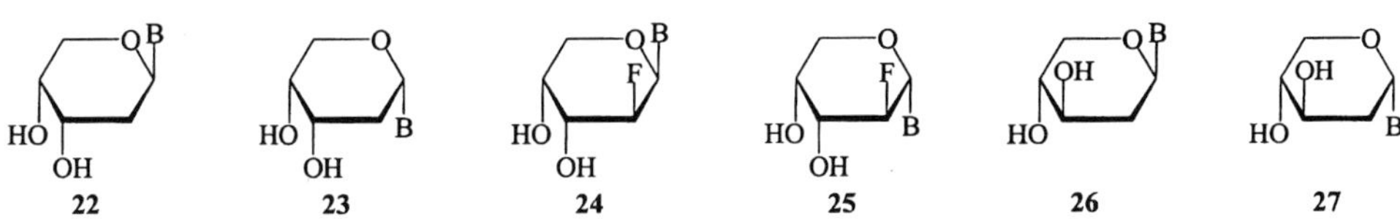

These 2-deoxy-D-*ribo*-pyranosyl nucleosides were previously synthesized from 3,4-di-*O*-acetyl-D-arabinal (Leutzinger *et al.*, 1968; Nagasawa *et al.*, 1967) by acid-catalyzed condensation with nucleobases, and, from 3,5-di-*O*-acyl-2-deoxy-D-*ribo*-pyranosyl chloride and silylated nucleobase in the presence of $AgClO_4$ (or $HgO/HgBr_2$) (Wittenburg, 1968a; Wittenburg, 1968b) or using the purine-HgCl complex (Zinner and Wittenburg, 1962). The 2-deoxy-D-*xylo*-pyranose analogues (**26**, **27**) are obtained using the same procedure (Wittenburg *et al.*, 1968).

X-ray analysis demonstrate that 1-(2,3-dideoxy-*erythro*-α-D-hexopyranosyl)thymine (De winter *et al.*, 1991a) 1-(2,3-dideoxy-*erythro*-β-D-hexopyranosyl)thymine (De Winter *et al.*, 1991a), 1-(2,3-dideoxy-*erythro*-β-D-hexopyranosyl)cytosine (De Winter *et al.*, 1992), 1-(2-deoxy-β-D-*ribo*-pyranosyl)-5-iodouracil (De Winter *et al.*, 1991b), 1-(2-deoxy-2-fluoro-α-D-*arabino*-pyranosyl)-5-iodouracil (De Winter *et al.*, 1991c), 1-(2-deoxy-2-fluoro-β-D-*arabino*-pyranosyl)thymine (De Winter *et al.*, 1991d) and 1-(2-deoxy-2-fluoro-β-D-*arabino*-pyranosyl)-5-ethyluracil (De Winter *et al.*, 1991c) all have an equatorial oriented base moiety.

9-β-D-Fucopyranosyladenine and its α-analogue (**28**, **29**) are obtained from tetra-*O*-acetyl-α-D-fucopyranose and 6-benzamidochloromercuripurine in the presence of titanium tetrachloride (Lerner, 1971). 1-β-D-Fucopyranosyl nucleosides with an uracil, thymine, cytosine, guanine and adenine base moiety (**28a-e**) were obtained from α-D-fucopyranose tetraacetate and silylated bases with $SnCl_4$ as catalyst (Lerner *et al.*, 1993).

The reaction of silylated purines with peracylglucoses in the presence of $SnCl_4$ gives 9-β-D-nucleosides in 60-70% yield (Lichtenthaler *et al.*, 1974a). The yield of the coupling reaction is given in Scheme 10. This reaction makes also 4'-amino-4'-deoxy-β-D-*gluco*-pyranosyl adenine (**30**) readily available.

A lot of work on ketopyranosyl nucleosides has been carried out by the group of K. Antonakis (1984). This work has been reviewed and therefore not included here.

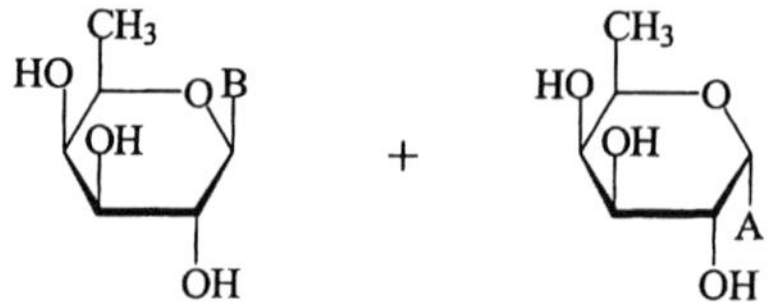

28a   B= adenin-9-yl
28b   B= uracil-1-yl
28c   B= thymin-1-yl
28d   B= cytosin-1-yl
28e   B= guanin-9-yl

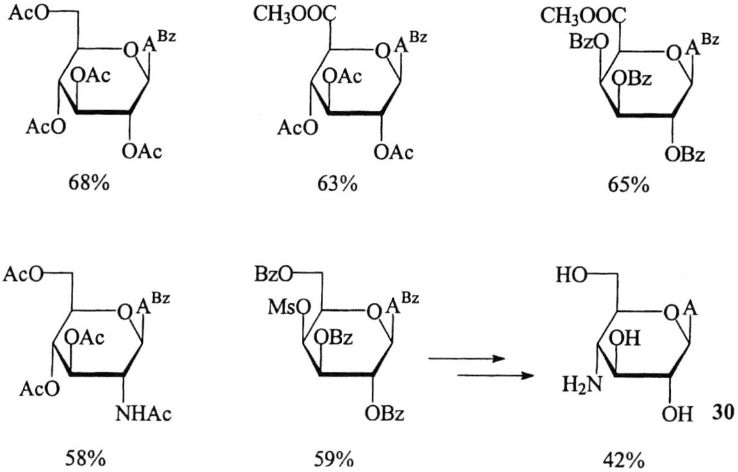

Scheme 10.

3.   Amino nucleosides

Amino nucleosides can be prepared by condensation of aminosugars with the appropriate bases. In 1954, Baker *et al.* described the reaction of 2,3,4,6-tetra-O-acetyl-α-chloro-D-glucosamine with chloromercuri-2-methylmercapto-6-dimethylaminopurine in toluene followed by desulfurization to give 6-dimethylamino-9-(2-amino-D-gluco-pyranosyl)purine (**31**) (Baker *et al.*, 1954).

31

A series of glucosamine pyrimidine nucleosides were synthesized from 1-chloro-3,4,6-tri-*O*-acetyl-2-deoxy-2-acetamido-, 2-carbo-benzyloxyamino- and 2-carbomethoxyamino-D-*gluco*-pyranose by the Hilbert-Johnson and the acetylcytosine-mercury procedures. These molecules were then further converted to 1-(2-deoxy-2-dimethylamino-β-D-*gluco*-pyranosyl)cytosine (**32**) (Stevens and Nagarajan, 1962).

Wolfrom *et al.* (1965) described the condensation of 3,4,6-tri-*O*-acetyl-2-deoxy-2-(2,4-dinitroanilino)-α-D-*gluco*-pyranosyl bromide with 6-acetamido-9-chloromercuripurine leading to the synthesis of α- and β-9-(2-amino-2-deoxy-β-D-*gluco*-pyranosyl)adenine (**33**, **34**). 1-(2-Amino-2-deoxy-D-*gluco*-pyranosyl)thymine (**35**) was obtained from 3,4,6-tri-*O*-acetyl-2-deoxy-2-trifluoroacetamido-α-D-*gluco*-pyranosyl bromide and bis(trimethylsilyl)-thymine by the fusion procedure. The protecting groups were removed with HCl in MeOH (Wolfrom and Bhat, 1967).

32   **33**   **34**   **35**

An elegant synthesis of 3'-amino-3'-deoxyhexopyranosyl nucleosides with a pyrimidine base was described by J. Fox (Watanabe *et al.*, 1965). Uridine was oxidized with sodium metaperiodate and the resulting dialdehyde was condensed with nitromethane in the presence of base. Hydrogenation with Raney nickel yielded the 3-amino-3-deoxy-β-D-*gluco*-pyranosyl nucleoside, which was further converted to 1-β-D-*allo*-pyranosyluracil (Scheme 11).

In fact, the reaction of the uridine 2',3'-dialdehyde with nitromethane followed by neutralization in non-aqueous conditions gives a mixture of the D-*galacto*-pyranosyl and D-*gluco*-pyranosyl isomer (Watanabe *et al.*, 1965). The protected D-*gluco*-pyranosyl compound was converted in several aminohexopyranosyl nucleosides using the Schmidt-Rutz dehydration as key reaction (Scheme 12) (Matsuda and Watanabe, 1996).

Similarly, 1-(3-amino-3-deoxy-β-D-*gluco*-pyranosyl)cytosine and 1-(3-amino-3-deoxy-β-D-*manno*-pyrano-syl)cytosine were obtained by the periodate-nitromethane-hydrogenation procedure (Scheme 13).

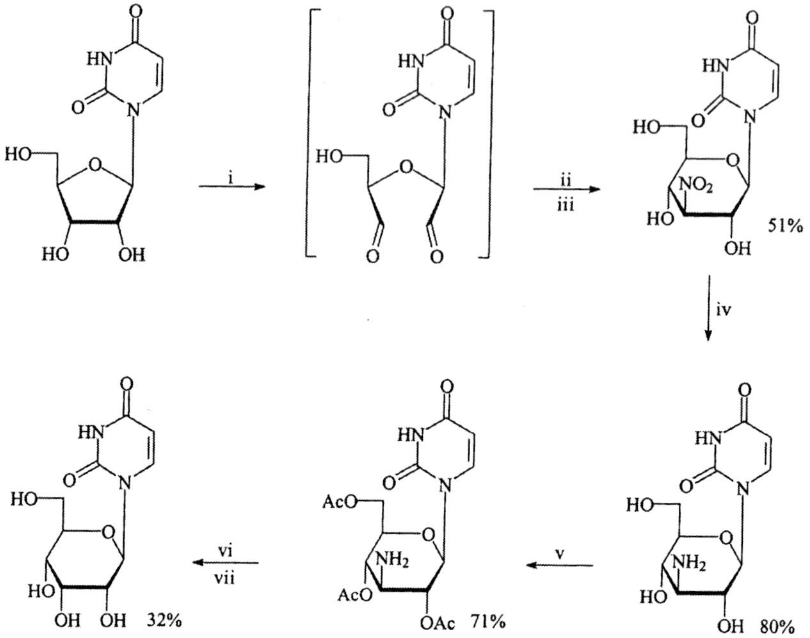

Scheme 11.

i: NaIO$_4$, H$_2$O; ii: CH$_3$NO$_2$, NaOCH$_3$, EtOH; iii: Dowex50 (H$^+$); iv: Raney Ni, MeOH, H$_2$O; v: AcOH, AcCl;
vi: AcOH, H$_2$O, C$_2$H$_5$ONO, EtOH; vii: MH$_3$, EtOH

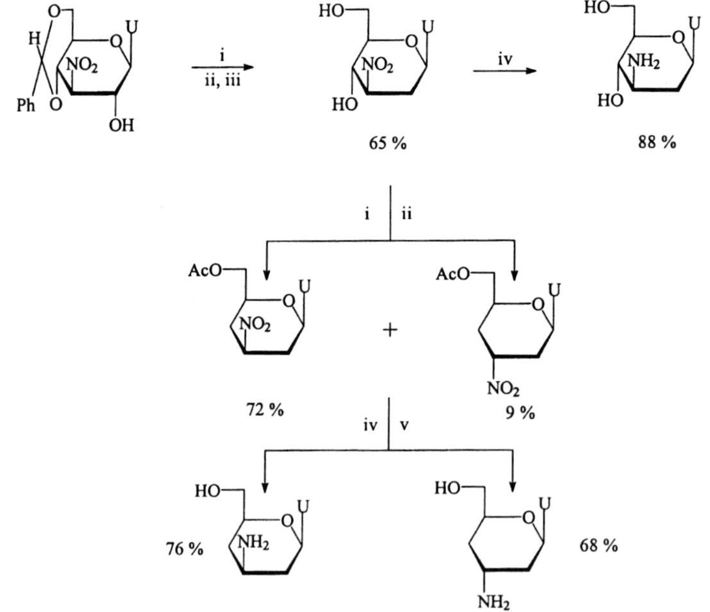

Scheme 12.

i: Ac$_2$O, DMAP, EtOAc; ii: NaBH$_4$, MeOH; iii: 90% CF$_3$COOH; iv: H$_2$, Raney Ni, EtOH, H$_2$O; v: NH$_3$, MeOH.

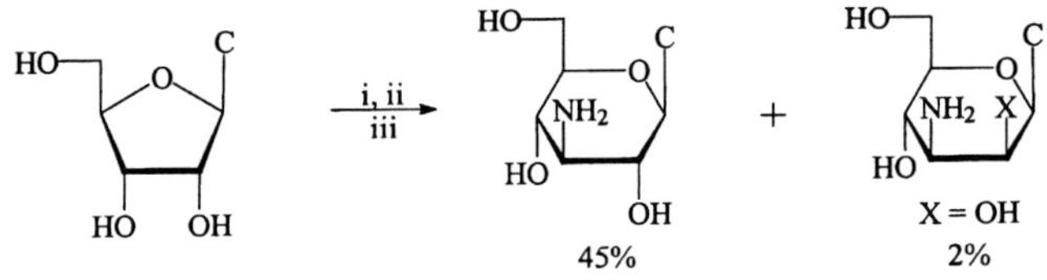

45% 2%

Scheme 13

i: $NaIO_4$, H_2O; ii: CH_3NO_2, NaOH, H_2O; iii: H_2, Raney-Ni, MeOH, H_2O. (yield given are those from reaction iii).

The 1-(3-Amino-3-deoxy-β-D-*gluco*-pyranosyl)uracil (**36**) was also further converted into 1-(3-amino-3-deoxy-β-D-*manno*pyranosyl)uracil (**37**) and 1-(3-amino-3-deoxy-β-D-*galacto*-pyranosyl)uracil (**38**) (Watanabe and Fox, 1966). The latter nucleoside can also be obtained directly from uridine by the periodate-nitromethane procedure.

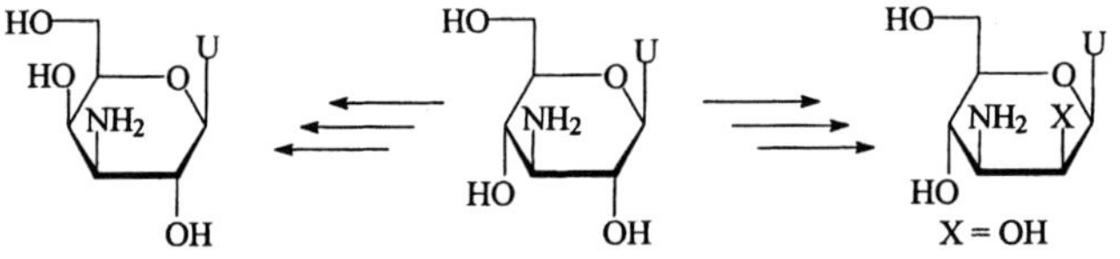

38   **36**   **37**

The reaction works as well with purine nucleosides. Treatment of adenosine with metaperiodate followed by condensation with nitromethane and reduction yielded a mixture of 9-(3-amino-3-deoxy-β-D-*manno*-pyranosyl)adenine, 9-(3-amino-3-deoxy-β-D-*galacto*-pyranosyl)adenine and 9-(3-amino-3-deoxy-β-D-*gluco*-pyranosyl)adenine (Scheme 14) (Beránek *et al.*, 1965).

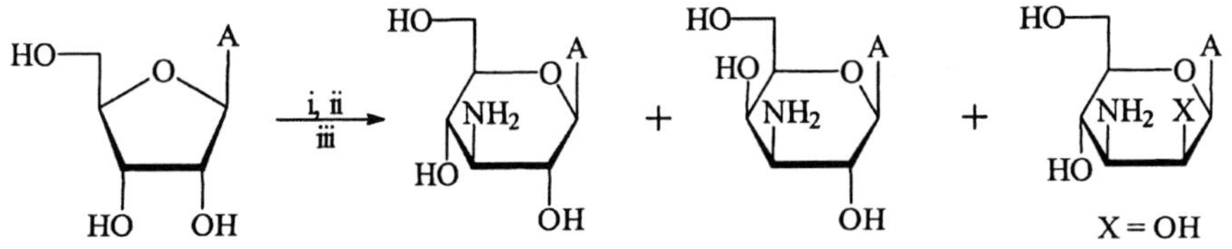

Scheme 14.

i: $NaIO_4$, H_2O; ii: NaOH, CH_3NO_2, EtOH; iii: H_2, Raney-Ni, MeOH.

This reaction sequence has become a classical method for the synthesis of 3-amino nucleosides. For example, 3-amino-3-deoxy-β-D-*gluco*-pyranose nucleosides with a uracil and hypoxanthine base moiety were synthesized by the periodate-nitromethane cyclisation method follow by reduction of the nitro group, by Lichtenthaler and Albrecht (1968).

Periodate oxidation of 6-(dimethylamino)-9-(β-D-*ribo*-furanosyl)purine followed by cyclization with nitromethane in the presence of sodium methoxide and hydrogenation, affords 9-(3-amino-3-deoxy-β-D-*gluco*-pyranosyl)-6-(dimethylamino)purine (**39**) (Lichtenthaler and Albrecht, 1968).

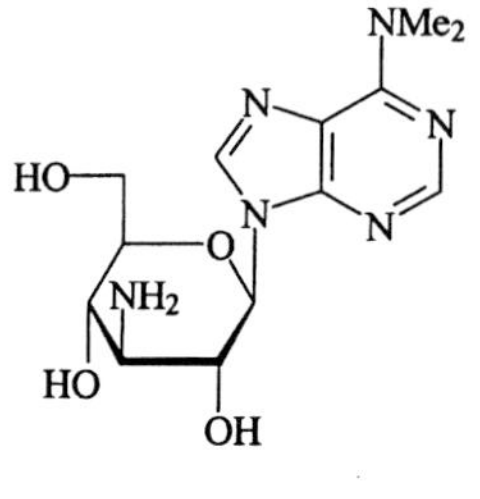

39

1-(4,6-*O*-Benzylidene-3-deoxy-3-nitro-β-D-*gluco*-pyranosyl)uracil was explored as Michael acceptor after dehydration, to synthesize 2-substituted –2,3-dideoxy-3-nitro-β-D-*gluco*-pyranosyl nucleosides (Scheme 15) (Ohta *et al.*, 1996; Tsuboike *et al.*, 1998).

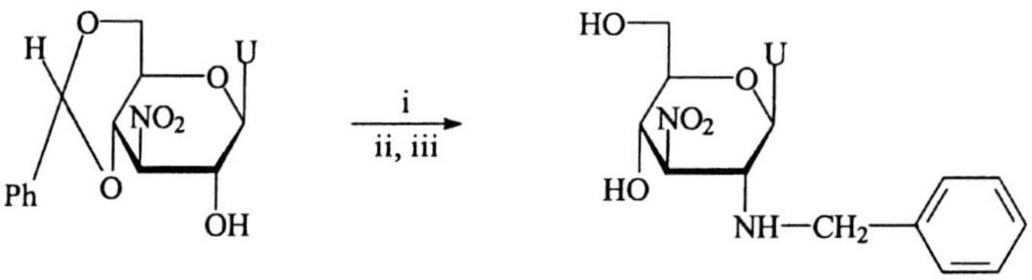

51 %

Scheme 15.

i: acetone, MsCl, Et₃N; ii: benzylamine; AcOH, MeOH; iii: 90% CF₃COOH

Polyamino nucleosides are also obtained from their nitro-precursors, based on this reaction. Reaction of 1-(2,4,6-tri-*O*-acetyl-3-nitro-3-deoxy-β-D-*gluco*-pyranosyl)uracil with ammonia followed by acetylation, reduction and acetylation gives 1-(2,3,4-triacetamido-6-*O*-acetyl-2,3,4-trideoxy-β-D-*gluco*-pyranosyl)uracil in 25% yield (Scheme 16). A similar reaction was performed with the hypoxanthin base (Lichtenthaler, 1969).

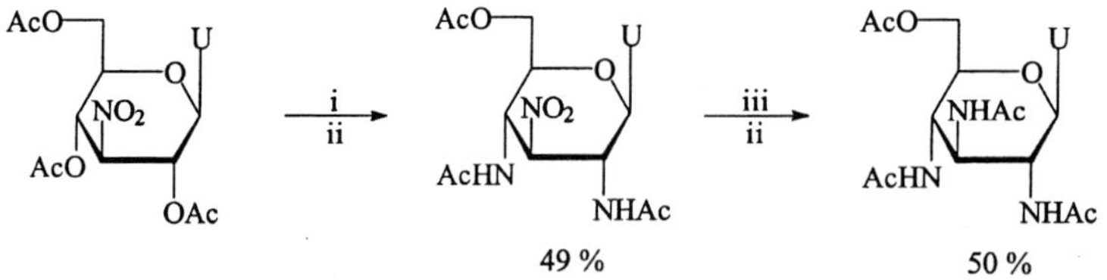

49 %　　　50 %

Scheme 16.

i: NH₃; ii: Ac₂O, MeOH; iii: H₂, Raney-Ni, MeOH, H₂O;

9-(4,6-*O*-benzylidene-2-*O*-acetyl-3-nitro-3-deoxy-β-D-*gluco*-pyranosyl)hypoxanthine can be converted into 9-(2,3-diacetamino-2,3-dideoxy-β-D-*gluco*-pyranosyl)hypoxanthine by reaction with benzylamine followed by hydrogenation and acetylation (Scheme 17) (Lichtenthaler, 1969).

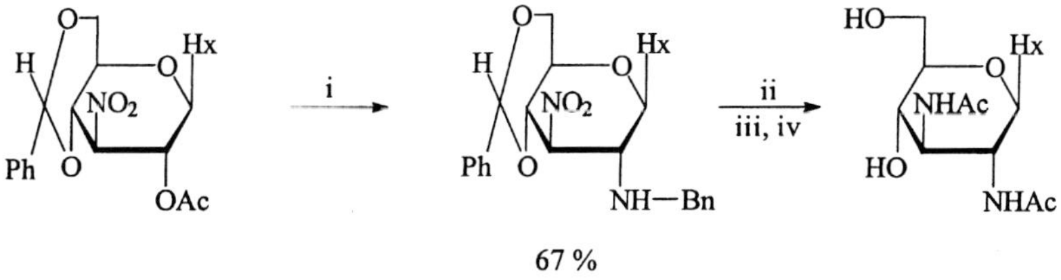

Scheme 17.

i: benzylamine, dioxane; ii: H_2 Pd/C; iii: Ac_2O; iv: HOAc. Hx: hypoxanthine

9-(6-Amino-6-deoxy-β-D-*gluco*-pyranosyl)adenine is obtained from 2,3,4-tri-*O*-acetyl-6-deoxy-6-nitro-α-D-*gluco*-pyranosyl bromide and chloromercuri 6-benzamido-purine followed by reduction (Scheme 18) (Baer and Bayer, 1971).

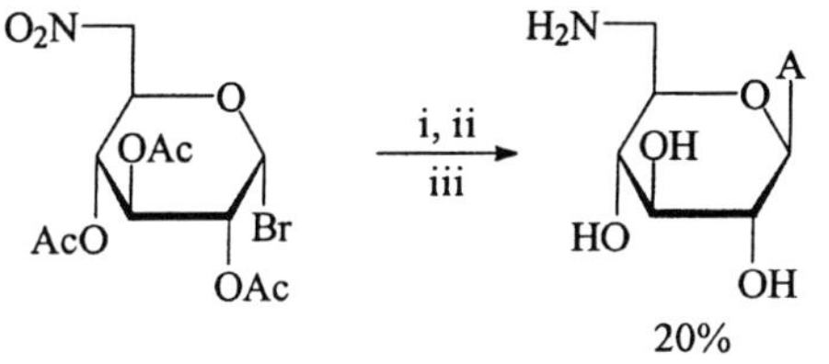

Scheme 18.

i: chloromercuri-6-benzamidopurine, toluene; ii: NaOMe MeOH; iii: H_2, Pd/C, EtOH; picric acid; EtOH, H_2O.

Nucleophilic substitution reactions have been used either to convert aminonucleosides into one another or to introduce an additional amino group via the azido functionality. 9-(2-Amino-2-deoxy-β-D-*allo*-pyranosyl)-6-dimethylaminopurine was obtained from 9-(2-acetamido-2-deoxy-β-D-*gluco*-pyranosyl)-6-dimethylamino-2-methylmercaptopurine via inversion of configuration at the 3-position (Scheme 19).

N-acetylated β-D-glucosamine nucleosides can be converted into 2-acetamido-2-deoxy-β-D-*allo*-pyranosyl nucleosides using an oxidation-reduction step for inversion of the configuration of C-3 (Scheme 20) (Al-Massoudi and Al-Atoom, 1995).

4'-Amino-4'-deoxy-β-D-*galacto*-pyranosylcytosine was synthesized from 1-*O*-acetyl-2,3,6-tri-*O*-benzoyl-4-*O*-mesyl-α-D-glucose and bis(trimethylsilyl)-N⁴-acetylcytosine involving a $SnCl_4$ catalysed nucleosidation reaction, azidolysis and reduction (Scheme

21) (Lichtenthaler *et al.*, 1974b). The uracil analogue was obtained in a multistep approach from β-D-*gluco*syluracil.

40%

20%

27%

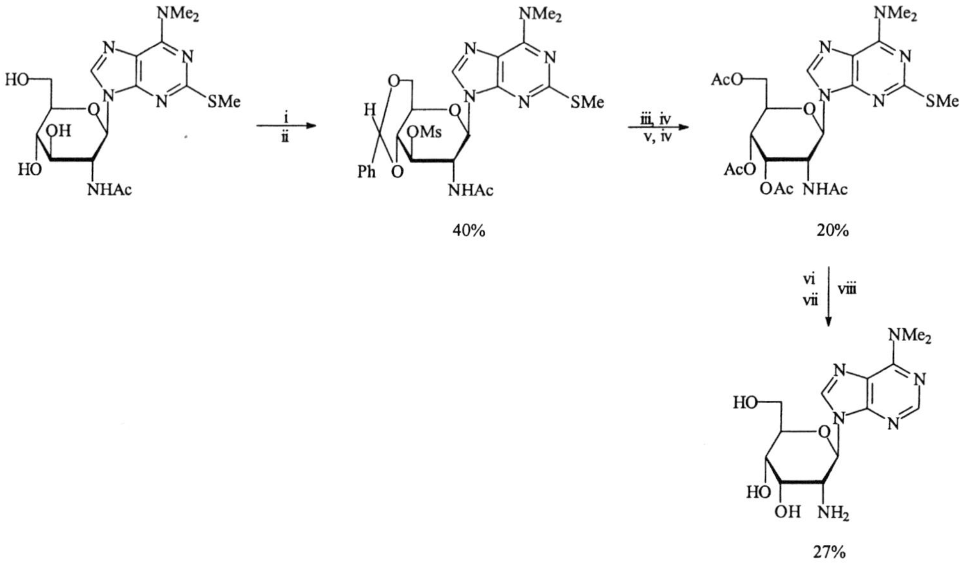

Scheme 19.

i: ZnCl$_2$, PhCH=O; ii: MsCl, pyridine; iii: NaOAc, MeOCH$_2$CH$_2$OH, Δ, iv: Ac$_2$O, pyridine; v: HCl, MeOH; vi: Raney-Ni, EtOH; vii: NaOH, MeOH; viii: Ba(OH)$_2$.

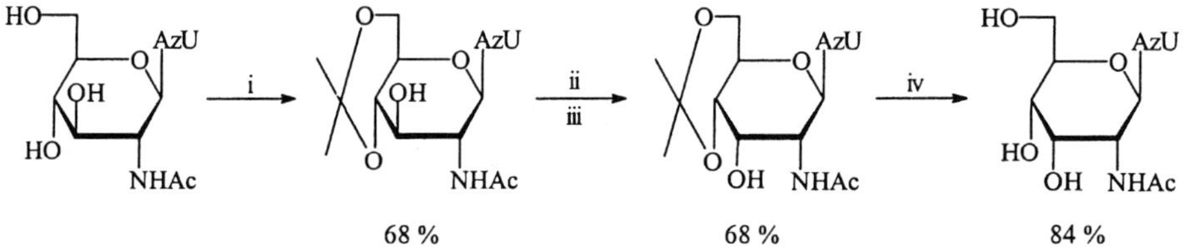

68 %   68 %   84 %

Scheme 20.

i: Me$_2$CO, MeC(OMe)$_2$Me, H$^+$; ii: CrO$_3$, pyridine, Ac$_2$O, CH$_2$Cl$_2$; iii: NaBH$_4$, EtOH, H$_2$O; iv: 80% HOAc

AzU: 6-azauracil

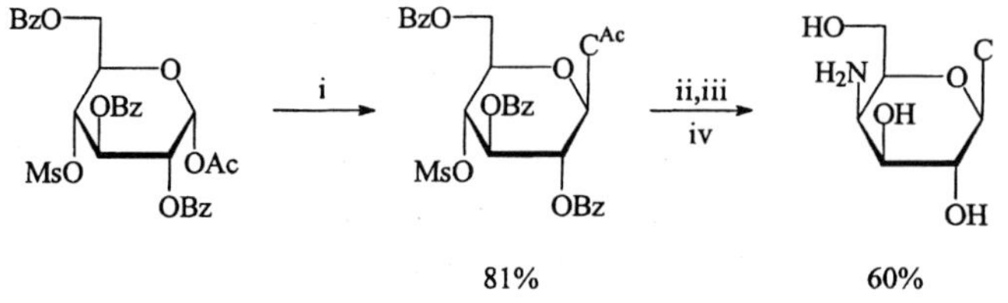

81%   60%

Scheme 21.

i: (Me$_3$Si)$_2$ N^{4}AcC, SnCl$_4$, dichloroethane; ii: NaN$_3$, HMPA; iii: NaOMe, MeOH; iv: H$_2$, Pd/C, H$_2$O.

6-Dimethylamino-9-(3-amino-3,4-dideoxy-β-D-L-*erythro*-pentopyranosyl)purine was synthesized as intermediate for the preparation of pentopyranose analogues of puromycin (Scheme 22) (Carret *et al.*, 1983).

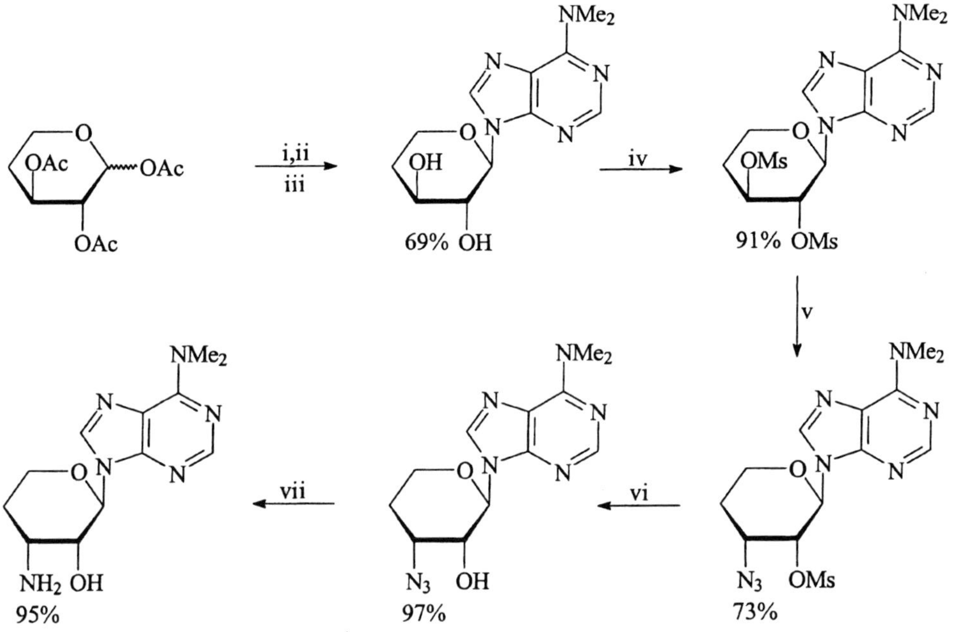

Scheme 22.

i : CH_3COCl, Et_2O, HCl; ii : CH_3NO_2, $Hg(CN)_2$, MS, 6-chloropurine; iii : Me_2N, MeOH; iv : MsCl, pyridine;
v : NaN_3, DMF; vi : NaOH, MeOH; vii : H_2, Pd/C, MeOH.

4. Pyranose nucleosides

The synthesis of 2,4-dideoxy-β-D-*erythro*-hexopyranosyl nucleosides was accomplished by two different synthetic routes from commercially available carbohydrate precursors (Augustyns *et al.*, 1993; Augustyns *et al.*, 1992). The most efficient route starts with tri-*O*-acetyl-D-glucal, via a methoxy mercuration reaction and a regioselective reductive opening of an epoxide (Scheme 23).

During sugar-base condensation reaction (and when no anchimeric assistance is involved) for the synthesis of hexopyranose, the thermodynamically favoured compound formed is that where the base moiety and the CH_2OR group are both equatorially oriented (Augustyns *et al.*, 1994).

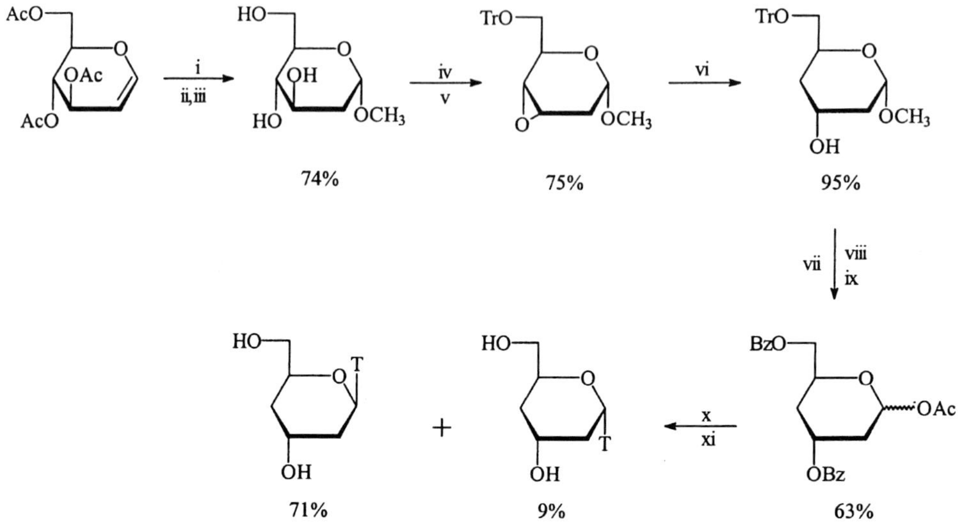

Scheme 23.
i: NaOMe, MeOH; ii: $Hg(OAc)_2$, MeOH; iii: NaCl, MeOH; $NaBH_4$, iPrOH; iv: TrCl, pyridine; v) HMPT, NaH, $(iPr)_3PhSO_2$imidazole, THF; vi: $LiAlH_4$, Et_2O, 95%; vii: pTsOH, MeOH; viii: BzCl, pyridine; ix: 80%, HOAc; Ac_2O-pyridine; x: $(Me_3Si)_2T$, TMSOTf, $C_2H_4Cl_2$; xi: NH_3, MeOH.

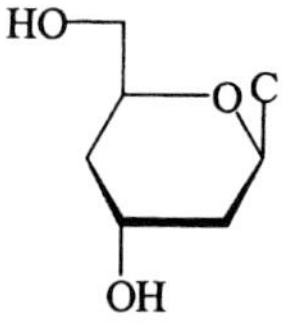

40

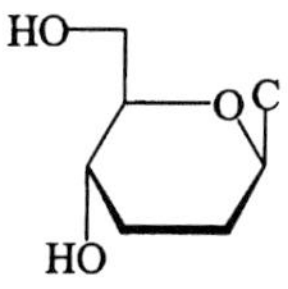

41

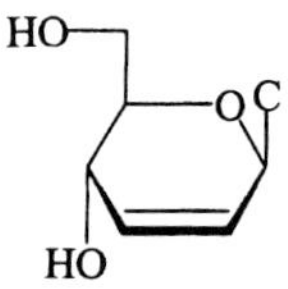

42

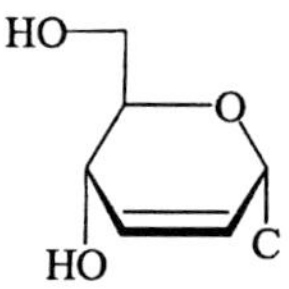

43

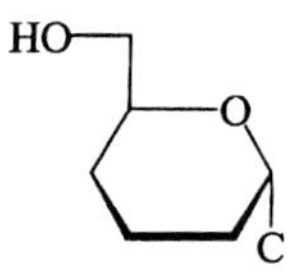

44

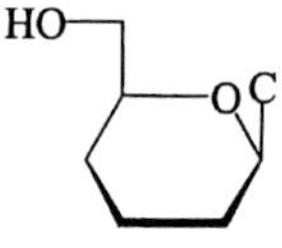

45

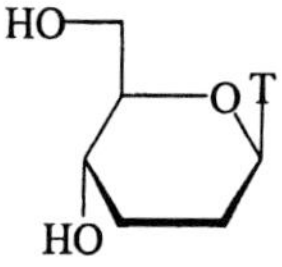

46

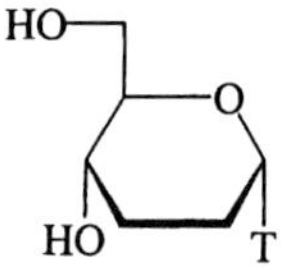

47

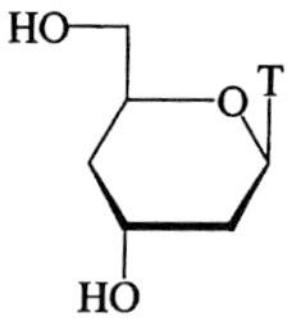

48

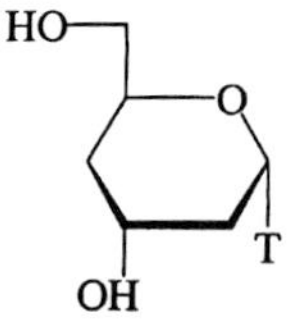

49

In acidic medium, **40** and **41** are about twice as stable as dCyd. The presence of a C_2-C_3 double bond markedly reduce acid stability. The β-anomers are much more stable than the α-anomers, (**42** versus **43**; **45** versus **44**; **46** versus **47**; **48** versus **49**) (Van Schepdael *et al.*, 1994; Thioti *et al.*, 2000).

Selective benzoylation of methyl α-D-*gluco*-pyranoside at position 2 and 6 is the key reaction for the synthesis of 3,4-dideoxy-β-D-*erythro*-hexopyranosyl nucleosides (Scheme 24) (Augustyns *et al.*, 1992c).

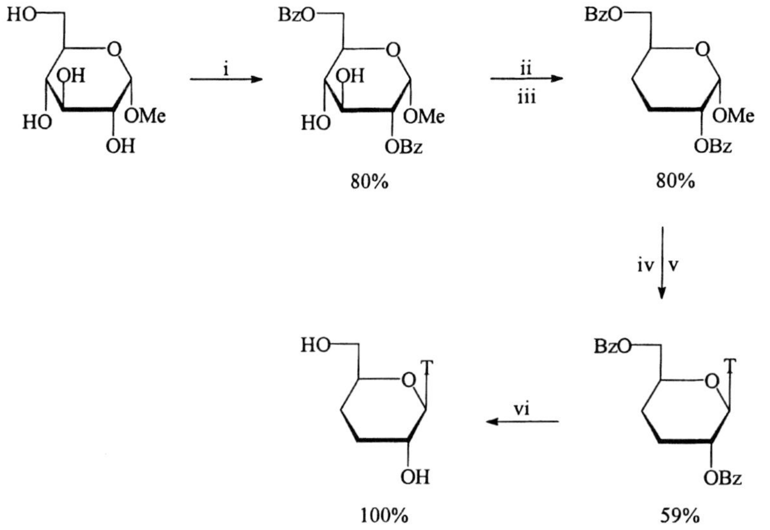

Scheme 24.

i: $(Bu_3Sn)_2O$, toluene, BzCl; ii: Ph_3P, ImH, triiodoimidazole, toluene; iii: H_2, Pd/C, Et_3N, EtOH; iv: AcOH, Ac_2O, H_2SO_4; v: $(Me_3Si)_2T$, $SnCl_4$, CH_3CN; vi: NH_3, MeOH.

Methyl 2,3-dideoxy-D-*erythro*-hexopyranoside (Scheme 25), obtained by Ferrier rearrangement of tri-*O*-acetyl-D-glucal in the presence of $BF_3.OEt_2$ was converted into a thioglycoside. The nucleobase was introduced at primary hydroxyl group, which makes a stereoselective intramolecular glycosylation reaction possible (Sugimura and Sujino, 1998). The compound was further converted into 2',3',6'-trideoxy-β-D-*erythro*-hexopyranosyl nucleosides, useful for the synthesis of amicetin analogues.

Hanessian described the use of 2-methoxypyridyloxy (MOP) and 2-thiopyridylcarbonate (TOPCAT) as activators of O-benzylated hexopyranoses (Scheme 26) (Hanessian *et al.*, 1980). The MOP group is activated by the catalyst (TMSOTf) and the reaction proceed via an 1,2-oxonium-triflate ion pair. Starting from 2,3,4,6-tetra-*O*-benzyl-β-D-*galacto*-pyranosyl MOP and silylated uracil they obtained a high yield of 1,2-trans-β-D-*galacto*-pyranosyl nucleoside (Hanessian *et al.*, 1996).

The protected 7-(3-deoxy-3-fluoro-β-D-*ribo*-hexopyranosyl-2-ulose)theophylline was converted into 7-(3,4-dideoxy-3-fluoro-β-D-*glycero*-hex-3-enopyranosyl-2-ulose)theophylline by an oxidation-elimination procedure (Scheme 29).

A series of unsaturated halogenoketonucleosides were synthesized by the group of K. Antonakis. As a representative example, the synthesis of 7-(3-deoxy-3-fluoro-β-D-*glycero*-hex-2-enopyranosyl-4-ulose)theophylline is described (Scheme 30) (Leclercq *et al.*, 1992).

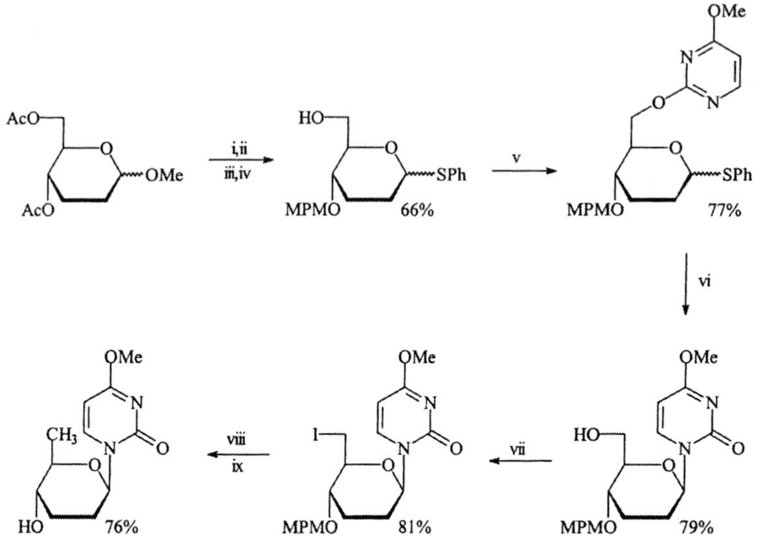

Scheme 25.

i: PhSSiMe$_3$, BF$_3$.Et$_2$O, CH$_2$Cl$_2$; ii: NaOMe, MeOH; iii: *p*-anisaldehyde dimethylacetal, HBF$_4$-OEt$_2$; iv: NaBH$_3$CN, CH$_3$CN; Me$_3$SiCl, CH$_3$CN: v: 2-chloro-4-methoxypyrimidine, NaH, DMF; vi: Me$_2$S(SMe)BF$_4$, M.S. 4Å, CH$_3$CN; NaOH, 1 M; vii: I$_2$, Ph$_3$P, pyridine; Na$_2$S$_2$O$_3$; viii: Bu$_3$SnH, AIBN, toluene; ix: DDQ, CH$_2$Cl$_2$, H$_2$O (MPM: 4-*O*-*p*-methoxybenzyl).

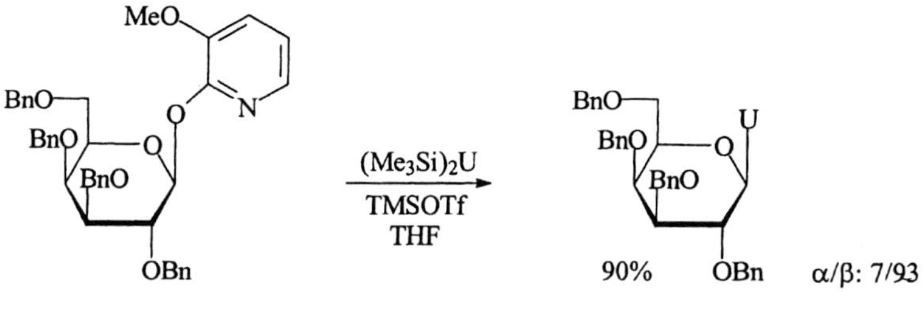

Scheme 26.

A solvent participation may influence the stereochemical outcome of the reaction. Reaction of 2,3,4,6-tetra-*O*-benzyl-β-D-*gluco*-pyranosyl TOPCAT with silylated N^6-benzoylcytosine and silver triflate in toluene gives predominantly the β-nucleoside. When the reaction is performed in THF, the α-nucleoside is the main compound (Scheme 27) (Hanessian *et al.*, 1996).

Direct fluorination at position 3',4'- and 6'- of β-D-*gluco*-pyranosyl-theophylline was accomplished using diethylaminosulfur trifluoride (Scheme 28) (Leclercq and Antonakis, 1989).

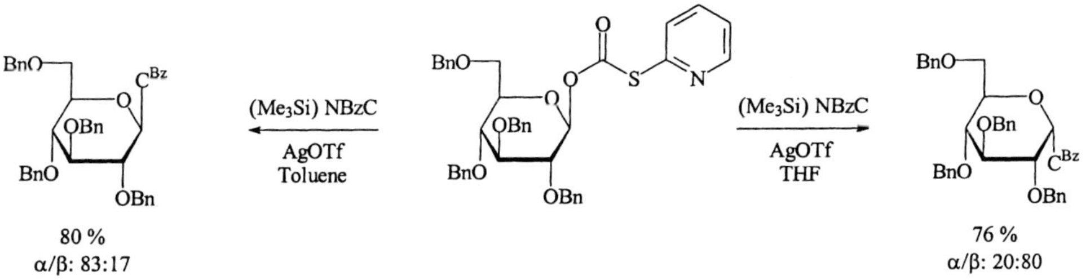

Scheme 27.

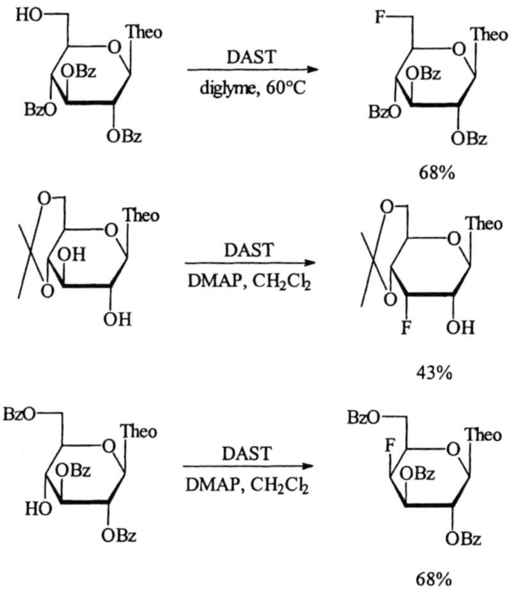

Scheme 28.

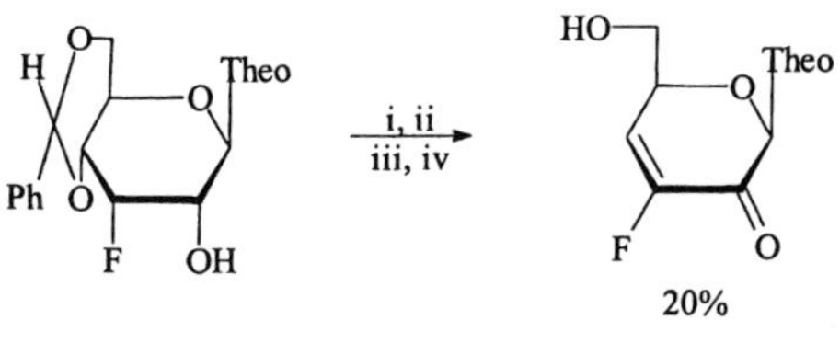

Scheme 29.

i: DMSO, oxalylchloride, CH_2Cl_2; ii: Amberlite IR-120 (H+), MeOH; iii: Ac_2O, pyridine; iv: HCl, MeOH, CH_2Cl_2 Theo: theophylline.

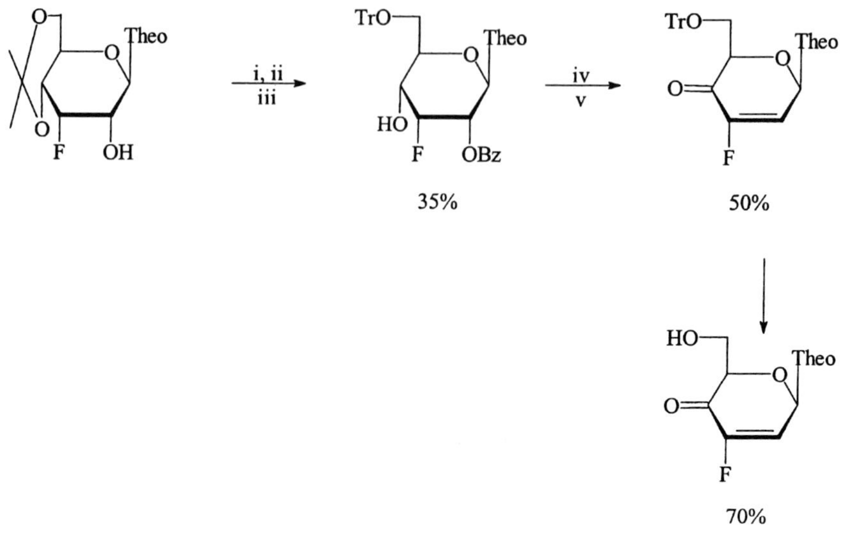

Scheme 30.

i: BzCl, pyridine, ii: Amberlite IR-120 (H⁺), MeOH; iii: TrCl, pyridine, DMAP; iv: pyridinium dichromate, M.S., AcOH, CH_2Cl_2; v: Ac_2O, pyridine; vi: 70% HOAc.

Extension of the reaction whereby "nucleoside dialdehydes" are cyclized by nitromethane to nitroethane produces branched nucleosides, examplified by the synthesis of a mixture of 3'-deoxy-3'-C-methyl-3'-nitro-pyranosyl nucleosides from uridine dialdehyde (Lichtenthaler and Zinke 1966).

For example, the C-methyl-branched 3'-nitro β-D-hexopyranosyl and 3'-amino β-D-hexopyranosyl nucleosides are obtained when the base-catalyzed cyclization reaction is carried out with nitroethane (Scheme 31) (Lichtenthaler and Zinke 1972).

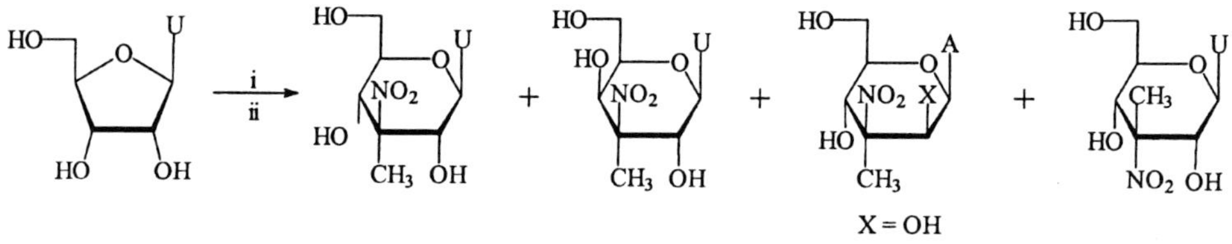

Scheme 31.

i: $NaIO_4$, H_2O; ii: $CH_3CH_2NO_2$, MeOH, NaOMe

1-(2,3,4-trideoxy-4-C-hydroxymethyl-β-D-*erythro*-hexopyranosyl) nucleosides were synthesized form (2S, 3R)-1-*O*-(4-bromobenzyl)-3-(2-propenyl)-1,2,4-butanetriol, itself obtained from the epoxide precursor and allylmagnesium bromide (Scheme 32) (Björnse *et al.*, 1993).

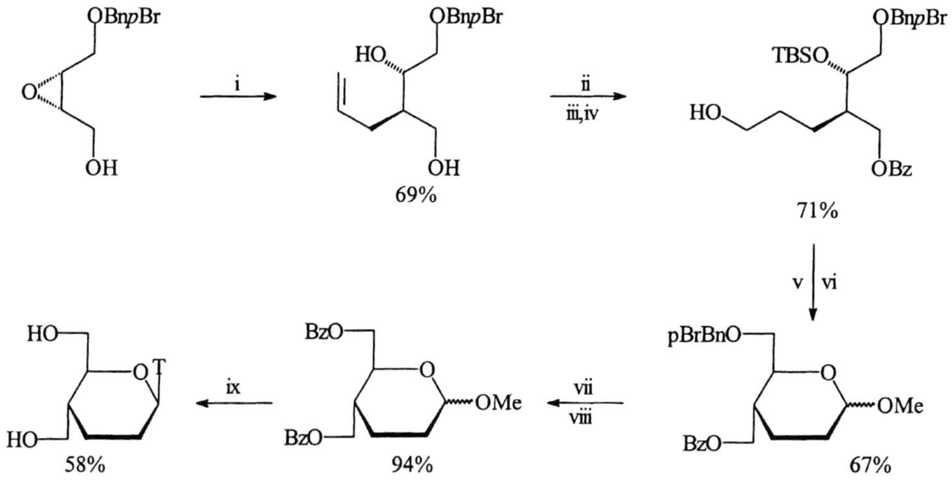

Scheme 32.

i: CH_2=CH-CH_2MgBr, Et_2O; ii: BzCl, CH_2Cl_2, pyridine; iii: TBSCl, ImH, DMF; iv: BH_3SMe_2,THF; $NaBO_3$; v: PDC, Ac_2O, CH_2Cl_2, DMF; vi: HCl, MeOH; vii: H_2, Pd/C, $NaHCO_3$, H_2O, EtOAc; viii: BzCl; CH_2Cl_2, pyridine; ix: $(Me_3Si)_2T$, TBPSTf, CH_2Cl_2, CH_3CN; x: NaOMe, MeOH.

3',4'-C-bishydroxymethyl-2',3',4'-trideoxy-β-L-*threo*-pentopyranosyl nucleosides were obtained from *trans*-(3S, 4S)-bis(methoxycarbonyl)cyclopentanone (Scheme 33) (Lundquist *et al.*, 1995). As could be expected, nucleobase is oriented in the β-position, giving a compound with all substituent in equatorial positions.

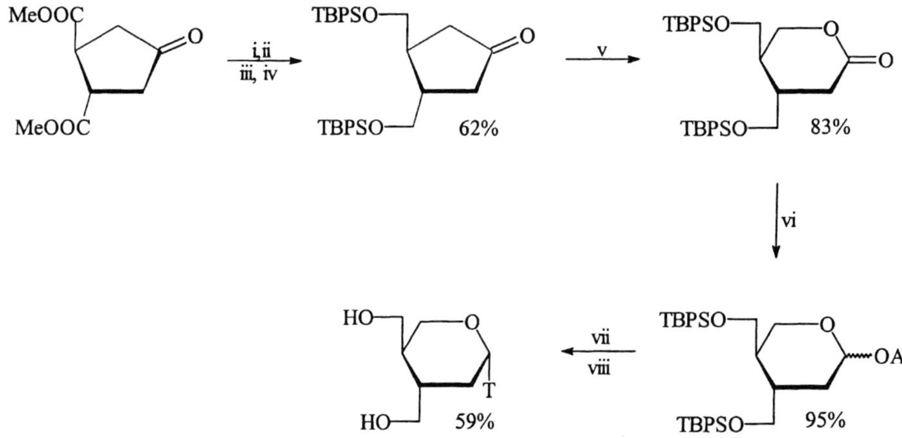

Scheme 33.

i: ethyleneglycol, H^+, toluene; ii: $LiAlH_4$, Et_2O, iii: TBPSCl, pyridine; iv: 80% HOAc, MeOH, acetone, v: mCPBA, CH_2Cl_2; vi: DIBAL-H, toluene; Ac_2O, pyridine; vii: $(Me_3Si)_2T$, TMSOTf, CH_2Cl_2; viii: TBAF, THF. (TBPS: tert-butyldiphenylsilyl).

The synthesis of 1-[4-deoxy-4-*C*-hydroxymethyl-α-L-*lyxo*-pyranosyl]thymine has been accomplished by two synthetic routes both starting from methyl 2,3-*O*-isopropylidene-β-D-ribopyranose (Doboszewski and Herdewijn, 1996a). The first route (Scheme 34) makes use of a ring opening, ring closure reaction sequence to increase the proportion of the desired L-isomers. The second route (Scheme 35) utilizes the soft nucleophilic character of malonyl anions and ozonolytic cleavage of enol ether to introduce the branched chain.

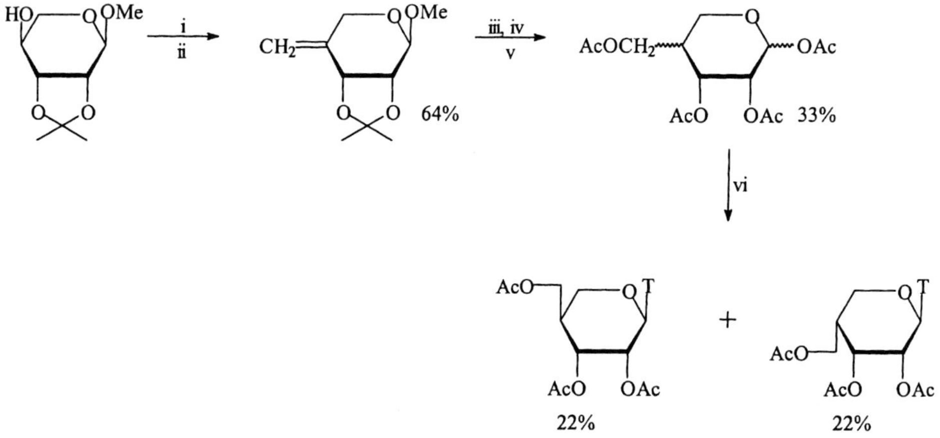

Scheme 34.

i: CrO_3-pyridine-Ac_2O, CH_2Cl_2; ii: CH_3PPh_3Br, THF BuLi; iii: diborane, THF, NaOH, H_2O_2; iv: CF_3COOH, Ac_2O-pyridine; v: Ac_2O, AcOH, H_2SO_4; vi: $(Me_3Si)_2T$; TMSOTf, $C_2H_4Cl_2$.

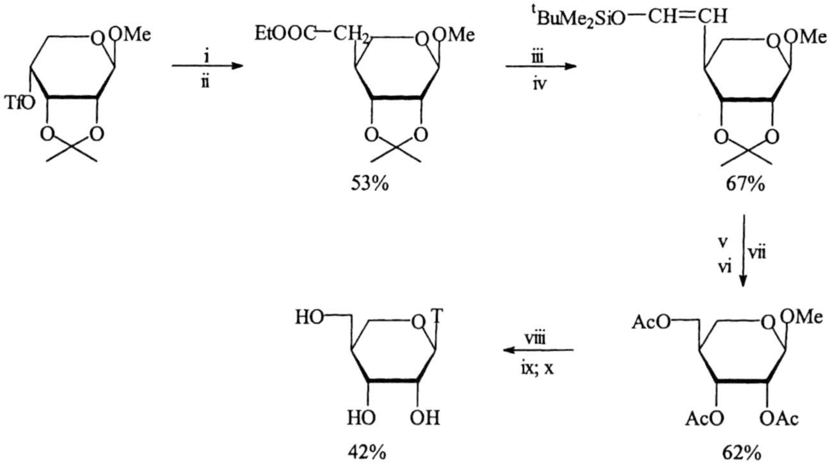

Scheme 35.

i: $NaCH(COOEt)_2$, DMF; ii: DMSO, H_2O, LiCl, Δ; iii: DIBAL, CH_2Cl_2; iv: $^tBuMe_2Si$-OTf, Et_3N, CH_2Cl_2; v: O_3, MeOH; $NaBH_4$; vi: CF_3COOH 90%; vii: Ac_2O-pyridine; viii: AcOH, Ac_2O; H_2SO_4; ix: $(Me_3Si)_2T$, TMSOTf, $C_2H_4Cl_2$; x: NaOMe, MeOH.

The introduction of a hydroxymethyl moiety into the C-4 carbon of a pyranose sugar as key reaction for the synthesis of 2,4-dideoxy-4-*C*-hydroxymethyl-α-L-*lyxo*pyranosyl nucleosides (Scheme 36), has also been accomplished by reaction of the higher order cuprate of trimethylsilylmethyllithium on α-L-methyl-3,4-anhydropentopyranose (Turner and Herdewijn, 1998).

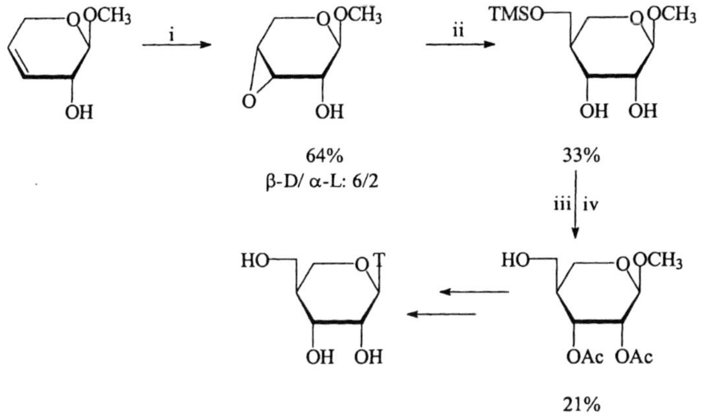

Scheme 36.

i: mCPBA, CH_2Cl_2; ii: $[(Me)_3SiCH_2]Cu(CN)Li_2$, $Ti(OiPr)_4$, THF, H_2O; iii: Ac_2O, pyridine; iv: KBr, AcOOH; AcOH, $HClO_4$.

1-[2,3-Dideoxy-3-*C*-(hydroxymethyl)-α-D-*erythro*-pentopyranosyl]thymine is synthesized from 3,4-di-*O*-acetyl-D-xylal using two stereoselective steps: introduction of the hydroxymethyl group by a radical cyclization of the (bromomethyl)dimethylsilyl ether of an allylic alcohol and Vorbruggen sugar-base condensation (Scheme 37) (Augustyns *et al.*, 1994).

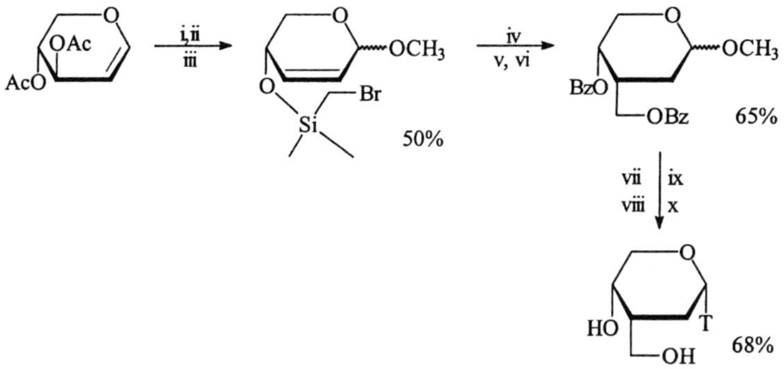

Scheme 37.

i: MeOH, $BF_3.Et_2O$; ii: NaOMe, MeOH; iii: $ClSi(CH_2Br)Me_2$, DMAP, Et_3N, CH_2Cl_2; iv: Bu_3SnH, AIBN, toluene; v: KF, $KHCO_3$, H_2O_2 MeOH, THF; vi: BzCl pyridine; vii: HOAc 80%; viii: Ac_2O, pyridine; ix: thymine, BSA, TMSOTf, CH_2Cl_2; x: NaOMe, MeOH.

The 3-deoxy-3-hydroxymethyl aldopentopyranosyl nucleoside in β-configuration is obtained from the 2,3-dideoxy-β-D-*glycero*-pent-2-enopyranosyl precursor via 2,3-dideoxy-3-*C*-hydroxymethyl-β-D-*erythro*-pentopyranosyl nucleosides (Scheme 38) (Doboszewski *et al.*, 1995a; Doboszewski *et al.*, 1995b).

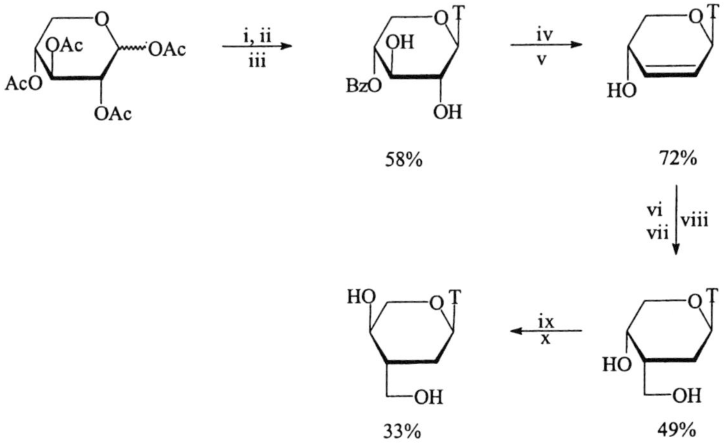

58%   72%

33%   49%

Scheme 38.

i:$(Me_3Si)_2T$; TMSOTf, $C_2H_4Cl_2$; ii: NaOMe, MeOH; iii: Bu_2SnO, MeOH; DMF, dioxane, BzCl; iv: ImH, Ph_3P, I_2, CH_3CN, toluene; Zn; v: NH_3 MeOH; vi: ImH, DMF, $(BrCH_2)SiMe_2Cl$; vii: Bu_3SnH, AIBN, toluene; viii: KF, $KHCO_3$, H_2O_2, DMF; ix: benzoic acid, Ph_3P, DEAD, dioxane; x: NH_3, MeOH.

The starting β-configured 2',3'-unsaturated pentopyranosyl nucleosides are more easily available using a fusion process between di-nitrobenzoyl protected D-xylal and heterocyclic bases. This method is superior over the acid mediated Ferrier rearrangements, which gives in addition to the anomeric nucleosides, 3'-substituted byproducts (Doboszewski *et al.*, 1995c; Doboszewski *et al.*, 1995d). Conformational analysis of these 2',3'-unsaturated pentopyranosyl nucleosides shows that purine bases are preferentially oriented pseudoaxially while pyrimidine bases prefer the pseudo equatorial position (Doboszewski *et al.*, 1995a). This two-state dynamic conformational equilibrium ($^6H_5 \Leftrightarrow {}^5E/^5S_4$) of β- and α-D-*glycero*-pent-2'-enopyranosyl nucleosides is driven by the tuning of anomeric effect, gauche effect, steric effect and the interaction between the π-system of the double bond with the heterocyclic aglycon and the 4'-OH in the allylic position (Polak *et al.*, 1997). The strenght of the π→σ* interaction increases in the following order: cytosine < uracil < thymine < adenine < guanine.

The synthesis of 3-deoxy-3-C-hydroxymethyl--L-lyxopyranosylthymine makes use of a furanose → pyranose conversion and of the formation of both furanose and pyranose nucleosides during Vörbruggen sugar-base condensation reaction starting from tetra-*O*-acetyl-3-deoxy-3-*C*-hydroxymethyl-L-lyxo-furanose (Scheme 39) (Doboszewski and Herdewijn 1996b).

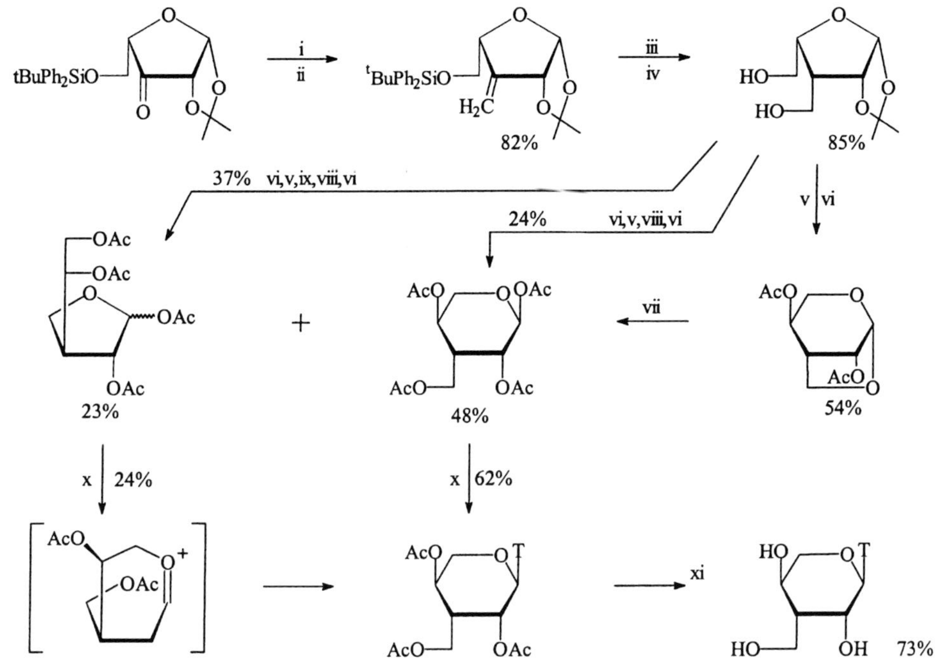

Scheme 39.

i: Me_3SiCH_2Li, CH_2Cl_2; ii: NaH, THF; iii: B_2H_6, THF; NaOH, H_2O_2; iv: TBAF, THF; v: 90% CF_3COOH; vi: Ac_2O, pyridine; vii: Ac_2O, AcOH, H_2SO_4; viii: NaOMe, MeOH; ix: CF_3COOH, DMF; x: $(Me_3Si)_2T$, TMSOTf, $C_2H_4Cl_2$; xi: NaOMe, MeOH.

5. Isonucleosides

Synthesis of isonucleosides by Michael addition of the purine base on an α,β-unsaturated sugar was first described by J. Carbon (1964). Isonucleosides are obtained as side compounds during the previously mentioned acid catalysed fusion reaction using glycals. The Michael type addition reaction between a silylated pyrimidine base and an unsaturated pyran-3(6H)-one is the key reaction for a synthetic procedure leading to pyranosyl isonucleosides (Scheme 40) (Prévost and Rouessac 1997).

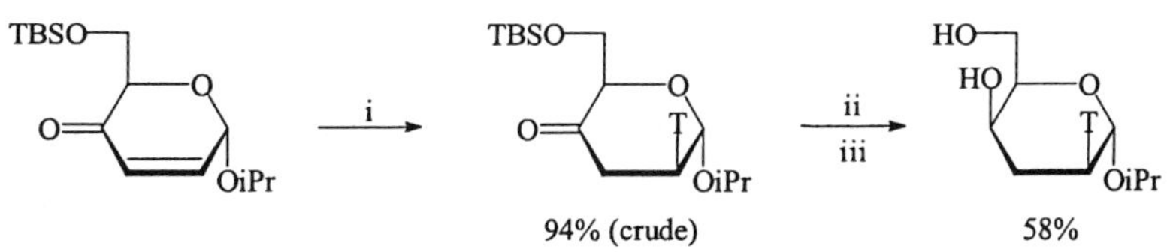

Scheme 40.

i: $(Me_3Si)_2T$, TMSOTf, CH_3CN; ii: $NaBH_4$, EtOH; iii: TBAF, THF

Reaction of methyl 2,3-anhydro-4,6-*O*-benzylidene-α-D-*allo*-pyranoside with adenine gives the 2-deoxy-D-*altro*-pyranoid nucleoside (Scheme 41) (Ohrui *et al.*, 1993). The corresponding D-*manno* derivative was synthesized in the same way starting from methyl 3-*O*-benzoyl-4,6-benzylidene-2-*O*-(trifluoromethylsulphonyl)-α-D-*gluco*-pyranoside (Ohrui *et al.*, 1993).

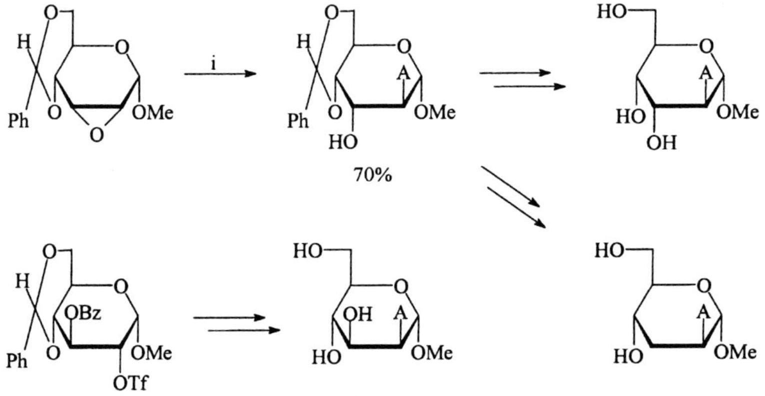

Scheme 41.

i: adenine, K$_2$CO$_3$, 18-Crown-6, DMF, Δ,

The synthesis of 1,5-anhydro-2,3-dideoxy-D-*arabino*-hexitol nucleosides starts from D-glucose. The compounds were first prepared by conversion of D-glucose in 3-deoxy-1,5-anhydro-D-hexitol, which was coupled at its 2-position to heterocyclic bases either by nucleophilic displacement or under Mitsunobu reaction conditions (Scheme 42) (Verheggen *et al.*, 1993; Verheggen *et al.*, 1995a; De Bouvere *et al.*, 1997; ostrowski *et al.*, 1998; Boudou *et al.*, 1999; Nandanan *et al.* 2000). Later, a more efficient approach starting from diacetone-D-glucose was published (Scheme 43) (Andersen *et al.*1996), which was also used to prepare the corresponding L-nucleoside analogues.

The synthesis of 1,5-anhydro-2,4-dideoxy-D-mannitol nucleosides, likewise, start with 1,5-anhydro-4,6-*O*-benzylidene-D-glucitol (Scheme 44). The compounds were obtained using the same principles as for the synthesis of 1,5-anhydro-2-deoxy-D-mannitol nucleosides except for the deoxygenation procedure which was performed before introduction of the base moiety (Hossain *et al.*, 1999).

1,5:2,3-Dianhydro-4,5-*O*-benzylidene-D-allitol can be prepared from commercially available tetraacetyl-α-D-bromoglucose. Nucleophilic opening of the epoxide yielded 1,5-anhydro-2-deoxy-D-*altro*-hexitol nucleosides (Scheme 45) (Allart *et al.* 2000; Verheggen *et al.*, 1995b; Pérez-Pérez *et al.*, 1996).

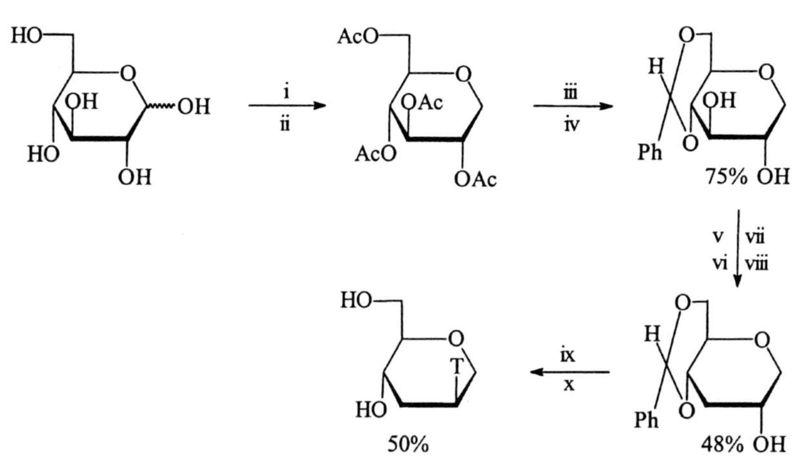

Scheme 42;

i : Ac$_2$O, HBr/HOAc; ii : Bu$_3$SnH, Et$_2$O; KF, H$_2$O; iii : NaOMe, MeOH; iv : C$_6$H$_5$CHO, ZnCl$_2$, toluene;

v : Bu$_2$SnO, benzene; CH$_3$C$_6$H$_4$COCl, dioxane; vi : CSCl$_2$, DMAP, CH$_2$Cl$_2$, 2,4-Cl$_2$C$_3$H$_3$OH, CH$_2$Cl$_2$;

vii : Bu$_3$SnH, AIBN, toluene; viii : NaOMe, MeOH; ix : N^3-benzoylT, Ph$_3$P, DEAD, dioxane; NH$_3$, MeOH;

x : 80% HOAc.

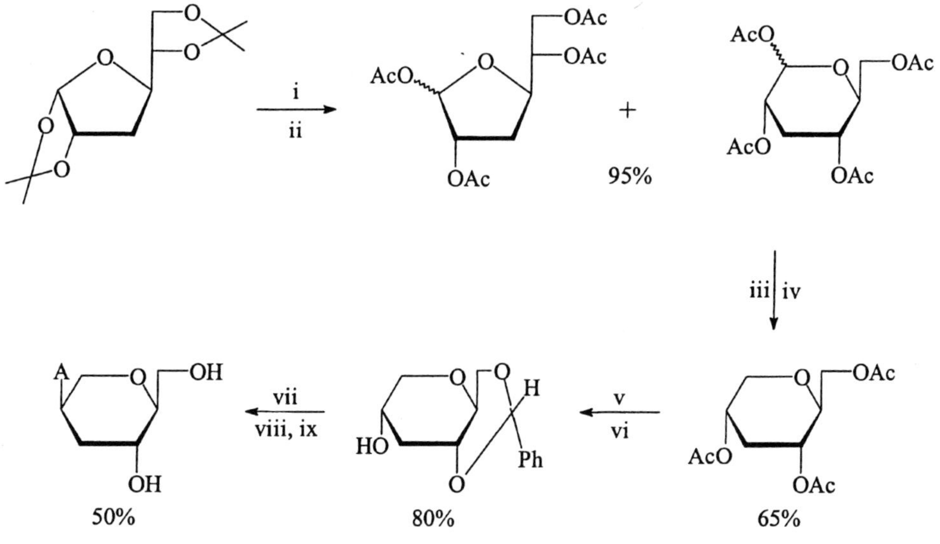

Scheme 43.

i : IRA-120(H$^+$); EtOH, H$_2$O, Δ; ii : Ac$_2$O, pyridine; iii : HBr, HOAc; iv : Bu$_3$SnH, Et$_2$O; v : NaOMe, MeOH;

vi : C$_6$H$_5$CH(OMe)$_2$, dioxane; vii : 6-chloropurine, Ph$_3$P, DEAD; viii : 80% HOAc; ix : NH$_3$.

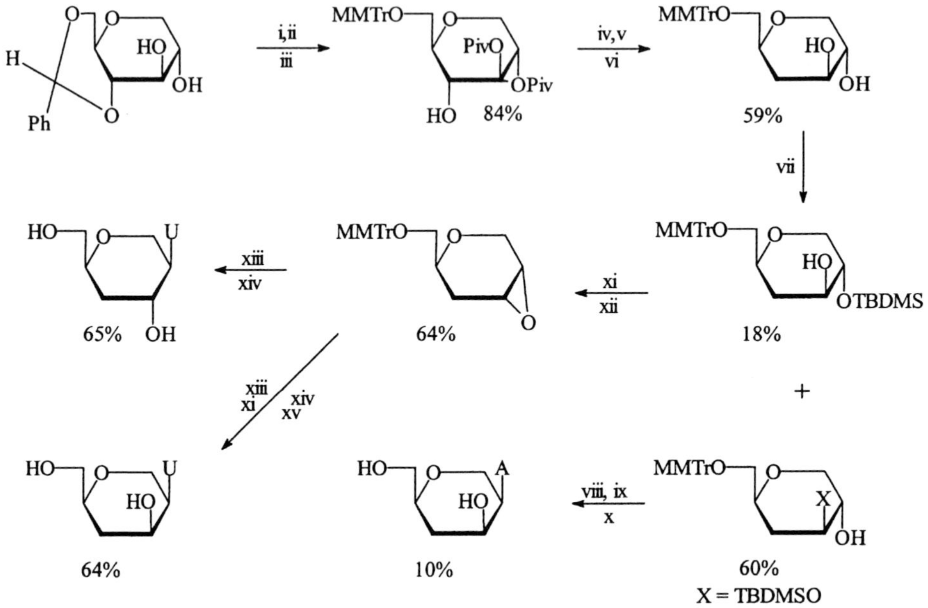

Scheme 44.

i: PivCl, pyridine; ii: 80% CF$_3$COOH, CH$_2$Cl$_2$; iii: MMTrCl, pyridine; iv: PhOC(S)Cl, 0DMAP, CH$_3$CN; v: Bu$_3$SnH, AIBN, toluene; vi: NaOH, H$_2$O, dioxane; vii: TBDMSCl, ImH, DMF; viii: (TFl)$_2$O, pyridine, CH$_2$Cl$_2$; ix: Bu$_4$N$^+$ adenine$^-$, CH$_2$Cl$_2$; x: CF$_3$COOH, H$_2$O; xi: MsCl, Et$_3$N, CH$_2$Cl$_2$; xii: TBAF, THF; xiii: Uracil, NaH, DMF; xiv: 80% HOAc; xv: NaOH, EtOH.

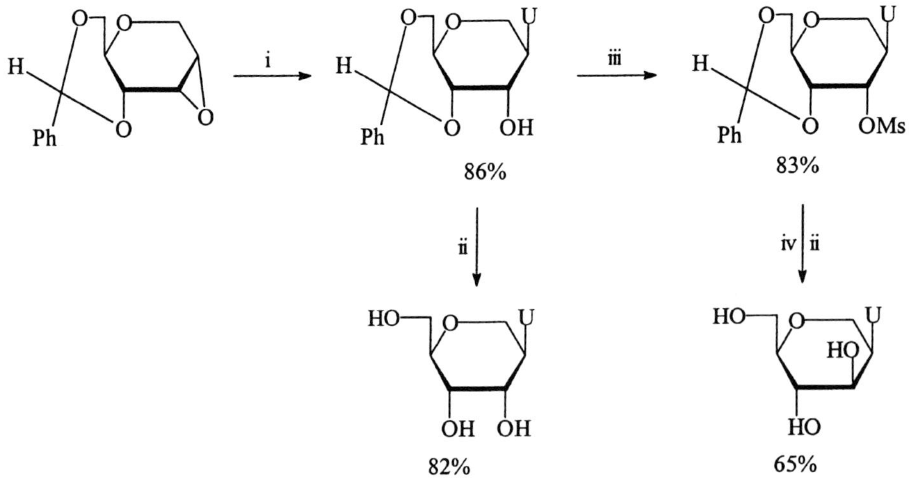

Scheme 45.

i: uracil, NaH, DMF; ii: 80%, HOAc; iii: MsCl, DMAP, pyridine; iv: NaOH, EtOH, Δ

The corresponding *manno*-hexitol nucleosides in the pyrimidine series can be obtained via O^2,3'-anhydro formation and "in situ" opening of the newly formed heterocyclic ring (Scheme 45) (Pérez-Pérez *et al.*, 1996). For the synthesis of the *manno*-hexitol nucleosides with a purine base moiety a stepwise protecting group approach is needed, starting from 1,5-anhydro-4,6-*O*-benzylidene-D-glucitol (Scheme 46) (Hossain *et al.*, 1998a).

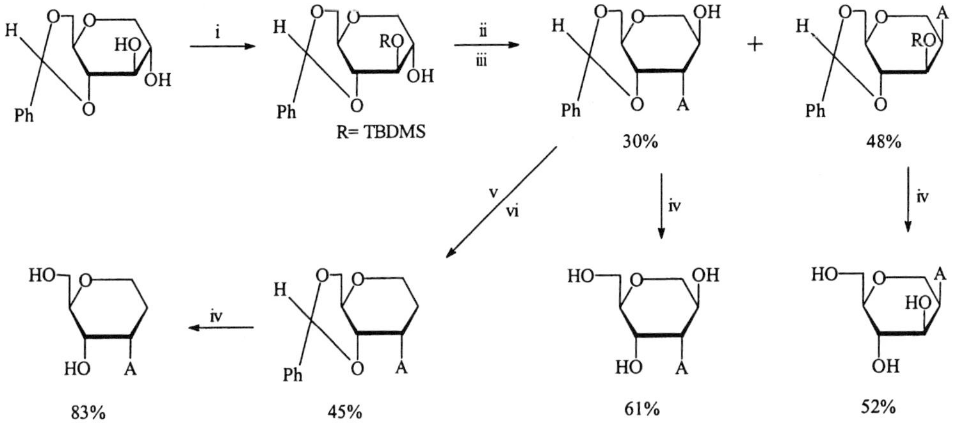

R= TBDMS

30%

48%

83%

45%

61%

52%

Scheme 46.

i: TBDMSCl, ImH, DMF; ii: TFl$_2$O, pyridine, CH$_2$Cl$_2$; iii: Bu$_4$N$^+$adenine$^-$, CH$_2$Cl$_2$; iv: CF$_3$COOH, H$_2$O; v: NaH, CS$_2$, MeI, THF; vi: Bu$_3$SnH, AIBN, toluene

The 1,5-anhydro-4,6-*O*-benzylidene-2-*O*-*p*-toluoyl-D-glucitol intermediate was used as starting material for the preparation of β-D-*threo*-hex-3-enopyranosyl nucleosides (Scheme 47). Unsaturation was introduced using the chlorodiphenylphosphine/iodine/imidazole system (Luyten and Herdewijn 1996).

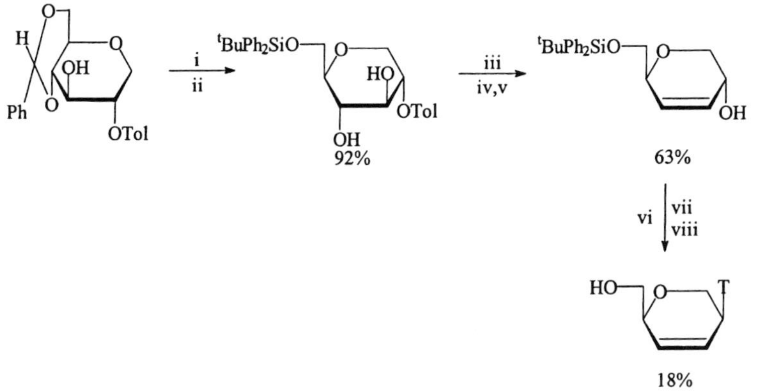

OTol

92%

63%

18%

Scheme 47.

i: 80% HOAc; ii: ^{t}BuPh$_2$SiCl, ImH, CH$_2$Cl$_2$; iii: Ph$_2$PCl$_2$, I$_2$, ImH, CH$_3$CN, toluene; iv: Zn dust; v: NaOMe, MeOH; vi: N^{3}BzT, Ph$_3$P, DEAD, dioxane; vii: NH$_3$, MeOH; viii: TBAF, THF.

The 1,5-anhydro-2,3,4-trideoxy-D-*threo*-hexitol nucleoside was synthesized starting from levoglucosenone via several routes (Scheme 48) (Jung and Kiankarimi 1998), one of them is given below.

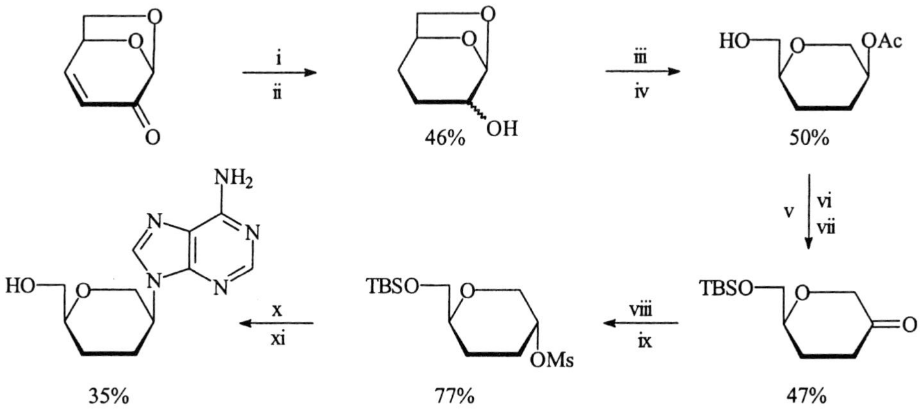

Scheme 48.

i: H_2 Pd/BaSO$_4$, EtOAc: ii: LiAlH$_4$, Et$_2$O; iii: AcCl, pyrimidine, CH$_2$Cl$_2$: iv: Et$_3$SiH, TMSOTf, CH$_3$CN; v: TBSCl, ImH, DMF; vi: K$_2$CO$_3$, MeOH; viii: (COCl)$_2$, DMSO, CH$_2$Cl$_2$, Et$_3$N; viii: LiAlH$_4$, Et$_2$O; ix: MsCl, Et$_3$N, DMAP, CH$_2$Cl$_2$; x: adenine, K$_2$CO$_3$, 18-crown-6, DMF, 100 °C; xi: TBAF, THF.

The α- as the β-analogues of 1,5-anhydrohexitol nucleosides could, likewise, be obtained from acyclic nucleoside precursors (Scheme 49) (Hossain *et al.*, 1997).

3-C-branched 1,5-anhydrohexitol nucleosides in the 3-(S) configuration was obtained via conversion of pentofuranose to nitrohexitol, addition of formaldehyde and removal of the nitro group using Bu$_3$SnH reduction (Scheme 50) (Hossain *et al.*, 1998b).

6.   Phosphonate nucleosides

Condensation of protected D-xylal with heterocyclic bases leads to two isomeric nucleosides which were converted to their respective phosphonate analogues (Scheme 51). The ring expanded D4T phosphonate analogue was also obtained from a longer route starting from peracylated D-xylose (Pérez-Pérez *et al.*, 1994; Pérez-Pérez *et al.*, 1995a).

The key step for the synthesis of the 2,5-*cis*-substituted dihydro-2H-pyranosyl nucleosides is the introduction of the phosphonomethoxy moiety on pentopyranosyl glycals through an acid catalyzed Ferrier-type rearrangement. The attack of the alcohol function occurs preferentially anti to the C-4-substituent (Scheme 52) (Pérez-Pérez *et al.*, 1995b). The compounds were also converted into their saturated congeners.

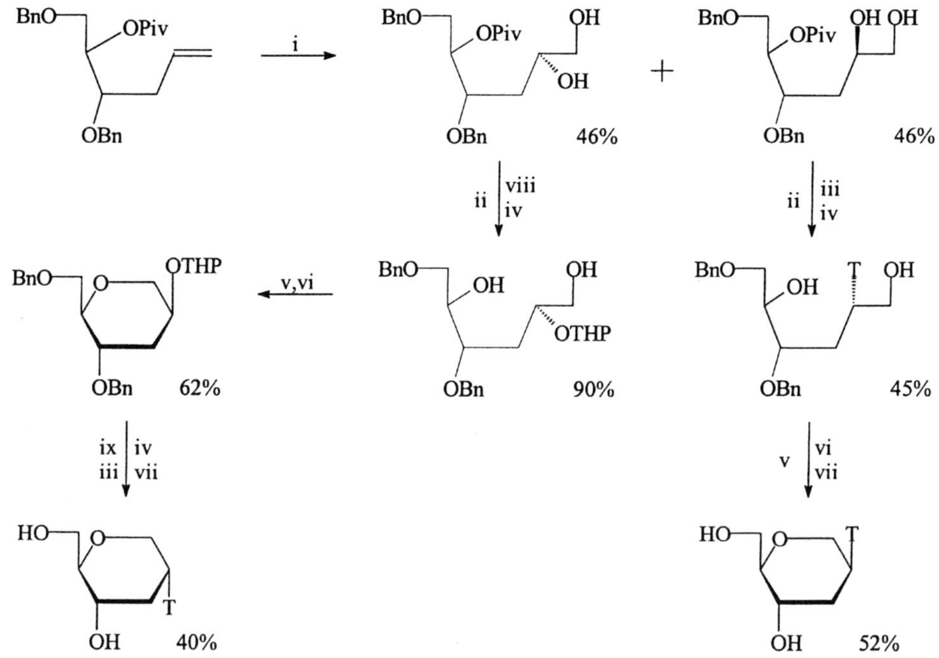

Scheme 49.

i : ADmixβ, 2-methyl-2-propanol, H_2O; ii : PivCl, pyridine; iii : N^3BzT, Ph_3P, DEAD, dioxane : iv : NaOH, H_2O, dioxane; v : TsCl, pyridine; vi : NaH, DMF; vii : $Pd(OH)_2/C$, cyclohexene, MeOH; viii : dihydropyran, H^+, CH_2Cl_2; ix : TsOH, MeOH.

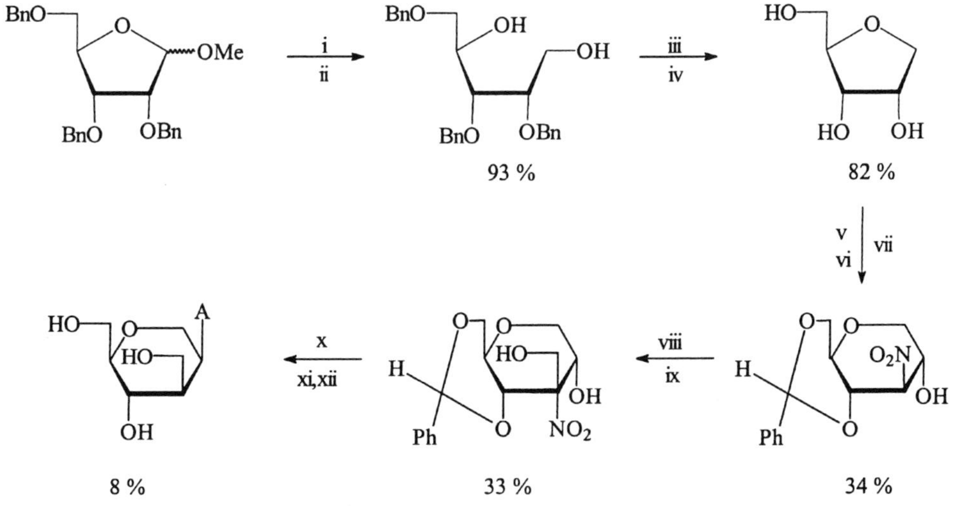

Scheme 50.

i: AcOH, H_2O; ii: $NaBH_4$, MeOH; iii: Ph_3P, DEAD, dioxane; iv: $Pd(OH)_2/C$ MeOH, cyclohexene; v: $NaIO_4$, H_2O; vi: CH_3NO_2, NaOMe, MeOH; vii: C_6H_5CHO, $ZnCl_2$; viii: CH_2O, TMG, CH_3CN; ix: MMTrCl, pyridine; x: Bu_3SnH, AIBN, toluene; xi: adenine, Ph_3P, DEAD, dioxane; xii: CF_3COOH, H_2O.

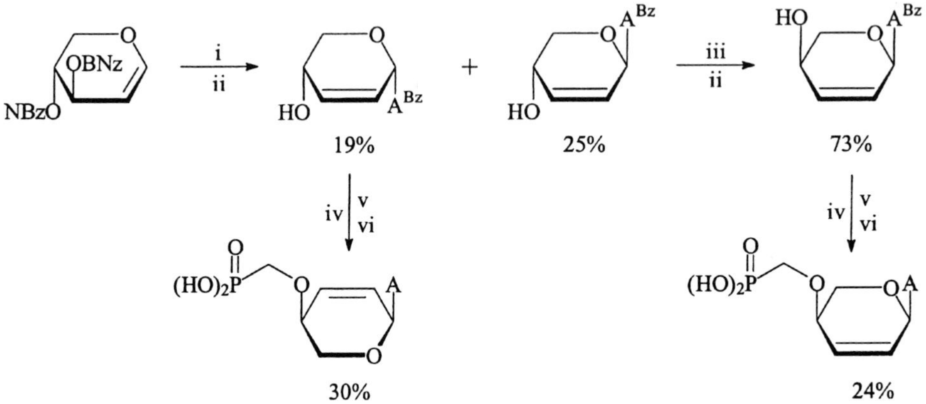

272   *P. Herdewijn*

19%   25%   73%

30%   24%

Scheme 51.

i : N⁶BzA, DMF; ii : NaOMe, MeOH; iii : pNBzOH, Ph₃P, DEAD, dioxane; iv : NaH, (iPrO)₂P(O)CH₂OTs, DMF;

v : NaOH, MeOH; vi : TMSiBr, DMF, lutidine, NH₃, H₂O.

50%   18%

55%   16%

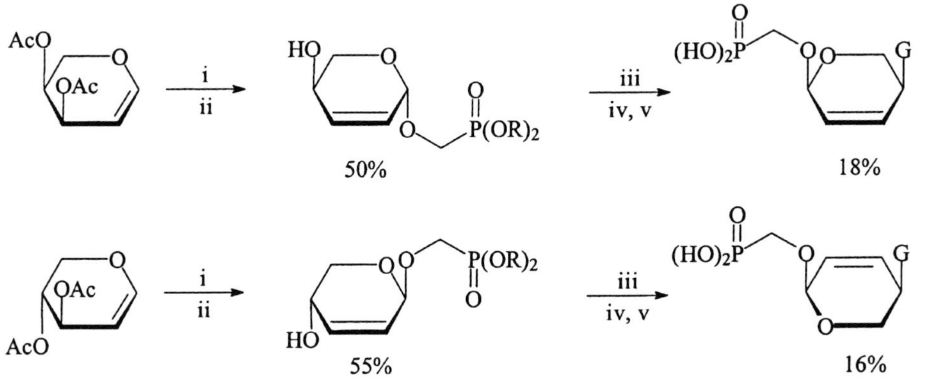

Scheme 52.

i : (iPrO)₂P(O)CH₂OH, TMSOTf, CH₃CN; ii : NH₃, MeOH; iii : 2-amino-6-chloropurine, Ph₃P, DEAD, dioxane;

iv : Me₃N, H₂O, DBU; v : TMSiBr, DMF, lutidine, NH₃, H₂O.

Addition of triisopropyl phosphite to a glycal delivers α- and β-2-enopyranosyl-phosphonates via a Ferrier rearrangement, which were substituted with the nucleobase under Mitsunobu conditions followed by deprotection (Scheme 53). This method allowed the synthesis of a series of 2',3'-unsaturated isonucleotide analogues. In case of the α-phosphonates, deprotection reaction induced double-bond migration leading to 1',2'-unsaturated compounds (Alexander *et al.*, 1996).

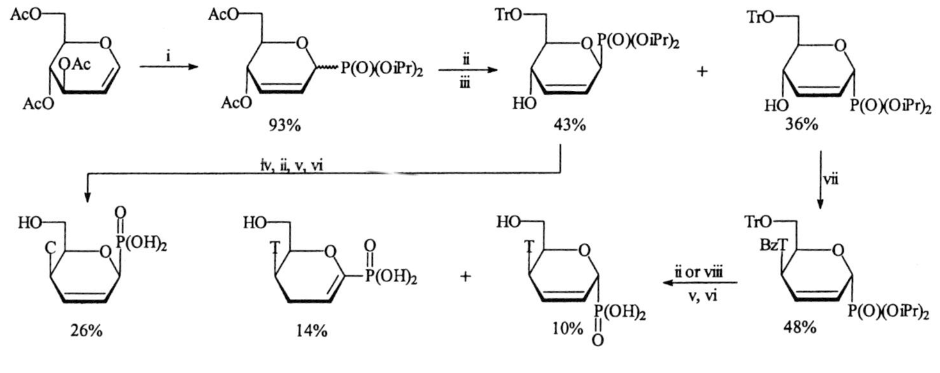

Scheme 53.

i : P(OiPr)$_3$, BF$_3$.Et$_2$O, toluene; ii : NH$_3$, MeOH; iii : TrCl, pyridine; iv : NBzC, DEAD, Ph$_3$P, THF; v : HOAc, H$_2$O; vi : TMSBr, lutidine, CH$_3$CN; vii : N$_3$-benzoylthymine, DEAD, Ph$_3$P, THF; viii : Et$_3$N, MeOH.

7. Carbocyclic nucleosides

The synthesis of cyclohexyl nucleosides was pioneered by H. Schaeffer (Schaeffer *et al.*, 1964a; Schaeffer *et al.*, 1964b; Schaeffer *et al.*, 1964c) and aimed at discovering new adenosine deaminase inhibitors. The compounds were synthesized by a gradual build up of the base moiety starting from the appropriate amino alcohol (Scheme 54).

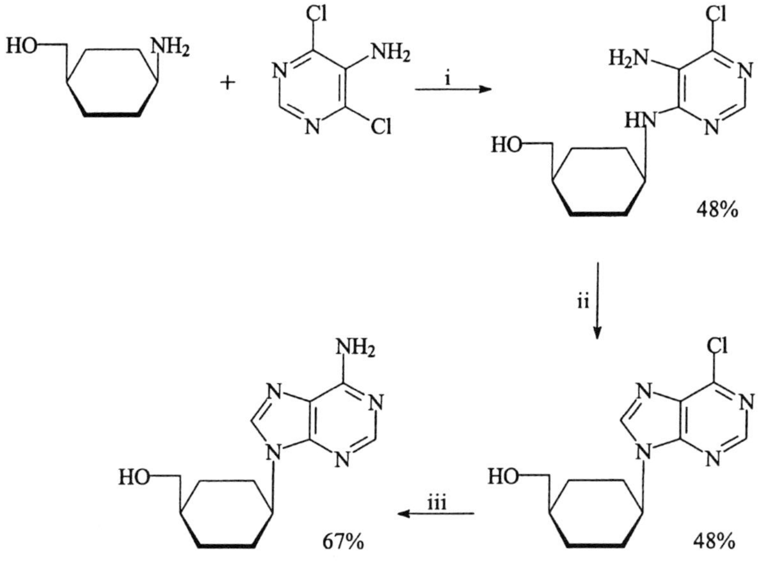

Scheme 54.

i : Et$_3$N, n-Butanol, reflux; ii : (EtO)$_3$CH, reflux; iii : NH$_3$, 55°C.

A common way to obtain cyclohexanyl nucleoside is by epoxide opening. The coupling of triethylamine-activated 6-chloropurine with 2,3-anhydro-1,5,6-tri-*O*-(methanesulfonyl)-*epi*-inositol gives the 6-chloropurine substituted *muco*-inositol (Scheme 55) (Aguilar *et al.*, 1992).

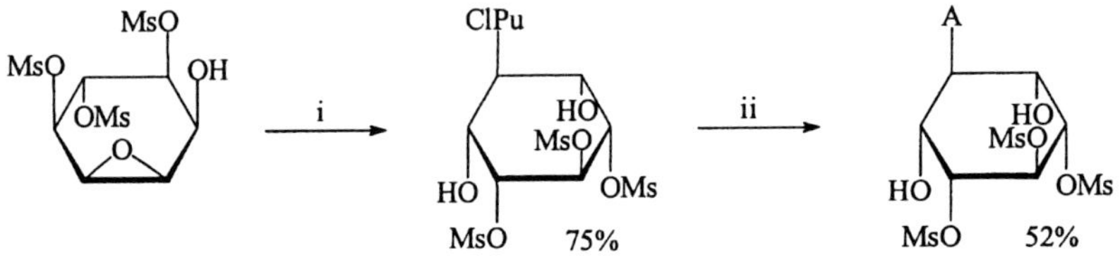

Scheme 55.

i : 6-chloropurine, Et$_3$N, DMF, Δ; ii : NH$_3$, 25%, Δ.

Nucleophilic ring opening of a *cis*-benzyloxy epoxide or a *trans*-benzyloxy epoxide using the salts of nucleobases yielded cyclohexanyl nucleosides with a simple substitution pattern (Scheme 56) (Calvani *et al.*, 1995).

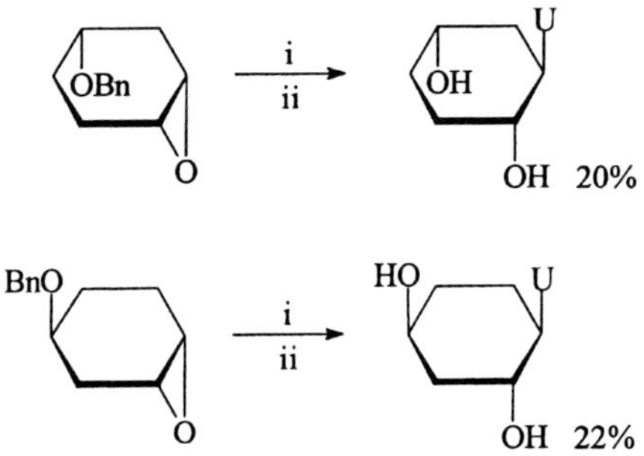

Scheme 56.

i : Uracil, BuLi, hexane, LiClO$_4$, DMF; ii : H$_2$, Pd/C, H$^+$, MeOH

This reaction was also used for the synthesis of dihydroxymethyl cyclohexane nucleosides (Scheme 57) (Mikhailov *et al.*, 1996).

A second method to obtain cyclohexanyl nucleosides is by using a Michael reaction. 9-*Pseudo*-β-D-glucopyranosyladenine and 9-*pseudo*-β-L-idopyranosyladenine, were obtained using as key reaction the Michael-type addition of adenine to nitro-cyclohexene derivatives (Scheme 58) (Kitagawa *et al.*, 1989).

Condensation of 2-amino-6-chloropurine with 2-cyclohexen-1-one in the presence of DBU gives, after reduction, 3-hydroxycyclohexyl nucleosides (Scheme 59) (Halazy *et al.*, 1992).

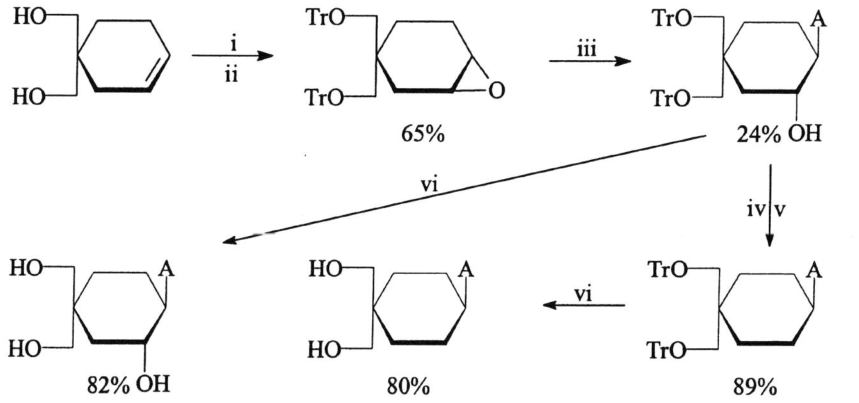

Scheme 57.

i : TrCl, pyridine; ii : MCPBA, CH_2Cl_2; iii : adenine, NaH, DMF; iv : PhOC(S)Cl, DMAP, CH_2Cl_2; v :
Bu_3SnH, AIBN, toluene; vi : AcOH 90%

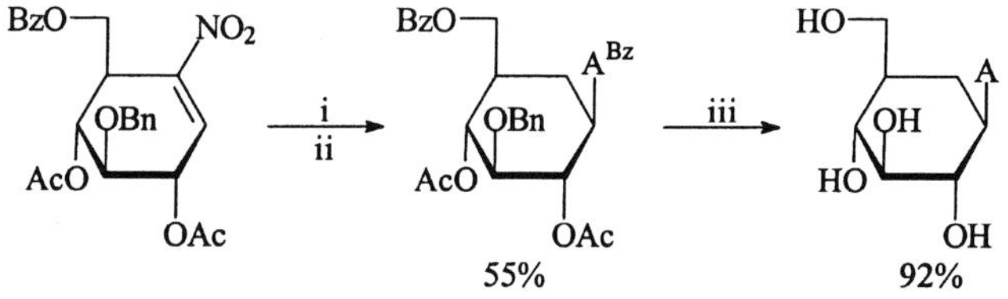

Scheme 58.

i : N⁶BzA, KF, 18-Crown-6, DMF; ii : Bu_3SnH, AIBN, benzene; iii : 1% NaOMe, MeOH; Na, liq.
NH_3, THF.

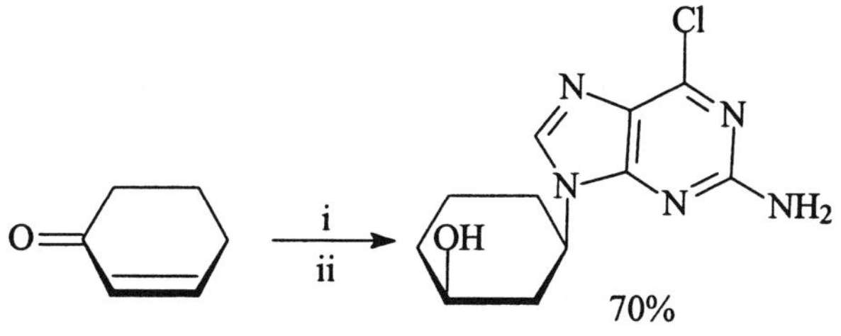

Scheme 59.

i) 2-amino-6-chloropurine, DBU, DMF; ii) $NaBH_4$, EtOH

Conjugate addition of heterocyclic bases to methyl 1,3-cyclohexadine-1-carboxylate leads to 4-hydroxymethyl-3-cyclohexenyl nucleosides (Scheme 60) (Arango *et al.*, 1993).

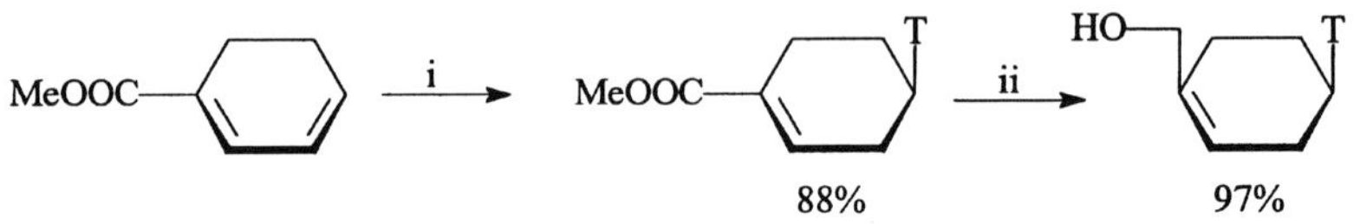

Scheme 60.
i) Thymine, DBU; DMF; ii) LiAlH$_4$, THF

This method was used to synthesize the carbocyclic analogues of the anhydrohexitol nucleosides (Scheme 61) (Maurinsh *et al.*, 1997). The enantiomers were separated as 3'-*O*-(R)-methylmandelic esters (Maurinsh *et al.*, 1999).

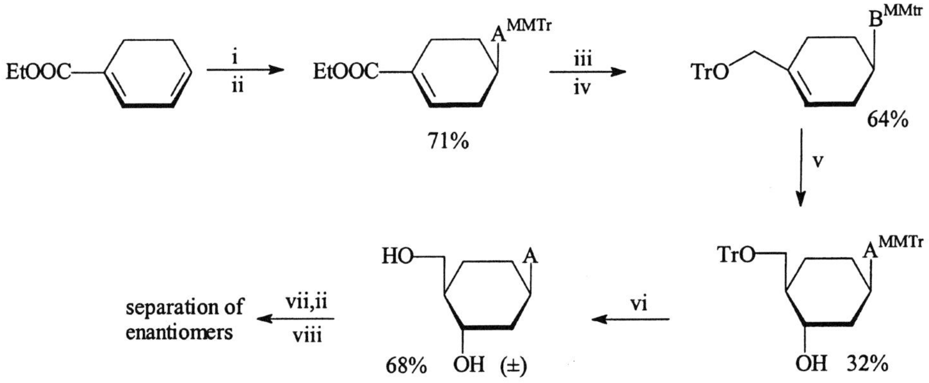

Scheme 61.
i) adenine, DBU, DMF, Δ; ii) MMTrCl, pyridine, Δ; iii) DIBAL, CH$_2$Cl$_2$; iv) TrCl, pyridine; v) BH$_3$, THF; NaOH, H$_2$O$_2$, vi) AcOH 80%; vii) TMSCl, pyridine, BzCl, NH$_4$OH; viii) (R)-(-)-methylmandelic acid, DCC, DMAP, CH$_2$Cl$_2$

A Pd(0)-catalyzed addition of adenine to cyclohexene epoxide afforded a 1,2-*cis* addition product, while the reaction in the absence of a Pd(0) catalyst afforded the *trans*-1,2-ring opened compound (Scheme 62) (Ramesh *et al.*, 1992). The obtained compounds were converted to various di- and trihydroxylated cyclohexenyl- and cyclohexanyl adenines.

Pd(0)-catalyzed addition of adenine to 3,4-epoxycyclohexene afforded 9-(4-hydroxy-2-cyclo-hexenyl)adenine, which was converted in several steps to 9-[(1'R,2'R,3'S)-2',3'-dihydroxy-cyclohexanyl]adenine (Scheme 63).

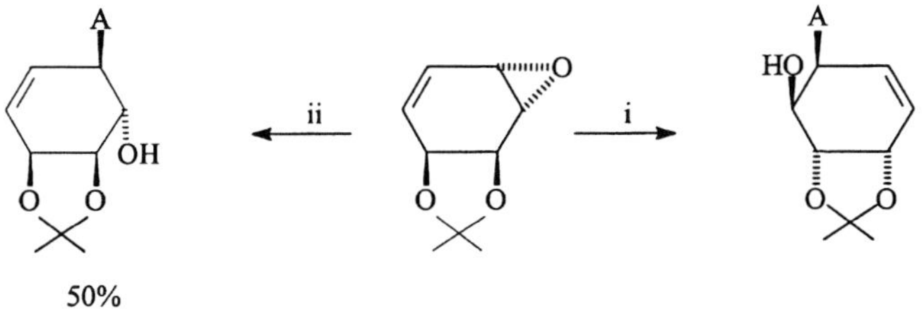

Scheme 62.

i) adenine, [(i-C$_3$H$_7$O)$_3$P]$_4$ Pd, THF, DMSO; ii) adenine, K$_2$CO$_3$ DMAC, Δ

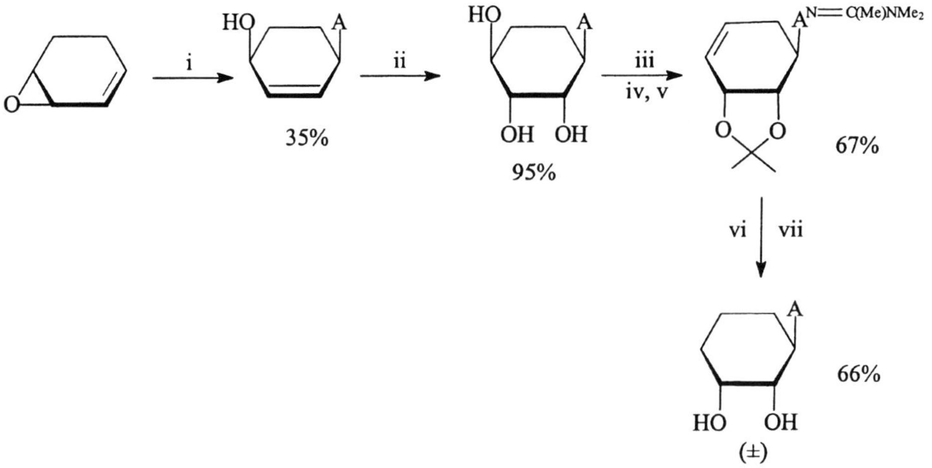

Scheme 63.

i) [(i-C$_3$H$_7$O)$_3$P]$_4$ Pd, adenine, THF, DMSO; ii) OsO$_4$, NMO, acetone; iii) DMP, HClO$_4$, acetone; iv) N,N-dimethylacetamide, dimethylacetal, dioxane; v) DAST, CH$_2$Cl$_2$; vi) a) aqueous NH$_4$OH, b) aqueous HCl, c) Dowex-50W(H$^+$); vii) Pd-C/H$_2$

Addition of thymine to 3,4-epoxycyclohexene in the presence of (Ph$_3$P)$_4$Pd gives cis-1-(4-hydroxy-2-cyclohexenyl)thymine and cis-1-(2-hydroxy-5-cyclohexenyl)thymine. Uncatalyzed nucleophilic ring-opening of the epoxide gives trans-1-(2-hdyroxy-5-cyclohexenyl)thymine (Scheme 64) (Arango *et al.*, 1993).

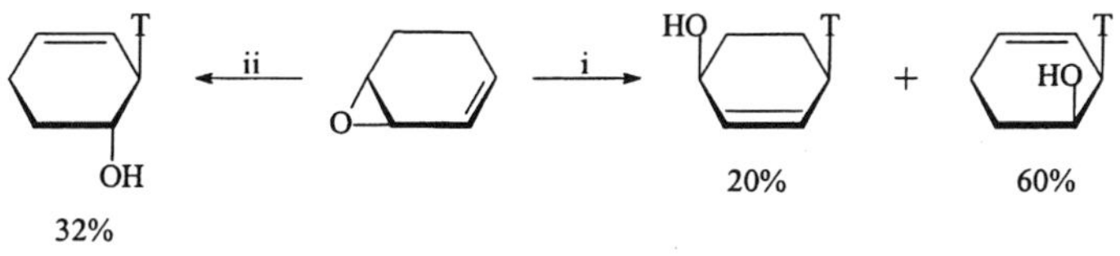

Scheme 64.

i) thymine, DMSO, (Ph$_3$)$_4$Pd, THF; ii) Thymine, DBU, DMF, Δ

Direct introduction of 6-chloropurine on (±)-*cis*-5-(*tert*-butyldimethylsilyloxymethyl)-2-cyclohexenol under Mitsunobu reaction conditions is low yielded (8%) (Konkel and Vince 1996a). Therefore first the *trans* configurated cyclohexenol was prepared, converted to the carbonate and reacted with 6-chloropurine under palladium coupling conditions (Scheme 65) (Konkel and Vince 1996a).

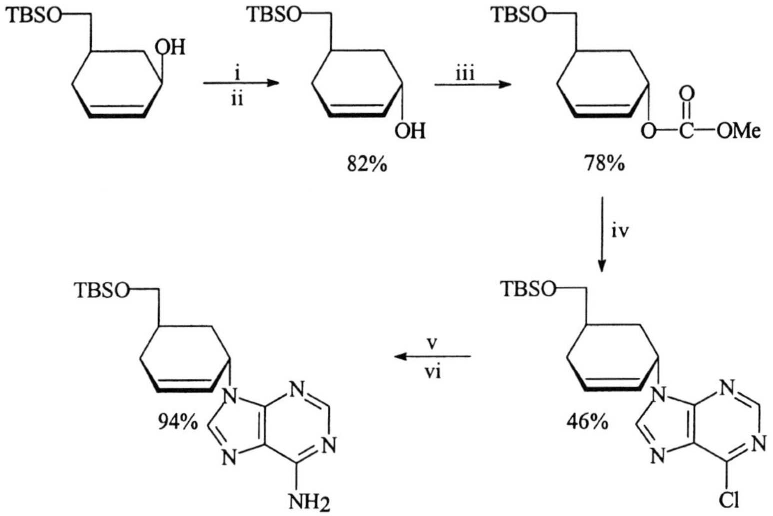

Scheme 65.

i : AcOH, DEAD, Ph_3P, THF; ii : K_2CO_3, MeOH; iii : $(MeO_2C)O$, DMAP, THF; iv : 6-chloropurine, NaH, $(Ph_3P)_4Pd$, DMF; v : TBAF, THF, AcOH; vi : NH_3, MeOH.

Reduction of 2-oxabicyclo[2.2.2.]oct-5-en-3-one with lithium aluminium hydride gives a diol which is reacted with dimethylpyrocarbonate. The purine moiety was introduced using palladium coupling methods (Scheme 66) (Konkel and Vince 1996b).

(±) 9-(4-β-hydroxymethylcyclohexene-2-en-1β-yl)-9H-adenine was also obtained from (±) *cis*-4-(hydroxymethyl)cyclohex-2-en-1-ylamine (Scheme 67) (Katagiri *et al.*, 1996).

Racemic *cis*[3-(adenin-9-yl)-4-cyclohexenyl]carbinol could be obtained starting from 5-azido-3-cyclohexene carboxylic acid (Scheme 68) (Konkel and Vince 1995). Reduction with $LiAlH_4$ yielded *cis*(3-amino-4-cyclohexenyl)carbinol from which the adenine base is gradually built up in three reaction (Konkel and Vince 1995).

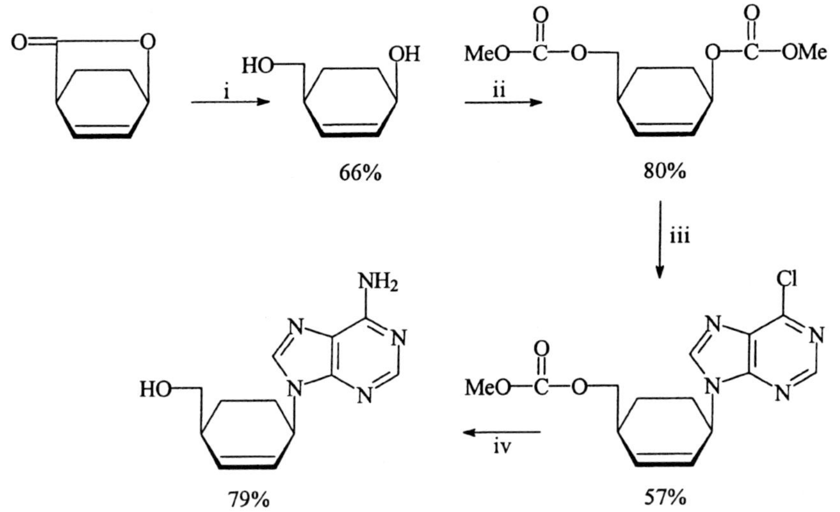

Scheme 66.

i : LiAlH$_4$, Et$_2$O; ii : (MeO$_2$C)$_2$O, DMAP, THF; iii : 6-chloropurine, NaH, (Ph$_3$P)$_4$Pd, DMF; iv : NH$_3$, MeOH; 10% NaOH, MeOH.

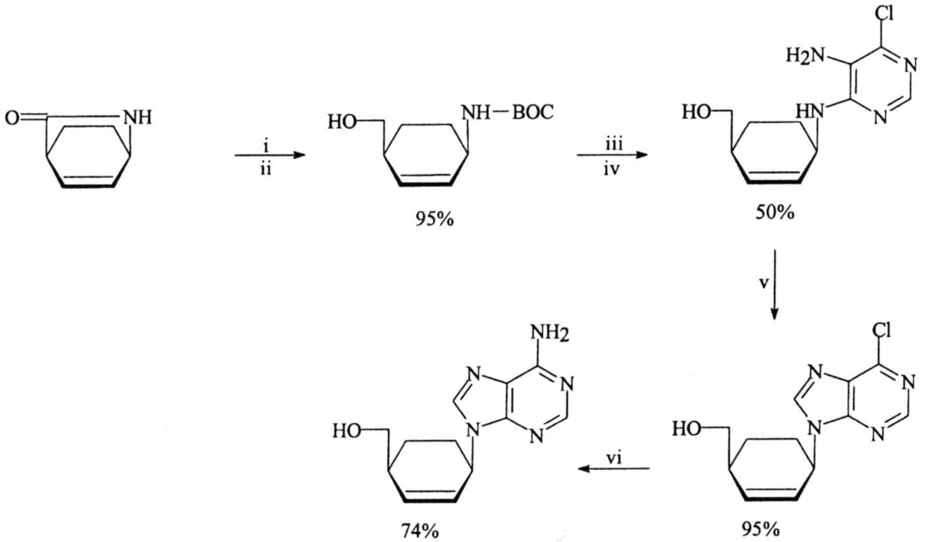

Scheme 67.

i : di-*tert*-butyl dicarbonate, Et$_3$N, DMAP, CH$_2$Cl$_2$; ii : NaBH$_4$, MeOH; iii : CF$_3$COOH; iv : 5-amino-4,6-dichloropyrimidine, BuOH, DIEA; v : (EtO)$_3$CH, HCl; vi : NH$_3$, MeOH.

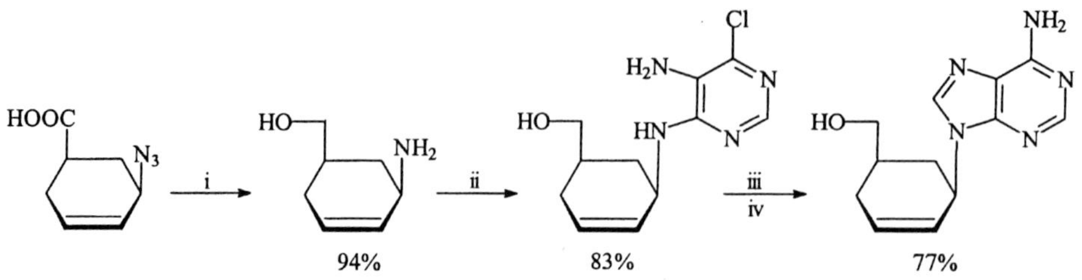

Scheme 68.

i : LiAlH$_4$, Et$_2$O; ii : 5-amino-4,6-dichloropyrimidine, Et$_3$N, BuOH; iii : (EtO)$_3$CH, HCl, DMF; iv : NH$_3$, MeOH;
HCl 2N.

A Pd(0) catalyzed alkylation of heterocyclic bases by allylic epoxide was low yielded (15-20%) (Pérez-Pérez *et al.*, 1995c). Therefore, a Mitsunobu-type condensation of nucleoside bases with a cyclohexenol was used for the synthesis of carbocyclic phosphonate nucleosides (Scheme 69) (Pérez-Pérez *et al.*, 1995c). The phosphonomethoxy moiety is introduced prior to coupling with the base to avoid protection or undesired alkylation of the base.

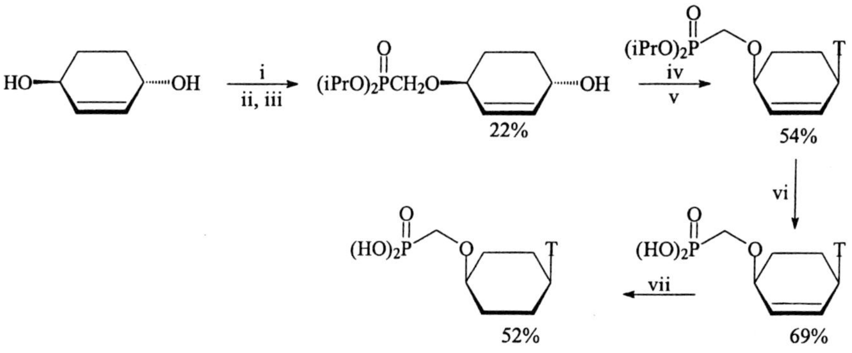

Scheme 69.

i) TrCl, DMAP, Et$_3$N, CH$_2$Cl$_2$; ii) (iPrO)$_2$P(O)CH$_2$Ts, NaH, DMF; iii) HOAc 80%; iv N^{3}BzT, Ph$_3$P, DEAD, dioxane; v) NH$_3$, MeOH; vi) TMSBr, DMF, NH$_4$OH; vii) H$_2$, Pd/C, EtOH

(R)-(-)-carvone was used as starting material for the synthesis of 2-(hydroxymethyl)-cyclohexane-1,3-diol nucleosides. The enantioselective precursors of the nucleoside analogues were obtained via a stereo- and regioselective hydroboration reaction (Scheme 70) (Wang *et al.*, 1998).

The cyclohexene nucleosides are obtained from the same precursors. Dependent on the protecting group strategy both the D- and L-analogue can be obtained (Schemes 71-72). Introduction of the base moiety via Mitsunobu reaction proceeded regio- and stereoselectively and with good chemical yield, while the Pd-coupling approach failed (Wang and Herdewijn 1999; Wang *et al.* 2000).

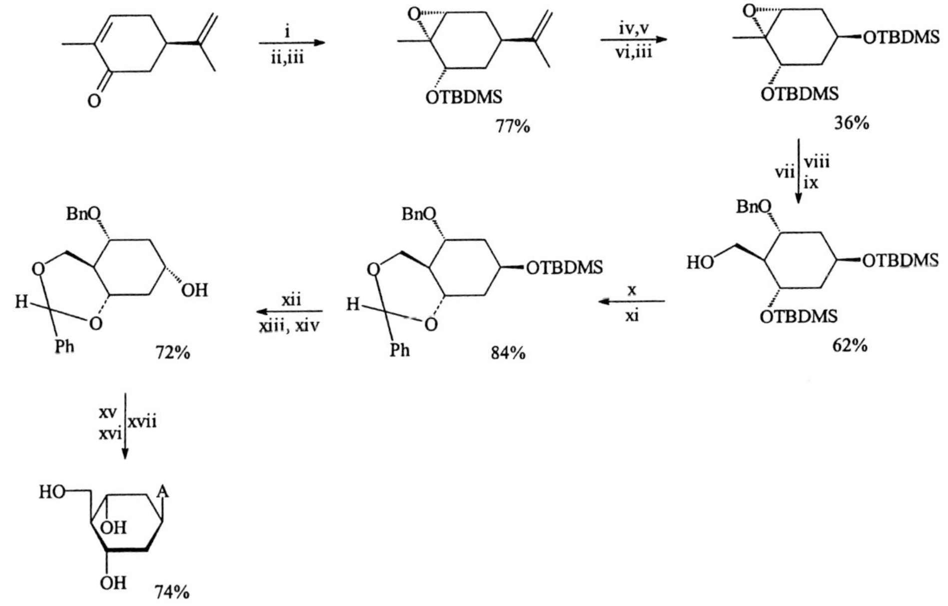

Scheme 70.

i : H_2O_2/NaOH, MeOH; ii : L-selectride, THF; iii : TBDMSCl, ImH, DMF; iv : 1% OsO_4, KIO_4, THF, H_2O; v) MCPBA, pH 8; vi : K_2CO_3, MeOH; vii : LiTMP/Et_2AlCl, benzene; viii : BnBr, NaH, TBAI; ix : 9-BBN, THF; H_2O_2, NaOH; x : 1 eq. TBAF, THF; xi : C_6H_5CH(OMe)$_2$, PTSA, dioxane; xii : TBAF, THF; xiii : C_6H_5COOH, Ph_3P, DEAD, dioxane; xiv : K_2CO_3, MeOH; xv : adenine, Ph_3P, DEAD, dioxane; xvi : 80% HOAc; xvii : Pd(OH)$_2$, cyclohexene, MeOH.

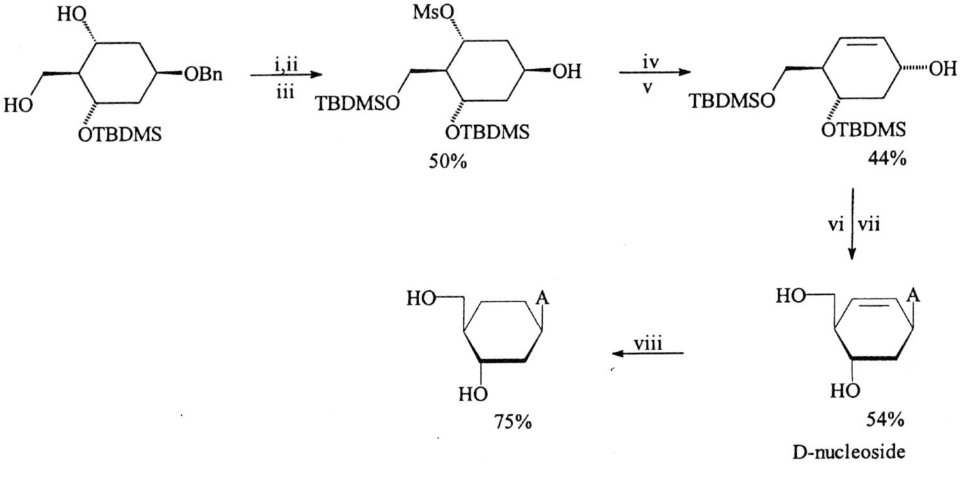

Scheme 71.

i : TBDMSCl, ImH, DMF; ii : MsCl, Et_3N, CH_2Cl_2; iii : 10% Pd/C, HCOONH$_4$, MeOH; iv : MnO_2, CH_2Cl_2 (47% recovery of starting material); v : NaBH$_4$, CeCl$_3$·7H$_2$O, MeOH; vi : adenine, Ph_3P, DEAD, dioxane; vii : CF$_3$COOH, H_2O; viii : 10% Pd/C, H_2, MeOH.

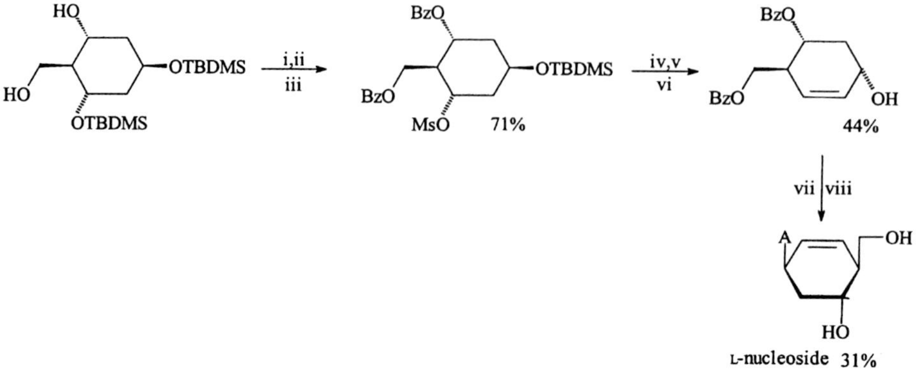

Scheme 72.

i : Bz$_2$O, DMAP, CH$_2$Cl$_2$; ii : 1 equiv. TBAF, THF; iii : MsCl, Et$_3$N, CH$_2$Cl$_2$; iv : TBAF, THF; v : PDC, CH$_2$Cl$_2$; vi : NaBH$_4$, CeCl$_3$.7H$_2$O, MeOH; vii : adenine, Ph$_3$P, DEAD, dioxane; viii : K$_2$CO$_3$, MeOH.

8.   Conclusion

With the exception of nucleoside antibiotics, nucleosides with a six-membered carbohydrate moiety were, for long time, considered as exotic structures. However, as mentioned in the introduction, the first nucleoside, that has ever been synthesized, was a pyranosyl nucleoside. The efforts to obtain six-membered nucleosides has increased considerably during last decade and these efforts will continue. Important reasons for that are the discovery of the antiviral activity of several six-membered nucleosides, the use of six-membered nucleosides as building blocks for oligonucleotides, and the renewed interest in antibiotic research. Still, many synthetic and biological aspects of this class of nucleosides remain to be explored.

9.   References

Aguilar, G.J., Gelpi, M.E. and Cadenas, R.A. (1992) Nucleocyclitols. Synthesis of Substituted Purinyl-*muco*-inositol Derivatives, J. Heterocyclic Chem. 29, 401-405.

Alexander, P., Krishnamurthy, V.V. and Prisbe, E.J. (1996) Synthesis and Antiviral Activity of Pyranosylphosphonic Acid Nucleotide Analogues, J. Med. Chem. 39, 1321-1330.

Allart, B., Busson, R., Rozenski, J., Van Aerschot, A. and Herdewijn, P. (1999) Synthesis of Protected D-Altritol Nucleosides as Building Blocks for Oligonucleotide Synthesis, Tetrahedron 55, 6527-6546.

Al-Masoudi, N.A. and Al-Atoom, A.A. (1995) Synthesis and Reactions of some Glycosylamine Derivatives of 6-Azauracil Nucleosides, Nucleosides Nucleotides 14, 1341-1348.

Al-Masoudi, N.A., Issa, F.B., Pfleiderer, W. and Lazrek, H.B. (1995) Synthesis and Biological Activity of some 5-Substituted-6-azauracil-N-1-nucleosides of 2-acetamido-2-deoxy-D-glucose, Nucleosides Nucleotides 14, 1693-1702.

Andersen, M.W., Daluge, S.M., Kerremans, L. and Herdewijn, P. (1996) The Synthesis of Modified D- and L-Anhydrohexitol Nucleosides, Tetrahedron Lett. 37, 8147-8150.

Antonakis, K. (1984) Ketonucleotides, Advances Carbocycl. Chem. Biochem. 42, 227-264.

Arango, J.H., Geer, A., Rodriguez, J., Younf, P.E. and Scheiner, P. (1993) Cyclohexenyl Nucleosides and Related Compounds, Nucleosides Nucleotides 12, 773-784.

Attia, A.M.E. and Elgemeie, G.E.H. (1995) Nucleic Acid Related Compounds: A Convenient Synthesis of 3-Deazauridine Analogues, Nucleosides Nucleotides, 1211-1218.

Attia, A.M.E., Ibrahim, E.I., Hay, F.E.A., Abbasi, M.M.A. and Mansour, H.A.E. (1995) Synthesis and Antiviral Activity of Some *N*-Pentopyranosyl-2-Pyridinethiones, Nucleosides Nucleotides, 14, 1581-1590.

Augustyns, K., Van Aerschot, A. and Herdewijn, P. (1992) Synthesis of 1-(24-Dideoxy-β-D-erythro-hexopyranosyl)thymine and its Incorporation into Oligonucleotides, Bioorg. Med. Chem. Lett. 2, 945-948. Augustyns, K., Van Aerschot, A., Urbanke, C. and Herdewijn, P. (1992) Influence of the Incorporation of 1-(2,3-dideoxy-β-D-erythro-hexopyranosyl)-thymine on the Enzymatic Stability and Base-pairing Properties of Oligodeoxynucleotides, Bull. Soc. Chim. Belg. 101, 119-129.

Augustyns, K., Vandendriesssche, F., Van Aerschot, A., Busson, R., Urbanke, C. and Herdewijn, P. (1992) Incorporation of Hexose Nucleoside Analogues into Oligonucleotides: Synthesis, Base-pairing Properties and Enzymatic Stability, Nucleic Acids Res. 20, 4711-4716.

Augustyns, K., Rozenski, J., Van Aerschot, A., Janssen, G. and Herdewijn, P. (1993) Synthesis of 2,4-Dideoxy-β-D-erythro-hexopyranosyl Nucleosides, J. Org. Chem. 58, 2977-2982.

Augustyns, K., Rozenski, J., Van Aerschot, A., Busson, R., Claes, P. and Herdewijn, P. (1994) Synthesis of a new branched-chain hexopyranosyl nucleoside : 1-[2',3'-dideoxy-3'-C-(hydroxymethyl)-α-D-erythro-pentopyranosyl]-thymine, Tetrahedron 50, 1189-1198.

Baddiley, J., Lythgoe, B. and Todd, A.R. (1943) Experiments on the Synthesis of Purine Nucleosides – Part III. 4-Glycosidaminopyrimidines, J. Chem. Soc. 43, 571-574.

Baer, H.H. and Bayer, M. (1971) A New Approach to 3'-Amino-3'-deoxynucleosides. Synthesis of 9-(3-Amino-3-deoxy-α-L-ribofuranosyl)adenine, Can. J. Chem. 49, 568-573.

Baker, B.R., Joseph, J.P., Schaub, R.E. and Williams, J.H. (1954) Puromycin, Synthetic Studies. V. 6-Dimethylamino-9-(2'-acetamino-β-D-gluco-pyranosyl)purine, J. Org. Chem. 19, 1786-1792.

Beránek, J., Friedman, H.A., Watanabe, K.A. and Fox, J.J. (1965) Nucleosides XXVIII. 3'-Amino-3'-deoxyhexopyranosyl Nucleosides. Part III. Synthesis of 3'-Amino-3'-deoxyhexopyranosyl Adenines, J. Heterocyclic Chem. 2, 188-191.

Björnse, M., Classon, B., Kvarnström, I. and Samuelsson, B. (1993) Synthesis of 4-C-Hydroxymethyl Hexopyranosyl Nucleosides as Potential Inhibitors of HIV, Tetrahedron 49, 8637-8644.

Böhringer, M., Roth, H.-J., Hunziker, J., Göbel, M., Krishnan, R., Giger, A., Schweizer, B., Schreiber, J., Leumann, C. and Eschenmoser, A. (1992) Warum Pentose- und nicht Hexose-Nucleinsäuren? Oligonucleotide aus 2',3'-Dideoxy-β-D-glucopyranosyl-Bausteinen ('Homo-DNS'): Herstellung, Helv. Chim. Acta 75, 1416-1477.

Boudou, V., Kerremans, L., De Bouvere, B., Lescrinier, E., Schepers, G., Busson, R., Van Aerschot, A and Herdewijn, P. (1999) Base Pairing of Anhydrohexitol Nucleosides with 2,6-Diaminopurine, 5-Methylcytosine and Uracil as Base Moiety, Nucleic Acids Res. 27, 1450-1456.

Bowles, W.A. and Robins, R.K. (1964) The Direct Utilization of Glycals for the Preparation of Purine Deoxynucleosides, J. Am. Chem. Soc. 86, 1252-1253.

Buchanan, J.G., Stoddart, J. and Wightman, R.H. (1994) Synthesis of the Indole Nucleoside Antibiotics Neosidomycin and SF-2140, J. Chem. Soc. Perkin Trans. 1, 1417-1426.

Calvani, F., Macchia, M., Rossello, A., Gismondo, M.R., Drago, L., Fassina, M.C., Cisternino, M. and Domiano, P. (1995) Synthesis and Antiviral Activity of Dihydroxycyclohexyl Pyrimidine and Purine Carbocyclic Nucleosides, Bioorg. Med. Chem. Lett. 5, 2567-2572.

Carbon, J.A. (1964) Direct Condensation of 2-Deoxy-D-ribose with Purines. Structure of the Products, J. Am. Chem. Soc. 6, 720-725.

Carret, G., Sarda, N., Abou-Assali, M., Anker, D. and Pachéco, H. (1983) Voies d'accès à un analogue pyrannique de la puromycine, J. Heterocyclic Chem. 20, 697-702.

Czernecki, S., Franco, S. and Valery, J.M. (1997) Further study toward amipurimycin: Synthesis of the northern part, J. Org. Chem. 62, 4845-4847.

Davoll, J. and Lythgoe, B. (1949) Deoxyribonucleosides and related compounds. Part I. Synthetic applications of some 1-halogeno 2-deoxy-sugar derivatives, J. Am. Chem. Soc., 2526-2531.

De Bouvere, B., Kerremans, L., Rozenski, J., Janssen, G., Van Aerschot, A., Claes, P., Busson, R., and Herdewijn, P. (1997) Improved Synthesis of Anhydrohexitol Building Blocks for Oligonucleotide Synthesis, Liebigs Ann./Recueil., 1453-1461.

De Fina, G.M., Varela, O. and de Lederkremer, R.M. (1994) Synthesis of 3,4-Dideoxyhexopyranosyl- and Hex-3-enopyranosid-2-ulose-pyrimidine Isonucleosides from 2-Acyloxyglycals, J. Chem. Research (S), 26-27.

De Winter, H.L., Blaton, N.M., Peeters, O.M., De Ranter, C.J., Van Aerschot, A. and Herdewijn, P. (1991a) Structures of 1-(2,3-Dideoxy-erythro-α-D-hexopyranosyl)thymine and the 1-(2,3-Dideoxy-erythro-β-D-hexopyranosyl)thymine.Dioxane Complex, Acta Cryst. C47, 838-842.

De Winter, H.L., Blaton, N.M., Peeters, O.M., De Ranter, C.J., Van Aerschot, A. and Herdewijn, P. (1991b) Structure of 1-(2-Deoxy-β-D-ribopyranosyl)-5-iodouracil, Acta Cryst. C47, 835-837.

De Winter, H.L., Blaton, N.M.,Peeters, O.M., De Ranter, C.J., Van Aerschot, A. and Herdewijn, P. (1991c) Structure of 1-(2-Deoxy-2-fluoro-α-D-arabinopyranosyl)5-iodouracil, Acta Cryst. C47, 2245-2247.

De Winter, H.L., Blaton, N.M., Peeters, O.M., De Ranter, C.J., Van Aerschot, A. and Herdewijn, P. (1991d) Structures of the 1-(2-Deoxy-2-fluoro-β-D-arabinopyranosyl)thymine – Water Complex and 1-(2-Deoxy-2-fluoro-β-D-arabinopyranosyl)-5-ethyluracil, Acta Cryst. C47, 1693-1697.

De Winter H L, De Ranter C J, Blaton N M, Peeters O M, Van Aerschot A and Herdewijn P (1992) 1-(2,3-Dideoxy-erythro-β-D-hexopyranosyl)cytosine: an Example of the Conformational and Stacking Properties of Pyranosyl Pyrimidine Nucleosides. A Crystallographic and Computational Approach, Acta Cryst. B48, 95-103.

Doboszewski B, Blaton N and Herdewijn P (1995a) Easy synthesis and different conformational behavior of purine and pyrimidine β-D-glycero-pent-2'-enopyranosyl nucleosides, J. Org. Chem. 60, 7909-7919.

Doboszewski, B., Blaton, N. and Herdewijn, P. (1995b) Synthesis of β-configured 2',3'-unsaturated pentopyranosyl nucleosides, Tetrahedron Lett. 36, 1321-1324.

Doboszewski, B., Blaton, N., Rozenski, J., De Bruyn, A. and Herdewijn, P. (1995c) 3'-deoxy-3'-hydroxymethyl-aldopentopyranosyl nucleoside synthesis, Tetrahedron 51, 5381-5396.

Doboszewski, B., De Winter, H., Van Aerschot, A. and Herdewijn, P. (1995d) Synthesis of 3'-deoxy-3'-C-hydroxymethyl aldopentopyranosyl nucleosides and their incorporation in oligonucleotides, Tetrahedron, 12319-12336.

Doboszewski, B. and Herdewijn, P. (1996a) Synthesis of 4-deoxy-4-C-hydroxymethyl-α-L-lyxopyranosyl thymine. Nucleosides Nucleotides 15, 1495-1518.

Doboszewski, B. and Herdewijn, P. (1996b) Branched-chain nucleosides: Synthesis of 3'-deoxy-3'-C-hydroxymethyl-α-L-lyxopyranosyl thymine and 3'-deoxy-3'-C-hydroxymethyl-α-L-threofuranosyl thymine, Tetrahedron 52, 1651-1668.

Ennifar, S., Das, B.C., Nash, S.M. and Nagarajan, R. (1977) Structural identity of anthelmycin with hikizimycin, J. Chem. Soc. Chem. Commun. 2, 41-42.

Etzold, G. and Langen, P. (1965) Pyrimidin-nucleoside der 2-Deoxy-D-glucose, Chem. Ber. 98, 1988-1997.

Etzold, G. and Langen, P. (1966) Pyranoide Isomere des α- und β-Thymidins, Naturwissenschaften 53, 178.

Etzold, G.V., Hintsche, R. and Langen, P. (1967) Pyranoide Thymin-O^2-3'-cylclonucleoside, Tetrahedron Lett. 48, 4827-4830.

Ferrier, R.J. and Ponpipom, M.M. (1971a) Unsaturated Carbohydrates. XIV. The Isomerisation of Hex-2-enopyranosylpurine Nucleoside Derivatives, J. Chem. Soc. (C), 553-559.

Ferrier, R.J. and Ponpipom, M.M. (1971b) Unsaturated Carbohydrates. Part XV. The Synthesis of (3'-Deoxyhex-2'-cnopyranosyl)purine Derivatives, J. Chem. Soc.(C), 560-562.

Fisher, E. and Helferich, B. (1914) Synthetische Glucoside der Purine, Chem. Ber., 210-25.

Fisher, L.V., Lee, W.W. and Goodman, L. (1969) Some 6-Substituted-9-(β-L-fucopyranosyl)purines, J. Heterocycl. Chem. 6, 949-951.

Fox, J.J. and Goodman, I. (1951) The synthesis of nucleosides of cytosine and 5-methylcytosine, J. Am. Chem. Soc. 73, 3256-3258.

Fox, J.J., Cavalieri, L.F. and Chang, N. (1953) The identification of cytidine acids *a* and *b* by spectrophotometric methods, J. Am. Chem. Soc. 75, 4315-4317.

Fox, J.J., Watanabe, K.A. and Bloch, A. (1966) Nucleoside antibiotics, Prog. Nucleic Acids Res Mol. Biol., 5, 251-313.

Fuertes, M., Garcia-Muñoz, R., Madroñero, R. and Stud, M. (1970) Heterocyclic N-Glycosides, *J. Chem. Soc.(c)*, Synthesis of Unsaturated N-Glycosides from 6-Chloropurine and Derivatives of D-Xylal and L-Arabinal. A. Conformational NMR Study. Tetrahedron 26, 4823-4837.

Halazy, S., Kenny, M., Dulworth, J. and Eggenspiller, A. (1992) Synthesis and Antiviral Properties of New Cycloalkanol Derivatives of Guanine, Nucleosides Nucleotides 11, 1595-1606.

Hanessian, S., Bacquet, C. and Lehong, N. (1980) Chemistry of the Glycosidic Linkage – Exceptionally Fast and Efficient Formation of Glycosides by Remote Activation, Carbohydr. Res. 80, C17-C22.

Hanessian, S., Conde, J.J., Khai, H.H. and Lou, B. (1996) The Stereocontrolled Synthesis of 1,2-trans Hexopyranosyl Nucleosides via a Novel Anomeric Activation, Tetrahedron 52, 10827-10834.

Harada, S., Mizuta, E. and Kishi, T. (1978) Structure of mildiomycin, a new antifungal nucleoside antibiotic. J. Am. Chem. Soc. 100, 4895-4897.

Herdewijn, P. and Van Aerschot, A. (1990) Synthesis of trideoxyhexopyranosylated and hexenopyranosylated nucleoside analogues as potential anit-HIV agents, Bull. Soc. Chim. Belg. 99, 895-901.

Herdewijn, P., Van Aerschot, A., Busson, R., Claes, P. and De Clercq, E. (1991) Synthesis of 2'-Deoxy-2'-fluoro-D-arabinopyranosyl Nucleosides and their 3',4'-Seco Analogues, Nucleosides Nucleotides 10, 1525-1549.

Herscovici, J., Montserret, R. and Antonakis; K. (1988) An improved method for the preparation of 2',3'-unsaturated nucleosides: synthesis of stereospecifically labeled ketonucleosides. Carbohydr. Res. 176, 219-229.

Hilbert, G.E. and Johnson, T.B. (1930) Researches on pyrimidines. CXVII. A method for the synthesis of nucleosides, J. Am. Chem. Soc. 52, 4489-4494.

Hilbert, G.E. and Jansen, E.F. (1936) Synthesis of 1-D-glucosidocytosine, J. Am. Chem. Soc. 58, 60-62.

Hilbert, G.E. (1937) Synthetic nucleosides – some 1-glycosidouracils, J. Am. Chem. Soc. 59, 330-333.

Hossain, N., Rozenski, J., De Clercq, E. and Herdewijn, P. (1997) Synthesis and Antiviral Activity of the α-Analogues of 1,5-Anhydrohexitol Nucleosides (1,5-Anhydro-2,3-dideoxy-D-ribohexitol Nucleosides, J. Org. Chem. 62, 2442-2247.

Hossain, N., Wroblowski, B., Van Aerschot, A., Rozenski, J., De Bruyn, A. and Herdewijn, P. (1998a) Oligonucleotides Composed of 2'-Deoxy-1,5-anhydro-D-mannitol Nucleosides with a Purine Base Moiety, J. Org. Chem. 63, 1574-1582.

Hossain, N., van Halbeek, H., De Clercq, E. and Herdewijn, P. (1998b) Synthesis of 3'-*C*-branched 1',5'-anhdyromannitol Nucleosides as new Antiherpes Agents, Tetrahedron 54, 2209-2226.

Hossain, N., Luyten, I., Rothenbacher, K., Busson, R. and Herdewijn, P. (1999) Synthesis and conformational analysis of 1,5-anhydro-2,4-dideoxy-D-mannitol nucleosides, Nucleosides Nucleotides 18, 161-180.

Jung, M.E. and Kiankarimi,M. (1998) Synthesis of methylene-expanded 2',3'-dideoxyribonucleosides, J. Org. Chem., 63, 8133-8144.

Katagiri, N., Ito, Y., Shiraishi, T., Maruyama, T., Sato, Y. and Kaneko, C. (1996) Deamination of 9-(Hydroxymethylated Cycloalkyl)-*9H*-adenines (Carbocyclic Adenine Nucleosides) by Adenosine Deaminase : Effect of High-Pressure upon Deamination Rate and Enantioselectivity, Nucleosides Nucleotides 15, 631-647.

Khodair, A.I., Ibrahim, E.E. and El Ashry, E.S.H. (1997) Glycosylation of 2-Thiouracil Derivatives. A Synthetic Approach to 3-Glycosyl-2,4-dioxypyrimidines, Nucleosides Nucleotides 16, 433-444.

Kitagawa, I., Cha, B.C., Nakae, T., Okaichi, Y., Takinami, Y. and Yoshikawa, M.(1989) A New Approach to the Synthesis of Optically Active Cyclohexane Analogs of Nucleoside Using a Michael-type Addition Reaction to Nitro-cyclohexenes as a Key Reaction, Chem. Pharm. Bull. 37, 542-544.

Kondo, T., Nakai, H., Goto, T. (1971) A Direct Synthesis of Unsaturated Uracil Nucleosides from Silylated Uracil and Triacetylglucal. Agr. Biol. Chem. 35, 1990-1991.

Konkel, M.J. and Vince, R. (1995) Synthesis and Biological Activity of Cyclohexenyl Nucleosides. Cis-5-(*9H*-Purin-9-yl)-3-cyclohexenyl Carbinols and Their 8-Azapurinyl Analogs, Nucleosides Nucleotides 14, 2061-2077.

Konkel, M.J. and Vince, R. (1996a) Cyclohexenyl Nucleosides: Synthesis of cis-4-(*9H*-Purin-9-yl)-2-cyclohexenylcarbinols, Tetrahedron 52, 799-808.

Konkel, M.J. and Vince, R. (1996b) Cyclohexenyl Nucleosides : Synthesis and Biological Activity of trans-3-(Purin-9-yl)-4-cyclohexenylcarbinols, Tetrahedron 52, 8969-8978.

Leclercq, F. and Antonakis, K. (1989) Direct Fluorination at Positions 3',4' and 6' of β-D-glucopyranosyltheophylline, Carbohydr. Res. 193, 307-313.

Leclercq, F., Egron, M.-J., Antonakis, K., Bennani-Baiti, M.I. and Frayssinet, C. (1992) Synthesis and Biological Activity of a Fluoroketonucleoside: 7-(3-Deoxy-3-fluoro-β-D-glycero-hex-2-enopyranosyl-4-ulose)-theophylline, Carbohydr. Res. 228, 95-102.

Lerner, L.M. and Kohn, P. (1964) The Preparation of Nucleosides from Allose, Altrose, Gulose, Talose, and Mannose, J. Med. Chem. 7, 655-658.

Lerner, L.M. (1971) Synthesis of 9-α- and 9-β-D-fucopyranosyladenine, Carbohydr. Res. 19, 225-258.

Lerner, L.M., Sheid, B. and Gaetjens, E. (1987) Preparation and antileukemic screening of some new 6'-deoxyhexopyranosyladenine nucleosides, J. Med. Chem. 30, 1521-1525.

Lerner, L.M., Mennitt, G., Gaetjens, E. and Sheid, B. (1993) Synthesis and antileikemic activity of certain D-fucopyranosyl nucleosides, Carbohydr. Res. 244, 285-294.

Leutzinger, E.E., Robins, R.K. and Townsend, L.B. (1968a) The Direct Utilization of Unsaturated Sugars in Nucleoside Syntheses. An Approach to the Preparation of Analogs of Blasticidin S. Synthesis of 9-(2',3'-didehydro-2',3'-dideoxy-D-*erythro*-hexopyranosyl) guanine, Tetrahedron Lett. 43, 4475-4478.

Leutzinger, E.E., Bowles, W.A., Robins, R.K. and Townsend, L.B. (1968b) Purine Nucleosides. XVIII. The Direct Utilization of Unsaturated Sugars in Nucleoside Syntheses. The Conformation and Structure of Certain 9-(2'-deoxyribopyranosyl)purines Prepared From D-Arabinal, J. Am. Chem. Soc. 90, 127-136.

Leutzinger, E.E., Robins, R.K. and Townsend, L.B. (1970) The direct utilization of unsaturated sugars in nucleoside synthesis. 3-deoxy-3-(6-chloro-2-methylthio-9-purinyl)-D-erythro-hex-1-enopyranose, a new and novel type of purine nucleoside, Tetrahedron Lett., 3751-3753.

Leutzinger, E.E., Meguro, T., Townsend, L.B., Shuman, D.A., Schweizer, M.P., Stewart, C.M. and Robins, R.K. (1972) The Direct Utilization of Unsaturated Sugars in Nucleoside Syntheses. The Synthesis, Configuration, and Conformation of Certain Hex-1-enitol-3-yl-, Hex-2-enopyranosyl-, and Hexopyranosylpurines. The Preparation of 9-(1,5-Anhydro-2,3-dideoxy-D-*arabino*-hex-1-enitol-3-yl)adenine and 9-(2,3-dideoxy-β-D-*erythro*-hex-2-enopyranosyl)adenine from D-Glucal, J. Org. Chem. 37, 3695-3703

Lichtenthaler F.W. and Albrecht, H.P. (1966) Nitromethan-Kondensation mit Dialdehyden, V. Hypoxanthin- und Uracil-Nucleoside der 3-Nitro- und 3-Amino-3-desoxy-gluco-pyranose, Chem. Ber. 99, 575-585.

Lichtenthaler, F.W. and Zinke, H. (1966) Nucleosides, XIII. Synthesis of *C*-Methyl Branched Uracil Nucleosides, Angew. Chem. Int. Ed. 5, 737.

Lichtenthaler F.W. and Albrecht, H.P. (1966) Nitromethan-Kondensation mit Dialdehyden, V. Hypoxanthin- und Uracil-Nucleoside der 3-Nitro- und 3-Amino-3-desoxy-gluco-pyranose, Chem. Ber. 99, 575-585.

Lichtenthaler, F.W., Trummlitz, G. and Zinke, H. (1969) Nucleoside IX. Synthese von Diamino- und Triamino-Zucker-Nucleosiden, Tetrahedron Lett. 16, 1213-1217.

Lichtenthaler, F.W. and Zinke, H. (1972) Nucleosides XIII. Synthesis and Interconversions of *C*-Methyl-Branched 1-(3-Amino-3-deoxy-β-D-hexopyranosyl)uracils. An Emperical Method for Configurational Assignments at the Branch Point by Nuclear Magnetic Resonance, J. Org. Chem. 37, 1612-1621.

Lichtenthaler, F.W., Voss, P. and Heerd, A. (1974) Nucleosides. XX. Stannic chloride catalyzed glycosidations of silylated purines with fully acylated sugars, Tetrahedron. Lett. 2141-2144.

Lichtenthaler, F.W., Ueno, T. and Voss, P. (1974) Nucleosides XXII. Pyrimidine Nucleosides of 4-Amino-4-deoxy-β-D-galactopyranose, Bull. Chem. Soc. Jpn. 47, 2304-2310.

Lundquist, Å., Kvarnström, I., Svensson, S.C.T., Classon, B. and Samuelsson, B. (1995) Synthesis of 3',4'-*C*-bishydroxymethyl-2',3',4'-trideoxy-β-L-threo-Pentopyranosyl Nucleosides as Potential Inhibitors of HIV, Nucleosides Nucleotides 14, 1493-1502.

Luyten, I. and Herdewijn P. (1996) Synthesis and Conformational Behavior of Purine and Pyrimidine β-D-threo-Hex-3'-enopyranosyl Nucleosides, Tetrahedron 52, 9249-9262.

Lythgoe, B., Smith, H. and Todd, A.R. (1947) Experiments on the Synthesis of Purine Nucleosides. Part XVI. 9-β-D-*Manno*-pyranosidoadenine. A Proof of the Location of the Sugar Residue in Adenosine, J.Chem.Soc., 355-357.

Mansour, A.K., Ibrahim, Y.A. and Khalil, N.S.A.M. (1999) Selective Synthesis and Structure of 6-Arylvinyl-2- and 4-Glucosyl-1,2,4-triazines of Expected Interesting Biological Activity, Nucleosides Nucleotides 18, 2265-2283.

Matsuda, A. and Watanabe, K.A. (1996) Polydeoxyaminohexopyranosylnucleosides. Synthesis of 1-(2,3,4-Trideoxy-3-nitro-β-D-erythro- and threo-hexopyranosyl)uracils from Uridine, Nucleosides Nucleotides 15, 205-217.

Maurinsh, Y., Schraml, J., De Winter, H., Blaton, N., Peeters, O., Lescrinier, E., Rozenski, J., Van Aerschot, A., De Clercq, E., Busson, R. and Herdewijn, P. (1997) Synthesis and Conformational Study of 3-Hydroxy-4-(Hydroxymethyl)-1-Cylcohexanyl Purines and Pyrimidines, J. Org. Chem. 62, 2861-2871.

Maurinsh, Y., Rosemeyer, H., Esnouf, R., Medvedovici, A., Wang, J., Ceulemans, G., Lescrinier, E., Hendrix, C., Busson, R., Sandra, P., Seela, F., Van Aerschot, A. and Herdewijn, P. (1999) Synthesis and Pairing Properties of Oligonucleotides Containing 3-Hydroxy-4-hydroxymethyl-1-cyclohexanyl Nucleosides,

Chem. Eur. J. 5, 2139-2149.

Mikhailov, S.N., Blaton, N., Rozenski, J., Balzarini, J., De Clercq, E. and Herdewijn P. (1996) Use of Cyclohexene Epoxides in the Preparation of Carbocyclic Nucleosides, Nucleosides Nucleotides 15, 867-878

Nagasawa, N., Kumashiro, I. and Takenishi, T. (1967) Synthesis of 9-(2'-Deoxy-β-D-ribopyranosyl)adenine, J. Org. Chem. 32, 251-252.

Nakamura, S. and Kondo, H. (1977) Brief review of nucleoside antibiotics, Heterocycles 8, 583-607.

Nandanan, E., Jang, S.-Y., Moro, S., Kim, H.O., Siddique, M.A., Russ, P., Marquez, V.E., Busson, R., Herdewijn, P., Harden, T.K., Boyer, J.L. and Jacobson, K.A. (2000) Synthesis, Biological Activity, and Molecular Modeling of Ribose-Modified Deoxyadenosine Bisphosphate Analogues as $P2Y_1$ Receptor Ligands, J. Med. Chem. 43, 829-842.

Nord, L.D., Dalley, N.K., McKernan, P.A. and Robins, R.K. (1987) Synthesis, Structure, and Biological Activity of Certain 2-Deoxy-β-D-ribo-hexopyranosyl Nucleosides and Nucleotides, J. Med. Chem. 30, 1044-1054.

Novák, J.J.K. and Šorm, F. (1962) Nucleic Acid Components and their Analogues. Synthesis of Anomeric 9-(2-deoxy-1-D-glucosyl)adenines and of their anhydroderivatives, Collection Czechoslov. Chem. Commun. 27, 902-905.

Ohrui, H., Waga,T. and Meguro,H. (1993) Synthesis of conceptually new, potentially antiviral, low- toxicity, acid- and enzyme-stable furanoid and pyranoid nucleosides, Biosci. Biotech. Biochem. 57, 1040-1041.

Ohta, N., Minamoto, K., Yamamoto, T., Koide, N. and Sakoda, R. (1996) Stereoselective Reactions of 1-(4,6-O-Benzylidene-2,3-didehydro-2,3-dideoxy-3-nitro-β-D-hexopyranosyl)uracil with some Nucleo-philes, Nucleosides Nucleotides, 15, 833-855.

Onodera, K., Hirano, S., Masuda, F. and Yajima, T. (1968) Unsaturated Rhamnosyltheophylline produced in the Fusion of 1,2,3,4-Tetra-O-acetyl-L-rhamnopyranose with Theophylline, Chem. Commun. 1538.

Onuma, S., Nawata, Y. and Saito, Y. (1966) An X-Ray Analysis of Blasticidin S Monohydrobromide, J. Chem. Soc. Jpn. 39, 1091.

Ostrowski, T., Wroblowski, B., Busson, R., Rozenski, J., De Clercq, E., Bennett, M.S., Champness, J.N., Summers, W.C., Sanderson, M.R. and Herdewijn, P. (1998) 5-Substituted Pyrimidines with a 1,5-Anhydro-2,3-dideoxy-D-arabino-hexitol Moiety at N-1 : Synthesis, Antiviral Activity, Conformational Analysis, and Interaction with Viral Thymidine Kinase, J. Med. Chem., 41, 4343-4353.

Pérez-Pérez, M.-J., Rozenski, J. and Herdewijn, P. (1994) Stereospecific Synthesis of a Pentopyranosyl Analogue of D4T Monophosphate, Bioorg. Med. Chem. Lett. 4, 1199-1202.

Pérez-Pérez, M.-J., Doboszewski, B., De Clercq, E. and Herdewijn, P. (1995a) Phosphonates Derivatives of 2',3'-Dideoxy-2',3'-didehydro-pentopyranosyl Nucleosides, Nucleosides Nucleotides 14, 707-710.

Pérez-Pérez, M.-J., Doboszewski, B., Rozenski, J. and Herdewijn, P. (1995b) Stereocontrolled Synthesis of Phosphonate Derivatives of Tetrahydro- and Dihydro-2H-Pyranyl Nucleosides: The Selectivity of the Ferrier Rearrangement, Tetrahedron Assymmetry 6, 973-984.

Pérez-Pérez, M.-J., Rozenski, J., Busson, R. and Herdewijn, P. (1995c) Application of the Mitsunoby-type Condensation Reaction to the Synthesis of Phosphonate Derivatives of Cyclohexenyl and Cyclohexanyl Nucleosides, J. Org. Chem. 60, 1531-1537.

Pérez-Pérez, M.-J., De Clercq, E. and Herdewijn, P. (1996) Synthesis and Antiviral Activity of 2-deoxy-1,5-anhydro-D-mannitol Nucleosides Containing a Pyrimidine Base Moiety, Bioorg. Med. Chem. Lett. 6, 1457-1460.

Polak, M., Doboszewski, B., Herdewijn, P. and Plavec, J. (1997) Conformational Studies of 2',3'-Unsaturated Pentopyranosyl Nucleosides by 1H NMR Spectroscopy. Impact of π–σ* Interactions on the Axial

Preference of the Purine versus Pyrimidine Nucleobase, J. Am. Chem. Soc. 119, 9782-9792.

Prévost,N. and Rouessac,F. (1997) A novel route to pyrimidine isodideoxynucleosides via Michael-type addition on unsaturated modified sugars, Synthetic Commun. 27, 2325-2335.

Ramesh, K., Wolfe, M.S., Lee, Y., Vander Velde, D. and Borchardt, R.T. (1992) Synthesis of Hydroxylated Cyclohexenyl- and Cyclohexanyladenines as Potential Inhibitors of *S*-Adenosylhomocysteine Hydrolase, J. Org. Chem. 57, 5861-5868.

Robins, R.K., Godefroi, E.F., Taylor, E.C., Lewis, L.R. and Jackson, A. (1961) Purine Nucleosides. I. The synthesis of Certain 6-Substituted-9-(tetrahydro-2-pyranyl)-purines as Models of Purine Deoxy-nucleosides. J. Am. Chem. Soc. 83, 2574-2579.

Schaeffer, H.J., Kaistha, K.K. and Chakraborti, S.K. (1964a) Enzyme Inhibitors II. Synthesis of trans-3-(6-Substituted-9-purinyl)cyclohexanols as Adenosine Deaminase Inhibitors, J. Pharm. Sci. 53, 1371-1374.

Schaeffer, H.J., Godse, D.D. and Liu, G. (1964b) Enzyme Inhibitors III. Syntheses of cis-(6-Substituted-9-purinyl)cycloalkylcarbinols as Adenosine Deaminase Inhibitors, J. Pharm.Sci. 53, 1510-1515.

Schaeffer, H.J., Marathe, S. and Alks, V. (1964c) Enzyme Inhibitors I. Inhibition of Adenosine Deaminase by Isosteric Nucleosides, J. Pharm. Sci 53, 1368-1370.

Seto, H., Koyoma, M., Ogino, H., Tsuruoka, T., Inouye, S. and Otake, N. (1983) The structures of novel nucleoside antibiotics, miharamycin-A an miharamycin-B, Tetrahedron Lett. 24, 1805-1808.

Stevens, C.L. and Nagarajan, K. (1962) Synthesis of Aminosugar Nucleosides, J. Med. Chem. 5, 1124-1147.

Sugimura, H. and Sujino, K. (1998) Synthesis of 2'3',6'-Trideoxy-β-D-erythro-hexopyranosyl Nucleosides Employing Intramolecular Glycosylation as a Key Step, Nucleosides Nucleotides 17, 53-63.

Thoithi, G.N., Van Schepdael, A., Busson, R., Janssen, G., Van Aerschot, A., Herdewijn, P., Roets, E. and Hoogmartens, J. Investigation of the Kinetics of Degradation of Hexopyranosylated Cytosine Nucleosides Using Liquid Chromatography, Nucleosides. Nucleotides 19, 189-203.

Tsuboike, K., Minamoto, K., Mizuno, G. and Yanagihara, K. (1998) Stereoselective synthesis of some 3-Nitroglucopyranosyladenine Analogues via a Nitroolefin Intermediate as Potential Therapeutic Agents. Nucleosides Nucleotides 17, 745-758.

Turner, S.J. and Herdewijn, P. (1998) Synthetic strategies towards the synthesis of 1-(2,4-Dideoxy-4-C-hydroxymethyl-α-L-lyxopyranosyl-base nucleosides, Nucleosides Nucleotides 17, 2085-2086.

Ueda, T. and Watanabe, S.-i. (1985) Synthesis and Optical Properties of 2,3-Dideoxy-D-*erythro*-hex-2-enopyranosyl Nucleosides. (Nucleosides and Nucleotides. LXII*)* Chem. Pharm. Bull. 33, 3689-3695.

Ulbricht, T.L.V. and Rogers, G.T. (1965) Nucleosides. Part I. O-glucosides of cytosine and their rearrangement to N-glucosides, J. Chem. Soc., 6125-6129.

Van Schepdael, A., Smets; K., Vandendriessche, F., Van Aerschot, A., Herdewijn, P., Roets, E. and Hoogmartens, J. (1994) Comparative Stability Study of Thymidine and (Dideoxy-D-*erythro*-hexopyranosyl)thymine Analogues Monitored by Capillary Electrophoresis, J. Chromatogr. A 687, 167-173.

Verheggen, I., Van Aerschot, A., Toppet, S., Snoeck, R., Janssen, G., Balzarini, J., De Clercq, E. and Herdewijn, P. (1993) Synthesis and Antiherpes Virus Activity of 1,5-Anhydrohexitol Nucleosides, J. Med. Chem. 36, 2033-2040.

Verheggen, I., Van Aerschot, A., Van Meervelt, L., Rozenski, J., Wiebe, L., Snoeck, R., Andrei, G., Balzarini, J., Claes, P., De Clercq, E. and Herdewijn, P. (1995a) Synthesis, Biological Evaluation, and Structure Analysis of a Series of New 1,5-Anhydrohexitol Nucleosides, J. Med. Chem. 38, 826-835.

Verheggen, I., Van Aerschot, A., Pillet, N., van der Wenden, E.M., Ijzerman, A. and Herdewijn, P. (1995b) Synthesis of 1,5-Anhdyro-2-(N^6-cyclopentyladenin-9-yl)-2-deoxy-D-altrohexitol, Nucleosides Nucleotides 14, 321-324.

After several less than meaningful results with the racemic guanosine analogs **4** [Patil and Schneller, 1991] and **5** [Patil *et al.*, 1992a] and adenosine (±)-**3** [Patil *et al.*, 1992b], an enantiomeric synthesis of (-)-**3** was accomplished [Siddiqi *et al.*, 1993a]. This derivative was found to possess significant activity towards human cytomegalovirus (HCMV), a herpes virus, and to be potent inhibitor of *S*-adenosyl-L-homocysteine (AdoHcy) hydrolase [Siddiqi *et al.*, 1994].

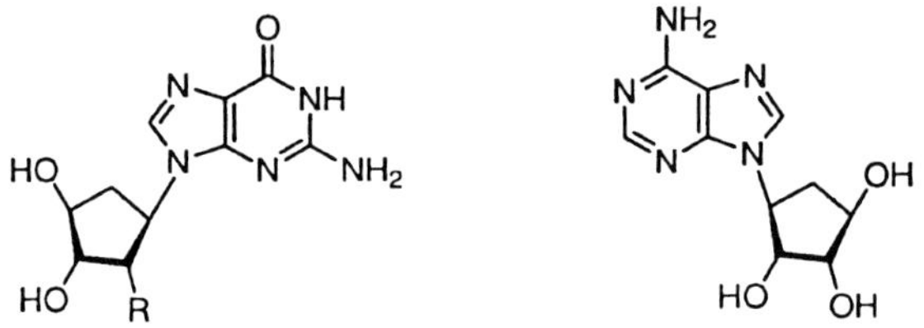

Figure 1. Early 5'-norcarbanucleosides.

In exploring structural variations of (-)-**3** for improved HCMV activity, its enantiomer (+)-**3** was prepared (Scheme 2) [Siddiqi *et al.*, 1994; Seley *et al.*, 1997a]. While (+)-**3** was 10-15 times less active towards HCMV, it did show inhibition of hepatitis B virus (HBV), which was not observed with (-)-**3**, at a concentration significantly below the 50% cytotoxic concentration [Seley *et al.*, 1997a].

In view of the need for developing effective chemotherapeutic agents for treating HBV [De Clercq, 1999] and thinking of (+)-**3** as resembling, structurally, L-nucleosides, which, as a group, have displayed promising antiviral properties (including versus HBV) [Wang *et al.*, 1998], efforts in our laboratory have sought more potent anti-HBV agents based on (+)-**3**.

2.   Results

The first two such derivatives considered replacing the 4'-hydroxyl of (+)-**3** with a thiol and an amino functionality (targets **8**, Schemes 2 and 3, and **9**, Scheme 2, respectively). In seeking **8**, the readily available allylic acetate **10** was subjected to a palladium (0) catalyzed coupling with potassium thioacetate as a means of forming the new carbon-sulfur bond necessary for **8**. The **11** obtained in this reaction provided the sulfur in a deactivated form; thus, avoiding its oxidation in the subsequent osmium tetroxide treatment to glycol **12**. However, attempts at debenzoylation and deacetylation of **12** under a variety of different conditions consistently led to decomposition. (It should be noted that the potassium thioacetate/palladium reaction could not be carried out on the derivative of **10** lacking the N-6 benzoyl group.)

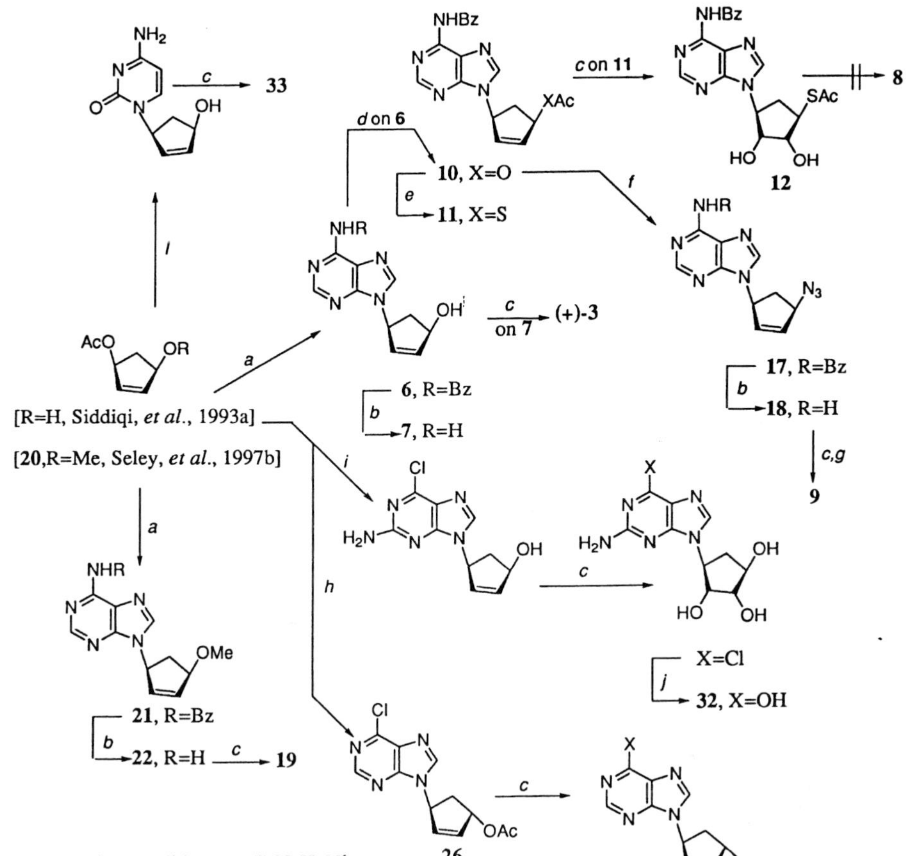

Reaction conditions: *a*, (i) NaH, N^6-benzoyladenine; (ii) $(Ph_3P)_4Pd/PPh_3$; *b*, NH_4OH in MeOH; *c*, $OsO_4/60\%$ aq. 4-methylmorpholine *N*-oxide; *d*, Ac_2O; *e*, KSAc, 5 mol % $(Ph_3P)_4Pd$, 15 mol % PPh_3; *f*, NaN_3, $Pd_2(dba)_3 \cdot (CHCl_3)$, 1,3-bis(diphenyl)-phosphinopropane (dppp); *g*, H_2, Pd/C; *h*, 6-chloropurine, PPh_3, DIAD; *i*, NaH, 2-amino-6-chloropurine; (ii) $(Ph_3P)_4Pd/PPh_3$; *j*, 1 N HCl; *k*, NH_3 in MeOH; *l*, Na salt of cytosine, $Pd_2(dba)_3 \cdot (CHCl_3)$, dppp

Scheme 2.

These problems were circumvented by taking a different approach (Scheme 3) beginning with the Michael addition of 4-methoxy-α-toluenethiol to the enone **13**. Reduction of the resultant **14** occurred favorably [Siddiqi *et al.*, 1993b] to give the 1,4-*trans* product **15**. Coupling **15** with 6-chloropurine under Mitsunobu conditions followed by reaction with ammonia yielded **16**. Acidic removal of the isopropylidene and 4-methoxybenzyl protecting groups produced **8**.

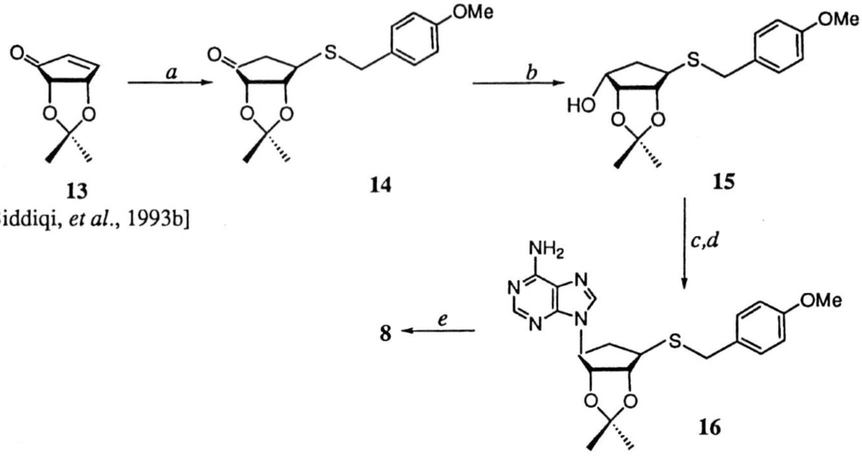

[Siddiqi, *et al.*, 1993b]

13   **14**   **15**

16

Reaction conditions. *a*, 4-methoxy-α-toluenethiol, K_2CO_3; *b*, BH_3•THF; *c*, DIAD, Ph_3P, 6-chloropurine; *d*, NH_3 in MeOH; *e*, TFA, PhOH

Scheme 3.

Contrary to the problem of achieving **8** from **10** via Scheme 2, a similar plan with sodium azide as the nucleophilic source in the initial palladium catalyzed reaction (path f, Scheme 2) led in a straightforward manner to **9**. This route involved standard procedures: debenzoylation of **17** to **18** followed by glycolization and catalytic hydrogenation to **9**.

To determine the role the C-4' hydroxyl hydrogen may play in anti-HBV agent design, the methyl derivative **19** (Scheme 2) was synthesized. Again, as is so often the case in this research, a palladium stimulated coupling involving a cyclopentenyl allylic acetate (here **20**) was the starting point. The product of this reaction with N^6-benzoyladenine (**21**) was then debenzoylated (to **22**) followed by oxidation to **19**.

Next to be considered were the epimer (**23**), deoxy (**24**), and deoxy-ene (**25**) forms of (+)-**3** (Schemes 2 and 4). Using the Mitsunobu reaction of (-)-(1S,4R)-4-hydroxy-2-cyclopenten-1-yl acetate with 6-chloropurine led to the inverted product **26**. Oxidation of **26** to **27** followed by ammonolysis gave **23**.

Another Mitsunobu process with 6-chloropurine using **28** led to **29** (Scheme 4). Ammonolysis of **29** followed by removal of the 2',3'-isopropylidene protecting unit provided **25**. Hydrogenation of **25** availed **24**.

Inspection of Table 1 shows that compounds **24** and **25** became the first derivatives of (+)-**3** in this study with greater anti-HBV potential clearly indicating the C-4' hydroxyl of (+)-**3** is unnecessary. To determine if this property was unique to carbanucleosides, the L-erythrofuranose analog **30** was prepared via an adaptation of a reported route [Lerner, 1969] and found to be inactive towards HBV.

Several additional derivatives have been prepared including the 7-deaza (**31**, Scheme 4), guanine (**32**, Scheme 2), and cytosine (**33**, Scheme 2). These compounds were inactive.

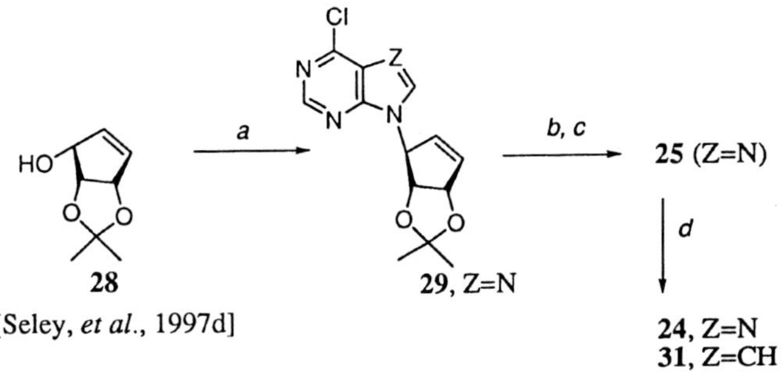

Reaction conditions: *a*, 6-chloropurine or 4-chloropyrrolo[2,3-*d*]pyrimidine, PPh$_3$, DIAD; *b*, NH$_3$ in MeOH; *c*, Dowex 50 x 8 acidic resin, MeOH; *d*, H$_2$, PtO$_2$

Scheme 4.

Table 1.   Inhibition of hepatitis B Virus by 5'-Noraristeromycin derivatives

Compound	CC50 (μM)	EC50 (μM)	EC90 (μM)	SI(CC50/EC90)
(+)-3	446 ± 20	1.4 ± 0.1	9.6 ± 0.8	46
23	1883 ± 101	>10	>10	-
24	93 ± 7.4	0.120 ± 0.016	0.978 ± 0.077	95
25	325 ± 17	0.145 ± 0.015	1.4 ± 0.2	232
3TC	1884 ± 123	0.070 ± 0.008	0.209 ± 0.018	9014

3.   Conclusion

Carbocyclic nucleosides (with adenine as the base) in the L-like configuration and lacking the C-5' methylene and C-5' hydroxymethylene have provided a new lead into anti-HBV agents. Efforts are now underway to pursue variations of **24** and **25**.

4.   Acknowledgments

This research was supported by funds from the Department of Health and Human Services (AI31718 and AI48495), which is appreciated. We are also grateful to Dr. Brent Korba of Georgetown University for providing the HBV assays.

5.   References

De Clercq, E. (1994) Antiviral activity spectrum and target of action of different classes of nucleoside analogues, Nucleosides Nucleotides 13, 1271-1295.

De Clercq, E. (1999) Perspectives for the treatment of hepatitis B virus infections, Int. J. Antimicrob. Agents 12, 81-95.

Hegde, V.R., Seley, K.S., Schneller, S.W., Elder, T.J.J. (1998) 5'-Amino-5'-deoxy-5'-noraristeromycin, J. Org. Chem. 63, 7092-7093.

Kim, C.U., Luh, B.Y., Martin, J.C. (1991) Regiospecific and highly stereoselective electrophilic addition to furanoid glycals: synthesis of phosphonate nucleotide analogues with potent activity against HIV, J. Org.Chem. 56, 2642-2647.

Koga, M., Schneller, S.W. (1990) The synthesis of two 2'-deoxy carbocyclic purine nucleosides lacking the 5'-methylene, Tetrahedron Lett. 31, 5861-5864.

Koga, M., Schneller, S.W. (1993) Oligonucleotides of carbocyclic 5'-nor 2'- and 3'-deoxyadenosine, Nucleic Acids Symposium Series number 29, 63-65.

Koga, M., Abe, K., Ozaki, S., Schneller, S.W. (1994) Synthesis and properties of carbocyclic oligonucleotides lacking the 5'-methylene, Nucleic Acids Symposium Series number 31, 65-66.

Lerner, L.M. (1969) Preparation of nucleosides via isopropylidene sugar derivatives. IV. synthesis of 9-α- and 9-β-erythrofuranosyladenine, J. Org. Chem. 34, 101-103.

Marquez, V. (1996) Carbocyclic nucleosides, Advances in Antiviral Drug Design 2, 89-146.

Patil, S.D., Schneller, S.W. (1991) (±)-5'-Nor ribofuranoside carbocyclic guanosine, J. Heterocycl. Chem. 28, 823-824.

Patil, S.D., Koga, M., Schneller, S.W., Snoeck, R., De Clercq, E. (1992a) (±)-Carbocyclic 5'-nor-2'-deoxy-guanosine and related purine derivatives: synthesis and antiviral properties, J. Med. Chem. 35, 2190-2195.

Patil, S.D., Schneller, S.W., Hosoya, M., Snoeck, R., Andrei, G., Balzarini, J., De Clercq, E. (1992b) Synthesis and antiviral properties of (±)-5'-noraristeromycin and related purine carbocyclic nucleosides. A new lead for anti-human cytomegalovirus agent design, J. Med. Chem. 35, 3372-3377.

Seley, K.L., Schneller, S.W., Korba, B. (1997a) A 5'-noraristeromycin enantiomer with activity towards hepatitis B virus, Nucleosides Nucleotides 1997, 2095-2099.

Seley, K.L., Schneller, S.W., De Clercq, E. (1997b) A methylated derivative of 5'-noraristeromycin, J. Org. Chem. 62, 5645-5646.

Seley, K.L., Schneller, S.W., Rattendi, D., Bacchi, C.J. (1997c) (+)-7-Deaza-5'-noraristeromycin as an anti-trypanosomal agents, J. Med. Chem. 40, 622-624.

Seley, K.L., Schneller, S.W., Rattendi, D., Lane, S., Bacchi, C.J. (1997d) Synthesis and anti-trypanosomal activity of various 8-aza-7-deaza-5'-noraristeromycin derivatives, J. Med. Chem. 40, 625-629,

Siddiqi, S.M., Chen, X., Schneller, S.W. (1993a) Enantiospecific synthesis of 5'-noraristeromycin and its 7-deaza derivative and a formal synthesis of (-)-5'-homoaristeromycin, Nucleosides Nucleotides 12, 267-278.

Siddiqi, S.M., Schneller, S.W., Ikeda, S., Snoeck, R., Andrei, G., Balzarini, J., De Clercq, E. (1993b) *S*-Adenosyl-L-homocysteine hydrolase inhibitors as antiviral agents: 5'-deoxyaristeromyicn, Nucleosides Nucleotides 12, 185-198.

Siddiqi, S.M., Chen, X., Schneller, S.W., Ikeda, S., Snoeck, R., Andrei, G., Balzarini, J., De Clercq, E. (1994) Antiviral enantiomeric preference for 5'-noraristeromycin, J. Med. Chem. 37, 551-554.

Wang, P., Hong, J.H., Cooperwood, J.S., Chu, C.K. (1998) Recent advances in L-nucleosides: chemistry and biology, Antiviral Res. 40, 19-44.

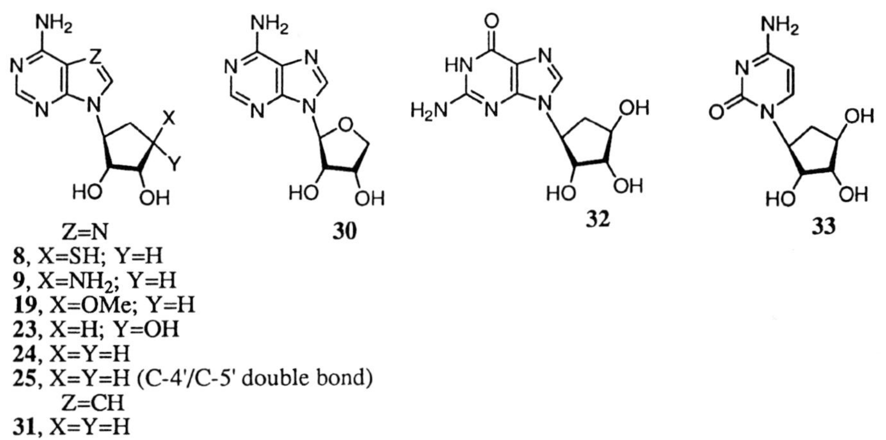

Z=N
8, X=SH; Y=H
9, X=NH$_2$; Y=H
19, X=OMe; Y=H
23, X=H; Y=OH
24, X=Y=H
25, X=Y=H (C-4'/C-5' double bond)
    Z=CH
31, X=Y=H

Figure 2; Target Analogs.

TRICYCLIC NUCLEOSIDES REVISITED

KATHERINE L. SELEY

School of Chemistry and Biochemistry, Georgia Institute of Technology, Atlanta, Georgia 30332-0400 USA

1.   Introduction

In the early 70's, Nelson Leonard introduced the nucleoside community to the concept of an expanded purine ring system, whereby a spacer ring (or rings) was inserted in between the two rings of the parent bicyclic purine moiety (**1-3**, Figure 1).

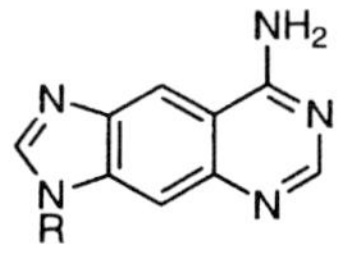

1, R=H
4, R=β-D-ribofuranosyl
5, R=2'-deoxy-D-ribofuranosyl

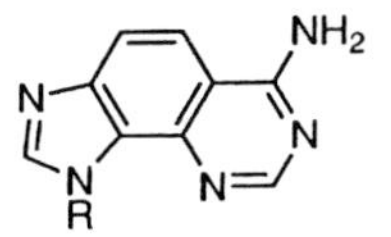

2, R=H
6, R=β–D-ribofuranosyl
7, R=2'-deoxy-D-ribofuranosyl

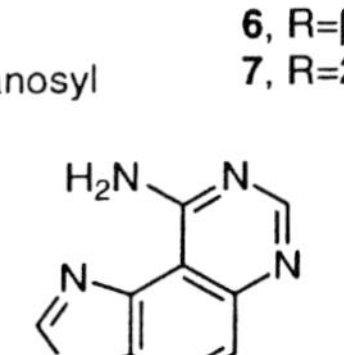

3, R=H
8, R=β–D-ribofuranosyl
9, R=2'-deoxy-D-ribofuranosyl

Figure 1.   "Expanded" tricyclic nucleosides and heterobases.

Leonard employed these stretched out purine analogues as dimensional probes for investigating enzyme/coenzyme binding sites. This led to a variety of shape-altered purines, which in turn, then led to the introduction of his expanded nucleosides, the *dist-*, *prox-* and *lin-*benzoadenosines (**4-9**, Figure 1). These analogues contained a benzene spacer ring separating the two components of the purine parent, adenosine. Leonard (Leonard and Hiremath, 1986; Leonard, 1982) [and references contained therein] and others [most notably, Nair (Nair *et al.*, 1984b; Nair *et al.*, 1984a; Nair and Offerman, 1985),

Recent Advances in Nucleosides: Chemistry and Chemotherapy, Ed. by C.K. Chu. 299 — 326

Schneller (Schneller and Clough, 1975; Schneller and Christ, 1982; Schneller and Ibay, 1986) and Townsend (Klimke *et al.*, 1979; Chung *et al.*, 1980b; Chung *et al.*, 1980a; Chung *et al.*, 1980c)] subsequently synthesized a large array of expanded analogues (nucleosides and nucleobases) testing the boundaries of numerous enzymes and enzyme binding sites. Many of these analogues showed potent biological activity, which fueled the interest in tricyclic nucleosides.

One significant physical characteristic of the tricyclic nucleosides is their intense fluorescence and therefore many of these analogues have been used to probe their environments. Protein-DNA interactions, base-base stacking interactions, neighboring group effects, hydrogen bonding interactions, as well as interactions with $NADP^+$ and ATP, have all been studied using the tricyclic nucleosides.

Leonard has written several excellent reviews of these dimensional probes (Leonard and Hiremath, 1986; Leonard, 1982) and the related etheno-bridged nucleotides, (Leonard, 1992; Leonard, 1993) therefore the tricyclics covered in those will not be included here.[1] Since the time of those reviews, however, there have been several interesting entries to the extended nucleoside family, and it is those analogues that this review will focus on.

The tricyclic nucleosides reviewed herein have been divided into two basic structural types; "expanded" and "extended" nucleosides.[2] As mentioned above, it is possible (in the case of the purines) to insert a spacer ring in-between the two rings of the bicyclic parent ring system, thereby "expanding" the purine. It is also possible, however, to add an additional ring onto a purine or pyrimidine ring system, thereby "extending" the heterocyclic moiety. There are, of course, many parent ring systems that have been used to construct the tricyclic nucleosides therefore the analogues covered in this review are classified by the parent heterocycle as well.

2.   "Expanded" dideoxyadenosine nucleosides

Related to Leonard's *lin*-benzo analogues, Nair *et al.* synthesized *lin*-benzo-dideoxy-adenosine **10** and its monophosphate **11** (Figure 2) (Zhang and Nair, 1997). Following on the significant biological activity shown by the isonucleosides, Nair also synthesized some *lin*-benzo isodideoxyadenosine analogues (**12-15** in Figure 2) (Zhang and Nair, 1997, Zhang *et al.*, 1998). The monophosphate analogues were found to have moderate to good inhibition of the viral-encoded enzyme HIV integrase. These results suggest that the nucleotide-binding site of HIV integrase, in contrast to the binding site of HIV reverse transcriptase, can accommodate major structural modifications in the nucleobase.

---

[1]   Much has also been written about the tricyclic wyosines, therefore they have been omitted from this review as well.

[2]   The triciribines, which are tricyclic as well, are triangular in shape, rather than "linear" or "bent". The author prefers to concentrate upon the later two categories, therefore the triciribines have been omitted from this review.

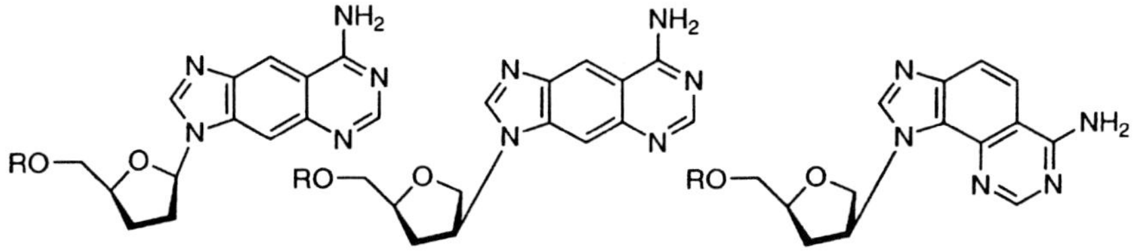

10, R=H
11, R=PO_3Na_2

12, R=H
13, R=PO_3Na_2

14, R=H
15, R=PO_3Na_2

Figure 2. "Expanded" tricyclic dideoxy- and isodideoxyadenosine analogues.

As shown in Scheme 1, the synthesis of **10** and **11** began with coupling persilylated **16**, the same base used by Leonard to construct his nucleosides. Coupling **16** to the protected diol **17** using TMSOTf[3] in dichloroethane provided **18**, which following standard deprotection with TBAF, and subsequent ammonolysis formed nucleoside **10**. Standard phosphorylation conditions afforded monophosphate **11**. Synthesis of the isodideoxy analogues is outlined in Scheme 2. Coupling the furan derivative **19** with the tricyclic base **16**, or its angular analogue **20**, yielded the protected isodideoxy nucleosides **21** or **22**, respectively. As before, ammonolysis gave the nucleosides **12** and **14**, and subsequent phosphorylation resulted in monophosphates **13** and **15**.

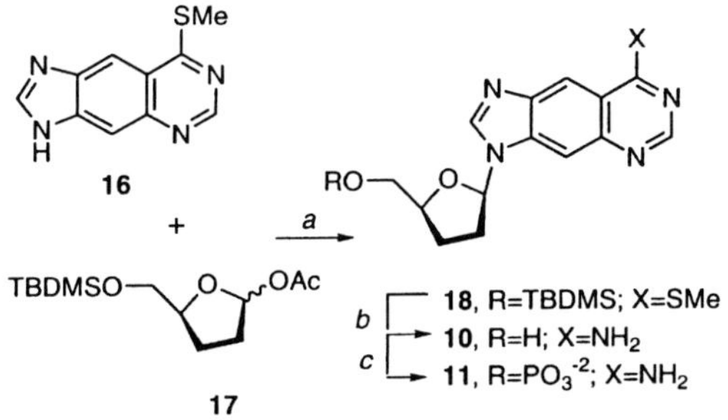

Reaction conditions: *a*, (i) TMSCl, HDMS, reflux; (ii) TMSOTf, $ClCH_2CH_2Cl$; *b*, (i) TBAF, THF; (ii) NH_3, MeOH, 100 °C; *c*, $POCl_3$, $(EtO)_3PO$, 0 °C

Scheme 1.

[3] Abbreviations used: TMSOTf, trimethylsilyl triflate; TBAF, tetrabutyl ammonium fluoride; TBAR, 1-*O*-acetyl-2,3,5-tri-*O*-benzoyl-β-D-ribofuranose; BSA, *N,O*-bis(trimethylsilyl)acetamide; HMDS, 1,1,1,3,3,3-hexamethyldisilazane; TMSCl, trimethylsilyl chloride; DEAD, diethyl azodicarboxylate; HCMV, human cytomegalovirus; TBDMSCl, *tert*-butyldimethylsilyl chloride; DMTr, dimethoxytrityl; AIBN, 2,2'-azobisisobutyronitrile; CBz, benzyloxycarbonyl; DMSO, dimethylsulfoxide; DBU, 1,8-diaza-bicyclo[5.4.0]undec-7-ene; t-BOC, *tert*-butoxycarbonyl; BnBr, benzyl bromide; TBAI, tetrabutyl ammonium iodide; t-BuONO, *tert*-butyl nitride; BSTFA, bis (trimethyl sityl) trifluoroacetamide; DMAP, dimethylamino pyridine

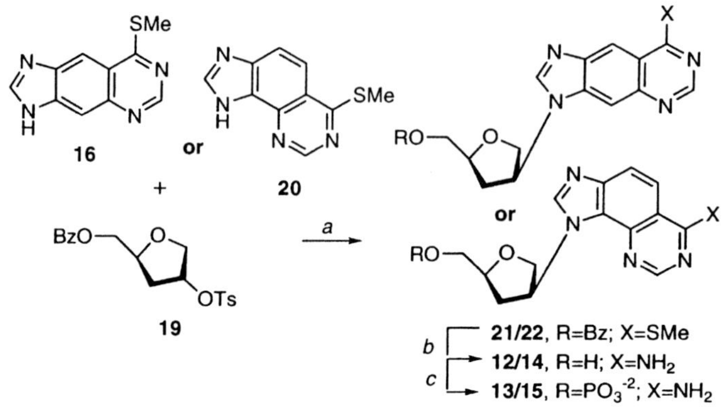
SMe
16
or
20
+
BzO
OTs
19
a
RO
X
SMe
b
c
21/22, R=Bz; X=SMe
12/14, R=H; X=NH2
13/15, R=PO3-2; X=NH2

Reaction conditions: *a*, K_2CO_3, 18-Crown-6, 95 °C; *b*, NH_3, MeOH, 100 °C; *c*, $POCl_3$, $(EtO)_3PO$, 0 °C

Scheme 2.

3. "Expanded" purine nucleosides

Drawing on Leonard's *lin*-benzoadenosine, recent synthetic efforts from our laboratories include the synthesis of three thieno-separated purines (**23-25** as shown in Figure 3). The heterocyclic bascs of these nucleosides have shown preliminary anticancer activity in colorectal cell lines (Seley *et al.*, 2000) and further anticancer and antiviral testing of the bases and of the nucleosides are presently underway.

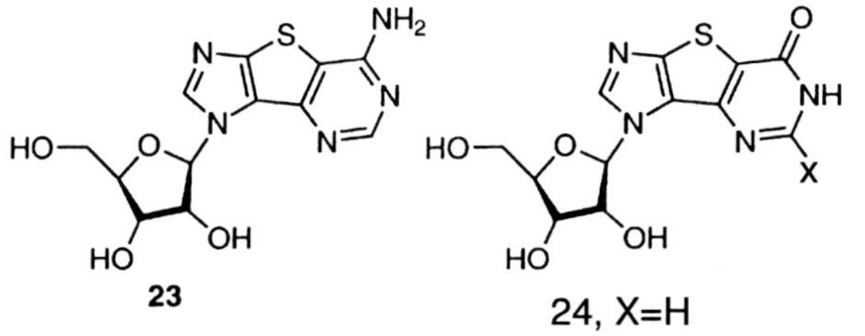
NH2
HO
OH
HO
23
O
NH
X

24, X=H
25, X=NH_2

Figure 3. "Expanded" purine tricyclic analogues.

The synthesis (Seley *et al.*, 2001, Seley *et al.*, submitted, Seley *et al.*, 2000) of the adenosine and inosine analogues **23** and **24** is outlined in Scheme 3 and was initiated with coupling persilylated dibromoimidazole 26 with 1,2,3,5-tetra-*O*-acetyl-ß-D-ribofuranose, using BSA and TMSOTf. Removal of the acetate groups of the resulting nuleoside **27** and reprotection with benzyl groups provided 28. Manipulation

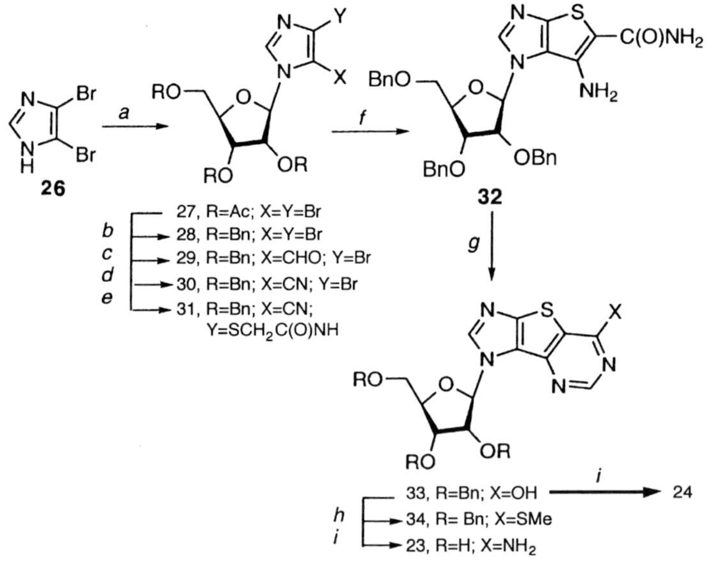

Reaction conditions; *a*, (i) BSA, CH2CN, 1,2,3,5,-tetra-*O*-acetyl-β-D-ribofuranose; (ii) TMTSOTf; *b*, (i) NH4OH, MeOH; (ii) NaH, BnBr, TBAL; *c*, EtMgBr, DMF; *d*, (i) NH2OH-HCl, NaHCO3; (ii) Ac2O; *e*, K2CO3, NH2C(O)CH2SH; *f*, NaOEt, EtOH; *g*, (EtO)3CH, Ac2O; *h*, (i) P2S-pyridine; (iii) K2CO3, McI; *i*, (i) NH3, BuOH; (ii), BF3-OEl2, EtSH.

Scheme 3.

of the bromine moieties was the undertaken utilizing a Grignard reaction to replace the C-5 bromide with a formyl group to give **29**. Conversion to the hydroxylamine and sybsequent dehydration with acetic antihydride provided nitrile **30**. Displacement of the C-4 bromide to form 31 was accomplished with thioglyolcamide.

The bicyclic nucleoside **32** was formed after treatment of **31** with sodium ethoxide. All of the tricyclic targets were then constructed from **32**. Treatment of **32** with triethylorthoformate afforded **33**, whose manipulation of the carbonyl with phosphorous pentasulfide followed by immediate conversion of the resulting thiocarbonyl with methyl iodide, gave thiomethyl **34**. Finally, ammonollysis using standard conditions, and removal of the benzyl protecting groups with boron trifluoride etherate, resulted in the desired adenoside analogue **23**. The inosine analogue **24** was available directly from **32** following deprotection.

The guanosine analogue **25** was also realized from **32**, albeit from a less tedious route (Scheme 4). Ring closure of **32** was accomplished with sequential treatment with sodium hydroxide in methanol, followed by carbon disulfide with heating. Next, addition of hydrogen peroxide, and then heating with ammonia in methanol preovided **35**. Finally, deprotection of the benzyl groups as before, gave **25**. All three nucleosides are presently undergoing broad screen biological testing for potential anticancer and antiviral activity. The results of these studies will be forthcoming elswhere.In a similar

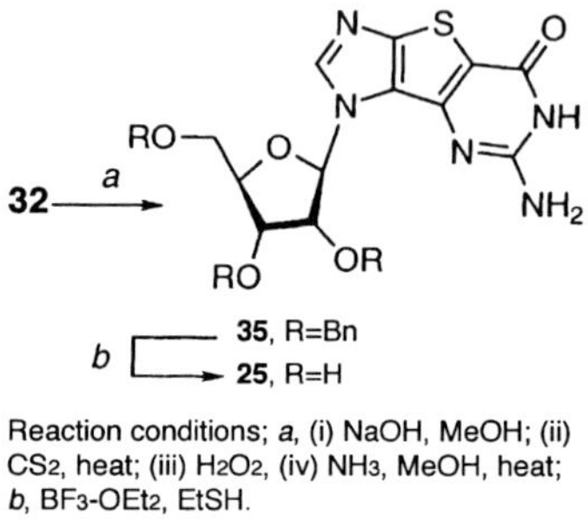

304   *K. L. Seley*

32 $\xrightarrow{a}$

Reaction conditions; *a*, (i) NaOH, MeOH; (ii) CS₂, heat; (iii) H₂O₂, (iv) NH₃, MeOH, heat; *b*, BF₃-OEt₂, EtSH.

b — **35**, R=Bn
— **25**, R=H

Scheme 4.

vein, Humphries and Ramsden have recently reported (Humphries and Ramsden, 1999; Humphries and Ramsden, 1995) the synthesis of three expanded tricyclics employing a pyridine spacer ring as shown in Figure 4; an expanded analogue of adenosine (**36**), a dithione analogue (**37**) and an inosine analogue (**38**). The syntheses are shown in Scheme 5. Starting with 5-amino-1-(β-D-ribofuranosyl)imidazole (AIRs) (**39**), reaction with ethoxymethylene malononitrile gave the 4-(2',2'-dicyanovinyl) derivative **40**, which underwent cyclization with aqueous sodium bicarbonate to give the bicyclic intermediate **41**. Treatment of **41** with diethoxymethyl acetate, followed by reaction with methanolic ammonia at room temperature afforded **36**. Refluxing **41** with carbon disulfide in pyridine, however, gave the dithione target **37**. Finally, reaction of **41** with ammonium hydroxide and hydrogen peroxide afforded amide **42**, which, following treatment with sodium ethoxide provided the inosine analogue **38**.

The synthesis of another expanded inosine analogue **43** (Figure 4) is outlined in Scheme 6 (O'Hara Dempcy and Skibo, 1991). Persilylation of **44** with HMDS and TMSCl for 36 hours, followed by ribosylation with 1,2,3,5-tetra-*O*-acetyl-β-D-ribofuranose and tin (IV) chloride, gave the desired N-3 intermediate **45** (Only trace amounts of the N-7 and N-1 were formed, however, the authors found that by altering

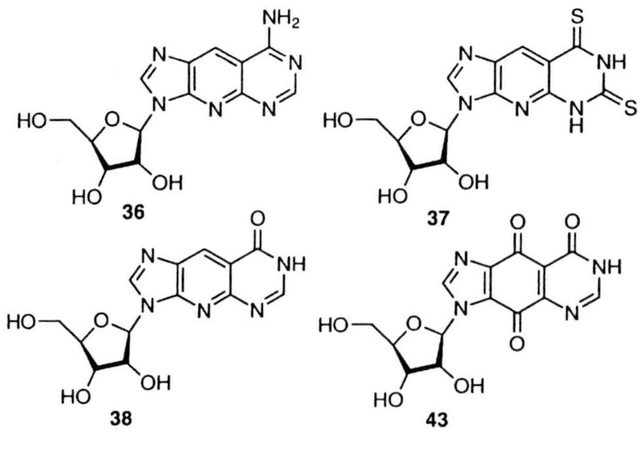

Figure 4. "Expanded" tricyclic purine analogues.

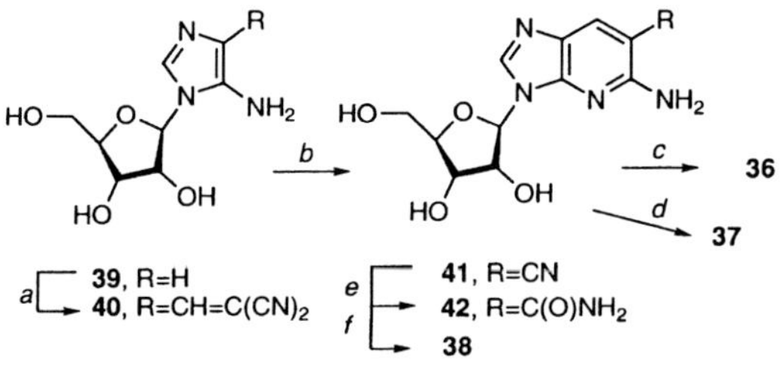

Reaction conditions: *a*, EtOCH=C(CN)$_2$; *b*, 5% aq. NaHCO$_3$, EtOH, 80 °C; *c*, (i) MeCO$_2$CH(OEt)$_2$; (ii) NH$_3$, MeOH; *d*, CS$_2$, pyridine, reflux; *e*, NH$_4$OH, H$_2$O$_2$, rt; *f*, NaOEt, HCO$_2$Et, EtOH, reflux.

Scheme 5.

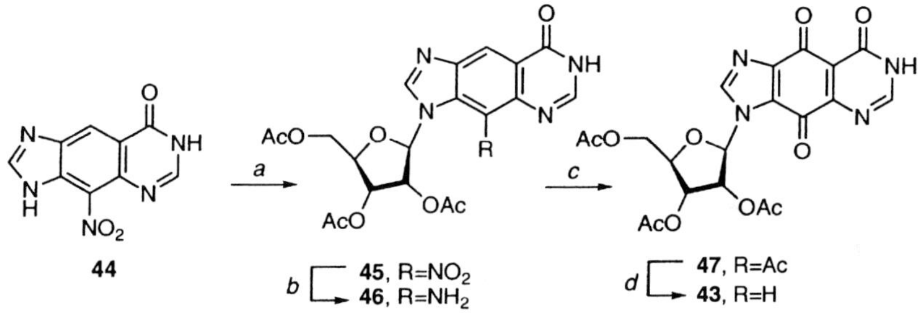

Reaction conditions: *a*, (i) HMDS, TMSCl, reflux, 36 hr; (ii) SnCl$_4$, 1,2,3,5-tetra-*O*-acetyl-β-D-ribofuranose, CH$_2$Cl$_2$, 30 min; *b*, sodium dithionite; *c*, Fremy oxidation; *d*, NaOH, MeOH.

Scheme 6.

the conditions they were, however, able to form those in larger quantities). Reduction of the nitro functionality of **45** was accomplished with sodium dithionite to give **46**, which was then converted to the quinone **47** by Fremy oxidation. Finally, removal of the acetyl groups resulted in the desired **43**. The substrate activity of **43** and the N-1 and N-7 analogues were assessed against xanthine oxidase. As anticipated, only **43** was found to be a substrate for the enzyme, thereby confirming its purine-like nature.

4.   "Expanded" carbocyclic purine nucleosides

Shown in Figure 5 are some very recent entries from the laboratories of Schneller *et al*; *lin*-benzoaristeromycin (**48**) (Rajappan and Schneller, 2001a) the carbocyclic analogue of Leonard's *lin*-benzoadenosine (**4**, Figure 1), *lin*-benzo-5'-noraristeromycin

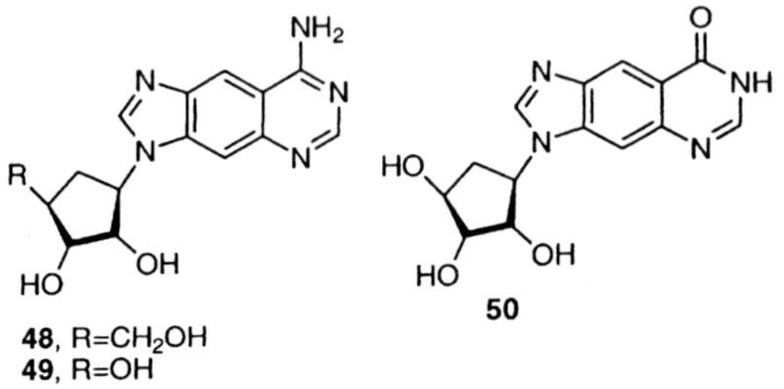

48, R=CH$_2$OH
49, R=OH

50

Figure 5. "Expanded" carbocyclic tricyclic purine analogues.

(**49**) (Rajappan and Schneller, 2001b) an expanded analogue of the potent anti-HCMV compound 5'-noraristeromycin, and carbocyclic *lin*-benzo-5'-norinosine (**50**) (Rajappan and Schneller, 2001b). These targets were designed as dimensional probes for *S*-adenosylhomocysteine hydrolase based on the significant inhibitory activity of aristeromycin and its 5'-nor analogue against the enzyme.

Using the previously employed thiomethyl *lin*-benzo intermediate **16** (Scheme 7) as a starting point, coupling **16** to triflate **51** (Wang *et al.*, 1999) using sodium hydride and 18-Crown-6 gave both the N-9 and N-7 coupled intermediates (**52a** and **b**, respectively), which were separable by chromatography. Treatment of **52a** with methanolic ammonia gave the amino intermediate **53**, which, following deprotection with trifluoroacetic acid, afforded the desired *lin*-benzoaristeromycin **48**.

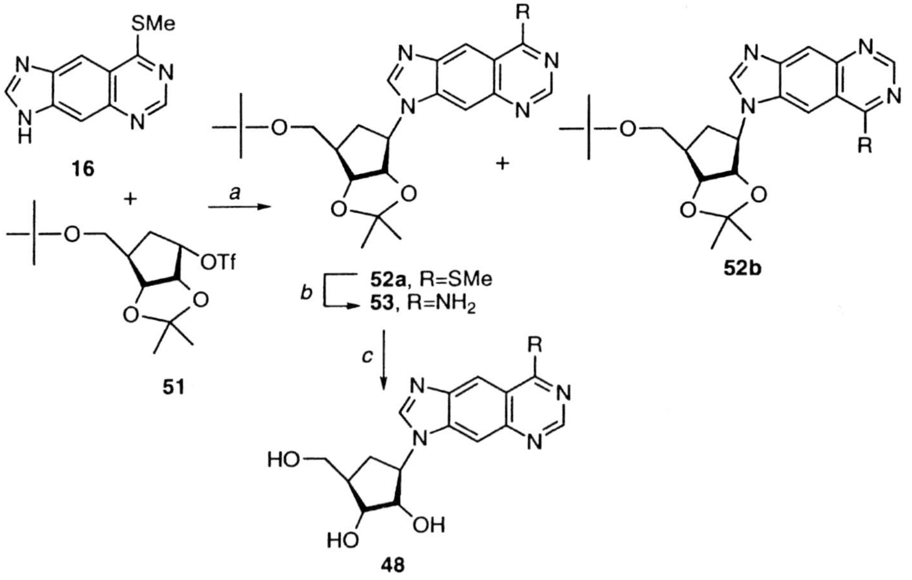

Reaction conditions: *a*, NaH, 18-Crown-6, DMF, rt; *b*, NH$_3$, MeOH; *c*, TFA, H$_2$O.

Scheme 7.

The *lin*-benzo-5'-noraristeromycin analogue **49** was synthesized as outlined in Scheme 8 (Rajappan and Schneller, submitted) starting with the standard palladium catalyzed coupling method used extensively in the synthesis of the 5'-nor analogues (Seley *et al.*, 1997a; Seley *et al.*, 1997b). Heterocycle **16** (Leonard *et al.*, 1975; Leonard *et al.*, 1987) was once again utilized, this time however, coupled to monoacetate **54** using tetrakis (triphenylphosphine)palladium and sodium hydride, to result in **55** and its N-7 analogue **56**. Separation by column chromatography, followed by dihydroxylation using standard conditions of osmium tetroxide and 4-methylmorpholine-*N*-oxide produced **57** (and similarly, **58**). Ammonolysis of **57** gave the desired **49**.

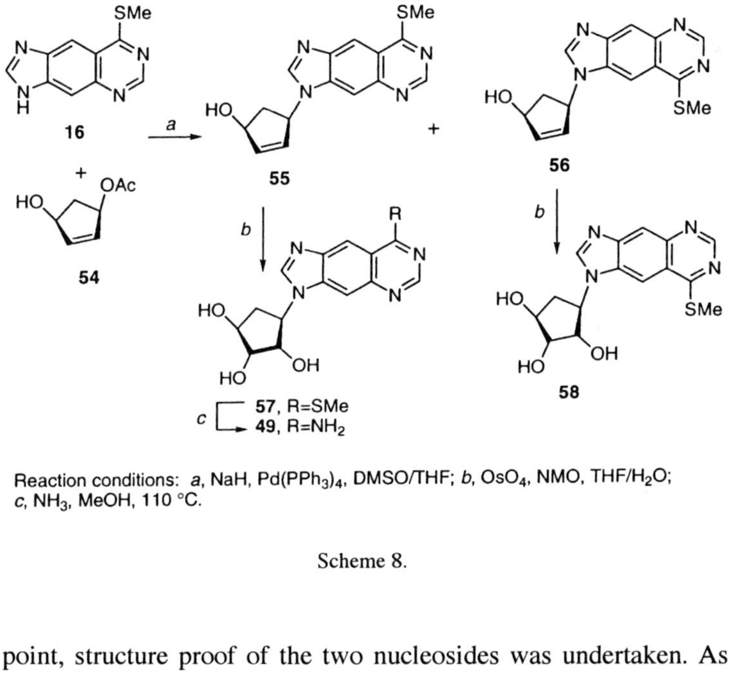

Reaction conditions: *a*, NaH, Pd(PPh₃)₄, DMSO/THF; *b*, OsO₄, NMO, THF/H₂O; *c*, NH₃, MeOH, 110 °C.

Scheme 8.

At this point, structure proof of the two nucleosides was undertaken. As shown in Scheme 9, hydrolysis of **57** (and similarly, **58**) yielded the inosine analogue **50** (and its N-7 analogue for **58**). An identical sample was also synthesized from coupling quinazoline **59** with the 2,3-*O*-isopropylidine-protected cyclopentanetriol **60** (Rajappan and Schneller, submitted). Following the coupling reaction of **59** and **60** to form the nitro intermediate **61**, reduction of the nitro functionality afforded **62**, which, after ring closure with triethyl orthoformate and subsequent hydrolysis, provided **50**.

5.   "Extended" ethenoadenosine nucleosides

Ethenoadenosine (**63** in Figure 6) and its analogues have been studied extensively. These analogues occur due to the reaction of nucleosides with chloro- or bromoacetaldehyde, vinyl chloride or chloroethylene oxide. New work in this area focuses

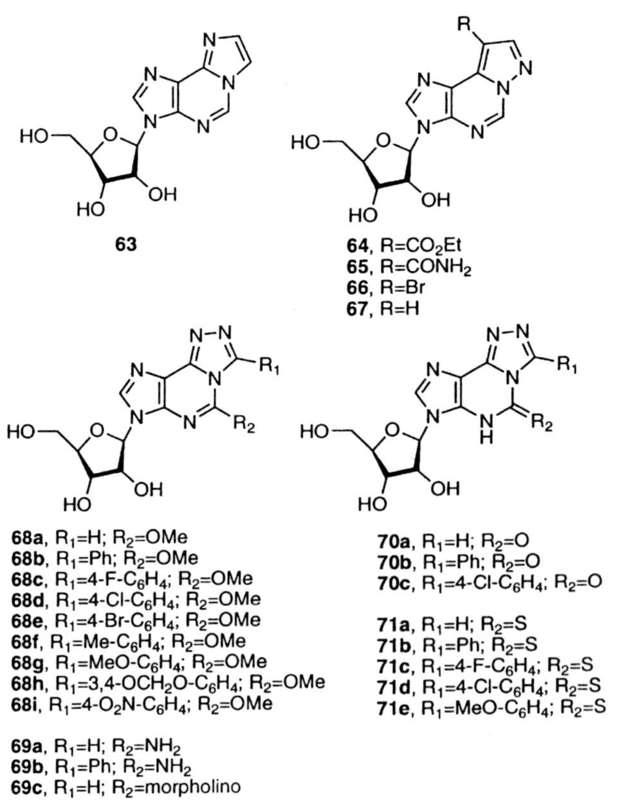

Reaction conditions: *a*, 1-BuOH, Et$_3$N, reflux; *b*, Pd/C, H$_2$, 45 psi; *c*, (i) EtO$_3$CH, HCl, rt; (ii) H$_2$O, 80 °C; *d*, 1N HCl, reflux.

Scheme 9.

68a, R$_1$=H; R$_2$=OMe
68b, R$_1$=Ph; R$_2$=OMe
68c, R$_1$=4-F-C$_6$H$_4$; R$_2$=OMe
68d, R$_1$=4-Cl-C$_6$H$_4$; R$_2$=OMe
68e, R$_1$=4-Br-C$_6$H$_4$; R$_2$=OMe
68f, R$_1$=Me-C$_6$H$_4$; R$_2$=OMe
68g, R$_1$=MeO-C$_6$H$_4$; R$_2$=OMe
68h, R$_1$=3,4-OCH$_2$O-C$_6$H$_4$; R$_2$=OMe
68i, R$_1$=4-O$_2$N-C$_6$H$_4$; R$_2$=OMe

69a, R$_1$=H; R$_2$=NH$_2$
69b, R$_1$=Ph; R$_2$=NH$_2$
69c, R$_1$=H; R$_2$=morpholino

70a, R$_1$=H; R$_2$=O
70b, R$_1$=Ph; R$_2$=O
70c, R$_1$=4-Cl-C$_6$H$_4$; R$_2$=O

71a, R$_1$=H; R$_2$=S
71b, R$_1$=Ph; R$_2$=S
71c, R$_1$=4-F-C$_6$H$_4$; R$_2$=S
71d, R$_1$=4-Cl-C$_6$H$_4$; R$_2$=S
71e, R$_1$=MeO-C$_6$H$_4$; R$_2$=S

Figure 6.   "Extended" ethenoadenosine tricyclic analogues.

on the syntheses of the pyrazolo analogues (**64-67** in Figure 6) by Hamamichi *et al.* (Hamamichi, 1991, Hamamichi and Miyasaka, 1994) and the triazolo series by Nagamatsu *et al.* (**68-71** in Figure 6) (Nagamatsu *et al.*, 1999).

The synthesis of the pyrazolo targets started with the protected 6-chloropurine nucleoside shown in Scheme 10 (Hamamichi, 1991, Hamamichi and Miyasaka, 1994). Transformation to the enamino ester **73** was accomplished by reaction of the protected 6-chloronucleoside **72** with methyl cyanoacetate, and subsequent hydrogenation. Reaction of **73** with hydrazine gave ester **74**, which can be converted to amide **75** with ammonia. Deprotection of **74** and **75** resulted in **64** and **65**, respectively. Alternatively, **75** was converted to the protected bromo analogue **76** with bromine at pH 6.9, which was then deprotected to give **66**, or reduced to give **77**. Compound **77** was then deprotected to give the unsubstituted parent compound **67**.

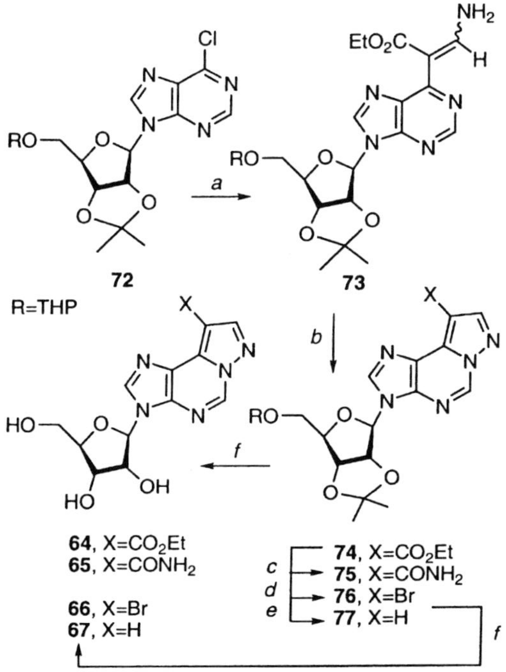

Reaction conditions: *a*, (i) CH$_2$(CN)CO$_2$Me, NaH; (ii) 5% Pd-C, H$_2$; *b*, NH$_2$NH$_2$; *c*, sat. NH$_3$, MeOH; *d*, Br$_2$, EtOAc-potassium phosphate buffer (pH 6.9); *e*, 5% Pd-C, H$_2$; *f*, aq. CF$_3$COOH.

Scheme 10.

The parent pyrazolo ethenoadenosine **67** showed moderate cytotoxic activity against mouse leukemia L5178Y cells, but even more significantly, the bromo analogue **66** proved to be extremely potent, while the others were found to be inactive. Based on these findings, it appears that the position 9 of these analogues is important for cytotoxic activity (Hamamichi, 1991; Hamamichi and Miyasaka, 1994).

The triazolo series was available from key intermediate **81**, and the syntheses are outlined in Schemes 11 and 12 (Nagamatsu *et al.*, 1999). First, treatment of the protected 6-chloro-2-iodonucleoside **78** with hydrazine afforded **79**, which was then either converted directly to the hydrazones **81a-j** by Methods A or B, or, alternatively, converted first to the hydrazino intermediates **80** using the appropriate aldehyde and DEAD, followed by conversion to the hydrazone intermediates **81a-j**, by Methods C or D (see Scheme 11 for specific conditions). Interestingly, most of the desired targets were available by both routes, with **81a** being the only exception. Method A was used for **81d, f, i**, and **j** and Method B for **81b, c, e, g**, and **h**. Method C was used to form intermediates **81a** and **b**, and Method D was used for intermediates **81c-j**.

Intermediates **81a-j** were then converted to the various final targets **68-71**, again by four general methods. Treatment of **81a** and **81c-j** with sodium methoxide formed the methoxy analogues **68a-i**. Treatment of **81a** or **81c** with ammonia gave **69a** and **b**, respectively, while treatment of **81a** with morpholine and potassium carbonate gave **69c**. To achieve the carbonyl or the thiocarbonyl targets, intermediates **81a, c**, and **e** were treated with 5% aqueous potassium hydroxide to give **70a-c**, while intermediates **81a, c, d, e** or **h** were treated with thiourea to afford **71a-e**. The authors report that some of the analogues showed more potent inhibitory activity than allopurinol against xanthine oxidase, however more extensive studies are underway to fully determine their potential.

6.   "Extended" benzimidazole nucleosides

Another type of tricyclic nucleoside that has been explored is the extended benzimidazole as shown in Figure 7 (Zhu and Townsend, 1996; Zhu *et al.*, 1996; Zhu *et al.*, 1998a; Zhu *et al.*, 1998c; Zhu *et al.*, 1998b; Zhu *et al.*, 1999; Zhu *et al.*, 2000). These include the naphtho[2,3-*d*]imidazoles (**82a-f**), the imidazo[4,5-*b*]quinoxalines (**83a-f**), the N-1 imidazo[4,5-*b*]quinolines (**84a-f**) and the N-3 imidazo[4,5-*b*]quinolines (**85a-f**). Townsend designed these nucleosides based on the lead provided by the potent biological activity of 2,5,6-trichloro-1-(β-D-ribofuranosyl)benzimidazole (TCRB) and its 2-bromo analogue BDCRB. Both TCRB and BDCRB have been found to be potent and selective inhibitors of HCMV involving a unique mechanism of action, which does not involve inhibition of viral DNA processing ADDIN ENRfu (Underwood *et al.*, 1993; Underwood *et al.*, 1994). It has been noted that two viral genes, UL56 and UL89 mutate to give a drug-resistant virus, so the proteins that are encoded by these genes must be considered as potential targets for TCRB and related analogues. Unfortunately, little is known about these proteins, and nothing at all is known about their overall three-dimensional structure or of the benzimidazole binding pocket(s) on the proteins. Based on the dearth of available structural information, Townsend designed these extended nucleosides as dimensional probes for the purpose of exploring the binding site(s) on the proteins.

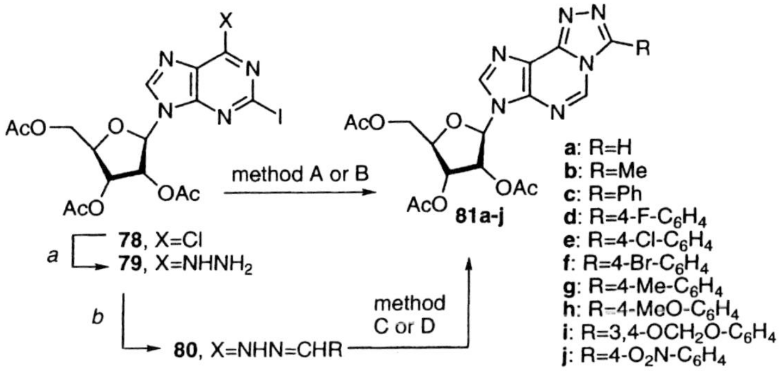

78, X=Cl
79, X=NHNH$_2$

a

b

80, X=NHN=CHR

method A or B

method C or D

81a-j

a: R=H
b: R=Me
c: R=Ph
d: R=4-F-C$_6$H$_4$
e: R=4-Cl-C$_6$H$_4$
f: R=4-Br-C$_6$H$_4$
g: R=4-Me-C$_6$H$_4$
h: R=4-MeO-C$_6$H$_4$
i: R=3,4-OCH$_2$O-C$_6$H$_4$
j: R=4-O$_2$N-C$_6$H$_4$

Reaction conditions: *a*, anh. NH$_2$NH$_2$, CH$_3$CN, rt, 2 hr; Method A: RC(OEt)$_3$, AcOH, 55-80 °C , 4-7 h; Method B: RCHO, CH$_3$CN, rt, 5 h, then DEAD, reflux, 5-24 h; *b*, RCHO, CH$_3$CN; Method C: DEAD, CH$_3$CN, reflux 5-10 h; Method D: Pb(OAc)$_4$, dioxane, reflux 1-15 h.

Scheme 11.

Scheme 12

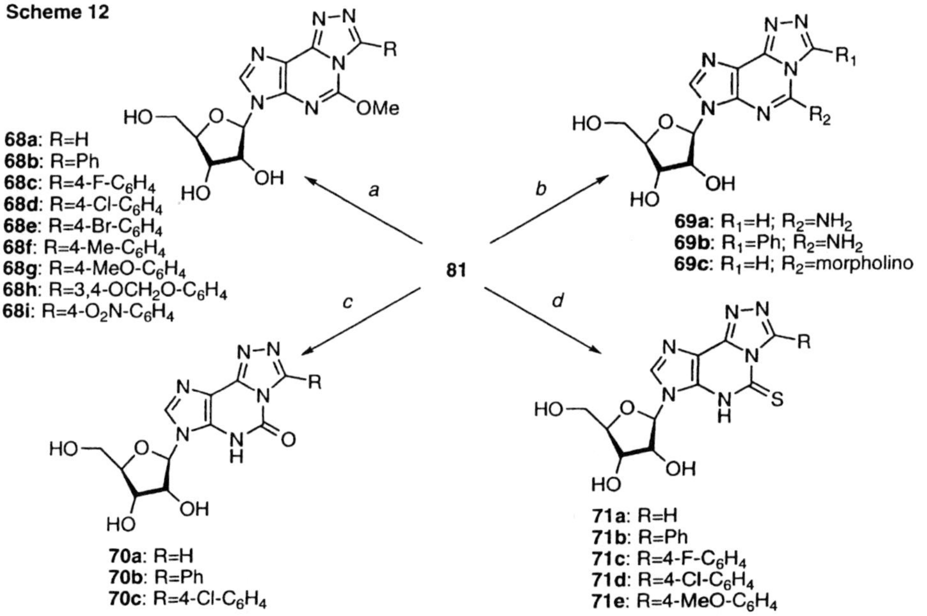

68a: R=H
68b: R=Ph
68c: R=4-F-C$_6$H$_4$
68d: R=4-Cl-C$_6$H$_4$
68e: R=4-Br-C$_6$H$_4$
68f: R=4-Me-C$_6$H$_4$
68g: R=4-MeO-C$_6$H$_4$
68h: R=3,4-OCH$_2$O-C$_6$H$_4$
68i: R=4-O$_2$N-C$_6$H$_4$

81

a

b

c

d

69a: R$_1$=H; R$_2$=NH$_2$
69b: R$_1$=Ph; R$_2$=NH$_2$
69c: R$_1$=H; R$_2$=morpholino

70a: R=H
70b: R=Ph
70c: R=4-Cl-C$_6$H$_4$

71a: R=H
71b: R=Ph
71c: R=4-F-C$_6$H$_4$
71d: R=4-Cl-C$_6$H$_4$
71e: R=4-MeO-C$_6$H$_4$

Reaction conditions: *a*, NaOMe, MeOH, 0-5 °C, 2 h, then 10% HCl; *b*, liquid NH$_3$, rt, 6 h or morpholine, K$_2$CO$_3$, dioxane, reflux, 20 h; *c*, 5% aq KOH, rt, 5-10 h, then 10% HCl; *d*, (H$_2$N)$_2$C=S, EtOH, 75 °C, 4 h, then 10% aq NaOH and 10% HCl.

Scheme 12.

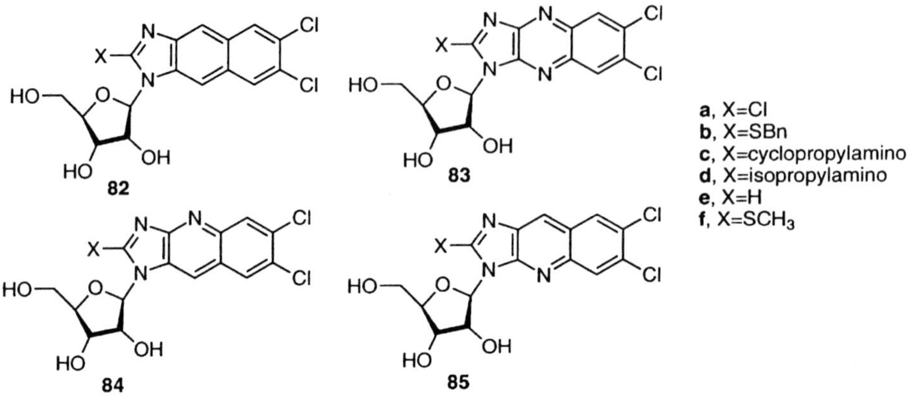

Figure 7. "Extended" benzimidazole tricyclic analogues.

The synthesis (Zhu *et al.*, 1998b; Zhu *et al.*, 2000) of the first series is outlined in Scheme 13, and begins with 2,3-diamino-6,7-dichloronaphthalene (**86**) which, following treatment with (thiocarbonyl)diimidazole, gave the tricyclic intermediate **87**. Compound **87** was then treated with either methyl iodide to give **88**, or, with benzyl bromide to give **89**. Using BSA, coupling of **88** or **89** with TBAR afforded **90** and **91**, respectively, which following deprotection, gave **82f** and **82b**. Reduction of **82f** with Raney nickel provided the unsubstituted analogue **82e**, while treatment of **82f** with chlorine in methanol at -78 °C provided the chloro analogue **82a**. Finally, reaction of **82a** with cyclopropylamine or isopropylamine gave **82c** and **82d**, respectively.

The imidazo[4,5-*b*]quinoxalines (**83a-f**) were synthesized in a similar fashion, starting with 2,3,6,7-tetrachloroquinoxoline (**92**) as shown in Scheme 14 (Zhu *et al.*, 1998c). Selective conversion to the 2,3-diamino **93** was achieved with methanolic ammonia. Ring-closure of **93** to the tricyclic **94** was accomplished with carbonyldiimidazole. Coupling of **94** with TBAR using BSA provided the nucleoside **95**, which was then treated with triazole to give intermediate **96**. Conversion to the thiocarbonyl using standard conditions gave **97**, which following deprotection gave **98**.

As shown in Scheme 15, intermediate **98** was then converted directly to the thiobenzyl analogue **83b** by reaction with benzyl bromide, or to the unsubstituted **83e** with Raney nickel. Conversely, the triazole intermediate **98** afforded the cyclopropyl and isopropyl analogues **83c** and **83d** following treatment with the appropriate amine, and subsequent deprotection with methanolic ammonia.

The remaining two series, the N-1 imidazo[4,5-*b*]quinolines (**84a-f**) and the N-3 imidazo[4,5-*b*]quinolines (**85a-f**) were realized from the same heterocyclic base, but employing different coupling conditions (Zhu and Townsend, 1996; Zhu *et al.*, 1996; Zhu *et al.*, 1999; Zhu *et al.*, 2000). Starting with the tri-substituted aniline **99** as shown in Scheme 16, conversion of the amino group of **99** to the nitrile **100** was accomplished using a non-aqueous diazotization procedure with boron trifluoride etherate and *tert*-butyl nitrite, followed by reaction with cuprous cyanide and sodium cyanide. Reaction

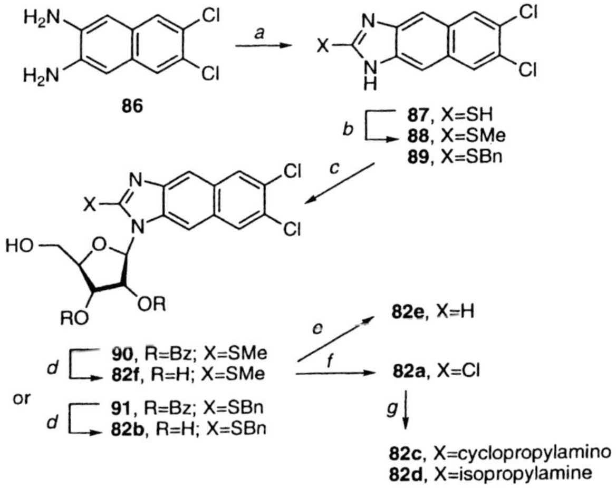

Reaction conditions: *a*, (thiocarbonyl)diimidazole, benzene, reflux; *b*, DMF, MeI, or, BnBr, DMF, rt; *c*, (i) BSA, ClCH$_2$CH$_2$Cl, rt; (ii) TBAR, TMSOTf, 50 °C; *d*, NH$_3$, MeOH, rt; *e*, Raney Ni, EtOH; *f*, Cl$_2$, MeOH, -78 °C; *g*, isopropylamine or cyclopropylamine.

Scheme 13.

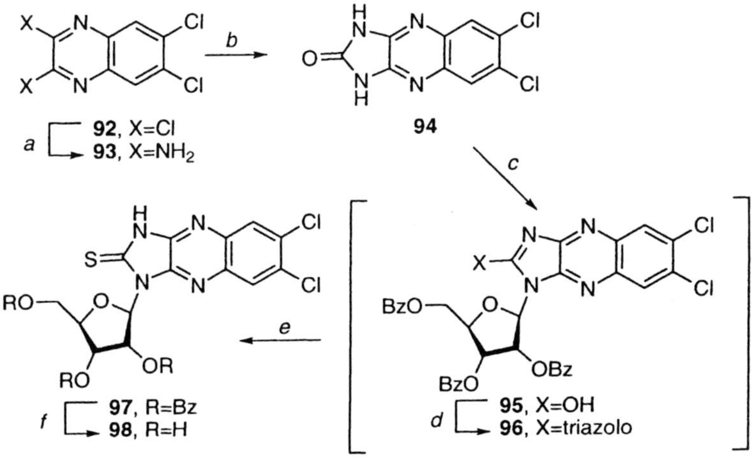

Reaction conditions: *a*, methanolic NH$_3$, 90 °C ; *b*, carbonyldiimidazole, toluene, reflux; *c*, (i) BSA, ClCH$_2$CH$_2$Cl, rt; (ii) TBAR, TMSOTf, 50 °C; *d*, POCl$_3$, triazole, Et$_3$N, rt; *e*, H$_2$S; *f*, NH$_3$, MeOH, rt.

Scheme 14.

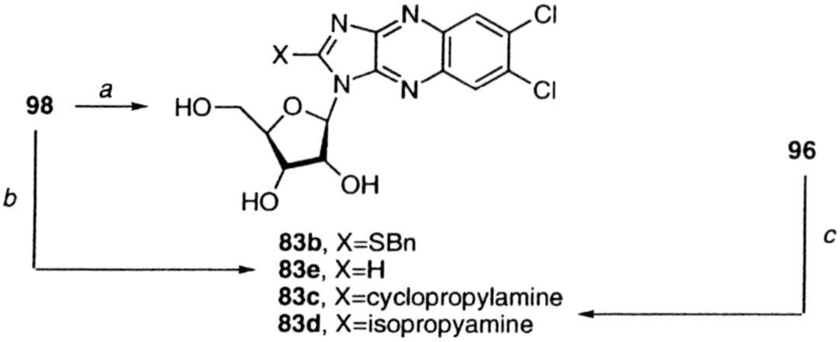

Reaction conditions: *a*, BnBr, NH$_4$OH ; *b*, Raney Ni, MeOH, reflux; *c*, (i) cyclopropylamine or isopropylamine; (ii) NH$_3$, MeOH.

Scheme 15.

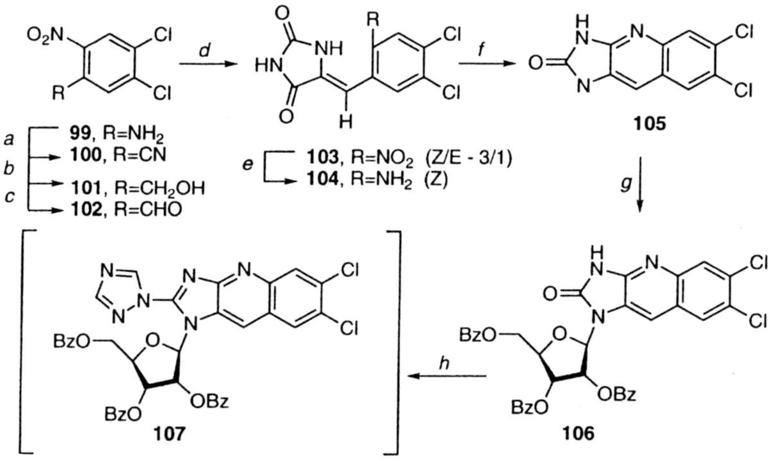

Reaction conditions: *a*, (i) t-BuONO, BF$_3$• OEt$_2$; (ii) NaCN, CuCN; *b*, (i) BF$_3$• OEt$_2$; (ii) NaNO$_2$, AcOH, H$_2$O; (iii) NaOH, H$_2$O; *c*, PCC, CH$_2$Cl$_2$; *d*, diethyl 2,4-dioxoimidazolidine-5-phosphonate, Et$_3$N, CH$_3$CN; *e*, (i) NaOH, MeOH; (ii) 10% HCl; (iii) Fe, FeSO$_4$, MeOH, H$_2$O, reflux; *f*, *hv*, AcOH, rt; *g*, TBAR, TMSOTf, ClCH$_2$CH$_2$Cl, CH$_3$CN; *h*, POCl$_3$, triazole, Et$_3$N, CH$_3$CN.

Scheme 16.

of **100** with boron trifluoride etherate, followed by sodium nitrite and acetic acid, then aqueous sodium hydroxide, provided alcohol **101**. Oxidation with pyridinium chlorochromate afforded aldehyde **102**.

Coupling **102** with diethyl 2,4-dioxoimidazoline-5-phosphate gave the linked intermediate **103** in a 3:1 mixture of the *Z* and *E* isomers. The mixture of the isomers is then isomerized to the *Z* isomer using alkaline conditions. Reduction of the nitro group with iron and iron (II) sulfate then resulted in amine **104**, which following a photoassisted ring closure (Zhu *et al.*, 1999; Zhu *et al.*, 1996) gave the desired heterocyclic intermediate **105**. Coupling **105** with TBAR using TMSOTf in 1,2-dichloroethane and acetonitrile provided the N-1 intermediate **106**. Treatment of **106** with triazole and phosphorous oxychloride gave **107**.

As shown in Scheme 17, **107** was then subjected to reaction with hydrogen sulfide to give thiocarbonyl **108**. Deprotection of **108** subsequently resulted in **109**, which was then converted either to the unsubstituted **84e** with Raney nickel, or to the thiobenzyl analogue **84b** with benzyl bromide. Alternatively, **108** was treated with chlorine in methanol to afford **110**, which following deprotection, resulted in **84a**. The remaining two targets **84c** and **84d** were realized from **84a** by reaction with either cyclopropylamine or isopropylamine, respectively.

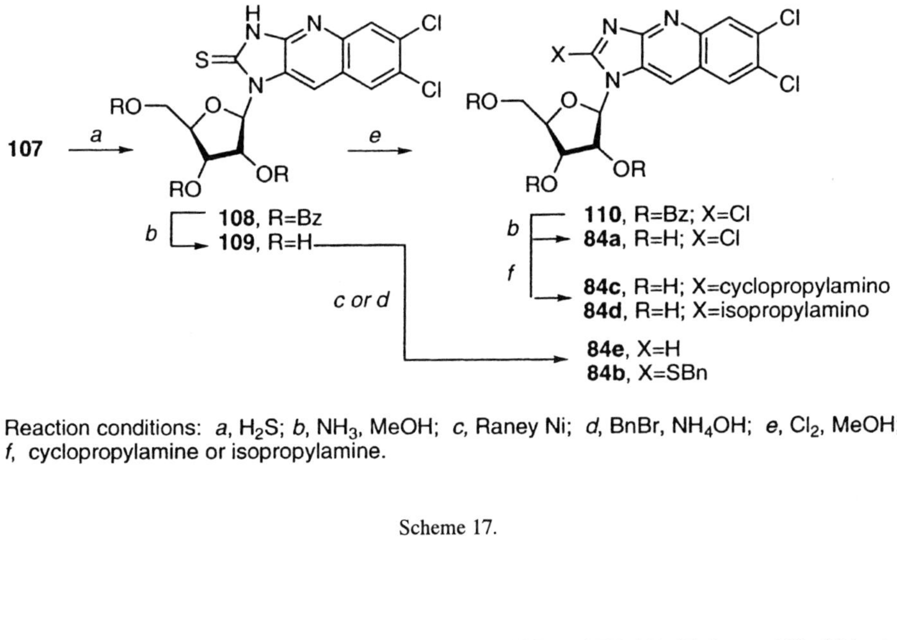

Reaction conditions: *a*, H_2S; *b*, NH_3, MeOH; *c*, Raney Ni; *d*, BnBr, NH_4OH; *e*, Cl_2, MeOH; *f*, cyclopropylamine or isopropylamine.

Scheme 17.

The final series once again employed coupling **105** and TBAR (Scheme 18). This time however, using tin (IV) chloride in acetonitrile at 50 °C to provide nucleoside **111**. Conversion of **111** to **112** followed by transformation to **85a-e** was as described for **84a-e**.

All of the tricyclic analogues were evaluated against HCMV and HSV-1. The chloro and thiobenzyl N-1 analogues were found to be nearly as active as TCRB in the HCMV assay, but both compounds proved to be more cytotoxic to HFF and KB cells than was TCRB or the thiobenzyl analogue of TCRB. In the N-3 series, only the trichloro

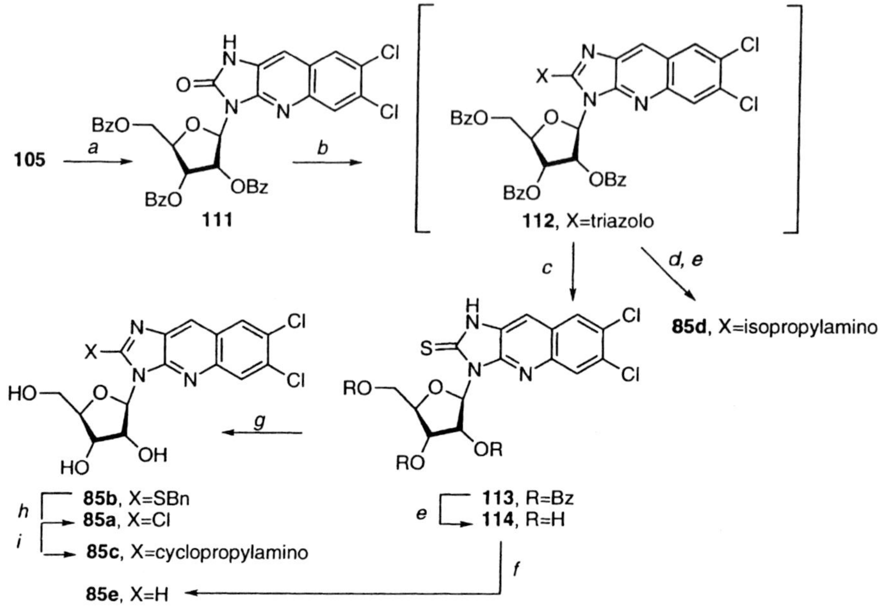

Reaction conditions: *a*, TBAR, SnCl$_4$, CH$_3$CN, 50 °C; *b*, POCl$_3$, triazole, Et$_3$N, CH$_3$CN; *c*, H$_2$S; *d*, isopropylamine; *e*, NH$_3$, MeOH; *f*, Raney Ni; *g*, BnBr, NH$_4$OH; *h*, Cl$_2$, MeOH; *i*, cyclopropylamine.

Scheme 18.

analogue was active against HCMV and HSV-1, but only at concentrations that were cytotoxic to HFF and KB cells. The data implies that the N-1 analogues interact with a greater affinity than the N-3 analogues, but given the increased cytotoxicity, additional modifications are necessary.

7. "Extended" benzoquinazoline nucleosides

Similar in structure to the extended benzimidazoles are the extended benzoquinazolines (Figure 8). Moreau *et al.* have synthesized a series of these tricyclic nucleosides, which have the typical fluorescent properties, as probes for duplex- and triplex-forming oligonucleotides (Godde *et al.*, 1998b; Godde *et al.*, 1998a; Godde *et al.*, 2000; Arzumanov *et al.*, 2000). These analogues of cytidine and thymidine have also been used to probe protonation sites and the binding of proteins to RNA. The series includes the 2'-deoxy (**115**), the ribose (**116**) and the 2'-methoxy (**117**) analogues, as well as the cytidine analogue **118** and the bent analogue **119**.

The synthesis of **115-117** begins with the coupling of benzo[*g*]quinazoline-2,4-(1*H*,3*H*)-dione **120** as shown in Scheme 19 (Godde *et al.*, 2000). Treatment of **120** with

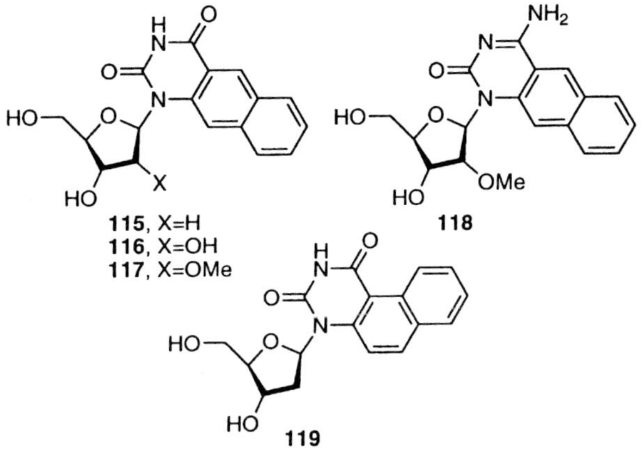

Figure 8. "Extended" benzoquinazoline tricyclic analogues.

ammonium sulfate, acetamide, TMSCl and HMDS for 3 days, followed by addition of TBAR and TMSOTf in a mixture of acetonitrile and toluene gave **121**. Deprotection of **121** with sodium methoxide afforded **116** directly. Reaction of **116** with diphenyl carbonate and a catalytic amount of sodium bicarbonate gave the $O^2,2'$-anhydro analogue **122** in excellent yield. Treatment of **122** with trimethyl borate, trimethyl orthoformate and traces of sodium bicarbonate in hot methanol gave the methoxy target **117**.

The thymidine analogue **118** is available from **117**. Protection of the 3'- and 5'-hydroxyls with TBDMSCl provided **123** (Scheme 19) which, following treatment with triazole, gave **124** as shown in Scheme 20. Conversion to the benzamide **125** was accomplished with benzamide and sodium hydride. Finally, removal of the silyl protecting groups with standard conditions, and subsequent hydrolysis of the resulting unprotected amide **126** afforded **118**.

The 2'-deoxy analogue **115** was realized from coupling **120** with 2-deoxy-3,5-di-*O*-*p*-toluoyl-D-pentofuranosyl chloride. First, persilylation of **120** with HMDS, ammonium sulfate and acetamide, followed by addition of the sugar and cuprous iodide in dry acetonitrile provided the nucleoside **127** as an inseparable mixture of anomers (Godde *et al.*, 1998b). However, using a deprotection/reprotection strategy, the mixture proved to be separable. First, removal of the toluoyl groups gave **128**, which, after reprotection of the hydroxyls with DMTrCl, resulted in a mixture of anomers (**129**) which proved to be separable by column chromatography. Finally, deprotection of the DMTr group of the β-anomer of **129** provided the desired 2'-deoxy analogue **115**.

The final compound in the series is the angular or "bent" analogue **119**. This was available from an analogous series of reactions as was employed for **115**, but

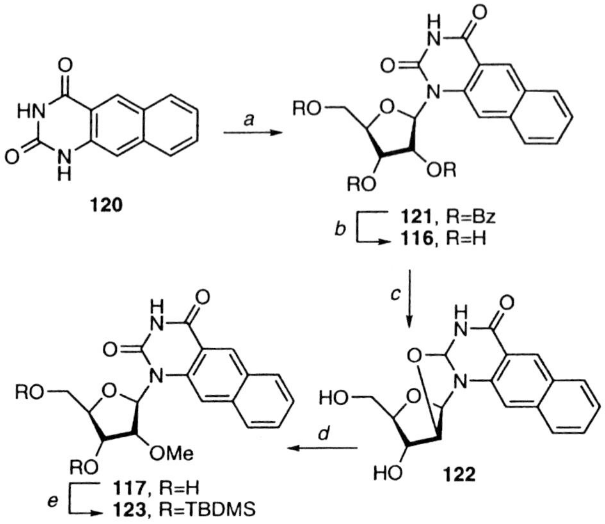

120

a →

121, R=Bz
116, R=H

b

c

122

d

117, R=H
123, R=TBDMS

e

Reaction conditions: *a*, (i) ammonium sulfate, acetamide, TMSCl, HMDS, 3 days; (ii) TBAR, CH_3CN, toluene, TMSOTf; *b*, NaOMe, MeOH; *c*, $(PhO)_2CO$, $NaHCO_3$, DMF, reflux; *d*, $B(OCH_3)_3$, $HC(OCH_3)_3$, $NaHCO_3$, MeOH, 50 °C; *e*, TBDMSCl, imidazole, DMF.

Scheme 19.

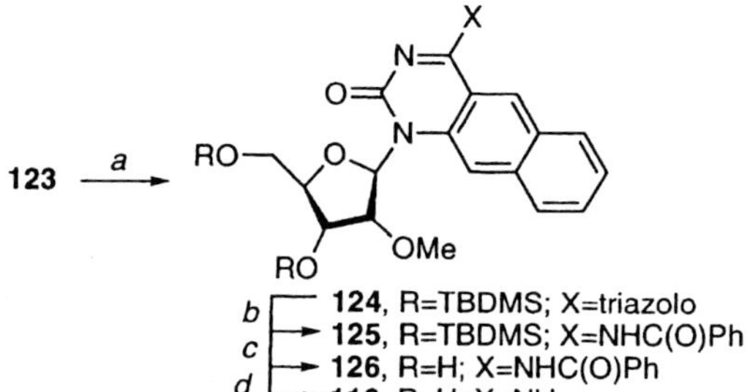

123 $\xrightarrow{a}$

124, R=TBDMS; X=triazolo
125, R=TBDMS; X=NHC(O)Ph
126, R=H; X=NHC(O)Ph
118, R=H; $X=NH_2$

b
c
d

Reaction conditions: *a*, $POCl_3$, TEA, triazole; *b*, NaH, benzamide, 1,4-dioxane; *c*, TBAF, THF; *d*, NH_4OH, DMF.

Scheme 20.

utilizing heterocycle **130** with the toluoyl protected sugar (Godde *et al.*, 1998b). Coupling once again gave an inseparable mixture of anomers (**131**), but the same deprotection/protection separation strategy was used to realize **119**.

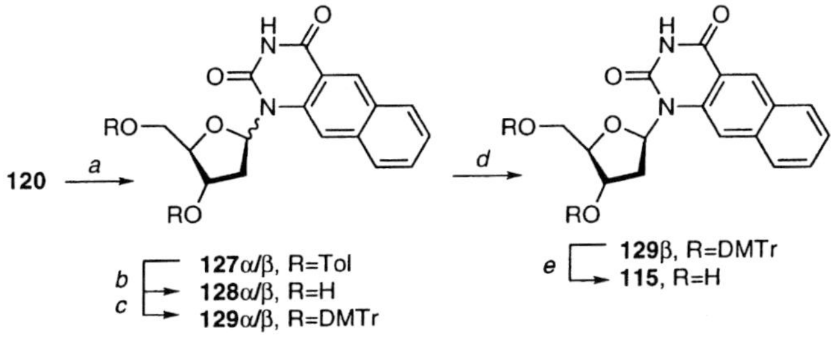

120 $\xrightarrow{a}$

b, *c*:
- **127α/β**, R=Tol
- **128α/β**, R=H
- **129α/β**, R=DMTr

$\xrightarrow{d}$

e:
- **129β**, R=DMTr
- **115**, R=H

Reaction conditions: *a*, (i) HMDS, $(NH_4)_2SO_4$, acetamide, 2 days; (ii) 2-deoxy-3,5-di-*O*-*p*-toluoyl-D-pentofuranosyl chloride, CuI, dry CH_3CN, 2 days; *b*, NaOMe, MeOH; *c*, DMTrCl, pyridine; *d*, separate by flash chromatography; *e*, AcOH, H_2O.

Scheme 21.

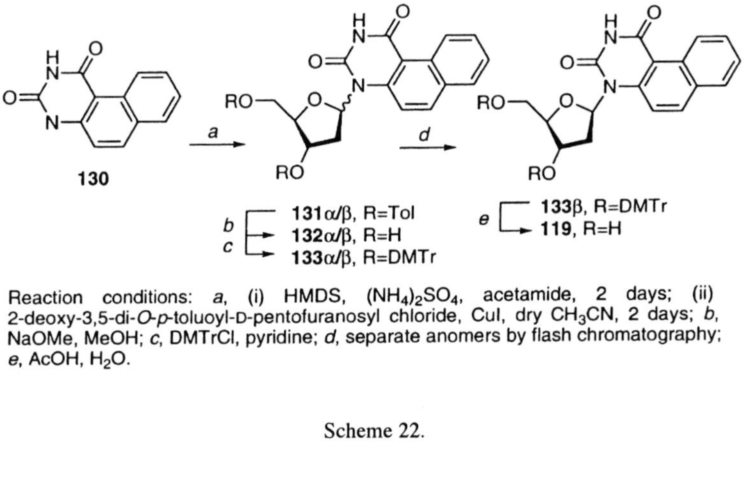

130 $\xrightarrow{a}$

b, *c*:
- **131α/β**, R=Tol
- **132α/β**, R=H
- **133α/β**, R=DMTr

$\xrightarrow{d}$

e:
- **133β**, R=DMTr
- **119**, R=H

Reaction conditions: *a*, (i) HMDS, $(NH_4)_2SO_4$, acetamide, 2 days; (ii) 2-deoxy-3,5-di-*O*-*p*-toluoyl-D-pentofuranosyl chloride, CuI, dry CH_3CN, 2 days; *b*, NaOMe, MeOH; *c*, DMTrCl, pyridine; *d*, separate anomers by flash chromatography; *e*, AcOH, H_2O.

Scheme 22.

8. "Extended" alloxazine nucleosides

Another of the tricyclic nucleoside groups, the alloxazines **134-137** (Figure 9) have been used extensively as fluorescent probes (Hawkins *et al.*, 1997, Rösler and Pfleiderer, 1997). Since the introduction of these nucleosides by Pfleiderer (Ienaga and Pfleiderer, 1977) several new analogues have been synthesized (Wang and Rizzo, 2000). These analogues can be viewed as potentially more hydrophobic analogues of thymidine, which would impart increased stabilization to the oligonucleotides that contained them (Lin *et al.*, 1995).

The synthesis of these analogues, along with a more efficient synthesis of the previously reported analogues, is outlined in Schemes 23 and 24 (Wang and Rizzo, 2000).

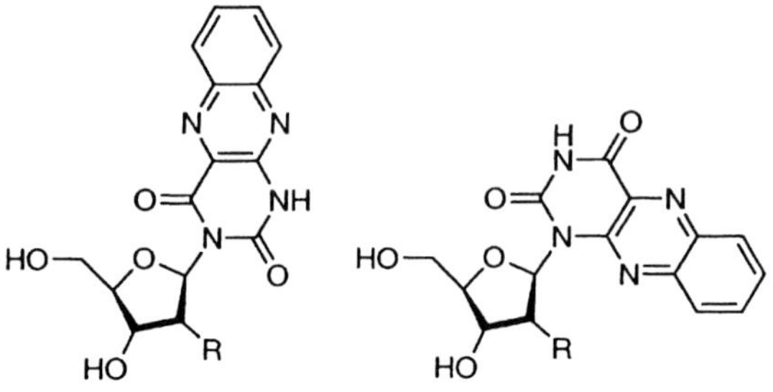

Figure 9. "Extended" alloxazine tricyclic analogues.

Coupling of alloxazine (**138**) has traditionally produced both the N-1 and N-3 products, however Rizzo *et al.* recently reported (Wang and Rizzo, 2000) they were able to obtain each of the isomers exclusively in excellent yield by carefully altering the solvent, the temperature and the length of time of the reaction. As shown in Scheme 23, coupling persilylated **138** to TBAR with tin (IV) chloride in dichloromethane at -20 °C for 15 minutes, followed by deprotection, gave the N-3 target **139** exclusively, while coupling the persilylated **138** to TBAR in acetonitrile at room temperature for 2 hours gave only the N-1 target **136** after deprotection.

The 2'-deoxy N-1 analogue was synthesized by Pfleiderer (Rösler and Pfleiderer, 1997) and employed the BSA coupling method. Coupling **138** to the 2-deoxy sugar gave **141**, which, following deprotection, provided **137**. Unfortunately, the 2'-deoxy N-1 analogue required a more tedious route (Scheme 24) (Wang and Rizzo, 2000). Starting with the unprotected **134**, selective protection of the 3'- and 5'-hydroxyls using tetraisopropyldisiloxane-1,3-diyl provided **142**, which was then subjected to Mitsunobu conditions to give the 2,2'-anhydronucleoside **143**. Using bromine as a nucleophile, **143** was opened to give the 2'-bromo intermediate **144**, which then underwent reduction with tributyltin hydride and AIBN to yield **145**. Removal of the silyl group of **145** then gave the desired 2'-hydroxy target **135**. Work is underway to determine the extent of the hydrogen bonding properties of these analogues.

9. "Extended" 2'-deoxycytidine nucleosides

As mentioned in the preceding section, oligonucleotides containing more hydrophobic pyrimidine analogues have been shown to stabilize DNA/DNA and DNA/RNA helices (Lin *et al.*, 1995). A series of tricyclic nucleosides **146-151** (Figure 10) which resemble cytidine has been synthesized by Matteucci *et al.* (Lin and Matteucci, 1998, Lin *et al.*, 1995). These phenoxazine tricyclics show increased stacking interactions and stability, as well as enhanced binding properties. More significantly, nucleoside **148** has been termed a "G-clamp" because it can form an additional hydrogen bond to

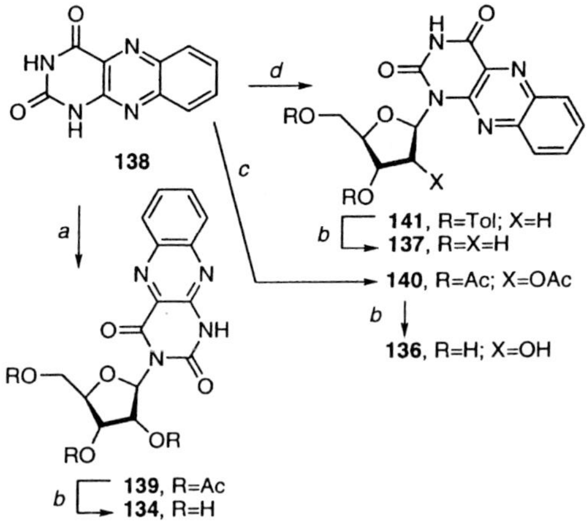

Reaction conditions: Reaction conditions: *a*, (i) TMSCl, benzene; (ii) TBAR, $SnCl_4$, CH_2Cl_2, -20 °C, 15 min; *b*, NaOMe, MeOH; *c*, (i) TMSCl, benzene; (ii) TBAR, $SnCl_4$, CH_3CN, rt, 2 h; *d*, (i) BSA, CH_2Cl_2, reflux, 15 min; (ii) 2-deoxy-3,5-di-*O-p*-toluoyl-D-pentofuranosyl chloride, dry $CHCl_3$; (iii) MeOH.

Scheme 23.

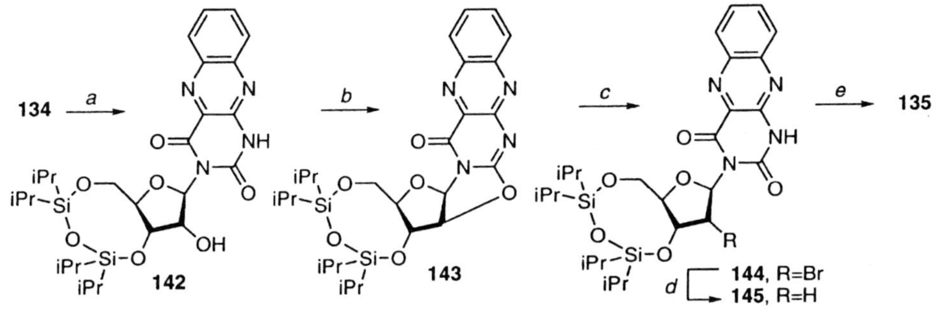

Reaction conditions: *a*, (*i*Pr$_2$SiCl)$_2$O, pyridine; *b*, DIAD, Ph$_3$P; *c*, HBr; *d*, *n*Bu$_3$SnH, AIBN, benzene, reflux; *e*, TBAF.

Scheme 24.

guanosine (Flanagan *et al.*, 1999b; Lin and Matteucci, 1998). The G-clamp base modification simultaneously recognizes both Watson-Crick and Hoogsteen faces of the complementary guanine within a helix, and this results in dramatically enhanced

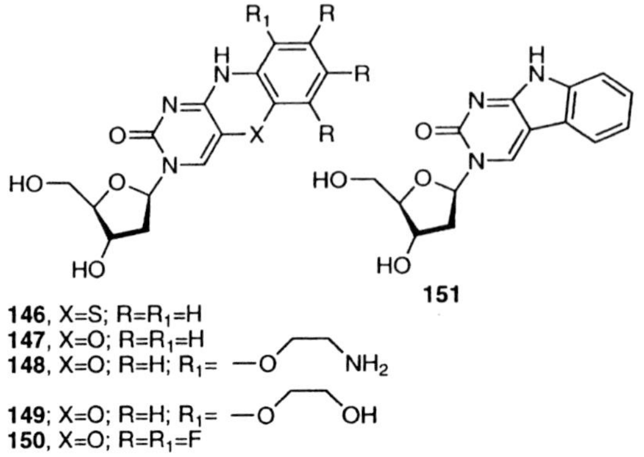

Figure 10.   "Extended" 2-deoxycytidine tricyclic analogues.

stability. When the G-clamp was incorporated into a previously optimized antisense oligonucleotide, the potency of the antisense oligonucleotide was enhanced 25-fold (Flanagan *et al.*, 1999b; Flanagan *et al.*, 1999a).

The synthesis of these targets began with manipulation of the halogen-substituted uridines **152a** and **b** as shown in Scheme 25 (Lin *et al.*, 1995; Lin and Matteucci, 1998). First, protection of the hydroxyl groups with acetic anhydride, followed by treatment with 2-mesitylenesulfonyl chloride and subsequent reaction with either 2-aminothiophenol or 2-aminophenol provided **153a** and **b**, respectively. Ring closure to form the tricyclic system and concurrent deprotection gave **146** and **147**. Alternatively, activation of the acetyl protected **152b** with triphenylphosphine and carbon tetrachloride, followed by reaction with 2-aminoresorcinol provided the dihydroxyl **154**. Mitsunobu alkylation of **154** with either CBz-N-protected ethanolamine or trityl-protected ethylene glycol produced **155** and **156** respectively, which, following deprotection, gave **148** and **149**.

The tetrafluoroderivative **150** was synthesized as outlined in Scheme 26 (Wang *et al.*, 1998). Using a double nucleophilic aromatic substitution reaction, bromination of 2'-deoxycytidine hydrochloride **157**, followed by treatment with diisopropylethylamine gave **158**. Subsequent treatment of **158** with potassium carbonate and hexafluorobenzene in DMSO gave the desired tetrafluoro derivative **150**.

The final target in this series possesses a central carbazole ring for its tricyclic system rather than the three six-membered analogues. This "bent" system would be a perturbation of the linear system, which as mentioned previously, showed increased (i) stability, (ii) stacking interactions and (iii) hydrogen bonding interactions. Matteucci *et al.* found that this analogue paired specifically with guanine and showed elevated Tm's, especially when the carbazole-containing nucleosides clustered (Matteucci and von Krosigk, 1996).

Stille coupling of **152a** with *t*-Boc-protected 2-trimethylstannyl aniline resulted in **159** (Scheme 27). Selective protection of the 5'-hydroxyl of **159** with DMTrCl,

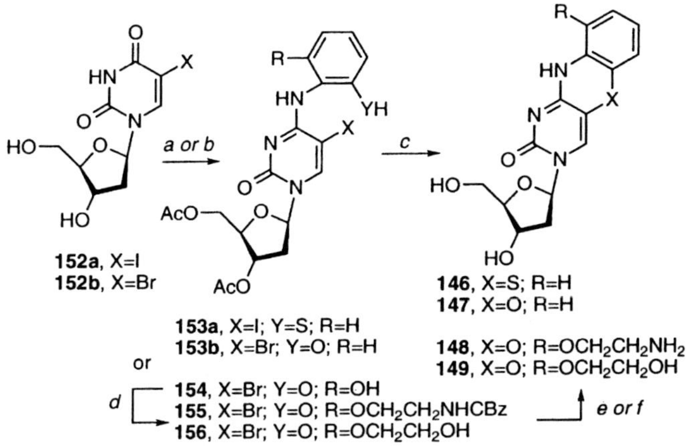

Reaction conditions: *a*, (i) Ac₂O in pyridine, (ii) 2-mesitylenesulfonyl chloride, TEA, then 2-aminophenol or 2-aminothiophenol, DBU, rt; *b*, (i) Ac₂O, pyridine (ii) Ph₃P, CCl₄, CH₂Cl₂; (iii) 2-aminoresorcinol, DBU, rt; *c*, for **146**: *t*-BuOK in ethanol, reflux; for **147**: NH₃, MeOH, rt; *d*, for **155**: HOCH₂CH₂NHCBz, DEAD, CH₂Cl₂, rt; for **156**: HOCH₂CH₂OTr, Ph₃P, DEAD, CH₂Cl₂; *e*, for **148**: H₂, 10% Pd on C; *f*, for **149**: NH₃, MeOH

Scheme 25.

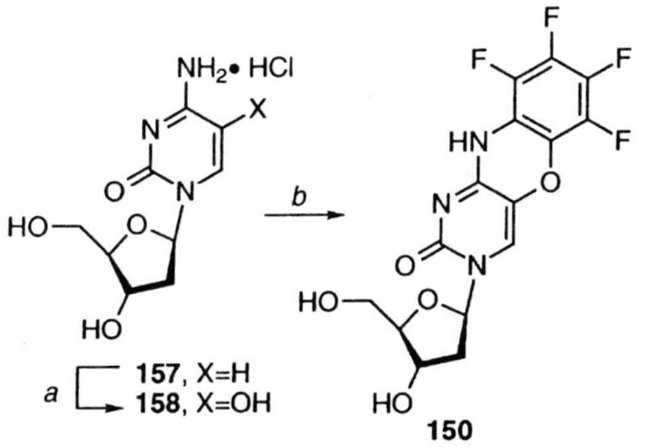

Reaction conditions: *a*, (i) bromine, H₂O, rt; (ii) *N,N*-diisopropylethylamine; *b*, K₂CO₃, hexafluorobenzene, DMSO, 50 °C.

Scheme 26.

and subsequent protection of the 3'-hydroxyl with a trimethylsilyl group, followed by activation of the C-4 of the uracil moiety with mesyl chloride, DBU-mediated ring closure and subsequent removal of the BOC-protecting group provided **160**. Finally, removal of the DMTr protecting group resulted in the carbazole target **151**.

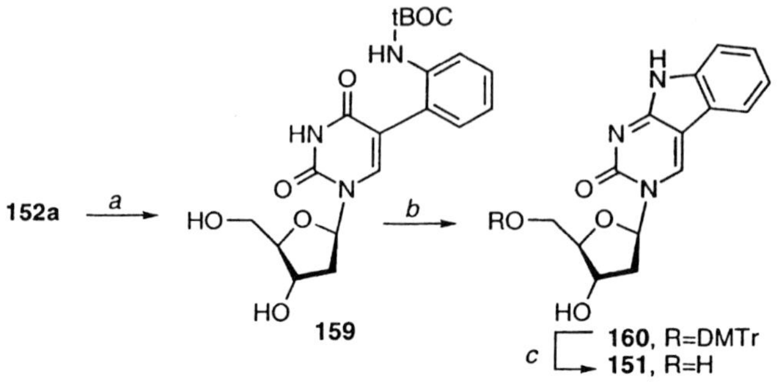

Reaction conditions: *a*, *t*-BOC protected 2-trimethylstannyl aniline (4 eq.), (Ph₃P)₂PdCl₂ (0.1 eq.), DMF, 80 °C. *b*, (i) DMTrCl, pyridine, rt; (ii) BSTFA (1.2 eq.), CH₂Cl₂, rt; (iii) MsCl (2 eq.), DMAP (0.1 eq.), Et₃N (10 eq.), CH₂Cl₂, rt; (iv) DBU (4 eq.), CH₂Cl₂, rt; (v) *t*-BuNH₂/MeOH (1:1), rt; *c*, AcOH, H₂O (4:1).

Scheme 27.

10.   Acknowledgements

This review was intended to provide an updated look at the advances that have been made in the area of tricyclic nucleosides since the time of Leonard's reviews. The author apologizes for any work that was accidentally overlooked, it was unintentional.

The author would like to express her gratitude to Professor Stewart W. Schneller, Professor Vasu Nair and Dr. Balkrishnen Bhat for the insightful conversations on tricyclic nucleosides; their guidance is greatly appreciated. The author would also like to thank Samer Salim, Liang Zhang and Peter O' Daniel for their excellent editorial assistance; this is also appreciated.

11.   References

Arzumanov, A., Godde, F., Moreau, S., Weeds, A. and Gait, M. J. (2000) *Helvetica Chimica Acta,* 83, 1424-1436.

Chung, F.-L., Earl, R. A. and Townsend, L. B. (1980a) *Tetrahedron Letters,* 21, 1599-1602.

Chung, F.-L., Earl, R. A. and Townsend, L. B. (1980b) *Journal of Organic Chemistry,* 45, 2532-2535.

Chung, F.-L., H., S. K., Panzica, R. P., Earl, R. A., Wotring, L. L. and Townsend, L. B. (1980c) *Journal of Medicinal Chemistry,* 23, 1158-1166.

Flanagan, W. M., Wagner, R. W., Grant, D., Lin, K.-Y. and Matteucci, M. D. (1999a) *Nature Biotechnology,* 17, 48-52.

Flanagan, W. M., Wolf, J. J., Olson, P., Grant, D., Lin, K.-Y., Wagner, R. W. and Matteucci, M. (1999b) *Proceedings of the National Academy of Sciences, U.S.A.,* 96, 3513-3518.

Godde, F., Aupeix, K., Moreau, S. and Toulmé, J.-J. (1998a) *Antisense & Nucleic Acid Drug Development,* **8,** 469-476.

Godde, F., Toulmé, J.-J. and Moreau, S. (1998b) *Biochemistry,* **37,** 13765-13775.

Godde, F., Toulmé, J.-J. and Moreau, S. (2000) *Nucleic Acids Research,* **28,** 2977-2985.

Hamamichi, N. (1991) *Tetrahedron Letters,* **32,** 7415-7418.

Hamamichi, N. and Miyasaka, T. (1994) *Journal of Organic Chemistry,* **59,** 1525-1531.

Hawkins, M. E., Pfleiderer, W., Balis, F. M., Porter, D. and Knutson, J. R. (1997) *Analytical Biochemistry,* **244,** 86-95.

Humphries, M. J. and Ramsden, C. A. (1995) *Synlett,,* 203-204.

Humphries, M. J. and Ramsden, C. A. (1999) *Synthesis,* **6,** 985-992.

Ienaga, K. and Pfleiderer, W. (1977) *Chem. Ber.,* **110,** 2449-3455.

Klimke, G., Ludemann, H.-D. and Townsend, L. B. (1979) *Z. Naturforsch,* **34c,** 653-657.

Leonard, N. J. (1982) *Accounts of Chemical Research,* **15,** 128-135.

Leonard, N. J. (1992) *CHEMTRACTS-Biochemistry and Molecular Biology,* **3,** 273-297.

Leonard, N. J. (1993) *CHEMTRACTS-Biochemistry and Molecular Biology,* **4,** 251-284.

Leonard, N. J. and Hiremath, S. P. (1986) *Tetrahedron,* **42,** 1917-1961.

Leonard, N. J., Kazmierczak, F. and Rykowski, A. (1987) *Journal of Organic Chemistry,* **52,** 2933-2935.

Leonard, N. J., Morrice, A. G. and Sprecker, M. A. (1975) *Journal of Organic Chemistry,* **40,** 356-366.

Lin, K.-Y., Jones, R. J. and Matteucci, M. D. (1995) *Journal of the American Chemical Society,* **117,** 3873-3874.

Lin, K.-Y. and Matteucci, M. D. (1998) *Journal of the American Chemical Society,* **120,** 8531-8532.

Matteucci, M. D. and von Krosigk, U. (1996) *Tetrahedron Letters,* **37,** 5057-5060.

Nagamatsu, T., Yamasaki, H., Akiyama, T., Hara, S., Mori, K. and Hitoshi, K. (1999) *Synthesis,* **4,** 655-663.

Nair, V. and Offerman, R. J. (1985) *Journal of Organic Chemistry,* **50,** 5627-5631.

Nair, V., Offerman, R. J. and Turner, G. A. (1984a) *Journal of Organic Chemistry,* **49,** 4021-4025.

Nair, V., Turner, G. A. and Offerman, R. J. (1984b) *Journal of the American Chemical Society,* **106,** 3370-3371.

O'Hara Dempcy, R. and Skibo, E. B. (1991) *Journal of Organic Chemistry,* **56,** 776-785.

Rajappan, V. P. and Schneller, S. W. (2001a) *Nucleosides, Nucleotides and Nucleic Acids,* **20,** 1117-1121.

Rajappan, V. P. and Schneller, S. W. (2001b) *Journal of Organic Chemistry,* **57,** 9044-9053.

Rösler, A. and Pfleiderer, W. (1997) *Helvetica Chimica Acta,* **80,** 1869-1881.

Schneller, S. W. and Christ, W. J. (1982) *Journal of Heterocyclic Chemistry,* **19,** S139-S161.

Schneller, S. W. and Clough, F. W. (1975) *Journal of Heterocyclic Chemistry,* **12,** 513-516.

Schneller, S. W. and Ibay, A. C. (1986) *Journal of Organic Chemistry,* **51,** 4067-4070.

Seley, K. L., Januszczyk, P., Hagos, A., Zhang, L. and Dransfield, D. T. (2000) *Journal of Medicinal Chemistry,* **43,** 4877-4883.

Seley, K. L., Schneller, S. W. and Korba, B. (1997a) *Nucleosides & Nucleotides,* **16,** 2095-2099.

Seley, K. L., Schneller, S. W., Rattendi, D. and Bacchi, C. J. (1997b) *Journal of Medicinal Chemistry,* **40,** 622-624.

Seley, K. L., Zhang, L., Hagos, A. (2001) Organic Letters, **3,** 3209-3210.

Seley, K. L., Zhang, L., Hagos, A., and Quirk, S. (submitted) *Journal of Organic Chemistry.*

Underwood, M. R., Biron, K. K., Hemphill, M. L., Miller, T. J., Stanat, S. C., Dornsife, R. E., Drach, J. C., Townsend, L. B., Edwards, C. A. and Harvey, R. J. Presented at the Herpesvirus Workshop, Pittsburgh, PA (1993).

Underwood, M. R., Stanat, S. C., Drach, J. C., Harvey, C. M. and Biron, K. K. Presented at the Herpesvirus

Workshop,Vancouver, BC (1994).

Wang, J., Lin, K.-Y. and Matteucci, M. D. (1998) *Tetrahedron Letters,* **39,** 8385-8388.

Wang, P., Agrofoglio, L. A., Newton, M. G. and Chu, C. K. (1999) *Journal of Organic Chemistry,* **64,** 4173-4178.

Wang, Z. and Rizzo, C. J. (2000) *Organic Letters,* **2,** 227-230.

Zhang, J. and Nair, V. (1997) *Nucleosides & Nucleotides,* **16,** 1091-1094.

Zhang, J., Neamati, N., Pommier, Y. and Nair, V. (1998) *Bioorganic & Medicinal Chemistry Letters,* **8,** 1887-1890.

Zhu, Z., Drach, J. C. and Townsend, L., B. (1998a) *Journal of Organic Chemistry,* **63,** 977-983.

Zhu, Z., Drach, J. C. and Townsend, L. B. (1998b) *Journal of Organic Chemistry,* **63,** 977-983.

Zhu, Z., Lippa, B. S., Drach, J. C. and Townsend, L. B. (2000) *Journal of Medicinal Chemistry,* **43,** 2430-2437.

Zhu, Z., Lippa, B. S. and Townsend, L. B. (1996) *Tetrahedron Letters,* **37,** 1937-1940.

Zhu, Z., Lippa, B. S. and Townsend, L. B. (1999) *Journal of Organic Chemistry,* **64,** 4159-4168.

Zhu, Z., Saluja, S., Drach, J. C. and Townsend, L. B. (1998c) *Journal of the Chinese Chemical Society,* **45,** 465-474.

Zhu, Z. and Townsend, L. B. (1996) *Tetrahedron Letters,* **37,** 3263-3266.

UNUSUAL ANALOGUES OF NUCLEOSIDES: CHEMISTRY AND BIOLOGICAL ACTIVITY

JIRI ZEMLICKA

Department of Chemistry, Barbara Ann Karmanos Cancer Institute, Prentis Bldg., Wayne State University School of Medicine, 110 E. Warren Ave., Detroit, Michigan 48201-1379 USA

Dedicated to Dr. Jack J. Fox on the occasion of his 84th birthday

1.   Introduction

In the late eighties we developed a new class of analogues of 2',3'-dideoxyribonucleosides (Zemlicka, 1993 and 1997). The structures of these compounds comprise an allene moiety which serves as a linker between the nucleic acid base residue and hydroxymethyl group (Chart 1, formula **1**). Adenallene (**1a**) and cytallene (**1b**) have a potent anti-HIV effect and analogue **1b** is also active against HBV. During investigations of the structure-activity relationships in this series, it became clear that replacement of one or both cummulated double bonds in structure **1** with a function that will not significantly change the overall size of the molecule could lead to new analogues of biological interest. It was reasoned that cyclopropane ring could fulfill this criterion. In fact, similarity of a double bond and cyclopropane ring has been long recognized in organic chemistry (March, 1992).

1

1a: B = Ade, **1b**: B = Cyt

2

3

2a, 3a: B = Ade

4

5

6

7

8

9

proximal   *medial-syn*   *medial-anti*   *distal*

10

11

6a - 9a: B= Ade, **6b - 9b**: B = Gua

B = nucleic acid base

Recent Advances in Nucleosides: Chemistry and Chemotherapy, Ed. by C.K. Chu. 327 — 357

In such a fashion, three basic types of new analogues were designed. Thus, replacement of the double bond of allene **1** *distal* from the nucleic acid base led to the Z- and E-isomers **2** and **3** (Chart 1). A similar operation at the *proximal* double bond gave the Z- and E-isomers **4** and **5** with interchanged base and hydroxymethyl functions. Finally, a more complex situation arose when both double bonds of allene **1** were replaced with cyclopropane rings. Thus, four isomeric forms are possible (Gajewski and Burka, 1970): *proximal* **6**, *medial-syn* **7**, *medial-anti* **8** and *distal* **9**. Because a strong antiviral activity was found in the Z-isomeric series **2** and, to some extent, also with the E-isomers **3** (see Section 3.1.1), the Z- and E-methylenecyclobutane analogues **10** and **11** were later included as a part of structure-activity relationships studies. The intermolecular distances between the nucleic acid base and hydroxymethyl group are closer in structures **1, 2, 4, 7, 8** and **10** than in **3, 5, 6, 9** and **11** (Table 1).

The synthesis, chemistry and biological activity of nucleoside analogues **2 - 11** are the subject of this chapter.

Table 1.   Comparison of intramolecular distances between the heterocyclic moiety (N^9 of adenine) and hydroxymethyl group of analogues **1a - 3a, 4, 5, 6a - 9a, 10** and **11**.

Analogue	N^9 --- $CH_2(OH)$ /Å/	Analogue	N^9 --- $CH_2(OH)$ /Å/
1a	4.24	**6a** (*proximal*)	3.54
2a (Z)	4.04	**7a** (*medial-syn*)	4.40
3a (E)	4.76	**8a** (*medial-anti*)	4.20
4 (Z)	3.93	**9a** (*distal*)	5.01
5 (E)	4.75	**10** (Z)	3.76
		11 (E)	4.63

2.   Chemistry

2.1.   Synthesis of analogues **2 - 11**

2.1.1.   *Methylenecyclopropane analogues of the type* **2** *and* **3**

Initially, the synthesis of analogues **2** and **3** was based on an alkylation-elimination of nucleic bases or suitable precursors with ethyl (*E,Z*)-2-bromo-2-(bromomethyl)cyclopropane-1-carboxylates (**12**) at an elevated temperature (Qiu *et al.*, 1998b,c, Scheme 1). The latter mixture of isomers served as an agent for a direct introduction of methylenecyclopropane residue into nucleic acid bases or suitable precursors. The reagent **12** was readily obtained by addition of carbene generated from ethyl diazoacetate (**13**) to 2,3-dibromopropene (**14**). Reaction of **12** with adenine or 2-amino-6-chloropurine

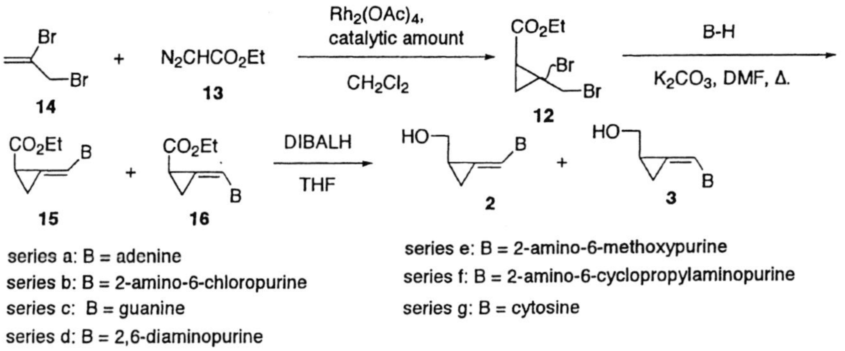

Rh$_2$(OAc)$_4$, catalytic amount / CH$_2$Cl$_2$

N$_2$CHCO$_2$Et **13**

14   **12**

B-H / K$_2$CO$_3$, DMF, Δ.

15 + **16**   DIBALH / THF   **2** + **3**

series a: B = adenine
series b: B = 2-amino-6-chloropurine
series c: B = guanine
series d: B = 2,6-diaminopurine

series e: B = 2-amino-6-methoxypurine
series f: B = 2-amino-6-cyclopropylaminopurine
series g: B = cytosine

Scheme 1.

afforded a mixture of the *Z*- and *E*-isomers **15a +16a** or **15b + 16b** which, in turn, were reduced with diisobutylaluminum hydride (DIBALH) to the corresponding carbinols **2a + 3a** and **2b + 3b**.

Although alkylation-elimination approach using reagent **12** is generally applicable for synthesis of methylenecyclopropane analogues **2** and **3**, in some cases, derivatization was necessary to separate the resultant *Z*- and *E*-isomers by chromatography on silica gel. Thus, the mixture of **2a + 3a** was derivatized with an N[6]-dimethylaminomethylene function (Qiu *et al.*, 1998b) prior to a chromatographic separation. The individual isomers were deprotected to afford pure synadenol[a] (**2a**) and the *E*-isomer **3a**. In contrast, 2-amino-6-chloropurine isomers **2b** and **3b** were separated without derivatization. Compound **2b** was conveniently used for synthesis of a series of the (*Z*)-2-aminopurine analogues Qiu *et al.*, 1998b,c, 1999b) **2c** - **2f** using standard protocols (Scheme 2). Thus, hydrolysis of **2b** with formic acid gave synguanol[a] (**2c**), ammonolysis led to 2,6-diaminopurine analogue **2d** whereas substitution with methoxy group furnished compound **2e**. Finally, reaction of **2b** with cyclopropylamine afforded 2-amino-6-cyclopropylamino derivative **2f**.

The procedure described in Scheme 1 was also applied for the synthesis of methylene-cyclopropane analogues in the pyrimidine series (Qiu *et al.*, 1998c). The N[4]-acetylcytosine served as a starting material for cytosine analogues. In this case, however, conversion to N[4]-benzoyl derivatives was obligatory to achieve separation of isomers. Syncytol[a] (**2g**) and the *E*-isomer **3g** were then obtained by a routine N-debenzoylation.

It is usually more convenient to perform alkylation and elimination as a one-pot procedure but separation of both steps is possible (Qiu *et al.*, 1998b,c). Alkylation of thymine is a typical example. The 2,4-bis-*O*-trimethylsilyl-5-methylpyrimidine (**17**) was alkylated with reagent **12** to give a mixture of the *Z*- and *E*-bromo derivatives (Qiu *et al.*, 1998c) **18** (Scheme 3). Although the latter are separable by chromatography, using individual isomers in a subsequent elimination does not offer any advantage (Qiu *et al.*, 1998b). Therefore, isomeric mixture **18** was subjected to elimination to give esters **19** and **20**. Reduction then furnished synthymol[a] (**21**) and the *E*-isomer **22** which were separated by chromatography.

---

[a]   Nomenclature (Qiu et al., 1998b) synadenol (2a), synguanol (2c), syncytol (2g) and synthymol (21) will be used throughout this Chapter.

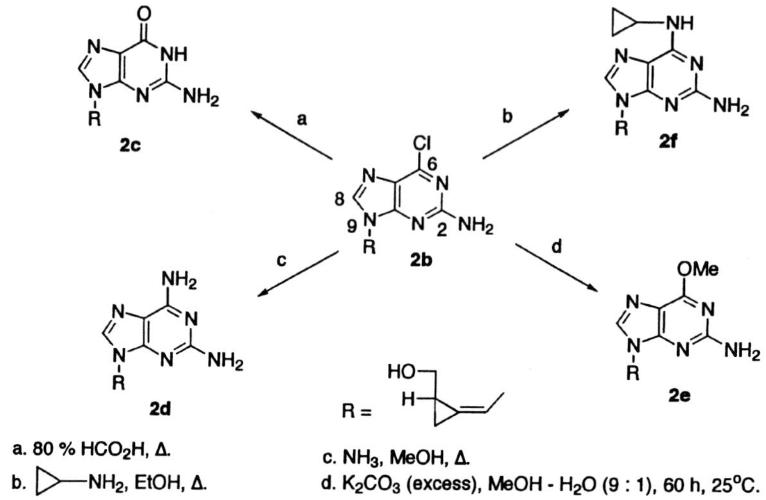

Scheme 2.

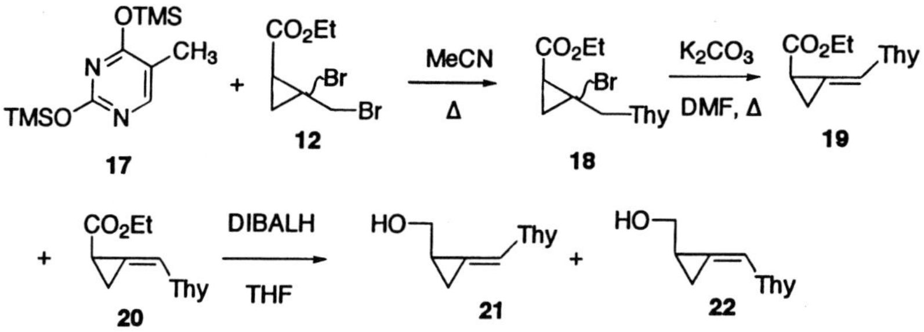

Scheme 3.

Although the procedures outlined above served well for obtaining sufficient materials for initial biological testing in vitro, we have sought a more straightforward method which would (i) eliminate reduction of ester intermediates carrying a nucleic acid base, e. g., **15b** + **16b** (Scheme 1) and (ii) derivatization necessary in some cases for separation of the *Z*- and *E*-isomers, e. g., **2a** and **3a**. Reduction of reagent **12** with DIBALH smoothly afforded (Qiu and Zemlicka, 1998a) carbinol **23** (Scheme 4). Acetylation then gave acetate **24** which was used for alkylation-elimination of 2-amino-6-chloropurine. The resultant *Z*- and *E*-isomeric mixture of **25** and **26** was deprotected under strictly controlled reaction conditions (to avoid substitution of the 6-chlorine atom) to give compounds **2b** and **3b**. In a similar fashion, carbinol **23** was protected with the methoxymethyl (MOM) group to furnish intermediate **27** which was

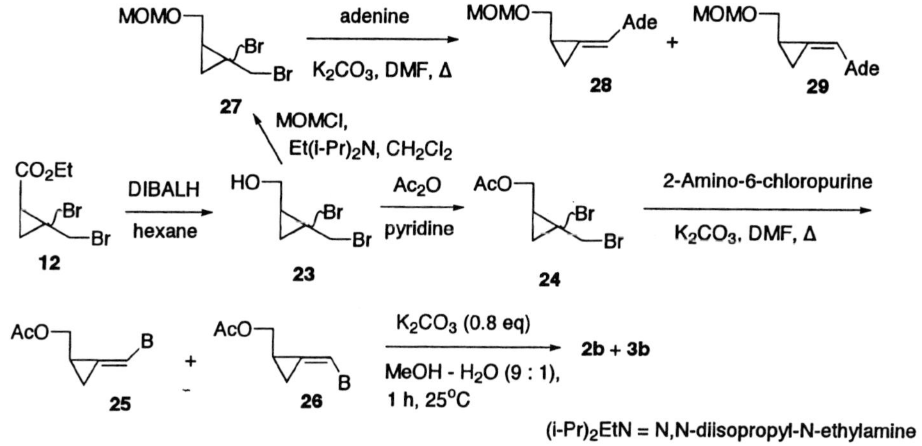

Scheme 4.

then used to transform adenine into a mixture of the *Z*- and *E*-isomers **28** and **29** separable by chromatography on silica gel. This result has shown that manipulation of the protecting groups can substantially simplify the separation of isomers.

A method similar to that described in Scheme 1 was adopted by another laboratory (Cheng *et al.*, 1998) for the synthesis of **2a** and **3a**. The alkylating agent **12** was prepared by addition of bromine to ethyl methylenecyclopropane carboxylate, the *Z*- and *E*-isomeric esters **15a** + **16a** were reduced with LiAlH$_4$ and the target analogues **2a** and **3a** were separated by a combination of silica gel chromatography and reversed phase HPLC. Likewise, the hydroxymethyl analogue of **2a** or **3a** (compound **30**) was prepared (Cheng *et al.*, 1997) as follows (Scheme 5). Dibromide **31** was obtained in a low (12 %) yield as a by-product of dibromocarbene addition to the dibenzyl ether of diol **32** readily accessible from Feist's acid (Hsiao and Hannick, 1990). Alkylation and elimination procedure (two steps) using **31** and adenine led to a protected intermediate **33** which, after deprotection gave analogue **30**. More recently, a similar approach was also exploited in the synthesis of an imidazo[4,5-e][1,3]diazepine methylenecyclopropane nucleoside analogue with a guanidinocarbamoyl function instead of the hydroxymethyl group (Chen and Hosmane, 2000). The obtained single isomer was formulated as *Z* although the presented NOE and molecular modeling data clearly favor the *E*-configuration.

Potent antiviral effect of the purine *Z*-isomeric analogues **2** (see Section 3.1.1) has prompted the investigation of approaches for obtaining *R*- and *S*-enantiomers.

Resolution of racemic synadenol **2a** was achieved by HPLC on a chiral column (Qiu *et al.*, 1998a; Cheng *et al.*, 1998). This method is generally suitable (Qiu *et al.*, 2000a) for resolution of the racemic *Z*- and *E*-isomeric purine methylenecyclopropane analogues (Table 2). It forms the basis for determination of the optical purity (enantiomeric excess)

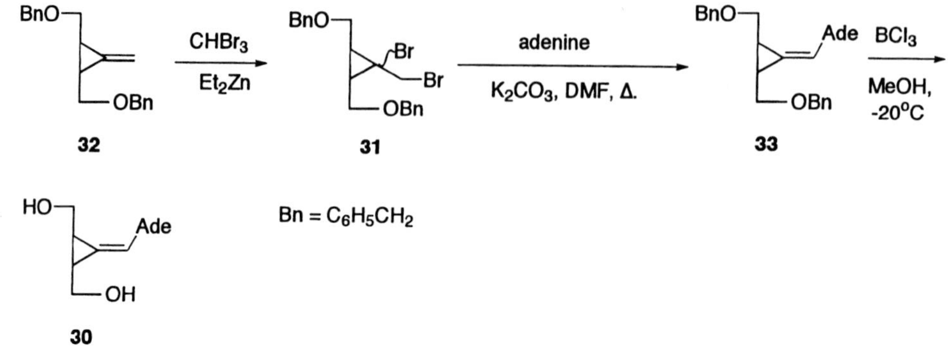

Scheme 5.

Table 2.   Chiral HPLC of enantiomeric methylenecyclopropane analogues of nucleosides[a].

	Z-Isomers:		E-Isomers:	
	34	**35**	**36**	**37**

series a: B = adenine

series b: B = 2-amino-6-chloropurine

series c: B = guanine

series d: B = 2,6-diaminopurine

series e: B = 2-amino-6-methoxypurine

series f: B = 2-amino-6-cyclopropylaminopurine

Enantiomer	Configuration	t_R (min)	Enantiomer	Configuration	t_R (min)
Z-isomers:					
34a	R	10.99[b]	**34e**	R	24.80
35a	S	10.25[b]	**35e**	S	5.32
34b	R	27.23	**34f**	R	13.68
35b	S	5.48	**35f**	S	6.80
34c	R	6.28	E-isomers:		
35c	S	3.85	**36a**	R	8.45[b,c]
34d	R	7.00	**37a**	S	9.91[b,c]
35d	S	5.35	**36b**	R	8.64
			37b	S	9.74

[a]   Chiralpak AD column, 10 μ, 4.6 x 250 mm, room temperature, methanol, flow rate 1 mL/min. t_R = retention time. Data from Qiu *et al.*, 2000a.

[b]   Data from Qiu *et al.*, 1998a. Cheng *et al.*, 1998 give t_R 42.3 (**34a**) and 33.1 min (**35a**) for hexane - 2-propanol (7 : 1), flow rate 0.7 mL/min.

[c]   Determined with a racemic mixture.

of the obtained enantiomers. Significant differences in retention times of *R*- and *S*-enantiomers **34b** and **35b** suggested a possibility of resolution of key racemic intermediate **2b** (Scheme 1) on a larger scale. However, because of prohibitive costs, other methods for synthesis of enantiomers have been sought.

Previously, enantiomers of anti-HIV agent adenallene **1a** were resolved by the action of adenosine deaminase (Megati *et al.*, 1992). Although synadenol (**2a**) is a weaker substrate for this enzyme, preliminary experiments (Qiu *et al.*, 1998b) indicated feasibility of such an approach. Digestion of racemic **2a** with adenosine deaminase from calf intestine gave the (+)-hypoxanthine analogue **36** and (-)-synadenol (**34a**, Scheme 6) that was resistant to further deamination (Qiu *et al.*, 1998a, 1999a). The absolute configuration (*R*) of **34a** was determined by X-ray diffraction. This finding corrected an erroneous assignment (Cheng *et al.*, 1998) of the *S* configuration to (-)-synadenol (**34a**). The deaminated product **36** was converted to (*S*)-(+)-synadenol (**35a**) as follows. Acetylation gave acetate **37** which, in turn, was transformed to 6-chloropurine intermediate **38**. Ammonolysis then afforded the desired (*S*)-(+)-enantiomer **35a**. It is worth of noting that methylenecyclopropane derivative **38** is significantly more stable toward ammonolysis than allene analogue **1** (B = 6-chloropurine, Megati *et al.*, 1992). A major disadvantage of the method described above is a lack of more general application. It is limited to synadenol (**2a**) given the fact that 2,6-diaminopurine analogue **2d** is completely resistant (Qiu *et al.*, 1998c) toward adenosine deaminase in contrast to 2-aminoadenallene (**1**, B = 2,6-diamino purine, Phadtare *et al.*, 1991). Another drawback is a necessity of converting the deaminated enantiomer back to the adenine derivative.

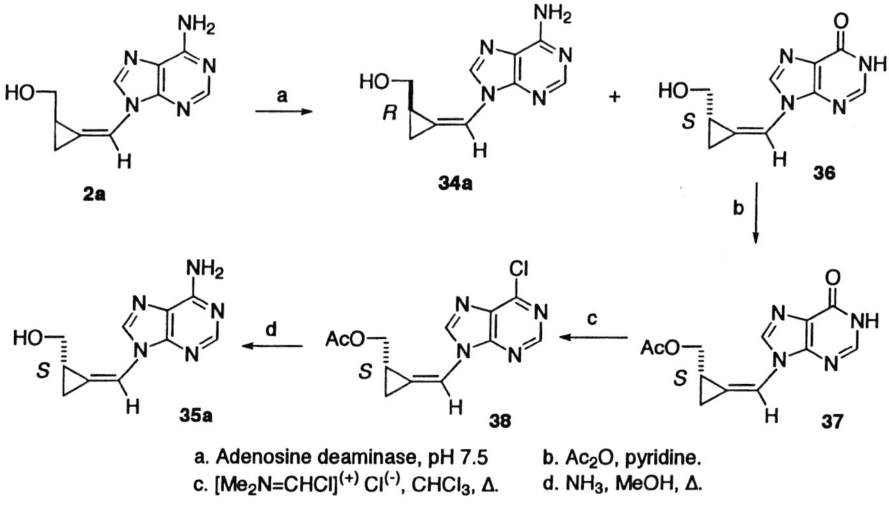

a. Adenosine deaminase, pH 7.5   b. Ac$_2$O, pyridine.
c. [Me$_2$N=CHCl]$^{(+)}$ Cl$^{(-)}$, CHCl$_3$, Δ.   d. NH$_3$, MeOH, Δ.

Scheme 6.

A general and non-destructive method (Qiu *et al.*, 2000a) started from the racemic methylenecyclopropane carboxylic acid (**39**) which was resolved as described (Lai *et al.*, 1991) via the diastereoisomeric (*R*)-2-phenylglycinol amides **40** and **41** (Scheme 7). The latter were separated by column chromatography on silica gel. Acid hydrolysis of **40** and **41** afforded the individual *R*- and *S*-enantiomers of methylenecyclopropane carboxylic acids **42** and **43**. The reagent suitable for alkylation-elimination of nucleic acid bases (see racemic form **24** in Scheme 4) was obtained as shown for the (*S*)-methylenecyclopropane carboxylic acid **43**. Esterification gave the ethyl ester **44**. Addition of bromine led to diastereoisomeric dibromo esters **45**. Reduction furnished the *S*-carbinol **46** which was, in turn, converted to acetate **47**. The latter reagent was used for alkylation-elimination of 2-amino-6-chloropurine to give a mixture of the *Z*- and *E*-isomers **48** + **49**. Deprotection afforded the *S*-enantiomers **35b** and **37b**. The *R*-enantiomers **34b** and **36b** were obtained by a similar procedure starting from the (*R*)-methylenecyclopropane carboxylic acid (**42**). The *R,Z*- and *S,Z*-enantiomers (Qiu *et al.*, 2000a) **34c** - **34f** and **35c** - **35f** were obtained as described in Scheme 2 for the respective racemic products. The procedure is also applicable for synthesis of adenine enantiomers **34a**, **35a**, **36a** and **37a**. The optical purity of all enantiomeric analogues was 94 - 100 %. The most important determinant of optical purity is the efficiency of chromatographic separation of the diastereoisomeric amides **40** and **41**.

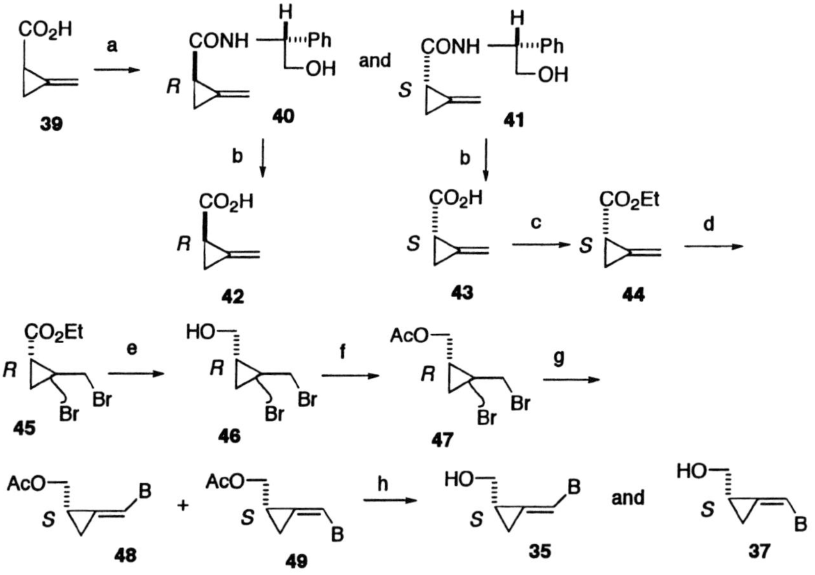

series a: B = adenine, series b: B = 2-amino-6-chloropurine

a. 1. IBuOCOCl, NEt$_3$, (*R*)-2-phenylglycinol; 2. Chromatography.   b. 1 M H$_2$SO$_4$, THF.
c. HCl, EtOH.   d. Br$_2$, CCl$_4$.   e. DIBALH, THF.   f. Ac$_2$O, pyridine.
g. 2-Amino-6-chloropurine, K$_2$CO$_3$, DMF, Δ.
h. 1. K$_2$CO$_3$ (0.8 eq.), MeOH - H$_2$O (9 : 1), 1 h, 25°C. 2. Chromatography.

Scheme 7.

2.1.2. *Methylenecyclopropane analogues* **4** *and* **5**

Analogues **4** and **5** are isomers of **2a** and **3a** derived by an interchange of the adenine and hydroxymethyl functions. Alkylation-elimination was again a method of choice. The synthesis (Qiu *et al.*, 2000b) commenced with 3-bromo-3-buten-1-ol (**50**) which was transformed to methoxymethyl (MOM) derivative **51** (Scheme 8). Addition of dibromocarbene generated from $CHBr_3$ under phase-transfer conditions afforded tribromocyclopropane **52**. Selective reduction with titanium tetraisopropoxide and ethylmagnesium bromide gave the *(E,Z)*-dibromocyclopropanes **53** which were used for an alkylation-elimination procedure with adenine. A mixture of the resultant *Z*- and *E*-intermediates **54** was deprotected to furnish analogues **4** and **5**. The latter were, unlike isomers **2a** and **3a**, readily separated by chromatography on silica gel.

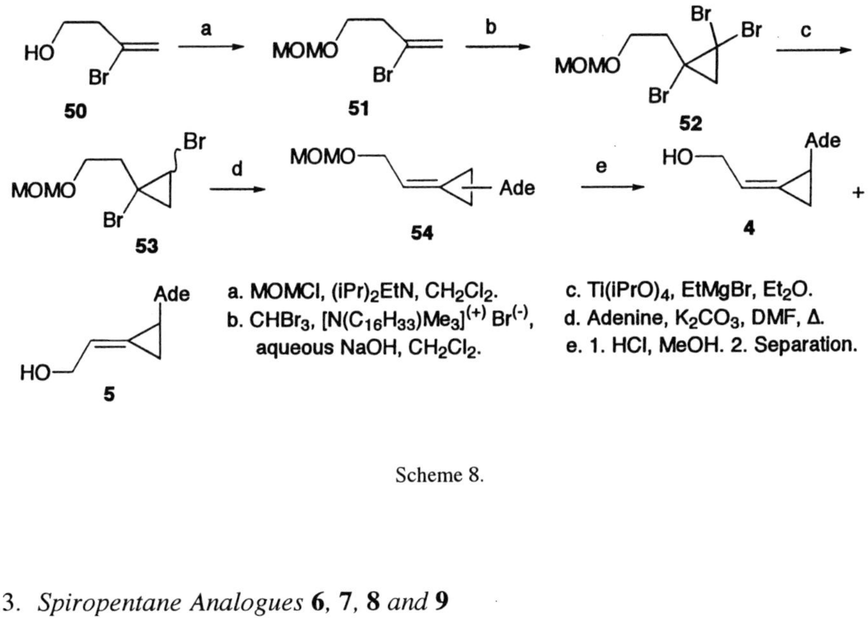

HO
Br
50
a
MOMO
Br
51
b
Br
Br
MOMO
Br
52
c
MOMO
Br
Br
53
d
MOMO
Ade
54
e
HO
Ade
4
+
Ade
HO
5
a. MOMCl, (iPr)2EtN, CH2Cl2.
b. CHBr3, [N(C16H33)Me3](+) Br(-), aqueous NaOH, CH2Cl2.
c. Ti(iPrO)4, EtMgBr, Et2O.
d. Adenine, K2CO3, DMF, Δ.
e. 1. HCl, MeOH. 2. Separation.

Scheme 8.

2.1.3. *Spiropentane Analogues* **6**, **7**, **8** *and* **9**

The synthesis of this type of analogues was the most difficult in view of the isomer complexity and a general lack of preparatively feasible stereoselective approaches in the spiro pentane series (Gajewski and Burka, 1970; Lukin and Zefirov, 1995). Because all four stereoisomeric types were necessary for antiviral evaluation, a combinatorial-like approach seemed the most appropriate. At the outset, synthesis of a mixture of all four possible isomers was contemplated from which all desired analogues could be generated in later stages. The success of this approach then depended to a large extent on the separation of at least some of the isomers in an early stage of the synthetic sequence. Following these principles, the syntheses of two complete sets of isomeric adenine and guanine spiropentanes, **6a** - **9a** and **6b** - **9b** were accomplished (Guan *et al.*, 2000a, Scheme 9).

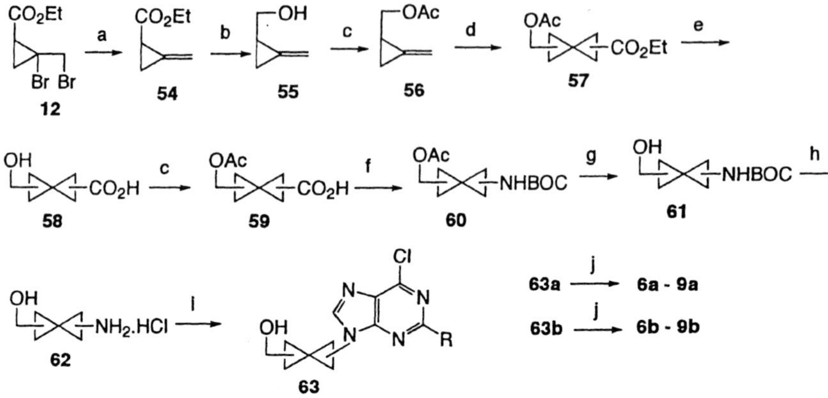

63a: R = H   63b: R = NHAc

a. Zn, AcOH - Et_2O. b. $LiAlH_4$, Et_2O. c. Ac_2O, pyridine. d. N_2CHCO_2Et, $Rh_2(OAc)_4$, CH_2Cl_2.
e. 1. NaOH, aq. MeOH. 2. Separation of *proximal + medial syn* and *medial-anti + distal* isomers.
f. 1.$(PhO)_2P(O)N_3$, NEt_3, tBuOH, Δ. g. 1. K_2CO_3, aq. MeOH. 2. Separation of all isomers.
h. HCl, MeOH. i. 1. 4,6-dichloro-5-nitropyrimidine or 2-acetamino-4,6-dichloro-5-nitropyrimidine,
NEt_3, EtOH. 2. $SnCl_2$, $CH(OEt)_3$. j. NH_3, MeOH, Δ.

Scheme 9.

Dibromo ester **12** (Scheme 1) was debrominated to give ethyl methylenecyclopropane carboxylate (**54**). Reduction afforded carbinol **55** which was then acetylated to furnish acetate **56**.

Addition of carbene generated from ethyl diazoacetate under catalysis with $Rh_2(OAc)_4$ led to a mixture of all four possible isomeric spiropentanes **57**. Hydrolysis afforded hydroxy carboxylic acids **58** which were resolved into two mixtures of *proximal + medial syn* and *medial anti + distal* isomers by chromatography. The next steps were perfomed separately with these portions. Reacetylation gave acetates **59** which, in turn, were subjected to a Curtius rearrangement following the activation with diphenyl phosphoryl azide in *tert*-butyl alcohol. The resultant *tert*-butoxycarbonyl (BOC) derivatives **60** were deacetylated to give intermediates **61**. At this stage, the two mixtures of *proximal + medial syn* and *medial anti + distal isomers* were resolved into individual components by chromatography. Deprotection with HCl in methanol then afforded hydrochlorides of all four isomeric aminospiropentanes **62**.

The adenine and guanine rings were introduced into individual spiropentanes **62** by stepwise procedures used routinely in nucleoside chemistry with some important exceptions. Aminospiropentanes **62** have limited stability as free bases and, therefore, procedures requiring higher temperatures were precluded. Therefore, 4,6-dichloro-5-nitropyrimidine (adenine series) and the respective 2-acetamino derivative (guanine series) with more reactive chlorine atoms were used for alkylation of aminospiropentanes **62** at room temperature. Reduction of the nitro group and imidazole ring closure were perfomed by a one-pot procedure using triethyl orthoformate and $SnCl_2$ to give 6-chloropurine derivatives **63a** or **63b**. In the final step, ammonolysis of **63a** gave adenine analogues **6a** - **9a** whereas a similar reaction with **63b** led to guanine spiropentanes **6b** - **9b**.

2.1.4. *Methylenecyclobutane analogues 10 and 11*

Two methods were elaborated for synthesis (Guan *et al.*, 2000b) of analogues **10** and **11**. The 2-(benzyloxymethyl)cyclobutanone **64** which was used for synthesis of methylenecyclobutane **65**, a key intermediate in both methods, was prepared by a modification of the procedure described (Scheme 10) for the corresponding *O*-benzoate (Lee-Ruff *et al.*, 1996). A 2 + 2 cycloaddition of ethyl acrylate (**66**) and ketene dimethyl thioacetal (**67**) gave ethyl 2,2-(bismethylthio)-1-cyclobutanecarboxylate (**68**) which was reduced with LiAlH$_4$ to carbinol **69**. Benzylation gave the *O*-benzyl thioacetal **70** and subsequent hydrolysis led to cyclobutanone **64**. Methylenecyclobutane **65** was then obtained by a Wittig methylenation. Addition of bromine using pyridinium perbromide (elemental bromine led to a contraction of the cyclobutane ring) gave dibromocyclobutane **71**. Alkylation-elimination of adenine with **71** gave the *E,Z*-bromo-methylenecyclobutanes **72** resulting from a simple β-elimination as major components. The *Z*- + *E*-isomers **73** + **74** (*E/Z* ratio 1 : 1) and an isomerization product thereof, cyclobutene **75**, were obtained only in low yields (5 and 4 %, respectively). This contrasts with the success of alkylation-elimination method in the methylenecyclopropane series (see Scheme 1, 3 and 4).

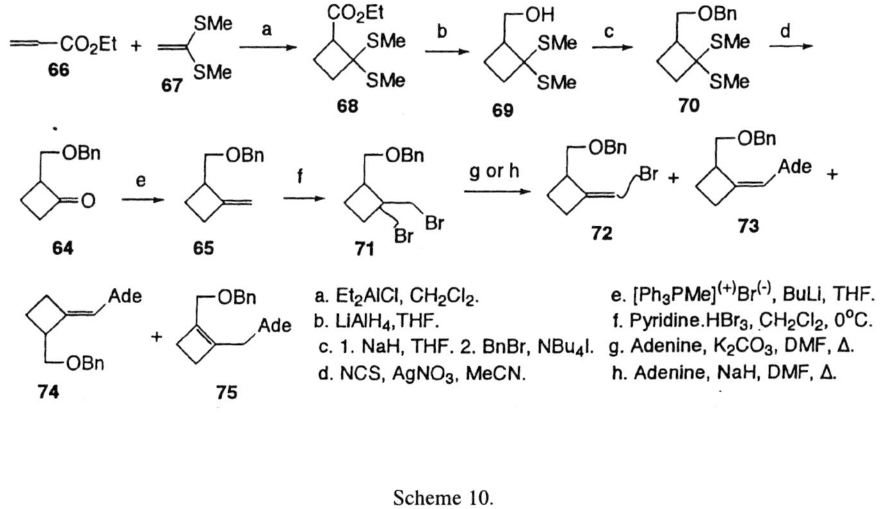

a. Et$_2$AlCl, CH$_2$Cl$_2$.
b. LiAlH$_4$,THF.
c. 1. NaH, THF. 2. BnBr, NBu$_4$I.
d. NCS, AgNO$_3$, MeCN.
e. [Ph$_3$PMe]$^{(+)}$Br$^{(-)}$, BuLi, THF.
f. Pyridine.HBr$_3$, CH$_2$Cl$_2$, 0°C.
g. Adenine, K$_2$CO$_3$, DMF, Δ.
h. Adenine, NaH, DMF, Δ.

Scheme 10.

An alternate approach made also use of methylenecyclobutane **65** (Scheme 11). Epoxidation gave a mixture of *E,Z*-spiro-oxiranes **76** which were used for alkylation of adenine to afford *E,Z*-hydroxycyclobutanes **77**. Reaction with methylsulfonyl chloride (MsCl) furnished *E,Z*-methylsulfonates **78** which, in turn, were transformed to *Z*- + *E*-

338   *J. Zemlicka*

methylenecyclobutanes **73** + **74** using potassium *tert*-butoxide (tBuOK). Deprotection furnished the target analogues **10** and **11** which were separated by chromatography on alumina. The *E*-isomer **11** was the predominant product. Using an excess of tBuOK, a double elimination occurred with the formation of diene **79**. Hydroboration of **79** followed by reaction with H_2O_2 gave the *E*-isomer **11**.

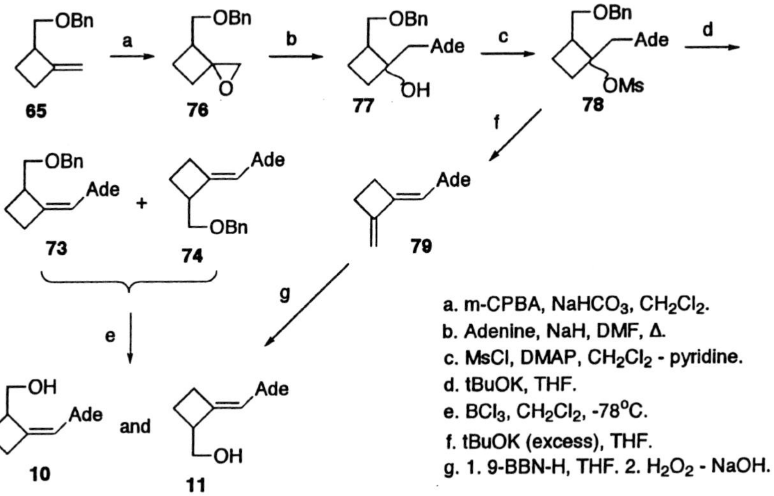

a. m-CPBA, NaHCO$_3$, CH$_2$Cl$_2$.
b. Adenine, NaH, DMF, Δ.
c. MsCl, DMAP, CH$_2$Cl$_2$ - pyridine.
d. tBuOK, THF.
e. BCl$_3$, CH$_2$Cl$_2$, -78°C.
f. tBuOK (excess), THF.
g. 1. 9-BBN-H, THF. 2. H$_2$O$_2$ - NaOH.

In Scheme 11, Bn = C$_6$H$_5$CH$_2$, NCS = N-chlorosuccinimide, Ms = CH$_3$SO$_2$, DMAP, 4-N,N-dimethylaminopyridine, m-CPBA = m-chloroperoxybenzoic acid, 9-BBN-H = 9-borabicyclo[3.3.1]nonane.

Scheme 11.

2.2. Assignment of isomeric and enantiomeric structure of the analogues

2.2.1. *Z,E-Isomerism*

Synthetic approaches leading to unsaturated analogues almost always led to both *Z*- and *E*-isomers (compounds of type **2, 4** and **10** vs. **3, 5** and **11**) which had to be separated and their geometrical isomerism assigned. Even more complex was the situation with analogues **6 - 9** where a total of four isomeric compounds was obtained. The NMR spectroscopy proved to be an indispensable tool for structural assignments in all these cases. With analogues **2a, 3a, 10** and **11** it was possible to deduce tentatively the *Z* or *E* configuration from the differences in H$_8$ chemical shifts: The H$_8$ of the *Z*-isomer was located downfield from that of the *E*-isomer (Table 3).

Follmann and Gremels (1974) have shown that the differences of the H$_8$ and H$_2$ chemical shifts δH$_8$ - δH$_2$ = Δδ can serve as indicators of intramolecular ribose-base interactions in adenine nucleosides. Thus, compounds with an *anti*-conformation of the

Table 3.   The H_8 and H_2 chemical shifts of adenine analogues in DMSO-d_6 solution[a].

Compound	H_8 (δ)	H_2 (δ)	$\Delta\delta$	Reference
1a	8.17	8.17	0	Phadtare and Zemlicka, 1989
2a	8.74	8.17	0.57	Qiu *et al.*, 1998b
3a	8.48	8.17	0.31	"
4	7.98	8.14	-0.16	Qiu *et al.*, 2000b
5	8.01	8.13	-0.12	"
6a	8.11	8.11	0	Guan *et al.*, 2000a
7a	8.11	8.12	-0.01	"
8a	8.10	8.11	-0.01	"
9a	8.09	8.10	-0.01	"
10	8.38	8.14	0.24	Guan *et al.*, 2000b
11	8.14	8.14	0	
80	8.21	8.07	0.14	Jones *et al.*, 1995
Adenosine	8.34	8.15	0.19	Follmann and Gremels, 1974

[a]   For purine ring numbering see formula **2b** in Scheme 2.

base and spatial orientation of the CH_2OH leading to a juxtaposition of the oxygen atom and H_8 proton have higher $\Delta\delta$ values in comparison with analogues lacking the CH_2OH. An absence or decrease of $\Delta\delta$ can also be expected in cases where the rotation of the base is not restricted or where the location or conformation of the CH_2OH does not allow an efficient deshielding of the H_8. These factors may be responsible for the fact that $\Delta\delta$ values of adenallene (**1a**), *E*-isomers **3a**, **11** and spiropentanes **6a** - **9a** are negligible (Table 3). Of interest is also a different behavior of another analogue with axial dissymmetry, compound **80** which shows a significant $\Delta\delta$ of 0.14. This is indicative of a deshielding of the H_8. The $\Delta\delta$ values of an isomeric pair **4** and **5** are difficult to interpret because the H_2 is shifted downfield from H_8. Generally, purine *Z*-isomers **2b** - **2f** have the H_8 at lower field (Qiu *et al.*, 1998b,c) from that of the *E*-isomers **3b** - **3f** and a similar trend was found for the H_6 chemical shifts of the pyrimidine analogues **2g**, **3g**, **21** and **22** (Qiu *et al.*, 1998c).

High $\Delta\delta$'s of the *Z*-isomer **10** (0.24) and, particularly, **2a** (0.57) that exceed that of adenosine (0.19) are indicative of an interaction of CH_2OH with the H_8. In fact, formation (Qiu and Zemlicka, 1998b) of an anhydronucleoside analogue **81** during attempted phosphorylation of synadenol (**2a**) with $POCl_3$ is also indicative of a proximity of the $CH_2(O)$ and adenine moiety although the reaction requires a *syn*-conformation of the base (Scheme 12). Cyclic structures comprising a methylenecyclopropane system are known (Brandi and Goti, 1998).

Although analysis of the H_8 chemical shifts of analogues of the type **2**, **3**, **10** and **11** (Table 3) made possible a preliminary *E,Z*-isomeric assignment, confirmation was necessary. In addition, this approach failed in cases of analogues with small differences

340   *J. Zemlicka*

in the H_8 shifts of the *Z*- and *E*-isomers such as **4** and **5**. Final solution of this problem came from the nuclear Overhauser effect (NOE) experiments and in case of **2a** also from the formation of anhydronucleoside analogue **81** (Scheme 12). Of particular importance were protons of the N^9-substituent located in the proximity of the H_8 of purine ring (Qiu *et al.*, 1998b and 2000b; Guan *et al.*, 2000b). The NOE was also instrumental for isomeric assignment of spiropentane analogues **6a** - **9a** (Guan *et al.*, 2000a).

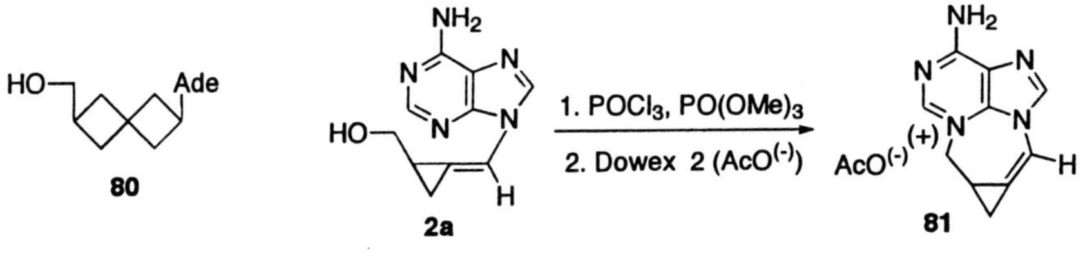

Scheme 12.

2.2.2. *R- and S-enantiomers*

A similarity of circular dichroism (CD) spectra with those of (*R*)-(-)- and (*S*)-(+)-adenallene (**82** and **83**, Megati *et al.*, 1992) gave an indication that (-)-synadenol (**34a**) has an *R* and the (+)-enantiomer **35a** *S* configuration (Figure 1). Nevertheless, an opposite assignment was reported in the literature (Cheng *et al.*, 1998). To confirm the assignment, two lines of evidence were obtained for absolute configuration of the *R*- and *S*-enantiomers of **2a** - **2f**. The *R* configuration of the (-)-enantiomer **34a** resultant from deamination (Scheme 6) of racemic **2a** with adenosine deaminase (Qiu *et al.*, 1998a, 1999a) was determined by X-ray diffraction of a single crystal of the hydrochloride (Figure 2). Both molecules of **34a** have adenine base in an *anti*-like conformation. An independent confirmation of the absolute configuration for all purine analogues was based on the fact that enantiomers **35a** - **35f** derived from the *(S)*-methylenecyclopropane carboxylic acid (Lai *et al.*, 1991) (**43**, Scheme 7) have the (*S*)-(+) configuration. Conversely, enantiomers **34a** - **34f** generated from the *R*-enantiomer **42** are of the (*R*)-(-) configuration (Qiu *et al.*, 2000a). It is also noteworthy that in the *Z*-isomeric series the chiral HPLC retention times (t_R) of the *R*-enantiomers were always higher than those of the *S*-enantiomers whereas an opposite trend was noted with the *E*-isomers (Table 2).

3.   Biological activity

Analogues of the type **2** - **11** were tested for antiviral activity against the following viruses: Human and murine cytomegalovirus (HCMV and MCMV), herpes simplex virus 1 and 2 (HSV-1 and -2), Epstein-Barr virus (EBV), varicella zoster virus (VZV),

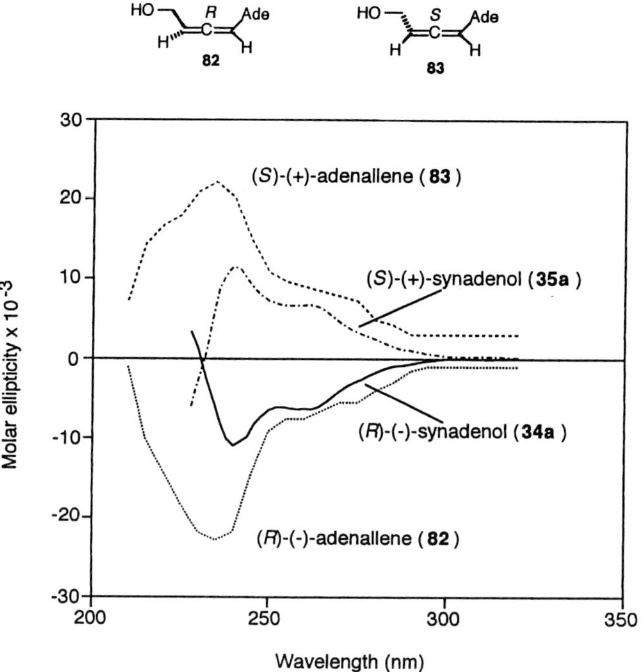

Figure 1.

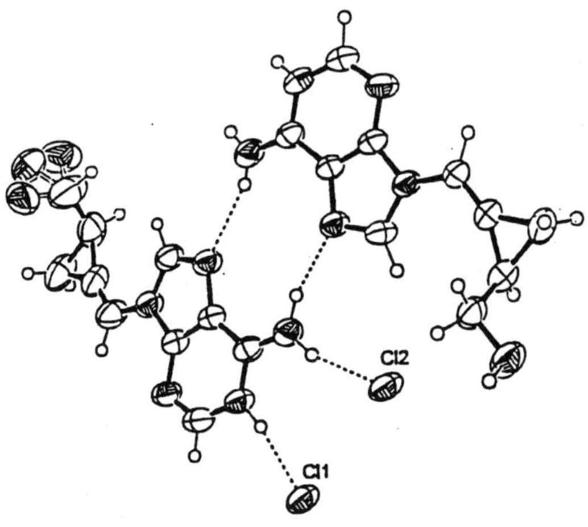

Figure 2.

hepatitis B virus (HBV) and human immunodeficiency virus (HIV-1). In selected cases, antitumor activity was also examined and adenine analogues were tested for substrate activity toward adenosine deaminase.

3.1. Antiviral activity

3.1.1. *Methylenecyclopropane analogues of type 2 and 3*

In terms of antiviral activity, the Z-isomers **2** are of most interest because of a strong and broad-spectrum effect (Qiu *et al.*, 1998b,c and 1999b; Rybak *et al.*, 1999 and 2000; Uchida *et al.*, 1999). The results with two strains of HCMV, Towne and AD 169, as well as murine virus are summarized in Table 4. It is clear that potency and cytotoxicity parameters of the Z-isomers **2a** - **2c**, **2e** and **2f** compare favorably with those of the current anti-HCMV drug ganciclovir (Qiu *et al.*, 1998b,c and 1999b; Rybak *et al.*, 1999 and 2000). In MCMV assays, activity of **2b** - **2f** was significantly higher. The 2,6-diaminopurine analogue **2d** exhibited a lower potency against both strains of HCMV but it was about equally effective against MCMV as the rest of 2-aminopurine analogues **2a** - **2c**, **2e** and **2f**. Of special importance is the fact that 2-aminopurine analogues **2b**, **2e** and **2f** derived from unnatural nucleic acid bases (with a modified substituent at the position 6 of purine ring) are all effective anti-CMV agents. The Z-isomers of pyrimidine derivatives lack a significant anti-CMV potency; syncytol (**2g**) was

Table 4.   Inhibition of human and murine cytomegalovirus (HCMV and MCMV) by racemic (Z)-methylene-cyclopropane analogues of nucleosides (EC_{50}/IC_{50}, μM)[a].

Compound	HCMV/HFF[b]		MCMV/MEF[b]
	Towne	AD 169	
2a	2.1/>100 (1.1)[c]	1.3/>460	2.1/>460
2b	5.0/>100 (0.8)[c]	6.9/185	0.27/260
2c	2.1/>100 (1.8)[c]	1.2/>429	0.3/>429
2d	23.7/100 (18)[c]	15.9/>431	0.6/>293
2e	3.1/>100	5.3/>404	0.4/260
2f	0.4/100 (1)[c]	2.4/327	0.37/220
2g	28.5/>100 (16)[c]	9.8/>518	52.3/>518
21	>100/>100	>480/>480	not tested
Ganciclovir	3.4/>100	2.3/>391	2.0/>192

[a]Data are compiled from Qiu *et al.*, 1998b,c, 1999b and Rybak *et al.*, 2000 which also list results of cytopathic effect inhibition assays. Cytotoxicities reported for HFF cells (Qiu *et al.*, 1998c) as IC_{50} 222 μM are actually >100 μM.

[b]   Plaque reduction.

[c]   Yield reduction, IC_{90}.

moderately effective whereas synthymol (**21**) was devoid of activity. The *E*-isomers **3a** - **3d**, **3g** and **22** were also ineffective.

Overall, the purine *Z*-isomers **2a** - **2f** are less effective against HSV-1 and HSV-2 (Table 5). The cytopathic inhibition effect (CPE) of synadenol (**2a**) showed a surprisingly low value of $EC_{50} < 0.14$ μM but the plaque reduction failed to show any potent activity (Qiu *et al.*, 1998b; Rybak *et al.*, 2000). The assay of **2a** in BSC-1 (ELISA) and Vero cells showed only moderate effects. The 6-cyclopropylaminopurine **2f** was somewhat more active. In the pyrimidine series, synthymol (**21**) was most potent with EC_{50} 2 μM in ELISA assay but this result was not duplicated in Vero cells. Syncytol (**2g**) was inactive and so were the *E*-isomers. None of the tested analogues had any appreciable activity against HSV-2 with the exception of synadenol (**2a**) whose potency in the CPE assay (EC_{50} 10.6 μM, Qiu *et al.*, 1998b; Rybak *et al.*, 2000) was close to acyclovir control but, again, only a moderate effect was seen in plaque reduction.

The methylenecyclopropane analogues, especially the *Z*-isomers, exhibited also strong potency against another class of herpesvirus, EBV (Qiu *et al.*, 1998b,c; Rybak *et al.*, 2000). The extent of this effect (Table 6) was roughly comparable with that observed against cytomegaloviruses (Table 4). The purine *Z*-isomers **2a** - **2d** exhibited

Table 5.　　Inhibition of herpes simplex virus type 1 and 2 (HSV-1 and HSV-2) replication by racemic (Z)-methylenecyclopropane analogues of nucleosides[a].

Compound	Antiviral activity, EC_{50} (μM)			Cytotoxicity, IC_{50} (μM)
	HSV-1		HSV-2	
	BSC-1 ELISA	Vero Plaque reduction	Vero Plaque reduction	CEM
2a	33	28[b]	59[b]	>100
2b	31	52	55	80
2c	100	85	>100	>100
2d	>100	>100	>100	>100
2e	33	>50	>50	>100
2f	15	31	41	>100
2g	>100	>100	>100	>100
21	2.0	>100	70	not tested
Control	3.5±2.1[c]	2.6±2.0[d]	9[d]	>20[d]

[a]　Data are compiled from Qiu *et al.*, 1999b,c, 1999b. These references and Rybak *et al.*, 2000 give also results of cytopathic effect (CPE) inhibition assays in HFF cells.

[b]　EC_{50}'s <0.14 and 10.6 μM were observed by CPE inhibition assays for HSV-1 and HSV-2 in HFF cells. In plaque reduction assays the EC_{50}'s were 140 and 47.9 μM, respectively (Rybak *et al.*, 2000).

[c]　Ganciclovir.

[d]　Acyclovir.

a strong anti-EBV effect in both H-1 and Daudi cells. Analogues **2e** and **2f** although potent in H-1 cells were somewhat less effective in Daudi culture. Interesting is a strong activity of syncytol (**2g**) in both types of host cells. The corresponding *E*-isomer **3g** was equipotent to **2g** in Daudi culture but it showed no effect in H-1 culture. Conversely, adenine analogue **3a** was active in H-1 but much less in Daudi cells. These results have indicated that the *E*-isomers of methylenecyclopropanes can also exhibit antiviral effects.

Table 6.   Inhibition of Epstein-Barr (EBV), varicella zoster (VZV) and hepatitis B virus (HBV) replication by racemic methylenecyclopropane analogues of nucleosides (EC$_{50}$/IC$_{50}$, μM)[a].

Compound	EBV		VZV[b]	HBV[c]	mtDNA[d]
	H-1[e]	Daudi[f]	HFF	2.2.15	
2a	0.2/>50	3.2/368	2.5	2	>50
3a	3/50	71.4/>229	90.2	>10	>150
2b	0.7/>50	1.2/>199	not tested	>10	>50
2c	0.3/>50	5.6/>230	61	10	>100
2d	1.5/>50	6.9/>215	93	10	>100
2e	1.6/>50	17.6/>202	23.9	>10	not tested
2f	4/>100[g]	11.8/184	13.2[h]	2	>100
2g	2.5/>50	<0.41/518	3.6	>10	not tested
3g	>50/>50	<0.41/518	>518	>10	not tested
21	>10/50	1.3/>241	3.6[h]	>10	>50
Control	5/75[i]	5.3/222[j]	9.8[j]	1.4[k]	0.07[k]

[a]   Data are compiled from Qiu *et al.*, 1998b,c, 1999b and Rybak *et al.*, 2000.

[b]   Plaque reduction, EC$_{50}$ data only. For cytotoxicity in HFF cells see Table 4.

[c]   EC$_{50}$ data only. For cytotoxicity in CEM cells see Table 5.

[d]   Inhibition of mitochondrial DNA synthesis (IC$_{50}$, μM).

[e]   DNA hybridization assay.

[f]   Viral capsid antigen immunofluorescence (VCA-IF) assay.

[g]   Cheng Y-C, unpublished data. Cytotoxicity was determined in CEM cells.

[h]   Kern ER, unpublished data.

[i]   Ganciclovir.

[j]   Acyclovir.

[k]   Zalcitabine (ddC).

Strong activity against another herpesvirus, VZV, was restricted to synadenol (**2a**) and, interestingly, to pyrimidine analogues syncytol (**2g**) and synthymol (**21**, Qiu *et al.*, 1998b,c; Rybak *et al.*, 2000). From the 2-aminopurine analogues, 6-cyclopropylaminopurine **2f** was somewhat active but potency of **2c**, **2d** and **2e** was lower. The *E*-isomers were devoid of anti-VZV effect. Against HBV, only synadenol (**2a**) and 2-amino-6-cyclopropylaminopurine analogue **2f** reached the activity level of zalcitabine (ddC) with IC_{50}'s of 2 μM. Nevertheless, in contrast to ddC, none of the tested analogues exhibited a long term toxicity measured as inhibition of mitochondrial DNA synthesis.

Three types of assays were employed for testing the activity against HIV-1 reverse transcriptase (RT) assay (Qiu *et al.*, 1998b,c, 1999b) in supernatant of CEM-SS cells infected with HIV_{IIIB}, cytopathic effect inhibition of HIV-1$_{LAI}$ and inhibition of gag p24 protein production in MT-2 cells (Table 7, Uchida *et al.*, 1999). All tested analogues showed anti-HIV potency at low multiplicity of infection (moi) in RT assay. However, this effect disappeared at high moi; only synadenol (**2a**) showed some activity. In MT-2 cells, analogue **2a** was again the most potent in both types of assays. A weaker efficacy of **2a** noted for MT-2 cells by other investigators (Cheng *et al.*, 1998) is perhaps related to a different strain of virus used. The 2,6-diaminopurine analogue **2d** was also somewhat active whereas the rest of the tested group was ineffective. Hydroxymethyl derivative of synadenol (**2a**), compound **30**, was inactive as an anti-HIV agent (Cheng *et al.*, 1997).

Table 7. Inhibition of human immunodeficiency virus (HIV-1) replication by racemic methylenecyclopropane analogues of nucleosides[a].

Compound	Antiviral activity, EC_{50} (μM)				Cytotoxicity, IC_{50} (μM)	
	HIV_{IIIB} RT[b]		HIV-1$_{LAI}$[c] MT-2	p24[d]	CEM-SS	MT-2
	low[e] moi	high[f] moi				
2a	0.8	20	0.75±0.35[g]	0.54±0.13	>100	32±11
3a	14	>100	>100	not tested	>100	>100
2c	1.1	>30	>70	not tested	>100	70±10
3c	1.9	>30	>100	not tested	>100	>100
2d	16	>100	12±1.5	4.1±2.0	>100	>100
2g	0.8	>30	64±18	38±9.3	>100	>100
3g	1.7	>100	>100	not tested	>100	>100
AZT	0.003	0.17	0.019±0.006	0.020±0.008	-	>100

[a] Data for reverse transcriptase assay (HIV_{IIIB} RT) and cytotoxicity in CEM-SS cells are from Qiu *et al.*, 1998b,c, 1999b and Rybak *et al.*, 2000. The rest of data are from Uchida *et al.*, 1999. The IC_{50} values of cytotoxicity in CEM-SS cells listed as 222 μM in Qiu *et al.*, 1998c are actually 100 μM.

[b] Reverse transcriptase assay.

[c] Cytopathic effect inhibition assay.

[d] Inhibition of p24 gag protein production.

[e] ~0.1 pfu/cell.

[f] ~1 pfu/cell.

[g] Cheng *et al.*, 1998 give EC_{50} 26 μM.

Potent in vitro activity of purine Z-methylenecyclopropane analogues against MCMV (Table 4) indicated a possibility of a therapeutic effect in murine model of cytomegalovirus infection. In the first set of experiments, synadenol (**2a**) and synguanol (**2c**) were administered intraperitoneally (i. p.) once daily at 50, 16.7 and 5.6 mg/kg for 5 days beginning 6, 24 and 48 h after the infection (Rybak *et al.*, 1999). At 50 and 16.7 mg/kg, both analogues displayed a protective effect similar to ganciclovir. At 5.6 mg/kg ganciclovir was effective whereas compounds **2a** and **2c** were not. In the second set of experiments, mice were treated orally with 100 mg/kg once daily or 50 mg/kg twice daily for 5 days with analogues **2a** and **2c**. The protection was demonstrated with synguanol (**2c**) with a dose of 100 mg/kg but not 50 mg/kg. Ganciclovir was effective at both concentrations but synadenol (**2a**) was ineffective. Oral evaluation of 2-amino-6-methoxypurine analogue **2e** showed protective effect at only 60 mg/kg whereas ganciclovir was also active at 20 mg/kg. In contrast, the efficacies of orally administered 2-amino-6-cyclopropylaminopurine derivative **2f** and ganciclovir at 60 and 20 mg/kg once daily for 5 days coincided. The effect of the most potent analogue **2f** on MCMV pathogenesis was next investigated. Compound **2f** reduced viral replication in target organs to the extent observed with ganciclovir. It should be also stressed that all analogues tested in vivo were racemic and, therefore, a comparison with achiral ganciclovir is biased. At any rate, analogue **2f** appears presently as the best candidate for further preclinical studies.

Efficacy of two analogues effective in vivo, synguanol (**2c**) and compound **2f**, against different strains of non-human cytomegalovirus was then compared (Table 8). Both analogues were significantly more potent than ganciclovir in all strains tested (Rybak *et al.*, 1999). Cytotoxicity of analogues **2c** and **2f** in proliferating cells (Table 9) was also of interest (Rybak *et al.*, 2000). Only in Daudi cells was the cytotoxicity of ganciclovir significantly lower than that of **2c** or **2f**. The situation was reversed in BFU-E cells where both analogues were superior to ganciclovir. Smaller differences, in favor of analogues **2c** and **2f**, were observed in proliferating HFF and CFU-GM cells.

Compounds **2c** and **2f** were also tested against a series of laboratory, clinical and ganciclovir-resistant strains of HCMV (Rybak *et al.*, 2000). In most of the laboratory and clinical strains investigated (Table 10), the anti-HCMV activities of both analogues were comparable to ganciclovir. The experiments with isolates resistant to ganciclovir indicated a cross-resistance pattern of **2c** and **2f** only in C8706/13-1-1. A limited cross-resistance of synguanol (**2c**) was noted in C8914-6 and C8805/37-1-1 strains. It appears that a tendency toward cross-resistance of analogue **2f** that lacks the guanine base of ganciclovir is smaller than that of synguanol (**2c**). Interestingly, the HCMV strain resistant to ganciclovir owing to a mutation of the UL97 gene was slightly resistant to synadenol (**2a**) but not to synguanol (**2c**). The strain with a mutated DNA polymerase gene was sensitive to both analogues (Drach *et al.*, 1997). These results indicate that 2-aminopurine (Z)-methylenecyclopropane analogues and, particularly, 6-cyclopropylamino derivative **2f**, are good candidates for a further development as potential anti-HCMV drugs.

Potent activity of purine (Z)-methylenecyclopropane analogues in culture and in vivo provided a strong motivation to study the enantioselectivity of antiviral effect. In some cases, antiviral efficacy of nucleoside analogues is associated with a particular

Table 8. Activity of synguanol (**2c**) and 6-cyclopropylamino analogue **2f** against non-human strains of cytomegalovirus (EC_{50}, μM)[a].

Strain	Synguanol (**2c**)	Compound **2f**	Ganciclovir
Murine	0.30	0.37	4.7±0.39
Rat	4.7±3.0	10.7±0.037	57.2±32.1
Guinea pig	54.5±0.86	7.7±0.74	217.0±33.3
Rhesus	11.6±0.02	2.2	23.9±10.6

[a] Plaque reduction assay. Modified from Rybak *et al.*, 1999.

Table 9. Inhibition of cell proliferation by synguanol (**2c**) and 6-cyclopropylamino analogue **2f** (IC_{50}, μM)[a].

Cells	Synguanol (**2c**)	Compound **2f**	Ganciclovir
Daudi	12.9	27.9	196
HFF	278	255	157
Bone marrow (CFU-GM)	55.7	77.5	35.3±9.7
Bone marrow (BFU-E)	162	130	6.6±1.0

[a] Modified from Rybak *et al.*, 2000.

enantiomer but in other it is not (Zemlicka, 2000). This may not depend only on the type of the analogue but also on the type of the virus. For example, the hallmark of adenallene (**1a**) and cytallene (**1b**) is a strict *R* selectivity of anti-HIV and, in case of cytallene, also anti-HBV effect (Zemlicka, 1997). Despite of a structural resemblance, allenic analogues **1** and methylenecyclopropanes **2** have different profiles of antiviral activity. Therefore, comparison of enantioselectivity patterns of both groups of analogs is also of interest.

Detailed investigations revealed surprisingly complex patterns of enantioselectivity of the (Z)-methylenecyclopropanes (Qiu *et al.*, 1998a, 2000a) rarely seen in any other group of nucleoside analogues (Zemlicka, 2000). The differences in activity between enantiomers vary from a virus to virus and reversals of enantioselectivity also occur. Type of host cells and, in some cases, even the assay can play a significant role. It is likely that these effects are related to changes in the mechanism of action among different analogues, viruses and host cells.

Table 10.   Antiviral activity of synguanol (**2c**) and 6-cyclopropylamino analogue **2f** against laboratory, clinical and ganciclovir-resistant strains of HCMV (EC_{50}, μM)[a].

Strain	Synguanol (**2c**)	Compound **2f**	Ganciclovir
A. Laboratory and clinical isolates:			
AD 169 (control)	1.2	2.4	2.3
Toledo	8.6±3.0	5.5±1.3	9.8±3.9
Davis	2.5±0.09	2.2±0.62	5.3±1.2
Towne	11.6	10.6	5.1
EC	5.1±4.3	2.2±0.6	3.9±0.8
CH	3.1	2.7	4.3
Coffman	6.9±6.0	3.1±0.8	4.7+0.6
C8708/17-1-1	9.4±2.6	4.4±3.3	5.9±3.8
C9208/3-3-1	6.9	2.2	4.7
C9208/5-4-2	9.4	4.0	3.8
B. Ganciclovir-resistant isolates:			
C8404/9-1-4	2.7±0.2	1.7±1.3	32.5±15.3
C8805/37-1-1	48.9±5.1	10.3±4.8	77.6±59.9
C8914-6	10.3±0.6	4.4±2.7	32.9±11.4
C8706/13-1-1	12	7.3	15.3
C9209/1-4-4	10.3	6.6	323
759[R]D100	23.2	16.9	194

[a]   Data were taken from Rybak *et al.*, 2000.

The results with HCMV and MCMV are summarized in Table 11. Enantiomers of synadenol **34a** and **35a** were equipotent against Towne strain of HCMV whereas the *S*-enantiomer **35a** was slightly better inhibitor in AD 169 strain. The most suprising was an abrupt change of enantioselectivity toward MCMV where the *S*-enantiomer **35a** was 100 times more potent than the *R*-enantiomer **34a**. The fact that enantioselectivity of an antiviral agent differs toward the same virus from two species can probably be related to different phosphorylation capability of the host cells (HFF vs. MEF). In contrast, all tested 2-aminopurine analogues were clearly *S*-selective toward both HCMV and MCMV and their activity pattern followed that of racemic compounds (Table 4). In Towne strain of virus, the enantioselectivity of enantiomers **34b** and **35b** was somewhat decreased. With one exception (**34b**) in a single assay, no significant cytotoxicity was observed with any of the enantiomers.

As already mentioned, racemic (Z)-methylenecyclopropane analogues of nucleosides are moderately effective against HSV-1 and HSV-2 (Table 5). The (*S*)-synadenol (**35a**) was the most active of all tested enantiomers. Interestingly, the *S*-enantiomers **35a**, **35e** and **35f** were equally effective toward wild HSV-1 and a strain deficient in thymidine kinase (Table 12).

Table 11.   Inhibition of HCMV and MCMV replication by enantiomers of (Z)-methylenecyclopropane analogues of purine nucleosides (EC$_{50}$, μM)[a].

Enantiomer		HCMV/HFF[b]		MCMV/MEF[b]
		Towne	AD 169	
34a	*R*	2.5 (1.3)[c]	6.9	55
35a	*S*	2.4 (0.5)[c]	1.9	0.55
34b	*R*	30[d]	256	not tested
35b	*S*	1.8	1.9	not tested
34c	*R*	>100	>429	>429
35c	*S*	2.5	2.6	0.39
34d	*R*	>100	>431	>86
35d	*S*	21	18	0.65
34e	*R*	>100	252	32
35e	*S*	2.5	2.0	0.24
34f	*R*	>100	>74	44
35f	*S*	1.8	1.9	<0.11

[a]   Data were taken from Qiu *et al.*, 1998a and 2000a.

[b]   Plaque reduction. The IC$_{50}$ cytotoxicity values were 100 μM or higher unless stated otherwise. For control see Table 4.

[c]   Yield reduction, EC$_{90}$.

[d]   IC$_{50}$ 32 μM.

This indicates that the latter enzyme is not involved in phosphorylation of these analogues.

Enantioselectivity of methylenecyclopropane analogues toward EBV has presented a complex pattern (Table 13). Thus, (*R*)-synadenol (**34a**) was the more effective enantiomer in all three EBV assays and it was separated from the *S*-enantiomer **35a** by a factor of approximately 3 to 7. Clearly, both enantiomers are effective anti-EBV agents resembling thus the situation found with HCMV (Table 11). Some similarity to CMV is also seen in the fact that the 2-aminopurine analogues exhibit also an *S* preference although of lower magnitude. Thus, the *S*-enantiomers **35b**, **35c**, **35e** (in two assays) and **35f** are superior to *R*-enantiomers **34b**, **34c**, **34e** and **34f** but is some assays a lower enantioselectivity was observed. The 2,6-diaminopurine enantiomers **34d** and **35d** are equipotent in H-1 cells but they exhibit a surpringly opposite preferences in Daudi cells with *R* (**34d**) being more effective in DNA hybridization and *S* (**35d**) in the VCA-IF assay. It is recognized that enantiomeric patterns of antiviral activity may depend on the virus strain or host cells used (Zemlicka, 2000) but variations in different assays with a single type of cells (Daudi) are more difficult to explain.

Table 12.   Inhibition of HSV-1, HSV-2 and VZV replication by enantiomers of (Z)-methylenecyclopropane analogues of purine nucleosides (EC_{50}, μM)[a].

Enantiomer	HSV-1		HSV-2	VZV	CEM[b]
	BSC-1/ELISA	Vero/Plaque	Vero/Plaque	HFF/Plaque	
34a *R*	37	>50	50	22[c]	>100
35a *S*	8.8	16 (33)[d]	35	56	65
34b *R*	>100	>50	>50	24	70
35b *S*	45	>50	20	12	96
34c *R*	>100	>50	>50	>429	>100
35c *S*	45	50	>50	55	90
34d *R*	>100	>50	>50	>431	>100
35d *S*	>100	>50	>50	68	>100
34e *R*	>100	>50	>50	>404	>100
35e *S*	20	23 (23)[d]	16	10	98
34f *R*	>100	>50	>50	291	95
35f *S*	20	15 (17)[d]	22	12	70

[a]   Data were taken from Qiu *et al.*, 1998a and 2000a.
[b]   Cytotoxicity, IC_{50} (μM). For controls see Table 5 and 6.
[c]   The EC_{50} values of both enantiomers were 1.5 μM in HEL cells.
[d]   HSV-1 strain deficient in thymidine kinase.

Against HBV, enantiomers of synadenol **34a** and **35a** were almost equipotent (Table 13). A similar order of potency was also seen with enantiomeric 2,6-diaminopurines **34d** and **35d**. Only the *R*-enantiomer **34a** was effective against HIV-1. A report (Cheng *et al.*, 1998) that the *S*-enantiomer **35a** is active against HIV-1 was based on an erroneous assignment of the absolute configuration as shown by Qiu *et al.* (1998a). Interestingly, an *R*-selective anti-HIV effect was also observed with adenallene (**1a**) although at a somewhat higher activity level (Megati *et al.*, 1992). Nevertheless, the latter analogue was ineffective against HBV (Zhu *et al.*, 1997). The rest of tested enantiomers were inactive against both HBV and HIV-1.

3.1.2. *Analogues 4, 5, 6a - 9a, 6b - 9b, 10 and 11*

Adenine analogues **4**, **5**, **10** and **11** where the distance between the nucleic acid base and hydroxymethyl group (Table 1) is similar to that in the *(Z)*-methylenecyclopropane analogues **2** or *E*-isomers **3**, were ineffective in any of the antiviral assays mentioned

Table 13.　Inhibition of EBV, HBV and HIV-1 replication by enantiomers of (Z)-methylenecyclopropane analogues of purine nucleosides (EC$_{50}$, μM)[a].

Enantiomer		H-1	EBV Daudi[b]	VCA-IF[d]	HBV 2.2.15	HIV-1 CEM-SS[c]
		DNA hybridization				
34a	*R*	0.09	<0.37	3	16	11[e]
35a	*S*	0.63	1	7.8	9	>100[e]
34b	*R*	27	6.4	5.2	>10	>100
35b	*S*	1.1	0.76[f]	1.6	>10	>100
34c	*R*	41	29	40	>10	>100
35c	*S*	0.47	0.34	2.3	>10	>100
34d	*R*	2.3	12	12	10	>100
35d	*S*	1.3	60	0.69	5	>100
34e	*R*	15	5.3	1.3	>20	>100
35e	*S*	0.44	1.0	3.8	>10	42
34f	*R*	49	1.3	61	>10	>100
35f	*S*	0.66	0.85	6.2	>10	65

[a]　Data were taken from Qiu *et al.*, 1998a and 2000a.

[b]　Cytotoxicities (IC$_{50}$) in stationary Daudi cells were >100 μM unless stated otherwise. For cytotoxicities in CEM cells see Table 12. For controls see Table 6 and 7.

[c]　Reverse transcriptase assay (see Table 7).

[d]　Viral capsid antigen immunofluorescence assay.

[e]　Cheng *et al.*, 1998 give EC$_{50}$ 13 (**34a**) and 50 μM (**35a**) in MT-2 cells. The assignment of absolute configuration of **34a** and **35a** is in error as shown by Qiu *et al.*, 1998a.

[f]　IC$_{50}$ 35 μM.

above. Nevertheless, only two pairs of the *Z,E*-isomers comprising an adenine ring were compared in each case and examination of analogues with other nucleic acid bases may be worthwhile. In contrast, antiviral activity was observed (Guan *et al.*, 2000a; Zemlicka *et al.*, 2000) with some spiropentane analogues (Table 14). Thus, *proximal* and *medial-syn* isomers **6a** and **7a** were effective against HCMV or MCMV but at significantly lower level than synadenol (**2a**, Table 4). Analogue **6a** was also effective against EBV but some cytotoxicity was apparent. A potent anti-EBV activity of the *distal* analogue **9b** well separated from cytotoxicity is worthy of note. This was the only guanine analogue with antiviral potency in the spiropentane series. It is interesting to note that a significant anti-EBV activity was also observed with the *E*-isomers **3a** and **3g** where the N^9 --- CH$_2$OH distance approximates that of **9b** (Table 1). However, adenine analogue **9a** was devoid of any antiviral effect.

3.2. Antitumor activity

Several of the methylenecyclopropane analogues were tested as potential antitumor agents (Qiu *et al.*, 1998b,c). In a clonogenic assay (Phadtare *et al.*, 1991), synadenol (**2a**) and synguanol (**2c**) inhibited murine leukemia L 1210 culture with IC_{50} 50 and 40 μM, respectively. Syncytol (**2g**) was also moderately effective (IC_{50} 50 μM). The less active *E*-isomers **3a**, **3c** and **3g** exhibited IC_{50} >100, 55 and 85 μM. In a disk diffusion assay with mouse solid tumors C-38, M-17/Adr and human tumor H-116 none of the analogues was active at levels lower than 500 μg/disk. Inhibition of KB cells, a human cell line derived from epidermoid oral carcinoma, was used as one of the cytotoxicity control tests in antiviral assays. From the tested analogues, only synadenol (**2a**, Qiu *et al.*, 1998b) and the *S*-enantiomer (**35a**, Qiu *et al.*, 2000a) had IC_{50} values lower than 100 (78 and 80 μM, respectively).

3.3. Substrate activity toward adenosine deaminase

Adenine analogues **2a**, **3a**, **4**, **5**, **6a**, **7a**, **8a**, **9a**, **10** and **11** were examined as substrates for adenosine deaminase from calf intestine (Qiu *et al.*, 1998b,c, 2000b; Guan *et al.*, 2000a,b). The use of enzymic deamination for resolution of synadenol (**2a**) was already described in Section 2.1.1 (Scheme 6). At best, all these analogues are moderate substrates. The *Z*(cis)-isomers are less reactive than *E*(trans)-isomers confirming thus the previously observed trend (Phadtare *et al.*, 1991; Xu *et al.*, 1995) which is opposite to that of antiviral activities. Thus, the order of the substrate activities is as follows: **1a** > **3a** > **5** > **2a** > **8a**. The *Z*-isomer **4**, spiropentane analogues **6a**, **7a**, **9a** and methylenecyclobutanes **10**, **11** are inactive. As already mentioned, the 2,6-diaminopurine analogue **2d** is not a substrate (Qiu *et al.*, 1998c) although the corresponding allene derivative is deaminated (Phadtare *et al.*, 1991). Enantioselectivity of deamination and anti-HIV effect (Qiu *et al.*, 1998a and 1999a) of synadenol (**2a**) follow the pattern found for adenallene (**1a**) (Megati *et al.*, 1992). Thus, both *S*-enantiomers **35a** and **83** are preferentially deaminated but the *R*-enantiomers **34a** and **82** are anti-HIV agents.

Table 14.   Antiviral activity of spiropentane analogues of nucleosides (EC_{50}/IC_{50}, μM)[a].

Compound		HCMV/HFF[b]	EBV/Daudi[c]
6a	(*proximal*)	28/>100	4.8/15 (0.95)
7a	(*medial-syn*)	20/>100[d]	22/>202 (0.61)
9b	(*distal*)	>100/>100	6.0/>199 (12)

[a]   Data were taken from Guan *et al.*, 2000a. Inactive compounds and viruses where no antiviral effect was seen are not listed. For controls see Table 4 and 6.

[b]   Towne strain of virus, plaque reduction assay.

[c]   VCA-IF or ELISA assay. Data in parentheses refer to inhibition of EBV DNA synthesis (IC_{50} values).

[d] EC_{50}/IC_{50} 40/>404 in AD 169 strain and 10.5/>404 in MCMV/MEF.

4. Structure-activity relationships

A set of analogues comprising all four DNA bases was investigated only in case of allenes **1** (Zemlicka, 1993 and 1997) and methylenecyclopropanes **2** and **3**. The antiviral effect of spiropentane adenine and guanine analogues **6a** - **9a** and **6b** - **9b** was also studied, whereas investigations of "reversed" methylenecyclopropanes **4**, **5** and methylenecyclobutanes **10**, **11** have been thus far limited to adenine derivatives. Inspite of this, some conclusions about the structure-activity relationships can be drawn.

The original design (Phadtare and Zemlicka, 1989) of allenic analogues was based on a similarity of intramolecular distances between the nucleic acid base and hydroxymethyl group of nucleosides in the $C_{3'}$-endo conformation (4.24 Å, Table 1). A similar distance (4.04 Å) is found in the methylenecyclopropane analogue **2a** derived by replacement of the "distal" double bond of **1a** with a cyclopropane ring. In the (Z)-methylenecycloalkane analogues **4**, **10** and spiropentanes **7a**, **8a** these distances vary between 3.76 and 4.40 Å. All these values correspond to the N^9 --- $C_{5'}$ distance (4.25 Å) found for the $C_{3'}$-endo conformation (Van Roey and Chu, 1993) of nucleosides. The N^9 --- $CH_2(OH)$ distance of the *E*-analogues **3a**, **5**, **11** and *distal* isomer **9a** are appreciably longer (4.63 - 5.01 Å) whereas that of the *proximal* spiropentane **6a** is significantly shorter (3.54 Å) than that observed in the Z-isomers. The latter distance coincides with that found for the $C_{3'}$-exo conformation of nucleosides. The distance factor satisfactorily explains the strong antiviral potency of allenes **1a**, **1b** and that of a series of the (Z)-methylenecyclopropanes **2a** - **2f** (regardless of the virus and base selectivity) and a general lack of activity of the *E*-isomers. Nevertheless, some caution must be exercised because analogues **3a** and **3g** exhibit quite a potent anti-EBV and -VZV effect (Table 6) although not in all assays. A total lack of potency of the "reversed" (Z)-methylenecyclopropane **4** and methylenecyclobutane **10** indicate that the distance criterion alone is not satisfactory for explaining the antiviral activity in this series of analogues. It appears that relatively minor structural variations can have detrimental effect on the antiviral potency. Thus, expansion of the *distal* cyclopropane ring of **2a** to cyclobutane in analogue **10** as well as replacement of the *proximal* double bong of **1a** with a cyclopropane ring (analogue **4**) led to a complete loss of antiviral potency. However, analogues carrying DNA bases other than adenine have not been studied. Interestingly, replacement of both double bonds in adenallene (**1a**) gave analogues with antiviral activity such as the *proximal* and *medial-syn* isomers **6a** and **7a**. Probably the most surprising is a potent anti-EBV effect (Table 14) of the *distal* guanine isomer **9b** where the N^9 --- $CH_2(OH)$ distance exceeds that of all *E*-isomers in Table 1.

Rotational restrictions of the nucleic acid base may also play a role in the antiviral activity of analogues **1a** and **2a** - **2f**. Crystal structures of (*R*)-(-)-adenallene (**82**, Megati *et al.*, 1992) and (*R*)-(-)-synadenol (**34a**, Figure 2) indicate an *anti*-like conformation of the base in both cases. The NMR spectra (Table 3) suggest an impairment of rotation of adenine base in synadenol (**2a**) but less so in adenallene (**1a**). In this aspect, analogue **2a** is more "nucleoside-like" than allene **1a**. It is possible that the latter factor may at least partly be responsible for a significantly broader antiviral effect of **2a**. Nevertheless, the NMR spectra suggest a similar rotational limitation in (inactive) methylenecyclobutane **10**.

H. Kumamoto, et al.

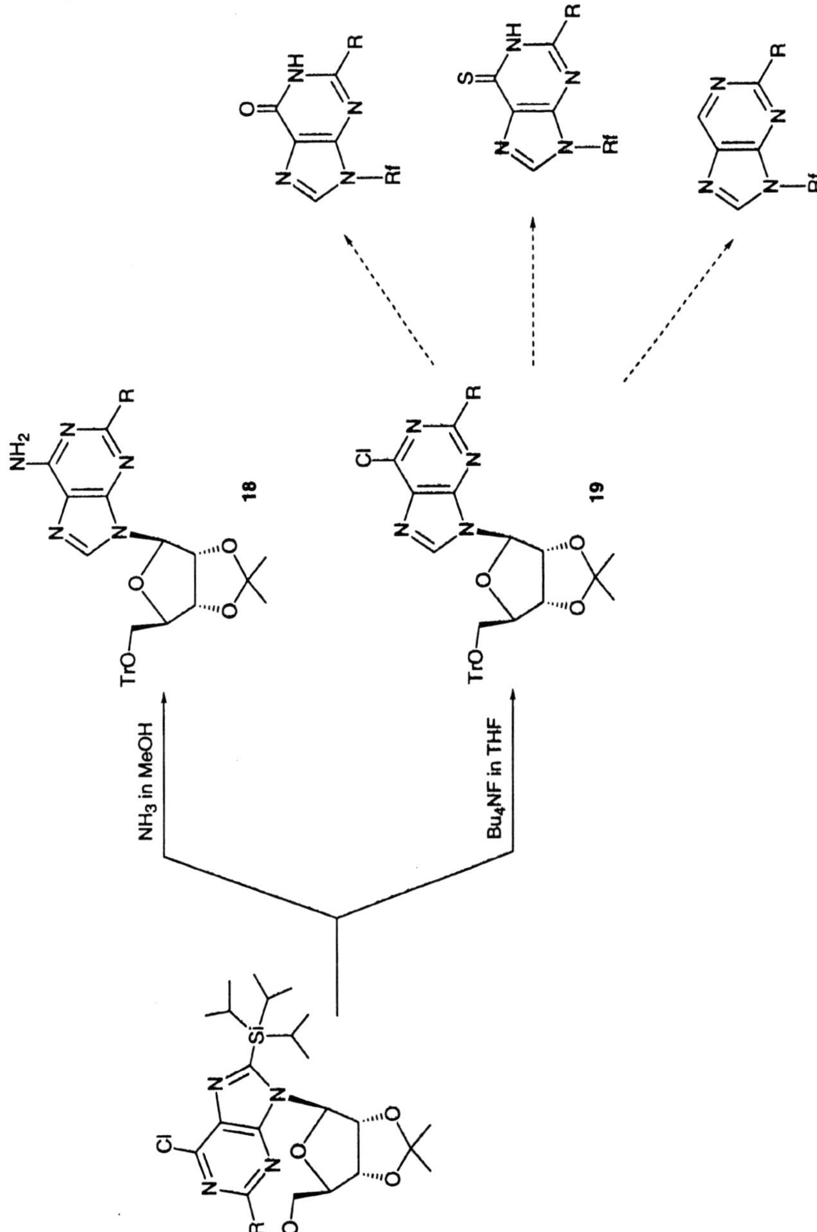

Figure 8. Removal of the 8-TIPS group (Rf = 1-β-D-ribofuranosyl or protected 1-β-D-ribofuranosyl).

lithiated species was reacted with Bu_3SnCl or Me_3SiCl, the expected 6-substituted derivative **21** was accompanied by the formation of the 2',6-bis-substituted product **22** (Figure 9). Although a precedent suggests that lithiation of the β-position of alkyl vinyl ethers occurs only when the anion is stabilized by an adjacent halogen atom (Lau *et al.*, 1978), the observed formation of **22** was initially considered simply as the consequence of further deprotonation at the 2'-position of **21** followed by reaction with the added electrophile.

The LDA lithiation-based deuteration of **21** (X = $SnBu_3$) was carried out to see if C2'-lithiation is actually involved in the formation of **22** (X = $SnBu_3$). Quite unexpectedly, this reaction gave the rearranged product **23** (X = $SnBu_3$), in which a 96% deuterium incorporation was observed at the 6-position (Kumamoto *et al.* 2000). The recovered **21** from this reaction, on the other hand, showed no incorporation of deuterium at the 2'-position. These results clearly suggest that deprotonation at the 2'-position of **21** was followed by an instantaneous migration of the stannnyl group from the 6-position. Under optimized reaction conditions wherein LTMP was used as a lithiating agent in the presence of HMPA, an 87% yield of the stannyl migration was accomplished.

That the silyl version of this anionic migration is also operative and that these migration reactions take place in an intramolecular manner were confirmed by the cross reaction between the 6-stannyl derivative (**21**) and the 6-silyl derivative (**24**) which have mutually different hydroxyl-protecting groups (Figure 10). Although the yield of the 2'-silyl derivative (**25**) was low due to desilylation during chromatography, there was no cross reaction product formed in this reaction.

7.   Synthesis of 2'-substituted 1',2'-unsaturated uridines

Manipulation of the 2'-stannyl group of 22 provides an entry to an hitherto unknown class of nucleosides, the 2'-substituted 1',2'-unsaturated uridines, as shown in Figure 11. Although fluorination of **22** under the reported conditions (Tuis 1995) gave a rather unsatisfactory yield of the product, other halogenation reactions proceeded in high yields. Introduction of carbon substituents can be effected by applying the Stille reaction (Mitchell 1992) to **22**. In cases where organotin reagents are available, the Stille reaction of the 2'-iodo derivative (**26**) is recommended.

It was found that UV absorption maxima of these 2'-substituted derivatives (λ_{max} in MeOH: 251-255 nm) appeared uniformly at much shorter wavelength (λ_{max} in MeOH: 251-255 nm) than that of the 2'-unsubstituted 1',2'-unsaturated uridine (**20**, λ_{max} in MeOH: 276 nm). Since the λ_{max} of **20** indicates some degree of its coplanar disposition between the base and the glycal portion, the above finding of the 2'-substituted derivatives may be rationalized by considering that the 2'-substituent, which is *ortho* to the N1-C1' pivot bond, would impede such coplanarity. If the reported instability of **20** during deprotection originates from its coplanar conformation, which might render the molecule to undergo aromatization at the glycal portion, then these 2'-substituted derivatives would be expected to be stable under desilylation conditions.

The above expectation turned out to be the case. When desilylation of the 2'-substituted derivatives was carried out with NH_4F in refluxing MeOH (Zhang *et al.*,

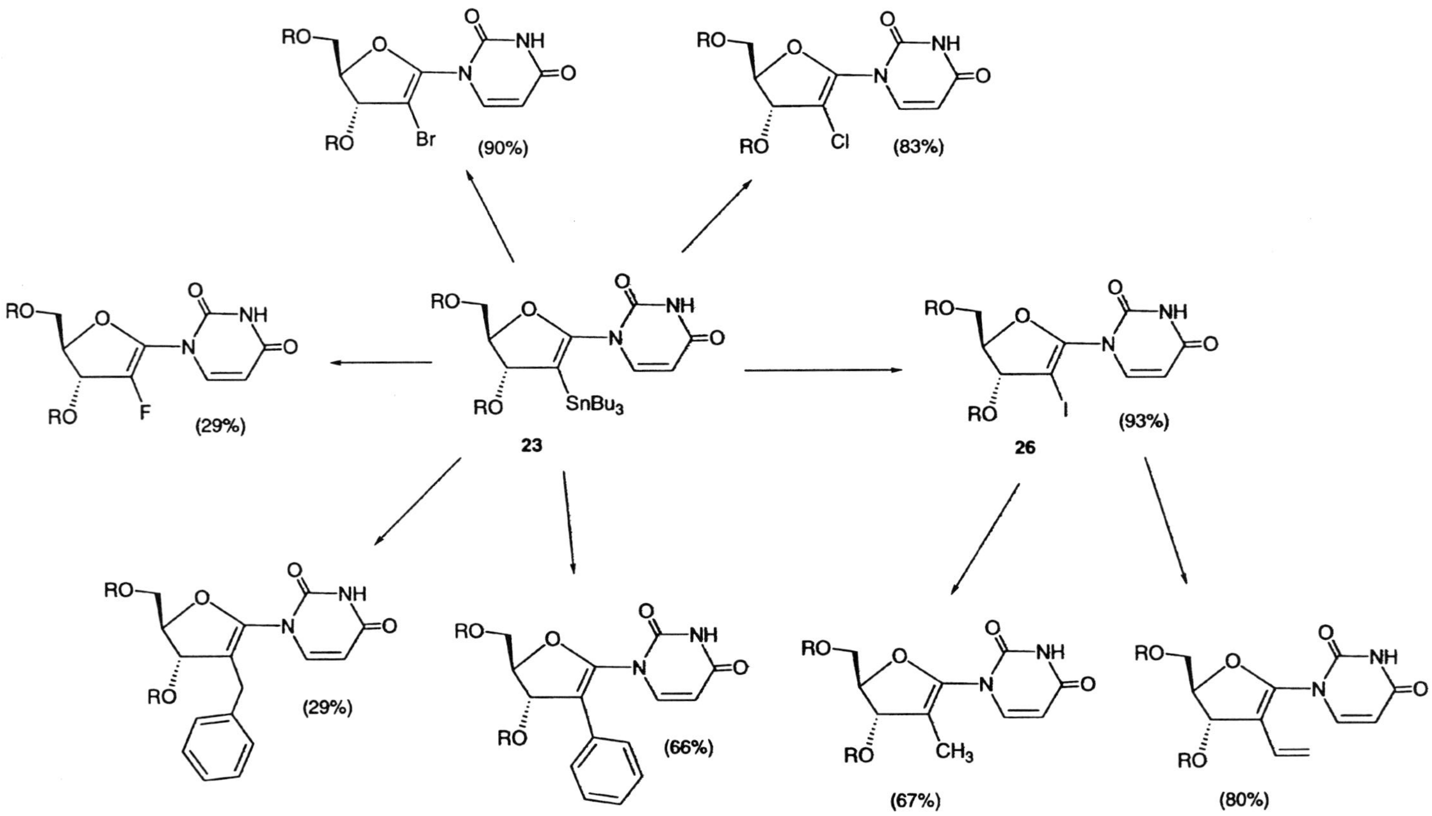

Figure 11. Synthesis of 2'-substituted 1',2'-unsaturated uridines (R = *tert*-butyldimethylsilyl). Yields of products are given in parentheses.

1992), the respective diols were formed uniformly in high yields. Formation of uracil or furan derivatives was not detected in any appreciable extent. On the other hand, under these desilylation conditions, the 2'-unsubstituted compound **20** formed none of the free nucleoside but gave the corresponding furan derivative and uracil.

8.   Conclusion

As described in this review, use of silicon- and tin-electrophiles in the lithiation of nucleosides has offered new aspects to the lithiation chemistry in this field. Thus, in LTMP-lithiation of 9-(2,3,5-tris-*O*-*tert*-butyldimethylsilyl-β-D-ribofuranosyl)-6-chloropurine (**1**), the initially introduced 8-tributylstannyl group enabled generation of the C2-lithiated species, and it was eventually transferred to the 2'-position. The resulting 2-stannylated product (**8**) and its adenosine counterpart (**14**) serve as useful intermediates for the preparation of 2-halogeno- and some 2-carbon-substituted derivatives.

An alternative lithiation-based C2-substitution of 6-chloropurine riboside has become feasible by introducing a triisopropylsilyl group to the 8-position of 9-(2,3-*O*-isopropylidene-5-*O*-trityl-β-D-ribofuranosyl)-6-chloropurine. Lithiation of this substrate (**17**) constitutes the first evident generation of C2-lithiated species of purine nucleosides. Addition of various types of electrophiles to the lithiated species of **17** has provided a more general entry to 2-substituted derivatives. Removal of the 8-triisopropylsilyl group can be effected either with NH_3 in MeOH, which gives the corresponding adenosine derivative, or with Bu_4NF. Since the latter method leaves the 6-chlorine atom intact, it would be possible by further displacement at the 6-position to prepared 2-substituted analogues of inosine, nebularine, and 6-mercaptopurine derivatives.

An anionic migration of silyl and stannyl groups was also observed in the lithiation of 1-[3,5-bis-*O*-(*tert*-butyldimethylsilyl)-2-deoxy-D-*erythro*-pent-1-enofuranosyl]uracil (**20**), this time from the 6-position of the base to the 2'-position of its glycal portion. The observed migration is again due to prior deprotonation of the more acidic H-6 and subsequent silylation or stannylation. The presence of the 6-silyl and stannyl group enabled generation of the 2'-lithiated species, which undergoes instantaneous migration from the 6-position. Manipulation of the 2'-stannyl group in **23** disclosed an entry to the 2'-halogeno- and 2'-carbon-substituted analogues for the first time.

9.   References

Adah SA, Nair V. (1995) Truflic enolates in the palladium-mediated synthesis of complex ethynyl adenosines. Tetrahedron Lett.; 36: 6371-6372.

Barton DHR, Hedgecock CJR, Lederer E, Motherwell WB. (1979) A direct method for the alkylation of adenosine nucleosides at position 8. Tetrahedron Lett.; 279-280.

Fraser RR, Baignée A, Bresse M, Hata K. (1982) Lithium dialkylamides. ^{13}C-Parameters and slow proton transfer. Tetrahedron Lett.; 23: 4195-4198.

Gschwend HG, Rodriguez HR, (1979) Heteroatom-facilitated lithiations. In: Dauben WG editor-in-chief. Organic Reactions. New York: John Wiley and Sons, Inc.; Vol. 26, p. 1-360.

Hayakawa H, Tanaka H, Maruyama Y, Miyasaka T. (1985) Regioselectivity in the lithiation of uridine. Effect of the sugar protecting groups. Chem. Lett.; 1401-1404.

Hayakawa H, Tanaka H, Obi K, Itoh M, Miyasaka T. (1987a) A simple entry to 5-substituted uridines based on the regioselective lithiation controlled by a protecting group in the sugar moiety. Tetrahedron Lett.; 28: 87-90.

Hayakawa H, Haraguchi K, Tanaka H, Miyasaka T. (1987b) Direct C-8 lithiation of naturally occurring purine nucleosides. A simple method for the synthesis of 8-carbon-substituted purine nucleosides. Chem. Pharm. Bull.; 35: 72-79.

Itoh Y, Haraguchi K, Tanaka H, Gen E, Miyasaka T. (1995) Divergent and stereocontrolled approach to the synthesis of uracil nucleosides branched at the anomeric position. J. Org. Chem.; 60: 656-662.

Kato K, Hayakawa H, Tanaka H, Kumamoto H, Shindoh S, Shuto S, Miyasaka T. (1997) A new entry to 2-substituted purine nucleosides based on lithiation-mediated stannyl transfer of 6-chloropurine nucleosides. J. Org. Chem.; 62, 6833-6841.

Kelly TR, Lang F. (1995) Synthesis of thiazole compounds via lithiation. An unexpected rearrangement. Tetrahedron Lett.; 36: 9293-9296.

Kumamoto H, Hayakawa H, Tanaka H, Shindoh S, Kato K, Miyasaka T, Endo K, Machida H, Matsuda A. (1998) Synthesis of 2-alkynylcordycepins and evaluation of their vasodilating activity. Nucleoside Nucleotides; 17: 15-27.

Kumamoto H, Tanaka H, Tsukioka R, Ishida Y, Nakamura A, Kimura S, Hayakawa H, Kato K, Miyasaka T. (1999) First evident generation of purin-2-yllithium. Lithiation of an 8-silyl-protected 6-chloropurine riboside as a key step for the synthesis of 2-carbon-substituted adenosines. J. Org. Chem.; 64: 7773-7780.

Kumamoto H, Shindoh S, Tanaka H, Itoh Y, Haraguchi K, Gen E, Kittaka A, Miyasaka T, Kondo M, Nakamura KT. (2000) An intramolecular migration of a stannyl group from the 6-position of 1-(2-deoxy-D-erythro-pent-1-enofuranosyl)uracil to the 2'-position. Synthesis of 2'-substituted 1',2'-undsaturated uridines. Tetrahedron; 56: 5363-5371.

Lau KSY, Schlosser M. (1978) (Z)-2-Ethoxyvinyllithium. A remarkably stable and synthetically useful 1,2-counterpolarized species. J. Org. Chem.; 43: 1595-1598.

Maeda M, Nushi K, Kawazoe Y. (1974) *C*-Alkylation of purine bases through free radical process catalyzed by ferrous ion. Tetrahedron; 30: 2677-2682.

Matsuda A, Nomoto Y, Ueda T. (1979) Synthesis of 2- and 8-cyanoadenosines and their derivatives. Chem. Pharm. Bull.; 27: 183-192.

Matsuda A, Shinozaki M, Miyasaka T, Machida H, Abiru T. (1985) Palladium-catalyzed cross-coupling of 2-iodoadenosine with terminal alkynes. Synthesis and biological activities of 2-alkynyladenosines. Chem. Pharm. Bull.; 33: 1766-1769.

Matsuda A, Shinozaki M, Yamaguchi T, Homma H, Nomoto R, Miyasaka T, Watanabe Y, Abiru T. (1992) 2-Alkynyladenosine, a novel class of selective adenosine A2-receptor agonists with potent antihypertensive effects. J. Med. Chem.; 35: 241-252.

Mitchell TN. (1992) Palladium-catalyzed reactions of organotin compounds. Synthesis; 803-815.

Nair V, Richardson SG. (1982) Modification of nucleic acid bases via radical intermediates. Synthesis of dihalogenated purine nucleosides. Synthesis; 670-672.

Nair V, Young DA. (1985) A new synthesis of isoguanosine. J. Org. Chem.; 50: 406-408.

Nair V, Lyons AG. (1990) Hypoxanthine nucleoside counterparts of the antibiotic cordycepin. Tetrahedron; 46: 7677-7692.

Nair V, Purdy DF. (1991) Synthetic approaches to new doubly modified nucleoside. Congeners of cordycepin and related 2'-deoxyadenosine. Tetrahedron; 47; 365-382.

Pereyre M, Quintard J-P, Rahm A. (1987) Tin in Organic Synthesis. London: Butterworth.

Pichat L, Godbillon J, Herbert M. (1973) Lithiations directes par le *n*-butyllithium d'uracile et thymine nucléosides isilyés. Méthylation de ces lithiens. Préparation de thymine-(D-6). Bull. Chim. Soc. Fr.; 2715-2719.

Robins MJ, Trip EM. (1974) 1-(2-Deoxy-D-*erythro*-pent-1-enofuranosyl)uracil. Synthesis of the first 1',2'-unsaturated pyrimidine nucleoside, a furanoid *N,O*-ketene acetal. Tetrahedron Lett.; 3369-3372.

Saenger W. (1984) Principles of Nucleic Acid Structure. New York: Springer-Verlag.

Sakamoto T, Uchiyama D, Kondo Y, Yamanaka H. (1993) 1,3-Dipolar cycloaddition reaction of substituted trimethylstannylacetylenes with nitrile oxides. Chem. Pharm. Bull.; 41: 478-480.

Sarma RH, Lee C-H, Evans FE, Yathindra N, Sundaralingam M. (1974) Probing the interrelation between the glycosyl torsion, sugar pucker, and backbone conformation in C(8)-substituted adenine nucleotides by 1H and 1H-31P fast Fourier transform nuclear magnetic resonance methods and conformational energy calculations. J. Am. Chem. Soc.; 96: 7337-7348.

Shimizu M, Tanaka H, Hayakawa H, Miyasaka T. (1990) Dynamic aspect in the LDA lithiation of an arabinofuranosyl derivative of 4-ethoxy-2-pyrimidinone. Regioselective entry to both C-5 and C-6 substitutions. Tetrahedron Lett.; 31: 1295-1298.

Srivastava PC, Robins RK, Meyer Jr. RB. (1988) Synthesis and properties of purine nucleosides. In: Townsend LB, editor. Chemistry of Nucleosides and Nucleotides. New York: Plenum Press; Vol. 1, p. 113-281.

Suzuki M, Tanaka H, Miyasaka T. (1987) Synthesis of 5-carbon-substituted 1-β-D-ribofuranosylimidazole-4-carboxamides via lithiation of a primary carboxamide. Chem. Pharm. Bull.; 35: 4056-4063.

Tanaka H, Nasu I, Miyasaka T. (1979) Regiospecific *C*-alkylation of uridine. A simple route to 6-alkyluridines. Tetrahedron Lett.; 4755-4758.

Tanaka H, Hayakawa H, Miyasaka T. (1982) Umpolung of reactivity at the C-6 position of uridine. A simple and general method for 6-substituted uridines. Tetrahedron; 38: 2635-2642.

Tanaka H, Uchida Y, Shinozaki M, Hayakawa H, Matsuda A, Miyasaka T. (1983a) A simplified synthesis of 8-substituted purine nucleosides via lithiation of 6-chloro-9-(2,3-*O*-isopropylidene-β-D-ribofuranosyl)purine. Chem. Pharm. Bull.; 31: 787-790.

Tanaka H, Matsuda A, Iijima S, Hayakawa H, Miyasaka T. (1983b) Synthesis and biological activities of 5-substituted 6-phenylthio and 6-iodouridines, a new class of antileukemic nucleosides. Chem. Pharm. Bull.; 31: 2164-2167.

Tanaka H, Hayakawa H, Iijima S, Haraguchi K, Miyasaka T. (1985) Lithiation of 3',5'-(tetraisopropyldisiloxane-1,3-diyl)-2'-deoxyuridine. Synthesis of 6-substituted 2'-deoxyuridines. Tetrahedron; 41: 861-866.

Tanaka H, Hirayama M, Suzuki M, Miyasaka T, Matsuda A, Ueda T. (1986a) A lithiation route to C-5 substitution of an imidazole nucleoside and its application to the synthesis of 3-deazaguanosine. Tetrahedron; 42: 1971-1980.

Tanaka H, Hayakawa H, Obi K, Miyasaka T. (1986b) Synthetic route to 5-substituted uridines *via* a new type of desulfurizative stannylation. Tetrahedron; 42: 4187-4195.

Tanaka H, Hayakawa H, Shibata S, Haraguchi K, Miyasaka T. (1992) Synthesis of 6-methyluridine via palladium-catalyzed cross-coupling between a 6-iodouridine derivative and tetramethylstannane. Nucleosides Nucleotides; 11: 319-328.

Tanji K, Kato H, Higasino T. (1991) Halogen-metal exchange reaction of 5-halo-3*H*-1,2,3-triazolo[4,5-d]pyrimidines with butyllithium. Chem. Pharm. Bull.; 39: 3037-3040.

Tuis M. (1995) Xenon difluoride in synthesis. Tetrahedron; 51: 6605-6634.

Ueda T. (1988) Synthesis and reaction of pyrimidine nucleosides. In: Townsend LB, editor. Chemistry of Nucleosides and Nucleotides. New York: Plenum Press; Vol. 1, p. 1-112.

Zhang W, Robins MJ. (1992) Removal of silyl protecting groups from hydroxyl functions with ammonium fluoride in methanol. Tetrahedron Lett.; 33: 1177-1180.

PURINE METABOLISM IN PARASITES: POTENTIAL TARGETS FOR CHEMOTHERAPY

MAHMOUD H. EL KOUNI

Department of Pharmacology and Toxicology,
Comprehensive Cancer Center, Center for AIDS Research,
University of Alabama at Birmingham, Birmingham, AL 35294, USA

1. Introduction

Parasitic diseases are the foremost worldwide health problem today, particularly in the under developed countries. It is estimated that the global prevalence of some of these diseases already exceeds 60% among the more than three billion people living in parasite endemic areas. Parasitic diseases are not confined to humans but also affect many domestic and wild animals causing an enormous economical blight to already poor countries and societies. In spite of the present alarming health and economic proportions of parasitic infections, these diseases are still on the rise largely because of poor sanitation and health education, inadequate control of measures, greater use of irrigation for agricultural development, increase and redistribution of world population, increased world travel, and development of resistance to drugs used for chemotherapy or control of vectors. In addition, with the recent advent of AIDS, several parasitic diseases which previously did not constitute a major threat to human health emerged as causative agents of lethal opportunistic infections (e.g. toxoplasmosis, cryptosporidiosis). Some of the parasitic diseases such as malaria cannot be ignored because of the high mortality rate. Malaria causes the death of over two million children every year. Most parasitic diseases, however, like *Ascaris* or *Ancylostoma* infections, remain neglected because their effects on human health are more subtle.

At the present time, chemotherapy is still the main stay to control most parasitic diseases, since antiparasitic vaccines are not yet available. Nevertheless, the need for new drugs is crucial to prevent or combat some major parasitic infections, such as trypanosomiasis, as there is still no single effective way of controlling this disease, or because some serious parasitic infections, such as malaria, developed resistance to presently available drugs.

Most of the currently available antiparasitic drugs have been discovered empirically by screening of large number of compounds for efficacy against parasites in animal models. Few of these drugs have been rationally designed. This is largely because, until recently, little was known about the basic biochemistry, physiology, and molecular biology of parasites and of their interactions with their hosts. The rational design of a drug is usually based on biochemical and physiological differences between pathogens and host. Some of the most striking differences between parasites and their mammalian

Recent Advances in Nucleosides: Chemistry and Chemotherapy, Ed. by C.K. Chu. 377 — 415

host are found in purine metabolism. Therefore, purine metabolism constitutes an excellent potential target for the rational design of antiparasitic chemotherapeutic regimens.

2.   Purine metabolism

Purines are of vital importance to all living organisms. They are essential for the synthesis of nucleic acids, proteins, other metabolites as well as for energy requiring reactions. In view of their high rate of replication or oviproduction, parasites require very active nucleic acid synthesis which necessitates large supplies of the indispensable purine nucleotides. In general, purine nucleotides can be synthesized by the *de novo* and/or the so called "salvage" pathways. The *de novo* pathway utilizes simple compounds for the synthesis of the various purine nucleotides. The salvage pathways, on the other hand, are reutilization routes by which the cell can satisfy its purine requirements from endogenous and/or exogenous sources of preformed purines. In a striking contrast to their host, all studied parasites, with the exception of the nematodes *Angiostrongylus cantonensis* [Wong and Ko, 1979] and *Metastrongylus apri* [Wong and Yeung, 1981], as well as the protozoa *Acanthamoeba polyphaga* [Ogbunude and Baer, 1993a], *A. astronyxis* and *A. castellani* [Hassan and Coombs, 1986a], are incapable of *de novo* purine biosynthesis and depend on the salvage pathways to meet their purine requirements [Perrotto *et al.*, 1971; Marr *et al.*, 1978; Boonlayangoor *et al.*, 1980; Berens *et al.*, 1981; Wang and Simashkevich, 1981; Gutteridge and Davies, 1981; Senft and Crabtree, 1983; Fish *et al.*, 1982a and b; Schwartzman and Pfefferkorn, 1982; Miller and Lindstead, 1983; Wang *et al.*, 1983; Wang CC and Aldritt, 1983; LaFon and Nelson, 1985; Dovey *et al.*, 1984]. In addition, because of the great phylogenic separation between the host and the parasite, there are in some cases sufficient distinction between corresponding enzymes of the purine salvage from the host and the parasite that can be exploited to design specific inhibitors or "subversive substrates" for the parasitic enzymes. Furthermore, the specificities of purine transport, the first step in purine salvage, have been shown to diverge significantly between parasites and their mammalian host. Therefore, transporters and enzymes responsible for the salvage of purines in parasites constitute excellent potential targets for chemotherapy against these organisms.

Parasites can be selectively deprived of purines by blocking or disrupting their purine uptake and salvage with one or more of the numerous available purine analogues, synthesized for the treatment of cancer and viral infections. Analogues which particularly inhibit transporters or enzymes involved in purine salvage metabolism in parasites are potential chemotherapeutic agents with a high degree of selectivity. Alternatively, analogues which are selectively transported and/or preferentially metabolized to toxic nucleotides in the parasite would act as "subversive substrates" and exhibit selective toxicity against parasites. In both cases, the lack of *de novo* purine biosynthesis and the total dependancy on the salvage of purines, would render the parasite specially sensitive to either strategy and would result in selective toxicity against parasites. However, few attempts were made to exploit these striking differences between parasites and their host. Nevertheless, the feasibility of using one or more of the already chemotherapeutic purine analogues which perturb nucleotide metabolism as potential antiparasitic agents was demonstrated in the treatment of some parasitic diseases including malaria [Gati *et al.*, 1987; Gero *et al.*, 1988 and 1989], schistosomiasis [Jaffe, 1975; Senft and Crabtree, 1983; el Kouni *et al.*, 1983, 1985, 1987 and 1989; Baer *et al.*, 1988; el Kouni, 1991], trypanosomiasis [Avila and Avila, 1981; Ogbunude and Ikediobi, 1982b; Avila *et al.*, 1983; Berens *et al.*, 1984], leishmaniasis [Berman *et al.*, 1983; Peters *et al.*, 1980; Walton *et al.*, 1983] and toxoplasmosis [Pfefferkorn and Pfefferkorn, 1976; Luft, 1986; el Kouni *et al.*, 1999].

In this review we will attempt to discuss the broad aspects of the differences between the specificities of the transport, metabolic pathways and enzymes involved in the metabolism of purines in parasites as compared to their mammalian host. Elucidation of differences between purine metabolism in parasites and their host not only contributes to the general knowledge of metabolism in the parasite under study, but may also assist in revealing potential targets for the treatment of parasitic diseases with one or more of the already available chemotherapeutic purine analogues. The importance of differences in transport and salvage metabolism of purine between the parasite and its hosts was highlighted by the discovery that potent nucleoside transport inhibitors of mammalian systems do not significantly inhibit the uptake of nucleosides in Schistosomes [el Kouni *et al.*, 1983 and el Kouni and Cha, 1987]. Based on these findings, a successful antischistosomal chemotherapeutic regimen was developed. The highly toxic purine analogues, tubercidin (7-deazaadenosine) and nebularine were made selectively toxic against *Schistosoma mansoni*, *S. japonicum* and *S. haematobium* by simultaneous administration of a nucleoside transport inhibitor as an antidote for adverse toxic effects to the host but not the parasite [el Kouni *et al.*, 1983, 1985, 1987 and 1989; Baer *et al.*, 1988; el Kouni, 1991].

The general pathways of purine salvage and the enzymes responsible for the different steps as currently known are shown in Figure 1. Because of the absence of *de novo* biosynthesis, most parasites depend mainly on one or two enzymes of the purine salvage pathways to satisfy their purine requirements. In the following are examples of crucial differences in the salvage metabolism of purines between parasites and their host that could serve as potential targets for the development of antiparasitic chemotherapeutic regimens.

3. Enzymes of purine salvage

3.1. Purine phosphoribosyltransferases (EC 2.4.2.-)

The phosphoribosyltransferase reactions involve the ribophosphorylation in one step of purine nucleobases such as hypoxanthine, guanine, adenine, xanthine and their analogues to their respective nucleoside 5'-monophosphate as follows:

Purine nucleobase + PRPP $\rightarrow$ Purine nucleoside 5'-monophosphate + PP_i

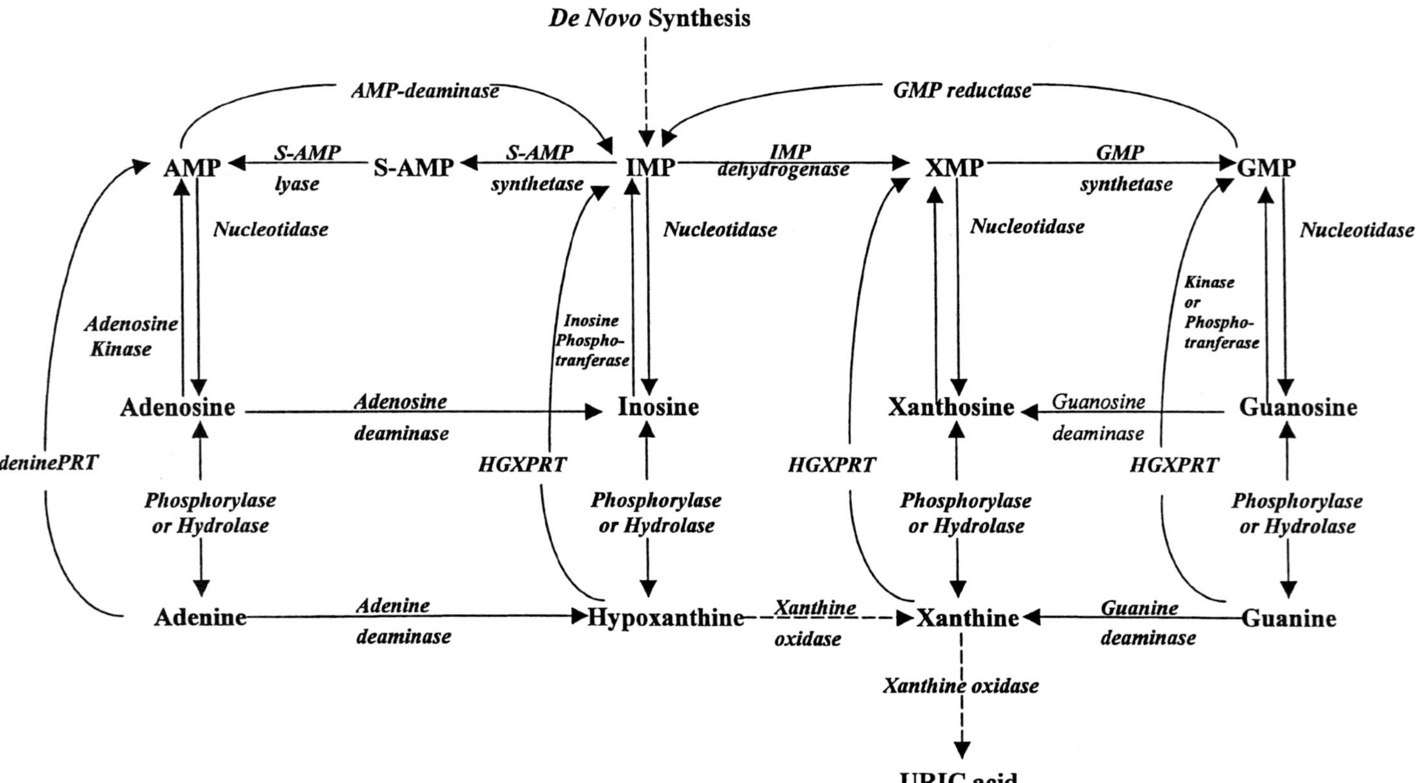

Figure 1.   Pathways of purine salvage in parasites. Dashed lines indicate reactions that are absent or could not be established.

Purine phosphoribosyltransferases are found in various parasites. However, as shown in Table 1, the specificity of these enzymes differs among the different species. In all cases studied, activity towards adenine is distinct from that towards hypoxanthine, guanine or xanthine. Therefore the enzyme is called adenine phosphoribosyltransferase (APRT). When activities towards hypoxanthine and guanine are attributed to a single enzyme, the enzyme is named hypoxanthine-guanine phosphoribosyltransferase (HGPRT) as in *Schistosoma mansoni* [Senft and Crabtree, 1983; Dovey *et al.*, 1984]. In other parasites the activity towards xanthine is attributed to one enzyme specific to xanthine (XPRT) as in *Leishmania donovani* [Tuttle and Krenitsky, 1979] or associated with the HGPRT activity and the enzyme is called hypoxanthine- guanine-xanthine phosphoribosyltransferase (HGXPRT) as in *Toxoplasma gondii* [Pfefferkorn and Borotz, 1994; Naguib *et al.*, 1995; Donald *et al.*, 1996]. It should be noted, however, that it is not yet clear how many xanthine phosphoribosyltransferases toxoplasma has. In contrast to other investigations [Pfefferkorn and Borotz, 1994; Donald *et al.*, 1996], kinetic and structure-activity relationship studies [Naguib *et al.*, 1995] suggest that the phosphoribosyltransferase activity towards xanthine in *Toxoplasma gondii* may be carried out by two different isoenzymes. The discrepancy between the results from the different studies could be due to post-translational modifications. The definitive proof of this suggestion must wait until the isolation and characterization of purine phosphoribosyltransferases from *T. gondii* are accomplished.

Trichomonas vaginalis [Heyworth *et al.*, 1982] appears to lack all purine phosphoribosyltransferases, as they depend on the purine nucleoside phosphorylase/kinase pathway for the salvage of purines (Figure 1) while *Entamoeba histolytica* seems to lack phosphoribosyltransferase activity towards guanine [Lo and Wang, 1985; Hassan and Coombs, 1986b].

In several of the parasites that are naturally (e.g *Tritrichomonas foetus* [Wang *et al.*, 1983)] or artificially (e.g. *Leishmania donovani* [Iovannisci *et al.*, 1984]) deficient in APRT, adenine is deaminated by adenine deaminase [EC 3.5.4.2] to hypoxanthine which is then anabolized to the nucleotide level by HGXPRT or HGPRT, as illustrated in Figure 1, depending on the species of the parasite. On the other hand, parasites which have APRT (e.g *Trypanosomes* [Berens *et al.*, 1981; Fish *et al.*, 1982a; Gutteridge and Davies, 1981]) usually lack adenine deaminase activity.

The potential of the parasitic purine phosphoribosyltransferases as a target for chemotherapy stems from the crucial role these enzymes play in the salvage of purines in parasites. For example in *Toxoplasma gondii,* all four purine nucleobases (i.e. xanthine, hypoxanthine, guanine, and adenine) and nucleosides other than adenosine (i.e. guanosine, inosine and xanthosine) are incorporated into the nucleotide pool *via* the phosphoribosyltransferase reactions. Neither kinase nor nucleoside phosphotransferase activities (Figure 1) were detected towards inosine, guanosine or xanthosine [Krug *et al.*, 1989]. Instead, all three nucleosides are first cleaved to their respective nucleobases by purine nucleoside phosphorylases [Krug *et al.*, 1989] or hydrolases [F.N.M. Naguib and M.H. el Kouni, unpublished results] and then converted to nucleotides by the phosphoribosyltransferases [Krug *et al.*, 1989]. Another example of the importance of phosphoribosyltransferases in the salvage of purines can be illustrated by enzymatic studies in *Plasmodium falciparum.* When the optimum assay conditions

Table 1.   Nomenclature and substrate specificity of purine phosphoribosyltransferases in parasites

Enzyme	EC Number	Substrate specificity	Parasite	Reference
APRT	2.4.2.7	Adenine	*Babesia divergens*	Hassan *et al.* (1987)
			Eimeria tenella	Wang and Simashkevich (1981)
			Entamoeba histolytica	Hassan and Coombs (1986b)
			Giardia Lamblia	Wang and Aldritt (1983)
			Leishmania donovani	Tuttle and Krenitsky (1979), Allen *et al.* (1989)
			Leishmania mexicana	Hassan and Coombs (1986c)
			Leishmania tarentolae	Hassan and Coombs (1986c)
			Plasmodium falciparum	Reyes *et al.* (1982), Pollack *et al.* (1985)
			Schistosoma mansoni	Senft and Crabtree (1983), Dovey *et al.* (1984)
			Toxoplasma gondii	Krug *et al.* (1989), Naguib *et al.* (1995)
			Trypanosoma brucei	Ogbunude and Ikediobi (1983), Hassan and Coombs (1986c)
			Trypanosoma congolense	Ogbunude and Ikediobi (1983)
			Trypanosoma cruzi	Berens *et al.* (1981), Gutteridge and Davies (1981)
			Trypanosoma vivax	Ogbunude and Ikediobi (1983)
GPRT	2.4.2.-	Guanine	*Giardia Lamblia*	Wang and Aldritt (1983), Aldritt and Wang (1985)
XPRT	2.4.2.22	Xanthine	*Entamoeba histolytica*	Hassan and Coombs (1986b)
			Leishmania donovani	Tuttle and Krenitsky (1979)
			Trypanosoma brucei	Fish *et al.* (1982a and b)
HGPRT	2.4.2.8	Hypoxanthine, Guanine	*Babesia divergens*	Hassan *et al.* (1987)
			Leishmania donovani	Tuttle and Krenitsky (1979), Allen *et al.* (1989)
			Schistosoma mansoni	Senft and Crabtree (1983), Dovey *et al.* (1984)
			Trypanosoma brucei	Fish *et al.* (1982a and b), Allen and Ullman (1993)
			Trypanosoma cruzi	Gutteridge and Davies (1981)

Table 1. continued.

Enzyme	EC Number	Substrate specificity	Parasite	Reference
HGXPRT	2.4.2.8	Hypoxanthine, Guanine, Xanthine	*Cryptosporidium parvum*	Doyle *et al.* (1998)
			Eimeria tenella	Aldritt and Wang (1985)
			Giardia Lamblia	Wang and Aldritt (1983), Aldritt and Wang (1985)
			Plasmodium falciparum	Reyes *et al.* (1982); Queen *et al.* (1988)
			Toxoplasma gondii	Krug *et al.* (1989), Pfefferkorn and Borotz (1994), Naguib *et al.* (1995), Donald *et al.* (1996)
			Tritrichomonas foetus	Wang *et al.* (1983)

for measurement of the major reactions of the salvage pathways were established, kinetic studies were performed to determine the efficiency of these major enzymes. The K_m, V_{max}, and enzyme efficiency (V_{max}/K_m) are shown in Table 2. The results in Table 2 indicate that the major enzyme reactions in the purine salvage pathway in *P. falciparum* are HGXPRT (V_{max}/K_m = 0.24) and purine nucleoside phosphorylase (V_{max}/K_m = 0.041). These results support the suggestion [Reyes *et al.*, 1982] that the major route of purine salvage in *P. falciparum* is *via* phosphoribosyltransferases.

Table 2.   Kinetic parameters of major enzymes of purine salvage pathways in *Plasmodium falciparum*.

Enzyme	Substrate	K_m [M]	V_{max} (nmol/min/mg protein)	V_{max}/K_m (min^{-1})
Phosphoribosyltransferases	Xanthine	315. ± 71.4	74.6 ± 13.4	0.240
	adenine	35.5 ± 9.56	0.12 ± 0.01	0.003
nucleoside phosphorylase	Inosine	266. ± 8.68	10.8 ± 0.25	0.041
	Adenosine	73.0 ± 0.01	1.06 ± 0.05	0.015
Adenosine deaminase	Adenosine	51.2 ± 11.1	0.70 ± 0.06	0.014
Adenosine kinase	Adenosine	<5	0.05 ± 0.01	<.01

An important factor in considering the phosphoribosyltransferase reaction towards xanthine as a potential target for chemotherapy is the fact that xanthine is not a substrate for the mammalian host phosphoribosyltransferases [Reyes *et al.*, 1982; F.N.M. Naguib and M.H. el Kouni MH, unpublished results], but is a good substrate for phosphoribosyltransferases from several parasites including *Entamoeba histolytica* [Hassan and Coombs, 1986b], *Toxoplasma gondii* [Pfefferkorn and Borotz, 1994; Naguib *et al.*, 1995; Donald *et al.*, 1996], *Eimeria tenella* [Wang and Simashkevich, 1981], *Leishmania donovani* [Marr *et al.*, 1978; Tuttle and Krenitsky, 1979; Hwang and Ullman, 1997], *Plasmodium falciparum* [Reyes *et al.*, 1982; Queen *et al.*, 1988], *Trypanosoma cruzi* [Berens *et al.*, 1981], *T. brucei* [Fish *et al.*, 1982a and b] and *Tritrichomonas foetus* [Wang *et al.*, 1983]. In addition, it should be noted that the conversion of xanthine to its nucleotide may be the only metabolic fate of this compound in some parasites. It has been shown that, unlike its host, *Entamoeba histolytica* [Hassan and Coombs, 1986b], *Toxoplasma gondii* [Krug *et al.*, 1989], *Eimeria tenella* [LaFon and Nelson, 1985], *Leishmania donovani* [Tuttle and Krenitsky, 1979; Allen and Ullman, 1994], *Plasmodium falciparum* [Reyes *et al.*, 1982], *Trypanosoma cruzi* [Berens *et al.*, 1981; Hammond and Gutteridge, 1984], *Trichomonas vaginalis*

[Miller and Lindstead, 1983], and probably *Tritrichomonas foetus* [Wang and Aldritt, 1983] have no xanthine oxidase (EC 1.2.3.2) activity towards xanthine or hypoxanthine (see Figure 1). This characteristic of xanthine metabolism is more likely not confined to these parasites but shared by other pathogenic protozoa.

These characteristics of xanthine metabolism in parasite can be exploited to develop antiparasitic drugs. Xanthine analogues may inhibit the parasitic phosphoribosyltransferases selectively and interfere with the purine salvage in the parasite but not the host. The selective disruption of the purine salvage pathways in such parasites by analogues that interfere specifically with XPRTase would be enhanced further by the common lack of *de novo* purine biosynthesis in these organisms. However, mutants deficient in HGXPRT are not lethal [Pfefferkorn and Borotz, 1994; Donald *et al.*, 1996]. This indicates that the parasites can utilize purines by other routes (e.g. kinase reaction, Figure 1). Therefore, the use of xanthine analogues as "subversive substrates" activated to toxic nucleotides only by the parasites would be a preferable strategy than inhibition of the enzyme.

It should also be mentioned that HGPRT and HGXPRT from Leishmania and trypanosomes [Tuttle and Krenitsky, 1979; Marr, 1983; Fish *et al.*, 1985] as well as from *Eimeria tenella* [Wang and Simashkevich, 1981] differ from the host HGPRT in their substrate specificities as they use several pyrazolopyrimidines (7-deaza-8-azapurines) efficiently as substrates and ultimately incorporate them into their nucleic acids causing the selective death of the parasite [Marr, 1983; Fish *et al.*, 1985]. Among the active pyrazolopyrimidine substrates, depicted in Figure 2, allopurinol was tested against visceral leishmaniasis and showed moderate results [Kager *et al.*, 1981]. Unfortunately, subsequent clinical trials were inconclusive.

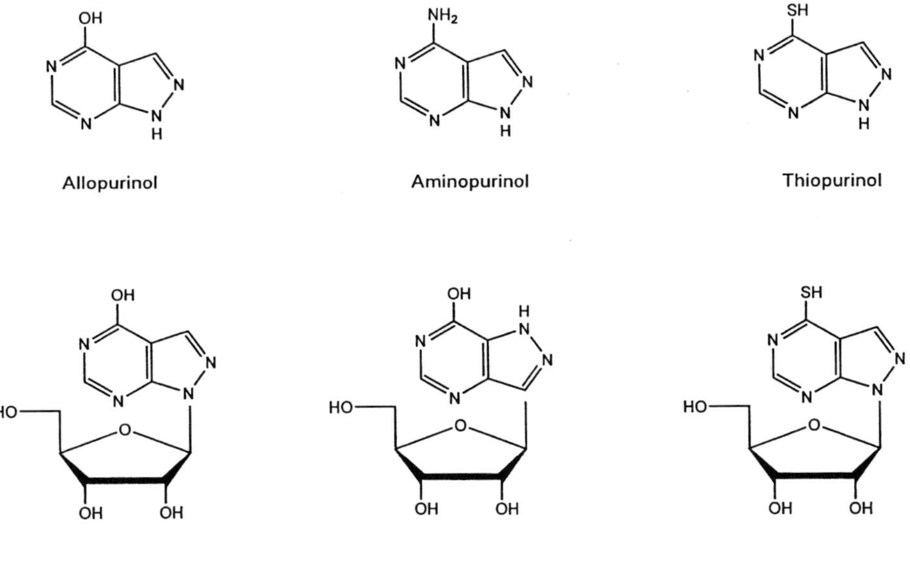

Figure 2.　Chemical structures of various pyrazolopyrimidines.

Because of the potential of the purine phosphoribosyltransferases as a target for chemotherapy in parasites, these enzymes were purified, cloned, expressed, crystallized and studied from different parasites including *Toxoplasma gondii* [Donald *et al.*, 1996; Schumacher *et al.*, 1996], *Eimeria tenella* [Wang and Simashkevich, 1981], *Leishmania donovani* [Allen *et al.*, 1989, 1995a and b; Jardim *et al.*, 1999; Phillips *et al.*, 1999], *Trypanosoma brucei* [Allen and Ullman, 1994], *T. cruzi* [Eakin *et al.*, 1997], *Plasmodium falciparum* [Pollack *et al.*, 1985; Reyes *et al.*, 1982; Queen *et al.*, 1988; Vasanthkumar *et al.*, 1990], *Tritrichomonas foetus* [Beck and Wang, 1993; Chin and Wang, 1994; Kanaaneh *et al.*, 1996 and 1997; Munagala *et al.*, 1998a and b; Somoza *et al.*, 1998; Pitera *et al.*, 1999; Aronov *et al.*, 2000], *Giardia lamblia* [Aldritt and Wang, 1985; Page *et al.*, 1999; Sommer *et al.*, 1999; Shi *et al.*, 2000], *Schistosoma mansoni* [Kanaaneh *et al.*, 1997; Craig III *et al.*, 1991; Yuan *et al.*, 1993; Kanaaneh *et al.*, 1994 and 1995] for the detailed characterization of the enzyme. The ultimate long term goal of these studies is to use molecular modeling so that ideal ligands can be designed as recently has been accomplished for the enzyme from *Tritrichomonas foetus* [Somoza *et al.*, 1998; Aronov *et al.*, 2000].

3.2. Adenosine kinase (EC 2.7.1.20)

This enzyme phosphorylate adenosine in the presence of ATP to adenosine 5'-monophosphate as follows:

$$\text{Adenosine} + \text{ATP} \rightarrow \text{AMP} + \text{ADP}$$

Adenosine kinase is found in *Eimeria tenella* [Wang and Simashkevich, 1981; Miller *et al.*, 1982], *E. acervulina, E. brunetii* [Miller *et al.*, 1982], *Entamoeba histolytica* [Lobee-Rich and Reeves, 1983; Hassan and Coombs, 1986b], *Leishmania donovani* [Iovannisci and Ullman, 1984; Datta *et al.*, 1987], *L. mexicana, L. tarentolae* [Hassan and Coombs, 1986b], *Babesia divergens* [Hassan *et al.*, 1987], *Metastrongylus cantonensis* [Wong and Ko, 1980], *Plasmodium falciparum* [Reyes *et al.*, 1982 and Table 2], *Schistosoma mansoni* [Senft and Crabtree, 1983; Dovey *et al.*, 1984; el Kouni *et al.*, 1983; el Kouni and Cha, 1987] *Trichomonas vaginalis* [Miller and Lindstead, 1983], *Tritrichomonas foetus* [Wang *et al.*, 1983], *Toxoplasma gondii* [Pfefferkorn and Pfefferkorn, 1978; Krug *et al.*, 1989; Iltzsch *et al.*, 1995; el Kouni *et al.*, 1999], *Trypanosoma cruzi* [Gutteridge and Davies, 1981], *T. congolense* and *T. vivax* [Ogbunude and Ikediobi, 1983], but absent or displaying very low activity in *T. brucei* [Ogbunude and Ikediobi, 1983] and *Giardia lamblia* [Berens and Marr. 1986].

The relative activity and importance of adenosine kinase as compared to other enzymes in the salvage pathways differs between the various parasites. Its activity is low in *Schistosoma mansoni* [Senft and Crabtree, 1983] but very high in *Toxoplasma gondii* [Pfefferkorn and Pfefferkorn, 1978; Krug *et al.*, 1987]. It is present in *Plasmodium falciparum* but does not have the major role in the salvage of adenosine by this parasite [Reyes *et al.*, 1982, and Table 2]. Adenosine kinase activity seems also related to the developmental stage of the parasite. The enzyme is 50-fold more active in *Leishmania donovani* amastigotes than in promastigotes [Looker *et al.*, 1983].

Like with other enzymes, the prospect of using adenosine kinase as a target for chemotherapy in parasites depends on its role in the salvage of purines, the activity of other enzymes which may interfere with that role, as well as how its substrate specificity differs from that of its counterpart in the host. A good example is the nature of *Toxoplasma gondii* adenosine kinase and its role in purine metabolism in this parasitic protozoa.

The adenosine kinase reaction is the main route of adenosine metabolism in *T. gondii* [Pfefferkorn and Pfefferkorn, 1978; Schwartzman and Pfefferkorn, 1982; Krug *et al.*, 1987]. It is the most active enzyme in the purine salvage pathways in this parasite. This results in preferential incorporation of adenosine into adenine nucleotides by at least a 10-fold higher rate than that of any other purine precursor tested [Krug *et al.*, 1987]. This also contrasts sharply with most mammalian cells, where adenosine is predominantly deaminated to inosine, which is then cleaved to hypoxanthine by adenosine deaminase (EC 3.5.4.4) and purine nucleoside phosphorylase (EC 2.42.1), respectively. Neither of these two enzymes is significantly active in toxoplasma [Krug *et al.*, 1987]. The high activity of adenosine kinase along with the unique characteristics of adenosine metabolism in toxoplasma make adenosine kinase an excellent target for chemotherapy. However, deficiency of adenosine kinase was shown not to be lethal to the parasites [Pfefferkorn and Pfefferkorn, 1978], indicating that inhibition of the enzyme will not lead to toxicity in toxoplasma. Therefore, attempts were made to search for analogues that can be used as "subversive" substrates and activated to toxic nucleotides selectively by *T. gondii,* but not the host, adenosine kinase. Preferential metabolism of such subversive substrates to toxic nucleotides in the parasites should lead to selective toxicity against toxoplasma.

Therefore, a complete structure-activity relationships study was carried out to identify compounds which could be substrates for adenosine kinase from *T. gondii* [Iltzsch *et al.*, 1986]. Various 6-substituted 9-β-D-ribofuranosylpurines were among the best ligands that bind to *T. gondii* adenosine kinase [Iltzsch *et al.*, 1986]. This finding is quite unusual since the compounds are not known to be active substrates for mammalian adenosine kinase. Furthermore, studies with toxoplasma in culture indicated that some of these compounds, the structure of which are shown in Figure 3, are selectively toxic to the parasite but not their host cells [el Kouni *et al.*, 1999 and Table 3]. Biochemical, metabolic and genetic studies on the fate of two of these 6-substituted 9-β-D-ribofuranosylpurines, namely nitrobenzylthioinosine (NBMPR) and 6-benzylthioinosine, demonstrated that toxoplasma can metabolize NBMPR and 6-benzylthioinosine to the nucleotide level by adenosine kinase [el Kouni *et al.*, 1999 and Table 3]. In the case of NBMPR, its nucleotide, NBMPR 5'-monophosphate appears to be the toxic agent since it is the only unique metabolite found in wild type parasites and mutants deficient in HGXPRT but not in adenosine kinase deficient mutants [el Kouni *et al.*, 1999]. Table 3 also shows that 6-benzylthioinosine is only toxic against parasites which contains adenosine kinase. These results emphasize the important role of adenosine kinase in purine metabolism in toxoplasma and clearly demonstrate that *T. gondii* adenosine kinase is different from its counter part in the host. Therefore, *T. gondii* adenosine kinase represents an excellent potential target for chemotherapy and for the development of chemotherapeutic regimens for the treatment of toxoplasmosis

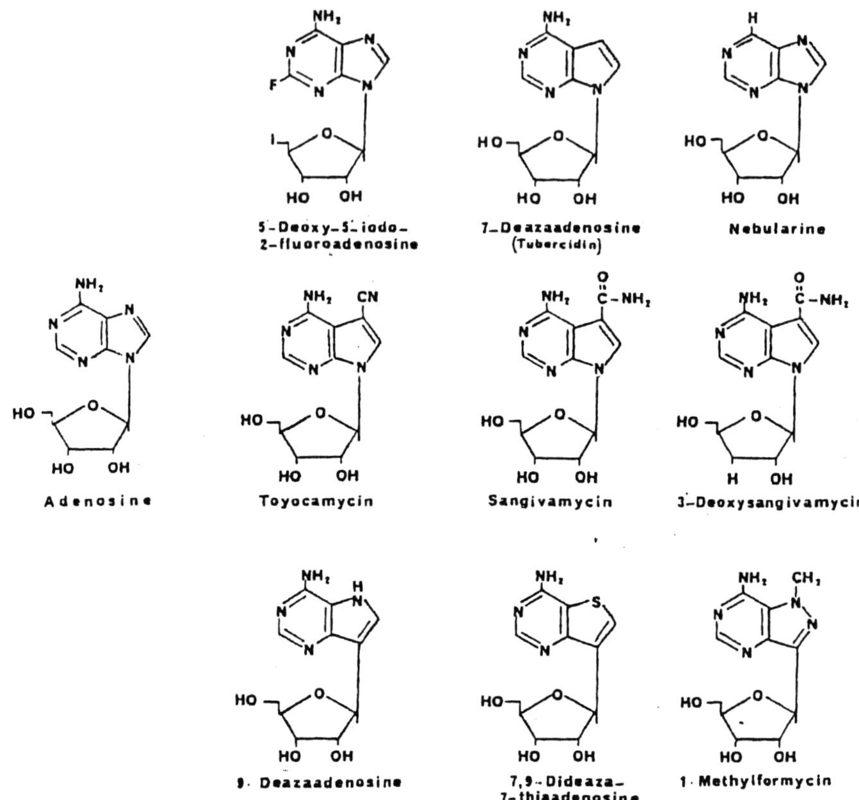

Figure 4. Chemical structures of adenosine and various analogues.

Previous experience dictates that similar analogues may not be metabolized in the same fashion. For example, it was found that schistosomes metabolize and incorporate tubercidin, 9-deazaadenosine, 7,9-dideaza-9-thiaadenosine, 2-fluoro-5'-deoxy- 5'-iodoadenosine, toyocamycin and nebularine but not 1-methylformycin, sangivamycin or 3'-deoxysangivamycin into their nucleotide pool, in spite of the fact that all these adenosine analogues are taken by the parasites [el Kouni et al., 1983; el Kouni and Cha, 1987]. The selectivity of schistosomes adenosine kinase in metabolizing these adenosine analogues irrespective of the similarity in their structures as shown in Figure 4, demonstrates the necessity of testing and evaluating the metabolism of each of the promising nucleoside analogues before a general conclusion can be made.

3.3. Purine nucleoside phosphotransferases (EC 2.7.1.77)

These enzymes are similar to the kinases as they phosphorylate nucleosides to their respective nucleosides 5'-monophosphate. The difference is that the kinase reaction require a high energy phosphate ester (i.e. ATP) while the phosphotransferase reaction utilizes a low energy phosphate ester (e.g. AMP, *p*-nitrophenylphosphate, etc.) as follows:

$$\text{Purine nucleoside}_1 + \text{Purine nucleoside}_2 \text{ 5'-monophosphate} \rightarrow$$
$$\text{Purine nucleoside}_1 \text{ 5'-monophosphate} + \text{Purine nucleoside}_2$$

Purine nucleoside phosphotransferases are found in *Tritrichomonas foetus* [Wang *et al.*, 1983], *Giardia lamblia* [Berens and Marr, 1986] and Leishmania species [Nelson *et al.*, 1979] but absent from *Trichomonas vaginalis* [Miller and Lindstead, 1983]. The enzyme from *Tritrichomonas Foetus* appears to be specific to guanosine [Wang *et al.*, 1983]. The *Leishmania donovani* and *Giardia lamblia* phosphotransferases also differ from one another with regard to the specificity of the phosphate donor. The Leishmania phosphotransferase uses *p*- nitrophenylphosphate better than either AMP or GMP while the reverse is true for the Giardial enzyme [Nelson *et al.*, 1979; Carson and Chang, 1981]. The Giardial and Leishmania nucleoside phosphotransferases are also unique in their substrate specificities. The Giardial enzyme cannot phosphorylate adenosine but can phosphorylate its 1-, 7- and 9-deaza- as well as 8-aza-analogues [Berens and Marr, 1986]. The Leishmania nucleoside phosphotransferase phosphorylates natural purine nucleosides as well as several pyrazolopyrimidine (7-deaza-8-azapurines) nucleoside analogues such as allopurinol riboside and formycin B (shown in Figure 2) to their respective pyrazolopyrimidine 5'-monophosphates [Nelson *et al.*, 1979; Carson and Chang, 1981]. These nucleotides are either converted to the triphosphate level and incorporated into nucleic acids of Leishmania or act as inhibitors of other important enzymes of the purine salvage pathways leading to the selective death of the parasites. Therefore, allopurinol riboside was tested against cutaneous leishmaniasis [Saenz *et al.*, 1989]. However, as the case with allopurinol against visceral leishmaniasis [Kager *et al.*, 1981], the results were unimpressive.

Attempts were later made to design new subversive substrates of the Leishmania nucleoside phosphotransferase by modification of the pentose moiety of the nucleoside. Carbocyclic-, 3'-deoxy- and 3'-deoxy-3'-flouroinosine analogues (Figure 5) were synthesized and proved to be substrates for Leishmania nucleoside phosphotransferase [Hiraoka *et al.*, 1986]. Test of these compounds showed selective antileishmanial activity *in vitro* and in animal models [Hiraoka *et al.*, 1986; Morshige *et al.*, 1995].

3.4.　5'-Methylthioadenosine/adenosine phosphorylase (MTA/Ado phosphorylase, EC 2.4.2.28)

This enzyme phosphorolytically cleaves the glycosidic bond of adenine containing nucleosides in the following manner:

$$\text{(5'-Deoxy- 5'-methylthio)adenosine} + P_i \rightarrow \text{Adenine} +$$
$$\text{(5'-Deoxy-5'-methylthio)--\text{D}-ribose-1-}P$$

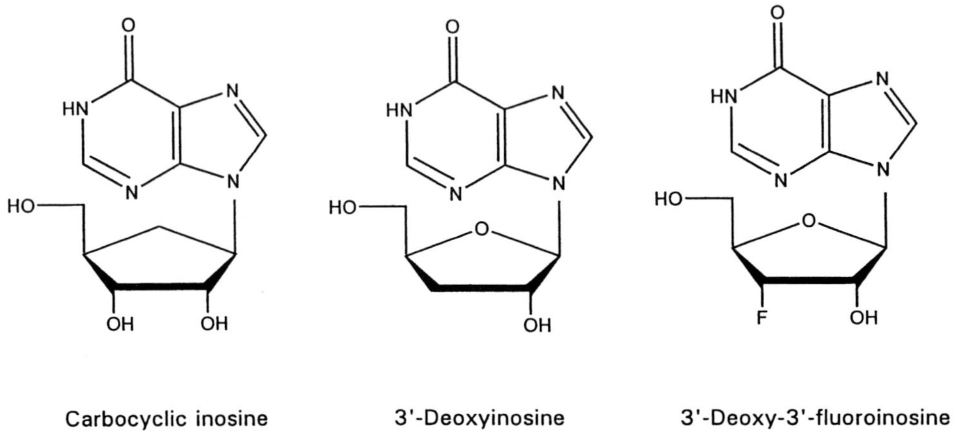

Carbocyclic inosine   3'-Deoxyinosine   3'-Deoxy-3'-fluoroinosine

Figure 5.   Chemical structures of various inosine analogues which are substrates of phosphotransferase from *Trypanosoma cruzi*

The enzyme is found in *Leishmania donovani* [Koszalka and Krenitsky, 1986], *Schistosoma mansoni* [Savarese *et al.*, 1989], *Trypanosoma cruzi* [Miller *et al.*, 1987] and *T. brucei* [Ghoda *et al.*, 1988] but absent from *Giardia lamblia*, *Plasmodium falciparum* and *Entamoeba invadens* [Riscoe *et al.*, 1989].

MTA/Ado phosphorylase from the parasites shares some characteristics with the mammalian enzyme but has some important differences. In common with the mammalian enzyme, the parasite MTA/Ado phosphorylase has a low K_m for the preferred substrate 5'-deoxy- 5'-methylthioadenosine, and tolerates a wide range of substitutions in the 5'-position of the nucleoside. However, unlike the mammalian MTA/Ado phosphorylase, the parasite enzyme has a relatively low K_m for adenosine and analogues with 2'-deoxy- or 2',3'-dideoxyribose moiety, and to a lesser extent with 3'-deoxyadenosine analogues [Ghoda *et al.*, 1988]. It can also use adenosine analogues with the 6-amino group substituted with a methyl group [Ghoda *et al.*, 1988]. Therefore, it is assumed that toxic purine nucleoside analogues containing for example 2',3'-dideoxyribose moieties can be designed as selectively toxic subversive substrates against parasites which have MTA/Ado phosphorylase (e.g. trypanosomes). Such subversive substrates would release the toxic purine nucleobase which can be then metabolized to the nucleotide level by APRT. In this respect, it should be noted that in species such as *Trypanosoma brucei* the rate of adenine uptake is the highest among all purines tested [Fish *et al.*, 1982a]. However, it has to be kept in mind that for this strategy to work and achieve selective toxicity, the designed adenosine analogue should neither be a substrate for the adenosine kinase nor inactivated by adenosine deaminase in the host (Figure 1). In addition, the released adenine analogue should be a good substrate for the parasite APRT and its nucleotide toxic to the parasite. It was proposed that 2',3'-dideoxy-2-flouroadenosine may fulfill these requirements and act as an antitrypanosomal [Ghoda *et al.*, 1988] or antischistosomal [Savarese *et al.*, 1989] agent.

It should also be mentioned here that 5'-deoxy-5'-methylthioadenosine, the preferred substrate of MTA/Ado phosphorylase, is a byproduct of polyamine biosynthesis as shown in Figure 6. Hence, inhibiting polyamine biosynthesis would block the generation of 5'-deoxy-5'-methylthioadenosine and subsequently enhance the utilization and toxicity of the subversive substrates of MTA/Ado phosphorylase. In this context, African trypanosomes are very sensitive to the inhibition of polyamine biosynthesis by the ornithine decarboxylase [EC 4.1.1.17] inhibitor D,L--difluoromethylornithine [Bacchi *et al.*, 1980, 1983 and 1987] and reduction of putrescine and spermidine synthesis. Therefore, the use of MTA/Ado phosphorylase subversive substrates in combination with D,L--difluoromethylornithine may prove synergistic against African trypanosomes [Ghoda *et al.*, 1988].

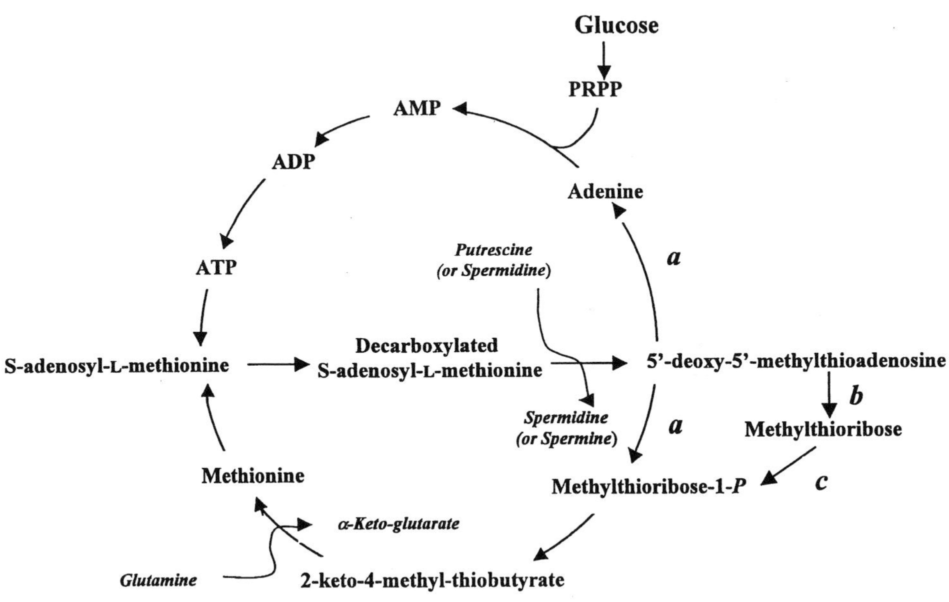

Figure 6.   Pathways of 5'-deoxy-5'-methylthioadenosine metabolism in parasites. *a*, 5'- methylthioadenosine/ adenosine phosphorylase; *b*, 5'-deoxy-5'-methylthioadenosine hydrolase; *c*, 5-methylthioribose kinase.

It is also known that 5'-deoxy-5'-methylthioribose-1-*P* is recycled to methionine and is an important source for methionine salvage [Schlenk and Ehninger, 1964; Sugimoto *et al.*, 1976; Shapiro and Schlenk, 1980; Shapiro and Barrett, 1981; Backlund Jr and Smith, 1981; Wang *et al.*, 1982; Yung *et al.*, 1982; Savarese *et al.*, 1983;]. Therefore, substrate analogues of 5'-deoxy- 5'-methylthioadenosine that release toxic methionine analogues (e.g. ethionine) may also be useful as antiparasitic drugs [Fitchen *et al.*, 1988; Riscoe *et al.*, 1989; Bacchi *et al.*, 1997].

not transport or metabolize inosine or its analogue formycin B [Boonlayangoor *et al.*, 1980; Lo and Wang, 1985; Das *et al.*, 1997].

Promastigotes of *Leishmania donovani* [Ogbunude *et al.*, 1981; Iovannisci *et al.*, 1984; Aronow *et al.*, 1987] and *L. major* [Baer *et al.*, 1992] have two non-overlapping substrate specific nucleoside transporters. The first is an adenosine/pyrimidine nucleoside transporter (ldNT1) and the second is an inosine/guanosine transporter (ldNT2) which are not inhibited by NBMPR or dipyridamole [Iovannisci and Ullman, 1984; Aronow *et al.*, 1987]. These two transporters have been cloned [Vasudevan *et al.*, 1998; Carter *et al.*, 2000]. In the amastigotes stage only ldNT1 is present while ldNT2 is replaced with a third stage-specific adenosine/purine nucleoside transporter (ldNT3*)* [Ghosh and Mukherjee, 2000]. This stage-specific expression of adenosine transporters is correlated with the increased requirement for adenosine by the amastigotes as evident by the 50-fold increase in their adenosine kinase activity over the promastigotes [Lookeret al., 1983]. This developmental regulation of the capacity and specificity of purine transporters is not particular to *Leishmania* as it is now well documented in trypanosomes [de Koning *et al.*, 2000] and *Plasmodium falciparum* [Carter *et al.*, 2000].

Adenosine is the best purine nucleoside taken up by *Trypanosoma congolense* and *T. brucei.* [James and Born, 1980]. The procyclic form of *T. brucei brucei* has a single adenosine/inosine carrier [de Koning *et al.*, 1998]. The blood form of *T. brucei* [Carter and Fairlamb, 1993; de Koning and Jarvis, 1999] and *T. equiperdum* [Barrett *et al.*, 1995], on the other hand, have two high affinity adenosine transporters (P1 and P2). P1 is an adenosine/inosine carrier and P2 is adenosine/adenine carrier [Carter and Fairlamb, 1993; de Koning and Jarvis, 1999]. P1 seems to be identical to the transporter of the procyclic form. Therefore, it is possible that *P1* is constitutively expressed in both forms of the parasite while *P2* is expressed only in the blood form [de Koning *et al.*, 1998]. The activity of these transporters seems to be regulated by purine availability and growth cycle [de Koning *et al.*, 2000].

An important difference between purine nucleoside transporters in trypanosomes and their counterparts in mammalian cells is that they are protonmotive force-driven transporters and are not inhibited by NBMPR, dilazep or dipyridamole at the concentration required to inhibit NBMPR-sensitive mammalian nucleoside transporters [de Koning *et al.*, 1998; de Koning and Jarvis, 1999]. Furthermore, the trypanosome nucleoside transporters, specially P1 has high affinity for toxic purine nucleoside analogues (e.g. tubercidin, ribavirin, Formycin A) which are poor ligands for the mammalian transporters [de Koning and Jarvis, 1999].

Trichomonas vaginalis has two separate nucleoside transporters. One is non-specific as it takes all purine and pyrimidine nucleosides [Harris *et al.*, 1988]. This carrier has a binding site for adenosine and pyrimidine nucleosides and a separate site for purine nucleosides. The second transporter takes adenosine, guanosine and uridine and has one binding site for adenosine and uridine, and a second separate binding site for guanosine. Neither of the two transporters is inhibited completely by NBMPR or dilazep at concentrations which would completely inhibit nucleoside transport in mammalian cells. It is interesting to note that the inhibition by NBMPR was competitive with similar inhibition constant to that of inhibitory nucleosides [Harris *et al.*, 1988]. This may suggest that NBMPR is transported by *Trichomonas vaginalis* as was observed in *Toxoplasma gondii* [el Kouni *et al.*, 1999; V. Guarcello and M.H. el Kouni, unpublished results]. Dilazep, on the other hand, displayed non-competitive inhibition suggesting that it binds to different site than the nucleoside binding site [Harris *et al.*, 1988].

In *Giardia lamblia* there is a single broad-specificity transporter for purine nucleosides which can also transport pyrimidine nucleosides [Davey *et al.*, 1992; Baum *et al.*, 1993]. The transporter requires the presence of 3'- but not 2'- or 5'-hydroxyl group for nucleoside transport [Davey *et al.*, 1992]. The transporter is not inhibited by NBMPR but by high concentrations of dipyridamole [Baum *et al.*, 1993].

Three adenosine transporters are reported in *Schistosoma mansoni* [Levy and Read, 1975]. The first transporter is specific for adenosine, the second has a high affinity for uridine but can also transport adenosine. The third transporter has high affinity for adenine but transports adenosine [Levy and Read, 1975]. Nucleoside transport in schistosomes is not inhibited by NBMPR, dilazep and dipyridamole [el Kouni *et al.*, 1983, 1985, 1987,1989; el Kouni and Cha, 1987; Baer *et al.*, 1988; Baer, 1989; el Kouni, 1991].

Toxoplasma gondii appears to have two purine transporters [Schwab *et al.*, 1995; Chiang *et al.*, 1999]. The first is a distinct inosine transporter that has not been fully characterized but seems to transport certain purine nucleobases as well as nucleosides [Schwab *et al.*, 1995]. The second is an adenosine/purine nucleoside transporter (TgAT) with adenosine being the preferred substrate. TgAT differs from the mammalian equilibrative and concentrative transporters in three notable characteristics. First, TgAT exhibits lack of stereospecificity, transporting both D-adenosine [Chiang *et al.*, 1999] and L-adenosine [V. Guarcello and M.H. el Kouni, unpublished results], whereas the mammalian transporters are stereospecific for the D-enantiomers (Figure 8).

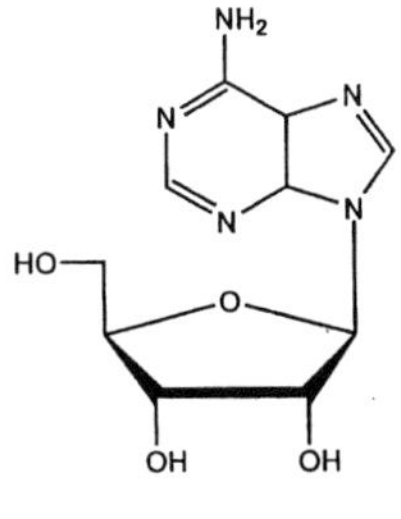

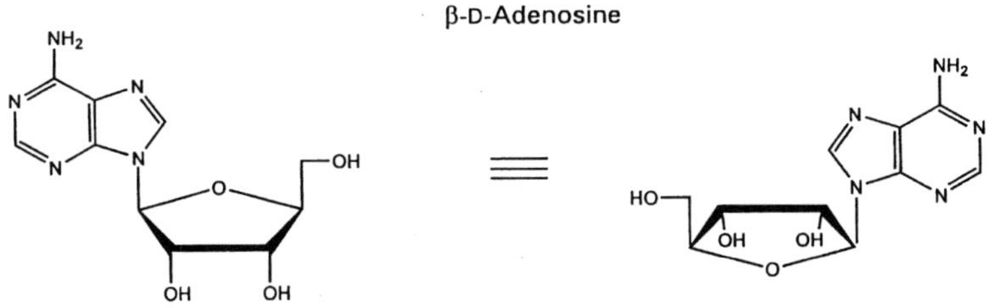

Figure 8. Chemical structures of the D- and L-enantiomers of adenosine.

Second, unlike mammalian transporters, TgAT is not inhibited by NBMPR but by high concentrations of dipyridamole. Third, NBMPR appears to be a permeant for this carrier as toxoplasma transports [V. Guarcello and M.H. el Kouni, unpublished results] and metabolizes [el Kouni *et al.*, 1999] NBMPR. The *TgAT* locus has been cloned and expressed [Chiang *et al.*, 1999]. Inactivation of the *TgAT* locus eliminates virtually all adenosine transport. However, this genetic lesion is not fatal [Chiang *et al.*, 1999], indicating that the parasites remain able to transport other purine nucleosides and/or nucleobases to satisfy their purine requirements.

Plasmodium falciparum has a high affinity equilibrative purine/pyrimidine nucleoside transporter (PfNT1) with adenosine being the preferred substrate. The locus *PfNT1* has been cloned and expressed [Carter *et al.*, 2000]. Like Leishmania [Ghosh and Mukherjee, 2000] and trypanosomes [de Koning *et al.*, 2000], the expression of PfNT1 is strongly regulated during the multiplication of the parasite erythrocytic stage. The increase in *PfNT1* expression during the life cycle is associated with the increased requirements and salvage of purines as well as the incorporation of adenosine into nucleic acids [Carter *et al.*, 2000]. The substrate specificity and inhibition profile of PfNT1 are unlike the mammalian transporters, but similar to those of TgAT in toxoplasma [Chiang *et al.*, 1999; V. Guarcello and M.H. el Kouni, unpublished results]. PfNT1 is also not entirely stereospecific as it is capable of transporting both the L- and D-enantiomers of purine and pyrimidine nucleosides and is not inhibited by NBMPR [Carter *et al.*, 2000].

4.2. Purine nucleobase transport

Virtually nothing was known about the mechanism of nucleobase transport in parasites until the recent studies on *Trypanosoma brucei brucei* [de Koning and Jarvis, 1997a and b]. It is apparent from these investigations that purine nucleobase transport mechanism may also be distinct from those in mammalian systems.

T. brucei brucei procyclic cells have a single (H1) high-affinity specific carrier for purine nucleobase as it does not take pyrimidine nucleobases or purine and pyrimidine nucleosides [de Koning and Jarvis, 1997a]. On the other hand, the bloodstream form has two purine nucleobase transporter; one is guanosine-sensitive (H2) and the other is guanosine-insensitive (H3) [de Koning and Jarvis, 1997b]. H3 resembles H1 in affinity and substrate specificity. H2 is less specialized as it binds guanosine and some pyrimidine nucleobases. As is the case with their purine nucleoside transport, the activity of purine nucleobase transporters in *T. brucei brucei* seems to be regulated by purine availability and growth cycle [de Koning *et al.*, 2000]. In contrast to mammalian cells, the *T. brucei brucei* purine nucleobase transporters like their nucleoside transporters [de Koning *et al.*, 1998 and 2000; de Koning and Jarvis, 1999], are not inhibited by either NBMPR, papaverine, dilazep or dipyridamole and are not sodium ion dependent transporter [de Koning and Jarvis, 1997a and b]. H1 and H3 are nucleobase/proton symporter as they are linked to protonmotive force [de Koning and Jarvis, 1997a and b]. Whether or not H2 is also a nucleobase/proton symporter remains to be determined.

Leishmania braziliensis also has one high affinity nucleobase transporter. However, the *L. braziliensis* transporter has a lower affinity for adenine [Hansen *et al.*, 1982] than that from *Trypanosoma brucei brucei* [de Koning and Jarvis, 1997a and b]. In addition, the specificity of *Leishmania braziliensis* nucleobase transporter with regard to pyrimidine nucleobases as permeants is not yet determined [Hansen *et al.*, 1982].

The high affinity and selectivity of the nucleobase transporter in *Trypanosoma brucei brucei* [de Koning and Jarvis, 1997a and b] and probably *Leishmania braziliensis* [Hansen *et al.*, 1982] appear uncommon among the nucleobase transporters in parasitic protozoa. For example, the affinities of the purine nucleobase carriers in *Tritrichomonas foetus* [Hedstorm and Wang, 1989] and *Giardia lamblia* [Ey *et al.*, 1992] are low in the millimolar range. Furthermore, the *G. lamblia* nucleobase transporter has broad-specificity as it permeates both purine and pyrimidine nucleobases [Ey *et al.*, 1992; Baum *et al.*, 1993]. This transporter is not inhibited by NBMPR or dipyridamole. [Ey *et al.*, 1992]. *Tritrichomonas foetus* [Hedstorm and Wang, 1989], on the other hand, has two nucleobase carriers, one for hypoxanthine as well as guanine, and the second for xanthine but can also transport pyrimidine nucleobases. Adenine enters by passive diffusion.

In *Schistosoma mansoni* hypoxanthine and guanine are assumed to be transported by one specific transporter. Adenine seems to be transported by two carriers; one is specific for adenine, the second has a high affinity for adenine but can also transport adenosine [Levy and Read, 1975]. In the cestode *Hymenolepis diminuta,* purine nucleobases are transported by at least one mediated transporter [McInnis *et al.*, 1965].

4.3. Purine transport in parasitized host cells

Intracellular parasites (e.g. toxoplasma, plasmodia, etc.) reside within specialized membrane surrounded vacuoles named parasitophorous vacuoles. Therefore, these intracellular parasites are cut off from vesicular membrane traffic within the host cells. Nevertheless, parasites within the parasitophorous vacuole rapidly multiply and obtain their nutritional requirements from the host through the parasitophorous vacuole membrane which acts as a molecular sieve allowing bidirectional equilibrated diffusion of small molecules (<1,300 Dalton), including nucleosides and nucleotides, between the vacuolar space and the host cell cytoplasm [Schwab *et al.*, 1994; Bermudes *et al.*, 1994 and references therein]. In contrast, the host cell membrane has tightly restricted permeability. Host cell membrane does not transport nucleotides because they are charged and, hence, nucleotides have to be dephosphorylated to nucleosides before they can be transported into the cell. As mentioned above, nucleosides are transported into mammalian cells with specific nucleoside transporters for some of which NBMPR is an inhibitor. However, infection was shown to alter the permeation and metabolism of purines in parasite-infected cells [Gati *et al.*, 1987; Gero *et al.*, 1988; Matias *et al.*, 1990; Upston and Gero, 1995]. It is, nevertheless, clear that not all nutrient transport is altered by infection, indicating that the alteration in nucleoside transport of infected cells has a degree of selectivity [Upston and Gero, 1995].

There is a current debate on how nucleosides enter the parasitized cells. Analysis of nucleoside transport in *Plasmodium falciparum*-infected human [Gero *et al.*, 1988; Upston and Gero, 1995] and *P. yoelii*-infected mouse [Gati *et al.*, 1987] erythrocytes demonstrated that infection with malaria parasites induces into the parasitized host cell an NBMPR insensitive, stereo-nonspecific nucleoside transporter that is similar to the PfNT1 described by Carter *et al.* (2000). There are at least four possible pathways for the induced nucleoside transport to occur in the parasitized host cell. The first is *via* an equilibrative high affinity adenosine transport system [Gero *et al.*, 1988; Upston and Gero, 1995]. The second is by way of a concentrative ion dependent channel [Kirk *et al.*, 1994]. The third pathway proposed is through tubovesicular membranes (TVM) which are interconnected network extending from the parasitophorous vacuolar membrane to the periphery of the infected cell [Lauer *et al.*, 1997]. The fourth is a duct for the transport of macromolecules that bypasses the host cell membrane [Pouvelle *et al.*, 1991]. The recent studies on the characteristics of *Plasmodium falciparum* nucleoside transporter PfNT1 [Carter *et al.*, 2000] ruled out all proposed pathways except the induction of an equilibrative high affinity adenosine transport system. Similarly, studies on toxoplasma-infected cells excluded the possibility that plasma membrane proteins of host cells might form membrane channels or transporters from the parasitophorous vacuole membrane that surrounds the parasites within the host cells [Schwab *et al.*, 1994; Bermudes *et al.*, 1994 and references therein].

4.4. Nucleoside transport inhibitors

As mentioned above, in contrast to mammalian cells, nucleoside transport in parasites or parasite-infected cells is insensitive to inhibitors of mammalian nucleoside transport (Table 4). It was shown that the insensitivity to NBMPR in *Trypanosoma gambiense* and *Plasmodium falciparum*-infected erythrocytes is due to the absence (*Trypanosoma gambiense* [Ogbunude and Ikediobi, 1982a]) or disappearance of NBMPR sensitive sites and appearance of NBMPR insensitive carrier mechanism(s) (malaria-infected cells [Gati *et al.*, 1987; Gero *et al.*, 1988 and 1989]). This seems also the case in *Toxoplasma gondii*-infected cells as neither NBMPR nor dilazep protect infected cells from tubercidin toxicity [V. Guarcello and M.H. el Kouni, unpublished results].

Nucleoside transport inhibitors were also reported to have antimalarial [Gero *et al.*, 1989] and antitoxoplasmosis [el Kouni *et al.*, 1999] effects on their own and to significantly change the intracellular nucleotide profile of malaria-infected erythrocytes [Gero *et al.*, 1989]. It was suggested that infection allows NBMPR to permeate the infected cell membrane and undergo catabolism to liberate 6-mercaptopurine which interferes with purine interconversion [Gero *et al.*, 1989]. In toxoplasma, NBMPR was shown to be cleaved to the nucleobase, 6- nitrobenzylthiohypoxanthine [el Kouni *et al.*, 1999]. However, the toxicity of NBMPR appears to be the result of its phosphorylation to the nucleotide level *via* adenosine kinase [el Kouni *et al.*, 1999].

The difference between parasites or parasite-infected host cells and uninfected host cells in sensitivity and chemotherapeutic efficacy of mammalian nucleoside transport inhibitors can be exploited for selective treatment of parasite or parasite-infected cells. This difference could provide the basis for the search to identify a specific inhibitor of the parasite purine transport. Since parasites are dependent on purine salvage and their purine transport is different from that of the host, finding a selective inhibitor of purine transport in the parasites will selectively starve these organisms for vital purines even in host cells which have *de novo* purine biosynthesis.

Table 4.   Effect of inhibitors of nucleoside transport in mammalian cells on the transport of purine nucleosides in various parasites.

Parasite	Inhibitors			Reference
	NBMPR	Dilazep	Dipyridamole	
Babesia Bovis	-/+[a]		n.d.	n.d. Matias *et al.* (1990)
Entamoeba histolytica	-/+[a]	-	n.d.	Das *et al.* (1997)
Leishmania donovani	-	n.d.	-	Iovannisci and Ullman (1984), Das *et al.* (1997)
Giardia lamblia	-	n.d.	+[b]	Baum *et al.* (1993)
Plasmodium falciparum	-	-	-/+[b]	Gero *et al.* (1988 and 1989), Carter *et al.* (2000)
Plasmodium yoelii	-	n.d.	n.d.	Gati *et al.* (1987)
Schistosoma mansoni	-	-	-/+[b]	el Kouni *et al.* (1983 and 1987), Baer *et al.* (1988)
Schistosoma japonicum	-	n.d.	n.d.	el Kouni *et al.* (1985)
Schistosoma haematobium	-	n.d.	n.d.	Baer *et al.* (1988)
Toxoplasma gondii	-	-	-/+[b]	Schwab *et al.* (1994), el Kouni *et al.* (1999), V. Guarcello and M.H. el Kouni (unpublished results), Chiang *et al.* (1999).
Trichomonas vaginalis	-/+[a]	-/+	n.d.	Harris *et al.* (1988)
Trypanosoma brucei (procyclic)	-	n.d.	-/+[b]	de Koning *et al.* (1998)
Trypanosoma brucei (blood form)	-	-	-/+[b]	James and Born (1980), de Koning and Jarvis (1999)
Trypanosoma congolense	n.d.	n.d.	-/+[b]	James and Born (1980),
Trypanosoma gambiense	-	n.d.	n.d.	Ogbunude and Ikediobi (1982a and b)

-    No inhibition; +, Inhibition; -/+, Inhibit only at concentrations several times higher than the concentration required to inhibit NBMPR-sensitive mammalian nucleoside transporters; n.d., Not determined.

[a]   Could be due to competition for a common carrier rather than inhibition.

[b]   Inhibits only at the high concentration required to inhibit NBMPR-insensitive transporters

Alternatively, the difference in nucleoside transport between parasites or parasite-infected host cells and uninfected host cells can also be exploited to make nucleoside analogues selectively toxic to the parasite or parasite-infected cells. It has been demonstrated that by coadministering an inhibitor of mammalian nucleoside transport with toxic purine analogues, the nucleoside transport inhibitor acted as a protective agent to the host cells but not the parasites or parasite-infected host cells from the toxicity of the purine analogues [Ogbunude and Ikediobi, 1982b; el Kouni *et al.*, 1983, 1985, 1987 and 1989; Gati *et al.*, 1987; Baer *et al.*, 1988; Gero *et al.*, 1988 and 1989; Baer, 1989; el Kouni, 1991]. Thus, the analogues were made selectively toxic against the parasite as will be discussed below.

5. Purine analogues as potential antiparasitic agents and possible protection against host-toxicity by nucleoside transport inhibitors

The feasibility of using purine analogues which perturb nucleotide metabolism as potential antiparasitic agents was demonstrated in the treatment of malaria [Gati *et al.*, 1987; Gero *et al.*, 1988 and 1989] and other parasitic diseases including schistosomiasis [Jaffe, 1975; el Kouni *et al.*, 1983, 1985, 1987 and 1989; Baer *et al.*, 1988; el Kouni, 1991], trypanosomiasis [Avila and Avila, 1981; Ogbunude and Ikediobi, 1982b; Avila *et al.*,1983; Berens *et al.*, 1984;], leishmaniasis [Peters *et al.*, 1980; Berman *et al.*, 1983; Walton *et al.*, 1983; Morshige *et al.*, 1995] and toxoplasmosis [Luft, 1986; el Kouni *et al.*, 1999] in animal models.

The use of purine analogues in chemotherapy may be limited by host-toxicity. One of the most exciting developments in the past few years was the demonstration that even a potentially very toxic compound may still be used as an antiparasitic drug without any apparent host toxicity (el Kouni *et al.*, 1983, 1985, 1987 and 1989; Baer *et al.*, 1988; Baer, 1989; el Kouni, 1991). As illustrated in Figure 9, it was found that nucleoside transport inhibitors of mammalian systems (NBMPR-P, dilazep or dipyridamole), do not inhibit the uptake of toxic nucleoside analogues into *Schistosoma mansoni* [el Kouni *et al.*, 1983; el Kouni and Cha 1987]. This difference in nucleoside transport between the schistosomes and their host was successfully manipulated to design an effective chemotherapeutic regimen for the treatment of schistosomiasis [el Kouni *et al.*, 1983, 1985, 1987 and 1989; Baer *et al.*, 1988; Baer, 1989; el Kouni, 1991]. By coadministering a nucleoside transport inhibitor with highly toxic nucleoside analogues (tubercidin or nebularine) the host but not the parasite was protected from the toxicity of the analogue [el Kouni *et al.*, 1983, 1985, 1987 and 1989; Baer *et al.*, 1988; Baer, 1989; el Kouni, 1991]. Similar combinations were also effective against *Trypanosoma gambiense* [Ogbunude and Ikediobi, 1982b], *Plasmodium falciparum* [Gero *et al.*, 1988] and *P. yoelii* [Gati *et al.*, 1987] infections. In contrast to non-infected erythrocytes, *P. falciparum-* or *P. yoelii*-infected red cells are insensitive to inhibitors of mammalian nucleoside transport [Gati *et al.*, 1987; Gero *et al.*, 1988 and 1989]. Thus, even with intracellular parasites, this method of host-protection can still be effective.

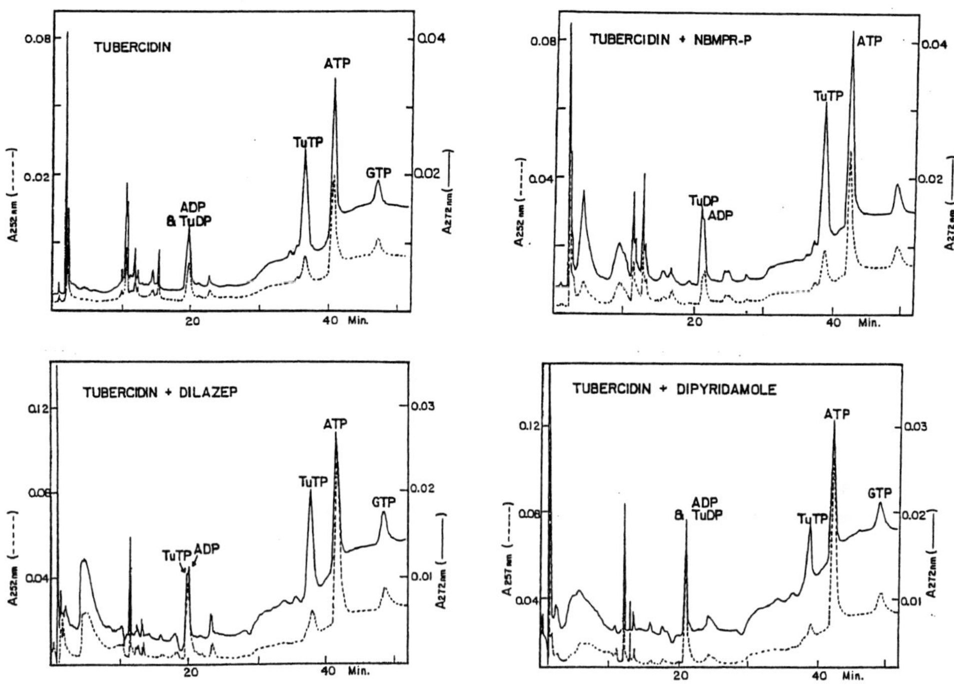

Figure 9. HPLC profiles showing the effect of the nucleoside transport inhibitors, NBMPR, dilazep and dipyridamole, on the incorporation of tubercidin into the nucleotide pools of *Schistosoma mansoni* after 4 hr incubation *in vitro*. The ratios of TuTP (tubercidin 5'- triphosphate)/ATP peak areas were 0.7,1.3, 0.9 and 0.5 for tubercidin alone and in the presence of NBMPR, dilazep and dipyridamole, respectively. The methodology was previously described by el Kouni *et al.* (1983) and el Kouni and Cha (1987).

It should also be noted that the use of NBMPR at 25 mg/kg/day × 4 for up to 3 weeks as an adjunct in the chemotherapy of schistosomiasis with purine analogues was without apparent host-toxicity [el Kouni *et al.*, 1983, 1985, 1987 and 1989; Baer *et al.*, 1988; Baer, 1989; el Kouni, 1991]. Such doses of NBMPR were non-toxic to the animals as judged from blood chemistry, hematological studies, and gross and histological examinations [el Kouni *et al.*, 1989]. No evidence for injury to the liver, kidney, spleen, pancreas, mesentery, or peritoneal mesothelium was observed. Furthermore, administration of NBMPR at the highest dose tested (100 mg/kg) showed no host toxicity [el Kouni *et al.*, 1999]. Hence, it can be stated that administration of NBMPR is safe at least up to the highest dose tested (100 mg/kg). Therefore, the mode of host-protection by coadministration of a nucleoside transport inhibitor may provide an attractive alternative for the treatment of this disease with currently available cytotoxic nucleoside analogues. In addition, dipyridamole and dilazep are already in clinical use as coronary vasodilator drugs (the latter in Europe and South America). The findings that they are similarly effective as NBMPR in improving the therapeutic index of nucleoside analogues against parasites [el Kouni *et al.*, 1987; el Kouni 1991] emphasizes the advantage of probable use of these drugs in man.

6. Conclusions and prospects

It is quite clear that parasites differ from their host in various aspects of purine metabolism including among other things: the lack of *de novo* purine biosynthesis, substrate specificity of various enzymes of the salvage pathway, the presence of special salvage enzymes exclusively in the parasites but not the host, nature and type of purine nucleobase and nucleoside transport and insensitivity of purine transport to inhibitors of mammalian host purine transport inhibitors. These various differences in purine metabolism between parasites and/or parasite infected cells and host cells present excellent targets for antiparasitic chemotherapeutic interventions as demonstrated by several studies. However, optimism in this endeavor should be accompanied by cautiousness. The existence of multiple independent pathways as well as transport systems for purine metabolism in parasites has clear implications for the design of new antiparasitic chemotherapeutic regimens aiming at blocking purine salvage, as no one purine enzyme or transporter seems to be essential. Therefore, strategies aiming at interfering with purine metabolism in parasites should be directed towards attacking two or more independent pathways to successfully overcome the ingenuity of the parasites in evading blocking their purine metabolism by a single agent.

7. Acknowledgments

Research work in the author's laboratory quoted in this review has been supported by grants AI-22219, AI-29848, AI-39550 and AI-42975 from the NIAID, DHHS and grants from the UNDP/World Bank/WHO Special Programme for Research and Training in Tropical Diseases. Thanks are due to Dr. Fardos N. M. Naguib and Dr. Omar N. Al Safarjalani for their help and valuable discussions during the preparation of this review.

8. References

Aldritt, S.M. and Wang, C.C. (1985) Purification and characterization of guanine phosphoribosyltransferase from *Giardia lamblia*. J. Biol. Chem. 261, 8528-8533.

Allen, T.E., Henschel, E.V., Coons, T., Cross, L., Conley, J. and Ullman, B. (1989) Purification and characterization of the adenine phosphoribosyltransferase and hypoxanthine-guanine phosphoribosyltransferase activities from *Leishmania donovani*. Molec. Biochem. Parasitol. 33, 273-282.

Allen, T.E., Hwang. H.Y., Jardim, A., Olafson, R. and Ullman, B. (1995) Cloning and expression of the hypoxanthine-guanine phosphoribosyltransferase from *Leishmania donovani*. Molec. Biochem. Parasitol. 73, 133-143.

Allen, T.E., Hwang, H.Y., Wilson, K., Hanson, S., Jardim, A. and Ullman, B. (1995) Cloning and expression of the adenine phosphoribosyltransferase gene from *Leishmania donovani*. Molec. Biochem. Parasitol. 74, 99-103.

Allen, T.E. and Ullman, B. (1993) Cloning and expression of the hypoxanthine-guanine phosphoribosyltransferase gene from *Trypanosoma brucei*. Nucleic Acid Res. 21, 5431-5438.

Allen, T.E. and Ullman, B. (1994) Molecular characterization and overexpression of the hypoxanthine-guanine phosphoribosyltransferase gene from *Trypanosoma cruzi*. Molec. Biochem. Parasitol. 65, 233-245.

Aronov, A.M., Munagala, N.R., Ortiz, De Montellano P.R., Kuntz, I.D. and Wang, C.C. (2000) Rational design of selective submicromolar inhibitors of *Tritrichomonas foetus* hypoxanthine-guanine-xanthine phosphoribosyltransferase. Biochemistry 39, 4684-4691.

Aronow, B., Kaur, K., McCartan, K. and Ullman, B. (1987) Two high affinity nucleoside transporters in *Leishmania donovani*. Molec. Biochem. Parasitol. 22, 29-37.

Avila, J.L. and Avila, A. (1981) *Trypanosoma cruzi*: Allopurinol in the treatment of mice with experimental acute chagas disease. Exptl. Parasitol. 51, 204-208.

Avila, J.L., Avila, A., Munoz, E. and Monzon, H. (1983) *Trypanosoma cruzi*: 4- Aminopyrazolopyrimidine in the treatment of experimental Chagas' disease. Exptl. Parasitol. 56, 236-240.

Bacchi, C.J., Garofalo, H., Mockenhaupt, D., McCann, P.P., Diekema, K.A., Pegg, A.E., Nathan, H.C., Mullaney, E.A., Chunosoff, L., Sjoerdsma, A. and Hutner, S.H. (1983) In vivo effects of -DL-difluoromethylornithine on the metabolism and morphology of *Trypanosoma brucei brucei*. Molec. Biochem. Parasitol. 7, 209-225.

Bacchi, C.J., Nathan, H.N., Clarkson, Jr. A.B., Biennen, E.J., Bitonti, A.J., McCann, P.P. and Sjoerdsma, A. (1987) Effects of the ornithine decarboxylase inhibitors DL-difluoromethylornithine and -DL-monofluoromethyldehydroornithine methyl ester alone and in combination with suramin against *Trypanosoma brucei brucei* central nervous system models. Am. J. Trop. Med. Hyg. 36, 48-54.

Bacchi, C.J., Nathan, H.C., Hutner, S.H., McCann, P.P. and Sjoerdsma, A. (1980) Polyamine metabolism: a potential therapeutic target in trypanosomes. Science 210, 332-334.

Bacchi, C.J., Sanabria, K., Spiess, A.J., Vargas, M., Marasco, C.J.Jr., Jimenez, L.M., Goldberg, B. and Sufrin, J.R. (1997) *In vivo* efficacies of 5'-methylthioadenosine analogs as trypanocides. Antimicrob. Agents Chemother. 41, 2108-2112.

Backlund, P.S.Jr. and Smith, R.A. (1981) Methionine synthesis from 5'-methylthioadenosine in rat liver. J. Biol. Chem. **256**, 1533-1535.

Baer, H.P. (1989) Cytotoxic nucleosides and parasitic diseases: A new therapeutic approach. Annals Saudi Med. 9, 570-575.

Baer, H.P., El-Soofi, A. and Selim, A. (1988) Treatment of *Schistosoma mansoni*- and *Schistosoma haematobium*-infected mice with a combination of tubercidin and nucleoside transport inhibitor. Med. Sci. Res. 16, 919.

Baer, H.P., Serignese, V., Ogbunude, P.O. and Dzimiri, M. (1992) Nucleosides transporters in *Leishmania major*: diversity in adenosine transporter expression or function in different strains. Am. J. Trop. Med. Hyg. 47, 87-91.

Barrett, M.P., Zhang, Z.Q., Denise, H., Giroud, C. and Baltz, T.A. (1995) diamidine-resistant *Trypanosoma equiperdum* clone contains a P2 purine transporter with reduced substrate affinity. Molec. Biochem. Parasitol. 73, 223-229.

Baum, K.F., Berens, R.L. and Marr, J.J. (1993) Purine nucleosides and nucleobase cell membrane transport in *Giardia lamblia*. J. Euk. Microbiol. 40, 643-649.

Beck, J.T. and Wang, C.C. (1993) The hypoxanthine-guanine-xanthine phosphoribosyltransferase from *Tritrichomonas foetus* has unique properties. Mol. Biochem. Parasitol. 60, 187-194.

Berens, R.L. and Marr, J.J. (1986) Adenosine analog metabolism in *Giardia lamblia*. Biochem. Pharmacol. 35, 4191-4197.

Berens, R.L., Marr, J.J., LaFon, S.W. and Nelson, D.J. (1981) Purine metabolism in *Trypanosoma cruzi*. Molec. Biochem. Parasitol. 3, 187-196

Berens, R.L., Marr, J.J., Looker, D.L., Nelson, D.J. and LaFon, S.W. (1984) Efficacy of pyrazolopyrimidine ribonucleosides against *Trypanosoma cruzi*: Studies *in vitro* and *in vivo* with sensitive and resistant strains. J. Infect. Dis. 150, 602-608.

Berman, J.D., Keenan, C.M., Lamb, S.R., Hanson, W.L. and Waits, V.B. (1983) *Leishmania donovani*: Oral efficacy and toxicity of Formycin B in the infected hamster. Exptl. Parasitol. 56, 215-221.

Bermudes, D., Peck, K.R., Afifi, M.A., Beckers, C.J.M. and Joiner, K.A. (1994) Tandemly repeated genes encode nucleoside triphosphate hydrolase isoforms secreted into parasitophorous vacuole of *Toxoplasma gondii*. J. Biol. Chem. 46, 29252-29260.

Bhaumik, D. and Datta. A.K. (1988) Reaction Kinetics and inhibition of adenosine kinase from *Leishmania donovani*. Molec. Biochem. Parasitol. 28, 181-188.

Bhaumik, D. and Datta, A.K. (1989) Immunochemical and catalytic characteristics of adenosine kinase from *Leishmania donovani*. J. Biol. Chem. 264,4356-4361.

Bhaumik, D. and Datta, A.K. (1992) Active site thiol(s) in *Leishmania donovani* adenosine kinase: comparison with hamster enzyme and evidence for the absence of regulatory adenosine binding site. Molec. Biochem. Parasitol. 52, 29-38.

Boonlayangoor, P., Albach, R.A. and Booden, T. (1980) Purine nucleotide synthesis in *Entamoeba histolytica*: A preliminary study. Arch. Invest. Med. (Mex.) 11, 83-88.

Buolamwini, J.K. (1997) Nucleoside transport inhibitors: Structure-activity relationships and potential therapeutic applications. Curr. Med. Chem. 4, 35-66.

Carter, N.S., Ben Mamoun, C., Liu, W., Silva, E.O., Landfear, S.M., Goldberg, D.E. and Ullman, B. (2000) Isolation and functional characterization of the PfNT1 nucleoside transporter gene from *Plasmodium falciparum*. J. Biol. Chem. 275, 10683-10691.

Carson, D.A. and Chang, K.P. (1981) Phosphorylation and anti leishmanial activity of Formycin B. Biochem. Biophys. Res. Commun. 100, 1377- 13.

Carter, N.S., Drew, M.E., Sanchez, M., Vasudevan, G., Landfear, S.M. and Ullman, B. (2000) Cloning of a novel inosine-guanosine transporter gene from *Leishmania donovani* by functional rescue of a transport-deficient mutant. J. Biol. Chem. 275, 20935-20941.

Carter, N.S. and Fairlamb, A.H. (1993) Arsenical-resistant trypanosomes lack an unusual adenosine transporter. Nature 361, 173-175.

Chiang, C.W., Carter, N., Sullivan, Jr. W.J., Donald R.G.K., Roos, D.S., Naguib, F.N.M., el Kouni, M.H., Ullman, B. and Wilson, C.M. (1999) The adenosine transporter of *Toxoplasma gondii*. Identification by insertional mutagenesis, cloning and recombinant expression. J. Biol. Chem. 274, 35255-35261.

Chin, M.S. and Wang, C.C. (1994) Isolation, sequencing and expression of the gene encoding hypoxanthine-guanine-xanthine phosphoribosyltransferase of *Tritrichomonas foetus*. Molec. Biochem. Parasitol. 63, 221-229.

Craig, S.P.III., Yuan, L., Kuntz, D.A., McKerrow, J.H. and Wang, C.C. (1991) High level expression in *Escherichia coli* of soluble enzymatically active schistosomal hypoxanthine/guanine phosphoribosyltransferase and trypanosomal ornithine decarboxylase. Proc. Natl. Acad. Sci., USA 88, 2500-2504.

Darling, J.A., Sullivan, Jr. W.J., Carter, D., Ullman, B. and Roos, D.S. (1999) Recombinant expression, purification, and characterization of *Toxoplasma gondii* adenosine kinase. Molec. Biochem. Parasitol. 103, 15-23.

Das, P., Das, S.R., Moorji, A. and Baer, H.P. (1997) Characterization of nucleoside uptake and transport in *Entamoeba histolytica*. Parasitol. Res. 83, 364-369.

Datta, A.K., Bhaumik, D. and Chatterjee, R. (1987) Isolation and characterization of adenosine kinase from *Leishmania donovani*. J. Biol. Chem. 262, 5515-5521.

Davey, R.A., Mayrhofer, G. and Ey, P. (1992) Identification of a broad-specificity transporter with affinity for the sugar moiety in *Giardia Intestinalis* trophozoites. Biochim. Biophys. Acta 1109, 172-178.

de Koning, H.P. and Jarvis, S.M. (1997a) Hypoxanthine uptake through a purine-selective nucleobase transporter in *Trypanosoma brucei brucei* procyclic cells is driven by protonmotive force. Eur. J. Biochem. 247, 1102-1110.

de Koning, H.P. and Jarvis, S.M. (1997b) Purine nucleobase transport in bloodstream forms of *Trypanosoma brucei brucei* is mediated by two novel transporters. Molec. Biochem. Parasitol. 89, 245-258.

de Koning, H.P. and Jarvis, S.M. (1999) Adenosine transporters in bloodstream forms of *Trypanosoma brucei brucei*: substrate recognition motifs and affinity for trypanocidal drugs. Molec. Pharmacol. 56, 1162-1170.

de Koning, H.P., Watson, C.J. and Jarvis, S.M. (1998) Characterization of a nucleoside/proton symporter in procyclic *Trypanosoma brucei brucei*. J. Biol. Chem. 273, 9486-9494.

de Koning, H.P., Watson, C.J., Sutcliffe, L. and Jarvis, S.M. (2000) Differential regulation of nucleoside and nucleobase transporters in *Crithidia fasciculata* and *Trypanosoma brucei brucei*. Molec. Biochem. Parasitol. 106, 93-107.

Dewey, V.C. and Kidder, G. (1973) Partial purification and properties of a nucleoside hydrolase from Crithidia. Arch. Biochem. Biophys. 157, 380-387.

Donald, R.G.K., Carter, D., Ullman, B. and Roos, D.S. (1996) Insertional tagging, cloning, and expression of the *Toxoplasma gondii* hypoxanthine-xanthine-guanine phosphoribosyltransferase gene. J. Biol. Chem. 271, 14010-14019.

Dovey, H.F., McKerrow, J.H. and Wang, C.C. (1984) Purine salvage in *Schistosoma mansoni* schistosomules. Molec. Biochem. Parasitol. 11, 157-167.

Doyle, P.S., Kanaani, J., Wang C.C., (1998) Hypoxanthine, guanine, xanthine phosphoribosyltransferase activity in *Cryptosporidium parvum*. Exptl. Parasitol. 89, 9-15, 1998.

Eakin, A.E., Gueraa, A., Focia, P.J., Torres-Martinez, J. and Craig III, S.P. (1997) Hypoxanthine phosphoribosyltransferase from *Trypanosoma cruzi* as a target for structure-based inhibitor design: Crystallization and inhibition studies with purine analogs. Antimicrob. Agents Chemother. 41, 1686-1692.

el Kouni, M.H. (1991) Efficacy of combination therapy with tubercidin and nitrobenzylthioinosine 5'-monophosphate against chronic and advanced stages of schistosomiasis. Biochem. Pharmacol. 41, 815-820.

el Kouni, M.H., and Cha, S. (1987) Metabolism of adenosine analogues by *Schistosoma mansoni* and the effect of nucleoside transport inhibitors. Biochem. Pharmacol. **36**, 1099-1106.

el Kouni, M.H., Diop, D. and Cha, S. (1983) Combination therapy of schistosomiasis by tubercidin and nitrobenzylthioinosine 5'-monophosphate. Proc. Natl. Acad. Sci. USA 80, 6667-6670.

el Kouni, M.H., Diop, D., O'Shea, P., Carlisle, R. and Sommadossi, J.-P. (1989) Prevention of tubercidin host toxicity by nitrobenzylthioinosine 5'-monophosphate for the treatment of schistosomiasis. Antimicrob. Agents Chemother. 33, 824-827.

el Kouni, M.H., Guarcello, V., Al Safarjalani, O.N. and Naguib, F.N.M. (1999) Metabolism and Selective Toxicity of 6-Nitrobenzylthioinosine in *Toxoplasma gondii*. Antimicrob. Agents Chemother. 43, 2437-2443.

el Kouni, M.H., Knopf, P.M. and Cha, S. (1985) Combination therapy of *Schistosoma japonicum* by tubercidin and nitrobenzylthioinosine 5'-monophosphate. Biochem. Pharmacol. 34, 3921-3923.

el Kouni, M.H., Messier, N.J. and Cha, S. (1987) Treatment of schistosomiasis by purine nucleoside analogues in combination with nucleoside transport inhibitors. Biochem. Pharmacol. 36, 3815-3821.

Ey, P.L., Davey, R.A. and Duffield G.A. (1992) A low affinity nucleobase transport in the protozoan parasite *Giardia intestinalis*. Biochim. Biophys Acta 1109, 179-186.

Fish, W.R., Looker, D.L., Marr, J.J. and Berens, R.L. (1982a) Purine metabolism in *Trypanosoma brucei gambiense*. Biochim. Biophys. Acta 714, 422-428.

Fish, W.R., Looker, D.L., Marr, J.J. and Berens, R.L. (1982b) Purine metabolism in the bloodstream forms of *Trypanosoma gambiense* and *Trypanosoma rhodesiense*. Biochim. Biophys. Acta 719, 223-231.

Fish, W.R., Marr, J.J., Berens, R.L., Looker, D.L., Nelson, D.J., LaFon, S.W. and Balber, A.E. (1985) Inosine analogs as chemotherapeutic agents for African trypanosomes: Metabolism in trypanosomes and efficacy in tissue culture. Antimicrob. Agents Chemother. 27, 33-36.

Fitchen, J.H., Riscoe, M.K. Ferro, A.J. (1988) Exploitation of methylthioribose kinase in the development of antiprotozoal drugs. Adv. Exptl. Med.Biol. 250,199-210.

Gati, W.P., Stoyke, A.F.W., Gero, A.M. and Paterson, A.R.P. (1987) Nucleoside permeation in mouse erythrocytes infected with *Plasmodium yoelii*. Biochem. Biophys. Res. Commun. 145, 1134-1141.

Gero, A.M., Bugledich, E.M.A., Paterson, A.R.P. and Jamieson, G.P. (1988) Stage-specific alteration of nucleoside membrane permeability and nitrobenzylthioinosine insensitivity in *Plasmodium falciparum* infected erythrocytes. Molec. Biochem. Parasitol. 27, 159-170.

Gero, A.M., and Kirk, K. (1994) Nutrient transport pathways in *Plasmodium*-infected erythrocytes: What and where are they? Parasitol. Today 10, 395-399.

Gero, A.M., Scott, H.V., O'Sullivan, W.J. and Christopherson, R.I. (1989) Antimalarial action of nitrobenzyl-thioinosine in combination with purine nucleoside antimetabolites. Molec. Biochem. Parasitol. 34, 87-98.

Ghoda, L.Y., Savarese, T.M., Northup, C.H., Parks, Jr. R.E., Garofalo, J., Katz, L., Ekkenbogen, B.B. and Bacchi, C.J. (1988) Substrate specificities of 5'-deoxy-5'-methylthioadenosine phosphorylase from *Trypanosoma brucei brucei* and mammalian cells. Molec. Biochem. Parasitol. 27, 109-118.

Ghosh, M. and Mukherjee, T. (2000) Stage-specific development of a novel adenosine transporter in *Leishmania donovani* amastigotes. Molec. Biochem. Parasitol. 108, 93-99.

Goldberg, B., Rattendib, D., Lloydd, D., Sufrine, J.R., Bacchi, C.J. (2001) In situ kinetic characterization of methylthioadenosine transport by the adenosine transporter (P2) of the African Trypanosoma brucei brucei and Trypanosoma brucei rhodesiense. Biochem. Pharmacol. 61, 449-457.

Griffith, D.A. and Jarvis, S.M. (1993) High affinity sodium-dependent nucleobase transport in cultured renal epithelial cells (LLC-PK). J. Biol. Chem. 268, 20085-20090.

Griffith, D.A. and Jarvis, S.M. (1996) Nucleoside and nucleobase transport systems in mammalian cells. Biochim. Biophys. Acta 1286, 153-181.

Gutteridge, W.E. and Davies, M.J. (1981) Enzymes of purine salvage in *Trypanosoma cruzi*. FEBS Letters 127, 211-214.

Hammond, D.J. and Gutteridge, W.E. (1984) Purine and pyrimidine metabolism in the Trypanosomatidae. Molec. Biochem. Parasitol. 13, 243-261.

Hansen, B.D., Perez-Arbelo, J., Walkony, J.F. and Hendricks, L.D. (1982) The specificity of purine base and nucleoside uptake in promastigotes of specificity of the *Leishmania braziliensis panamensis*. Parasitol. 85, 271-282.

Harris, D.I., Beechey, R.B., Linstead, D. and Barrett, J. (1988) Nucleoside uptake by *Trichomonas vaginalis*. Molec. Biochem. Parasitol. **29**, 105-116.

Hassan, H.F. and Coombs, G.H. (1986a) *De novo* synthesis of purines by *Acanthamoeba castellani* and *A. astronyxis*. IRCS Med. Sci. 14, 559-560.

Hassan, H.F. and Coombs, G.H. (1986b) Purine-metabolizing enzymes in *Entamoeba histolytica*. Molec. Biochem. Parasitol. 19, 19-25.

Hassan, H.F. and Coombs, G.H. (1986c) Comparative study of purine- and pyrimidine metabolizing enzymes of range of trypanosomatids. Comp. Biochem. Physiol. 84B, 217-223.

Hassan, H.F., Philips R.S. and Coombs, G.H. (1987) Purine-metabolizing enzymes in *Babesia divergens*. Parasitol. Res. 73, 121-125.

Hcdstorm, L. and Wang, C.C. (1989) Purine base transport in wild-type and mycophenolic acid-resistant *Tritrichomonas foetus*. Molec. Biochem. Parasitol. 35, 219-228.

Heyworth, P.G., Gutteridge, W.E. and Ginger, C.D. (1982) Purine metabolism in *Trichomonas vaginalis*. FEBS Letters 141, 106-110.

Hiraoka, O., Satake, H., Iguchi, S., Matsuda, A., Ueda, T. and Wataya, Y. (1986) Carbocyclic inosine as a potent anti-Leishmania agent: The metabolism and selective cytotoxic effects of carbocyclic inosine in promastigotes of *Leishmania tropica* and *Leishmania donovani*. Biochem. Biophys. Res. Comm. 134, 1114-1121.

Hwang, H.Y. and Ullman, B. (1997) Genetic analysis of purine metabolism in *Leishmania donovani*. J. Biol. Chem. 272, 19488-19496.

Iltzsch, M.H., Uber, S.S., Tankersly, K.O. and el Kouni, M.H. (1995) Structure-activity relationship of the binding of nucleoside ligands to adenosine kinase from *Toxoplasma gondii*. Biochem. Pharmacol. 49, 1501-1512.

Iovannisci, D.M., Goebel, D., Allen, K., Kaur, K. and Ullman, B. (1984) Genetic analysis of adenine metabolism in *Leishmania donovani* promastigotes. J. Biol. Chem. 259, 14617-14623.

Iovannisci, D.M. and Ullman, B. (1984) Characterization of a mutant *Leishmania donovani* deficient in adenosine kinase activity. Molec. Biochem. Parasitol. 12, 139-151.

Jaffe, J.J. (1975) Nucleoside analogs as antiparasitic Agents. Ann. N.Y. Acad. Sci. 255, 306-316.

James, D.M. and Born, G.V.R. (1980) Uptake of purine bases and nucleosides in African trypanosomes. Parasitol. 81, 383-393.

Jardim, A., Bergeson, S.E., Shih, S., Carter, N., Lucas, R.W., Merlin, G., Myler, P.J., Stuart, K. and Ullman, B. (1999) Xanthine phosphoribosyltransferase from *Leishmania donovani*. Molecular cloning, biochemical characterization, and genetic analysis. J. Biol. Chem. 274, 34403-34410.

Kager, P.A., Rees, P.H., Wellde, P.T., Hockmeyer, W.T. and Lyerly, W.H. (1981) Allopurinol in the treatment of visceral leishmaniasis. Trans. Roy. Soc. Trop. Med. Hyg. 75, 556-559.

Kanaaneh, J., Craig III, S.P. and Wang, C.C. (1994) Differential inhibitory effects of GMP-2',3'-dialdehyde on human and schistosomal hypoxanthine-guanine phosphoribosyltransferases. Eur. J. Biochem. 223, 595-601.

Kanaaneh, J., Maltby, D., Focia, P. and Wang, C.C. (1995) Identification of the active sites of human schistosomal hypoxanthine-guanine phosphoribosyltransferases by GMP-2',3'-dialdehyde affinity labeling. Biochemistry 34, 14987-14996.

Kanaaneh, J., Maltby, D., Somoza, J.R. and Wang, C.C. (1996) Probing the active site of *Trichomonas foetus* hypoxanthine-guanine-xanthine phosphoribosyltransferase using covalent modification of cysteine residues. Eur. J. Biochem. 239, 764-772.

Kanaaneh, J., Maltby, D., Somoza, J.R. and Wang, C.C. (1997) Inactivation of *Trichomonas foetus* and *Schistosoma mansoni* purine phosphoribosyltransferases by arginine-specific reagents. Eur. J. Biochem. 244, 810-817.

Kirk, K., Horner, H.A., Elford, B.C., Ellory, J.C. and Newbold, C.I. (1994) Transport of diverse substrates into malaria-infected erythrocytes via a pathway showing functional characteristics of a chloride channel. J. Biol. Chem. 269, 3339-3347.

Koszalka, G.W. and Krenitsky, T.A. (1979) Hydrolases in *Leishmania donovani*, J. Biol. Chem. 254, 8185-81937.

Koszalka, G.W. and Krenitsky, T.A. (1986) 5'-Methylthioadenosine (MeSAdo) phosphorylase from promastigotes of *Leishmania donovani*. In: W.L. Nyhan, L.F. Thompson and R.W.E. Wattseds (Eds.), Purine and Pyrimidine Metabolism in Man V pp. 559-563 Plenum, New York

Kraup, M. and Marz, R. (1995) Membrane transport of nucleobases: Interaction with inhibitors, Gen. Pharmacol. 26, 1185-1190.

Krug, E.C., Marr, J.J. and Berens, R.L. (1989) Purine metabolism in *Toxoplasma gondii*. J. Biol. Chem. 264, 10601-10607, 1989.

LaFon, S.W. and Nelson, D.J. (1985) Purine metabolism in th intact sporozoites and merozoites of *Eimeria tenella*. Molec. Biochem. Parasitol. 14, 11-22.

Lauer, S.A., Rathod, P.K., Ghori, N. and Haldar, K. (1997) A membrane network for nutrient import in red cells infected with malaria parasite. Science 276, 1122-1125.

Levy, M.G. and Read, C.P. (1975) Purine and pyrimidine transport in *Schistosoma mansoni*. J. Parasitol. 61, 627-632, 1975

Lo, H.-S. and Wang, C.C. (1985) Purine salvage in *Entamoeba histolytica*. J. Parasitol. 71, 662- 670.

Lobee-Rich, P.A. and Reeves, R.E. (1983) The partial purification and characterization of adenosine kinase from *Entamoeba histolytica,* Am. J. Trop. Med. Hyg. 32, 976-979.

Looker, D.L., Berens, R.L. and Marr, J.J. (1983) Purine metabolism in *Leishmania donovani* amastigotes and promastigotes. Molec. Biochem. Parasitol. 9, 15-28.

Luft, B.J. (1986) Potent in vivo activity of arprinocid, a purine analogue, against murine toxoplasmosis. J. Infect. Dis. 154, 692-693.

Marr, J.J. (1983) Pyrazolopyrimidine metabolism in Leishmania and Trypanosomes: Significant differences between host and parasite. J. Cell. Biochem. **22**, 187-196.

Marr, J.J. and Bernes, R.L. (1983) Pyrazolopyrimidine metabolism in the pathogenic Trypanosomatidae Molec. Biochem. Parasitol. 7, 339-356.

Marr, J.J., Bernes R.L. and Nelson, D.J. (1978) Purine metabolism in *Leishmania donovani* and *Leishmania braziliensis*. Biochim. Biophys. Acta. 544, 360-371.

Matias, C., Nott, S.E., Bagnara, A.S., O'Sullivan, W.J. and Gero, A.M. (1990) Purine salvage and metabolism in *Babesia bovis*. Parasitol. Res. 76, 207-213.

McInnis, A.J., Fisher, F.M. and Read, C.P. (1965) Membrane transport of purines and pyrimidines in a cestode. J. Parasitol. 51, 260-267.

Miller, R.L., Adamczyk, D.L., Rideout, J.L. and Krenitsky, T.A. (1982) Purification, characterization, substrate and inhibitor specificity of adenosine kinase from several *Eimeria* species. Molec. Biochem. Parasitol. 6, 209-223.

Miller, R.L. and Lindstead, D. (1983) Purine and pyrimidine metabolizing activities in *Trichomonas vaginalis* extracts. Molec. Biochem. Parasitol. **7**, 41-51.

Miller, R.L., Sabourin, C.L. and Krenitsky, T.A. (1987) *Trypanosoma cruzi* adenine nucleoside phosphorylase: purification and substrate specificity. Biochem. Pharmacol. 36, 553-559.

Miller, R.L., Sabourin, C.L.K., Krenitsky, T.A., Berens, R.L. and Marr, J.J. (1984) Nucleoside hydrolases from *Trypanosoma cruzi*. J. Biol. Chem. **259**, 5073-5077.

Morshige, K., Aji, T., Ishii, A., Yasuda, T. and Wataya, Y. (1995) *Leishmania donovani*: Pilot study for evaluation of therapeutic effect of inosine analogs against amastigotes *in vitro* and *in vivo*. Exptl. Parasitol. 80, 665-671.

Munagala, N.R., Chin, M.S. and Wang, C.C. (1998) Steady-state kinetics of the hypoxanthine-guanine-xanthine phosphoribosyltransferase from *Tritrichomonas foetus*: the role of threonine-47. Biochemistry 37, 4045-4051.

Munagala, N.R. and Wang, C.C. (1998) Altering the purine specificity of hypoxanthine-guanine-xanthine phosphoribosyltransferase from *Tritrichomonas foetus* by structure-based point mutations in the enzyme protein. Biochemistry 37, 16612-16619.

Naguib, F.N.M., Iltzsch, M.H., el Kouni, M.M., Panzica, R.P. and el Kouni, M.H. (1995) The binding of ligands to guanine and xanthine phosphoribosyltransferases from *Toxoplasma gondii*. Biochem. Pharmacol. 50, 1685-1693.

Nelson, D.J., LaFon, S.W., Tuttle, J.V., Miller, R.L., Miller, T.A., Krenitsky, T.A., Elion, G.B., Berens, R.L. and Marr, J.J. (1979) Allopurinol ribonucleoside as an antileishmanial agent. Biological effects, metabolism and enzymatic phosphorylation. *Leishmania donovani*, J. Biol. Chem. 254, 11544-115.

Ogbunude, P.O.J, al-Jaser, M.H. and Baer, H.P. (1991) *Leishmania donovani*: Characteristics of adenosine and inosine transporters in promastigotes of two different strains. Exptl. Parasitol. **73**, 369-375.

Ogbunude, P.O.J and Baer, H.P. (1993a) De novo purine biosynthesis in a free living protozoan parasite, *Acanthamoeba polyphaga*. Trop. Med. and Parasitol. 44, 19-22.

Ogbunude, P.O.J and Baer, H.P. (1993b) Nucleoside transport in parasite. Current status and methodological aspects. Int. J. Biochem. 25, 471-477.

Ogbunude P.O.J. and Ikediobi, C.O. (1982a) Interaction of a nucleoside transporter inhibitor nitrobenzylthioinosine with parasitic protozoa. IRCS Med. Sci. 10, 693-694.

Ogbunude P.O.J. and Ikediobi, C.O. (1982b) Effects of nitrobenzylthioinosinate on the toxicity of tubercidin and ethidium against *Trypanosoma gambiense*. Acta Tropica 39, 219-224.

Ogbunude P.O.J. and Ikediobi, C.O. (1983) Comparative aspects of purine metabolism in some African trypanosomes. Molec. Biochem. Parasitol. 9, 279-287.

Page, J.P., Munagala, N.R. and Wang, C.C. (1999) Point mutations in the guanine phosphoribosyltransferase from *Giardia lamblia* modulate pyrophosphate binding and enzyme catalysis. Eur. J. Biochem. 259, 565-571.

Parker, M.D., Hyde, R.J., Yao, S.Y., McRobert, L., Cass, C.E., Young, J.D., McConkey, G.A., Baldwin, S.A. (2000) Identification of a nucleoside/nucleobase transporter from Plasmodium falciparum, a novel target for anti-malarial chemotherapy. Biochem. J. 349, 67-75.

Parkin, D.W., Limberg, G., Tyler, P.C., Furneaux, Chen, X-Y. and Schramm, V.L. (1997) Isozyme-specific transition state inhibitors for the trypanosomal hydrolases. Biochemistry 36, 3528-3534.

Perrotto, J., Keister, D.B. and Gelderman, A.H. (1971) Incorporation of precursors into *Toxoplasma* DNA. J. Protozool. 18, 470-473.

Peters, W., Trotter, E.R. and Robinson, B.L. (1980) The experimental chemotherapy of leishmaniasis. Ann. Trop. Med. Parasitol. 74, 321-335.

Pfefferkorn, E.R. and Borotz, S.E. (1994) *Toxoplasma gondii*: Characterization of a mutant resistant to 6-thioxanthine. Exptl. Parasitol. 79, 374-382.

Pfefferkorn, E.R. and Pfefferkorn, L.C. (1976) Arabinosyl nucleosides inhibit *Toxoplasma gondii* and allow the selection of resistant mutants. J. Parasitol. 62, 993-999.

Pfefferkorn, E.R. and Pfefferkorn, L.C. (1978) The biochemical basis for resistance to adenine arabinoside in a mutant of *Toxoplasma gondii*. J. Parasitol. 64, 486-492.

Phillips, C.L., Ullman, B., Brennan, R.G. and Hill, C.P. (1999) Crystal structures of adenine phosphoribosyltransferase from *Leishmania donovani*. EMBO J. 18, 3533-3545.

Pitera, J.W., Munagala, N.R. Wang, C.C. and Kollman, P.A. (1999) Understanding substrate specificity in human and parasite phosphoribosyltransferases through calculation and experiment. Biochemistry 38, 10298-10306.

Pollack, Y., Shemer, R., Metzger, S., Spira, D.T. and Golenser (1985) APRT from *Plasmodium falciparum*: Expression of the adenine phosphoribosyltransferase gene in mouse cells. Exptl. Parasitol. 60, 270-275.

Pouvelle, B., Spiegel, R., Hsiao, L., Howard, R.J., Morris, R.L., Thomas, A.P. and Taraschi, T.F. (1991) Direct access to serum macromolecules by intraerythrocytes malaria parasites. Nature 353,73-75.

Queen, S.A., Jagt, D.V. and Reyes, P. (1988) Properties and substrate specificity of a purine phosphoribosyltransferase from the human malaria parasite *Plasmodium falciparum*. Molec. Biochem. Parasitol. 30, 123-134.

Rager, N., Mamoun, C.B., Carter, N.S., Goldberg, D.E., Ullman, B. (2001) Localization of the Plasmodium falciparum PfNT1 nucleoside transporter to the parasite plasma membrane. J. Biol. Chem. 276, 41095-41099.

Recacha, R., Talalaev, A., Delucas, L.J.and Chattopadhyay, D. (2000) *Toxoplasma gondii* adenosine kinase: expression, purification, characterization, crystallization and preliminary crystallographic analysis, Biological Crystal. 56,76-78.

Reyes, P., Rathod, P.K., Sanchez, D.J., Mrema, J.E.K., Rieckmann, K.H. and Heidrich, H.G. (1982) Enzymes of purine and pyrimidine metabolism from the human malaria parasite, *Plasmodium falciparum*. Molec. Biochem. Parasitol. 5, 275-290.

Riscoe, M.K., Ferro, A.J. and Fitchen, J.H. (1989) Methionine recycling as a target for antiprotozoal drug development. Parasitol. Today 5, 330-333.

Saenz, R.E., Paz, H.M., Johnson, C.M., Marr, J.J., Nelson, D.J., Pattishall, K.H. and Rogers, M.D. (1989) Treatment of American cutaneous leishmaniasis with orally administered allopurinol riboside. J. infect. Diseases 160, 153-158.

Savarese, T.M., Ghoda, L.Y., Dexter, D.L. and Parks, R.E.Jr. (1983) Conversion of 5'-deoxy-5'-methylthioadenosine and 5'-deoxy-5'-methylthioinosine to methionine in cultured human leukemic cells. Cancer Res. 43, 4699-4702.

Savarese, T.M., Harrington, S. and el Kouni, M.H. (1989) Adenine nucleoside phosphorylase and purine nucleoside phosphorylase in *Schistosoma mansoni*. J. Cell. Biol. 107, 396a.

Schlenk, F. and Ehninger, D.J. (1964) Observations on the metabolism of 5'-methylthioadenosine. Arch. Biochem. Biophys. 106, 95-100.

Schmidt, G., Walter, R.D. and Konigk, E. (1975) A purine nucleoside hydrolase in *Trypanosoma gambiense*, purification and properties. Tropenmed. Parasitol. 26, 19-26.

Schumacher, M.A., Carter, D., Roos, D.S., Ullman, B. and Brennan, R.G. (1996) Crystal structures of *Toxoplasma gondii* HGXPRTase reveal the catalytic role of a long flexible loop. Nature Structural Biol. 3, 881-887.

Schumacher, M.A., Scott, D.M., Mathews II, Ealick, S.E., Roos, D.S., Ullman, B. and Brennan, R.G. (2000) Crystal structures of *Toxoplasma gondii* adenosine kinase reveal a novel catalytic mechanism and prodrug binding. J. Molec. Biol. 296, 549-567.

Schwab, J.C., Afifi, M.A., Pizzorno, G., Handschumacher, R.E. and Joiner, K.A. (1995) *Toxoplasma gondii* tachyzoites possess an unusual plasma membrane adenosine transporter. Molec. Biochem. Parasitol. 70, 59-69.

Schwab, J.C., Beckers, C.J.M. and Joiner, K.A. (1994) The parasitophorous vacuole membrane surrounding intracellular *Toxoplasma gondii* as a membrane sieve. Proc. Natl. Acad. Sci. USA 91, 509-513.

Schwartzman, J.D. and Pfefferkorn, E.R. (1982) *Toxoplasma gondii*: purine synthesis and salvage in mutant host cells and parasites. Exptl. Parasitol. 53, 77-86.

Senft, A.W. and Crabtree, G.W. (1983) Purine metabolism in schistosomes: Potential Targets for chemotherapy. Pharmac. Ther. 20, 341-356.

Shapiro, S.K. and Barrett, A. (1981) 5-Methylthioribose as a precursor of the carbon chain of methionine. Biochem. Biophys. Res. Commun. 102, 302-307.

Shapiro, S.K. and Schlenk, F. (1980) Conversion of 5'-methylthioadenosine into S-adenosylmethionine by yeast cells. Biochim. Biophys. Acta. 633, 176-180.

Sherman, I.W. (1988) Mechanism of molecular trafficking in malaria. Parasitol. 96, S57-S81.

Shi, W., Munagala, N.R., Wang, C.C., Li, C.M., Tyler, P.C., Furneaux, R.H., Grubmeyer, C., Schramm, V.L. and Almo, S.C. (2000) Crystal structures of *Giardia lamblia* guanine phosphoribosyltransferase at 1.75 $. Biochemistry 39, 6781-6790.

Sommer, J.M., Ma, H. and Wang, C.C. (1999) Cloning, expression and characterization of an unusual guanine phosphoribosyltransferase form *Giardia lamblia*. Molec. Biochem. Parasitol. 103, 1-14.

Somoza, J.R., Skillman, A.G. Jr., Munagala, N.R., Oshiro, C.M., Knegtel, R.M., Mpoke, S., Fletterick, R.J., Kuntz, I.D. and Wang, C.C. (1998) Rational design of novel antimicrobials: blocking purine salvage in a parasitic protozoan. Biochemistry 37, 5344-5348.

Sugimoto, Y., Toraya, T. and Fukui, S. (1976) Studies on the metabolic role of 5'-methylthioadenosine in *Ochromonas malhamensis* and other microorganisms. Arch. Microbiol. 108, 175-182.

Sullivan, W.J.Jr., Chiang, C.W., Wilson, C.M., Naguib, F.N.M., el Kouni, M.H., Donald, R.G.K. and Roos, D.S. (1999) Insertional tagging of at least two loci associated with resistance to adenine arabinoside in *Toxoplasma gondii*, and cloning of the adenosine kinase locus. Molec. Biochem. Parasitol. 103, 1-14.

Tuttle, J.V. and Krenitsky, T.A. (1979) Purine phosphoribosyltransferase from *Leishmania donovani*. J. Biol. Chem. 255, 9098–916.

Upston, J.M. and Gero, A.M., (1995) Parasite induced permeation of nucleosides in *Plasmodium falciparum* malaria. Biochim. Biophys. Acta 1236, 249-258.

Vasanthkumar, G., Davis Jr., R.L., Sullivan, M.A. and Donahue, J.P. (1990) Cloning and expression in *Escherichia coli* of a hypoxanthine-guanosine phosphoribosyltransferase-encoding c-DNA from *Plasmodium falciparum*. Gene 91, 63-69.

Vasudevan, G., Carter, N.S., Drew, M.E., Beverley, S.M.., Sanchez, M.A., Seyfang, A., Ullman, B., Landfear, S.M. (1998) Cloning of Leishmania nucleoside transporter genes by rescue of a transport-deficient mutant. Proc. Natl. Acad. Sci. U.S.A. 95,9873-8.

Walton, B.C., Harper, J. and Neal, R.A. (1983) Effectiveness of allopurinol against *Leishmania braziliensis panamensis* in *Aotus trivirgatus*. Am. J. Trop. Med. Hyg. 32, 46-50.

Wang, C.C. and Aldritt, J. (1983) Purine salvage networks in *Giardia lamblia*. J. Exptl. Med. 185,1703-1712.

Wang, C.C. and Simashkevich, P.M. (1981) Purine metabolism in the protozoan parasite *Eimeria tenella*. Proc. Natl. Acad. Sci. USA 78, 6618-6622.

Wang, C.C., Verham, R., Rice, A. and Tzeng, S. (1983) Purine salvage by *Tritrichomonas foetus*. Molec. Biochem. Parasitol. 8, 325-337.

Wang, S.Y., Adams, D.O. and Lieberman, M. (1982) Recycling of 5'-methylthioadenosine-ribose carbon atoms into methionine in tomato tissue in relation to ethylene production. Plant Physiol. 70, 117-121.

Wong, P.C.L., and Ko, R.C. (1979) *De novo* purine ribonucleotide biosynthesis in adult *Angiostrongylus cantonensis* (Nematoda: Metastrongyloidea) Comp. Biochem. Physiol. 62B, 129-132.

Wong, P.C.L. and Ko, R.C. (1980) Purine ribonucleotide biosynthesis in the gravid *Metastrongylus cantonensis* (Nematoda: Metastrongyloidea) Comp. Biochem. Physiol. 65B, 303-308.

Wong, P.C.L. and Yeung, S.B. (1981) Pathways of purine ribonucleotide biosynthesis in the adult *Metastrongylus apri* (Nematoda: Metastrongyloidea) from pig lung. Molec. Biochem. Parasitol. 2, 285-293, 1981.

Yuan, L., Wu, C.S.C., Craig III, S.P., Liu, A.F. and Wang, C.C. (1993) Comparing the human and schistosomal hypoxanthine-guanine phosphoribosyltransferases by circular dichroism. Biochim. Biophys. Acta 1162, 10-16.

Yung, K.H., Yang, S.F. and Schlenk, F. (1982) Methionine synthesis from 5-methylthioribose in apple tissue. Biochem. Biophys. Res. Commun. 104, 771-774.

ANTIVIRAL β-L-NUCLEOSIDES SPECIFIC FOR HEPATITIS B VIRUS INFECTION

JEAN-PIERRE SOMMADOSSI

*Novirio Pharmaceuticals, Inc., 125 CambridgePark Dr.,
Cambridge, MA 02140, USA*

1.   Abstract

Three simple, related nucleosides, β-L-2'-deoxycytidine (L-dC), β-L-thymidine (L-dT), and β-L-2'-deoxyadenosine (L-dA), have been discovered to be potent, specific and selective inhibitors of the replication of hepatitis B virus (HBV) as well as the closely related duck and woodchuck hepatitis viruses (WHV). Structure-activity relationship analysis indicates that the 3'-OH group of the β-L-2'-deoxyribose of the β-L-2'-deoxynucleoside confers specific anti-hepadnavirus activity. These simple nucleosides had no effect on the replication of 15 other RNA and DNA viruses, did not inhibit human DNA polymerases (α, β, and γ) or compromise mitochondrial function. The nucleosides are efficiently converted intracellularly into active triphosphate metabolites that have a long half life. Once daily oral administration of these compounds in the woodchuck efficacy model of chronic HBV infection reduced viral load by as much as 10^8 genome equivalents/ml serum and there was no drug-related toxicity. In addition, a decline in WHV surface antigen (WHsAg) paralleled the decrease in viral load. This class of nucleosides displays an excellent overall safety profile. The first compound, L-dT, has already entered phase IIb clinical trials and L-dC, currently being developed as a prodrug, is in phase I/II studies. These compounds have the potential for use in combination therapy with the goal of achieving superior viral suppression and diminishing the onset of resistance.

2.   Introduction

This chapter summarizes the findings on a series of three structurally simple β-L-nucleosides, L-dA, L-dC, and L-dT, that were recently identified as potent, selective and highly specific inhibitors of HBV replication. Each compound has been shown to exhibit an excellent safety profile in preclinical testing. The first member of this series to reach clinical trials was L-dT. A dose escalation study in HBV infected patients is currently in progress under a US IND.

Recent Advances in Nucleosides: Chemistry and Chemotherapy, Ed. by C.K. Chu. 417 — 432

3.   Results and discussion

3.1.   The β-L-nucleosides, L-dA, L-dC and L-dT, are specific and selective inhibitors of hepadnaviruses

The structures of the β-L-nucleosides L-dA, L-dC, and L-dT, which are collectively known as the Novirio NV-02 series, are shown in Figure 1. These molecules are simple in structure and closely resemble the natural β-D-deoxynucleosides deoxyadenosine (dAdo), thymidine (thd) and deoxycytidine (dCyd). Unlike most nucleoside analogs, they exhibit no chemical modifications and differ from their natural nucleoside counterparts only with respect to the spatial relationship of their base and sugar moieties. They have an L-configuration *versus* the D-configuration of the natural deoxynucleosides.

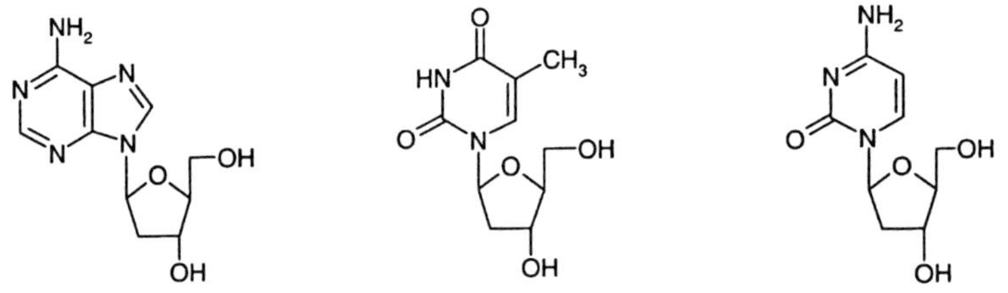

L-dA (NV-02A)   **L-dT (NV-02B)**   **L-dC (NV-02C)**

Figure 1.   Structures of β-L-nucleoside compounds.

An extensive structure-activity analysis of Novirio's nucleoside collection identified the NV-02 series molecules as the most potent, selective and specific inhibitors of hepatitis B virus replication in the HepG 2.2.15 tissue culture assay. The structure–activity relationships (SAR) established among the β-L-2'-deoxycytidine, -thymidine and -deoxyadenosine series are presented in Table 1, which compares the antiviral activity of these molecules against HBV and HIV.

The data reveal L-dC, L-dT and L-dA to be potent inhibitors of HBV replication (EC_{50}s in the 100 to 250 nM range) with excellent specificity as shown by their lack of activity against HIV. Closer examination of the SAR shows that the key to obtaining specific inhibitors of HBV is the hydroxyl (-OH) group in the 3'-position of the β-L-2'-deoxyribose sugar. This is most clearly seen in the L-dC series, where only compounds retaining the 3'-OH moiety (e.g., β-L-2'-deoxy-5-fluorocytidine, L-5-FdC and β-L-2'-deoxy-5-chlorocytidine, L-5-CldC) are specific inhibitors of HBV. Conversely, replacement of the 3'-OH (R3) on the deoxyribose sugar resulted in several instances in molecules with good activity against both HBV and HIV, reflecting a loss of antiviral specificity (e.g., β-L-2',3'-dideoxycytidine, L-ddC; β-L-2',3'-dideoxy-3'-thiacytidine, 3TC; β-L-2',3'-didehydro-2',3'-dideoxycytidine, L-d4C).

In the L-dA series, the activity of β-L-2',3'-didehydro-2',3'-dideoxyadinosine (L-d4A) against both HIV and HBV again shows that specificity is lost along with the R3 OH group.

Table 1.   Structure-activity relationship of β-L-2'-deoxynucleosides

	R1	R2	R3	X	anti-HBV (2.2.15 cells)	anti-HIV (PBM cells)
L-dC	H	H	OH	CH	0.24 0.08	>200
L-5-FdC	F	H	OH	CH	5	>100
L-5-CldC	Cl	H	OH	CH	10	>100
L-ddC	H	H	H	CH	0.1	0.26
3TC	H	H	–	S	0.05 0.01	0.002
L-3'-azido-5-FddC	F	H	N_3	CH	0.11 ± 0.09	0.05
L-3'-FddC	H	H	F	CH	0.5	82
FTC	F	H	–	S	0.04	0.008
L-5-ClddC	Cl	H	H	CH	10	>100
L-d4C	H	–	–	CH	<0.1	1.0
L-d4FC	F	–	–	CH	<0.1	0.034
L-3'-F-5-FddC	F	–	F	CH	4	>100
L-5-FddC	F	–	–	CH	0.10 ± 0.05	0.021
L-dT		H	OH		0.19 ± 0.09	>200
L-ddT		H	H		>10	>100
L-3'-FddT		H	F		>10	>100
L-3'-azido-ddT		H	N_3		>10	>100
L-3'-amino-ddT		H	NH_2		>10	>10
L-d4T		–	–		>10	>100
L-xylo-dT		OH	H		>10	>10
L-dA	H	H	OH		0.10 - 1.9	>10
L-2-CldA	Cl	H	OH		>10	>10
L-ddA	H	H	H		5	>10
L-d4A	H	–	–		0.80 ± 0.10	0.38
L-3'-azido-ddA	H	H	N_3		5	>10
L-3'-amino-ddA	H	H	NH_2		>10	>10
L-3'-fluoro-ddA	H	H	F		>10	>100
L-ddAMP-bis(tbutylSATE)	H	H	H		0.08 ± 0.03	0.002
L-3'-azido-d4A	H	–	N_3		>10	>100

[a] Antiviral 50% effective concentration (EC50). The greater than symbol (>) is used to indicate the highest concentration at which the compounds were tested. Values represent the means of at least three independent experiments. Anti-HIV data for L-ddC, 3TC, FTC, L-5-FddC, L-d4FC from references (Gosselin *et al.*, 1994, Schinazi *et al.*, 1992, Shi *et al.*, 1999). L-d4T, L-ddA and L-d4A data from references (Bolon *et al.*, 1996, Gosselin *et al.*, 1997).

Similarly, in the L-dT series, exclusive specificity for HBV is only seen in the presence of the 3'-OH group. In this case, only L-dT itself is active, while closely related molecules are not, suggesting that the 3'-OH group may play a role in determining the affinity of the molecule for the HBV polymerase.

To further assess their antiviral activity and specificity, L-dC, L-dT and L-dA were screened against 15 different RNA and DNA viruses (Table 2). The striking finding was that the β-L-2'-deoxynucleosides, L-dC, L-dT and L-dA, inhibited HBV replication as well as the replication of the closely related duck hepatitis B virus (DHBV). However, they had no activity against HIV-1, HSV-1, HSV-2, VZV, EBV, HCMV, adenovirus type-1, influenza A and B, measles virus, parainfluenza type-3, rhinovirus type-5 and RSV type-A at concentrations up to 100 μM. Potent antiviral activity against the woodchuck hepatitis B virus (WHV) using an *in vivo* model of chronic hepatitis B virus infection is described later in this chapter. Thus, the unmodified β-L-2'-deoxynucleosides L-dC, L-dT and L-dA, exhibit an unusual degree of specificity for inhibiting members of the small family of hepadnaviruses, HBV, DHBV, and WHV.

Table 2.   Antiviral Activity of L-dC, L-dT and L-dA

Virus[a]	Cell line	EC_{50} (μM)[b]			CC_{50} (μM)[b]		
		L-dC	L-dT	L-dA	L-dC	L-dT	L-dA
HBV	2.2.15	0.24	0.19	0.10	>2000	>2000	>1000
DHBV	PDH	0.87	0.18	0.15	ndc	nd	nd
HIV-1	PBMC	>100	>100	>100	>100	>100	>100
HSV-1	HFF	>20	>200	>100	>60	>200	>100
HSV-2	HFF	>100	>100	>100	>100	>100	>100
VZV	HFF	>100	45.2	>100	>100	18.6	>100
EBV	Daudi	>50	>50	5.7	>50	>50	23.1
HCMV	HFF	>100	>100	>100	>100	>100	>100
adenovirus type-1	A549	>100	nd	>100	>100	nd	>100
influenza A	MDCK	>100	>100	>100	>100	>100	>100
influenza B	MDCK	>100	>100	>100	>100	>100	>100
measles	CV-1	>100	>100	>100	>100	>100	>100
parainfluenza type-3	MA-104	>100	>100	>100	>100	>100	>100
rhinovirus type-5	KB	>100	nd	>100	>100	nd	>100
RSV type-A	MA-104	>100	>100	>100	>100	>100	>100

[a]   The specific antiviral activity of L-dC, L-dT and L-dA was confirmed using a panel of viruses tested by the NIH NIAID Antiviral Research and Antimicrobial Chemistry Program.

[b]   Antiviral 50% effective concentration (EC_{50}) and 50% cytotoxic concentration (CC_{50}). PDH, primary duck hepatocytes; PBMC, peripheral blood mononuclear cells; HFF, human foreskin fibroblast; Daudi, Burkitt's B-cell lymphoma; A549, human lung carcinoma; MDCK, canine kidney epithelial cells; CV-1, African green monkey kidney fibroblast cells; MA-104, Rhesus monkey kidney epithelial cells; KB, human nasopharyngeal carcinoma.

[c]   nd, not determined.

The majority of nucleoside analogs with antiviral activity inhibit the viral replication step via direct interaction of their 5'-triphosphate metabolites with the respective viral polymerase. Consistent with this idea, the 5'-triphosphates of the β-L-2'-deoxynucleosides, L-dC, L-dA and L-dA inhibit the WHV DNA polymerase in *in vitro* assays with 50% inhibitory concentration (IC_{50}) values of 0.24–1.82 μM (data not shown). By analogy with other nucleoside analogs, L-dC, L-dT and L-dA likely inhibit the reverse transcription of pregenomic RNA and/or the synthesis of HBV second-strand DNA. They may, however, have a different mechanism of action through inhibition of the unique HBV priming reaction. Additionally, it is possible that these compounds may inhibit other important activities of the polymerase (which include RNascH activity, the hepadnavirus-specific priming of reverse transcription and the co-ordination of intracellular virion assembly). Should the antiviral mechanism of action differ significantly from that of lamivudine, it is possible that the genotypes of variants selected under drug pressure will be fully sensitive to combination therapy.

In addition to antiviral specificity, the selectivity of antiviral drugs becomes a critical factor in determining whether they will ultimately be suitable for use in human patients. This is particularly true when long-term therapy is required, as is the case for chronic HBV infection. Toxic side effects, primarily related to non-selective interaction with cellular polymerases, have been a major limitation for the clinical use of several nucleoside analogs (Faulds and Brogden, 1992, Hurst and Noble, 1999, Whittington and Brogden, 1992, Wilde and Langtry, 1993).

When tested in *in vitro* polymerase assays using purified human DNA polymerases α, β and γ, the 5'-triphosphates of L-dC, L-dT and L-dA did not inhibit enzymatic activity at concentrations up to 100 μM (data not shown). Krayevsky and coworkers also reported that the 5'-triphosphates of L-dC and L-dT were not substrates for human DNA polymerases (Semizarov *et al.*, 1997). Thus, these compounds are highly selective for viral *versus* host cell polymerases. It has also been previously reported that the 5'-triphosphate of L-dT is inactive against the HIV reverse transcriptase (von Janta-Lipinski *et al.*, 1998).

Further evidence for the selectivity of these compounds comes from the lack of cytotoxicity seen for the NV-02 nucleosides (Table 2), implying a lack of effect on host cell functions. When tested against 10 different cell lines, L-dC, L-dT and L-dA showed little or no evidence of cytotoxicity at concentrations greater than 100 μM. In particular, L-dC, L-dT and L-dA had no cytotoxic effect on primary human peripheral blood mononuclear cells (PBMC), human foreskin fibroblasts (HFF), or other cell types of mammalian origin (Table 2). In addition, studies by Verri *et al.* (1997) demonstrated that L-dC was not cytotoxic toward lymphoblastoid T cells. Finally, these compounds were not cytotoxic in the human hepatoma cell line 2.2.15 (CC_{50} values > 2,500 μM).

4.   Intracellular activation, metabolism, and pharmacology

Metabolic pathways have been worked out for L-dT and L-dC based on extensive intracellular accumulation and decay data, and on competition experiments using the corresponding endogenous D-nucleosides. These pathways are summarized in Figure 2 for L-dT and Figure 3 for L-dC. L-dT is converted into the triphosphate (TP) form by redundant cellular nucleoside/nucleotide kinases, whereas formation of L-dCTP utilizes only deoxycytidine nucleoside/nucleotide kinases.

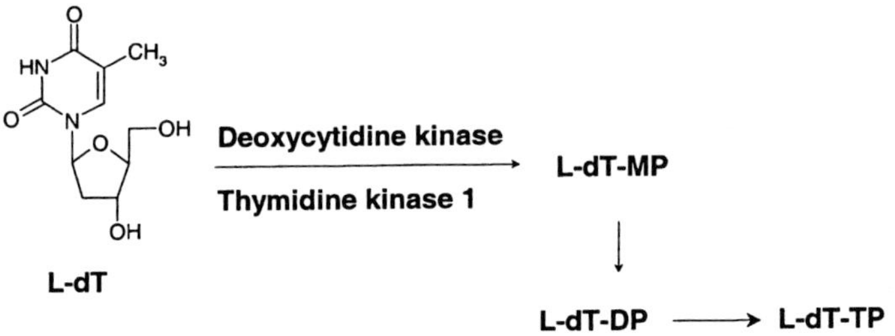

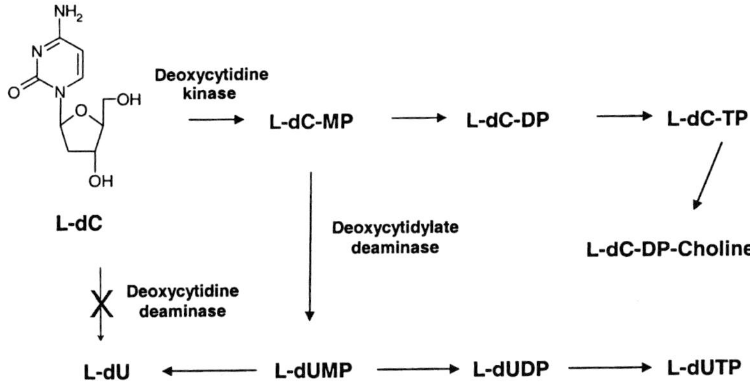

Figure 2.   Proposed metabolic pathway for L-dT.

Figure 3.   Proposed metabolic pathway for L-dC.

L-dC, L-dT and L-dA are metabolized (activated) efficiently to their respective 5'-triphosphate derivatives in HepG2 cells and human hepatocytes in primary culture (Placidi *et al.*, 1999). This is in contrast to earlier studies reporting limited intracellular activation of L-dT (Focher *et al.*, 1995, Spadari *et al.*, 1992). The metabolic profiles obtained after a 24-hour exposure of HepG2 cells to 10 μM [^{3}H]-L-dT and [^{3}H]-L-dC are shown in Figure 4. L-dT was efficiently converted into the active triphosphate form, which reached a peak concentration of just below 30 μM at 24 h. The mono- and

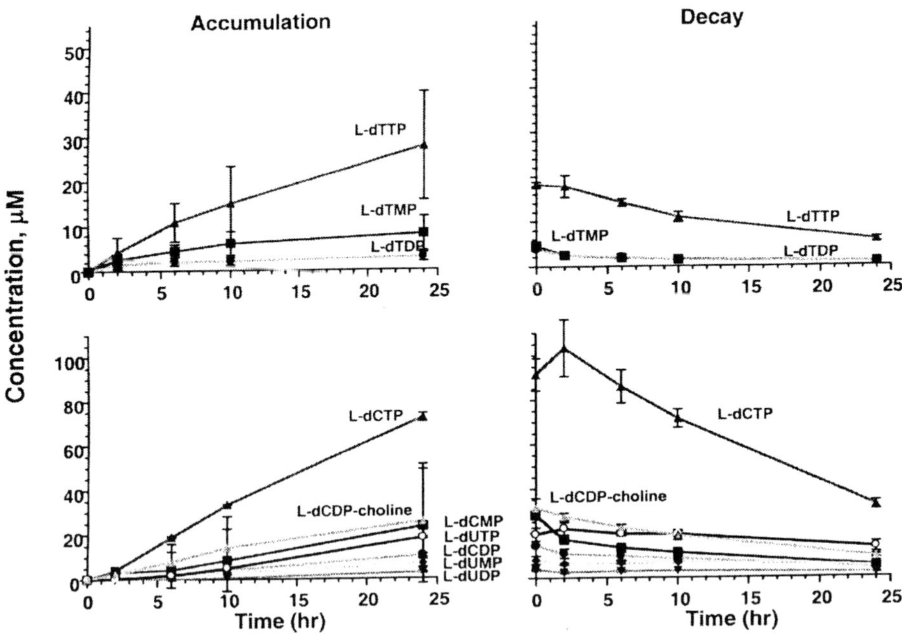

Figure 4.   Intracellular accumulation and decay of metabolites after 24 h exposure of HepG2 cells to 10 µM L-dT or L-dC.

diphosphate forms were present intracellularly at much lower levels. For L-dC, the triphosphate form again accumulated efficiently, reaching a maximal intracellular concentration of 70 µM, but the metabolic pathway is more complex. Along with L-dCTP, a second 5'-triphosphate derivative, corresponding to β-L-2'-deoxyuridine 5'-triphosphate (L-dUTP), was formed. Similar to β-L-cytidine analogs (Chang *et al.*, 1992, Furman *et al.*, 1992, Martin *et al.*, 1997, Verri *et al.*, 1997), L-dC was not a substrate for cytosolic cytidine deaminase; thus, deoxycytidylate deaminase acting on L-dC 5'-monophosphate is presumed to explain the formation of this metabolite. Another metabolite corresponding to a choline form of L-dCDP was also detected in HepG2 cells. However, the important point is that these metabolites are minor; their formation does not significantly diminish the concentration of the active L-dCTP species.

Metabolic decay experiments revealed the apparent intracellular half-lives of the L-dT-5'-TP and L-dC-5'-TP to be long, i.e., $\geq$ 14 hours. Thus, even after 24 h, the intracellular TP concentrations were well in excess of the estimated IC_{50} values (~0.24–1.82 µM for the WHV DNA polymerase) and remained above the IC_{90} values (~5 µM).

In summary, the efficient conversion of L-dT and L-dC into high concentrations of the respective triphosphate forms, coupled with the long half lives of the triphosphates, creates a favorable scenario for HBV antiviral therapy.

5.   Pharmacokinetic profiles

The pharmacokinetic profile of L-dT in the cynomologous monkey is presented in Figure 5. Following intravenous administration, plasma concentrations of L-dT declined in a bi-exponential manner and to undetectable levels after 8 hours. The observed terminal phase half-life was ~1.5 hr in monkeys and somewhat longer (~3.5 hr) in woodchucks. The total clearance was higher in monkeys (~0.60 l/hr/kg) than in woodchucks (~0.30 l/hr/kg). The apparent volume of distribution (Vd) indicated good tissue distribution in both species. Oral absorption of L-dT was slow in monkeys and in woodchucks, with peak concentrations occurring 1 to 4 hours after dosing. The absolute oral bioavailability (%F) for L-dT reached 68.6% in monkeys and 38.3% in woodchucks.

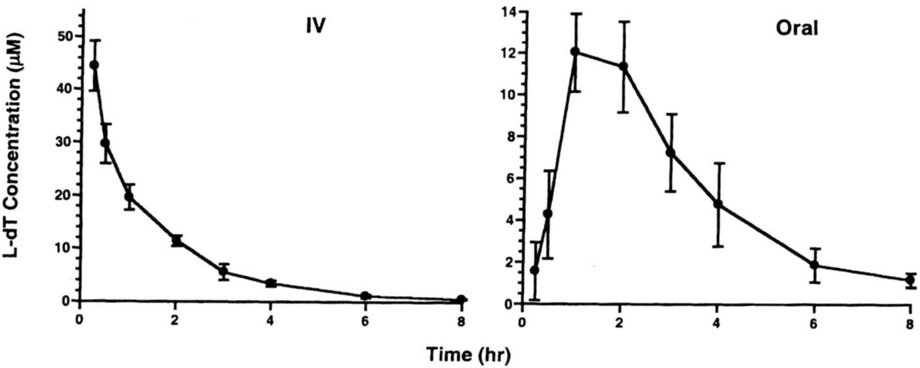

Figure 5.   Plasma concentration in monkeys after intravenous (IV) or oral administration of 10 mg/kg L-dT. The data are the mean (±SD) from three animals per group.

The oral bioavailability of L-dC was lower and more variable than L-dT in woodchucks (9.6%) and monkeys (16.4%). To improve oral absorption, a series of ester prodrugs was synthesized. The oral bioavailability of the 3', 5' valine ester prodrug of L-dC, increased at least 4-fold in monkeys compared to L-dC.

6.   Antiviral activity in the woodchuck chronic hepatitis model

Woodchucks chronically infected with WHV are widely accepted as a model of HBV infection and have proven useful in the evaluation of anti-HBV agents. This model has been shown to be a positive predictor of antiviral activity as well as safety for the treatment of human chronic HBV infection (Tennant *et al.*, 1998, Korba *et al.*, 1990, Korba *et al.*, 2000).

The study for assessing efficacy in this model involved 4 weeks of daily treatment (3 animals per group) with 10 mg/kg/day L-dT or L-dC (delivered by oral gavage) and 8 weeks of follow-up. The study included two control arms: a placebo arm and a 10 mg/kg/day lamivudine treatment arm. Serum levels of WHV DNA were determined throughout the study by DNA dot-blot hybridization (detection limit, approximately 10^7 genome equivalents/ml serum) and by quantitative PCR (detection limit, 300 genome equivalents/ml serum).

WHV DNA replication was significantly inhibited within the first few days of treatment with either L-dT or L-dC, whereas placebo levels remained unaffected (data not shown). Most notably, serum WHV DNA levels (WHV viremia) decreased up to 8 logs to below the limit of detection by PCR in the L-dT treated group (see below) and decreased by 4 to 6 logs in the L-dC treated animals (data not shown). WHV DNA levels rebounded to near pre-treatment levels by 8 weeks following drug withdrawal.

In contrast, the cytidine analog lamivudine (10 mg/kg/d) reduced the HBV genome equivalents/ml in serum by only 0.5 to 1.0 log. This limited effect is consistent with previous studies using similar doses of lamivudine (Genovesi *et al.*, 1998). Higher doses (40–200 mg/kg) of this drug are required to produce significant antiviral activity in this model (Mason *et al.*, 1998). The low activity of lamivudine in the woodchuck model has been ascribed in part to poor absorption and in part to the low conversion of lamivudine and other cytidine analogs to their active 5'-triphosphate forms seen in rodent/woodchuck liver compared to that in human liver. The oral bioavailability of lamivudine in woodchucks has been reported to be 18% – 54% *versus* 82% in humans (Rajagopalan *et al.*, 1996, van Leeuwen *et al.*, 1992). With these caveats in mind, the performance of L-dC in the woodchuck is surprisingly good, suggesting that the ester prodrug of L-dC, which had an oral bioavailability of four times that of L-dC in the monkey, should have good potency against HBV in human patients.

For the L-dT treated animals, which showed the most marked reduction in viral load, we also observed a decline in WHV surface antigen as measured using the method of Cote, *et al.* (Cote *et al.*, 1993). The data are summarized in Figure 6. The strength of the surface antigen response broadly paralleled the viral load response, but the onset of the surface antigen response was delayed by at least one week compared to the reduction in viral load. Surface antigen levels continued to fall for several weeks after drug removal before rebounding. This result is intriguing since a correlation has been demonstrated in this model between HBsAg reduction and the clearance of cccDNA from infected hepatocytes (Cote *et al.*, 1993).

In a separate 12 week study in the woodchuck, the combination of 1 mg/kg/day L-dT and 1 mg/kg/day L-dC reduced viral load to levels significantly lower than either agent alone. This combination of L-dT and L-dC (each at one-tenth the concentration of monotherapy) reduced viral load to the limit of detection (300 genome equivalents per ml serum). Following drug removal, the time to viral rebound was markedly prolonged when L-dT and L-dC were administered in combination. A dramatic decrease in hepatitis B surface antigen, as a marker of viral replication, was also seen (data not shown). In both the 28 day study and the 12 week study, no toxicity was seen at the highest dose tested.

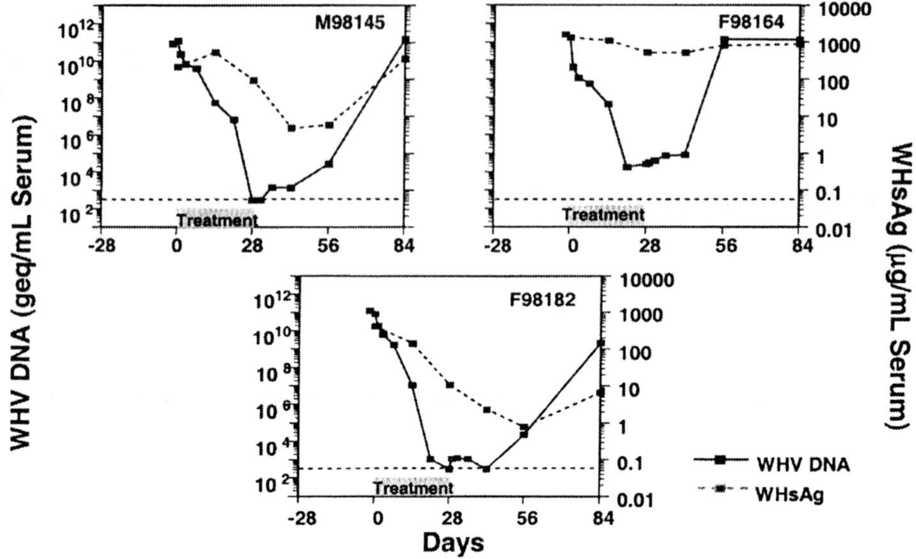

Figure 6.   Reduction of serum virus load and WHsAg in the Woodchuck chronic HBV model.

7.   Safety profile of L-dT, L-dC and L-dA

As discussed earlier in this chapter, long-term therapy is expected for chronic HBV infection. Thus, the nucleoside safety profile is a critical issue, particularly since clinically limiting side effects have been well documented for some nucleoside analogs (Faulds and Brogden, 1992, Hurst and Noble, 1999, Whittington and Brogden, 1992, Wilde and Langtry, 1993). The lack of inhibitory activity of the L-dT, L-dC and L-dA triphosphates has been discussed earlier along with the lack of cytotoxicity in a number of different mammalian cell lines. Described briefly below are additional safety studies that have been performed with these compounds and in particular with the lead compound, L-dT.

Human bone marrow stem cells in primary culture have been shown to be a good predictor of potential nucleoside analog-induced hematotoxicity in patients (Faraj *et al.*, 1994, Sommadossi *et al.*, 1989). Granulocyte-macrophage (CFU-GM) and erythroid (BFU-E) precursors exposed to L-dC, L-dT and L-dA in clonogenic assays, which routinely detect the cellular toxicity of zidovudine were not affected (Table 3). These results suggest that L-dC, L-dT and L-dA are highly selective and their phosphorylated forms will be non-toxic *in vivo*.

Nucleoside analogs used in AIDS therapy, such as zidovudine (AZT, β-D-3'-azido-3'-deoxythymidine), stavudine (d4T, β-L-2',3'-didehydro-2',3'-dideoxythymidine) didanosine

Table 3.   Human bone marrow toxicity of L-dA, L-dT and L-dC in granulocyte macrophage progenitor and erythrocyte precursor cells

Compound	CFU-GM[a] IC$_{50}$ (μM)	BFU-E[b] IC$_{50}$ (μM)
L-dA	> 40	> 10
L-dT	> 40	> 10
L-dC	> 40	> 10
ZDV	1.9 ± 1.2	0.6 ± 0.5

[a]   CFU-GM: Granulocyte macrophage progenitor cells, colony forming units.

[b]   BFU-E: Erythrocyte precursor cells, burst forming units.

(ddI, β-D-2',3'-dideoxyinosine) and zalcitabine (ddC, β-D-2',3'-dideoxycytidine), have shown clinically limiting delayed toxicities such as peripheral neuropathy, myopathy, and pancreatitis (Faulds and Brogden, 1992, Hurst and Noble, 1999, Whittington and Brogden, 1992, Wilde and Langtry, 1993). These adverse effects are attributable to decreased mitochondrial DNA (mtDNA) content and/or altered mitochondrial function leading to increased lactic acid production and hepatic steatosis (Chen and Cheng, 1989, Cui *et al.*, 1997, Cui *et al.*, 1996, Cui *et al.*, 1995, Dalakas *et al.*, 1990, Lewis *et al.*, 1992, Pan-Zhou *et al.*, 2000). Concomitant morphological changes in mitochondria (e.g., loss of cristae, matrix dissolution and swelling, and lipid droplet formation) can be observed with ultrastructrual analysis using transmission electron microscopy (Cui *et al.*, 1996, Lewis *et al.*, 1996, Pan-Zhou *et al.*, 1998). For example, fialuridine (1,2'-deoxy-2'-fluoro-1-β-D-arabinofuranosly-5-iodo-uracil, FIAU) toxicity was shown to be associated with an irreversible intracellular event that decreased mitochondrial respiratory function, resulting in decreased mitochondrial ATP production and fatty acid metabolism. This form of mitochondrial toxicity can be initially identified in cell culture by increased lactic acid production and intracellular lipid droplet formation. In HepG2 cells incubated with 10 μM FIAU, a substantial increase in lactic acid production was observed (Table 4). Electron micrographs of these cells showed the presence of enlarged mitochondria with morphological changes consistent with mitochondrial dysfunction (data not shown). Lamivudine (10 μM) did not affect mitochondrial structure or function. Using similar conditions, exposure of HepG2 cells to 10 μM L-dC, L-dT or L-dA for 14 days had no effect on lactic acid production, mitochondrial DNA content or morphology (Table 4).

In acute (50 to 2000 mg/kg single oral dose) and subacute (500 to 2000 mg/kg/day orally for 28 days) toxicology studies of L-dT in rats and monkeys there were no overt signs of toxicity, nor were there any L-dT related effects on body weight, food consumption, or clinical pathology parameters (hematology and serum chemistry). In addition, there were no macroscopic lesions observed at necropsy, nor were there any microscopic findings on histomorphological analysis attributable to L-dT. Based on

Table 4.   Effect of L-dC, L-dT and L-dA on mitochondria in HepG2 cells

Compound	Conc. (μM)	% of control			Lipid Droplet Formation	Mitochondrial Morphology
		Cell Density	L-Lactate	mtDNA		
Control		100	100	100	neg[a]	normal
L-dC	0.1	102 12	100 4	105 11	nd	nd
	1.0	100 ± 6	101 ± 6	99 ± 10	nd	nd
	10	101 ± 10	101 ± 2	107 ± 8	neg	normal
L-dT	0.1	103 ± 7	102 ± 2	103 ± 4	nd[b]	nd
	1.0	106 ± 8	99 ± 2	101 ± 7	nd	nd
	10	97 ± 7	105 ± 2	97 ± 4	neg	normal
L-dA	0.1	103 ± 14	99 ± 3	97 ± 14	nd	nd
	1.0	102 ± 14	102 ± 3	92 ± 8	nd	nd
	10	100 ± 14	103 ± 5	88 ± 18	neg	normal
Lamivudine[c]	0.1	101 ± 2	99 ± 5	107 ± 8	nd	nd
	1.0	99 ± 1	101 ± 3	96 ± 9	nd	nd
	10	99 ± 1	98 ± 3	98 ± 10	neg	normal
FIAU[c]	0.1	83 ± 6	119 ± 5	101 ± 2	nd	nd
	1.0	73 ± 9	134 ± 9	118 ± 5	nd	nd
	10	37 ± 10	203 ± 13	86 ± 4	positive	abnormal

HepG2 cells were treated with the indicated concentrations of L-dT, L-dC or L-dA for 14 days. Values are presented as means and standard deviations of three independent experiments.

[a]   neg, negative.

[b]   nd, not determined.

[c]   Data from reference (Wilde and Langtry, 1993, Lewis *et al.*, 1992).

the results of these studies, the no observed adverse effect level (NOAEL) for L-dT following a single oral dose, or repeated dosing for 28 days by oral gavage in the Sprague-Dawley rat and cynomologus monkey was 2000 mg/kg.

In normal healthy woodchucks or woodchucks chronically infected with hepatitis B virus, no toxicity was observed during acute (10 mg/kg single dose IV and PO) and subacute (28 days at 10 mg/kg/day orally and 12 weeks at 1 mg/kg/day orally) studies. There was no weight loss in the L-dT treatment groups compared to control animals, clinical pathology parameters (hematology and serum chemistry) were in the normal range and end of treatment liver biopsies in the 12-week study showed no evidence of fatty change (microvesicular steatosis).

Genotoxicity assays have been completed on L-dT and L-dC. Neither compound was mutagenic in the *S. typhimurium* or *E. coli* plate incorporation mutagenicity assay at concentrations up to a maximum of 5000 μg/plate tested. There was no evidence of chromosomal aberrations in the Chinese hamster ovary (CHO) assay after exposure to L-dT or L-dC at concentrations up to a maximum of 5000 μg/mL (20.6 mM). In the mouse micronucleus assay, L-dT and L-dC were not clastogenic to male or female animals (maximum dose tested 2000 mg/kg).

8. Summary

Three structurally simple β-L-2'-deoxynucleosides, L-dT, L-dC, L-dA, and have been identified as highly specific and selective inhibitors of HBV replication. The presence of a hydroxyl group in the 3'-position appears to be the key to the specificity of these nucleosides towards hepatitis B viruses. L-dT and L-dC are potent inhibitors of WHV replication in the woodchuck efficacy model of chronic hepatitis B virus infection, reducing serum viral titers by as much as eight logs. The excellent safety profile of the first member of this series, L-dT, coupled with its antiviral potency and selectivity, has led to its entry into a phase I/II clinical trial in HBV infected patients under a US IND recently followed by initiation of a phase IIb. An ester prodrug of L-dC is currently in clinical trials. It is also anticipated that these new nucleosides will be used in combination (i.e., L-dT and L-dC or L-dT and lamivudine) to further reduce chronic HBV replication and prevent the selection of resistant virus.

9. References

Bolon, P. J., Wang, P., Chu, C., Gosselin, G., Boudou, V., Pierra, C., Mathe, C., Imbach, J. L., Faraj, A., Alaoui, A., Sommadossi, J.-P., Pai, S. B., Zhu, Y. L., Lin, J. S., Cheng, Y. C. and Schinazi, R. F. (1996) Anti-human immunodeficiency and anti-hepatitis B virus activities of beta-L-2',3'-dideoxy purine nucleosides, Bioorg Med Chem Lett, 6, 1657-1662.

Chang, C. N., Doong, S. L., Zhou, J. H., Beach, J. W., Jeong, L. S., Chu, C. K., Tsai, C. H., Cheng, Y. C., Liotta, D. and Schinazi, R. (1992) Deoxycytidine deaminase-resistant stereoisomer is the active form of (+/-)-2',3'-dideoxy-3'-thiacytidine in the inhibition of hepatitis B virus replication, J Biol Chem, 267, 13938-42.

Chen, C. H. and Cheng, Y. C. (1989) Delayed cytotoxicity and selective loss of mitochondrial DNA in cells treated with the anti-human immunodeficiency virus compound 2',3'- dideoxycytidine, J Biol Chem, 264, 11934-7.

Cote, R. J., Roneker, C., Cass, K., Schodel, F., Peterson, D., Tennant, B., De Noronha, F. and Gerin, J. (1993) New enzyme immunoassays for the serologic detection of woodchuck hepatitis virus infection, Viral Immunol, 6, 161-9.

Cui, L., Locatelli, L., Xie, M. Y. and Sommadossi, J.-P. (1997) Mitochondrial DNA effect of nucleoside analogs on neurite regeneration and mitochondrial DNA synthesis in PC-12 cells, J Pharmacol Exp Ther, 280, 1228-1234.

Cui, L., Schinazi, R. F., Gosselin, G., Imbach, J.-L., Chu, C. K., Rando, R. F., Revankar, G. R. and Sommadossi, J.-P. (1996) Effect of β-enantiomeric and racemic nucleoside analogues on mitochondrial funtions in HepG2 cells, Biochem Pharmacol, 52, 1577-1584.

Cui, L., Yoon, S., Schinazi, R. F. and Sommadossi, J.-P. (1995) Cellular and molecular events leading to mitochondrial toxicity of 1-(2-deoxy-2-fluoro-1-□-D-arabinofuranoxyl)-5-iodouracil in human liver cells, J Clin Invest, 95, 555-563.

Dalakas, M. C., Illa, I., Pezeshkpour, G. H. and al., e. (1990) Mitochondrial myopathy caused by long-term zidovudine therapy, N Engl J Med, 322, 1098-1105.

Faraj, A., Fowler, D. A., Bridges, E. G. and Sommadossi, J.-P. (1994) Effects of 2',3'-dideoxynucleosides on proliferation and differentiation of human pluripotent progenitors in liquid culture and their effects on mitochondrial DNA synthesis, Antimicrob Agents Chemother, 38, 924-930.

Faulds, D. and Brogden, R. N. (1992) Didanosine. A review of its antiviral activity, pharmacokinetic properties and therapeutic potential in human immunodeficiency virus infection, Drugs, 44, 94-116.

Focher, F., Maga, G., Bendiscioli, A., Capobianco, M., Colonna, F., Garbesi, A. and Spadari, S. (1995) Stereospecificity of human DNA polymerases alpha, beta, gamma, delta and epsilon, HIV-reverse transcriptase, HSV-1 DNA polymerase, calf thymus terminal transferase and Escherichia coli DNA polymerase I in recognizing D- and L-thymidine 5'-triphosphate as substrate, Nucleic Acids Res, 23, 2840-7.

Furman, P. A., Davis, M., Liotta, D. C., Paff, M., Frick, L. W., Nelson, D. J., Dornsife, R. E., Wurster, J. A., Wilson, L. J., Fyfe, J. A., Tuttle, J. V., Miller, W. H., Condreay, L., Averett, D. R., Schinazi, R. F. and Painter, G. R. (1992) The anti-hepatitis B virus activities, cytotoxicities, and anabolic profiles of the (-) and (+) enantiomers of cis-5-fluoro-1-[2-(hydroxymethyl)-1,3-oxathiolan-5-yl]cytosine, Antimicrob Agents Chemother, 36, 2686-92.

Genovesi, E. V., Lamb, L., Medina, I., Taylor, D., Seifer, M., Innaimo, S., Colonno, R. J., Standring, D. N. and Clark, J. M. (1998) Efficacy of the carbocyclic 2'-deoxyguanosine nucleoside BMS-200475 in the woodchuck model of hepatitis B virus infection, Antimicrob Agents Chemother, 42, 3209-17.

Gosselin, G., Boudou, V., Griffon, J. F., Pavia, G., Pierra, C., Imbach, J. L., Aubertin, A. M., Schinazi, R. F., Faraj, A. and Sommadossi, J. P. (1997) New unnatural L-nucleoside enantiomers: From their stereoselective synthesis to their biological activities, Nucleosides Nucleotides, 16, 1389-98.

Gosselin, G., Mathe, C., Bergogne, M. C., Aubertin, A. M., Kirn, A., Schinazi, R. F., Sommadossi, J. P. and Imbach, J. L. (1994) Enantiomeric 2',3'-dideoxycytidine derivatives are potent human immunodeficiency virus inhibitors in cell culture, C R Acad Sci III, 317, 85-9.

Hurst, M. and Noble, S. (1999) Stavudine: an update of its use in the treatment of HIV infection, Drugs, 58, 919-49.

Korba, B. E., Cote, P., Hornbuckle, W., Tennant, B. C. and Gerin, J. L. (2000) Treatment of chronic woodchuck hepatitis virus infection in the eastern woodchuck (Marmota monax) with nucleoside analogues is predictive of therapy for chronic hepatitis B virus infection in humans, Hepatology, 31, 1165-75.

Korba, B. E., Cote, P. J., Tennant, B. C. and Gerin, J. L. (1990) Woodchuck hepatitis virus infection as a model for the development of antiviral therapies against HBV. In: Ed, Hollinger, F. B., Lemon, S.M., Margolis, H.) Viral hepatitis and liver disease, Williams and Wilkins, Baltimore, pp. 663-5.

Lewis, W., Gonzalez, B., Chomyn, A. and Papoian, T. (1992) Zidovudine induces molecular, biochemical, and ultrastructural changes in rat skeletal muscle mitochondria, J Clin Invest, 89, 1354-60.

Lewis, W., Levine, E. S., Griniuviene, B., Tankersly, K. O., Colacino, J. M., Sommadossi, J. P., Watanabe, K. A. and Perrino, F. W. (1996) Fialuridine and its metabolites inhibit DNA polymerase gamma at sites of multiple adjacent analog incorporation, decrease mtDNA abundance, and cause mitchondrial structural defects in cultured hepatoblasts, Proc Natl Acad Sci U S A, 93, 3592-7.

Martin, L. T., Faraj, A., Schinazi, R. F., Gosselin, G., Mathe, C., Imbach, J.-L. and Sommadossi, J.-P. (1997)

Effect of stereoisomerism on the cellular pharmacology of beta- enantiomers of cytidine analogs in Hep-G2 cells, Biochem Pharmacol, 53, 75-87.

Mason, W. S., Cullen, J., Moraleda, G., Saputelli, J., Aldrich, C. E., Miller, D. S., Tennant, B., Frick, L., Averett, D., Condreay, L. D. and Jilbert, A. R. (1998) Lamivudine therapy of WHV-infected woodchucks, Virology, 245, 18-32.

Pan-Zhou, X.-R., Cretton-Scott, E., Zhou, X.-J., Yang, M.-X., Lasker, J. M. and Sommadossi, J.-P. (1998) Role of human liver P450s and cytochrome b5 in the reductive metabolism of 3'-azido-3'-deoxythymidine (AZT) to 3'-amino3'-deoxythymidine, Biochem Pharmacol, 55, 757-766.

Pan-Zhou, X. R., Cui, L., Zhou, X. J., Sommadossi, J.-P. and Darley-Usmar, V. M. (2000) Differential effects of antiretroviral nucleoside analogs on mitochondrial function: dual inhibition of citrate synthase and cytochrome c oxidase by AZT, Antimicrob Agents Chemother, 44, 496-503.

Placidi, L., Hernández, B., Cretton-Scott, E., Faraj, A., Bryant, M., Imbach, J.-L., Gosselin, G., Pierra, C., Dukhan, D. and Sommadossi, J.-P. (1999) Cellular pharmacology of β-L-thymidine (L-dT, NV-02B) and β-L-2'-deoxycytidine (L-dC, NV-02C) in HepG2 cells and primary rat, monkey and human hepatocytes, Antivir Ther, 4, A122.

Rajagopalan, P., Boudinot, F. D., Chu, C. K., Tennant, B. C., Baldwin, B. H. and Schinazi, R. F. (1996) Pharmacokinetics of (-)-2'-3'-dideoxy-3'-thiacytidine in woodchucks, Antimicrob Agents Chemother, 40, 642-5.

Schinazi, R. F., McMillan, A., Cannon, D., Mathis, R., Lloyd, R. M., Peck, A., Sommadossi, J.-P., St. Clair, M., Wilson, J., Furman, P. A., Painter, G., Choi, W.-B. and Liotta, D. C. (1992) Selective inhibition of human immunodeficiency viruses by racemates and enantiomers of cis-5-fluoro-1-[2-hydroxymethyl)-1,3-oxatiolan-5-yl]cytosine, Antimicrob Agents Chemother, 36, 2423-31.

Semizarov, D. G., Arzumanov, A. A., Dyatkina, N. B., Meyer, A., Vichier-Guerre, S., Gosselin, G., Rayner, B., Imbach, J.-L. and Krayevsky, A. A. (1997) Stereoisomers of deoxynucleoside 5'-triphosphates as substrates for template-dependent and -independent DNA polymerases, J. Biol Chem, 272, 9556-60.

Shi, J., McAtee, J. J., Schlueter Wirtz, S., Tharnish, P., Juodawlkis, A., Liotta, D. C. and Schinazi, R. F. (1999) Synthesis and biological evaluation of 2',3'-didehydro-2',3'- dideoxy-5- fluorocytidine (D4FC) analogues: discovery of carbocyclic nucleoside triphosphates with potent inhibitory activity against HIV-1 reverse transcriptase, J Med Chem, 42, 859-67.

Sommadossi, J.-P., Carlisle, R. and Zhou, Z. (1989) Cellular pharmacology of 3'-azido-3'-deoxythymidine with evidence of incorporation into DNA of human bone marrow cells, Mol Pharmacol, 36, 9-14.

Spadari, S., Maga, G., Focher, F., Ciarrocchi, G., Manservigi, R., Arcamone, F., Capobianco, M., Carcuro, A., Colonna, F., Iotti, S. and et al. (1992) L-thymidine is phosphorylated by herpes simplex virus type 1 thymidine kinase and inhibits viral growth, J Med Chem, 35, 4214-20.

Tennant, B. C., Peek, S. F., Tochkov, I. A., Baldwin, B. H., Hornbuckle, W. E., Korba, B. E., Cote, P. J. and Gerin, J. L. (1998) The woodchuck in preclinical assessment of therapy for hepatitis B virus infection. In: (Eds, Schinazi, R. F., Sommadossi, J.-P. and Thomas, H. C.)Therapies for Viral Hepatitis, 171-176, International Medical Press, London, pp. 171-176.

van Leeuwen, R., Lange, J. M., Hussey, E. K., Donn, K. H., Hall, S. T., Harker, A. J., Jonker, P. and Danner, S. A. (1992) The safety and pharmacokinetics of a reverse transcriptase inhibitor, 3TC, in patients with HIV infection: a phase I study, AIDS, 6, 1471-5.

Verri, A., Focher, F., Priori, G., Gosselin, G., Imbach, J. L., Capobianco, M., Garbesi, A. and Spadari, S. (1997) Lack of enantiospecificity of human 2'-deoxycytidine kinase: relevance for the activation of beta-L-deoxycytidine analogs as antineoplastic and antiviral agents, Mol Pharmacol, 51, 132-8.

von Janta-Lipinski, M., Costisella, B., Ochs, H., Hubscher, U., Hafkemeyer, P. and Matthes, E. (1998)

Newly synthesized L-enantiomers of 3'-fluoro-modified β-2'-deoxyribonucleoside 5'-triphosphates inhibit hepatitis B DNA polymerases but not the five celluar DNA polymerases a,β,γ,δ, and ε nor HIV-1 reverse transcriptase, J Med Chem, 41, 2040-2046.

Whittington, R. and Brogden, R. N. (1992) Zalcitabine. A review of its pharmacology and clinical potential in acquired immunodeficiency syndrome (AIDS), Drugs, 44, 656-83.

Wilde, M. I. and Langtry, H. D. (1993) Zidovudine. An update of its pharmacodynamic and pharmacokinetic properties, and therapeutic efficacy, Drugs, 46, 515-78.

ANTIVIRAL ACTIVITY OF NUCLEOSIDE ANALOGUES: THE BVDU CONNECTION

ERIK DE CLERCQ

Division of Virology and Chemotherapy, Department of Microbiology and Immunology, Rega Institute for Medical Research, K.U.Leuven, Minderbroedersstraat 10, B-3000 Leuven, BELGIUM

1. Introduction

BVDU [(E)-5-(2-bromovinyl)-2'-deoxyuridine, brivudin] was originally synthesized in 1976 at the Chemistry Department of the University of Birmingham by P.J. Barr, A.S. Jones and R.T. Walker, as a potential radiation-sensitizing agent (assuming that it would be incorporated into DNA). Its potent and selective activity against herpes simplex virus type 1 (HSV-1) was first mentioned at the FEBS (Federation of European Biochemical Societies) Symposium on "Antimetabolites in Biochemistry, Biology and Medicine" (Prague, Czechoslovakia, 10-12 July 1978) (De Clercq *et al.*, 1979a) and the IVth Symposium on the Chemistry of Nucleic Acid Components (Bechyne Castle, Czechoslovakia, 3-10 September 1978) (Walker *et al.*, 1978). When it was discovered, BVDU, and its closely related congener, IVDU [(E)-5-(2-iodovinyl)-2'-deoxyuridine] proved more potent and more selective in their activity against HSV-1 than all other anti-herpes compounds (De Clercq *et al.*, 1979b), and this has virtually remained so, now more than 20 years later.

At the joint NATO Advanced Study Institute/FEBS Advanced Study Course held at Sogesta (near Urbino) in Italy (7-18 May 1979) on "Nucleoside Analogues: Chemistry, Biology and Medical Applications", P. Langen presented a long list of 5-substituted 2'-deoxyribopyrimidine nucleosides as anti-HSV-1 agents, the most active on the list being 5-(1-bromovinyl)-2'-deoxyuridine, a compound obtained by the selective bromination and subsequent dehydrobromination of 5-ethyl-2'-deoxyuridine (Bärwolff and Langen, 1975). As it turned out later, the compound thus synthesized was not the 5-(1-bromovinyl)- but 5-(2-bromovinyl)-2'-deoxyuridine, and thus the superiority of BVDU over other anti-HSV-1 agents (Reefschläger *et al.*, 1982) was confirmed in a truly blinded fashion.

The discovery of BVDU as a selective anti-herpesvirus agent came shortly after that of acyclovir [9-(2-hydroxyethoxymethyl)guanine] (Elion *et al.*, 1977; Schaeffer *et al.*, 1978). Shortly thereafter, the 2'-fluoro-2'-deoxyarabinofuranosylpyrimidine nucleosides, and particularly 2'-fluoro-5-iodoaracytosine (FIAC) were reported as potent and selective anti-herpesvirus agents (Watanabe *et al.*, 1979; Lopez *et al.*, 1980). For many years these three compounds (acyclovir, BVDU and FIAC) (Figure 1) would remain the

433

Recent Advances in Nucleosides: Chemistry and Chemotherapy, Ed. by C.K. Chu. 433 — 454

"gold standards" or reference compounds for the development of new, and potentially more effective and/or selective antiviral agents (De Clercq, 1985a).

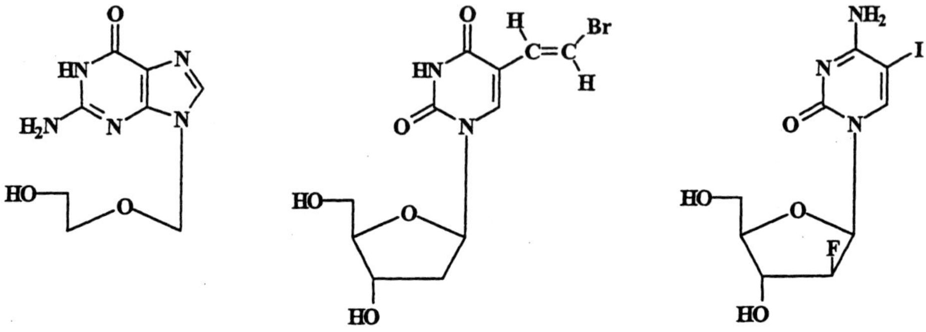

Figure 1. The "top three" anti-herpesvirus agents of the early eighthies: acyclovir (ACV), brivudin (BVDU) and fluoroiodoaracytosine (FIAC).

From a (clinical) therapeutic viewpoint, the three compounds fared quite differently. Acyclovir became (worldwide) the drug of choice for the treatment of HSV-1, HSV-2 and varicella-zoster virus (VZV) infections (although it has now been replaced by its prodrug, valaciclovir, for the oral treatment of these infections). BVDU has been used for many years, albeit at a limited scale, for the topical treatment of herpetic keratitis, and for the oral treatment of VZV infections (in particular, shingles in immunocompromised patients). FIAC, and its uracil counterpart FIAU, once considered for the treatment of hepatitis B virus (HBV) infections, are no longer pursued, but their L-conformer (L-FMAU) is still in (preclinical) development for the treatment of HBV infections.

The antiviral potency and selectivity of BVDU, its activity spectrum, mechanism of action, structure-function relationship relative to that of other 5-substituted 2'-deoxy-uridines, and clinical efficacy relative to that of acyclovir, FIAC and other anti-herpes agents have been reviewed repeatedly in the years that followed the initial discovery of BVDU as a selective HSV-1 inhibitor (De Clercq, 1980, 1982a,b, 1983a,b, 1984a,b,c, 1985b, 1986; De Clercq *et al.*, 1981, 1985; De Clercq and Walker, 1984, 1986). From these initial studies BVDU not only emerged as a potent and selective inhibitor of HSV-1 (De Clercq *et al.*, 1980a), but also of VZV replication. In fact, in a comparative study of various anti-herpes drugs against VZV, the EC50 (50% effective concentration) of BVDU was 0.0024 μg/ml, as compared to 4.64 μg/ml for acyclovir, attesting to a more than 103-fold superiority in potency of BVDU over acyclovir (Shigeta al., 1983).

Within two years after its discovery, BVDU was introduced in the clinic, in Belgium, for the oral treatment of severe herpes zoster (De Clercq *et al.*, 1980b) and topical treatment of herpes simplex keratitis (Maudgal *et al.*, 1981). All patients, whether adults (Wildiers and De Clercq, 1984) or children (Benoit *et al.*, 1985), who were treated with oral BVDU for zoster or varicella, responded promptly to the treatment without any adverse side effects. In the treatment of herpetic keratitis, BVDU proved successful where other antiviral drugs including 5-iodo-2'-deoxyuridine (idoxuridine), 5-trifluoromethyl-2'-deoxyuridine (trifluridine), 9-β-D-arabinofuranosyladenine (vidarabine) and 9-(2-hydroxyethoxymethyl)guanine (acyclovir) had failed (Maudgal and De Clercq, 1991). Independently from the clinical studies in Belgium, BVDU was also pursued in the former Deutsche Demokratische Republik (DDR), for the treatment of herpetic keratitis (Töpke *et al.*, 1984) and VZV infections in patients with malignancies (Wutzler *et al.*, 1988). In these studies, as in ours, the compound proved clearly efficacious, without any adverse side effects.

BVDU (Brivudin) has been licensed, first in the DDR, then Germany, for the treatment of severe VZV and HSV-1 infections in immunosuppressed patients: it is administered orally at 4 × 125 mg per day for 5 days, as tablets (Helpin®). This makes that there are three "standard" treatments available for varicella-zoster virus infections in Germany: brivudin (Helpin®), valaciclovir (Valtrex®, Zelitrex®) and the diacetyl ester of 9-(4-hydroxy-3-hydroxymethyl-but-1-yl)-6-deoxyguanine (famciclovir) (Famvir®).

The biological properties of BVDU are clearly dependent on the presence of its (E)-2-bromovinyl substituent, which, therefore, can be considered as the pharmacophore of the molecule. Here I will provide an update on the activity spectrum, mechanism of action, and other unique properties that is shared by BVDU and its congeners and that are all determined by the presence of the (E)-2-bromovinyl entity.

2. Antiviral activity spectrum

The antiviral activity spectrum of BVDU is not restricted to HSV-1 and VZV but also encompasses several other herpesviruses such as suid herpesvirus type 1 (SHV-1), bovid herpesvirus type 1 (BHV-1), simian varicella virus (SVV), herpesvirus saimiri, and herpesvirus platyrrhinae (Figure 2). Also, Epstein-Barr virus (EBV) is rather sensitive to BVDU, whereas HSV-2 and cytomegalovirus (CMV) are relatively resistant to the antiviral action of the compound (De Clercq, 1984d). Murine herpesvirus 68 (MHV-68), a murine gamma herpesvirus closely related to EBV, is also sensitive to BVDU, albeit to a lesser extent than EBV (Neyts and De Clercq, 1998). More sensitive are the bovine herpes mammillitis virus (Harkness *et al.*, 1986), the macropodid herpesvirus 1 (Smith and Whalley, 1998) and the macropodid herpesvirus 2 (Smith, 1996). The macropodid herpesviruses have been held responsible for the death of kangaroos and wallabies in European and North American zoos, and BVDU has been considered the drug of choice for experimental therapy of herpesvirus infections in captive macropodids (Harkness *et al.*, 1986; Smith, 1996). Like human CMV [which corresponds to human herpesvirus type 5 (HHV-5)], human herpesvirus type 6 (HHV-6) (Reymen *et al.*, 1995), human herpesvirus type 7 (HHV-7) (Zhang *et al.*, 1999) and human herpesvirus type 8 (HHV-8, or Kaposi's sarcoma-associated herpesvirus) (Neyts and De Clercq, 1997) showed little, if any, sensitivity to BVDU. The characteristic activity spectrum of BVDU (Figure 2)

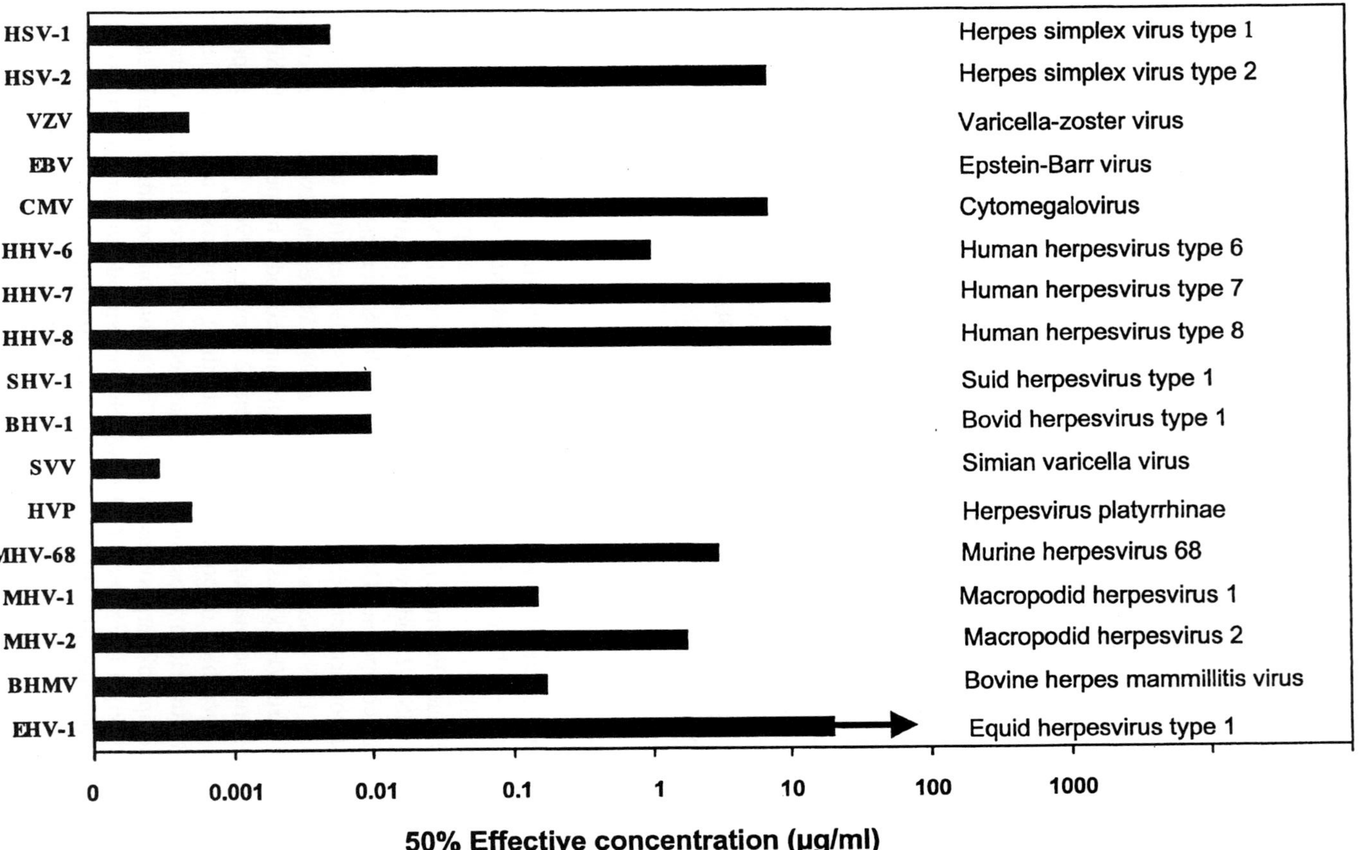

Figure 2. Antiviral activity spectrum of BVDU.

thus explains why, from a human clinical viewpoint, the compound has been primarily pursued for the treatment of HSV-1 and VZV infections.

3.　Mechanism of action

The exquisite potency of BVDU against HSV-1 and VZV, in comparison with the potency of other antiviral compounds, has been demonstrated with various clinical isolates of both HSV-1 (Andrei *et al.*, 1992) and VZV (Andrei *et al.*, 1995). The mechanism of action of BVDU against HSV-1 and VZV (Figure 3) depends on a specific phosphorylation by the virus-encoded thymidine kinase (TK), the HSV-1 TK and VZV TK, which converts BVDU to its 5'-monophosphate (BVDU-MP) and 5'-diphosphate (BVDU-DP) (Descamps and De Clercq, 1981). Upon further phosphorylation by cellular kinase(s), i.e. nucleoside 5'-diphosphate (NDP) kinase, BVDU 5'-triphosphate (BVDU-TP) can then interact with the viral DNA polymerase, either as a competitive inhibitor with respect to the natural substrate (dTTP) (Allaudeen *et al.*, 1981), or as an alternative substrate, allowing the incorporation of BVDU-TP (as BVDU-MP) into the growing DNA chain (Figure 3).

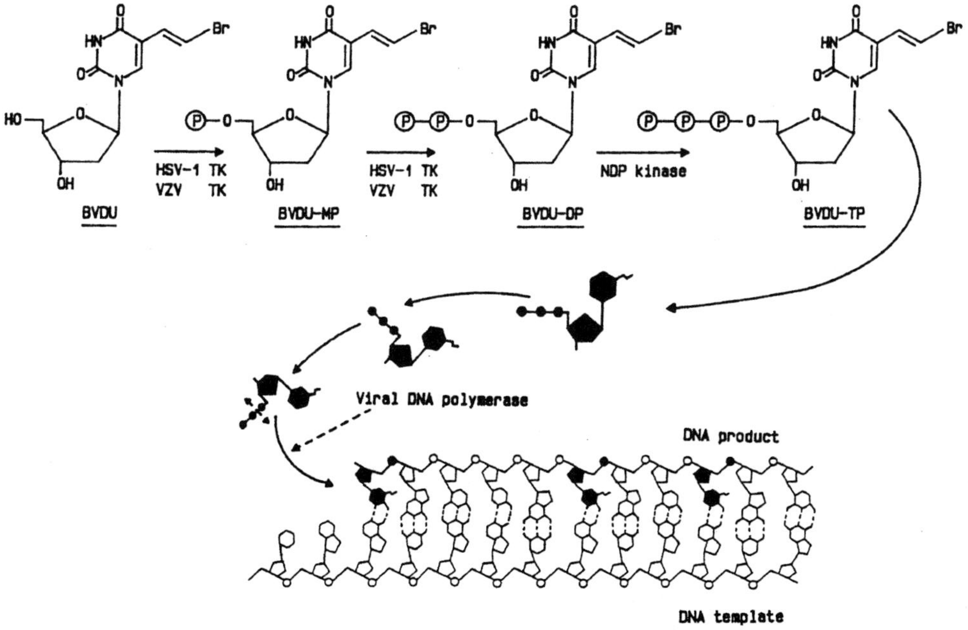

Figure 3.　Mechanism of action of BVDU. Following uptake by the (virus-infected) cells, BVDU is phosphorylated by the virus-encoded thymidine kinase (TK) to the 5'-monophosphate (BVDU-MP) and 5'-diphosphate (BVDU-DP), and further on to the 5'-triphosphate (BVDU-TP) by cellular kinases, i.e. nucleoside 5'-diphosphate (NDP) kinase. BVDU-TP can act as a competitive inhibitor/alternative substrate of the viral DNA polymerase, and as a substrate it can be incorporated internally (via internucleotide linkages) into the (growing) DNA chain.

This incorporation may affect both the stability and functioning of the DNA during the replication and transcription processes. In fact, a close correlation has been found between the incorporation of BVDU into HSV-1 DNA, DNA integrity and viral infectivity (Mancini *et al.*, 1983; Balzarini *et al.*, 1990b).

A remarkable feature in the antiviral specificity of BVDU is that it is a highly potent inhibitor of HSV-1 but not HSV-2, so that it can be used as a marker for differentiating HSV-2 from HSV-1 strains (De Clercq, 1984d). The reason for the differential sensitivity of HSV-1 and HSV-2 towards BVDU resides in the fact that the HSV-2-encoded, unlike its HSV-1 counterpart, together with its TK activity, also possesses dTMP kinase activity, but is unable to phosphorylate BVDU 5'-monophosphate onto BVDU 5'-diphosphate (Fyfe, 1982). This results in a substantial reduction in the supply of the active BVDU metabolite, BVDU-TP, in the HSV-2-infected cells (Ayisi *et al.*, 1984), and, thus, reduced ability to block viral DNA synthesis. BVDU-MP may interact as an alternate substrate (Barr *et al.*, 1983) or inhibitor (Yokota *et al.*, 1994) of thymidylate synthase, but it is questionable that the interference of BVDU-MP with dTMP synthase contributes to the antiviral potency that is eventually achieved by BVDU.

Thus, the predominant determinant in the antiviral activity of BVDU is the virus-encoded thymidine kinase (TK). The latter can apparently be reduced, independently from the TK activity, as is the case for HSV-2 and also some HSV-1 isolates that have been more recently described (Jennings Wilber and Docherty, 1994). One of these HSV-1 isolates had a single mutation (G → A at base position 502) that resulted in the substitution of threonine for alanine at amino acid position 168 in the viral TK: this led to a decreased dTMP kinase activity, concomitantly with a reduced sensitivity of the viral isolate towards BVDU.

4.   Clinical efficacy

BVDU and its arabinofuranosyl counterpart BVaraU belong to the most potent inhibitors of VZV that have ever been described: BVDU inhibits VZV replication in cell culture at an EC50 of 0.001-0.003 μg/ml, and BVaraU at an even 3-fold lower EC50 (Shigeta *et al.*, 1983; Andrei *et al.*, 1995). This exquisite potency has prompted the pursuit of both BVDU and BVaraU for the treatment of VZV infections in immunocompromised patients. The closest experimental model for VZV infections in humans is simian varicella virus (SVV) infection in monkeys, and BVDU was found effective in suppressing this disease when administered orally at either 15, 10 or 5, or even 1 mg/kg/day (Soike *et al.*, 1981). When given orally at 7.5 mg/kg/day (that is 4 × 125 mg-tablets per day), BVDU proved as efficacious as acyclovir given intravenously at 30 mg/kg/day, in the treatment of herpes zoster in immunocompromised patients (Wutzler *et al.*, 1995).

In another randomized double-blind controlled clinical trial (Gnann *et al.*, 1998), BVaraU (sorivudine) administered orally at 40 mg daily was compared with acyclovir, also given orally at 4 g (five times 800 mg) daily, both over a 10-day course, in the treatment of dermatomal herpes zoster in patients infected with human immunodeficiency virus (HIV): as shown in Figure 4, BVaraU effected a slightly faster cessation of new

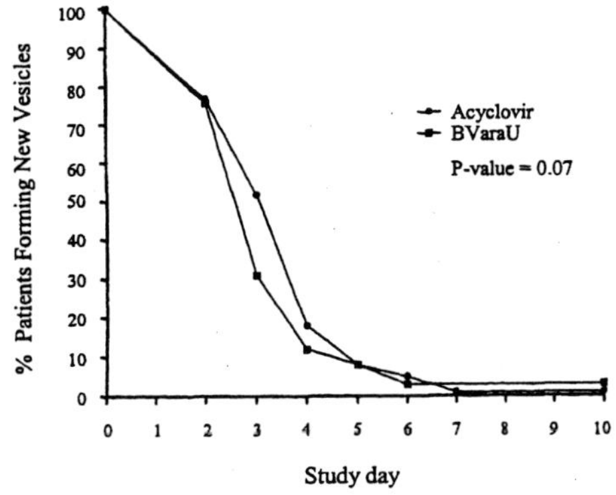

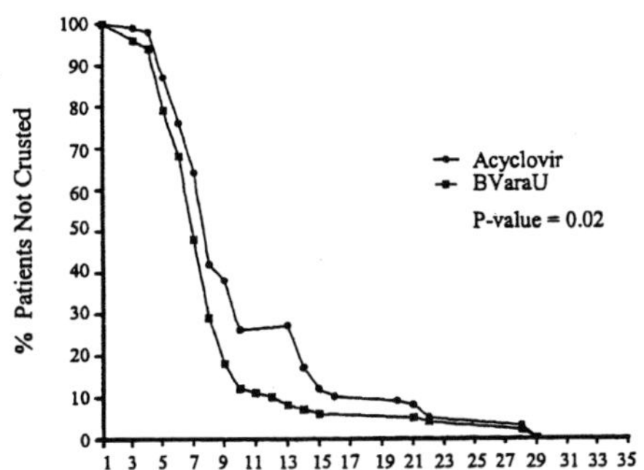

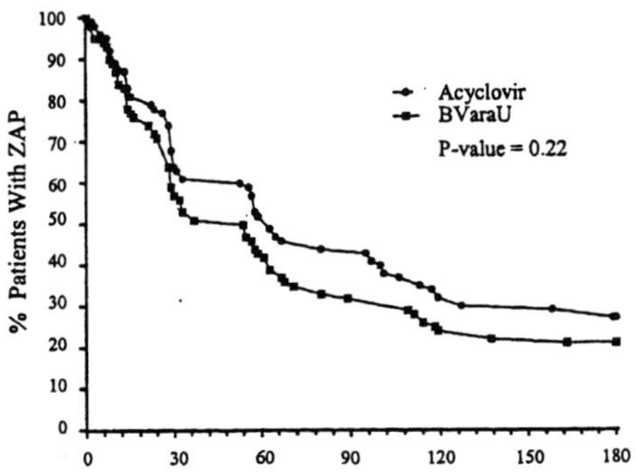

Figure 4.   Kaplan-Meier curves demonstrating (i) time to cessation of new vesicle formation (upper panel), (ii) time to total lesion crusting (middle panel), and (iii) time to resolution of zoster-associated pain (ZAP) (lower panel), for the acyclovir (N = 82) and BVaraU (N = 84) treatment groups. P values were determined by the log-rank test. Treatment existed of a 10-day course of orally administered BVaraU (40 mg

lesion formation (p = 0.07), a significantly accelerated cutaneous healing (crusting) (p = 0.02) and a similar resolution of zoster-associated pain (ZAP) (p = 0.22) when compared with acyclovir therapy. Thus, BVaraU could be considered as efficient a treatment for herpes zoster (in HIV-infected individuals) at a daily dose of 40 mg as acyclovir at a daily dose of 4 g.

The efficacy of BVDU in the treatment of herpes zoster has been examined in several European clinical centers, with BVDU given at various dosage schedules (i.e., 50 or 125 mg twice daily; and 31.25, 62.5 or 125 mg once daily) in comparison with 4 g (five times 800 mg) daily for acyclovir, for a 7-day treatment period. The results of this multicentered study with a large number of patients that have been followed up for an extended time period should be revealed in the near future.

As already mentioned (Maudgal and De Clercq, 1991), BVU has been used for many years in the topical treatment (as 0.1% eyedrops) of herpetic keratitis, since it is efficacious against various manifestations of this disease (dendritic and geographic corneal ulcers, and stromal keratitits), also when clinical resistant to other antiviral drugs such as idoxuridine, fluridine, vidarabine, or acyclovir (Maudgal and De Clercq, 1991).

BVDU as a 5% cream in Beeler base (15 g of cetylalcohol, 1 g of cera alba, 10 g of propylene glycol, 2 g of sodium lauryl sulfate, and enough water to make 100 g) has been used, with success, in the topical treatment of recurrent herpes labialis. This use has been based on the protective activity seen with BVDU in the topical treatment of intracutaneous HSV infection in hairless (hr/hr) mice (De Clercq, 1984e). When an entirely blinded protocol was followed to assess the efficacy of BVDU in this HSV-1 model infection [E. De Clercq, unpublished observations (1994)] BVDU, when formulated as an hydrogel cream at 5, 2 or 0.5%, completely suppressed all manifestations of the infection (i.e. skin lesions, paralysis of the hind legs, and mortality) (Figure 5). At all three concentration levels, topical BVDU treatment resulted in a 100% survival rate at the 20th day post infection. In contrast, all the placebo-treated mice developed lesions within 4-7 days after the infection and succumbed within 7-14 days after the infection. These observations provide unequivocal evidence for the effectiveness of topical BVDU (at 5, 2 and 0.5%) in the treatment of intracutaneous HSV-1 infection.

5.   Interaction with 5-fluorouracil

It has been known for more than 15 years (De Clercq, 1986) that BVDU is recognized as substrate by thymidine phosphorylase that converts BVDU to BVU [(E)-5-(2-bromovinyl)uracil] and 2-deoxyribose-1-phosphate (Figure 6). The resulting BVU can be re-converted to BVDU, both *in vitro* and *in vivo*, through a pentosyl transfer reaction with any 5-substituted 2'-deoxyuridine, including 2'-deoxythymidine, as the pentosyl donor (De Clercq, 1986). BVU itself is a potent inhibitor of dihydrothymine dehydrogenase (DPD), the enzyme that is responsible for the first step in the catabolic pathway of pyrimidines. As DPD is also needed for the degradation of 5-fluorouracil, BVU protects 5-fluorouracil against breakdown and significantly increases its half-life. This marked increase in the half-life of 5-fluorouracil has also been demonstrated in

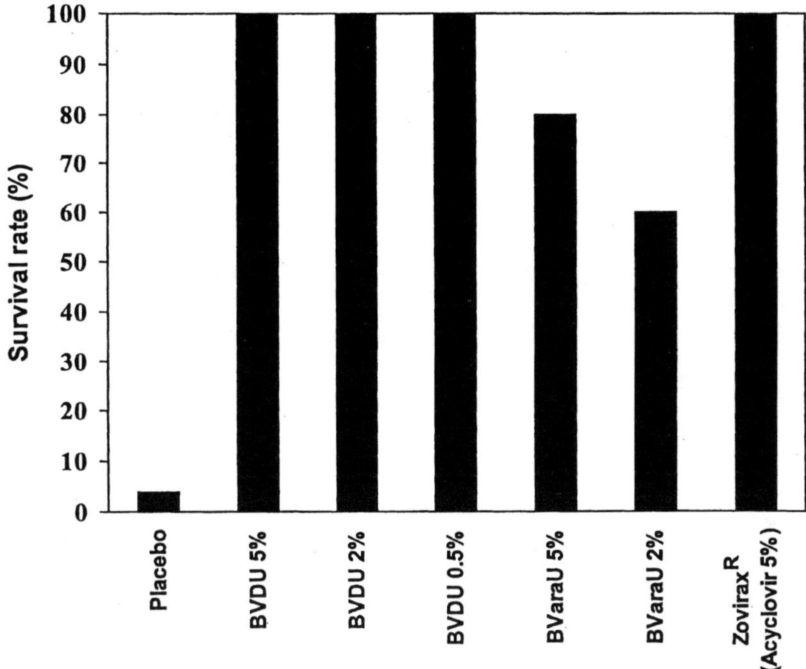

Figure 5. Survival rates of hairless mice inoculated intracutaneously with HSV-1 and treated topically with hydrogel creams containing BVDU or BVaraU, or acyclovir cream (Zovirax®). The KOS strain of HSV-1 was inoculated at $10^{4.7}$ PFU per 0.05 ml per mouse. The different creams were applied topically, four times daily, for 5 days only, starting immediately after virus infection, and the survival rates were estimated at the 20th day post infection.

cancer patients which had been given 5-fluorouracil (intravenously), concomitantly with BVDU (orally) (Keizer *et al.*, 1994). The combination of BVDU with 5-fluorouracil results in a significant enhancement of the antitumor activity of 5-fluorouracil, as has been clearly shown in different tumor models, i.e. adenocarcinoma 755 (Iigo *et al.*, 1988) and Lewis lung carcinoma (Iigo *et al.*, 1990), in mice. In fact, the combination of BVDU with 5-fluorouracil significantly enhanced the life-span of mice bearing liver metastases of Lewis lung carcinoma (Iigo *et al.*, 1990).

While the combination of BVDU (or BVU) with 5-fluorouracil may be endowed with enhanced antitumor activity, one should also beware of increased toxicity associated with the elevated plasma levels of 5-fluorouracil (Keizer *et al.*, 1994). The Pharmaceutical Affairs Bureau, Japanese Ministry of Health and Welfare, reported that in 1993 fifteen deaths occurred in Japanese patients following the co-administration of BVaraU (sorivudine) with a 5-fluorouracil prodrug, and this within 40 days after sorivudine was approved by the Japanese government and began to be used clinically. Before death, all of the patients had severe symptoms of toxicity, such as diarrhea with bloody flux and marked decreases in white blood cell and platelet counts.

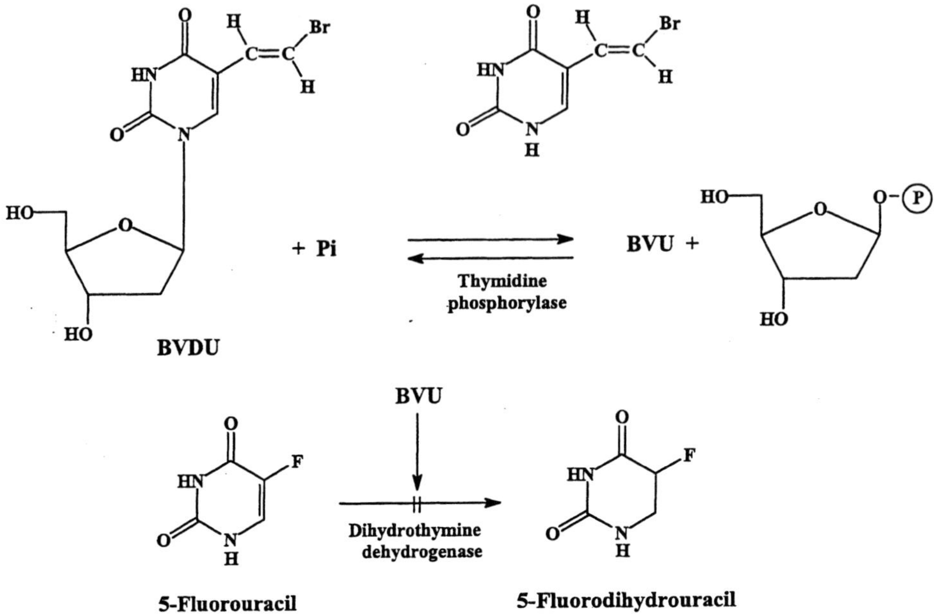

Figure 6.   Degradation of BVDU to BVU by thymidine phosphorylase and inhibition of the degradation of 5-fluorouracil by BVU.

Also, eight other Japanese patients that had received both drugs during this period had severe symptoms of gastrointestinal toxicity and myelotoxicity. Obviously, this severe toxicity could be attributed to the enhanced plasma and tissue levels of 5-fluorouracil, consequently to its retarded catabolism (Okuda *et al.*, 1997, 1998); the culprit being BVU that was found to irreversibly inhibit DPD, as originally shown by Desgranges *et al.* (1986), by covalent binding of a reduced form of BVU as a suicidal inactivator (Ogura *et al.*, 1998). The formation of BVU, after oral administration of BVaraU, could be due to the action of the thymidine phosphorylase(s) from enterobacteria, i.e. Klebsiella pneumoniae (Machida *et al.*, 1995), and/or the anaerobic Bacteroides species (Nakayama *et al.*, 1997).

Even at the 40 mg once-daily oral dosage regimen for 10 consecutive days (Gnann *et al.*, 1998), BVaraU leads to a profound depression of DPD activity, and recovery of DPD activity occurred only within 4 weeks of completion of BVaraU therapy (Yan *et al.*, 1997). This indicates that patients receiving sorivudine are not only at risk to develop potentially life-threatening toxicity from 5-fluorouracil or any of its prodrugs while receiving both drugs simultaneously but also for the next few weeks after the last dose of sorivudine (Diasio, 1998). More recent studies have addressed the mechanism-based inactivation of DPD by BVU (Kanamitsu *et al.*, 2000): in the presence of NADPH, the sulfhydryl group of Cys671 in the human DPD would interact with 5,6-dihydro-5-(2-bromoethyldenyl)uracil (BEDU), a putative allyl bromide type of reactive molecule, to form a sulfide bond with loss of hydrogen bromide (Nishiyama *et al.*, 2000).

6.   Combined gene therapy and chemotherapy

We were the first (De Clercq, 1986) to note that tumor cells, i.e. murine mammary FM3A carcinoma cells, when transformed by the HSV-1 thymidine kinase (TK) gene, acquire a dramatically (about 10,000 fold) increased sensitivity to the cytostatic action of BVDU and related compounds such as IVDU and BVDC, but not BVaraU (Figure 7) (Balzarini *et al.*, 1985a). Evidently, BVDU and its congeners must rely on phosphorylation by the HSV-1 TK to exert their cytostatic action (Balzarini *et al.*, 1985b). Thymidylate synthase was identified as the principal target enzyme for the cytostatic activity of BVDU in both HSV-1 and HSV-2 TK gene-transformed tumor cells (Balzarini *et al.*, 1987). Tumor cells transfected by the HSV-1 or HSV-2 TK genes also gain a substantial (100- to 1,000-fold) increase in sensitivity to acyclic guanosine analogues such as buciclovir, penciclovir and ganciclovir, but in these cases cytostatic activity seems to be related to an action targeted at the cellular DNA polymerase rather than thymidylate synthase (Balzarini *et al.*, 1993, 1994). From the latter study, BVDU, BVDC and the 4'-thio counterpart of BVDU, S-BVDU (Figure 8), emerged as particularly potent inhibitors of FM3A TK-/HSV-1 TK+ cells, with the notion that S-BVDU, because of its resistance to phosphorolysis by thymidine phosphorylase, should be a promising candidate for further investigation in the treatment of HSV TK gene-transfected tumors *in vivo* (Balzarini *et al.*, 1994).

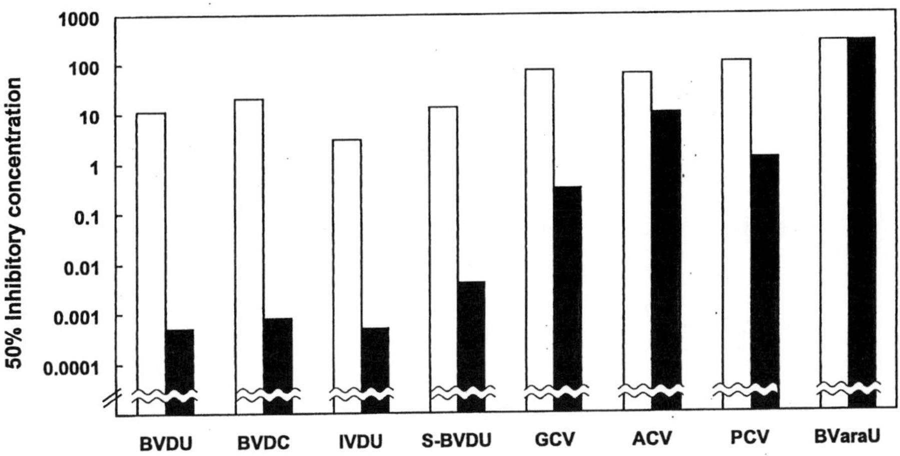

Figure 7.   Inhibitory effects of BVDU and other selected antiviral compounds [BVDC, IVDU, S-BVDU, GCV (ganciclovir), ACV (acyclovir), PCV (penciclovir), BVaraU] on the proliferation of murine mammary carcinoma (FM3A) cells transfected by the HSV-1 thymidine kinase (TK) gene.□: FM3A/0. ■: FM3A TK-/HSV-1 TK+.

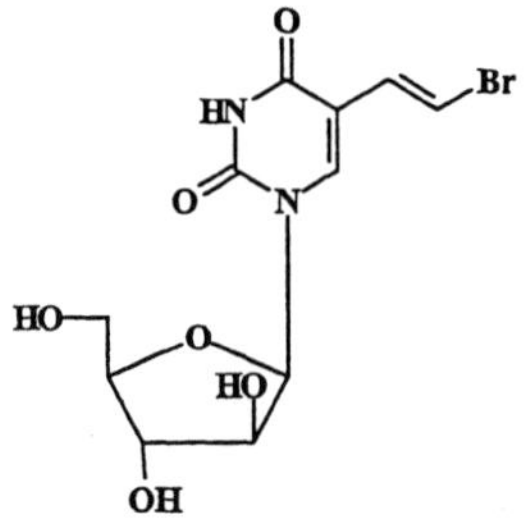

E. De Clercq

BVaraU

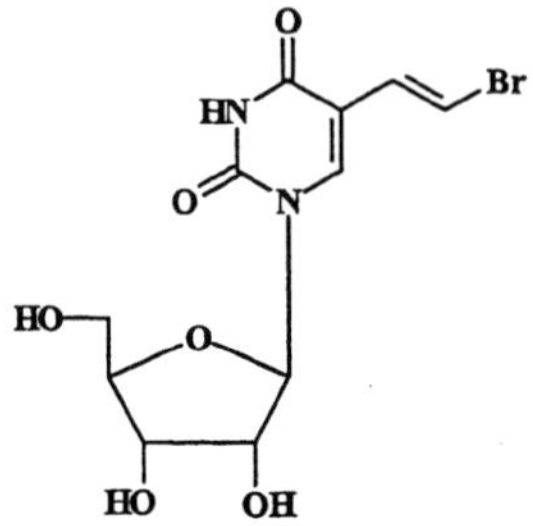

BVriboU

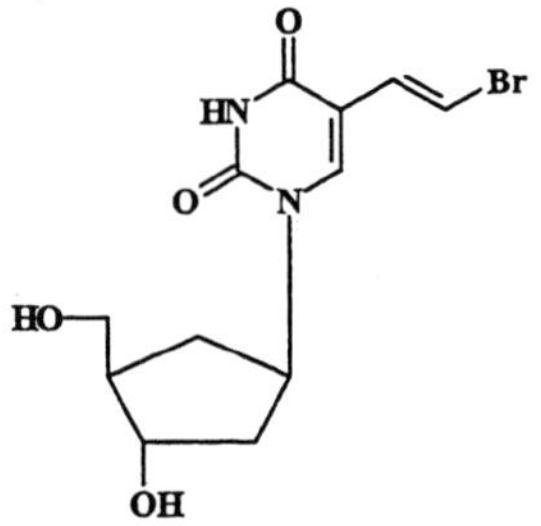

C-BVDU

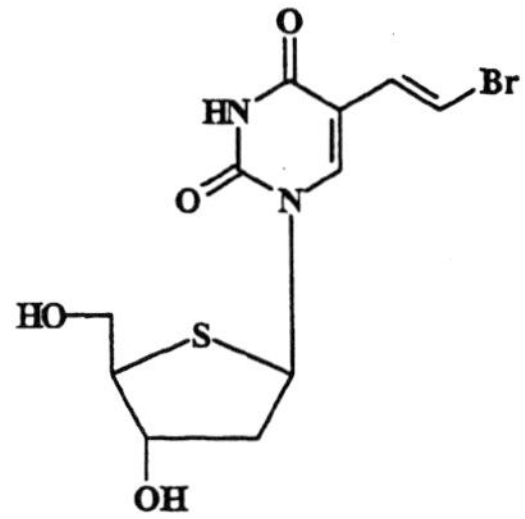

S-BVDU

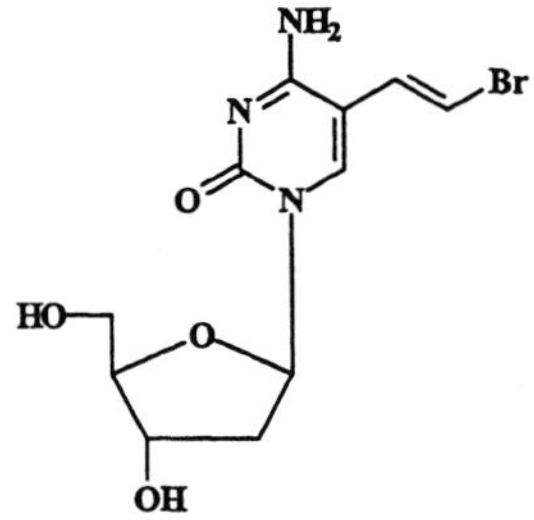

BVDC

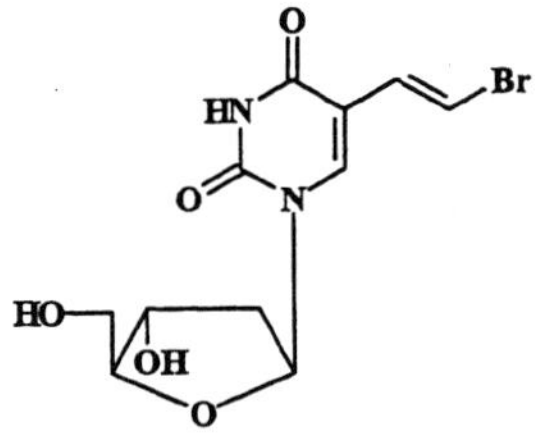

L-BVDU

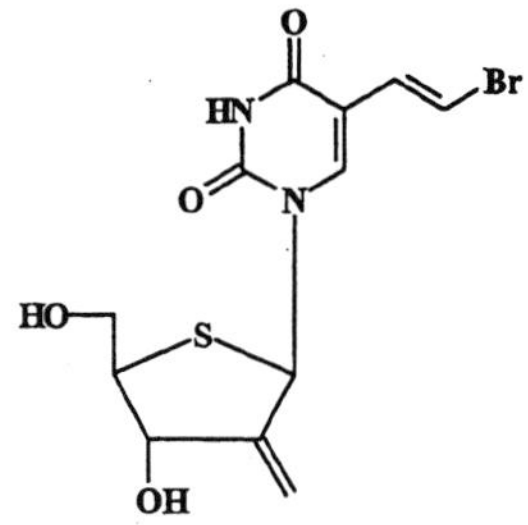

S-BVMDU

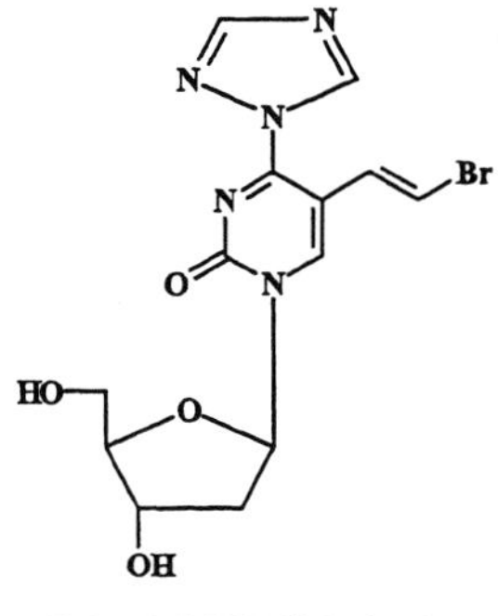

Triazolyl BVDC derivative

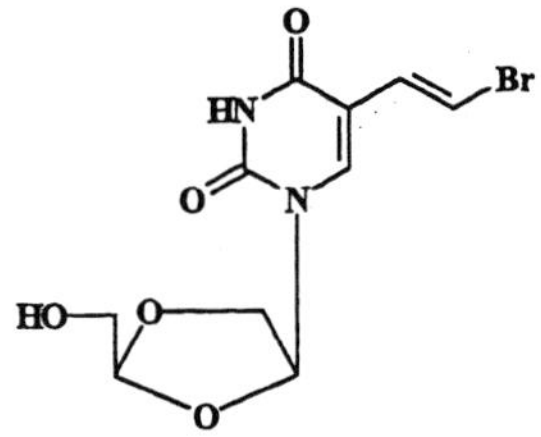

L-BVODDU

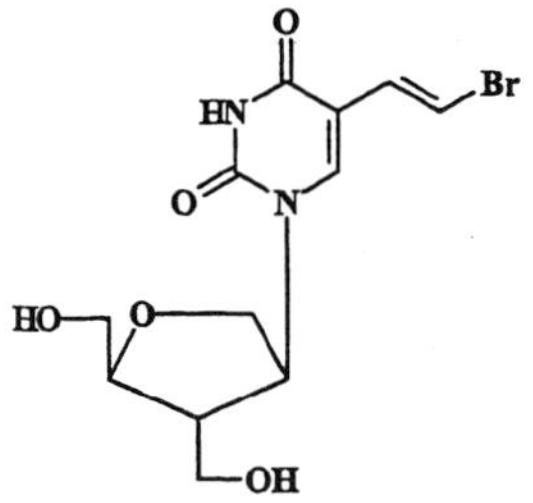

BMS-181165

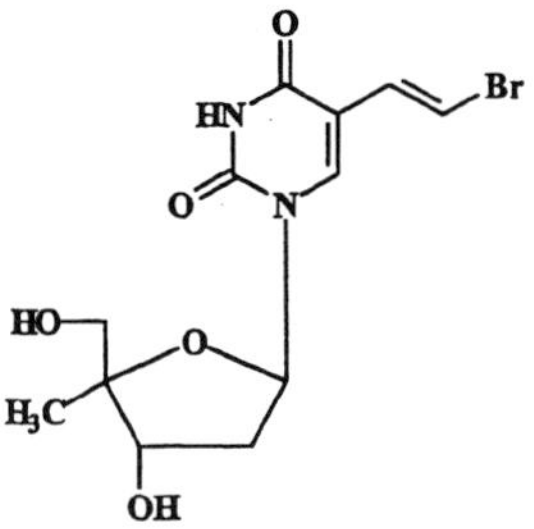

4'-methyl BVDU

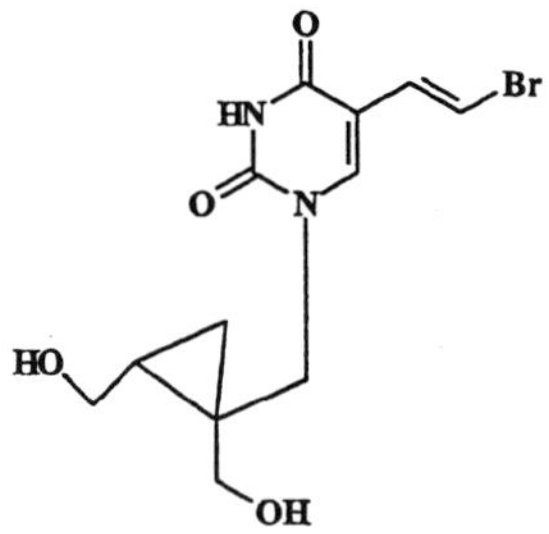

AV-100

Figure 8.   BVDU derivatives all based on the (E)-2-bromovinyl substituent as the pharmacophore: (E)-5-(2-bromovinyl)-1-β-D-arabino-pentofuranosyluracil (BVaraU), (E)-5-(2-bromovinyl)uridine (BVriboU), (±)-carbocyclic BVDU (C-BVDU), (E)-5-(2-bromovinyl)-2'-deoxy-4'-thiouridine (S-BVDU), (E)-5-(2-bromo-vinyl)-2'-deoxycytidine (BVDC), (E)-5-(2-bromovinyl)-2'-deoxy-L-uridine (L-BVDU), (E)-5-(2-bromovinyl)-1-(2-deoxy-2-C-methylene-4-thio-β-D-erythro-pentofuranosyl)uracil (S-BVMDU), the 4-(1,2,4-triazol-1-yl) derivative of BVDC, the L-dioxolane derivative of (E)-5-(2-bromovinyl)uracil (L-BVODDU), [3S-(3α,4β-5α)]-(E)-5-(2-bromovinyl)-1-[tetrahydro-4,5-bis(hydroxymethyl)-3-furanyl]-2,4(1H,3H)-pyrimidinedione (BMS-181165), 4'-C-methyl-BVDU (4'-methyl BVDU), and (1'S,2'R)-5-[(E)-2-bromovinyl]-[[1',2'-bis(hydroxymethyl)cycloprop-1'-yl]methyl-2,4-(1H,3H)-pyrimidinedione (AV-100).

From the parent compound (IVDU), two 2'-fluoro-substituted congeners (IVFRU, with the fluorine in the ribo configuration; and IVFAU, with the fluorine in the ara configuration) were derived: again, these compounds inhibited FM3A TK-/HSV-1 TK+ cells at a 1,000- to 10,000-fold lower concentration than wild-type FM3A cells, and, because of their resistance to phosphorolytic cleavage by thymidine phosphorylase, could be considered as promising candidate compounds for further evaluation for the combined gene/chemotherapy of HSV TK gene-transfected tumors in animal models (Balzarini *et al.*, 1995). It should be noticed, however, that purine nucleoside analogues, represented by ganciclovir, possess a stronger bystander killing effect than pyrimidine nucleoside analogues, represented by BVDU (Degrève *et al.*, 1999), at least *in vitro* in osteosarcoma cells. This might translate in greater potency for the purine nucleoside analogues in killing tumor cells *in vivo*. Yet, despite its potential "handicaps" (i.e., sensitivity to phosphorolytic cleavage by thymidine phosphorylase and little bystander cell killing), BVDU treatment (15 mg/kg × 3/day, intraperitoneally), added onto irradiation, has proven effective in increasing the median survival time of rats with HSV TK-transduced 9L gliosarcoma cells implanted in the brain (Kim *et al.*, 1995).

In addition to the HSV-1 and HSV-2 TK genes, VZV TK gene has also been used as a suicide gene in human breast cancer MDA-MB-435 cells (Grignet-Debrus and Calberg-Bacq, 1997) and human osteosarcoma (OST) cells (Degrève *et al.*, 1997). Again, BVDU and its congeners (now also including BVaraU, which, as shown in Figure 7, did not gain increased cytostatic activity towards FM3A TK-/HSV-1 TK+ cells) displayed cytostatic activity against OST TK-/VZV TK+ cells at drug concentrations that were 1,000- to 10,000-fold lower than those required to inhibit the corresponding non-transfected tumor cells (Degrève *et al.*, 1997). Also, BVDU gained a 2,000-fold increased cytotoxicity towards the VZV TK gene-transfected MDA-MB-435 cells (Grignet-Debrus and Calberg-Bacq, 1997). Moreover, when these cells were implanted subcutaneously in nude mice, their growth was significantly suppressed following BVDU treatment (intraperitoneally, at a dose of 80 mg/kg/day) (Grignet-Debrus and Calberg-Bacq, 1997). BVDU did not exert a bystander killing effect on mixed popula-tions of VZV TK+ and TK- MDA-MB-435 cells; however, an important bystander effect was observed when similar experiments were performed with 9L gliosarcoma cells (Grignet-Debrus and Calberg-Bacq, 1997).

BVDU has also been shown to effect greater cytotoxicity for Aedes albopictus cells transfected with HSV TK than for wild-type mosquito cells (Mazzacano and Fallon,

dihydropyrimidine dehydrogenase, the enzyme that initiates the degradative pathway of 5-fluorouracil. The prime determinant in the unique behavior of BVDU is its (E)-5-(2-bromovinyl) substituent. Numerous BVDU analogues have been described that, when equipped with this particular pharmacophore, demonstrate an activity spectrum characteristic of BVDU, including selective anti-VZV activity.

9.   Acknowledgments

Prof. Erik De Clercq holds the Professor P. De Somer Chair of Microbiology at the Katholieke Universiteit Leuven School of Medicine and thanks Christiane Callebaut for her invaluable editorial assistance.

10.   References

Allaudeen, H.S., Kozarich, J.W., Bertino, J.R. and De Clercq, E. (1981) On the mechanism of selective inhibition of herpesvirus replication by (E)-5-(2-bromovinyl)-2'-deoxyuridine. Proc. Natl. Acad. Sci. USA 78, 2698-2702.

Andrei, G., Snoeck, R., Goubau, P., Desmyter, J. and De Clercq E. (1992) Comparative activity of various compounds against clinical strains of herpes simplex virus. Eur. J. Clin. Microbiol. Infect. Dis. 11, 143-151.

Andrei, G., Snoeck, R., Reymen, D, Liesnard, C., Goubau, P., Desmyter, J. and De Clercq E. (1995) Comparative activity of selected antiviral compounds against clinical isolates of varicella-zoster virus. Eur. J. Clin. Microbiol. Infect. Dis. 14, 318-328.

Ayisi, N.K., De Clercq, E., Wall, R.A., Hughes, H. and Sacks, S.L. (1984) Metabolic fate of (E)-5-(2-bromovinyl)-2'-deoxyuridine in herpes simplex virus- and mock-infected cells. Antimicrob. Agents Chemother. 26, 762-765.

Balzarini, J., De Clercq, E., Ayusawa, D. and Seno, T. (1985a) Murine mammary FM3A carcinoma cells transformed with the herpes simplex virus type 1 thymidine kinase gene are highly sensitive to the growth-inhibitory properties of (E)-5-(2-bromovinyl)-2'-deoxyuridine and related compounds. FEBS Lett. 185, 95-100.

Balzarini, J., De Clercq, E., Verbruggen, A., Ayusawa, D. and Seno, T. (1985b) Highly selective cytostatic activity of (E)-5-(2-bromovinyl)-2'-deoxyuridine derivatives for murine mammary carcinoma (FM3A) cells transformed with the herpes simplex virus type 1 thymidine kinase gene. Mol. Pharmacol. 28, 581-587.

Balzarini, J., De Clercq, E., Verbruggen, A., Ayusawa, D., Shimizu, K. and Seno, T. (1987) Thymidylate synthase is the principal target enzyme for the cytostatic activity of (E)-5-(2-bromovinyl)-2'-deoxyuridine against murine mammary carcinoma (FM3A) cells transformed with the herpes simplex virus type 1 or type 2 thymidine kinase gene. Mol. Pharmacol. 32, 410-416.

Balzarini, J., De Clercq, E., Baumgartner, H., Bodenteich, M. and Griengl, H. (1990a) Carbocyclic 5-iodo-2'-deoxyuridine (C-IDU) and carbocyclic (E)-5-(2-bromovinyl)-2'-deoxyuridine (C-BVDU) as unique examples of chiral molecules where the two enantiomeric forms are biologically active: interaction of the (+) and (-)-enantiomers of C-IDU and C-BVDU with the thymidine kinase of herpes simplex virus type 1. Mol. Pharmacol. 37, 395-401.

Balzarini, J., Bernaerts, R., Verbruggen, A. and De Clercq, E. (1990b) Role of the incorporation of (E)-5-(2-iodovinyl)-2'-deoxyuridine and its carbocyclic analogue into DNA of herpes simplex virus type 1-infected cells in the antiviral effects of these compounds. Mol. Pharmacol. 37, 402-407.

Balzarini, J., Bohman, C. and De Clercq, E. (1993) Differential mechanism of cytostatic effect of (E)-5-(2-bromovinyl)-2'-deoxyuridine, 9-(1,3-dihydroxy-2-propoxymethyl)guanine, and other antiherpetic drugs on tumor cells transfected by the thymidine kinase gene of herpes simplex virus type 1 or type 2. J. Biol. Chem. 268, 6332-6337.

Balzarini, J., Bohman, C., Walker, R.T. and De Clercq, E. (1994) Comparative cytostatic activity of different antiherpetic drugs against herpes simplex virus thymidine kinase gene-transfected tumor cells. Mol. Pharmacol. 45, 1253-1258.

Balzarini, J., Morin, K.W., Knaus, E.E., Wiebe, L.I. and De Clercq, E. (1995) Novel (E)-5-(2-iodovinyl)-2'-deoxyuridine derivatives as potential cytostatic agents against herpes simplex virus thymidine kinase gene transfected tumors. Gene Ther. 2, 317-322.

Balzarini, J., Degrève, B., Hatse, S., De Clercq, E., Breuer, M., Johansson, M., Huybrechts, R. and Karlsson, A. (2000) The multifunctional deoxynucleoside kinase of insect cells is a target for the development of new insecticides. Mol. Pharmacol. 57, 811-819.

Bamford, M.J., Coe, P.L. and Walker, R.T. (1990) Synthesis and antiviral activity of 3'-deoxy-3'-C-hydroxymethyl nucleosides. J. Med. Chem. 33, 2494-2501.

Barr, P.J., Oppenheimer, N.J. and Santi, D.V. (1983) Thymidylate synthetase-catalyzed conversions of E-5-(2-bromovinyl)-2'-deoxyuridylate. J. Biol. Chem. 258, 13627-13631.

Bärwolff, D. and Langen, P. (1975) A selective bromination of thymidine. Nucleic Acids Res., special publication no 1, s29-s31.

Basnak, I., Otter, G.P., Duncombe, R.J., Westwood, N.B., Pietrarelli, M., Hardy, G.W., Mills, G., Rahim, S.G. and Walker, RT. (1998) Efficient syntheses of (E)-5-(2-bromovinyl)-2'-deoxy-4'-thiouridine. a nucleoside analogue with potent biological activity. Nucleosides Nucleotides 17, 29-38.

Benoit, Y., Laureys, G., Delbeke, M.-J. and De Clercq, E. (1985) Oral BVDU treatment of varicella and zoster in children with cancer. Eur. J. Pediatr. 143, 198-202.

Bernaerts, R., Desgranges, C. and De Clercq, E. (1989) (E)-5-(2-Bromovinyl)uridine requires phosphorylation by the herpes simplex virus (type 1)-induced thymidine kinase to express its antiviral activity. Biochem. Pharmacol. 38, 1955-1961.

Choi, Y.S., Li, L., Grill, S., Gullen, E., Lee, C.S., Gumina, G., Cheng, Y.-C. and Chu, C.K. (2000) Structure-activity relationships of (E)-5-(2-bromovinyl)uracil and related pyrimidine L-nucleosides as antiviral agents for varicella zoster virus. Abstracts of the Thirteenth International Conference on Antiviral Research, Baltimore, MD, USA, 16-21 April 2000. Antiviral Res. 46, A81, no 149.

De Clercq, E. (1980) Antiviral and antitumor activities of 5-substituted 2'-deoxyuridines. Meth. Find. Exptl. Clin. Pharmacol. 2, 253-267.

De Clercq, E. (1982a) Antiviral activity of pyrimidine nucleoside analogs: a structure-function analysis. Alderweireldt FC, Esmans EL, editors. Proceedings of the 4th International Round Table on Nucleosides, Nucleotides and their Biological Applications, Antwerp, 4-6 February 1981. Antwerp, The University of Antwerp, p. 25-45.

De Clercq, E. (1982b) Selective antiherpes agents. Trends Pharmacol. Sci. 3, 492-495.

De Clercq, E. (1983a) Antiviral activity of 5-substituted pyrimidine nucleoside analogues. Pure Appl. Chem. 55, 623-636.

De Clercq, E. (1983b) The chemotherapy of herpesvirus infections with reference to bromovinyldeoxyuridine and other antiviral compounds. In: Sir Stuart-Harris CH, Oxford J, editors. Problems of Antiviral

450   *E. De Clercq*

Therapy. London: Academic Press, p. 295-315.

De Clercq, E. (1984a) Bromovinyldeoxyuridine (BVDU): current status in antiviral therapy. In: Kurstak E, editor. Control of Virus Diseases. New York: Marcel Dekker, Inc., p. 443-458.

De Clercq, E. (1984b) Therapeutic potentials of bromovinyldeoxyuridine (BVDU) in the treatment of herpesvirus infections. I. Fundamental aspects. In: Rapp F, editor. Herpesvirus. New York: Alan R. Liss, Inc., p. 573-585.

De Clercq, E. (1984c) Therapeutic potentials of bromovinyldeoxyuridine (BVDU) in the treatment of herpesvirus infections. I. Clinical aspects. In: Rapp F, editor. Herpesvirus. New York: Alan R. Liss, Inc., p. 587-599.

De Clercq, E. (1984d) The antiviral spectrum of (E)-5-(2-bromovinyl)-2'-deoxyuridine. J. Antimicrob. Chemother. 14, suppl A, 85-95.

De Clercq, E. (1984e) Topical treatment of cutaneous herpes simplex virus infection in hairless mice with (E)-5-(2-bromovinyl)-2'-deoxyuridine and related compounds. Antimicrob. Agents Chemother. 26, 155-159.

De Clercq, E. (1985a) Antiviral agents. In: Greenwood D, O'Grady F, editors. Symposium of the Society for General Microbiology. Scientific Basis of Antimicrobial Chemotherapy. Cambridge: Cambridge University Press, p. 155-184.

De Clercq, E. (1985b) Synthetic pyrimidine nucleoside analogues. Harnden MR, editor. Approaches to Antiviral Agents. London: MacMillan Press, p. 57-99.

De Clercq, E. (1986) Towards a selective chemotherapy of virus infections. Development of bromo-vinyldeoxyuridine as a highly potent and selective antiherpetic drug. Verh. K. Acad. Geneesk. Belg. 48, 261-290.

De Clercq, E. and Walker, R.T. (1984) Synthesis and antiviral properties of 5-vinylpyrimidine nucleoside analogs. Pharmac. Ther. 26, 1-44.

De Clercq, E. and Walker, R.T. (1986) Chemotherapeutic agents for herpesvirus infections. In: Ellis GP, West GB, editors. Progress in Medicinal Chemistry, vol 23. Amsterdam: Elsevier Science Publishers, B.V., p. 187-218.

De Clercq, E., Descamps, J., Barr, P.J., Jones, A.S., Serafinowski, P., Walker, R.T., Huang, G.F., Torrence, P.F., Schmidt, C.L., Mertes, M.P., Kulikowski, T. and Shugar, D. (1979a) Comparative study of the potency and selectivity of anti-herpes compounds. In: Skoda J, Langen P, editors. Antimetabolites in Biochemistry, Biology and Medicine. Oxford: Pergamon Press, p. 275-285.

De Clercq, E., Descamps, J., De Somer, P., Barr, P.J., Jones, A.S. and Walker, R.T. (1979b) (E)-5-(2-Bromovinyl)-2'-deoxyuridine: a potent and selective anti-herpes agent. Proc. Natl. Acad. Sci. USA. 76, 2947-2951.

De Clercq, E., Descamps, J., Verhelst, G., Walker, R.T., Jones, A.S., Torrence, P.F. and Shugar, D. (1980a) Comparative efficacy of antiherpes drugs against different strains of herpes simplex virus. J. Infect. Dis. 141, 563-574.

De Clercq, E., Degreef, H., Wildiers, J., de Jonge, G., Drochmans, A., Descamps, J. and De Somer, P. (1980b) Oral (E)-5-(2-bromovinyl)-2'-deoxyuridine in severe herpes zoster. Br. Med. J. 281, 1178.

De Clercq, E., Verhelst, G., Descamps, J. and Bergstrom, D.E. (1981) Differential inhibition of herpes simplex viruses, type 1 (HSV-1) and type 2 (HSV-2), by (E)-5-(2-X-vinyl)-2'-deoxyuridines. Acta microbiol. Acad. Sci. hung. 28, 307-312.

De Clercq, E., Desgranges, C., Herdewijn, P., Sim, I.S. and Walker, R.T. (1985) Bromovinyluracil nucleoside analogues as antiherpes agents. Dahlbom R, Nilsson JLG, editors. Proceedings of the VIIIth International Symposium on Medicinal Chemistry, vol 1. Stockholm: Swedish Pharmaceutical Press, p. 198-210.

De Clercq, E., Desgranges, C., Herdewijn, P., Sim, I.S., Jones, A.S., McLean, M.J. and Walker, R.T. (1986) Synthesis and antiviral activity of (E)-5-(2-bromovinyl)uracil and (E)-5-(2-bromovinyl)uridine. J. Med. Chem. 29, 213-217.

De Clercq, E., Walker, R.T. and Whale, R.F. (1992) 3'-O-benzyl-(E)-5-(2-bromovinyl)-2'-deoxyuridine is active as an anti-herpes agent in vivo but not in vitro. Med. Chem. Res. 2, 111-118.

Degrève, B., Andrei, G., Izquierdo, M., Piette, J., Morin, K., Knaus, E.E., Wiebe, L.I., Basrah, I., Walker, R.T., De Clercq, E and Balzarini, J. (1997) Varicella-zoster virus thymidine kinase gene and antiherpetic pyrimidine nucleoside analogues in a combined gene/chemotherapy treatment for cancer. Gene Ther. 4, 1107-1114.

Degrève, B, De Clercq, E. and Balzarini, J. (1999) Bystander effect of purine nucleoside analogues in HSV-1tk suicide gene therapy is superior to that of pyrimidine nucleoside analogues. Gene Ther. 6, 162-170.

Descamps, J. and De Clercq, E. (1981) Specific phosphorylation of E-5-(2-iodovinyl)-2'-deoxyuridine by herpes simplex virus-infected cells. J. Biol. Chem. 256, 5973-5976.

Desgranges, C., Razaka, G., De Clercq, E, Herdewijn, P., Balzarini, J., Drouillet, F. and Bricaud, H. (1986) Effect of (E)-5-(2-bromovinyl)uraci on the catabolism and antitumor activity of 5-fluorouracil in rats and leukemic mice. Cancer Res. 46, 1094-1101.

Diasio, R.B. (1998) Sorivudine and 5-fluorouracil; a clinically significant drug-drug interaction due to inhibition of dihydropyrimidine dehydrogenase. Br. J. Clin. Pharmacol. 46, 1-4.

Elion, G.B., Furman, P.A., Fyfe, J.A., de Miranda, P., Beauchamp, L. and Schaeffer, H.J. (1977) Selectivity of action of an antiherpetic agent, 9-(2-hydroxyethoxymethyl)guanine. Proc. Natl. Acad. Sci. USA. 74, 5716-5720.

Farrow, S.N, Jones, A.S, Kumar, A., Walker, R.T., Balzarini, J. and De Clercq, E. (1990) Synthesis and biological properties of novel phosphotriesters: a new approach to the introduction of biologically active nucleotides into cells. J. Med. Chem. 33, 1400-1406.

Fyfe, J.A. (1982) Differential phosphorylation of (E)-5-(2-bromovinyl)-2'-deoxyuridine monophosphate by thymidylate kinases from herpes simplex viruses types 1 and 2 and varicella zoster virus. Mol. Pharmacol. 21, 432-437.

Gnann Jr, J.W., Crumpacker, C.S., Lalezari, J.P., Smith, J.A., Tyring, S.K., Baum, K.F., Borucki, M.J., Joseph, W.P., Mertz, G.J., Steigbigel, R.T., Cloud, G.A., Soong, S.-J., Sherrill, L.C., DeHertogh, D.A., Whitley, R.J. and the Collaborative Antiviral Study Group (CASG)/AIDS Clinical Trials Group (ACTG) Herpes Zoster Study Group. (1998) Sorivudine versus acyclovir for treatment of dermatomal herpes zoster in human immunodeficiency virus-infected patients: results from a randomized, controlled clinical trial. Antimicrob. Agents Chemother. 42, 1139-1145.

Grignet-Debrus, C. and Calberg-Bacq, C-M. (1997) Potential of varicella zoster virus thymidine kinase as a suicide gene in breast cancer cells. Gene Ther. 4, 560-569.

Harkness, J.W., Sands, J.J. and Sim I.S. (1986) Bovine herpes mammillitis therapy. Veterinary Record 118, 282.

Ichikawa, E., Yamamura, S. and Kato, K. (1999) Synthesis of 2',3'-dideoxy-3'-C-(hydroxymethyl)-4'-thiopentofuranosyl nucleosides as potential antiviral agent. Bioorg. Med. Chem. Lett. 9, 1113-1114.

Iigo, M., Araki, E., Nakajima, Y., Hoshi, A. and De Clercq, E. (1988) Enhancing effect of bromo-vinyldeoxyuridine on antitumor activity of 5-fluorouracil against adenocarcinoma 755 in mice. Biochem. Pharmacol. 37, 1609-1613.

Iigo, M., Nishikata, K.-i., Nakajima, Y., Hoshi, A. and De Clercq, E. (1990) Effect of (E)-5-(2-bromovinyl)-2'-deoxyuridine on life-span and 5-fluorouracil metabolism in mice with hepatic metastases. Eur. J.

452   *E. De Clercq*

Cancer. 26, 1089-1092.

Jennings Wilber, B.A. and Docherty, J.J. (1994) Analysis of the thymidine kinase of a herpes simplex virus type 1 isolate that exhibits resistance to (E)-5-(2-bromovinyl)-2'-deoxyuridine. J. Gen. Virol. 75, 1743-1747.

Jones, A.S., Rahim, S.G., Walker, R.T. and De Clercq, E. (1981) Synthesis and antiviral properties of (Z)-5-(2-bromovinyl)-2'-deoxyuridine. J. Med. Chem. 24, 759-760.

Jones, A.S., Sayers, J.R., Walker, R.T. and De Clercq, E. (1988) Synthesis and antiviral properties of (E)-5-(2-bromovinyl)-2'-deoxycytidine-related compounds. J. Med. Chem. 31, 268-271.

Kanamitsu, S.-I., Ito, K., Okuda, H., Ogura, K., Watabe, T., Muro, K. and Sugiyama, Y. (2000) Prediction of in vivo drug-drug interactions based on mechanism-based inhibition from in vitro data: inhibition of 5-fluorouracil metabolism by (E)-5-(2-bromovinyl)uracil. Drug Metab. Dispos. 28, 467-474.

Keizer, H.J., De Bruijn, E.A., Tjaden, U.R. and De Clercq, E. (1994) Inhibition of fluorouracil catabolism in cancer patients by the antiviral agent (E)-5-(2-bromovinyl)-2'-deoxyuridine. J. Cancer Res. Clin. Oncol. 120, 545-549.

Kim, J.H., Kim, S.H., Kolozsvary, A., Brown, S.L., Kim, O.B. and Freytag S.O. (1995) Selective enhancement of radiation response of herpes simplex virus thymidine kinase transduced 9L gliosarcoma cells in vitro and in vivo by antiviral agents. Int. J. Radiation Oncol. Biol. Phys. 33, 861-868.

Kitano, K., Machida, H., Miura, S. and Ohrui, H. (1999) Synthesis of novel 4'-C-methyl-pyrimidine nucleosides and their biological activities. Bioorg. Med. Chem. Lett. 9, 827-830.

Kumar, A., Coe, P.L., Jones, A.S., Walker, R.T., Balzarini, J. and De Clercq, E. (1990) Synthesis and biological evaluation of some cyclic phosphoramidate nucleoside derivatives. J. Med. Chem. 33, 2368-2375.

Lopez, C., Watanabe, K.A. and Fox, J.J. (1980) 2'-Fluoro-5-iodo-aracytosine, a potent and selective anti-herpesvirus agent. Antimicrob. Agents. Chemother. 17, 803-806.

Machida, H., Watanabe, Y., Kano, F., Sakata, S., Kumagai, M. and Yamaguchi, T. (1995) Deglycosylation of antiherpesviral 5-substituted arabinosyluracil derivatives by rat liver extract and enterobacteria cells. Biochem. Pharmacol. 49, 763-766.

Mancini, W.R., De Clercq, E. and Prusoff, W.H. (1983) The relationship between incorporation of E-5-(2-bromovinyl)-2'-deoxyuridine into herpes simplex virus type 1 DNA with virus infectivity and DNA integrity. J. Biol. Chem. 258, 792-795.

Maudgal, P.C. and De Clercq, E. (1991) Bromovinyldeoxyuridine treatment of herpetic keratitis clinically resistant to other antiviral agents. Curr. Eye Res. 10, suppl, 193-199.

Maudgal, P.C., Missotten, L., De Clercq, E., Descamps, J. and De Meuter, E. (1981) Efficacy of (E)-5-(2-bromovinyl)-2'-deoxyuridine in the topical treatment of herpes simplex keratitis. Albrecht von Graefes Arch. Klin. Ophthalmol. 216, 261-268.

Mazzacano, C.A. and Fallon, A.M. (1995) Evaluation of a viral thymidine kinase gene for suicide selection in transfected mosquito cells. Insect Mol. Biol. 4, 125-134.

Nakayama, H., Kinouchi, T., Kataoka, K., Akimoto, S;, Matsuda, Y. and Ohnishi, Y. (1997) Intestinal anaerobic bacteria hydrolyse sorivudine, producing the high blood concentration of 5-(E)-5-(2-bromovinyl)uracil that increases the level and toxicity of 5-fluorouracil. Pharmacogenetics 7, 35-43.

Neyts, J. and De Clercq, E. (1997) Antiviral drug susceptibility of human herpesvirus 8. Antimicrob. Agents Chemother. 41, 2754-2756.

Neyts, J. and De Clercq, E. (1998) In vitro and in vivo inhibition of murine gamma herpesvirus 68 replication by selected antiviral agents. Antimicrob. Agents Chemother. 42, 170-172.

Nishiyama, T., Ogura, K;, Okuda, H., Suda, K., Kato, A. and Watabe, T. (2000) Mechanism-based inactivation of human dihydropyrimidine dehydrogenase by (E)-5-(2-bromovinyl)uracil in the presence of NADPH. Mol. Pharmacol. 57, 899-905.

Ogura, K., Nishiyama, T., Takubo, H., Kato, A., Okuda, H., Arakawa, K., Fukushima, M., Nagayama, S., Kawaguchi, Y. and Watabe, T. (1998) Suicidal inactivation of human dihydropyrimidine dehydrogenase by (E)-5-(2-bromovinyl)uracil derived from the antiviral, sorivudine. Cancer Lett. 122, 107-113.

Okuda, H., Nishiyama, T., Ogura, Y., Nagayama, S., Ikeda, K., Yamaguchi, S., Nakamura, Y., Kawaguchi, K. and Watabe, T. (1997) Lethal drug interactions of sorivudine, a new antiviral drug, with oral 5-fluorouracil prodrugs. Drug Metab. Dispos. 25, 270-273.

Okuda, H., Ogura, K., Kato, A., Takubo, H. and Watabe, T. (1998) A possible mechanism of eighteen patient deaths caused by interactions of sorivudine, a new antiviral drug, with oral 5-fluorouracil prodrugs. J. Pharmacol. Exp. Ther. 287, 791-799.

Onishi, T., Mukai, C., Nakagawa, R., Sekiyama, T., Aoki, M., Suzuki, K., Nakazawa, H., Ono, N., Ohmura, Y., Iwayama, S., Okunishi, M. and Tsuji, T. (2000) Synthesis and antiviral activity of novel anti-VZV 5-substituted uracil nucleosides with a cyclopropane sugar moiety. J. Med. Chem. 43, 278-282.

Reefschläger, J., Bärwolff, D., Engelmann, P., Langen, P. and Rosenthal, H.A. (1982) Efficiency and selectivity of (E)-5-(2-bromovinyl)-2'-deoxyuridine and some other 5-substituted 2'-deoxypyrimidine nucleosides as anti-herpes agents. Antiviral Res. 2, 41-52.

Reymen, D., Naesens, L., Balzarini, J., Holý, A., Dvoraková, H. and De Clercq, E. (1995) Antiviral activity of selected acyclic nucleoside analogues against human herpesvirus 6. Antiviral Res. 28, 343-357.

Satoh, H., Yoshimura, Y;, Watanabe, M., Ashida, N., Ijichi, K., Sakata, S., Machida, H. and Matsuda, A. (1998) Synthesis and antiviral activities of 5-substituted 1-(2-deoxy-2-C-methylene-4-thio-β-D-erythro-pentofuranosyl)uracils. Nucleosides Nucleotides 17, 65-79.

Schaeffer, H.J., Beauchamp, L., de Miranda, P., Elion, G.B., Bauer, D.J. and Collins, P. (1978) 9-(2-Hydroxyethoxymethyl)guanine activity against viruses of the herpes group. Nature 272, 583-585.

Shigeta, S., Yokota, T., Iwabuchi, T., Baba, M., Konno, K., Ogata, M. and De Clercq, E. (1983) Comparative efficacy of antiherpes drugs against various strains of varicella-zoster virus. J. Infect. Dis. 147, 576-584.

Smith, G. (1996) In vitro sensitivity of macropodid herpesvirus 2 to selected anti-herpetic compounds. J. Wildlife Dis. 32, 117-120.

Smith, G.A. and Whalley, J.M. (1998) (E)-5-(2'-Bromovinyl)-2'-deoxyuridine inhibition of macropodid herpesvirus 1 in vitro. J. Zoo Wildlife Med. 29, 157-159.

Soike, K.F., Gibson, S. and Gerone P.J. (1981) Inhibition of simian varicella virus infection of African green monkeys by (E)-5-(2-bromovinyl)-2'-deoxyuridine (BVDU). Antiviral Res. 1, 325-337.

Soike, K.F., Huang, J.-L., Russell, J.W., Whiterock, V.J., Sundeen, J.E., Stratton, L.W. and Clark, J.M. (1994) Pharmacokinetics and antiviral activity of a novel isonucleoside, BMS-181165, against simian varicella virus infection in African green monkeys. Antiviral Res. 23, 219-224.

Spadari, S., Ciarrocchi, G., Focher, F., Verri, A., Maga, G., Arcamone, F., Iafrate, E., Manzini, S., Garbesi, A. and Tondelli, L. (1995) 5-Iodo-2'-deoxy-L-uridine and (E)-5-(2-bromovinyl)-2'-deoxy-L-uridine: selective phosphorylation by herpes simplex virus type 1 thymidine kinase, antiherpetic activity, and cytotoxicity studies. Mol. Pharmacol. 47, 1231-1238.

Tino, J.A., Clark, J.M., Field, A.K., Jacobs, G.A., Lis, K.A., Michalik, T.L., McGeever-Rubin, B., Slusarchyk, W.A., Spergel, S.H., Sundeen, J.E., Tuomari, A.V., Weaver, E.R., Young, M.G. and Zahler, R. (1993) Synthesis and antiviral activity of novel isonucleoside analogs. J. Med. Chem. 36, 1221-1229.

Töpke, H., Reefschläger, J., Bärwolff, D. and Halibrand, C. (1984) Erste Erfahrungen mit dem Antiherpetikum (E)-5-(2-Bromvinyl)-2'-Desoxyuridin (BrVUdR) bei der Behandlung des Herpes corneae. Folia ophthal. 9, 39-48.

Walker, R.T., Barr, P.J., De Clercq, E, Descamps, J., Jones, A.S. and Serafinowski, P. (1978) The synthesis and properties of some antiviral nucleosides. Nucleic Acids Res., special publication no 4, s103-s106.

Watanaben K.A., Reichman, U., Hirota, K., Lopez, C. and Fox, J.J. (1979) Nucleosides. 110. Synthesis and antiherpes virus activity of some 2'-fluoro-2'-deoxyarabinofuranosyl-pyrimidine nucleosides. J, Med. Chem. 22, 21-24.

Weber, O. (2000) Novel mouse models for the investigation of experimental drugs with activity against human varicella-zoster virus. Antiviral Chem. Chemother. 11, 283-290.

Wildiers, J. and De Clercq, E. (1984) Oral (E)-5-(2-bromovinyl)-2'-deoxyuridine treatment of severe herpes zoster in cancer patients. Eur. J. Cancer Clin. Oncol. 20, 471-476.

Wutzler, P., Wutke, K., Bärwolff, D. and Reefschläger, J. (1988) BVDU-Therapie von Zostererkrankungen bei Patienten mit malignem Grundleiden. Z. gesamte inn. Med. 43, 677-680.

Wutzler, P., De Clercq, E., Wutke, K. and Färber, I. (1995) Oral brivudin vs. intravenous acyclovir in the treatment of herpes zoster in immunocompromised patients: a randomized double-blind trial. J. Med. Virol. 46, 252-257.

Yan, J., Tyring, S.K., McCrary, M.M., Lee, P.C., Haworth, S., Raymond, R., Olsen, S.J. and Diasio, R.B. (1997) The effect of sorivudine on dihydropyrimidine dehydrogenase activity in patients with acute herpes zoster. Clin. Pharmacol. Ther. 61, 563-573.

Yokota, T., Konno, K. and Shigeta, S. (1994) Inhibition of thymidylate synthetase activity induced in varicella-zoster virus infected cells by (E)-5-(2-bromovinyl)-2'-deoxyuridine. Antiviral Chem. Chemother. 5, 191-194.

Zhang, Y., Schols, D. and De Clercq, E. (1999) Selective activity of various antiviral compounds against HHV-7 infection. Antiviral Res. 43, 23-35.

THE ROLE OF THE CELLULAR DEOXYNUCLEOSIDE KINASES IN ACTIVATION OF NUCLEOSIDE ANALOGS USED IN CHEMOTHERAPY

STAFFAN ERIKSSON and LIYA WANG

Department of Veterinary Medical Chemistry,
Swedish University of Agricultural Sciences,
The Biomedical Center, Box 575, S-751 23 Uppsala, SWEDEN

1.   Introduction

Deoxynucleoside analogs are cornerstones in the treatment of viral and cancer diseases and their importance as chemotherapeutic agents is increasing. The world market value of nucleoside drugs was in 1998 in the order of 30 billion $US and the projected growth to 2005 is about 50 billion $US. Therefore, the achievements that led to establishment of chemical methods for analysis and synthesis of this group of compounds deserve the recognition of the scientific community. Professor, Dr. Jack Fox is one of the pioneers in the field and has made major contributions, particularly in the area of pyrimidine nucleoside chemistry and this review is dedicated to Dr Fox.

Nucleoside analogs are pro-drugs that after uptake into cells need activation by phosphorylation to form nucleotides that serve as inhibitors for viral or cellular DNA or RNA synthesis. The initial 5′-phosphorylation, carried out by nucleoside kinases, is usually the rate limiting step in the activation of nucleoside analogs and this review attempts to summarize the major advancements in the biochemistry and molecular genetics related to the cellular deoxynucleoside kinases. There are some earlier and recent reviews relevant for this field and we refer to these for a more comprehensive overview of the literature (Arnér, 1996; Arnér and Eriksson, 1995; Balzarini, 1994; Eriksson and Wang, 1997; Johansson and Eriksson, 1996; Maury, 2000; Plunkett and Gandhi, 1996; van der Wilt and Peters, 1994; Wintersberger, 1997).

2.   Overview of the salvage pathways for deoxynucleosides

The major pathway for biosynthesis of DNA precursors is via the reduction of ribonucleotides to deoxyribonucleotides by the ribonucleotide reductase enzyme system (Jordan and Reichard, 1998) which occurs in all proliferating cells. However, as a complement to this *de novo* synthesis pathway all cells also express deoxynucleoside kinases that can phosphorylate deoxynucleosides to form nucleoside monophosphates using a nucleoside triphosphate as phosphate donor. There are four deoxynucleoside kinases in animal cells, the two cytosolic enzymes, deoxycytidine kinase (dCK) and

455

Recent Advances in Nucleosides: Chemistry and Chemotherapy, Ed. by C.K. Chu. 455 — 475

thymidine kinase (TK1) as well as two mitochondrial enzymes, thymidine kinase 2 (TK2) and deoxyguanosine kinase (dGK). These enzymes are expressed in different ways, so that TK1 is only found in proliferating cells, dCK is expressed predominantly in lymphoid tissues, but also in growing cells, and the mitochondrial kinases are expressed in all types of cells, but at low levels as will be discussed below.

The salvage pathway is initiated by the uptake of nucleosides and there are at least two different classes of nucleoside transporters in mammalian cell membranes, i.e. equilibrative transport proteins and sodium dependent concentrative transporters as reviewed by Cass *et al.* (1999). Some types of nucleoside analogs (e.g. 3′-azidothymidine) appears to rely on passive diffusion for uptake, but in all cases the uptake process is reversible and it is the phosphorylation step that traps the nucleosides intracellularly. Nucleoside monophosphates are usually not the active compounds, but require further phosphorylation by nucleoside monophosphate kinases and nucleoside diphosphate kinase to form nucleoside triphosphates. These metabolites compete with endogenous DNA or RNA precursors for incorporation into nucleic acids, leading to blocked proliferation or apoptosis i.e. programmed cell death (Balzarini, 1994; Leoni *et al.*, 1998; Plunkett and Gandhi, 1996).

There are several types of nucleoside monophosphate kinases for the various nucleosides monophosphates (Yan and Tsai, 1999), only in case of thymidylate kinase this step is specific for deoxyribonucleotides, and it has been shown to be rate limiting for the 3′-azido-TMP activation (Lavie *et al.*, 1997). The nucleoside diphosphate kinase reaction is usually not limiting in the anabolism of nucleosides, but some 3′-modified nucleoside analogs have shown much reduced rates of phosphorylation as compared to the natural compounds (Bourdais *et al.*, 1996), and this fact could be of importance in the accumulation of active metabolites. However, the initial 5′-phosphorylation step is a prerequisite for further activation of nucleosides as exemplified by the role of the herpes virus encoded thymidine kinases in the efficacy and selectivity of anti-herpes nucleoside analogs (Balzarini, 1994). Below follows an overview of the properties and regulation of the cellular deoxynucleoside kinases.

3.   Pharmacologically important deoxynucleoside analogs

There are currently more than 20 nucleoside analogs registered for use as antiviral or anticancer drugs. Table 1 lists the generic and systematic names of some of these analogs, indicates their clinical applications and identifies which cellular kinase is involved in their initial phosphorylation, as specified further in the section describing the enzymes. In some cases it is known that cytosolic 5′-nucleotidase (5′-NT) is responsible for the phosphorylation of purine nucleoside analogs using IMP as phosphate donor. Thus this enzyme has the capacity to phosphorylate nucleosides in addition to being a nucleotidase (Johnson and Fridland, 1989; Arnér and Eriksson, 1995 and references therein).

The nucleoside analogs known to be activated only by virus coded nucleoside kinases are not included and the first ten analogs are presented in the order of their importance as estimated by their approximate sales values 1998. The last six analogs are in clinical trials or experimental drugs and are not appearing in any priority order.

Table 1.   Pharmacologically active deoxynucleoside analogs

Generic name[a]	Systemic name	Activating dNK	Application[b]
Lamivudine (3TC)	(-) β-L-2′,3′-dideoxythiacytidine	Viral kinases dCK	HIV, HBV
Stavudine (D4T)	2′,3′-didehydro-2′,3'-dideoxy thymidine	TK1	HIV
Zidovudine (AZT)	3′-azido-2′,3′-dideoxythymidine	TK1, TK2	HIV
Gemcitabine (dFdC)	2′,2′-difluoro-2′-deoxycytidine	dCK, TK2	Pancreatic cancer Ovarian cancer Lung cancer
Didanosine (ddI)	2′,3′-dideoxyinosine	5'-NT, dGK	HIV
Fludarabine (FaraA)	2-fluoro-9-β-D-arabinofuranosyladenine	dCK, dGK	Leukemias
Cytarabine (AraC)	1-β-D-arabinofuranosylcytosine	dCK, TK2	Leukemias, Lymphomas
Penciclovir (PCV)	9-(4-hydroxy-3-hydroxymethylbutyl)guanine	Viral kinases dGK	HSV, VZV, EBV HBV
Cladribine (CdA)	2-chloro-2′-deoxyadenosine	dCK, dGK	Leukemias
Zalcitabine (ddC)	2′-3′-dideoxycytidine	dCK	HIV
Vidarabine (AraA)	9-β-D-arabinofuranosyladenine	viral kinases dCK, dGK	HSV, VZV, CMV, HBV
Floxuridine (FdUrd)	5-fluoro-2′-deoxyuridine	TK1, TK2	Breast cancer Gastrointestinal cancer
Fialuridine (FIAU)	2'-deoxy-2'-fluoro-5-iodo-1-β-D-arabinofuranosyluracil	viral kinase TK1, TK2	HSV, VZV, HBV
L-FMAU	2'-fluoro-5-methyl-β-L-arabinofuranosyluracil	viral kinases TK1, TK2, dCK	HBV, EBV
L-Fd4C	2′,3′-dideoxy-2′,3′-didehydro-β-L(-)-5-fluorocytidine	viral kinases, dCK	HSV, HBV
β-L-OddC	β-L-1,3-dioxolane-cytidine	dCK	Prostate cancer Colon cancer Liver cancer Leukemias
CdG	Carbocyclic 2'-deoxyguanosine	dCK, dGK, 5′-NT	HSV, CMV

a   Common abbreviations are in parenthesis.

b   Abbreviations; HSV, herpes simplex virus; HIV, human immunodeficiency virus; EBV, Epstein-Barr virus; CMV, human cytomegalovirus; VZV, Varicella-zoster virus; HBV, hepatitis B virus.

4.   Cytosolic thymidine kinase

TK1 (EC 2.7.1.21) is present in most organisms with few exceptions such as yeast. Almost all DNA viruses e.g., bacteriophages, vaccinia viruses, pox viruses and herpes viruses code for enzymes with thymidine kinase activity. However, the herpes virus thymidine kinases belong to the same enzyme family as dCK and not the TK1 family as described below. The expression of TK1 is cell cycle regulated and active enzyme is detected only in S-phase cells. The regulation of TK1 is complex and occurs both at the transcriptional, the translational and post-translational levels (Wintersberger, 1997). The biological role of TK1 is still unclear, since several cell lines and also recently transgenic mice, which lacking TK1 activity, showed only minimal alterations in phenotypic properties (Dobrovolsky *et al.*, 1999).

The human gene for TK is localized to chromosome 17q25 (Petty *et al.*, 1996) and it contains seven exons and six introns of different lengths, all together encompassing a 13 kb DNA fragment (Flemington *et al.*, 1987). The human TK1 mRNA is 1.5 kb and expressed only in proliferating cells (Bradshaw, 1983). A human TK1 cDNA of 1421 bp has been cloned and sequenced as well as a number of other TK1 cDNAs from various species. The cDNAs encodes a protein of 25.5 kDa with highly conserved regions typical for nucleoside kinases as shown in Figure 1. There is a phosphate-binding loop (A) in the N-terminal, one region (B) involved in Mg^{2+} binding, and the third region (C) in the center of the protein is most likely involved in thymidine binding (Folkers *et al.*, 1991). Valine 106 in the mouse TK is most likely involved in subunit interactions and the methionine106 in human TK is apparently a variant found only in special cases (Bradshaw, 1983). In the majority of human TKs amino acid 106 is a valine (Berenstein *et al.*, 2000). However, these structural features have been deduced by sequence comparisons with other kinases of known structure, since there is still no 3-D structure available representing the TK1 enzyme family.

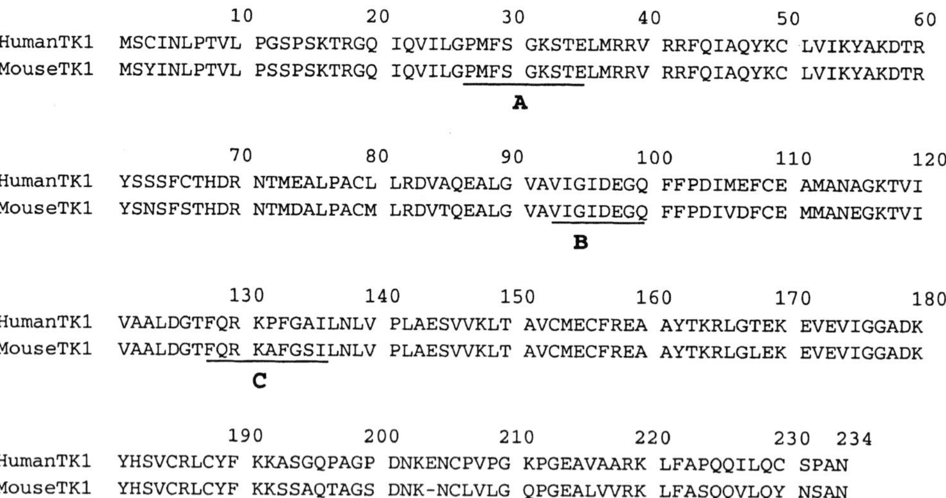

Figure 1.   Alignment of the amino acid sequences of human and mouse TK1 (GenBank accession numbers: K02581 and M68468). Regions involved in the function of the enzyme are underlined (Folkers *et al.*, 1991) as described in the text.

The cell cycle regulation of TK1 has been studied extensively and during recent years primarily with the mouse enzyme (Wintersberger, 1997; Sutterluety *et al.*, 1998; Posch *et al.*, 2000). Recently, it has been observed that TK1 mRNA levels are up-regulated in S-phase cells but that there appears to be a relatively high constitutive activity of the TK1 promoter also in resting cells. However in this type of cells the transcription of an antisense RNA originating in intron 3 of the mouse TK1 gene is induced and this leads to a drastic down regulation of total TK1 mRNA in non-growing cells (Sutterluety *et al.*, 1998). If the antisense mechanism of TK1 gene regulation is operating also in human cells remains to be determined.

The post-transcriptional modifications of TK1 occurs by M-phase specific phosphorylation of serine 13 in the N-terminal of human TK1 (Chang *et al.*, 1998), and by specific degradation of TK1 in growth arrested cells (Kauffman and Kelly, 1991; Posch *et al.*, 2000). This degradation process requires an intact C-terminal sequence and is inhibited by the presence of the substrate. Substrate binding to TK1 apparently block the cell cycle dependent proteolysis (Posch *et al.*, 2000).

ATP as well as other nucleoside- and deoxynucleoside triphosphates can be used as phosphate donors in the TK1 reaction, with the exception of CTP. The final end-product, dTTP, is an inhibitor, probably acting as a bisubstrate analog blocking the active site of TK1, a feed-back inhibitory mechanism that appears to be preserved in all deoxynucleoside kinases. ATP not only acts as a co-substrate but also induces the formation of the highly active TK1 tetramer (Munch-Petersen *et al.*, 1993). This is a slow process that leads to an enzyme with higher affinity for the nucleoside substrates, while the dimeric low activity form of TK1 is found when the enzyme is tested without pre-incubation with ATP. The methionine106 mutant TK1 mentioned above does not require ATP activation and is constantly in the tetrameric high affinity form. However, this enzyme variant is less stable than "natural" valine106 TK1. The kinetic behavior of these two TK1 variants differed so that TK1Met106 phosphorylation of dThd followed Michaelis-Menten kinetics, while TK1val106 showed negative co-operativity with dThd (Berenstein *et al.*, 2000). The origin of TK1Met106 is not clear but it is important that many of the earlier enzyme studies with recombinant TK1 have been done with the unusual TK1Met106.

In addition to the natural substrates, thymidine (dThd) and deoxyuridine (dUrd), TK1 also phosphorylates several clinically important nucleoside analogs (Table 1), for instance the anti-HIV compounds, 3′-azido-2′,3′-dideoxythymidine (AZT), 3′-fluoro-2′,3′-dideoxythymidine (FLT) and 2′,3′-dideoxy-2′,3′-didehydrothymidine (D4T) (Furman *et al.*, 1986; Matthes *et al.*, 1988; Wang *et al.*, 2000). The first two analogs are efficient substrates while D4T is active at a level only 1-5% of that of dThd. Several 5-substituted dUrd analogs are accepted by TK1 as substrates, e.g. 5-fluoro, 5-bromo, and 5-ethyl dUrd, while bulkier substitutions, such as 5-propenyl, 5-(2-chloroethyl) and 5-(2-bromovinyl) are not substrates, nor is arabinosylthymine (AraT) (Eriksson *et al.*, 1991; Johansson and Eriksson, 1996). The anti hepatitis B virus analogs 2′-deoxy-2′-fluoro-5-methyl-β-D-arabinofuranosyluracil (FMAU) and 2′-deoxy-2′-fluoro-5-iodo-β-D-arabinofuranosyluracil (FIAU) are efficient substrates for pure TK2 but showed only minimal activity with TK1 (Table 1) (Wang and Eriksson, 1996). The high capacity to activate FIAU in the mitochondria may be a contributing factor for the severe mitochondrial toxicity observed with this analog (Parker and Cheng, 1994). Unexpectedly, the L-enantiomer of FMAU was found to be a relatively good substrate for TK1 and it is also a substrate for dCK and TK2 (Liu *et al.*, 1998). This is the only L-nucleoside that has been shown to be phosphorylated by TK1 (Maury, 2000). A number of large substitutions at the N3 position, e.g. o-carboranylalkyl Thd, was recently shown to be good substrates for TK1 and this may be advantageous for development of analogs intended for boron neutron capture therapy against cancer (Lunato *et al.*, 1999).

There is also a role for TK1 in cancer diagnostics and monitoring, since determination of serum TK levels with a commercially available highly sensitive 125 I-dUrd assay have been used clinically for this purpose (Gronowitz *et al.*, 1983; Hallek *et al.*, 1999).

5.   Deoxycytidine kinase

dCK (NTP:deoxycytidine 5′-phosphotransferase, EC 2.7.1.74) is a cytosolic enzyme with a very broad substrate specificity. It is responsible for the phosphorylation of deoxynucleosides to form the corresponding monophosphates using ATP or UTP as phosphate donors. dCK phosphorylates dCyd most efficiently but also purine nucleosides such as dAdo and dGuo and in addition several of the pharmacologically most important antiviral and cytostatic deoxynucleosides (Table 1), e.g. L-2′3′-dideoxy-3′-thiacytidine (Lamivudine, 3TC), arabinosylcytosine (Cytosar, araC), 2-chlorodeoxy-adenosine (Cladribine, CdA) and 2′,2′-diflourdeoxycytidine (Gemcitabine, dFdC) (Arnér and Eriksson, 1995; Carson *et al.*, 1980; Furman *et al.*, 1986; Heinemann *et al.*, 1988; Matthes *et al.*, 1988; Plagemann *et al.*, 1978). These analogs are between 5-50% as efficient as dCyd as substrates for dCK and CdA is a much better substrate than dAdo. The apparent affinities for the nucleoside substrates decrease when UTP is used as phosphate donor instead of ATP and it is likely that UTP is the preferred phosphate donor in intact cells (Kraweic *et al.*, 1995; Shewach *et al.*, 1992). A further description of the specificity of dCK will be presented below but first some of the basic biochemical properties of the enzyme will be summarized.

Deoxycytidine kinase is composed of two identical polypeptides of 261 amino acids (Arnér and Eriksson, 1995). The enzymatic properties of recombinant dCK are similar but not identical to dCK purified from tissues (Arnér and Eriksson, 1995; Bohman and Eriksson, 1988; Chottiner *et al.*, 1991; Eriksson *et al.*, 1991; Ives and Durham, 1970; Kierdaszuk *et al.*, 1993). The enzyme shows sequence similarity with the herpes simplex type 1 virus thymidine kinase (HSV1-TK), as well as about 40% sequence identity to the mitochondrial thymidine kinase 2 (TK2) and deoxyguanosine kinase (dGK) (Figure 2) (Eriksson and Wang, 1997), and these enzymes constitute a family with many structural and kinetic properties in common. The structure determination of dGK has recently provided a model for the structure of dCK (Johansson *et al.*, 2001). The model predicts that dCK has 6 β-sheets and 7 α-helices and the N-terminal phosphate binding loop is in the centre of the active site. The 5′-OH group of the nucleoside held in place by a hydrogen bond to a conserved Glu and the base by hydrogen bonds to a conserved Gln. The phosphate donor binds in the opposite direction and the β and γ phosphates interacting with the Lys and Ser of the phosphate loop and two arginines in the LID region which is covering the active site.

The reaction kinetics of dCK is negative co-operativity both for the phosphate donor and acceptors, giving Hill coefficients <1 (Bohman and Eriksson, 1988; Ives and Durham, 1970; Kierdaszuk *et al.*, 1993). The reaction follows a random bi-bi pathway with ATP, while with UTP at low nucleoside concentration is ordered with the donor binding first (Hughes *et al.*, 1997; Ives and Durham, 1970; Kierdaszuk *et al.*, 1993; Turk *et al.*, 1999). Fluorescence quenching experiments demonstrate that substrate binding induces conformational changes (Kierdaszuk *et al.*, 1993; Turk *et al.*, 1999) which indicate that the enzyme may exist in different conformational states with different affinities for the substrates. Binding of substrates to one of the states could then stabilise this type of state in both subunits of the active dimeric enzyme. Feed-back inhibitors (e.g. dCTP) most likely function as bisubstrate analogs binding to both sites in the active cleft blocking the enzyme (Ikeda *et al.*, 1986).

The pharmacologically most important nucleosides are in addition to those mentioned above: 2-fluoro-arabinofuranosyladenine (Fludarabine, FaraA), 2′,3′-dideoxycytidine (Zalcitabine, ddC) and arabinosyladenine (Vidarabine, araA). The two latter analogs are active against HIV and HBV, both of which require cellular enzymes for activation of nucleoside prodrugs. ddC is about 10-20% as efficient substrate as dCyd, while the two other arabinosyl analogs are much less active, and dGK is probably also of importance for the phosphorylation of these nucleosides (see Section 7). A carbocyclic deoxyguanosine analog has shown broad spectrum antiviral activity and is phosphorylated by dCK but also by dGK and 5′-nucleotidase (Table 1) (Bennett *et al.*, 1998; Wang *et al.*, 1993).

dCK does not discriminate between the enantiomeric forms of its substrates and in several cases shows preferential phosphorylation of L-nucleosides (Chang *et al.*, 1992; Gaubert *et al.*, 1999; Maury, 2000; Shewach *et al.*, 1993; Wang J. *et al.*, 1999). This has led to the development of several L-nucleosides which are substrates for dCK and which have been or will be tested both for antiviral and anticancer activity (Grove *et al.*, 1995; Kierdaszuk *et al.*, 1999; Maury, 2000; Verri *et al.*, 1997). Two of these new L-nucleosides deserve to be mentioned specifically (Table 1), i.e. L-FMAU and L-OddC, which showed relatively high activity with dCK. The former has broad and impressive anti viral activities and is now in final clinical testing, and L-OddC is the first L-nucleoside that shows good promise as an anti tumor agent (Grove *et al.*, 1995; Liu *et al.*, 1998). Some didehydro deoxynucleoside analogs are substrates for dCK although with low efficiency. One new L-analog (L-Fd4C) of this type has shown good antiviral activity and is a substrate both for dCK and dGK in addition to 5′-NT and thus has several possible routes of activation (Table 1) (Zhu *et al.*, 1998).

The surprising fact that α-ddC is a more efficient substrate for dCK than β-ddC (Wang *et al.*, 1999) indicates that dCK prefers nucleosides with the sugar in the S-conformation (C2′-endo-C3′-exo), since α-ddC adopts that conformation preferentially. This conclusion was recently verified by direct determination of the structure of the bound nucleoside using NMR studies with complexes of $^{13}C/^{2}H$ double labeled dCyd and dAdo and dCK (Maltseva *et al.*, 2000). These results are in agreement with molecular modeling studies with HSV-1 TK and conformationally restricted nucleoside analogs (Prota *et al.*, 2000) and should be considered in future design of nucleoside analogs.

Mouse dCK has a lower capacity to activate certain deoxynucleosides, e.g. CdA and ddC as compared with human dCK, which is one of the reasons that these analogs are less toxic to mouse cells than to human cells. The amino acid sequence has 93% identity between mouse and human dCK, but the enzymes have different substrate specificity with ATP or UTP used as phosphate donors. The amino acids in the LID region (aa175-aa185) differ in several important positions between the mouse and human dCK (Usova and Eriksson, 2002), which apparently change the specificity of the kinases. These results are of importance for the choice of animal models used in drug development of deoxynucleoside therapeutics.

The gene for human dCK is localized to chromosome band 4q13.3-q21.1 and is a single copy gene of 34 kb (Song *et al.*, 1993; Stegmann *et al.*, 1993). Despite initial reports of higher transcriptional activity of the dCK gene in T- as compared to B-lymphocytes the more detailed analysis of the transcriptional regulation showed no difference between these two types of lymphocytes or resting and actively growing cells (Chen *et al.*, 1995; Hengstschläger *et al.*, 1993). The expression of dCK mRNA is tissue specific and shows high levels in lymphocytic tissues, intermediate levels in proliferating epithelial cells and very low levels in terminally differentiated tissues such as liver, kidney and pancreas (Chottiner *et al.*, 1991; Johansson and Karlsson, 1997). The activity of the enzyme varied between 2-10 folds in extracts from cells of different cell cycle phases while the mRNA levels were relatively constant, which suggest that the enzyme is regulated by post transcriptional processes (Hengstschläger *et al.*, 1993).

dCK was found in the cytosolic fraction during biochemical isolation procedures, but recently it was found that the enzyme has a nuclear localisation signal in the N-terminal region and overexpression of dCK cDNA resulted in nuclear localization of dCK protein in the transfected cells (Johansson *et al.*, 1997). However, endogenous dCK in normal cells was found only in the cytosol using a dCK peptide antibody (Hatzis *et al.*, 1998). The role of the nuclear localisation signal and the possible regulation of the subcellular localisation of dCK remains to be determined, but it is clear that native dCK is a cytosolic enzyme.

Cell lines lacking dCK are resistant to several nucleoside analogs and dCK deficiency was shown to be a reason for araC, dFdC and CdA resistance in cultured cells (Ruiz van Haperen *et al.*, 1994; Owens *et al.*, 1992). However, when malignant cells were isolated from acute myeloic leukemia (AML) patients who were resistant to AraC, only in one out of 16 cases was dCK deficiency detected (Flasshove *et al.*, 1994). In another case of AraC resistant AML cells, several shorter dCK mRNA variants were detected and these truncated forms of dCK mRNA's lacked one or several exons (Veuger *et al.*, 2000). The alternatively spliced forms of dCK could be involved in the process of nucleoside analog resistance.

In some studies a correlation between the sensitivity of cells to nucleoside analogs and the level of dCK was observed, e.g., patients responding to CdA had a somewhat higher level of dCK than non-responders (Arnér *et al.*, 1994; Kawasaki *et al.*, 1992). However, the correlation was weak and it is not clear if determinations of dCK levels in the clinical situation are of diagnostic or prognostic value. Transfection experiments with dCK gene constructs has shown that higher cellular levels of dCK give higher sensitivity of the cells to cytotoxic nucleoside which serves as substrates for the enzyme (Hapke *et al.*, 1996). Therefore, dCK gene transfer to tumour cells could be a method to increase the sensitivity for a nucleoside analog. Transfection of glioma cells with vectors containing the dCK gene gave higher sensitivity of the transfected cells to AraC and when tested in animal models it lead to tumour regression in response to AraC chemotherapy (Manome *et al.*, 1996).

Synergistic effect of certain combinations of nucleosides e.g. AraC and FAraA (Plunkett and Gandhi, 1996) has been shown to at least in part be due to activation of dCK and thereby more efficient activation of the nucleosides by the treatment of cells with DNA inhibitors (Spasokoukotskaja *et al.*, 1999). Incubation of cells with several of the nucleosides activated by dCK but also with unrelated agents such as the topoisomerase inhibitor etoposide leads to higher dCK activity without a concomitant increase in dCK mRNA or protein. The molecular nature of this post transcriptional activation process is not defined but the effect is of considerable clinical importance and could be utilised to improve anticancer and antiviral chemotherapy.

6.   Mitochondrial thymidine kinase

TK2 catalyses the transfer of a gamma phosphate group from ATP to the 5'-hydroxyl group of dThd, the same substrate as for TK1. The level of TK2 is low as compared to TK1 in proliferating cells, but it is significant in resting or terminally differentiated cells where TK1 activity is undetectable. TK2 is constitutively expressed and the activity levels are most likely correlated to the mitochondria content of the tissues.

The open reading frame of human TK2 cDNA codes for 233 amino acids with calculated molecular weight of 28 kDa and recombinant human TK2 showed similar enzyme activity as the native enzyme (Wang L. *et al.*, 1999). Recent studies on recombinant mouse TK2 have shown that the active form of TK2 is a dimer (Wang and Eriksson, 2000).

Human TK2 utilizes dThd, dCyd and dUrd as substrates but with different efficiency and kinetic mechanisms. TK2 phosphorylates dThd with negative co-operativity, which means that the affinity for the substrate decreased with increasing substrate concentrations. The Km value for dThd was 16 μM at concentrations above 8 μM, and was 0.2-0.4 μM at dThd concentrations below 8 μM. The phosphorylation of dCyd and dUrd showed normal Michaelis-Menten mechanism with Km value of 36 μM and 6 μM, respectively. Both dTTP and dCTP inhibit the enzyme and ATP and CTP can be used as phosphate donors by TK2 (Munch-Petersen *et al.*, 1991; Wang L. *et al.*, 1999).

dThd analogs with modification on the sugar moiety such as AZT, AraT, FLT and ribothymine could be phosphorylated, but with low efficiency (Table 1). TK2 showed about 5-10% of the activity with AZT as compared to that with dThd. The ability of TK2 to phosphorylate AZT may explain the anti HIV activity of AZT in resting cells such as macrophages (Arnér *et al.*, 1992). A large number of dUrd analogs can be phosphorylated by TK2, including 5-substitutions such as halogen, amino, ethyl, 5-(2-bromovinyl), and some bulky groups and also modifications on sugar moiety e.g. arabinosyluracil (AraU), 2′,2′-difluoro-2′-deoxyuridine (dFdU), FIAU and FMAU, (Table 1) (Eriksson *et al.*, 1991, 1995; Munch-Petersen *et al.*, 1991; Wang and Eriksson, 1996; Wang J. *et al.*, 1999). Uridine (Urd) itself was not a substrate for TK2 but 5-substituted analogs such as 5-methyl-Urd, 5-(2-bromovinyl)-Urd and 5-iodo-Urd could be phosphorylated by TK2 (Balzarini *et al.*, 2000). dCyd analogs with 5-substitution of halogen, amino or even bulky groups, such as 5-(2-chloroethyl) and 5-(2-bromovinyl) showed activity with TK2. Modifications of the sugar moiety such as dFdC and AraC were also acceptable but with low efficiency (Table 1) (Eriksson *et al.*, 1991, 1995; Wang L. *et al.*, 1999). Interestingly some of the 5-aryl substituted deoxycytidine analogues were phosphorylated by TK2 but not by dCK.

TK2, in contrast to TK1, showed relaxed enantioselectivity. L-Thd and L-dCyd were efficiently phosphorylated by TK2 and the phosphorylation of L-Thd was shown to follow Michaelis-Menten type kinetics. Several other L- nucleosides such as L-BvdU, L-FMAU and L-5-iodo-dUrd were also substrates for TK2 (Table 1) (Liu *et al.*, 1998; Maury, 2000; Verri *et al.*, 1997; Wang J. *et al.*, 1999).

The human TK2 gene is localized to chromosome 16q22 (Johansson and Karlsson, 1997; Willecke *et al.*, 1977). The expression of TK2 mRNA is complex and tissue specific. In most tissues there are two transcripts of 4.0 and 2.2 kb while in proliferating tissues, like testis, ovary and thymus, two additional transcripts were found, which indicates a complex transcriptional regulation of TK2 expression (Johansson and Karlsson, 1997; Wang *et al.*, 1999). The cloned cDNA sequences of human TK2 do not include a mitochondrial targeting sequence thereby they are not complete (Johansson and Karlsson, 1997; Wang L. *et al.*, 1999). Mouse TK2 cDNA has recently been cloned and the open reading frame codes for a polypeptide containing both a mitochondrial targeting signal and a catalytic domain, which shows about 80% sequence identity to human TK2. *In vitro* translation and import experiment showed that the N-terminal sequence directs the import of precursor protein into isolated mitochondria (Wang and Eriksson, 2000). Therefore, it is clear now that TK2 is localized in the mitochondria.

The amino acid sequence of TK2 shows approximately 40% identity to dCK and dGK (Figure 2). The key amino acid residues in substrate recognition in dGK sequence are also conserved in both dCK and TK2 sequence and this may explain the overlapping substrate specificity of these three enzymes.

In cells, dThd is either degraded to thymine or salvaged to dTMP by TK1 and TK2. TK2 phosphorylates both dThd and dCyd with similar efficiency but with different kinetic mechanisms. At normal metabolic conditions dThd and dCyd phosphorylation by TK2 may be regulated by the availability of the substrate in the tissues and by feed back regulation of the end products, e.g. dTTP and dCTP. In this way balanced dNTP pools in the mitochondria can probably be maintained. However, dThd inhibits efficiently the dCyd phosphorylation carried out by TK2, while dCyd is a poor inhibitor of the dThd phosphorylation (Wang L. *et al.*, 1999). Mitochondrial neurogastrointestinal encephalomyopathy (MNGIE) is an autosomal recessive disease with multiple deletions and depletion of mitochondrial DNA in skeletal muscle due to mutations in the gene encoding thymidine phosphorylase (TP), an enzyme that catabolizes thymidine to thymine and 2-deoxy-D-ribose 1-phosphate. Lack of TP activity led to a 50-fold increase in plasma dThd level in MNGIE patients (Nishino *et al.*, 1999; Nishino *et al.*, 2000).

In vivo this high dThd level will lead to a increase in the dTMP pool and a decrease in the dCMP pool due to inhibition of dCyd phosphorylation by dThd. The resulting large dTMP pool and small dCMP pool will lead to altered dTTP and dCTP pools. Such pool imbalances may induce mutations or deletions in newly synthesized mtDNA, suggesting that TK2 plays a role in certain forms of mitochondrial diseases (Wang and Eriksson, 2000).

7. Deoxyguanosine kinase

dGK (NTP: deoxyguanosine 5'-phosphotransferase, EC 2.7.1.113) catalyzes the phosphorylation of purine deoxynucleosides and their analogs, using a nucleoside triphosphate as phosphate donor. dGK is directly involved in providing DNA precursors for mtDNA synthesis. dGK is constitutively expressed and the dGK activity is found in most tissues, including lymphoid tissues, spleen, skin, liver and brain. In certain brain tumor extracts and also cytomegalovirus infected cell extracts the levels of dGK were found to be elevated (Lewis *et al.*, 1985; Wang *et al.*, 1993).

Human dGK has a molecular mass of 60 kDa, and consists of two identical subunits of 30 kDa. dGK purified from various tissues/or species had similar properties but relatively low specific activity, probably due to the inactivation during purification. Recombinant dGK showed a broad substrate specificity and a much higher specific activity when purified in the presence of ATP and Triton X-100, and thus may represents the true physiological state of the enzyme. The amino acid sequence of dGK shows ≈ 40% identity to dCK and TK2 and several important sequence motifs are highly conserved (Figure2).

The structure of dGK has recently been solved with ATP bound to the active site (Johansson *et al.*, 2000). The dGK structure consists of an αβ structure made up of 9 α-helices and five β-sheets (Figure 2). The five stranded parallel β-sheet, as part of the core of the molecule, together with 7 α-helices forms the active site cleft, which exhibits a high degree of homology with HSV1-TK. The active site consists of substrate nucleoside-binding pocket and the ATP binding loop. Similar to HSV1-TK and adenylate kinase, the ATP binding loop is formed by a strand-turn-helix (residues 40-63) that constitute a large anion hole, which coordinates the phosphate groups of ATP. Arg 206 and Arg208 from helix 7 (the LID region) are also interacting with phosphate groups and assist in phosphoryl transfer. The base of the substrate nucleoside stack between residues Trp75, Phe110 and Phe151 of helix 2, 4 and 5 while the deoxyribose 5'-OH group is forming a hydrogen bond with Glu70 and the 3'-OH group interacts with Tyr100 and Glu211. Additional residues that surround the substrate may also participate in catalysis. The structure of dGK showed high overall similarity to that of HSV1-TK and Drosophila melanogaster-deoxynucleoside kinase (Johansson *et al.*, 2001).

Several nucleoside triphosphates can be used as phosphate donors with ATP and UTP as the most efficient ones. However dATP and dGTP are feedback inhibitors of the enzyme. A random Bi Bi kinetic mechanism, which may include the formation of a dead-end complex, was proposed for bovine dGK. The end product dGTP, which

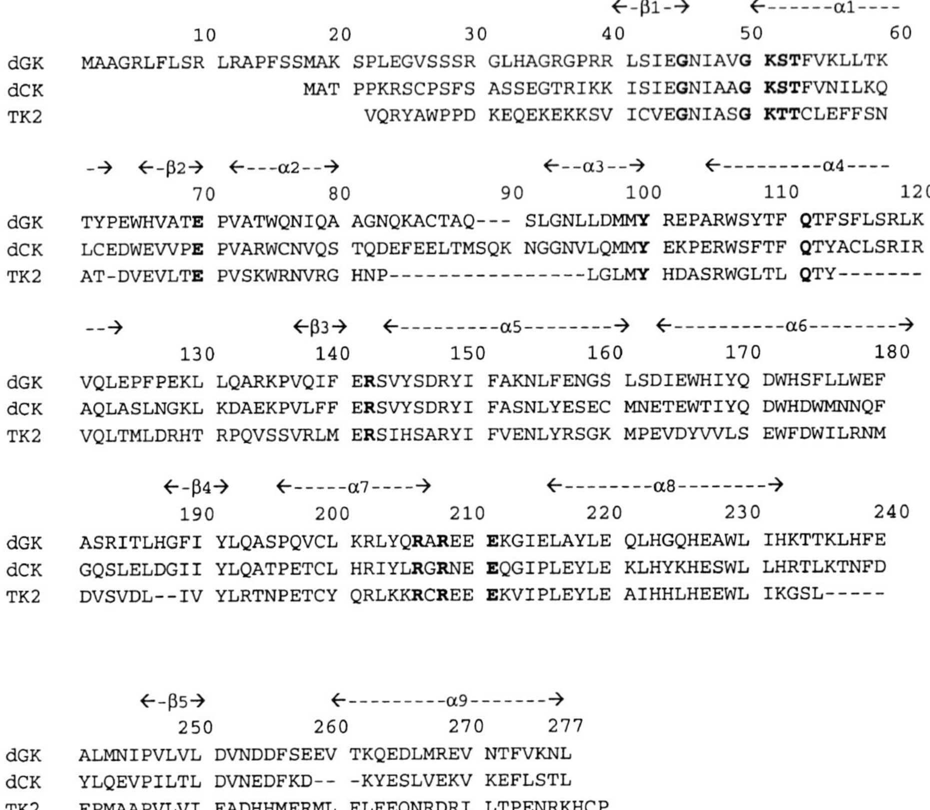

Figure 2. Alignment of the amino acid sequences of human deoxycytidine kinase and deoxyguanosine kinase and TK2 showing the conserved amino acids involved in substrate binding (bold) as described in the text. The predicted secondary structure of dCK (top line) is presented above the sequences based on the observed structure of dGK (Johansson *et al.*, 2000). The accession numbers: dGK: X97386 and U41668, dCK: M60527, and TK2: Y10498 and U77088. On the second line is the amino acid numbering of dGK.

competes very efficiently with ATP (Ki = 0.03 μM), but is non-competitive towards dGuo, behaved as a bisubstrate analog, most likely interacting with its nucleoside moiety towards the dGuo site, and the phosphate groups of dGTP interacting with the ATP binding site. This is apparently a general mechanism for the feed-back inhibition of deoxyribonucleoside kinases (Park and Ives, 1995).

Natural purine deoxynucleosides, i.e. dGuo, dAdo and deoxyinosine (dIno) as well as a number of purine nucleoside analogs are substrates for dGK. At high concentration the pyrimidine nucleoside dCyd showed some activity with human dGK. Nucleosides with modifications on either the sugar or base can be efficiently phosphorylated but with restrictions as to the substitution at 2' position of deoxyribose. dGK activates several antiviral and anticancer nucleoside analogs (Table 1), including AraG (9-β-D-arabinofuranosylguanine), dFdG (2',2'-difluorodeoxyguanosine), CdA, CAFdA (2-chloro-2'-arabinofluoro-2'-deoxyadenosine), AraA, F-AraA, and to some extent ddI, 5-FdC (5-fluoro-2'-deoxycytidine), dFdC, and to a low extent Ganciclovir (9-(1,3-dihydroxy-2-propoxymethyl)-guanine) and Penciclovir (Vectavir, 9-(2-hydroxy-1-hydroxymethyl-ethoxymethyl)-guanine) (Table 1) (Herrström Sjöberg *et al.*, 1998, 2001; Wang *et al.*, 1993). dGK exhibits a relaxed enantioselectivity, accepting both D- and L- enantiomers of β-dGuo with similar efficiencies, but with low efficiency for α-L-dGuo. β-L-dAdo and β-L-dCyd were also substrates, but with low activity (Wang J. *et al.*, 1999).

The human dGK gene is localized to chromosome band 2q13 (Johansson *et al.*, 1996), a region involved in translocations observed in some patients with lymphoproliferative disorders, such as acute and chronic lymphatic leukemia (Berkowicz *et al.*, 1995). The structure of the gene is quite similar to the dCK gene with 7 exons, ranging in size from 100 to 237 base pairs, spread on a 35 kb fragment. A major 1.35 kb dGK mRNA transcript was found in most tissues by northern blot analysis. The cDNA for human dGK codes for a 277 amino acid polypeptide with an N-terminal 17 amino acid sequence characteristic of a mitochondrial import signal, which is able to direct the precursor protein into the mitochondria. Biochemical studies of subcellular fractions as well as in situ immunohistochemical experiment using a dGK antibody clearly showed that the dGK protein is localized in the mitochondrial matrix (Johansson and Karlsson, 1996; Jüllig and Eriksson, 2000; Wang *et al.*, 1996).

Mouse dGK cDNA showed 75% sequence identity to human dGK, and two mRNA transcripts were identified, one full length form, representing the mitochondrial form of enzyme, and an N-terminal truncated form, lacking the mitochondrial targeting signal. The shorter mRNA may represent a cytosolic form of mouse dGK. Recombinant mouse dGK showed similar specific activity and substrate specificity as compared to the human enzyme (Petrakis *et al.*, 1999).

Mitochondria play a critical role in apoptosis in many cell types with the release of cytochrome c from the intermembrane space activating the caspases that initiate apoptosis (Green and Reed, 1998). It was observed that dGK translocated from mito-chondria to cytosol during the course of apoptosis using western blotting of subcellular fractions as well as immunofluorescence in situ studies of apoptotic 293 cells and/or Molt4 cells. What role dGK plays in the apoptosis process is not clear, but dGK may be selectively transported into the cytosol to assist in amplifying the apoptotic cascade (Jüllig and Eriksson, 2001).

The intracellular level of dGTP is usually the limiting DNA precursor in almost all cells studied so far and thus the synthesis of dGTP is of vital importance. A high dGTP level is cytotoxic and dGK is directly involved in the regulation of mitochondrial dGTP levels. Evidence from a recent study with radioactive nucleosides, araG and CdA, which are dGK substrates, showed that they were selectively incorporated into mtDNA via intramitochondrial phosphorylation (Zhu *et al.*, 2000). Purine nucleoside phosphorylase (PNP) deficiency in man, an inherited disease, is associated with severe T-cell immunodeficiency and variable neurological manifestations including spasticity and behavioral disorders. The lack of PNP activity resulted in a high intracellular level of dGuo, one of PNP substrates, in affected patients. The phosphorylation of dGuo by dGK will lead in turn to the accumulation of dGTP, which can inhibit ribonucleotide reductase activity, abrogating DNA synthesis or DNA repair, which are the cause of impaired T-cell maturation and differentiation (Cohen *et al.*, 1978).

PNP-deficient mice showed similar symptoms as in human patients, exhibiting T cell lymphopenia and abnormalities in T cell function. It was observed that the levels of intracellular dGTP were elevated and a marked increase of intramitochondrial dGTP levels was found in these mice. Consistent with the observation that incubation of mouse thymocytes with dGuo in the presence of PNP inhibitor led to a large accumulation of dGTP in the mitochondria. Therefore, it was proposed that the T cell abnormalities were due to the selective accumulation of dGTP in the mitochondria, leading to defective mtDNA and T cell apoptosis (Arpaia *et al.*, 2000). This is in accordance with the fact that in order to reduce the toxic level of dGTP a secondary loss of dGK activity was found in PNP deficient mice (Arpaia *et al.*, 2000; Snyder *et al.*, 1994). These results suggest that dGK plays a key role in the mechanism of inherited T-cell immunodeficiencies.

The dominant role of cytosolic dCK in the activation of nucleoside analogs used in chemotherapy has prevented an evaluation of the role of dGK in the cytotoxicity of nucleoside analogs. When both dGK and dCK are present, dCK is the major contributor in the activation of many nucleoside analogs, such as araC, CdA and dFdC etc. However, in tissues such as brain and liver that do not contain dCK, dGK is the only purine nucleoside phosphorylating enzyme and thus is responsible for the activation of nucleoside analogs.

Penciclovir is used in chemotherapy of chronic hepatitis B virus infection (Shaw and Locarnini, 1999). Unlike herpes viruses, hepatitis B virus does not code for a nucleoside kinase, therefore, the activation of Penciclovir is depending on cellular enzymes and dGK is responsible for the initial phosphorylation of this drug. ddI was recently shown to be substrate for dGK, and the observed side effect with ddI treatment might be because of incorporation of the analog metabolite into mtDNA (Herrström Sjöberg *et al.*, 1998).

8.   Conclusions

The genes for cellular deoxynucleoside kinases are now sequenced and the proteins can be expressed in recombinant form in relatively large amounts. This has enabled the determination of the dGK structure, which, in turn, provided good models for the structures of dCK and TK2 since these enzymes belong to the same enzyme family. In the near future, we will be in the position to use molecular modeling techniques, co-crystallization methods and *in vitro* mutagenesis to define the detailed structure function relationships within this family. This information will aid future rational drug design of nucleoside analogs that need to be activated by deoxynucleoside kinases. Although the active site structure is remarkably similar within this family, including the herpes virus thymidine kinase, detailed knowledge about the similarities and differences is needed for design of selective and highly active chemotherapeutic analogs. The regulation of dCK, TK2 and dGK at the transcriptional and post-transcriptional levels is largely unknown and remains to be clarified. Another major challenge in this field is elucidation of the biological role of these kinases, and for this purpose transgenic animal models, as well as microarray profiling techniques, should provide a starting point.

In case of TK1 the 3-D structure is eagerly need to clarify the differences and similarities to the other kinases. Much more is known about the regulation of TK1, which occurs with high precision and in many different ways in different cells and cell cycle stages. The lack of TK1 gives only a subtle phenotype in mice, but clearly the expression of this enzyme can be used as a marker for S-phase cells and as such it is useful in therapeutic and diagnostic anticancer approaches.

9.   Acknowledgement

Work performed in the authors laboratories have been funded by the Swedish University for Agricultural Sciences, The Swedish Medical Research Council, The Swedish Natural Science Research Council, The Swedish Board for Technical Development, The Swedish Strategic Research Foundation, the European Commission (BMH4-CT96-0479) and Medivir AB, Huddinge, Sweden.

10.   References

Arnér, E. (1996) On the phosphorylation of 2-chlorodeoxyadenosine (CdA) and its correlation with clinical response in leukemia treatment. Leuk. Lymphoma. 21, 225-231

Arnér, E., Spasokoukotskaja, T., Juliusson, G., Liliemark, J., and Eriksson, S. (1994) Phosphorylation of 2-chlorodeoxyadenosine (CdA) in extracts of peripheral blood mononuclear cells of leukaemic patients, Br. J. Haematol. 87, 715-718

Arnér, E.S., Valentin, A. and Eriksson, S. (1992) Thymidine and 3'-Azido-3'-deoxythymidine metabolism in Human Peripheral Blood Lymphocytes and Monocyte-derived macrophages, J. Biol. Chem. 267, 10968-10975.

Arnér, E.S.J. and Eriksson, S. (1995) Mammalian deoxyribonucleoside kinases, Pharmac. Ther. 67, 155-186.

Arpaia, E., Benveniste, P., Di Cristofano, A., Gu, Y., Dalal, I., Kelly, S., Hershfield, M., Pandolfi, P., Roifman, C. and Cohen, A. (2000) Mitochondrial basis for immune deficiency. Evidence from purine nucleoside phosphorylase-deficient mice, J. Exp. Med. 191, 2197-2208.

Balzarini, J. (1994) Metabolism and mechanism of antiretroviral action of purine and pyrimidine derivatives, Pharm. World Sci. 16, 113-126.

Balzarini, J., Zhu, C., De Clercq, E., Perez-Perez, M., Chamorro, C., Camarasa, M. and Karlsson, A. (2000) Novel ribofuranosylnucleoside lead compounds for potent and selective inhibitors of mitochondrial thymidine kinase-2, Biochem. J. 351, 167-171.

Bennett, L. J., Allan, P., Arnett, G., Shealy, Y., Shewach, D., Mason, W., Fourel, I. and Parker, W. (1998) Metabolism in human cells of the D and L enantiomers of the carbocyclic analog of 2'-deoxyguanosine: substrate activity with deoxycytidine kinase, mitochondrial deoxyguanosine kinase, and 5'-nucleotidase, Antimicrob. Agents Chemother. 42, 1045-1051.

Berenstein, D., Christensen, J., Kristensen, T., Hofbauer, R. and Munch-Petersen, B. (2000) Valine, Not

Methionine, Is Amino Acid 106 in Human Cytosolic Thymidine Kinase (TK1). Impact on oligomerization, stability, nad kinetic properties, J. Biol. Chem. 275, 32187-32192.

Berkowicz, M., Toren, A., Rosner, E., Biniaminov, M., Rosenthal, E., Gipsh, N., Berman, S., Hardan, I., Mandel, M. and Amariglio N. (1995) Translocation (2;14)(p13;q32) in CD10+;CD13+ acute lymphatic leukemia, Cancer Genet. Cytogenet. 83, 140-143.

Bohman, C. and Eriksson, S. (1988) Deoxycytidine Kinase from Human leukemic Spleen: Preparation and characterization of the Homogeneous Enzyme, Biochemistry 27, 4258-4265.

Bourdais, J., Biondi, R., Sarfati, S., Guerreiro, C., Lascu, I., Janin, J. and Veron, M. (1996) Cellular phosphorylation of anti-HIV nucleosides. Role of nucleoside diphosphate kinase, J. Biol. Chem. 271, 7887-7890.

Bradshaw, H. D. (1983) Molecular cloning and cell cycle-specific regulation of a functional human thymidine kinase gene, Proc. Natl. Acad. Sci. USA 80, 5588-5591.

Carson, D.A., Wasson, D.B., Kaye, J., Ullman, B., Martin, D.J., Robins, R.K. and Montgomery, J.A. (1980) Deoxycytidine Kinase-mediated toxicity of deoxyadenosine analogs toward malignant human lymphoblasts in vitro and toward murine L1210 leukemia in vivo, Proc. Natl. Acad. Sci. USA 77, 6865-6869.

Cass, C., Young, J., Baldwin, S., Cabrita, M., Graham, K., Griffiths, M., Jennings, L., Mackey, J., Ng, A., Ritzel, M., Vickers, M. and Yao, S. (1999) Nucleoside transporters of mammalian cells, Pharm. Biotechnol. 12, 313-352.

Chang, C., Skalski, V., Zhou, J. and Cheng, Y. (1992) Biochemical pharmacology of (+)- and (-)-2',3'-dideoxy-3'-thiacytidine as anti-hepatitis B virus agents, J. Biol. Chem. 267, 22414-22420.

Chang, Z., Huang, D. and Chi, L. (1998) Serine 13 is the site of mitotic phosphorylation of human thymidine kinase, J. Biol. Chem. 273, 12095-12100.

Chen, E.H., JohnsonII, E.E., Vetter, S.M. and Mitchell, B.S. (1995) Characterization of the deoxycytidine kinase promoter in human lymohoblast cell lines, J. Clin. Invest. 95, 1660-1668

Chottiner, E.G., Shewach, D.S., Datta, N.S., Ashcraft, E., Gribbin, D., Ginsburg, D., Fox, I.H. and Mitchell, B.S. (1991) Cloning and expression of human deoxycytidine kinase cDNA, Proc. Natl. Acad. Sci. USA 88, 1531-1535.

Cohen, A., Gudas, L., Ammann, A., Staal, G. and Martin, D.J. (1978) Deoxyguanosine triphosphate as a possible toxic metabolite in the immunodeficiency associated with purine nucleoside phosphorylase deficiency, J. Clin. Invest. 61, 1405-1409.

Dobrovolsky, V., Casciano, D. and Heflich, R. (1999) Tk[+/-] mouse model for detecting in vivo mutation in an endogenous, autosomal gene, Mutat. Res. 423, 125-136.

Eriksson, S., Kierdaszuk, B., Munch-Petersen, B., Öberg, B. and Johansson, N.G. (1991) Comparison of the substrate specificities of human thymidine kinase 1 and 2 and deoxycytidine kinase toward antiviral and cytostatic nucleoside analogs, Biochem. Biophys. Res. Commun. 176, 586-592.

Eriksson, S., Wang, J., Gronowitz, S. and Johansson, N.G. (1995) Substrate specificities of mitochondrial thymidine kinase and cytosolic deoxycytidine kinase against 5-aryl substituted pyrimidine 2'-deoxyribose analogues, Nuclsides & Nucleotides 14, 507-510.

Eriksson, S. and Wang, L. (1997) Substrate specificities, expression, and primary sequences of deoxynucleoside kinases; implications for chemotherapy, Nucleosides & Nucleotides 16, 653-659.

Flasshove, M., Strumberg, D., Ayscue, L., Mitchell, B. S., C. Tirier, W. H., Seeber, S. and Schutte, J. (1994) Structural analysis of the deoxycytidine kinase gene in patients with acute myeloid leukemia and resistance to cytosine arabinoside, Leukemia 8, 780-785.

Flemington, E., Bradshaw, H.D., Traina-Dorge, V., Slagel, V. and Deininger, P.L. (1987) Sequence, structure and promoter characterization of the human thymidine kinase gene, Gene 52, 267-277.

Folkers, G., Trumpp-Kallmeyer, S., Gutbrod, O., Krickl, S., Fetzer, J. and Keil, G. (1991) Computer-aided active-site-directed modeling of the herpes simplex virus 1 and human thymidine kinase, J. Comput. Aided Mol. Des. 5, 385-404.

Furman, P.A., Fyfe, J.A., Clair, M.H.S., Weinhold, K., Rideout, J.L., Freeman, G.A., Lehrma, S.N., Bolognesi, D.P., Broder, S., Mitsuya, H. and Barry, D.W. (1986) Phosphorylation of 3'-azido-3'-deoxythymidine and selective interaction of the 5'-triphosphate with human immunodeficiency virus reverse transcriptase, Proc. Natl. Acad. Sci. USA 83, 8333-8337.

Gaubert, G., Gosselin, G., Boudou, V., Imbach, J., Eriksson, S. and Maury, G. (1999) Low enantioselectivities of human deoxycytidine kinase and human deoxyguanosine kinase with respect to 2'-deoxyadenosine, 2'-deoxyguanosine and their analogs, Biochimie 81, 1041-1047.

Green, D. and Reed, J. (1998) Mitochondria and apoptosis, Science 281, 1309-1312.

Gronowitz, J., Hagberg, H., Kallander, C. and Simonsson, B. (1983) The use of serum deoxythymidine kinase as a prognostic marker, and in the monitoring of patients with non-Hodgkin's lymphoma., Br. J. Cancer 47, 487-495.

Grove, K., Guo, X., Liu, S., Gao, Z., Chu, C. and Cheng, Y. (1995) Anticancer activity of beta-L-dioxolane-cytidine, a novel nucleoside analogue with the unnatural L configuration, Cancer Res. 55, 3008-3011.

Hallek, M., Langenmayer, I., Nerl, C., Knauf, W., Dietzfelbinger, H., Adorf, D., Ostwald, M., Busch, R., Kuhn-Hallek, I., Thiel, E. and Emmerich, B. (1999) Elevated serum thymidine kinase levels identify a subgroup at high risk of disease progression in early, nonsmoldering chronic lymphocytic leukemia, Blood 93, 1732-1737.

Hapke, D.M., Stegmann, A.P.A. and Mitchell, B.S. (1996) Retroviral transfer of deoxycytidine kinase into tumor cell lines enhancec nucleoside toxicity, Cancer Res. 56, 2343-2347.

Hatzis, P., Al-Madhoon, A., Jüllig, M., Petrakis, T., Eriksson, S. and Talianidis, I. (1998) The intracellular localization of deoxycytidine kinase, J. Biol. Chem. 273, 30239-30243.

Heinemann, V., Hertel, L., Grindey, G. and Plunkett, W. (1988) Comparison of the cellular pharmacokinetics and toxicity of 2',2'-difluorodeoxycytidine and 1-beta-D-arabinofuranosylcytosine, Cancer Res. 48, 4024-4031.

Hengstschläger, M., Denk, C. and Wawra, E. (1993) Cell cycle regulation of deoxycytidine kinase-evidence for post-transcriptional control, FEBS lett. 321, 237-240.

Herrström Sjöberg, A., Wang, L. and Eriksson, S. (1998) Substrate specificity of human recombinant mitochondrial deoxyguanosine kinase with cytostatic and antiviral purine and pyrimidine analogs, Mol. Pharmacol. 53, 270-273.

Herrström Sjöberg, A., Wang, L. and Eriksson, S. (2001) Antiviral guanosine analogs as substrates for deoxyguanosne kinase: implications for chemotherapy, Antimicrob. Agents Chemother. (in press).

Hughes, T., Hahn, T., Reynolds, K. and Shewach, D. (1997) Kinetic analysis of human deoxycytidine kinase with the true phosphate donor uridine triphosphate, Biochemistry 36, 7540-7547.

Ikeda, S., Chakravarty, R. and Ives, D. (1986) Multisubstrate analogs for deoxynucleoside kinases. Triphosphate end products and synthetic bisubstrate analogs exhibit identical modes of binding and are useful probes for distinguishing kinetic mechanisms, J. Biol. Chem. 261, 15836-15843.

Ives, D. and Durham, J. (1970) Deoxycytidine kinase. 3. Kinetics and allosteric regulation of the calf thymus enzyme, J. Biol. Chem. 245, 2285-2294.

Johansson, K., Ramaswamy, S., Ljungcrantz, C., Knecht, W., Piskur, J., Munch-Petersen, B., Eriksson, S. and Eklund, H. (2001) Crystal structures of two eukaryotic deoxyribonucleoside kinases, Nat. Struct. Biol. 8, 616-620.

Johansson, M., Bajalica-Lagercrantz, S., Lagercrantz, J. and Karlsson, A. (1996) Localization of the human deoxyguanosine kinase gene (DGUOK) to chromosome 2p13, Genomics 38, 450-451.

Johansson, M., Brismar, S. and Karlsson, A. (1997) Human deoxycytidine kinase is located in the cell nucleus, Proc. Natl. Acad. Sci. USA 94, 11941-11945.

Johansson, M. and Karlsson, A. (1996) Cloning and expression of human deoxyguanosine kinase cDNA, Proc. Natl. Acad. Sci. USA 93, 7258-7262.

Johansson, M. and Karlsson, A. (1997) Cloning of the cDNA and chromosome localization of the gene for human thymidine kinase 2, J. Biol. Chem. 272, 8454-8458.

Johansson, N. and Eriksson, S. (1996) Structure-activity relationships for phosphorylation of nucleoside analogs to monophosphates by nucleoside kinases, Acta Biochim. Pol. 43, 143-160.

Johnson, W.A. and Fridland, A. (1989) Phosphorylation of 2',3'-Dideoxyinosine by Cytosolic 5'-Nucleotidase of Humam Lymphoid Cells, Mol. Pharmacol. 36, 291-295.

Jordan, A. and Reichard, P. (1998) Ribonucleotide reductases, Annu. Rev. Biochem. 67, 71-98.

Jüllig, M. and Eriksson, S. (2001) Apoptosis induces efflux of the mitochondrial matrix enzyme deoxyguanosine kinase, J. Biol. Chem. 276, 24000-24004.

Jüllig, M. and Eriksson, S. (2000) Mitochondrial and submitochondrial localization of human deoxyguanosine kinase, Eur. J. Biochem. 267, 5466-5472.

Kauffman, M.G. and Kelly, T.J. (1991) Cell cycle regulation of thymidine kinase: residues near the carboxyl terminus are essential for the specific degradation of the enzyme at mitosis, Mol. Cell Biol. 11, 2538-2546.

Kawasaki, H., Caricia, C.J., and Carson, D.A. (1992) Quantitative immunoassay of human deoxycytidine kinase in malignant cells. Anal. Biochem. 207, 193-196.

Kierdaszuk, B., Krawiec, K., Kazimierczuk, Z., Jacobsson, U., Johansson, N., Munch-Petersen, B., Eriksson, S. and Shugar, D. (1999) Substrate/inhibitor properties of human deoxycytidine kinase (dCK) and thymidine kinases (TK1 and TK2) towards the sugar moiety of nucleosides, including O'-alkyl analogues, Nucleosides & Nucleotides 18, 1883-1903.

Kierdaszuk, B., Rigler, R. and Eriksson, S. (1993) Binding of Substrates of Human Deoxycytidine Kinase Studied with Ligand-Dependt Quenching of Enzyme Instrinsic Fluorescence, Biochemistry 32, 1993.

Kraweic, K., Kierdaszuk, B., Eriksson, S., Munch-Petersen, B. and Shugar, D. (1995) Nucleoside triphosphate donors for nucleoside kinases: Donor properties of UTP with human deoxycytidine kinase, Biochem. Biophys. Res. Commun. 216, 42-48.

Lavie, A., Vetter, I., Konrad, M., Goody, R., Reinstein, J. and Schlichting, I. (1997) Structure of thymidylate kinase reveals the cause behind the limiting step in AZT activation, Nat. Struct. Biol. 4, 601-604.

Leoni, L., Chao, Q., Cottam, H., Genini, D., Rosenbach, M., Carrera, C., Budihardjo, I., Wang, X. and Carson, D. (1998) Induction of an apoptotic program in cell-free extracts by 2-chloro-2'-deoxyadenosine 5'-triphosphate and cytochrome c, Proc. Natl. Acad. Sci. USA 95, 9567-9571.

Lewis, R.A., Wakins, L. and Jeor, S.S. (1985) Enhancement of deoxyguanosine kinase activity in human lung fibroblast cells infected with human cytomegalovirus, Mol. Cell. Biochemistry 65, 67-71.

Liu, S., Grove, K. and Cheng, Y. (1998) Unique metabolism of a novel antiviral L-nucleoside analog, 2'-fluoro-5-methyl-beta-L-arabinofuranosyluracil: a substrate for both thymidine kinase and deoxycytidine kinase, Antimicrob. Agents Chemother. 42, 833-839.

Lunato, A., Wang, J., Woollard, J., Anisuzzaman, A., Ji, W., Rong, F., Ikeda, S., Soloway, A., Eriksson, S., Ives, D., Blue, T. and Tjarks, W. (1999) Synthesis of 5-(carboranylalkylmercapto)-2'-deoxyuridines and 3-(carboranylalkyl)thymidines and their evaluation as substrates for human thymidine kinases 1 and 2, J. Med. Chem. 42, 3378-3389.

Maltseva, T., Usova, E., Eriksson, S., Mileki, J., Földesi, A. and Chattopadhayaya, J. (2000) An NMR conformational study of the complexes of 13C/2H double-labelled 2´-deoxynucleosides and deoxycytidine kinase, J. Chem. Soc. Trans. Perkin 2, 2199-2207.

Manome, Y., Wen, P.Y., Dong, Y., Tanaka, T., Mitchell, B.S., Kufe, D.W. and Fine, H.A. (1996) Viral vector transduction of the human deoxycytidine kinase cDNA sensitizes glioma cells to the cytotoxic effects of cytosine arabinoside in vitro and in vivo, Nat. Med. 2, 567-573.

Matthes, E., Lehmann, C., Scholz, D., Rosenthal, H. A. and Langen, P. (1988) Phosphorylation, anti-HIV activity and cytotoxicity of 3'-fluorothymidine, Biochem. Biophys. Res. Commun. 153, 825-831.

Maury, G. (2000) The enantioselectivity of enzymes involved in current antiviral therapy using nucleoside analogues: a new strategy?, Antivir. Chem. Chemother. 11, 165-189.

Munch-Petersen, B., Cloos, L., Tyrsted, G. and Eriksson, S. (1991) Diverging substrate specificity of pure human thymidine kinases 1 and 2 against antiviral dideoxynucleosides, J. Biol. Chem. 266, 9032-9038.

Munch-Petersen, B., Tyrsted, G. and Cloos, L. (1993) Reversible ATP-dependent transition between two forms of human cytosolic thymidine kinase with different enzymatic properties, J. Biol. Chem. 268, 15621-15625.

Nishino, I., Spinazzola, A. and Hirano, M. (1999) Thymidine phosphorylase gene mutations in MNGIE, a human mitochondrial disorder, Science 283, 689-692.

Nishino, I., Spinazzola, A., Papadimitriou, A., Hammans, S., Steiner, I., Hahn, C., Connolly, A., Verloes, A., Guimaraes, J., Maillard, I., Hamano, H., Donati, M., Semrad, C., Russell, J., Andreu, A., Hadjigeorgiou, G., Vu, T., Tadesse, S., Nygaard, T., Nonaka, I., Hirano, I., Bonilla, E., Rowland, L., DiMauro, S. and Hirano, M. (2000) Mitochondrial neurogastrointestinal encephalomyopathy: an autosomal recessive disorder due to thymidine phosphorylase mutations, Ann. Neurol. 47, 792-800.

Owens, J.K., Shewach, D.S., Ullman, B. and Mitchell, B.S. (1992) Resistance to 1-β-D-arabinofuranosylcytosine in human T-lymphoblasts mediated by mutations within the deoxycytidine kinase gene, Cancer Res. 52, 2389-2393.

Park, I. and Ives, D. H. (1995) Kinetic mechanism and end-product regulation of deoxyguanosine kinase from beef liver mitochondria, J. Biol. chem. 117, 1058-1061.

Parker, W. B. and Cheng, Y. C. (1994) Mitochondrial toxicity of antiviral nucleoside analogs, J. NIH Res. 6, 57-61.

Petrakis, T., Ktistaki, E., Wang, L., Eriksson, S. and Talianidis, I. (1999) Cloning and characterization of mouse deoxyguanosine kinase. Evidence for a cytoplasmic isoform, J. Biol. Chem. 274, 24726-24730.

Petty, E., Miller, D., Grant, A., Collins, E., Glover, T. and Law, D. (1996) FISH localization of the soluble thymidine kinase gene (TK1) to human 17q25, a region of chromosomal loss in sporadic breast tumors, Cytogenet. Cell Genet. 72, 319-321.

Plagemann, P., Marz, R. and Wohlhueter, R. (1978) Transport and metabolism of deoxycytidine and 1-beta-D-arabinofuranosylcytosine into cultured Novikoff rat hepatoma cells, relationship to phosphorylation, and regulation of triphosphate synthesis, Cancer Res. 38, 978-989.

Plunkett, W. and Gandhi, V. (1996) Pharmacology of purine nucleoside analogues, Hematol. Cell Ther. 38 Suppl 2, S67-74.

Posch, M., Hauser, C. and Seiser, C. (2000) Substrate binding is a prerequisite for stabilisation of mouse thymidine kinase in proliferating fibroblasts, J. Mol. Biol. 300, 493-502.

Prota, A., Vogt, J., Pilger, B., Perozzo, R., Wurth, C., Marquez, V., Russ, P., Schulz, G., Folkers, G. and Scapozza, L. (2000) Kinetics and crystal structure of the wild-type and the engineered Y101F mutant of Herpes simplex virus type 1 thymidine kinase interacting with (North)-methanocarba-thymidine,

Biochemistry 39, 9597-9603.

Ruiz van Haperen, V.W.T., Veeman, G., Eriksson, S., Boven, E., Stegmann, A.P.A., Hermsen, M., Vermorken, J.B., Pinedo, H.M. and Peters, G.J. (1994) Development and molecular characterization of a 2', 2'-difluorodeoxycytidine resistant variant of the human ovarian carcinoma cell line A2780, Cancer Res. 54, 4138-4143.

Shaw, T. and Locarnini, S. (1999) Preclinical aspects of lamivudine and famciclovir against hepatitis B virus, J. Viral. Hepat. 6, 89-106.

Shewach, D., Liotta, D. and Schinazi, R. (1993) Affinity of the antiviral enantiomers of oxathiolane cytosine nucleosides for human 2'-deoxycytidine kinase, Biochem. Pharmacol. 45, 1540-1543.

Shewach, D.S., Reynolds, K.K. and Hertel, L. (1992) Nucleotide specificity of human deoxycytidine kinase, Mol. Pharmacol. 42, 518-524.

Snyder, F.F., Jenuth, J.P., Dilay, J.E., Fung, E., Lightfoot, T. and Mably, E.R. (1994) Secondary loss of deoxyguanosine kinase acticity in purine nucleoside phosphorylase deficient mice, Biochim. Biophys. Acta 1227, 33-40.

Song, J.J., Walker, S., Chen, E., Johnson, E.E., Spychala, J., Gribbin, T. and Mitchell, B.S. (1993) Genomic structure and chromosomal localization of the human deoxycytidine kinase gene, Proc. Natl. Acad Sci. USA 90, 431-434.

Spasokoukotskaja, T., Sasvari-Székely, M., Keszler, G., Albertioni, F., Eriksson, S. and Staub, M. (1999) Treatment of normal and malignant cells with nucleoside analogues and etoposide enhances deoxycytidine kinase activity, Eur. J. Cancer 35, 1862-1867.

Stegmann, A.P.A., Honders, M.W., Bolk, M.W.J., Wessels, J., Willemze, R. and Landegent, J.E. (1993) Assignment of the human deoxycytidine kinase (DCK) gene to chromosome 4 Band q13.3-q21.1, Genomics 17, 528-529.

Sutterluety, H., Bartl, S., Doetzlhofer, A., Khier, H., Wintersberger, E. and Seiser, C. (1998) Growth-regulated antisense transcription of the mouse thymidine kinase gene, Nucleic Acids Res. 26, 4989-4995.

Turk, B., Awad, R., Usova, E., Bjork, I. and Eriksson, S. (1999) A pre-steady-state kinetic analysis of substrate binding to human recombinant deoxycytidine kinase: a model for nucleoside kinase action, Biochemistry 38, 8555-8561.

Usova, E. and Eriksson, S. (2002) Mutational analysis of the amino acids in deoxycytidine kinase that determined differences in substrate specificity between the mouse and the human enzyme; implications for animal models in drug development, Mol. Pharmacol. (submitted)

van der Wilt, C. and Peters, G. (1994) New targets for pyrimidine antimetabolites in the treatment of solid tumours. 1: Thymidylate synthase. Pharm. World Sci. 16, 84-103.

Verri, A., Focher, F., Priori, G., Gosselin, G., Imbach, J., Capobianco, M., Garbesi, A. and Spadari, S. (1997) Lack of enantiospecificity of human 2'-deoxycytidine kinase: relevance for the activation of beta-L-deoxycytidine analogs as antineoplastic and antiviral agents, Mol. Pharmacol. 51, 132-138.

Verri, A., Priori, G., Spadari, S., Tondelli, L. and Focher, F. (1997) Relaxed enantioselectivity of human mitochondrial thymidine kinase and chemotherapeutic uses of L-nucleoside analogues., Biochem. J. 328, 317-320.

Veuger, M., Honders, M., Landegent, J., Willemze, R. and Barge, R. (2000) High incidence of alternatively spliced forms of deoxycytidine kinase in patients with resistant acute myeloid leukemia, Blood 96, 1517-1524.

Wang, J., Choudhury, D., Chattopadhyaya, J. and Eriksson, S. (1999) Stereoisomeric selectivity of human deoxyribonucleoside kinases, Biochemistry 38, 16993-16999.

Wang, J. and Eriksson, S. (1996) Phosphorylation of the anti-hepatitis B nucleoside analog 1-(2'-deoxy-2'-

fluoro-1-beta-D-arabinofuranosyl)-5-iodouracil(FIAU) by human cytosolic and mitochondrial thymidine kinase and implication for cytotoxicity, Antimicrob. Agents Chemother. 40, 1555-1557.

Wang, J., Su, C., Neuhard, J. and Eriksson, S. (2000) Expression of human mitochondrial thymidine kinase in Escherichia coli: correlation between the enzymatic activity of pyrimidine nucleoside analogues and their inhibitory effect on bacterial growth, Biochem. Pharmacol. 59, 1583-1588.

Wang, L. and Eriksson, S. (2000) Cloning and characterization of full length mouse thymidine kinase 2: the N-terminal sequence directs import of the precursor protein into mitochondria, Biochem. J. 351, 469-476.

Wang, L., Hellman, U. and Eriksson, S. (1996) Cloning and expression of human mitochondrial deoxyguanosine kinase cDNA, FEBS lett. 390, 39-43.

Wang, L., Karlsson, A., Arnér, E. S. J. and Eriksson, S. (1993) Substrate specificity of mitochondrial 2'-deoxyguanosine kinase Efficient phosphorylation of 2-chlorodeoxyadenosine, J. Biol. Chem. 268, 22847-22852.

Wang, L., Karlsson, A., Mathiesen, T. and Eriksson, S. (1993) 2-Chloro-2´-deoxyadenosine phosphorylation by deoxyguanosine kinase in crude extracts of malignant human brain tissue. Recent Advances in Chemotherapy. In: J. Einhorn, C.E., Nord and S.R. Norrby (eds), Proceedings of the 18th International Congress of Chemotherapy, American Society for Microbiology; pp. 919-921, Stockholm, Sweden.

Wang, L., Munch-Petersen, B., Herrström Sjöberg, A., Hellman, U., Bergman, T., Jörnvall, H. and Eriksson, S. (1999) Human thymidine kinase 2: molecular cloning and characterisation of the enzyme activity with antiviral and cytostatic nucleoside substrates, FEBS Lett. 443, 170-174.

Willecke, K., Teber, T., Kucherlapati, R. S. and Ruddle, F. H. (1977) Human mitochondrial thymidine kinase is coded for by a gene on chromosome16 of the nucleus, Somatic Cell Genet. 3, 237-245.

Wintersberger, E. (1997) Regulation and biological function of thymidine kinase, Biochem. Soc. Trans. 25, 303-308.

Yan, H. and Tsai, M.D. (1999) Nucleoside monophosphate kinase: Structure, mechanism and substrate specificity. Adv. Enzymol. Relat. Areas Mol. Biol. 73, 103-134.

Zhu, C., Johansson, M. and Karlsson, A. (2000) Incorporation of nucleoside analogs into nuclear or mitochondrial DNA is determined by the intracellular phosphorylation site, J. Biol. Chem. 275, 26727-26731.

Zhu, Y., Dutschman, D., Liu, S., Bridges, E. and Cheng, Y. (1998) Anti-hepatitis B virus activity and metabolism of 2',3'-dideoxy-2',3'-didehydro-beta-L(-)-5-fluorocytidine, Antimicrob. Agents Chemother. 42, 1805-1810.

CELLULAR TRANSPORT OF NUCLEOTIDE ANALOGS

TOMAS CIHLAR, DAMIAN MCCOLL
and NORBERT BISCHOFBERGER

Gilead Sciences, Foster City, California, USA

1. Introduction

Inhibitors of viral DNA polymerases and reverse transcriptases represent a large group of effective antiviral agents. Among them, acyclic nucleoside phosphonates (ANPs), a class of unique nucleotide analogs, are currently being clinically utilized or investigated for the treatment of diseases caused by various DNA viruses and retroviruses (De Clercq *et al.*, 1986; Naesens *et al.*, 1997). The most advanced antivirals of this class are cidofovir, adefovir, and tenofovir (Figure 1).

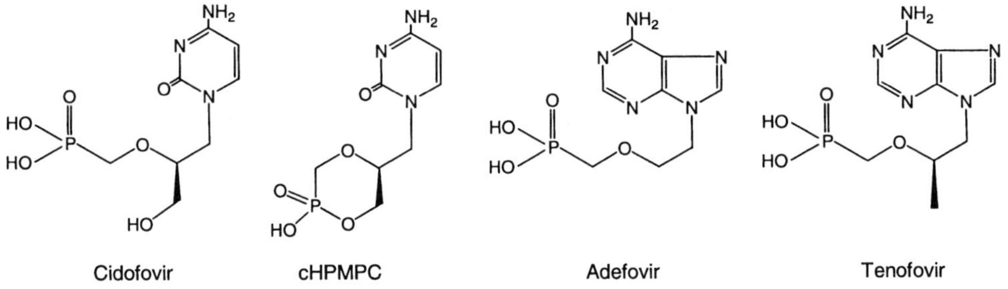

Figure 1. Acyclic nucleoside phosphonate analogs.

Cidofovir (HPMPC) effectively inhibits replication of numerous DNA viruses (Hitchcock *et al.*, 1996). It has been approved for the treatment of CMV retinitis in AIDS patients (Lalezari *et al.*, 1997) and has also been shown active in managing the papillomavirus-associated cutaneous diseases including cervical neoplasia (Snoeck *et al.*, 2000) and laryngeal papillomatosis (Snoeck *et al.*, 1998). In addition, HIV-infected patients with progressive multifocal leukoencephalopathy due to JC virus infection have been treated with cidofovir (Brambilla *et al.*, 1999). An intracellular cyclic prodrug of cidofovir (cHPMPC, Figure 1) has been designed and shown to have the same *in vitro* antiviral efficacy (Bischofbeger *et al.*, 1994) and improved *in vivo* toxicity profile compared to cidofovir (Hitchcock *et al.*, 1995). Unlike cidofovir, adefovir (PMEA) exhibits *in vitro* antiviral activity against both DNA viruses and retroviruses (reviewed

477

Recent Advances in Nucleosides: Chemistry and Chemotherapy, Ed. by C.K. Chu. 477 — 503

in Cihlar and Bischofberger, 1998). Adefovir dipivoxil [Bis(POM)PMEA], an oral prodrug of adefovir, has been extensively studied as an anti-HIV agent (Kahn *et al.*, 1999). Adefovir also exhibits potent anti-hepadnaviral activity (Ying *et al.*, 2000) and adefovir dipivoxil is being evaluated for the treatment of hepatitis B virus infections (Perrillo *et al.*, 2000). Tenofovir (PMPA), the third clinically advanced nucleotide analog, has shown potent antiretroviral (Balzarini *et al.*, 1993) as well as antihepadnaviral (Ying *et al.*, 2000) activity *in vitro*. Its orally bioavailable prodrug, tenofovir disoproxil [Bis(POC)PMPA], is recently approved as an once-daily oral agent for treatment of HIV-1 infection (Schooley *et al.*, 2000).

In addition to their antiviral activity, particular ANPs exhibit antiparasitic (Smeijsters *et al.*, 1999), immunomodulatory (Frankova *et al.*, 1999), and antitumor (Otova *et al.*, 1997) activity. As antivirals, ANPs possess a number of distinct characteristics including a unique resistance profile (Xiong *et al.*, 1998; Miller *et al.*, 2000). They have been designed to circumvent the first phosphorylation step in the intracellular nucleoside activation pathway by incorporating the non-hydrolyzable phosphonate moiety into their structure. Despite their highly polar nature, ANPs are taken up and accumulated by various mammalian cell types. In addition to other characteristics, the intracellular pharmacokinetics (i.e. uptake, metabolism, and efflux) of ANPs affects their antiviral potency, *in vivo* dosing as well as toxicity profile. Together with discussing the current knowledge about cellular transport of ANPs, this review summarizes membrane transporters, most of which have been identified only recently, that may potentially mediate cellular uptake and/or efflux of this class of important therapeutics.

2.   Mechanisms of cellular uptake of ANPs

Natural nucleotides are not efficiently imported across plasma membrane in their intact form. Instead, they are first dephosphorylated by the plasma membrane ectoenzymes to the respective nucleosides, which then are taken up via nucleoside transporters and re-phosphorylated by intracellular nucleoside kinases (Chiba *et al.*, 1984). 5'-Nucleotidase (Shah *et al.*, 1986; Edwards *et al.*, 1986) and alkaline phosphatase (Zekri *et al.*, 1989; Vorbrodt, 1979) are examples of the most common ecto-phosphatases. This process, however, does not apply to ANPs because they are refractory to these enzymes.

Several *in vitro* studies revealed differences in the mechanism of cellular uptake of ANPs in various cell types. In the established H9 cell line derived from human T-lymphocytes, adefovir uptake was insensitive to specific inhibitors of nucleoside trans-port and was characterized by slow kinetics and significant temperature-sensitivity sug-gesting that a nonspecific fluid-phase endocytosis is the main transport mechanism (Palu *et al.*, 1991). In the same study, identical transport characteristics were found for another nucleoside phosphonate HPMPA (adenine analog of cidofovir) (Palu *et al.*, 1991). Subsequently, the fluid-phase endocytosis was confirmed as the main mechanism of adefovir uptake into T-cells of human origin (Olsanska *et al.*, 1997). In this study, CEM T-lymphotropic cells were used to demonstrate a temperature-sensitive, non-concentrative, and non-saturable uptake of adefovir. The process was strictly ATP-dependent and insensitive to competitive inhibition by other ANPs. In addition, adefovir uptake into CEM cells exhibited identical kinetics to carboxyinulin, an established marker for fluid-phase endocytosis (Olsanska *et al.*, 1997).

In contrast to T-cells, uptake of adefovir into HeLa S3 epithelial cells has been identified as saturable process (Km = 0.39 μM) proceeding against the concentration gradient (Cihlar *et al.*, 1995). At extracellular adefovir concentrations < 1 μM, active transport resulted in at least a 10-fold higher intracellular concentration. The transport was specific with respect to the type of nucleobase and the acyclic sugar-like moiety since a few analogs of adefovir, e.g. PMEDAP (2,6-diaminopurine counterpart), but not cidofovir, tenofovir, or their analogs were able to act as competitive inhibitors. In addition, natural nucleotides competitively inhibited the transport process. Their effect was dependent on the nature of nucleobase and the position of the phosphate group indicating that the uptake of adefovir into HeLa cells is mediated by a highly specific membrane transporter. By using affinity chromatography, a putative 50-kDa membrane transporter was identified (Cihlar *et al.*, 1995).

Uptake of adefovir has also been characterized in Vero cells derived from the African green monkey kidney (Prus *et al.*, 1991). The uptake was found to be Na$^+$-dependent, but it was not experimentally established whether the coupling with Na$^+$ transport is direct or indirect (through other transporters). Similar to HeLa cells, the transport of adefovir in Vero cells was saturable (Km = 130 μM) and sensitive to inhibition by other ANPs and natural nucleotides. In addition, mersalyl acid and probenecid, two inhibitors of organic anion transport, reduced adefovir uptake into Vero cells suggesting that some form of the organic anion transport system expressed in renal cells may be involved (see Chapter 4).

Vero cells have also been used to characterize the uptake of cidofovir (Connelly *et al.*, 1993). Unlike adefovir, the uptake of cidofovir was not saturable across a wide concentration range and was insensitive to other ANPs and natural nucleotides. Kinetics and inhibitor susceptibility of cidofovir uptake was identical to that of sucrose, a probe for fluid-phase endocytosis, pointing to a difference between the mechanism of cidofovir and adefovir uptake into Vero cells. Alternatively, the observed difference may partly have been due to different experimental conditions in the two studies (Prus *et al.*, 1991; Connelly *et al.*, 1993).

3.   Cellular metabolism and efflux of ANPs

3.1. Metabolism

Following their entry into the cells, ANPs are activated by cellular enzymes to their diphosphorylated derivatives. The enzymes involved in phosphorylation of ANPs are distinct for the pyrimidine (cidofovir) and purine series (adefovir, tenofovir, HPMPA). For pyrimidine ANPs, pyrimidine nucleoside monophosphate (PNMP) kinase catalyzes the conversion of cidofovir to its monophosphate, which is further phosphorylated to diphosphate by several enzymes including pyruvate kinase, nucleoside diphosphate (NDP) kinase and creatine kinase (Ho *et al.*, 1992; Cihlar and Chen, 1996). There appear to be two potential pathways for phosphorylation of purine ANPs. AMP(dAMP) kinase from murine T-leukemia cells can phosphorylate adefovir and several other ANPs (HPMPA, FPMPA) to their diphosphates in a two-step process that proceeds via a monophosphate intermediate (Merta *et al.*, 1992). This phosphorylation is stereoselective and approximately three orders of magnitude less efficient than the phosphorylation of AMP. Further evidence for the role of AMP(dAMP) kinase in the activation of adefovir has come from studies on a human CEM T-lymphoid cell line and a derived adefovir-resistant cell line, CEM-r1 (Robbins *et al.*, 1995a). The CEM-r1 cells showed a 2-fold decrease in the activity of AMP kinase and interestingly, also had a 7-fold increased rate of adefovir efflux. Mitochondrial AMP kinase (AK2) phosphorylates adefovir more efficiently than the cytosolic counterpart (AK1) (Robbins *et al.*, 1995b). Adefovir monophosphate can be converted to diphosphate by creatine kinase (Merta *et al.*, 1992). This phosphorylation step can also be catalyzed by NDP kinase, which shows a broad substrate specificity for most NDP analogues, including acyclic molecules (Miller and Miller, 1982).

An alternative pathway involves the phosphorylation of adefovir directly to diphosphate by 5-phosphoribosyl-1-pyrophosphate (PRPP) synthetase from either bacterial or eukaryotic sources (Balzarini and De Clercq, 1991; Balzarini *et al.*, 1991a). PRPP synthetase does not display stereospecificity, recognizing both the (S)- and (R)-enantiomers of HMPMA and FPMPA (Balzarini *et al.*, 1991b). The relevance of this enzyme in the physiological activation of adefovir is unclear because of relatively low efficiency of phosphorylation.

A general feature of ANPs is the long intracellular half-life of the diphosphate metabolite, a property which may explain their long-lasting antiviral activity. In the case of cidofovir, intracellular levels of the mono- and diphosphate forms show a bi-phasic decline with half-lives of approximately 24h and 65h, respectively (Ho *et al.*, 1992; Aduma *et al.*, 1995). These long half-lives could be due to the formation of a cidofovir-phosphocholine metabolite, which is formed from cidofovir diphosphate by CTP: phosphorylcholine cytidylyl-transferase (Cihlar *et al.*, 1992). This metabolite has a half-life of approximately 85h and may act as a reservoir for subsequent conversion to cidofovir diphosphate. No equivalent reservoir has been demonstrated for adefovir or tenofovir. The half-life of adefovir diphosphate is 5-17 hours in different cell types (Aduma *et al.*, 1995; Balzarini *et al.*, 1991a). Tenofovir diphosphate exhibited a half-life of 10 hours and approximately 50 hours in activated and resting PBMCs, respectively (Robbins *et al.*, 1998).

3.2. Cellular efflux of ANPs

Recent studies have implicated two members of the ABC transporter family as being involved in the cellular efflux of adefovir. ABC transporters are a large family of ATP-dependent membrane-bound proteins found in all eukaryotes and are characterized by the presence of a 200-250 amino acid domain that contains two short, conserved peptide motifs (Klein *et al.*, 1999). These two motifs, designated the Walker A and Walker B motifs, are involved in ATP binding and are found in many other ATP-utilizing proteins. A third conserved sequence motif, the "C" motif, is located between the "A" and "B" motifs and is diagnostic for the ABC transporter family as a whole. ABC transporters can be subdivided into eight subfamilies based on sequence homology. They transport a wide variety of compounds including lipids, peptides, glutathione conjugates and anionic compounds. Multidrug resistance protein 1 (MDR1 or P-glycoprotein), the first ABC transporter to be described has been shown to be responsible for multidrug resistance in cancer chemotherapy (Juliano and Ling, 1976; Ambudkar *et al.*, 1999). Multidrug resistance is characterized by resistance to multiple structurally unrelated drugs and is often observed after exposure to only a single agent. ABC transporters belonging to the MRP sub-family, notably MRP1 and MRP2, have also been shown to mediate multidrug resistance (Cole *et al.*, 1992). MDR1 appears to be able to transport a wide variety of hydrophobic compounds. In contrast, MRP1, MRP2 and MRP3 mediate the ATP-dependent, unidirectional transport of lipophilic compounds in the form of their glutathione, glucuronate or sulfate conjugates (Borst *et al.*, 1999; Hipfner *et al.*, 1999; Konig *et al.*, 1999a). In addition to anionic conjugates, unconjugated amphiphilic anions can also serve as substrates for MRP1, MRP2 and possibly other members of the MRP subfamily.

MRP4 (ABCC4 or MOAT-B) (Robbins *et al.*, 1995a; Lee *et al.*, 1998; Borst *et al.*, 1999; Schuetz *et al.*, 1999) and MRP5 (ABCC5 or MOAT-C) (Kool *et al.*, 1997; Belinsky *et al.*, 1998; Borst *et al.*, 1999; Wijnholds *et al.*, 2000) have been identified as playing a role in the efflux of ANPs. Robbins *et al.* (1995a) isolated a human T-lymphoid cell line, CEM-r1 that was generated by selecting CEM cells in gradually increasing concentrations (up to 10 mM) of adefovir. The CEM-r1 cell line demonstrated resistance to the cytostatic effects of adefovir. When CEM-r1 cells were infected with HIV-1, a reduced antiviral effect of adefovir was found. CEM-r1 cells also showed some cross-resistance to the cytostatic effects of other ANPs, e.g. PMEDAP, PMEG, adefovir dipivoxil, as well as partial resistance to some purine nucleoside analogs. Accumulation of adefovir and its metabolites in CEM-r1 cells was greatly reduced compared to the parental cell line. This was due to a 7-fold increase in the rate of adefovir efflux, as well as a 2-fold decrease in AMP kinase activity. Subsequently, it was shown that the efflux phenotype was stable in the absence of drug and was ATP-dependent (Schuetz *et al.*, 1999). The efflux was partially selective, displaying a preference for adefovir and AZT-monophosphate. CEM-r1 cells also displayed a cross-resistance to the cytostatic and antiviral effects of other drugs such as AZT and d4T. The MDR1 antagonist verapamil had no effect on the efflux of adefovir from CEM-r1 cells. Indeed, MDR1 protein was not detectable in the parental or resistant cell line and the latter did not display resistance to the MDR1 substrates vinblastine and colchicine.

Further analysis of the CEM-r1 cells demonstrated the amplification of the MRP4 gene. Genes for other members of the MRP family, including MRP1 and MRP2, were not amplified. Rates of adefovir efflux from CEM-r1 correlated with levels of MRP4 protein expression, whereas MRP1 expression was unchanged. Transfer of high-level MRP4 expression via somatic cell fusion resulted in transfer of the resistance phenotype. These studies clearly link the overexpression of MRP4 with the adefovir resistance phenotype. It remains to be determined, however, whether adefovir is transported as such or as a conjugate.

A role for MRP5 in the transport of ANPs has also been reported. Winjholds *et al.*, (2000) used retroviral transduction to study MRP5 in 293 human embryonic kidney (HEK) cells and Madin-Darby canine kidney II (MDCKII) cells. In growth inhibition assays, transfected HEK cells expressing MRP5 showed 2- to 3-fold resistance to adefovir, 6-mercaptopurine, and thioguanine, but not to other analogs. Further characterization of MRP5-transfected 293 cells revealed a 2-3 fold decrease in the steady state intracellular accumulation of adefovir and its metabolites due to increased adefovir efflux. The observation that MRP5-transfected MDCKII cells can transport (S)-(2,4-dinitrophenyl)glutathione and that efflux of adefovir from 293/MRP5 cells was inhibited by sulfinpyrazone, supports the notion of MRP5 as a glutathione conjugate transporter. However, as for MRP4, it is unclear whether MRP5 transports nucleotide analogs directly or as conjugates. The role of other members of the MRP family in the transport of ANPs remains to be determined. MRP1, MRP2, and MRP3 are all glutathione-conjugate pumps with distinct substrate preferences whereas little is known about the function of MRP6 and MRP7 (Borst *et al.*, 1999; Hipfner *et al.*, 1999; Konig *et al.*, 1999a). Interestingly, both MRP4 and MRP5 lack an N-terminal transmembrane domain found in all other members of the MRP family, however, both retain the structural characteristics of glutathione-conjugate pumps found in MRP1.

A distinct efflux mechanism for adefovir has been identified in an erythroleukemia K562 cell line selected for resistance to the anti-proliferative effects of adefovir (Hatse *et al.*, 1998). Uptake of adefovir into K562/PMEA cells was unaltered; however, steady state accumulation of the drug and its metabolites was substantially reduced and correlated with increased efflux of adefovir into culture medium. Furthermore, adefovir phosphorylation was 25- to 50-fold less efficient compared to the parental line. Adefovir efflux in the resistant cells was shown to be both temperature- and ATP-dependent and was strongly inhibited by indomethacin, an anion channel blocker, implicating a mechanism distinct from MDR/MRP-mediated efflux. This efflux mechanism did not recognize other adenine nucleotides or AZT metabolites. Another unique efflux mechanism for adefovir dipivoxil operates in Caco-2 cell derived from human intestinal epithelium (Annaert *et al.*, 1998) (see also Section 5.4). Efflux in these cells shows two components; one component specifically affects adefovir dipivoxil and is sensitive to verapamil implicating an MDR1-like mechanism. The second component is sensitive to indomethacin, i.e. non-MDR/MRP, and affects mono(POM)PMEA and adefovir but not the intact prodrug.

An interesting counterpoint to the above mechanisms of adefovir efflux has been described in a murine leukemia L1210 cells selected in the presence of adefovir (L1210/PMEA) (Balzarini *et al.*, 1998). These cells display 300-fold resistance to adefovir's cytostatic effect but not to adefovir dipivoxil or other ANPs. L1210/PMEA cells retained resistance to adefovir in the presence of both indomethacin and verapamil, implicating a mechanism distinct from that of altered efflux. It is likely that this cell line has a highly specific defect in adefovir uptake. Thus, adefovir may be both exported via a variety of cellular mechanisms including upregulation of MRP-like proteins and/or prevented from entering the cell by down-regulation or mutation of transporters involved in the uptake. The potential clinical significance of these "cellular resistance" mechanisms is currently under investigation. Although much work has focused on their role in the molecular pharmacology of ANPs, it will be of considerable interest to examine whether other nucleoside-based antivirals, such as d4T, are also able to induce similar cellular responses.

4. ANPs and their renal transport

Both in animal models and in humans, cidofovir (Cundy *et al.*, 1995b), adefovir (Cundy *et al.*, 1995a), and tenofovir (Deeks *et al.*, 1998) are almost exclusively eliminated by the kidney. Their renal clearance rates exceed that of creatinine indicating that all three drugs undergo efficient tubular secretion. Recently, a human renal organic anion transporter 1 (hOAT1; hPAHT) has been identified which participates in the tubular secretion of ANPs (Cihlar *et al.*, 1999; Lu *et al.*, 1999; Hosoyamada *et al.*, 1999). When functionally expressed in *Xenopus laevis* oocytes, hOAT1 is capable of mediating efficient uptake of cidofovir, adefovir, and other ANPs (Cihlar *et al.*, 1999). Cidofovir and adefovir are also substrates for the rat renal organic anion transporter 1 (Oat1; ortholog of hOAT1), although their transport is noticeably less efficient in comparison with hOAT1 (Cihlar *et al.*, 1999). High-level expression of hOAT1 and Oat1 has been specifically localized to kidney with some low-level expression of hOAT1 detected in brain (see Chapters 5 and 6) (Cihlar *et al.*, 1999; Hosoyamada *et al.*, 1999). Detailed immunohistochemical analysis of Oat1 in kidney revealed its exclusive localization to the basolateral (antiluminal) membrane in the S2 segment of proximal tubules (Tojo *et al.*, 1999). This observation is consistent with the proposed role of Oat1 and hOAT1 in the active tubular secretion of ANPs.

Both hOAT1 and Oat1 function as organic anion/dicarboxylate exchangers with α-ketoglutarate (α-KG) being presumably the most efficient intracellular counter-ion (Figure 2) (Van Aubel *et al.*, 2000). The concentration gradient of α-KG generated partly by the intracellular metabolism and partly by other tubular transport systems (mainly NaDC-1 and SDCT2) (Sekine *et al.*, 1998a; Chen *et al.*, 1999) provides energy for the active hOAT1-mediated uptake of organic anions, including ANPs, from blood into proximal tubule epithelium. Several membrane channels/transporters have been identified in renal tubules from various species that presumably work in concert with hOAT1 (or its respective orthologs) and may mediate the efflux of ANPs from tubular epithelium into the lumen (Figure 2). They include mainly MRP2 (Schaub *et al.*, 1999), MRP6 (Kool *et al.*, 1999) and presumably also other members of MRP family (Kool *et al.*, 1997). In addition, a recent study indicated the ability of human inorganic phosphate transporter 1 (NTP1) to transport organic anions such as p-aminohippuric acid (PAH), β-lactams, and uric acid (Uchino *et al.*, 2000a; Uchino *et al.*, 2000b). It is believed that in addition to the re-absorption of Pi from glomerular filtrate, NPT1 can mediate the luminal efflux of organic anions.

Expression of a number of other organic anion transporters has been detected in kidney, e.g. rat Oatp1 (Jacquemin *et al.*, 1994), an ortholog of human OATP1 (Kullak-Ublick *et al.*, 1995), Oatp3 (Abe *et al.*, 1998), Oat-K1 (Masuda *et al.*, 1997), Oat-K2 (Masuda *et al.*, 1999a), Oat2 (Sekine *et al.*, 1998b), Oat3 (Kusuhara *et al.*, 1999), and Oat4 (Cha *et al.*, 2000) (Figure 2). These transport proteins, some of which have

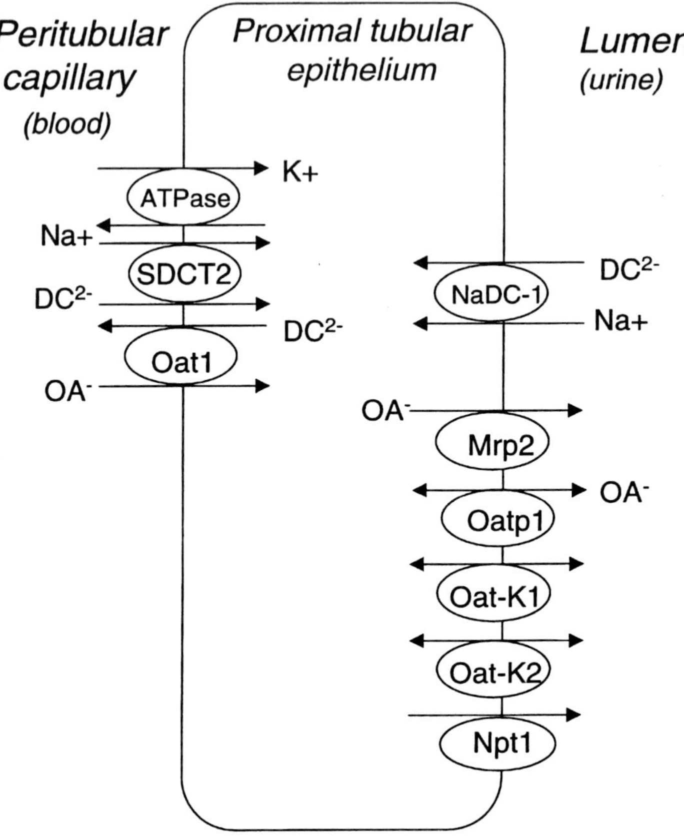

Figure 2.   Transport of organic anions in renal proximal tubules. In addition to the depicted transporters, expression of Oat2, Oat3, OAT4, Oatp2, and Oatp3 has been found in the kidney. However, their precise localization has not been determined yet. ATPase, Na^+/K^+ adenosine triphosphatase; Oat1, organic anion transporter 1; SDCT2, sodium-dicarboxylate transporter 2; NaDC-1, sodium-dicarboxylate transporter 1; Mrp2, multidrug resistance protein 2; Oatp1, organic anion transport polypeptide 1; Oat-K1 and Oat-K2, kidney-specific organic anion transporter 1 and 2; Npt1, sodium-phosphate transporter 1; OA-, organic anion; DC2-, dicarboxylate.

initially been identified in liver, may potentially contribute to the tubular uptake and/or secretion of ANPs. In some cases, the transporter localization in particular segment(s) of the nephron is not known. Table 2 (Chapter 6) provides a summary of renal organic anion transporters and their characteristics. In addition, several reviews discussing the molecular pharmacology of renal organic ion transporters have been recently published (Van Aubel *et al.*, 2000; Sweet and Pritchard, 1999; Ullrich, 1999).

In rabbits (Cundy *et al.*, 1996) and rats (Cundy *et al.*, 1996a), cidofovir accumulates in kidneys to much higher extent than in other tissues with the majority of drug deposited in the superficial renal cortex. This is presumably due to a difference between the rate of cidofovir basolateral uptake into the proximal tubular cells and the rate of its luminal efflux. The main clinical consequence of cidofovir renal accumulation is nephrotoxicity. (Lalezari *et al.*, 1997; Lacy *et al.*, 1998). Prolonged anti-HIV therapy with high doses of adefovir dipivoxil (60 or 120 mg) is also associated with nephrotoxicity (Kahn *et al.*, 1999). In both cases, the nephrotoxicity has primarily tubular character. Stable heterologous expression of hOAT1 in various cell types has been found to enhance the cytotoxicity of cidofovir and adefovir by 200- to 400-fold relative to the corresponding control cells (Ho *et al.*, 2000). In contrast, the cyclic prodrug of cidofovir (cHPMPC) showed only marginally increased cytototoxicity upon hOAT1 expression, which correlates with its reduced nephrotoxicity *in vivo* (Hitchcock *et al.*, 1995). Together with the tissue-specific localization of hOAT1, these observations indicate a direct involvement of this transporter in the etiology of nephrotoxicity associated with cidofovir and adefovir (Ho *et al.*, 2000).

A number of hOAT1 inhibitors, which may serve as nephroprotective agents, have been identified. Probenecid and betamipron (N-benzoyl-β-alanine) efficiently reduce the hOAT1-mediated uptake and cytotoxicity of cidofovir and adefovir in cells expressing hOAT1 (Ho *et al.*, 2000; Cihlar *et al.*, 1999). In cynomolgus monkeys receiving chronic i.v. cidofovir treatment, oral coadministration of probenecid decreased the drug-associated nephrotoxicity (Lacy *et al.*, 1998). Probenecid is also coadministered clinically to ameliorate the nephrotoxicity in patients treated with cidofovir. By using a recently developed microtiter fluorescent hOAT1 assay (Cihlar and Ho, 2000), a number of nonsteroidal anti-inflammatory drugs (NSAIDs) have been identified as potent inhibitors of hOAT1 with ketoprofen, flurbiprofen, and several other NSAIDs being more efficient than probenecid in reducing the hOAT1-mediated transport and cytotoxicity of cidofovir and adefovir (Mulato *et al.*, 2000).

In contrast to cidofovir and adefovir, prolonged administration of tenofovir disoproxil to patients does not adversely affect renal functions (Schooley *et al.*, 2000). Steady-state kinetic experiments revealed no significant difference in the efficiency of hOAT1-specific transport of tenofovir compared to that of cidofovir and adefovir (Table 1) (Ho *et al.*, 2000; Cihlar *et al.*, 2000). Further studies, however, demonstrated a minimal effect of tenofovir on the *in vitro* growth and viability of human renal proximal tubular epithelial cells as well as the *in vitro* integrity of renal proximal tubular epithelium (Table 1) suggesting that a lack of interference with essential cellular functions rather than the reduced renal uptake is the reason for the absence of tenofovir nephrotoxicity (Cihlar *et al.*, 2000). On the other hand, the reduced nephrotoxicity of cHPMPC can be explained by its markedly diminished hOAT1-mediated transport efficiency compared to the parental cidofovir (Ho *et al.*, 2000).

5.   Transport of ANPs in other tissues – Potential interactions

5.1. Liver

As mentioned in Chapter 4, ANPs are almost exclusively eliminated by the kidney with no significant hepatobiliary secretion (Cundy, 1999). Cidofovir administered

Table 1.   Profile of cidofovir, adefovir, and tenofovir in the in vitro models for renal proximal tubular toxicity.

In vitro Assay	Drug		
	Cidofovir	Adefovir	Tenofovir
Efficiency of hOAT1-mediated transport; Vmax/Km [pmol/10^6 cells.min.μM] [a]	1.77	1.93	3.26
Viability of RPTECs[b]; T1/2 [days]	9.5	21	>25
Inhibition of RPTECs growth; CC50 [μM]	260	495	> 2,000
Integrity of epithelium formed by RPTECs; CTER50[c] [μM]	113	1,100	> 3,000

[a]   Data generated by using the in vitro cell-based hOAT1 assay (Ho *et al.*, 2000; Cihlar *et al.*, 2000).

[b]   T1/2 of human renal proximal tubule epithelial cells (RPTECs) in the presence of 500 μM drug (Cihlar *et al.*, 2000)

[c]   CTER50 – concentration reducing the transepithelial resistance of RPTEC monolayer cultured on microporous membrane by 50% (Cihlar *et al.*, 2000).

intravenously to rats accumulates in liver to a level approximately 20-fold lower than that in kidney, but still higher than in other organs (Cundy *et al.*, 1996a). Similar results have been observed with HPMPA, the adenine analog of cidofovir (Bijsterbosch *et al.*, 1998). Preinjection with probenecid reduced both renal and hepatic uptake of HPMPA in rats by approximately 75% indicating that ANPs may be taken up into hepatocytes by an active transport mechanism. HPMPA has been shown to accumulate to a much higher extent in liver parenchymal cells than in endothelial or Kupffer cells (Bijsterbosch *et al.*, 1998). In addition, adefovir as well as tenofovir exhibit potent anti-HBV activity suggesting their efficient accumulation in hepatocytes. Similar to kidney, a number of liver organic anion transporters have been identified and characterized over the last few years, some of which may potentially interact with ANPs.

Liver anion transport systems are essential for the production of bile and for the biliary secretion of a wide range of anionic substances (Meier *et al.*, 1997). In order to achieve efficient secretion, the sinusoidal (basolateral) transporters mediating uptake of substrates from blood function in concert with the canalicular (apical) efflux pumps (Kullak-Ublick *et al.*, 2000). Figure 3 summarizes the organic anion transporters identified in hepatocytes. Oat2 (formerly NLT-1) and Oat3, both detected in rat liver at high levels, exhibit a number of characteristics similar to renal Oat1. They belong to the same family of transporters and show 42% and 49% amino acid identity, respectively, to Oat1 (Kusuhara *et al.*, 1999). Both Oat2 and Oat3 are capable of transporting PAH, a prototype substrate for Oat1, and are sensitive to probenecid (Sekine *et al.*, 1998b; Kusuhara *et al.*, 1999). Oat2 has been localized to the hepatocyte sinusoidal membrane (Simonson *et al.*, 1994). Thus, Oat2 and Oat3 may be candidates for mediating the uptake of ANPs into hepatocytes. In addition, an organic anion/dicarboxylate exchanger was

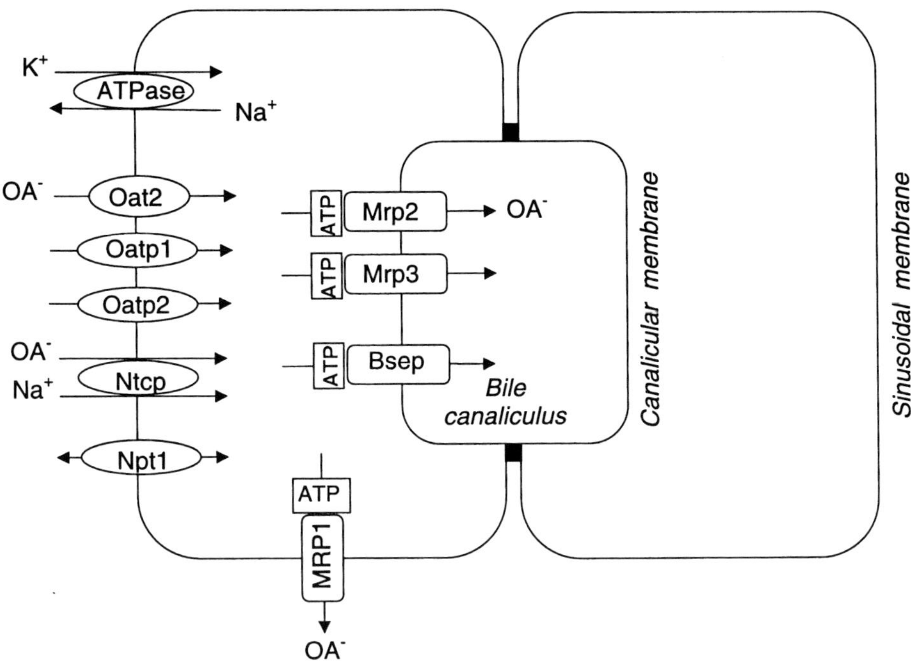

Figure 3. Transport of organic anions in hepatocytes. In addition to the depicted transporters, expression of Oat3, Oatp3, and MRP6 has been found in the liver. However, their precise localization has not been determined yet. ATPase, Na+/K+ adenosine triphosphatase; Oat2, organic anion transporter 2; Oatp1 and Oatp2, organic anion transport polypeptide 1 and 2; Ntcp, Na+-taurocholate cotransporting polypeptide; Npt1, sodium-phosphate transporter 1; Mrp1, Mrp2, and Mrp3, multidrug resistance protein 1, 2, and 3; Bsep, bile salt export pump; OA-, organic anion.

identified in rat hepatocytes by using functional transport experiments. This transporter recognizes cholate as a high-affinity substrate and similarly to Oat1 is inhibited by exogenous α-KG, PAH, and probenecid (Boelsterli *et al.*, 1995). Since the sequence of this transporter is not know, it is not clear if it is identical with Oat2 or Oat3, or if it represents a novel liver anion transporter.

A distinct family of organic anion transport polypeptides (oatps) has been identified in the liver. They include rat transporters Oatp1 (Jacquemin *et al.*, 1994), Oatp2 (Noe *et al.*, 1997), and Oatp3 (Abe *et al.*, 1998) as well as human homologues OATP (Kullak-Ublick *et al.*, 1995; Meier *et al.*, 1997), OATP2 (Konig *et al.*, 2000b), and OATP8 (Konig *et al.*, 2000a). Oatp1 and 2 are presumably bi-directional transporters with high level expression in sinusoidal membrane indicating that they could mediate both uptake and efflux of substrates from and into the blood. Oatp3 is expressed in the liver at lower levels and has not been precisely localized yet (Abe *et al.*, 1998). All three Oatps are capable of Na+-independent transport of bile acids and other anionic substrates. Importantly, they are insensitive to PAH and/or probenecid suggesting that they may not play a major role in the hepatic transport of ANPs (Kullak-Ublick *et al.*,

Table 2.   continued.

Transporter Name [a]	Species	Tissue Distribution (Membrane Localization)[b]	Transport Direction/Mechanism	Substrates [c]	Additional Characteristics	References
Npt1	Mouse	K (pt-am), L (sm)	Pi – uptake; OA – primarily efflux in kidney and uptake/efflux in liver	β-lactams, (penicillin, faropenem, foscarnet), mevalonic acid	Cl-sensitive; primary function is Na+-dependent resorption of Pi	Chong et al., 1995; Tenenhouse et al., 1998; Yabuuchi et al., 1998; Uchino et al., 2000a.
OATP	Human	B, K, L(sm), Lu	Uptake	BSP, bile acids, estrone-sulfate	Na+-independent	Kullak-Ublick et al., 1995; Meier et al., 1997.
OATP2 (LST-1)	Human	L(sm)	Uptake	glucuronosyl estradiol, cholyltaurine, glucuronosyl bilirubin	Na+-independent; 44% identity to OATP	Abe et al., 1999; Konig et al., 2000b.
OATP8	Human	L(sm)	Uptake	BSP, glucuronosyl estradiol	80% identity to OATP2	Konig et al., 2000a.
Oatp1	Rat	L(sm), K (pt(S3), am), SM, B, I, Lu	Bi-directional	GSH, BSP, bile acids, estrone-sulfate,	Ortholog of OATP; OA-/GSH exchanger ochratoxin A	Jacquemin et al., 1994; Bergwerk et al., 1996; Ballatori and Rebbeor, 1998; Meier et al., 1997.
Oatp2	Rat	B, L(sm), K[J], R	Bi-directional	GSH, GSH and estrogen conjugates, ouabain, digoxin, thyroid hormons	Anion exchanger	Noe et al., 1997; Abe et al., 1998; Kakyo et al., 1999; Li et al. 2000.
Oatp3	Rat	K, L, R	Uptake	taurocholate, thyroid hormons	Na+-independent	Abe et al., 1998.
NTCP	Human	L (sm)#	Uptake	Bile acids, thyroid hormons (rat Ntcp)	Na+-dependent; Rat Ntcp also identified	Meier et al., 1997; Hagenbuch and Meier, 1994; Stieger et al., 1994; Schroeder et al., 1998.
MRP1 (ABCC1)	Human Rat Mouse	Ubiquitous including: I, L, Lu, K (sm), B, Pl	ATP-dependent uni-directional efflux	GSH, glucoronate and sulphate conjugates of liphophilic compounds Leukotrienes, bilirubin, etoposide, MTX, vincristine	Confers high level multidrug resistance when over-expressed in tumors. Closest homology with MRP2, MRP3 and MRP6.	Borst, 1999; Hipfner et al., 1999; Konig et al., 1999a.

Table 2.   continued.

Transporter Name [a]	Species	Tissue Distribution (Membrane Localization)[b]	Transport Direction/ Mechanism	Substrates [c]	Additional Characteristics	References
MRP2 (ABCC2, cMOAT)	Human Rat Mouse	L, K (ap), I	ATP-dependent, uni-directional efflux	GSH, glucoronate and sulphate conjugates of liphophilic compounds Leukotrienes, bilirubin, estradiol MTX	Mutations associated with Dubin-Johnson syndrome. Identity to: MRP1 (49%) MRP4 (37%) MRP5 (35%)	Kool *et al.*, 1997; Borst *et al.*, 1999; Konig *et al.*, 1999a.
MRP3 (ABCC3, MOAT-D, cMOAT2)	Human Rat	L (sm), K (blm), I, Pa, Bl, Gb	ATP-dependent, uni-directional efflux	GSH, glucoronate and sulphate conjugates of lipophilic compounds. Leukotrienes, bilirubin, estradiol, MTX, dinitrobenzene	Identity to: MRP1 (58%) MRP4 (36%) MRP5 (33%)	Kool *et al.*, 1997; Borst *et al.*, 1999; Konig *et al.*, 1999a; Kool *et al.*, 1999.
MRP4 (ABCC4, MOAT-B)	Human	I, L, Lu, K, B, Gb, M, Pa, Pr, O, T Membrane location not defined	ATP-dependent uni-drectional efflux	Ability to efflux GSH conjugates unknown; ANPs (PMEA, PMEG), nucleoside monophos-phates (AZTM[p], D4TM[p])	Identity to: MRP1 (39%) MRP5 (36%) Lacks N-terminal extension found in MRP1, -2, -3 and -6. Closest structural homology to MRP5	Robbins *et al.*, 1995a; Kool *et al.*, 1997; Lee *et al.*, 1998; Borst *et al.*, 1999; Konig *et al.*, 1999a; Schuetz *et al.*, 1999.
MRP5 (ABCC5, MOAT-C)	Human Rat Mouse	I, L, Lu, K (blm), B, Pl, M, Gb, S, T.	ATP-dependent, uni-directional efflux	GSH conjugates, ANPs (PMEA), thiopurine nucleoside monophosphates	Identity to: MRP1 (34%) MRP4 (36%) Closest structural homology to MRP4.	Borst *et al.*, 1999; Konig *et al.*, 1999a; Wijnholds *et al.*, 2000.
MRP6 (ABCC6)	Human Rat Mouse	L(blm, cm), K, B	Presumably ATP-dependent, uni-directional efflux	Physiological substrates unknown. BQ-123 an anionic cyclopentapeptide.	Mutations associated with pseudoxanthoma elasticum. Identity to: MRP1 (45%) MRP4 (34%) MRP5 (31%)	Borst *et al.*, 1999; Konig *et al.*, 1999a; Madon *et al.*, 2000; Ringpfeil *et al.*, 2000.
MRP7 (ABCC10)	Human Mouse	unknown	unknown	unknown	unknown	partial Genbank sequences only
Bsep (spgp)	Rat	L(cm), B, I	Efflux	Bile acids	ATP-dependent	Gerloff *et al.*, 1998; Torok *et al.*, 1999.

Table 2.   continued.

[a]   hOAT1, human organic anion transporter 1; Oat1, Oat2, and Oat3, organic anion transporter 1, 2, and 3; OAT4, organic anion transporter 4; NLT, normal liver transporter; Oatp1, Oatp2, and Oatp3, organic anion transport polypeptide, Oat-K1 and Oat-K2, kidney-specific organic anion transporter 1 and 2; NPT1/Npt1, sodium-phosphate transporter 1; NTCP, sodium-taurocholate cotransporting polypeptide; MRP, multidrug resistance protein; ABCC, ATP-binding cassette tranporter sub-family C; MOAT, multi-specific organic anion transporter; Bsep - bile salt export pump; spgp – sister of P-glycoprotein; LST-1, liver-specific transporter 1.

[b]   I, Intestine; L, liver; Lu, lung; K, kidney; B, brain; Bl, Bladder; Gb, gall bladder; M, muscle; O, ovary, Pa, pancreas; Pl, placenta; Pr, prostate; R, retina; S, spleen; T, testis; blm, basolateral membrane; ap, apical membrane; sm, sinusoidal membrane; cm, canalicular membrane; cp, choroid plexus; pt, proximal tubule; ccd, cortical collecting ducts.

[c]   PAH, p-aminohippuric acid; OA, organic anion; GSH, glutathion; BSP, bromosulfophtalein; MTX, methotrexate.

[*]   Based on the apical expression after Oat-K2 gene transfection into MDCK cells (Masuda *et al.*, 1999a).

[¶]   Noe *et al.* (Noe *et al.*, 1997), but not Abe *et al.* (Abe *et al.*, 1998) detected the expression of Oatp2 in the kidney.

[#]   Based on the basolateral (sinusoidal) localization of the rat ortholog Ntcp (Stieger *et al.*, 1994).

infection, respectively, suggesting that inflammation of BCSFB or BBB may change the membrane permeability leading to an increased accumulation of cidofovir in CSF. Indeed, cidofovir was detected in CSF obtained by lumbar puncture from a patient with PML (Cundy, 1999).

5.3. Eye

The potent and long-lasting clinical effect of intravenous cidofovir in CMV retinitis patients suggests efficient penetration of the drug across the blood-retinal barrier (BRB; choroid) and its prolonged half-life in retinal tissue. An interesting observation has been made from the comparison of ocular pharmacokinetics of cidofovir and cHPMPC. When administered intravitreally to rabbits, cHPMPC was cleared from vitreous 2-times faster then cidofovir (Cundy *et al.*, 1996b). 24 hours post-dose, the total amount of cidofovir in retina was 10-fold higher than that of cHPMPC with a proportional difference in the phosphorylated metabolites. However, the mechanism of cidofovir and cHPMPC retinal clearance was not determined in this study.

Currently, only limited information is available on the transport of organic anions across BRB with the majority of experiments done with fluorescein or carboxyfluorescein, two fluorescent organic anion probes. Both compounds have been shown to move across BRB in the basolateral-to-apical direction (i.e. from blood to vitreous) by a passive non-saturable transport mechanism. In contrast, the transport of both compounds in the opposite direction appears to be active, saturable, and sensitive to probenecid indicating involvement of a specific organic anion transporter (Engler *et al.*, 1994; Yoshida *et al.*, 1992; Tsuboi and Pederson, 1986). Expression of Oatp2 and Oatp3 has recently been demonstrated in the retina (Abe *et al.*, 1998). However, it is not known if these transporters can transport fluorescein and/or carboxyfluorescein. As discussed above (Section 5.1.), these transporters may not be able to interact with ANPs. Localization of the vast majority of the other known organic anion transporters to eye has yet to be determined.

5.4. Intestine

The low oral bioavailability of ANPs in humans and animals (<10%) in conjunction with no first pass metabolic clearance of these compounds indicate that their penetration across intestinal epithelium is limited (Cundy, 1999). This may partly be due to the negative charge of ANPs and partly due to the apical efflux of ANPs back into the intestinal lumen. Several MRP isoforms including MRP4 have been detected in colon and/or small intestine (Kool *et al.*, 1997; Lee *et al.*, 1998; Konig *et al.*, 1999b), which may mediate efflux of ANPs back into the lumen. In order to efficiently deliver ANPs by an oral administration, lipophilic prodrugs of adefovir and tenofovir have been designed (reviewed in Cihlar and Bischofberger, 1998). In human intestinal epithelial cells (Caco-2) preloaded with adefovir dipivoxil, the basolateral efflux of adefovir and mono(POM)PMEA proceeds with approximately a 3- to 4-fold higher rate than the apical efflux suggesting that the efflux back into the lumen may play a minor role in the limited oral bioavailability of ANPs (Annaert *et al.*, 1998). Based on the susceptibility to verapamil, the same study has demonstrated that the intact adefovir dipivoxil is a substrate for MDR1-like efflux system in Caco-2 cells. Pharmacokinetic experiments with orally dosed adefovir dipivoxil in the wild-type and mdr1 -/- knock-out mice indicated that although mdr1-mediated efflux may impact the initial intestinal absorption rate of the prodrug, it does not appear to have a substantial effect on the resulting exposure to adefovir, based on the identical AUC of the drug in the wild-type and knock-out mice (Kearney, 2000).

6.   Summary – Organic anion transporters

Despite the negative charge on the phosphonate moiety, ANPs can penetrate host cells and exhibit potent antiviral effects. Various mechanisms of ANPs uptake have been identified in different cell types, which include both non-specific mechanisms such as fluid-phase endocytosis and highly specific receptor- or transporter-mediated processes. Recent studies have demonstrated the importance of tissue-specific uptake of ANPs in the etiology of some of their adverse effects. Together with the uptake and metabolism, the process of efflux has been shown to be a component of the cellular pharmacokinetics of ANPs affecting their intracellular levels.

During the last few years, remarkable progress has been made in the identification and molecular characterization of a large number of novel organic anion transporters. To date, only four of them have been shown to interact with ANPs and mediate their cellular uptake or efflux. They include hOAT1 and its ortholog Oat1 and two MRP efflux pumps, MRP4 and MRP5. However, based on the similarity in the structure, transport mechanism, or function, some other organic anion transporters are presumably also capable of transporting ANPs. Basic characteristics of the main organic anion transporters identified to date are given in Table 2. Further study of their interactions with ANPs will contribute to a better understanding of the efficacy, pharmacology, and toxicology of this clinically important class of antiviral nucleotide analogs.

7.   References

Abe, T., Kakyo, M., Sakagami, H., Tokui, T., Nishio, T., Tanemoto, M., Nomura, H., Hebert, S. C., Matsuno, S., Kondo, H., and Yawo, H. (1998). Molecular characterization and tissue distribution of a new organic anion transporter subtype (oatp3) that transports thyroid hormones and taurocholate and comparison with oatp2. J. Biol. Chem. 273, 22395-401.

Abe, T., Kakyo, M., Tokui, T., Nakagomi, R., Nishio, T., Nakai, D., Nomura, H., Unno, M., Suzuki, M., Naitoh, T., Matsuno, S., and Yawo, H. (1999). Identification of a novel gene family encoding human liver-specific organic anion transporter LST-1. J. Biol. Chem. 274, 17159-17163.

Aduma, P., Connelly, M. C., Srinivas, R. V., and Fridland, A. (1995). Metabolic diversity and antiviral activities of acyclic nucleoside phosphonates. Mol. Pharmacol. 47, 816-822.

Ambudkar, S. V., Dey, S., Hrycyna, C. A., Ramachandra, M., Pastan, I., and Gottesman, M. M. (1999). Biochemical, cellular, and pharmacological aspects of the multidrug transporter. Annu. Rev. Pharmacol. Toxicol. 39, 361-398.

Angeletti, R. H., Novikoff, P. M., Juvvadi, S. R., Fritschy, J. M., Meier, P. J., and Wolkoff, A. W. (1997). The choroid plexus epithelium is the site of the organic anion transport protein in the brain. Proc. Natl. Acad. Sci. U S A 94, 283-286.

Annaert, P., Van Gelder, J., Naesens, L., De Clercq, E., Van den Mooter, G., Kinget, R., and Augustijns, P. (1998). Carrier mechanisms involved in the transepithelial transport of bis(POM)PMEA and its metabolites across Caco-2 monolayers. Pharm. Res. 15, 1168-1173.

Apiwattanakul, N., Sekine, T., Chairoungdua, A., Kanai, Y., Nakajima, N., Sophasan, S., and Endou, H. (1999). Transport properties of nonsteroidal anti-inflammatory drugs by organic anion transporter 1 expressed in Xenopus laevis oocytes. Mol. Pharmacol. 55, 847-854.

Ballatori, N., and Rebbeor, J. (1998). Roles of MRP2 and oatp1 in hepatocellular export of reduced glutathione. Semin. Liver. Dis. 18, 377-387.

Balzarini, J., and De Clercq, E. (1991). 5-Phosphoribosyl-1-pyrophosphate synthetase converts the acyclic nucleoside phosphonates 9-(3-hydroxy-2-phosphonylmethoxypropyl)adenine and 9-(2-phosphonyl-methoxyethyl)adenine directly to their antivirally active diphosphate derivatives. J. Biol. Chem. 266, 8686-8689.

Balzarini, J., Hao, Z., Herdewijn, P., Johns, D. G., and De Clercq, E. (1991a). Intracellular metabolism and mechanism of anti-retrovirus action of 9-(2-phosphonylmethoxyethyl)adenine, a potent anti-human immunodeficiency virus compound. Proc. Natl. Acad. Sci. USA 88, 1499-1503.

Balzarini, J., Hatse, S., Naesens, L., and De Clercq, E. (1998). Selection and characterization of murine leukaemia L1210 cells with high-level resistance to the cytostatic activity of the acyclic nucleoside phosphonate 9-(2-phosphonylmethoxyethyl)adenine (PMEA). Biochim. Biophys. Acta 1402, 29-38.

Balzarini, J., Holy, A., Jindrich, J., Dvorakova, H., Hao, Z., Snoeck, R., Herdewijn, P., Johns, D. G., and De Clercq, E. (1991b). 9-[(2RS)-3-Fluoro-2-phosphonylmethoxypropyl]derivatives of purines: a class of highly selective antiretroviral agents in vitro and in vivo. Proc. Natl. Acad. Sci. USA 88, 4961-4965.

Balzarini, J., Holy, A., Jindrich, J., Naesens, L., Snoeck, R., Schols, D., and De Clercq, E. (1993). Differential antiherpes virus and antiretrovirus effects of the (S) and (R) enantiomers of acyclic nucleoside phosphonates: potent and selective in vitro and in vivo antiretrovirus activities of (R)-9-(2-phosphonomethoxypropyl)-2,6-diaminopurine. Antimicrobial Agents Chemother. 37, 332-338.

Belinsky, M. G., Bain, L. J., Balsara, B. B., Testa, J. R., and Kruh, G. D. (1998). Characterization of MOAT-C and MOAT-D, new members of the MRP/cMOAT subfamily of transporter proteins. J. Natl. Cancer Inst. 90, 1734-1741.

Bergwerk, A. J., Shi, X., Ford, A. C., Kanai, N., Jacquemin, E., Burk, R. D., Bai, S., Novikoff, P. M., Stieger, B., Meier, P. J., Schuster, V. L., and Wolkoff, A. W. (1996). Immunologic distribution of an organic anion transport protein in rat liver and kidney. Am. J. Physiol. 271, G231-G238.

Bijsterbosch, M., Smeijsters, L., and van Berkel, T. (1998). Disposition of the acyclic nucleoside phosphonate (S)-9-(3-hydroxy-2-phosphonylmethoxypropyl)adenine. Antimicrobial Agents Chemother. 42, 1146-1150.

Bischofbeger, N., Hitchcock, M.J.M., Chen, M. S., Barkhimer, D. B., Cundy, K. C., Kent, K. M., Lacy, S. A., Lee, W. A., Li, Z. H., and Mendel, D. B. *et al.* (1994). 1-[((S)-2-hydroxy-2-oxo-1,4,2-dioaphosphorinan-5-yl)methyl] cytosine, an intracellular prodrug for (S)-1-(3-hydroxy-2-phosphonylmehtoxypropyl)cytosine with improved therapeutic index in vivo. Antimicrobial Agents Chemother. 38, 2387-2391.

Boelsterli, U. A., Zimmerli, B., and Meier, P. J. (1995). Identification and characterization of a basolateral dicarboxylate/cholate antiport system in rat hepatocytes. Am. J. Physiol. 268, G797-805.

Borst, P., Evers, R., Kool, M., and Wijnholds, J. (1999). The multidrug resistance protein family. Biochim. Biophys. Acta 1461, 347-357.

Brambilla, A. M., Castagna, A., Novati, R., Clinque, P., Terreni, M. R., Mouoli, M. C., and Lazzarin, A. (1999). Remission of AIDS-associated progressive multifocal leukoencephalopathy after cidofovir therapy. J. Neurol. 246, 723-725.

Cha, S. H., Sekine, T., Kusuhara, H., Yu, E., Kim, J. Y., Kim, D. K., Sugiyama, Y., Kanai, Y., and Endou, H. (2000). Molecular cloning and characterization of multispecific organic anion transporter 4 expressed in the placenta. J. Biol. Chem. 275, 4507-12.

Chen, X., Tsukaguchi, H., Chen, X. Z., Berger, U., and Hediger, M. (1999). Molecular and functional analysis of SDCT2, a novel rat sodium-dependent dicarboxylate transporter. J. Clin. Invest. 103, 1159-1168.

Chiba, P., Kraupp, M., Rumpold, H., and Muller, M. M. (1984). Uptake of AMP into K 562 and HL-60 leukemic cell lines. Adv. Exp. Med. Biol. 165, 273-278.

Chong, S. S., Kozak, C. A., Liu, L., Kristjansson, K., Dunn, S. T., Bourdeau, J. E., and Hughes, M. R. (1995). Cloning, genetic mapping, and expression analysis of a mouse renal sodium-dependent phosphate cotransporter. Am. J. Physiol. 268, F1038-1045.

Chong, S. S., Krisjansson, K., Zoghbi, H. Y., and Hughes, M. R. (1993). Molecular cloning of the cDNA encoding a human renal sodium phosphate transport protein and its assignment to chromosome 6p21.3-p23. Genomics 18, 355-359.

Cihlar, T., and Chen, MS. (1996). Identification of enzymes catalysing two-step phosphorylaiton of cidofovir and the effect of cytomegalovirus infection on their activities in host cells. Mol. Pharmacol. 50, 1502-1510.

Cihlar, T., and Bischofberger, N. (1998). PMEA and PMPA: acyclic nucleoside phosphonates with potent anti-HIV activity. Trends in Drug Research II (H. Van der Goot, Ed.), pp 105-116. Elsevier, Amsterdam.

Cihlar, T., and Ho, E. S. (2000). Fluorescence-based assay for the interaction of small molecules with the human renal organic anion transporter 1. Anal. Biochem. 283, 49-55.

Cihlar, T., Lin, D. C., and Ho, E. S. (2000). Tenofovir exhibits a highly favorable profile in the experimental cell culture models for renal tubular toxicity. Antiviral Res. 46, A69.

Cihlar, T., Lin, D. C., Pritchard, J. B., Fuller, M. D., Mendel, D. B., and Sweet, D. H. (1999). The antiviral nucleotide analogs cidofovir and adefovir are novel substrates for human and rat renal organic anion transporter 1. Mol. Pharmacol. 56, 570-80.

Cihlar, T., Rosenberg, I., Votruba, I., and Holy, A. (1995). Transport of 9-(2-phosphonylmethoxyethyl)adenine across plasma membrane of HeLa S3 cells is protein mediated. Antimicrobial Agents Chemother. 39, 117-124.

Cihlar, T., Votruba, I., Horska, K., Liboska, R., Rosenberg, I., and Holy, A. (1992). Metabolism of 1-(S)-(3-hydroxy-2-phosphonomethoxypropyl)cytosine (HPMPC) in human embryonic lung cells. Collect. Czech. Chem. Commun. 57, 661-672.

Cole, S. P., Bhardwaj, G., Gerlach, J. H., Mackie, J. E., Grant, C.E., Almquist, K.C., Stewart, A. J., Kurz, E. U., Duncan, A. M., and Deeley, R.G. (1992). Overexpression of a transporter gene in a multidrug-resistant human lung cancer cell line. Science 258, 1650-1654.

Connelly, M. C., Robbins, B. L., and Fridland, A. (1993). Mechanism of uptake of the phosphonate analog (S)-1-(3-hydroxy-2-phosphonylmethoxypropyl)cytosine (HPMPC) in Vero cells. Biochem. Pharmacol. 46, 1053-1057.

Cundy, K. C. (1999). Clinical pharmacokinetics of the antiviral nucleotide analogues cidofovir and adefovir. Clin. Pharmacokinet. 36, 127-143.

Cundy, K. C., Barditch-Crovo, P., Walker, R. E., Collier, A. C., Ebeling, D., Toole, J., and Jaffe, H. S. (1995a). Clinical pharmacokinetics of adefovir in human immunodeficiency virus type 1-infected patients. Antimicrobial Agents Chemother. 39, 2401-2405.

Cundy, K. C., Bidgood, A. M., Lynch, G., Shaw, J. P., Griffin, L., and Lee, W. A. (1996a). Pharmacokinetics, bioavailability, metabolism, and tissue distribution of cidofovir (HPMPC) and cyclic HPMPC in rats. Drug Metab. Dispos. 24, 745-752.

Cundy, K. C., Li, Z. H., and Lee, W. A. (1996). Effect of probenecid on the distribution, metabolism, and excretion of cidofovir in rabbits. Drug Metab. Dispos. 24, 315-321.

Cundy, K. C., Lynch, G., Shaw, J. P., Hitchcock, M. J., and Lee, W. A. (1996b). Distribution and metabolism of intravitreal cidofovir and cyclic HMPMC in rabbits. Curr. Eye Res. 15, 569-576.

Cundy, K. C., Petty, B. G., Flaherty, J., Fisher, P. E., Polis, M. A., Wachsman, M., Lietman, P. S., Lalezari, J. P., Hitchcock, M. J. M., and Jaffe, H. S. (1995b). Clinical pharmacokinetics of cidofovir in human immunodeficiency virus-infected patients. Antimicrobial Agents Chemother. 39, 1247-1252.

De Clercq, E., Holy, A., Rosenberg, I., Sakuma, T., Balzarini, J., and Maudgal, P. C. (1986). A novel selective broad-spectrum anti-DNA virus agent. Nature 323, 464-467.

Deeks, S. G., Bardich-Provo, P., Lietman, P. S., Hwang, F., Cundy, K. C., Rooney, J. F., Hellmann, N. S., Safrin, S., and Kahn, J. O. (1998). Safety, pharmacokinetics, and antiretroviral activity of intravenous 9-[2-(R)-(phosphonomethoxy)propyl]adenine, a novel anti-human immmunodeficiency virus (HIV) therapy, in HIV-infected adults. Antimicrobial Agents Chemother. 42, 2380-2384.

Edwards, N. L., Zaytoun, A. M., and Renard, G. A. (1986). Seperate mechanisms for cellular uptake of purine nucleotides by B-and T-lymphoblasts. Adv. Exp. Med. Biol. 195, 463-465.

Engler, C. B., Sander, B., Larsen, M., Koefoed, P., Parving, H. H., and Lund-Anderson, H. (1994). Probenecid inhibition of the outward transport of flourescein across human blood-retina barrier. Acta Ophthalmol. 72, 663-667.

Frankova, D., Zidek, Z., Buchar, E., Jioieka, Z., and Holy, A. (1999). Antiinflammatory effects of phosphonomethoxyethyl analogues of adenine in a model of adjuvant arthritis. Nucleosides Nucleotides 18, 955-958.

Gerloff, T., Stieger, B., Hagenbuch, B., Madon, J., Landmann, L., Roth, J., Hofmann, A. F., and Meier, P. J. (1998). The sister of P-glycoprotein represents the canalicular bile salt export of mammalian liver. J. Biol. Chem. 273, 10046-10050.

Hagenbuch, B., and Meier, P. J. (1994). Molecular cloning, chromosomal localization, and functional characterization of a human liver Na^+/bile acid cotransporter. J. Clin. Invest. 93, 1326-1331.

Hakvoort, A., Haselbach, M., and Galla, H. J. (1998). Active transport properties of porcine choroid plexus cells in culture. Brain Res. 795, 247-256.

Hatse, S., De Clercq, E., and Balzarini, J. (1998). Enhanced 9-(2-phosphonylmethoxyethyl)adenine secretion by a specific, indomethacin-sensitive efflux pump in a mutant 9-(2-phosphonylmethoxyethyl)adenine-resistant human erythroleukemia K562 cell line. Mol. Pharmacol. 54, 907-917.

Hipfner, D. R., Deeley, R. G., and Cole, S. P. (1999). Structural, mechanistic and clinical aspects of MRP1. Biochim. Biophys. Acta 1461, 359-376.

Hitchcock, M. J. M., Jaffe, J. S., Martin, J. C., and Stagg, R. J. (1996). Cidofovir, a new agent with potent anti-herpesvirus activity. Antiviral Chemistry Chemother. 7, 115-127.

Hitchcock, M. J. M., Lacy, S. A., Lindsey, J. R., and Kern, E. R. (1995). The cyclic congener of cidofovir has reduced nephrotoxicity in three species. Antiviral Res. 26, A358.

Ho, E. S., Lin, D. C., Mendel, D. B., and Cihlar, T. (2000). Cytotoxicity of antiviral nucleotides adefovir and cidofovir is induced by the expression of human renal organic anion transporter 1. J. Am. Soc. Nephrol. 10, 383-393.

Ho, H. T., Woods, K. L., Bronson, J. J., De Boeck, H., Martin, J. C., and Hitchcock, M. J. M. (1992).

Intracellular metabolism of the antiherpes agent (S)-1-[3-hydroxy-2-(phosphonylmethoxy)propyl]cytosine. Mol. Pharmacol. 41, 197-202.

Holloway, L. S., and Cassin, S. (1972). In vitro uptake of PAH by choroid plexus from dogs of various ages. Am. J. Physiol. 223, 507-509.

Hosoyamada, M., Sekine, T., Kanai, Y., and Endou, H. (1999). Molecular cloning and functional expression of a multispecific organic anion transporter from human kidney. Am. J. Physiol. 276, F122-F128.

Jacquemin, E., Hagenbuch, B., Stieger, B., Wolkoff, A., and Meier, P. (1994). Expression cloning of a rat liver Na$^+$-independent organic anion transporter. Proc. Natl. Acad. Sci. USA 91, 133-137.

Jariyawat, S., Sekine, T., Takeda, M., Apiwattanakul, N., Kanai, Y., Sophasan, S., and Endou, H. (1999). The interaction and transport of beta-lactam antibiotics with the cloned rat renal organic anion transporter 1. J. Pharmacol. Exp. Ther. 290, 672-7.

Juliano, R. L., and Ling, V. (1976). A surface glycoprotein modulating drug permeability in Chinese hamster ovary cell mutants. Biochim. Biophys. Acta 455, 152-162.

Kahn, J., Lagakos, S., Wulfsohn, M., Cherng, D., Miller, M., Cherrington, J., Hardy, D., Beall, G., Cooper, R., Murphy, R., Basgoz, N., Ng, E., Deeks, S., Winslow, D., Toole, J. J., and Coakley, D. (1999). Efficacy and safety of adefovir dipivoxil with antiretroviral therapy: a randomized controlled trial. JAMA 282, 2305-2312.

Kakyo, M., Sakagami, H., Nishio, T., Nakai, D., Nakagomi, R., Tokui, T., Naitoh, T., Matsuno, S., Abe, T., and Yawo, H. (1999). Immunohistochemical distribution and functional characterization of an organic anion transporting polypeptide 2 (oatp2). FEBS Lett. 445, 343-346.

Kearney, B.P., Abbott, E., and Cundy, K.C. (2000). The effect of P-glycoprotein on the relative bioavailability and pharmacokinetics of adefovir dipivoxil (ADV) and tenofovir disoproxil fumarate (TDF) in mice. 1st International Workshop on Clinical Pharmacology of HIV Therapy, Noordwijk, The Netherlands.

Kim, R., Fromm, M., Wandel, C., Leake, B., Wood, A., Roden, D., and Wilkinson, G. (1998). The drug transporter P-glycoprotein limits oral absorption and brain entry of HIV-1 protease inhibitors. J. Clin. Invest. 101, 289-294.

Klein, I., Sarkadi, B., and Varadi, A. (1999). An inventory of the human ABC proteins. Biochim. Biophys. Acta 1461, 237-262.

Konig, J., Cui, Y., Nies, A. T., and Keppler, D. (2000a). Localization and genomic organization of a new hepatocellular organic anion transporting polypeptide. J. Biol. Chem. 275, 23161-8.

Konig, J., Cui, Y., Nies, A. T., and Keppler, D. (2000b). A novel human organic anion transporting polypeptide localized to the basolateral hepatocyte membrane. Am. J. Physiol. 278, G156-G164.

Konig, J., Nies, A. T., Cui, Y., Leier, I., and Keppler, D. (1999a). Conjugate export pumps of the multidrug resistance protein (MRP) family: localization, substrate specificity, and MRP2-mediated drug resistance. Biochim. Biophys. Acta. 1461, 377-394.

Konig, J., Rost, D., Cui, Y., and Keppler, D. (1999b). Characterization of the human multidrug resistance protein isoform MRP3 localized to the basolateral hepatocyte membrane. Hepatology 29, 1156-1163.

Kool, M., de Haas, M., Scheffer, G. L., Scheper, R. J., van Eijk, M. J., Juijn, J. A., Baas, F., and Borst, P. (1997). Analysis of expression of cMOAT (MRP2), MRP3, MRP4 and MRP5, homologues of the multidrug resistance-associated protein gene (MRP1), in human cancer cell lines. Cancer Res. 57, 3537-3547.

Kool, M., Van der Linden, M., de Haas, M., Baas, F., and Borst, P. (1999). Expression of Human MRP6, a homologue of the multidrug resistance protein gene MRP1, in tissues and cancer cells. Cancer Res. 59, 175-182.

Kullak-Ublick, G., Beuers, U., and Paumgartner, G. (2000). Hepatobiliary transport. J. Hepatology 32, 3-18.

Kullak-Ublick, G., Hagenbuch, B., Stieger, B., Schteingart, C., Hofmann, A., Wolkoff, A., and Meier, P. (1995). Molecular and functional characterization of an organic anion transporting polypeptide cloned from human liver. Gastroenterology 109, 1274-1282.

Kusuhara, H., Sekine, T., Utsunomiya-Tate, N., Tsuda, M., Kojima, R., Cha, S., Sugiyama, Y., Kanai, Y., and Endou, H. (1999). Molecular cloning and characterization of a new multispecific organic anion transporter from rat brain. J. Biol. Chem. 274, 13675-13680.

Kuze, K., Graves, P., Leahy, A., Wilson, P., Stuhlmann, H., and You, G. (1999). Heterologous expression and functional characterization of a mouse renal organic anion transporter in mammalian cells. J. Biol. Chem. 274, 1519-1524.

Lacy, S. A., Hitchcock, M. J. M., Lee, W. A., Tellier, P., and Cundy, K. C. (1998). Effect of oral probenecid coadministration on the chronic toxicity and pharmacokinetics of intravenous cidofovir in cynomolgus monkeys. Toxicol. Sci. 4, 97-106.

Lalezari, J. P., Stagg, R. J., Kuppermann, B. D., Holland, G. N., Kramer, F., Ives, D. V., Youle, M., Robinson, M. R., Drew, W. L., and Jaffe, H. S. (1997). Intravenous cidofovir for peripheral cytomegalovirus retinitis in patients with AIDS. A randomized, controlled trial. Ann. Intern. Med. 126, 257-263.

Lee, K., Belinsky, M. G., Bell, D. W., Testa, J. R., and Kruh, G. D. (1998). Isolation of MOAT-B, a widely expressed multidrug resistance-associated protein/canalicular multispecific organic anion transporter-related transporter. Cancer Res. 58, 2741-2747.

Li, L., Meier, P. J., and Ballatori, N. (2000). Oatp2 mediates bidirectional organic solute transport: a role for intracellular glutathione. Mol. Pharmacol. 58, 335-340.

Lin, D. C., Ho, E. S., Mendel, D. B., and Cihlar, T. (1999). Expression of the human renal organic anion transporter 1 (hOAT1) potentiates in vitro cytotoxicity of cephalosporin antibiotics. 39th ICAAC, San Francisco, CA.

Lopez-Nieto, C. E., You, G., Bush, K. T., Barros, E., Beier, D., and Nigam, S. (1997). Molecular cloning and characterization of NKT, a gene product related to the organic cation transporter family that is almost exclusively expressed in the kidney. J. Biol. Chem. 272, 6471-6478.

Lu, R., Chan, B. S., and Schuster, V. L. (1999). Cloning of the human kidney PAH transporter: narrow substrate specificity and regulation by protein kinase C. Am. J. Physiol. 276, F295-F303.

Madon, J., Hagenbuch, B., Landmann, L., Meier, P. J., and Stieger, B. (2000). Transport, function and hepatocellular localization of MRP6 in rat liver. Mol. Pharmacol. 57, 634-641.

Masuda, S., Hideyuki, S., Nonoguchi, H., Tomita, K., and Inui, K. (1997). mRNA distribution and membrane localization of the OAT-K1 organic anion transporter in rat renal tubules. FEBS Lett. 407, 127-131.

Masuda, S., Ibaramoto, K., Takeuchi, A., Saito, H., Hashimoto, Y., and Inui, K. I. (1999a). Cloning and functional characterization of a new multispecific organic anion transporter, OAT-K2, in rat kidney. Mol. Pharmacol. 55, 743-52.

Masuda, S., Saito, H., and Inui, K. I. (1997a). Interactions of nonsteroidal anti-inflammatory drugs with rat renal organic anion transporter, OAT-K1. J. Pharmacol. Exp. Ther. 283, 1039-42.

Masuda, S., Takeuchi, A., Saito, H., Hashimoto, Y., and Inui, K. (1999b). Functional analysis of rat renal organic anion transporter OAT-K1: bidirectional methotrexate transport in apical membrane. FEBS Lett. 459, 128-132.

McAleer, M. A., Breen, M. A., White, N. L., and Matthews, N. (1999). PABC11 (also known as MOAT-C and MRP5), a member of the ABC family of proteins, has anion transporter activity but does not confer multidrug resistance when overexpressed in human embryonic kidney 293 cells. J. Biol. Chem. 274, 23541-23548.

Meier, P., Eckhardt, U., Schroeder, A., Hagenbuch, B., and Stieger, B. (1997). Substrate specificity of sinusoidal bile acid and organic anion uptake systems in rat and human liver. Hepatology 26, 1667-1677.

Merta, A., Votruba, I., Jindrich, J., Holy, A., Cihlar, T., Rosenberg, I., Otmar, M., and Herve, T. Y. (1992). Phosphorylation of 9-(2-phosphonomethoxyethyl)adenine and 9-(S)-(3-hydroxy-2-phosphono-methoxypropyl)adenine by AMP(dAMP)kinase from L1210 cells. Biochem. Pharmacol. 44, 2067-2077.

Miller, M. D., Margot, N. A., Hertogs, K., Larder, B., and Miller, V. (2000). Anti-HIV activity profile of tenofovir (PMPA) against a panel of nucleoside-resistant clinical samples. 4th International Workshop on HIV Drug Resistance & Treatment Strategies Sitges, Spain.

Miller, T. B., and Ross, C. R. (1972). Transport of organic cations and anions by choroid plexus. J. Pharmacol. Exp. Ther. 196, 771-777.

Miller, W. H., and Miller, R. L. (1982). Phosphorylation of acyclovir diphosphate by cellular enzymes. Biochem. Pharmacol. 31, 3879-3884.

Mulato, A. S., Ho, E. S., and Cihlar, T. (2000). Nonsteroidal anti-inflammatory drugs efficiently reduce the transport and cytotoxicity of adefovir mediated by the human renal organic anion transporter 1. J. Pharmacol. Exp. Ther. 295, 10-15.

Naesens, L., Snoeck, R., Andrei, G., Balzarini, J., Neyts, J., and De Clercq, E. (1997). HPMPC (cidofovir), PMEA (adefovir) and related acyclic nucleoside phosphonate analogues: a review of their pharmacology and clinical potential in the treatment of viral infections. Antiviral Chemistry Chemother. 8, 1-23.

Noe, B., Hagenbuch, B., Stieger, B., and Meier, P. J. (1997). Isolation of a multispecific organic anion and cardiac glycoside transporter from rat brain. Proc. Natl. Acad. Sci. USA 94, 10346-50.

Olsanska, L., Cihlar, T., Votruba, I., and Holy, A. (1997). Transport of adefovir (PMEA) in human T-lymphoblastoid cells. Collect. Czech. Chem. Commun. 62, 821-828.

Otova, B., Zidek, Z., Holy, A., Vortruba, I., Sladka, M., Marinov, I., and Leskova, V. (1997). Antitumor activity of novel purine acyclic nucleotide analogs PMEA and PMEDAP. In Vivo 11, 163-167.

Palu, G., Stefanelli, S., Rassu, M., Parolin, C., Balzarini, J., and De Clercq, E. (1991). Cellular uptake of phosphonylmethoxyalkylpurine derivatives. Antiviral Res. 16, 115-119.

Perrillo, R., Schiff, E., Yoshida, E., Statler, A., Hirsch, K., Wright, T., Gutfreund, K., Lamy, P., and Murray, A. (2000). Adefovir dipivoxil for the treatment of lamivudine-resistant hepatitis B mutants. Hepatology 32, 129-134.

Pritchard, J. B., Sweet, D. H., Miller, D. S., and Walden, R. (1999). Mechanism of organic anion transport across the apical membrane of choroid plexus. J. Biol. Chem. 274, 33382-33387.

Prus, K. L., Hill, E. L., and Ellis, M. N. (1991). The transport of 9-(2-phosphonylmethoxyethyl)adenine (PMEA) into Vero cells. Antiviral Res. Suppl. 1, 144.

Rao, V., Dahlheimer, J., Bardgett, M., Snyder, A., Finch, R., Sartorelli, A., and Piwnica-Worms, D. (1999). Choroid plexus epithelial expression of MDR1 P glycoprotein and multidrug resistance-associated protein contribute to the blood-cerebrospinal-fluid drug-permeability barrier. Proc. Natl. Acad. Sci. 96, 3900-3905.

Ringpfeil, F., Lebwohl, M. G., Christiano, A. M., and Uitto, J. (2000). Pseudoanxthoma elasticum: mutations in the MRP6 gene encoding a transmembrane ATP-binding cassette (ABC) transporter. Proc. Natl. Acad. Sci. USA 97, 6001-6006.

Robbins, B. L., Connelly, M. C., Marshall, D. R., Srinivas, R. V., and Fridland, A. (1995a). A human T lymphoid cell variant resistant to the acyclic nucleoside phosphonate 9-(2-phosphonylmethoxyethyl)adenine shows a unique combination of phosphorylation defect and increase efflux of the agent. Mol. Pharmacol. 47, 391-397.

Robbins, B. L., Greenhaw, J., Connelly, M. C., and Fridland, A. (1995b). Metabolic pathways for activation of the antiviral agent 9-(2-phosphonylmethoxyethyl)adenine in human lymphoid cells. Antimicrobial Agents Chemother. 39, 2304-2308.

Robbins, B. L., Srinivas, R. V., Kim, C., Bischofberger, N., and Fridland, A. (1998). Anti-human immunodeficiency virus activity and cellular metabolism of a potential prodrug of the acyclic nucleoside phosphonate 9-R-(2-phosphonomethoxypropyl)adenine (PMPA), Bis(isopropyloxymethylcarbonyl)PMPA. Antimicrobial Agents Chemother. 42, 612-617.

Saito, H., Masuda, S., and Inui, K. (1996). Cloning and functional characterization of a novel rat organic anion transporter mediating basolateral uptake of methotrexate in the kidney. J. Biol. Chem. 271, 20719-20725.

Schaub, T., Kartenbeck, J., Konig, J., Spring, H., Dorsam, J., Staehler, G., Storkel, S., Thon, W., and Keppler, D. (1999). Expression of the MRP2 gene-encoded conjugate export pump in human kidney proximal tubules and in renal cell carcinoma. J. Am. Soc. Nephrol. 10, 1159-1169.

Schooley, R., Myers, R., Ruane, P., Beall, G., Lampiris, H., Miller, M., Mills, R., and McGowan, I. (2000). Tenofovir disoproxil fumarate (TDF) for the treatment of antiretroviral experienced patients. A double-blind placebo-controlled study. 40th ICAAC,Toronto, Canada.

Schroeder, A., Exkhardt, U., Stieger, B., Tynes, R., Schteingart, C. D., Hofmann, A. F., Meier, P. J., and Hagenbuch, B. (1998). Substrate specificity of the rat liver Na(+)-bile salt cotransporter in Xenopus laevis oocytes and in CHO cells. Am. J. Physiol. 274, G370-375.

Schuetz, J. D., Connelly, M. C., D., S., Paibir, S. G., Flynn, P. M., Srinivas, R.V.,

Kumar, A., and Fridland, A. (1999). MRP4: A previously unidentified factor in resistance to nucleoside-based antiviral drugs. Nature Med. 5, 1048-1051.

Sekine, T., Cha, S. H., Hosoyamada, M., Kanai, Y., Watanabe, N., Furuta, Y., Fukuda, K., Igarashi, T., and Endou, H. (1998a). Cloning, functional characterization, and localization of a rat renal Na$^+$-dicarboxylate transporter. Am. J. Physiol. 275, F298-F305.

Sekine, T., Cha, S. H., Tsuda, M., Apiwattanakul, N., Nakajima, N., Kanai, Y., and Endou, H. (1998b). Identification of multispecific organic anion transporter 2 expressed predominantly in the liver. FEBS Lett. 429, 179-82.

Sekine, T., Watanabe, N., Hosoyamada, M., Kainai, Y., and Endou, H. (1997). Expression cloning and characterization of a novel multispecific organic anion transporter. J. Biol. Chem. 272, 18526-18529.

Shah, T., Simpson, R. J., Webster, A. D., and T.J., P. (1986). Uptake of free adenosine and adenosine from adenosine monophosphate by human peripheral blood lymphocytes: possible kinetic role for ecto-5'-nucleotidase in the regulation of intracellular adenosine. Clin. Exp. Immunol. 66, 158-165.

Simonson, G. D., Vincent, A. C., Roberg, K. J., Huang, Y., and Iwanij, V. (1994). Molecular cloning and characterization of a novel liver-specific transport protein. J. Cell. Sci. 107, 1065-1072.

Smeijsters, L. J., Franssen, F. F., Naesens, L., de Vries, E., Holy, A., Balzarini, J., De Clercq, E., and Overdulve, J. P. (1999). Inhibition of the in vitro growth of Plasmodium falciparum by acyclic nucleoside phosphonates. Int. J. Antimicrob. Agents 12, 53-61.

Snocck, R., Noel, J. C., Muller, C., De Clercq, E., and Bossens, M. (2000). Cidofovir, a new approach for the treatment of cervix intraepithelial neoplasia grade III (CIN III). J. Med. Virol. 60, 205-209.

Snoeck, R., Wellens, W., Desloovere, C., Van Ranst, M., Naesens, L., De CLercq, E., and Feenstra, L. (1998). Treatment of severe laryngeal papillomatosis with intralesional injections of cidofovir [(S)-1-(3-hydroxy-2-phosphonylmethoxypropyl)cytosine]. J. Med. Virol. 54, 219-225.

Stieger, B., Hagenbuch, B., Landmann, L., Hochli, M., Schroeder, A., and Meier, P. J. (1994). In situ localization of the hepatocytic Na$^+$/taurocholate cotransporting polypeptide in rat liver. Gastroenterology 107, 1787-1787.

Sweet, D., Wolff, N., and Pritchard, J. (1997). Expression cloning and characterization of ROAT1. J. Biol. Chem. 272, 30088-30095.

Sweet, D. H., and Pritchard, J. B. (1999). The molecular biology of renal organic anion and organic cation transporters. Cell. Biochem. Biophys. 31, 89-118.

Takeuchi, A., Masuda, S., Saito, H., Hashimoto, Y., and Inui, K. (2000). Trans-stimulation effects of folic acid derivatives on mehtotrexate transport by rat renal organic anion transporter, OAT-K1. J. Pharmacol. Exp. Ther. 293, 1034-1039.

Tenenhouse, H., Roy, S., Martel, J., and Gauthier, C. (1998). Differential expression, abudance, and regulation of Na$^+$ -phospahte cotransporter genes in murine kidney. Am. J. Physiol. 275, F527-F534.

Tojo, A., Sekine, T., Nakajima, N., Hosoyamada, M., Kanai, Y., Kimura, K., and Endou, H. (1999). Immunohistochemical Localization of Multispecific Renal Organic Anion Transporter 1 in Rat Kidney. J. Am. Soc. Nephrol. 10, 464-471.

Torok, M., Gutmann, H., Fricker, G., and Drewe, J. (1999). Sister of P-glycoprotein expression in different tissues. Biochem. Pharmacol. 57, 833-835.

Tsuboi, S., and Pederson, J. E. (1986). Permeability of the isolated dog retinal pigment epithelium to carboxyfluorescein. Invest. Ophthalmol. Vis. Sci. 27, 1767-1770.

Uchino, H., Tamai, I., Yabuuchi, H., China, K., Miyamoto, K. I., Takeda, E., and Tsuji, A. (2000a). Faropenem transport across the renal epithelial luminal membrane via inorganic phosphate transporter Npt1. Antimicrobial Agents Chemother. 44, 574-577.

Uchino, H., Tamai, I., Yamashita, K., Minemoto, Y., Sai, Y., Yabuuchi, H., Miyamoto, K., Takeda, E., and Tsuji, A. (2000b). p-Aminohippuric acid transport at renal apical membrane mediated by human inorganic phosphate transporter NPT1. Biochem. Biophys. Res. Commun. 270, 254-259.

Ullrich, K. J. (1999). Affinity of drugs to the different renal transporters for organic anions and organic cations. Pharm. Biotechnol. 12, 159-179.

Van Aubel, R., Masereeuw, R., and Russel, F. (2000). Molecular pharmacology of renal organic anion transporters. Am. J. Physiol. 279, F216-F232.

Vorbrodt, A., and Borun, TW. (1979). Ultrastructural cytochemical studies of plasma membrane phosphatase activities during the HeLa S3 cell cycle. J. Histochem. Cytochem. 27, 1596-1603.

Wada, S., Tsuda, M., Sekine, T., Cha, S. H., Kimura, M., Kanai, Y., and Endou, H. (2000). Rat multispecific organic anion transporter 1 (rOAT1) transports zidovudine, acyclovir, and other antiviral nucleoside analogs. J. Pharmacol. Exp. Ther. 294, 844-849.

Wijnholds, J., Mol, C. A., van Deemter, L., de Haas, M., Scheffer, G. L., Baas, F., Beijnen, J. H., Scheper, R. J., Hatse, S., De Clercq, E., Balzarini, J., and Borst, P. (2000). Multidrug-resitance protein 5 is a multispecific organic anion transporter able to transport nucleotide analogs. Proc. Natl. Acad. Sci. USA 97, 7476-7481.

Xiong, X., Flores, C., Yang, H., Toole, J. J., and Gibbs, C. S. (1998). Mutations in hepatitis B DNA polymerase associated with resistance to lamivudine do not confer resistance to adefovir in vitro. Hepatology 28, 1669-1673.

Yabuuchi, H., Tamai, I., Morita, K., Kouda, T., Miyamoto, K. I., Takeda, E., and Tsuhi, A. (1998). Hepatic sinusoidal membrane transport of anionic drugs mediated by anion transporter Npt1. J. Pharmacol. Exp. Ther. 286, 1391-1396.

Ying, C., De Clercq, E., and Neyts, J. (2000). Lamivudine, adefovir and tenofovir exhibit long-lasting anti-hepatitis B virus activity in cell culture. J. Viral Hepat. 7, 79-83.

Yoshida, A., Ishiko, S., and Kojima, M. (1992). Outward permeability of the blood-retinal barrier. Graefes. Arch. Clin. Exp. Ophthalmol. 230, 78-83.

Zekri, M., Harb, J., Bernard, S., Poirier, G., Devaux, C., and Meflah, K. (1989). Differences in the release of 5'-nucleotidase and alkaline phosphatese from plasma membrane of several cell types by PI-PLC. Comp. Biochem. Physiol. B. 93, 673-679.

Zhang, Y., Han, H., Elmquist, W. F., and Miller, D. W. (2000). Expression of various multidrug resistance-associated protein (MRP) homologues in brain microvessel endothelial cells. Brain Res. 876, 148-153.

5-AZA-7-DEAZAPURINES: SYNTHESIS AND PROPERTIES OF NUCLEOSIDES AND OLIGONUCLEOTIDES

FRANK SEELA and HELMUT ROSEMEYER

Laboratorium für Organische und Bioorganische Chemie, Institut für Chemie, Universität Osnabrück, Barbarastr. 7, D-49069 Osnabrück, GERMANY

Dedicated to Prof. J. J. Fox

1.   Introduction

Naturally-occurring purine nucleosides have received considerable attention as lead structures in medicinal chemistry (G. R. Revankar and R. K. Robins, 1991). Their structural analogues have the potential either to emulate or to antagonize the function of the parent nucleosides. Moreover, base-modified purine nucleosides have gathered interest in the field of antisense technology (E. Uhlmann and A. Peyman, 1990). As substitutes of the regular nucleic acid constituents they can increase the affinity of a DNA-RNA hybrid and can stabilize their structure with regard to enzymatic degradation (U. Englisch and D. H. Gauss, 1991).

This article reports on base-modified purine nucleosides with a nitrogen atom in a bridgehead position (G. R. Revankar and R. K. Robins, 1991). Not many reports have appeared describing this class of molecules. Among these base-modified nucleosides the 5-aza-7-deazapurine {= imidazo[1,2-a]-*s*-triazine} nucleosides are of considerable interest as they display a similar shape as the parent purine compounds. Formally, these compounds can be constructed by the transposition of nitrogen-7 to the bridgehead position 5 leading to a 7-deazapurine structure which exhibits a very stable N-glycosylic bond (purine numbering is used throughout the manuscript) (Scheme 1).

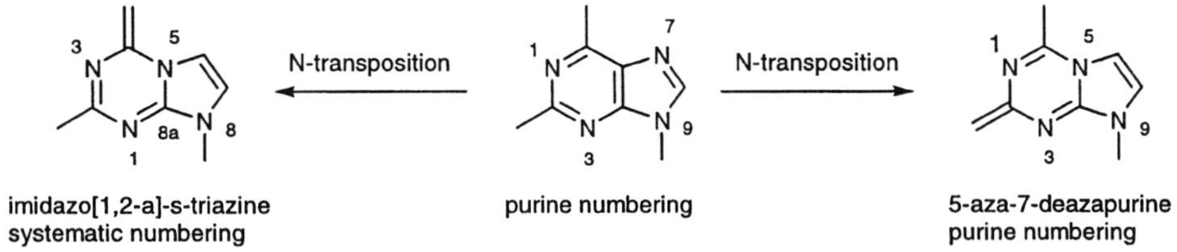

N-transposition

N-transposition

imidazo[1,2-a]-s-triazine
systematic numbering

purine numbering

5-aza-7-deazapurine
purine numbering

Scheme 1.

505

Recent Advances in Nucleosides: Chemistry and Chemotherapy, Ed. by C.K. Chu. 505 — 533

According to the structural modification of the heterocycle the pyrimidine ring of 5-aza-7-deazapurines must carry an exocyclic substituent with a double bond being present either in the positions 2 and/or 6 of the base. This is the structural requirement when the five-membered ring carries a residue at the nitrogen-9, e.g. a sugar moiety. Common exocyclic substituents are oxo or thioxo groups but also an imino group is conceivable. Several 5-aza-7-deazapurine nucleosides with an oxo or a thioxo group (**1-6**) are displayed in Scheme 2.

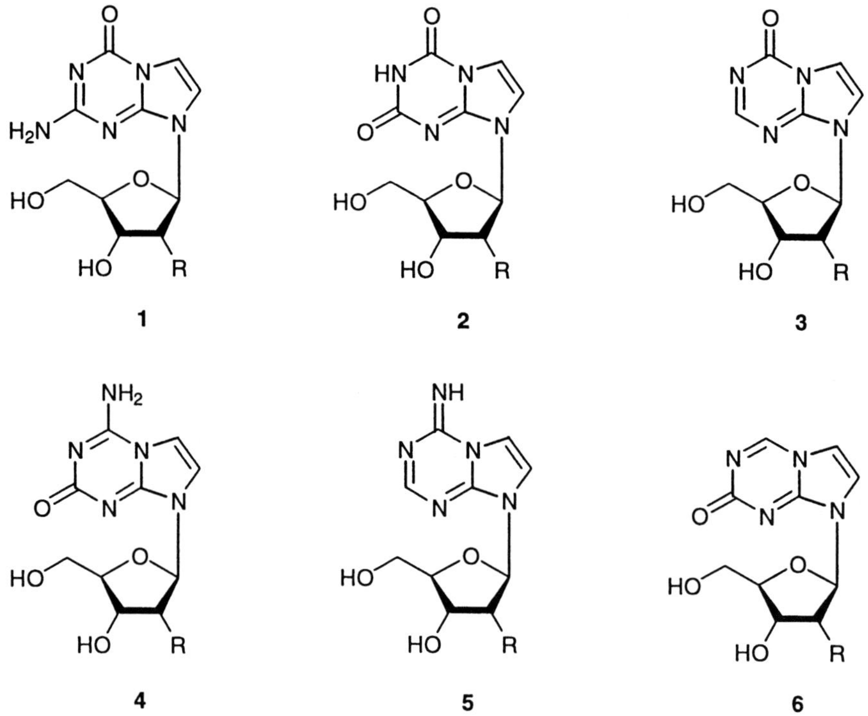

a: R = OH; b: R = H

Scheme 2.

The presence of a nitrogen in position-5 changes the donor-acceptor pattern of the base with regard to the parent purines. The 5-aza-7-deaza-2'-deoxyguanosine (**1b**), for example, is a structural analogue of 2'-deoxyguanosine (**8**) as well as of 7-deaza-2'-deoxyguanosine (**9**, c^{7}G$_d$). It displays an almost identical molecular shape of the parent purine nucleoside but shows an altered Watson-Crick recognition site; the proton at nitrogen-1 is absent and this position is now a proton acceptor site and not longer a proton donor. Compound **1b** is the purine counterfeit of 2'-deoxyisocytidine (**7**).

Nevertheless, protonation of **1b** at nitrogen-1 restores the donor-acceptor pattern of 2'-deoxyguanosine - or more precisely - of 7-deaza-2'-deoxyguanosine (**9**); the molecule is now a cation (**1b**-cation, Scheme 3).

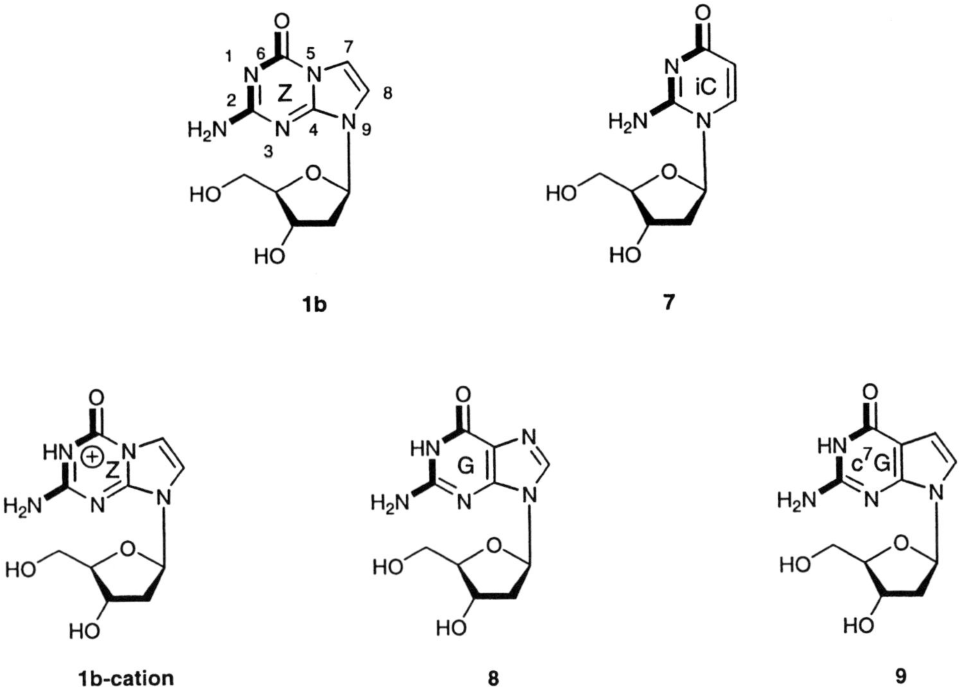

1b

7

1b-cation

8

9

Scheme 3.

2.   Synthesis of 5-aza-7-deazapurines

5-Aza-7-deazapurines can be synthesized by two different routes depending on the heterocyclic system which is used as educt. The first route (i) uses *sym*-triazine (*s*-triazine or [1,3,5]-triazine) derivatives as starting materials, the second (ii) imidazole precursors. For a review see A. R. Katritzky *et al.* (A. R. Katritzky *et al.*, 1996).

(i) *Closure of the imidazole ring.* - An early synthesis of the 5-aza-7-deazapurine ring system in the form of a 7,8-dihydro derivative (**12**) involved the rearrangement of 2,4-diamino-6-(1-aziridinyl)-*s*-triazine (**10**) with hydrochloric acid (F. C. Schaefer, 1955) (Scheme 4). The reaction proceeds most likely via a corresponding 2-(chloroethylamino) derivative **11**. Compound **12** was also prepared by condensation of potassium dicyano-guanidine with 2-chloroethylamine hydrochloride (F. C. Schaefer, 1955). The 7,8-dihydro derivative of 5-aza-7-deazapurine (**14**) was obtained on the thermolysis of 2-chloroe-thoxy-substituted *s*-triazine **13** (Scheme 5) (V. V. Dovlatyan *et al.*, 1977; 1980; 1981; 1985). Other examples of this synthetic route are reported by the same authors.

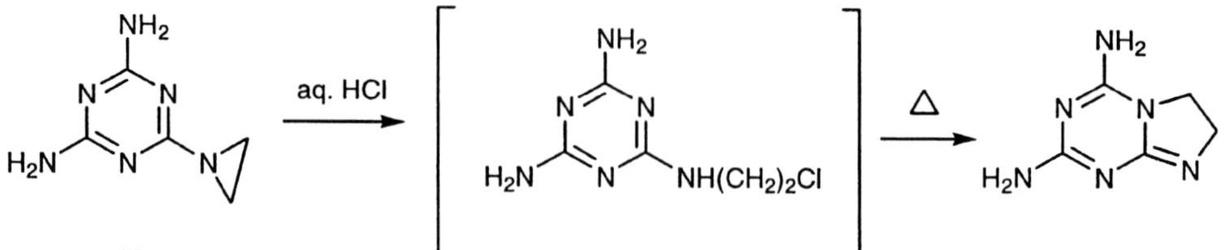

10 11 12

Scheme 4.

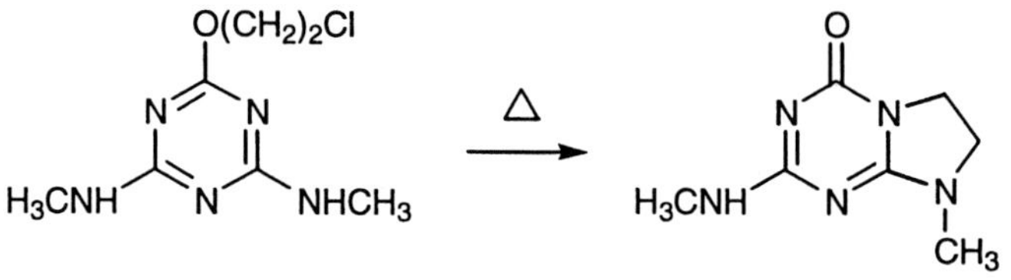

13 14

Scheme 5.

A similar route towards the synthesis of 7,8-dihydro derivatives of 5-aza-7-deazapurines (**16**) starting from hydroxyethylamino derivatives of *s*-triazine such as **15** was reported by T. Unishi and co-workers (T. Unishi *et al.*, 1987; 1988) (Scheme 6) as well as by R. Hinkens *et al.* (R. Hinkens *et al.*, 1961). More recently it was shown that thermal rearrangement of 2,4-dianilino-*s*-triazin-6-yl propargyl ethers such as **17** results in the formation of the 6-methyleno derivative **18** which isomerizes to the fully aromatic 5-aza-7-deazapurine derivative **19** (K. K. Balasubramanian *et al.*, 1980) (Scheme 7).

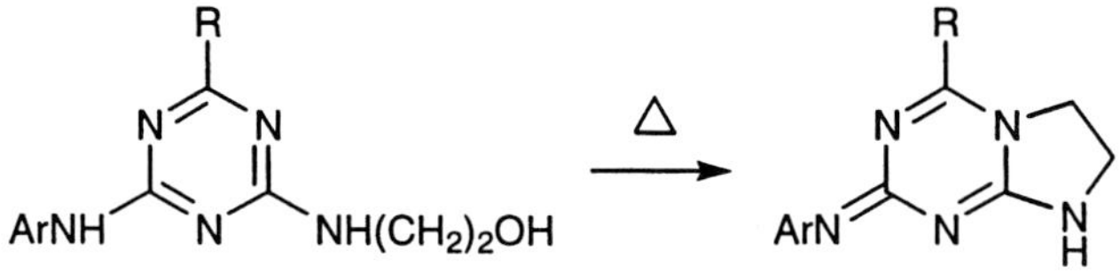

15 16

Scheme 6.

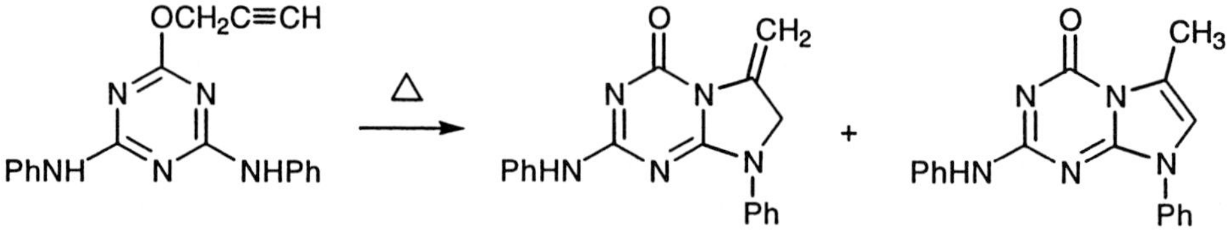

17 18 19

Scheme 7.

The reaction of 2-amino-4,6-bis(methylthio)-*s*-triazine (**20**) with bromoacetaldehyde in ethanol (Scheme 8) (J. Kobe *et al.*, 1970) furnished a mixture of the fully aromatic bases **21** and **22a**; the first was converted to the latter by heating in hot ethanol. The bis(methylthio) derivative **22a** gave the 7-bromo derivative **22b** upon treatment with N-bromosuccinimide.

A similar approach was used by V. Nair (V. Nair *et al.*, 1991) to synthesize 5-aza-7-deazahypoxanthine (**24**) in a one step reaction from 5-azacytosine (**23**) and chloroacetaldehyde in aqueous solution. After chromatographic work-up crystalline **24** was isolated in 86% yield which was used later for nucleoside synthesis (see Chapter 3) (Scheme 9).

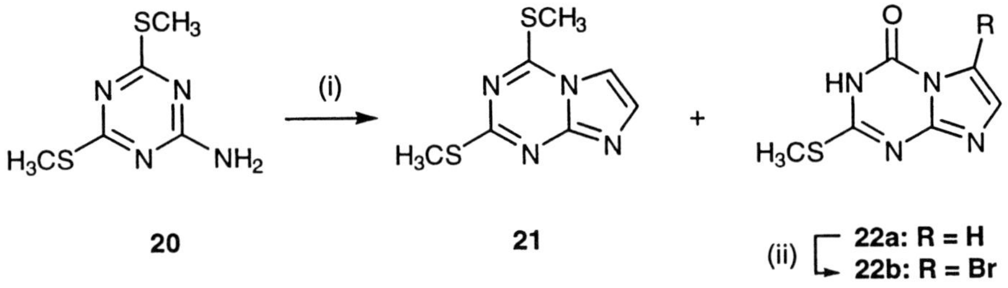

20   **21**   **22a: R = H**
(ii) **22b: R = Br**

(i) BrCH$_2$CHO, EtOH, 65-70°C; (ii) N-bromosuccinimide, CHCl$_3$, reflux, 5 min.

Scheme 8.

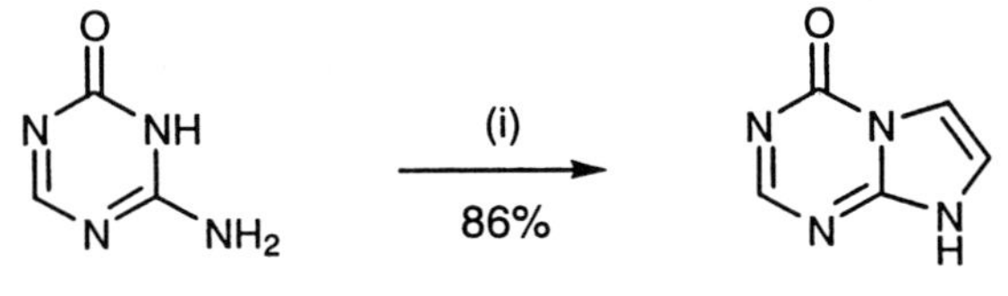

23   **24**

(i) ClCH$_2$CHO, H$_2$O, 45°C, 5d

Scheme 9.

The nowadays most often applied synthetic route towards the synthesis of 5-aza-7-deazapurines was developed by R. K. Robins and co-workers (S.-H. Kim *et al.*, 1978). It was employed during the synthesis of 5-aza-7-deazaguanine (**30**) and 5-aza-7-deazaxanthine (**31**) (Scheme 10).

As educt cyanuric chloride (**25**) was used, and the key intermediate **27** was obtained by selective and stepwise amination and/or hydroxylation reactions (**25** → **26** → **27**). Condensation of the latter with aminoacetaldehyde dimethyl acetal furnished compound **28** which was treated with 6N aq. HCl to afford the 2-amino-4-(2-hydroxy-vinylene-amino)-*s*-triazine (**29**). Ring closure of compound **29** with conc. sulphuric acid gave 5-aza-7-deazaguanine (**30**). Deamination of **30** with barium nitrite provided 5-aza-7-deazaxanthine (**31**). Compound **31** was alternatively prepared from 2-chloro-*s*-triazine-

4,6-dione (not shown). The formation of 5-aza-7-deazaisoguanine (**32**) which is also conceivable was, however, not observed.

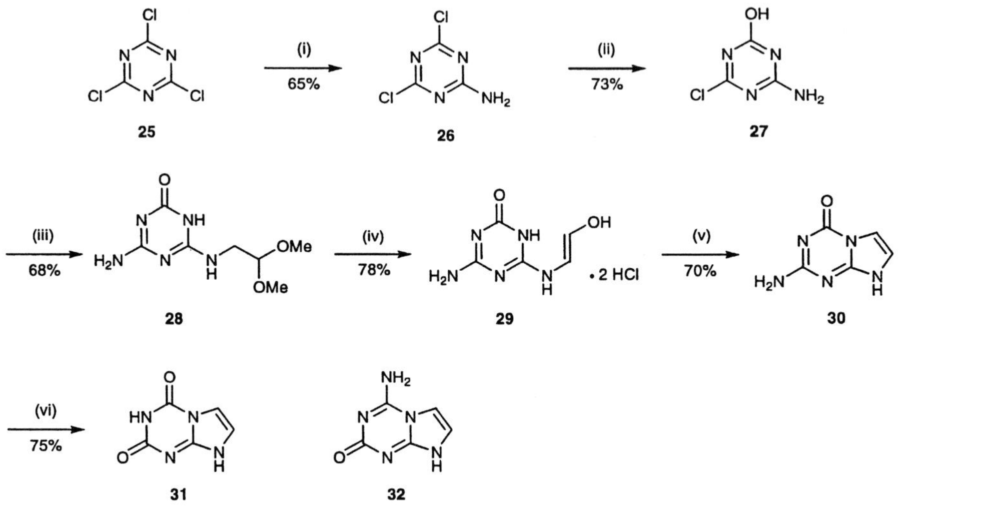

Scheme 10.

(i) NH₃ / -5 °C; (ii) aq. NaOH, r.t.; (iii) H₂NCH₂CH(OCH₃)₂, reflux; (iv) 6N HCl, reflux; (v) conc. H₂SO₄; (vi) Ba(NO₂)₂ / HOAc.

The versatility of this reaction sequence was demonstrated during the synthesis of the tricyclic derivative **31** (Scheme 11). The initial displacement of the chloro substituents of cyanuric chloride (**25**) was performed in 1M aq. NaHCO₃ (4h, r.t.) to give the sodium salt of 2,4-dichloro-6-hydroxy-*s*-triazine (**33**). Then, the other two halogens were displaced by the action of two equivalents of aminoacetaldehyde dimethyl acetal yielding compound **34**.

Hydrolysis furnished the aldehyde **35** which was cyclized (conc. sulphuric acid; 90°C, 1h) to give the hitherto unknown heterocyclic system imidazo[1,2-a]-imidazo[2,1-d]-*s*-triazin-5-one (**36**) (H. Rosemeyer and F. Seela, unpublished results, a). Related compounds were claimed to be useful as photographic couplers (K. Yamakawa and H. Naruse, 1991).

(ii) *Closure of the s-triazine ring.* - Appending an *s*-triazine ring to an imidazole precursor to produce 5-aza-7-deazapurines has been performed since 1971 and was reported by several groups (A. C. Veronese *et al.*, 1971; L. Capuano and H. J. Schrepfer, 1971; H. Staehle *et al.*, 1973; F. Saczewski and M. Gdaniec, 1987). Although this synthetic route towards the synthesis of 5-aza-7-deazapurine bases has not been used often, an important application is the synthesis of 5-aza-7-deazaisoguanine (**32**). The 2-amino-1*H*-imidazole (**37**) was used as starting material (J. J. Voegel *et al.*, 1993). This was reacted with (phenylthio)carbonyl isothiocyanate to give compound **38** which was converted to the dibenzyl derivative **39**. The benzylmercapto group was displaced by an amino function to yield compound **40**. The latter was reduced (Na / liquid NH₃) to

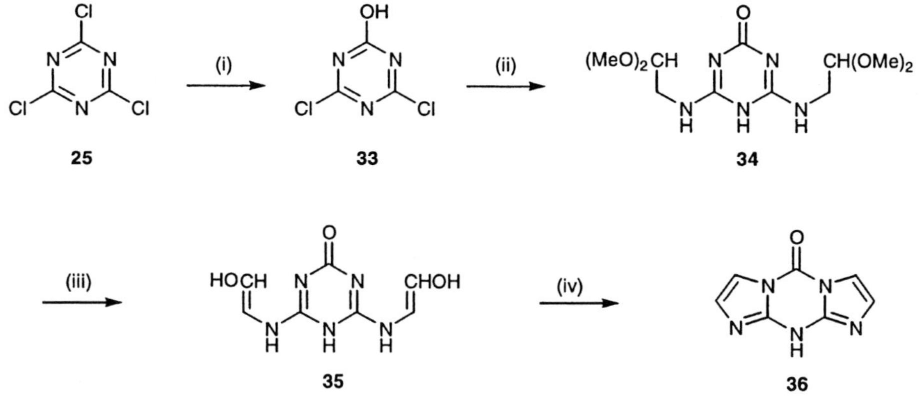

(i) 1M aq. NaHCO₃, r.t.; (ii) H₂NCH₂CH(OCH₃)₂, reflux; (iii) 6N HCl, reflux; (iv) conc. H₂SO₄.

Scheme 11.

furnish the desired 5-aza-7-deaza-1*H*-isoguanine (**32**) (Scheme 12). A similar approach has been applied for the linear synthesis of 5-aza-7-deazapurine nucleosides which will be described in the next chapter.

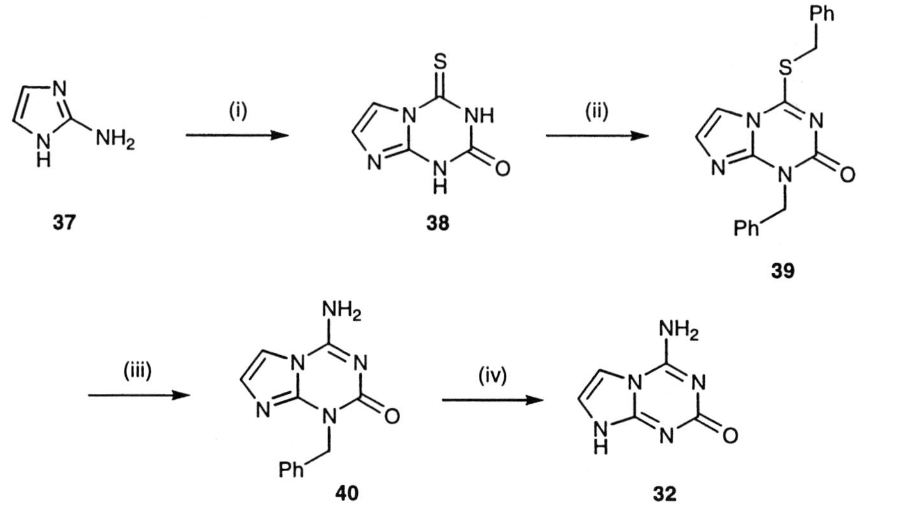

(i) PhSC(O)NCS, MeCN; (ii) PhCH₂Br, Na₂CO₃, DMF; (iii) NH₃ / MeOH; (iv) Na / NH₃ (l)

Scheme 12.

3.   Synthesis of 5-aza-7-deazapurine nucleosides

The synthesis of 5-aza-7-deazapurine nucleosides can be performed either by a linear (i) or a convergent (ii) route. The first uses imidazole nucleosides as starting materials and comprises a subsequent closure to a triazine ring. The second method uses preformed 5-aza-7-deazapurines as precursors.

(i) *Linear syntheses:* A series of 5-aza-7-deazapurine α-D- and β-D-ribofuranosides have been prepared on the linear route by Verheyden and co-workers (E. J. Prisbe *et al.*, 1978a). Condensation of 2-nitroimidazole (**41**) with 2,3,5-tri-O-benzoyl-β-D-ribofuranosyl bromide (**42**) in the presence of $Hg(CN)_2$ gave the β-D-ribonucleoside **43**. Reduction of **43** (Ni / H_2) (→ **44**), followed by condensation with S-phenylthiocarbonyl isothiocyanate furnished compound **45** which was subsequently methylated at the thioxo group (→ **46**). Treatment of **46** with methanolic ammonia afforded 5-aza-7-deazaisoguanosine (**4a**) (Scheme 13).

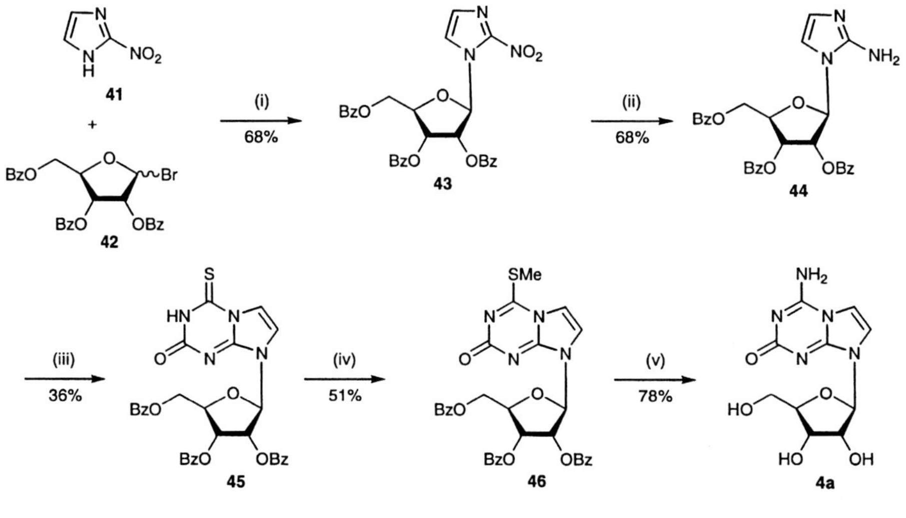

(i) $Hg(CN)_2$, MeCN, 60°C, 2.5 h; (ii) Ni / H_2, r.t., 1.5 h; (iii) PhSC(O)NCS, r.t., 2.5 h; (iv) CH_3I, r.t., 32 h; (v) $NH_3 / MeOH$, r.t., 24 h.

Scheme 13.

On the other hand (Scheme 14), the $SnCl_4$-catalyzed glycosylation of **41** with 1-O-acetyl-2,3,5-tri-O-benzoyl-β-D-ribofuranose (**47**) in acetonitrile gave the α-D-nucleoside **48** as the main product. The formation of the α-D anomer may be due to the presence of the nitro group. It can take part in an initial glycosylation step as nucleophile followed by an intramolecular S_N2-displacement reaction by a second inversion of configuration (F. Seela and W. Bourgeois, 1989) (Scheme 14, [I]). Subsequent reduction (**48** → **49**), followed by condensation with phenoxycarbonyl isocyanate and deprotection of the sugar moiety afforded the 5-aza-7-deazaxanthine α-D-ribofuranoside (**51**). Analogously, the β-D-configurated 5-aza-7-deazaxanthosine as well as the α-*D*-configurated 5-aza-7-deazaisoguanosine were synthesized by the same authors.

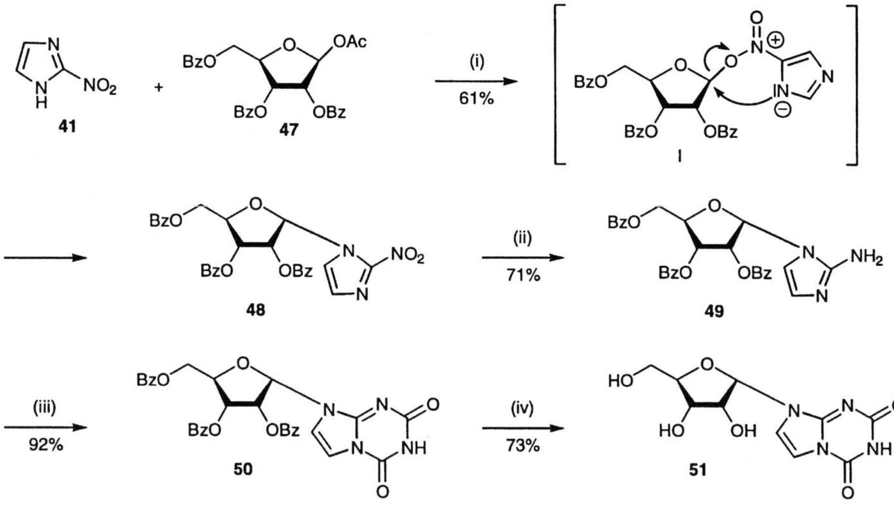

(i) $SnCl_4$, MeCN, 60°C, 75 min; (ii) Ni / H_2, r.t., 1.5 h; (iii) PhOC(O)NCO, r.t., 3.5 h; (iv) NH_3 / MeOH, r.t., 48 h.

Scheme 14.

In another reaction compound **44** was reacted with ethyl-N-cyanoformimidate to give the bis-adduct **52** (Scheme 15) which was cyclized to compound **53** (E. J. Prisbe *et al.*, 1978a). The latter was isolated as a hydrochloride which can be considered as 5-aza-7-deazaadenosine derivative. In a subsequent manuscript Verheyden (E. J. Prisbe *et al.*, 1978b) described the reaction of the 2-aminoimidazole (**37**) with 1-*O*-acetyl-2,3,5-tri-*O*-benzoyl-β-*D*-ribofuranose (**47**). Interestingly, the glycosylation site was the amino group and not the ring nitrogen (→ **54**). This key intermediate was then converted in N(3)-glycosylated derivatives of 5-aza-7-deazaxanthosine (**55**) and 5-aza-7-deazaguanosine (**56**) using various isocyanates or isothiocyanates (Scheme 16).

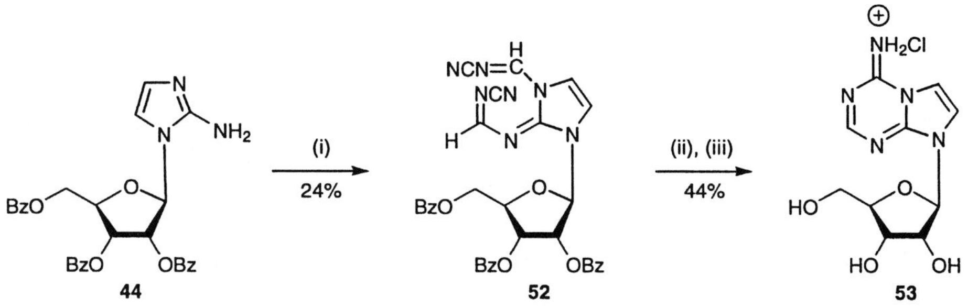

(i) NCN=CHOEt, r.t., 2.25 h; (ii) $NaOCH_3$ / MeOH, r.t., 2 h; (iii) 1N HCl

Scheme 15.

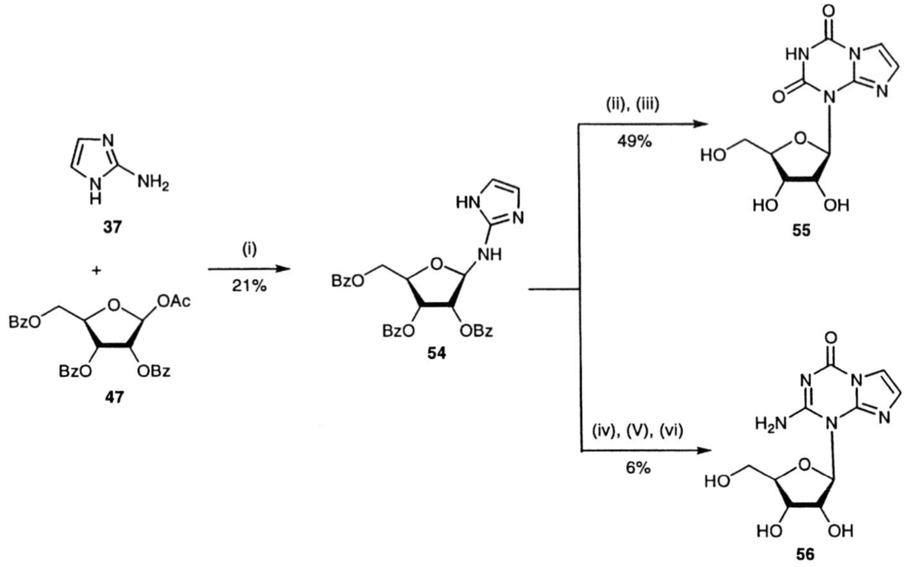

(i) $SnCl_4$, $Hg(CN)_2$ 60°C, 2 h; (ii) PhOC(O)NCO; (iii) NH_3 / MeOH; (iv) PhOC(S)NCS; (v) MeI; (vi) NH_3 / MeOH

Scheme 16.

(ii) *Convergent syntheses:* R. K. Robins and co-workers (S.-H. Kim *et al.*, 1978) have performed the convergent synthesis of 5-aza-7-deazaguanosine (**1a**) applying the glycosylation procedure of H. Vorbrüggen (U. Niedballa and H. Vorbrüggen, 1974). Silylation of 5-aza-7-deazaguanine (**30**) with hexamethyldisilazane (HMDS)/ammonium sulfate, (E. Wittenburg, 1964) followed by condensation with 1-*O*-acetyl-2,3,5-tri-*O*-benzoyl-β-D-ribofuranose (**47**) ($SnCl_4$, anhydr. 1,2-dichloroethane) and subsequent debenzoylation afforded the desired ribonucleoside **1a** (Scheme 17). This was deaminated with barium nitrite to give 5-aza-7-deazaxanthosine (**2a**). Also 5-aza-7-deazaisoguanine (**32**) was converted into its ribonucleoside (**4a**) by the reaction of the silylated base with 1-*O*-acetyl-2,3,5-tri-*O*-benzoyl-β-D-ribofuranose (**47**) ($SnCl_4$, anhydr. 1,2-dichloroethane) followed by debenzoylation (Scheme 18) (J. J. Voegel *et al.*, 1993).

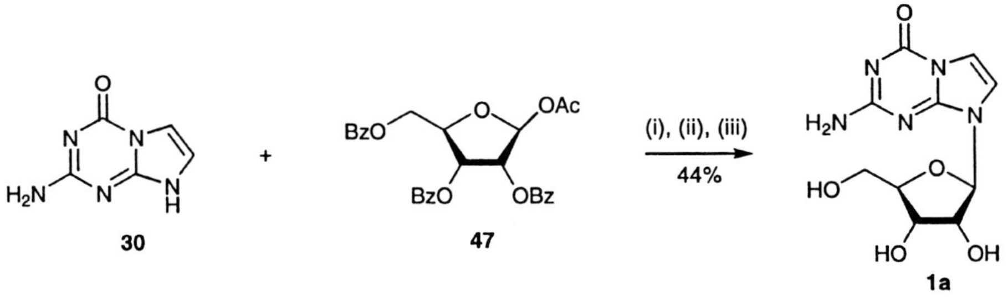

(i) HMDS, 15 h, reflux; (ii) $SnCl_4$, r.t., 30 h; (iii) NaOMe / MeOH.

Scheme 17.

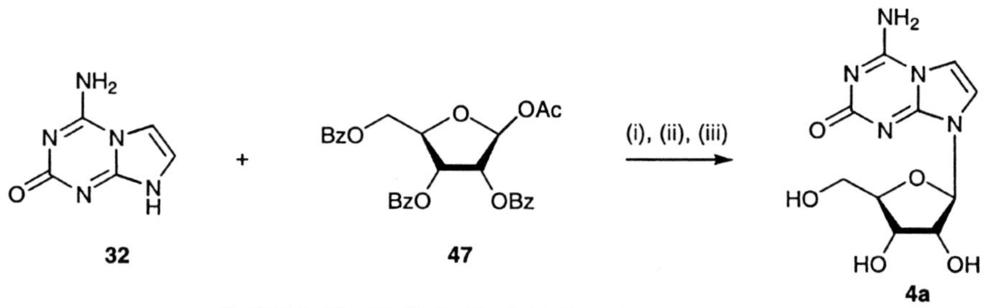

(i), (ii), (iii)

32 + **47**

(i) HMDS, NH₄Cl; (ii) SnCl₄, C₂H₄Cl₂; (iii) NH₃ / MeOH

4a

Scheme 18.

The synthesis of the anomeric 2'-deoxy-*D*-ribofuranosides of 5-aza-7-deazaguanine (**1b, 59**) has been reported (H. Rosemeyer and F. Seela, 1987) (Scheme 19). Solid-liquid as well as liquid-liquid phase-transfer glycosylation (PTG) of nucleobase anions with 3,5-di-toluoyl-2-deoxy-α-*D*-ribofuranosyl chloride (**58**) was employed. The liquid-liquid PTG (method A) made use of the unprotected nucleobase **30**; the N-isobutyryl derivative **57** was used for the solid-liquid PTG (method B) (Scheme 19). For comparison, also the NaH-mediated glycosylation of **57** was evaluated. All three methods gave anomeric mixtures of the toluoyl-protected nucleosides with different ratios of α-*D* vs. β-*D* anomers. The formation of anomeric mixtures could not be avoided due to an unfavourable partition of the nucleobase anion between the organic and the aqueous phase (liquid-liquid) and a prolonged reaction time, during which the halogenose **58** anomerized.

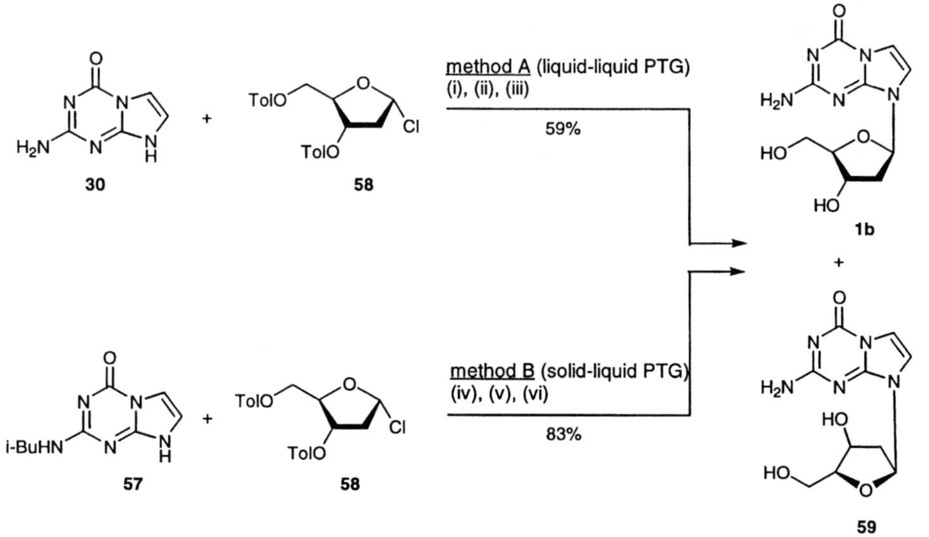

method A (liquid-liquid PTG)
(i), (ii), (iii)

59%

30 + **58**

1b

+

method B (solid-liquid PTG)
(iv), (v), (vi)

83%

57 + **58**

59

(i) 10% aq. K₂CO₃, Bu₄NHSO₄, CH₂Cl₂, 10 min, r.t.; (ii) 4% NH₃ (g) in MeOH, r.t., 48 h; (iii) chromatographic separation of anomers; (iv) K₂CO₃ (s), TDA-1, MeCN, 60 min, r.t.; (v) chromatographic separation of anomers; (vi) 4% NH₃ (g) in MeOH.

Scheme 19.

It has been found to be generally tedious to separate the fully unprotected anomers **1b/59** formed upon liquid-liquid phase-transfer glycosylation (method A, Scheme 19). However, this techniques afforded the highest β-*D*/α-*D* ratio of anomers as desired. Therefore, this mixture of anomers (**1b/59**) was submitted to transient 5'-O-silylation with *t*-butyldiphenylsilyl chloride (H. Rosemeyer, F. Seela, unpublished results, b). On the stage of the 5'-protected derivatives the anomers proved to be separable by silica gel chromatography (CHCl$_3$-MeOH, 9:1). The anomers **1b** and **59** were obtained after treatment of their silylated precursors with Bu$_4$NF in THF.

Our laboratory has demonstrated that the incubation of 5-aza-7-deazaguanine (**30**) with 2-deoxy-α-*D*-ribofuranose 1-phosphate and purine nucleoside phosphorylase (pH 7.5, 2h, r.t.) resulted in a stereoselective formation of the β-*D*-nucleoside **1b**. The reaction can be performed on preparative scale (H. Rosemeyer and F. Seela, 1987). Other groups (S. Benner and co-workers) have also used this technique to prepare 5-aza-7-deaza-2'-deoxyisoguanosine (**4b**) (J. J. Voegel *et al.*, 1993). In this case 2'-deoxy-7-methylguanosinium iodide was used as glycosyl donor. The reaction was optimised by repetition (8 reaction cycles; 24-36 h, each, with enzyme recovery) to give 830 mg of **4b** (59%) from 800 mg of the 5-aza-7-deazaisoguanine (**32**).

McGee and co-workers (D. P. C. McGee *et al.*, 1985) were the first who published the synthesis of an acyclic nucleoside analogue of 5-aza-7-deazaguanosine, namely of 2-amino-8-[(1,3-dihydroxy-2-propoxy)methyl]-imidazo[1,2-a]-*s*-triazin-4-one (**63**). Acid-catalyzed alkylation of 2-acetamido-imidazo[1,2-a]-s-triazin-4-one (**60**) with 1,3-dibenzyloxy-2-acetoxymethylglycerol (**61**) gave the protected compound **62** which was subsequently de-benzylated by catalytic hydrogenation and further deacetylation with methanolic ammonia (→ **63**). The author observed a partial reduction of **63** to the 7,8-dihydro derivative **64** (5-7%). (Scheme 20).

The anomeric 2',3'-dideoxyribonucleosides of 5-aza-7-deazaguanine (**69** and **73**) were prepared in our laboratory (F. Seela *et al.*, 1990; V. Nair *et al.*, 1991) (Scheme 21). The synthesis used the isobutyryl derivative **57** as base and an anomeric mixture of the halogenose **66** as starting materials. The halogenose **66** was prepared from its anomeric lactol precursor by chlorination (R. Appel, 1975; F. Seela *et al.*, 1990). The nucleobase anion glycosylation furnished the anomeric mixture of the glycosylation products **67/71**. The latter were separated chromatographically (**67**: 25%; **71**: 27%). Desilylation (1M Bu$_4$NF in THF) afforded compounds **68** and **72**. Deisobutyrylation (methanolic ammonia) furnished the dideoxynucleosides **69** and **73**. The resulting 2',3'-dideoxynucleosides as well as their precursors were assigned by ^{1}H- and ^{13}C-NMR spectroscopy. Their anomeric configuration was established by [1]H-NOE difference spectroscopy (H. Rosemeyer *et al.*, 1989). The β-*D*-anomer was converted into its 5'-triphosphate **70** (Scheme 21) by applying the one-pot phosphorylation method of J. Ludwig (J. Ludwig, 1981).

Similarly, V. Nair and co-workers prepared the anomeric 2',3'-dideoxynucleoside of 5-aza-7-deazahypoxanthine (**75** and **76**) (Scheme 22). Glycosylation of the silylated base with the protected dideoxy sugar derivative **74** in the presence of trimethylsilyl triflate furnished a mixture of the anomeric glycosylation products (1:1) in a total yield of 76% which was deprotected (methanolic ammonia) to yield the nucleosides **75** and **76**. The authors reported a remarkable N-glycosylic bond stability of these compounds.

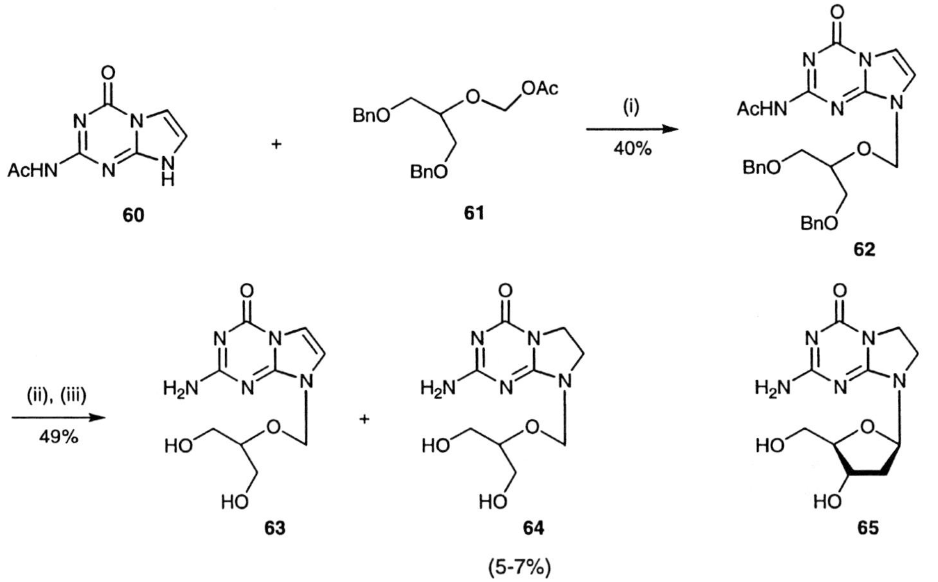

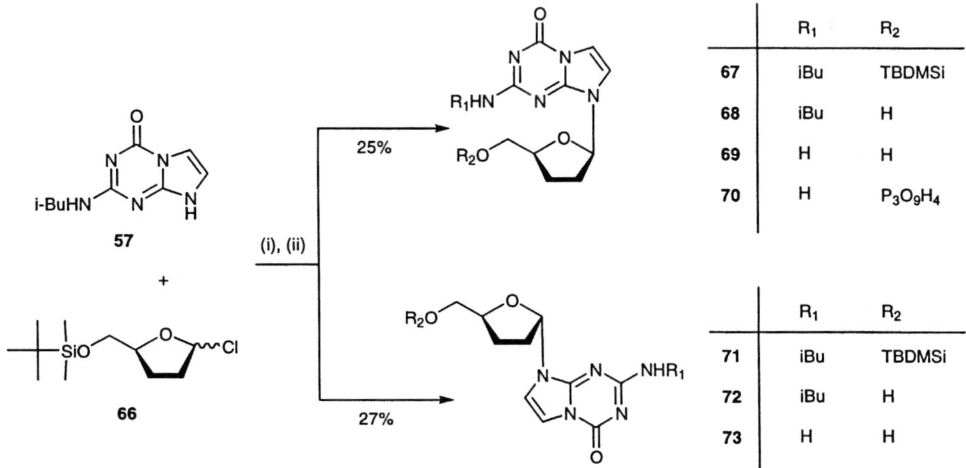

(i) H^+, Δ; (ii) $Pd(OH)_2$ / C; (iii) NH_3 / MeOH.

Scheme 20.

	R_1	R_2
67	iBu	TBDMSi
68	iBu	H
69	H	H
70	H	$P_3O_9H_4$

	R_1	R_2
71	iBu	TBDMSi
72	iBu	H
73	H	H

(i) K_2CO_3, TDA-1, MeCN, r.t.; (ii) chromatographic separation of anomers

Scheme 21.

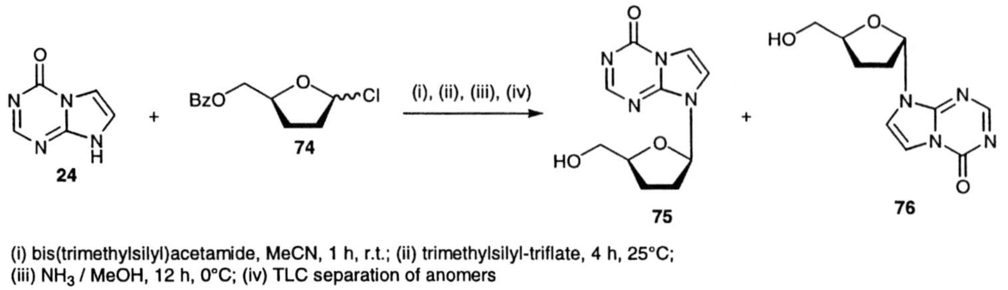

24
74
75
76
(i), (ii), (iii), (iv)
(i) bis(trimethylsilyl)acetamide, MeCN, 1 h, r.t.; (ii) trimethylsilyl-triflate, 4 h, 25°C;
(iii) NH3 / MeOH, 12 h, 0°C; (iv) TLC separation of anomers

Scheme 22.

4. Determination of the protonation sites, the position of glycosylation and the conformation of 5-aza-7-deaza-2'-deoxyguanosine (1b)

pK-Values and protonation site. - Early studies have shown that 5-aza-7-deaza-2'-deoxyguanosine (**1b**) as well as its base **30** are protonated on the s-triazine ring exhibiting a pK_{BH+} value of 3.7 (H. Rosemeyer and F. Seela, 1987). Later, the site of protonation was established using ^{13}C-NMR spectroscopy (F. Seela and A. Melenewski, 1999). Spectra were measured in D_2O within a pD range of 1-14. A significant downfield shift of the C(2) and C(6) resonances ($\Delta\delta = 8.4$ ppm, both) was observed when changing the pD from 5 to 1 while all other chemical shifts remained almost constant. These experiments prove N(1) as protonation site which is different from 2'-deoxyguanosine (**8**) ($pK_{BH+} = 3.5$) which is protonated at N(7) under simultaneous destabilization of its N-glycosylic bond. UV-Spectra show, that in solvents of different polarity 5-aza-7-deaza-2'-deoxyguanosine (**1b**) exists in different tautomeric forms; in dioxane e.g., **1b** displays a UV maximum of 266 nm, whereas in H_2O the maximum is shifted to 258 nm (F. Seela and A. Melenewski, 1999). ^{13}C-NMR spectroscopy revealed that in water compound **1b** exists in the tautomeric form **69**, while in DMSO **1b** is predominant (Scheme 23) (H. Rosemeyer and F. Seela, 1987).

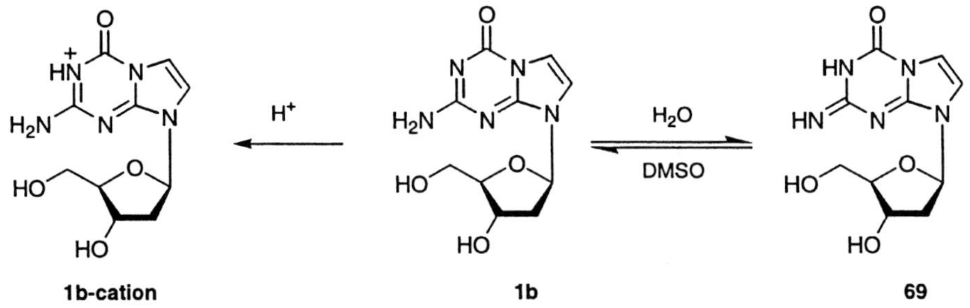

H+
H2O
DMSO
1b-cation
1b
69

Scheme 23.

Compound **1b** is extraordinarily stable at the N-glycosylic bond compared to 2'-deoxyguanosine (**8**): while the latter is hydrolysed in 0.5 N aq. HCl (r.t.) with a half-life value of 10.6 min, **1b** is stable under these conditions.

Conformation of 5-Aza-7-deaza-2'-deoxyguanosine (**1b**). - The various conformational states of a nucleoside are (i) the puckering of the pentofuranosyl moiety (N $\leftrightarrow$ S, $^{3'}T_{2'} \leftrightarrow {}_{3'}T^{2'}$), (ii) the rotational equilibrium about the C(4')-C(5') bond ($\alpha^{(+)g} \leftrightarrow \alpha^t \leftrightarrow \alpha^{(-)g}$), as well as (iii) the syn-anti equilibrium of the base about the N-glycosylic bond. These equilibria are interdependent, and the energy barriers between these conformational states are low in the case of regular nucleosides (H. Rosemeyer and F. Seela, 1997). The nucleobases linked to the anomeric sugar carbon are driving the two-state N $\leftrightarrow$ S pseudorotational equilibrium in nucleosides by two counteracting contributions (i) the anomeric effect [stereoelectronic interactions between O(4') and the nucleobase nitrogen at C(1')], which places the aglycone in the pseudoaxial orientation and (ii) the inherent steric effect of the nucleobase, which opposes the anomeric effect by its tendency to take up the pseudoequatorial position (Figure 1). The latter is sterically favoured in the S-type conformations.

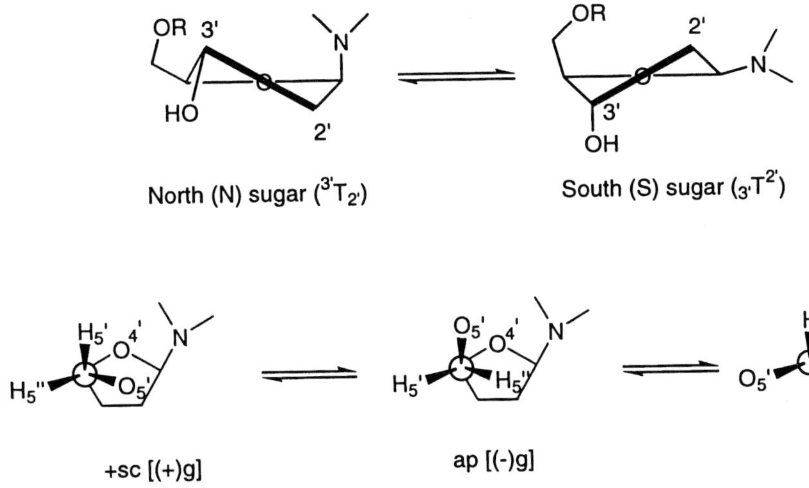

North (N) sugar ($^{3'}T_{2'}$)   South (S) sugar ($_{3'}T^{2'}$)

+sc [(+)g]   ap [(-)g]   -sc [t]

Figure 1.

The conformation of 5-aza-7-deaza-2'-deoxyguanosine (**1b**) with respect to the torsion at the N-glycosylic bond as well as to the sugar puckering in solution has been studied by H. Rosemeyer *et al.* (H. Rosemeyer *et al.*, 1990). Applying 1D-NOE difference spectroscopy with irradiation of H(8) a low $\eta(H(1'))$ of 1.8% indicated a significantly pronounced anti-conformation of the base. An analogous result was obtained for the α-D-anomer **59**. Also the 2',3'-dideoxy-D-ribofuranosides **69**, **73** exhibit a strongly predominant anti conformation as deduced from NOE experiments [**69**: $\eta(H-1') = 1.6\%$; **73**: $\eta(H-1') = 1.5\%$, both upon irradiation of H-8 resonances].

Moreover, the sugar conformation of 5-aza-7-deaza-2'-deoxyguanosine (**1b**) was studied (Figure 1) and compared with those of 2'-deoxyguanosine (**8**) as well as of 7-deaza-2'-deoxyguanosine (**9**) (Table 1) (H. Rosemeyer and F. Seela, 1997). For this purpose, the vicinal ^3J(H,H) coupling constants were determined from well-resolved ^{1}H NMR spectra measured in D_2O. Information on the preferred sugar puckering was obtained by using the *PSEUROT* program (version 6.2; J. van Wijk and C. Altona, 1993).

Table 1.   Conformer populations of 2'-deoxynucleosides at 303K[a].

Compound	%N	%S	%$\gamma^{(+)g}$	%γ^t	%$\gamma^{(-)g}$
dG, **8**	29	71	53	30	17
c^{7}G$_d$, **9**	28	72	43	33	24
z^{5}c^{7}G$_d$, **1b**	37	63	48	33	19

[a]   Solvent, D_2O; RMS, = 0.4 Hz; $|\Delta J_{max}| = 0.5$ Hz.

Using the ^{1}H,^{1}H-coupling constants 1',2', 1',2'', 2',3', 2'',3', and 3',4' the *N/S* conformer populations were determined to be 37% *N* and 63% *S*. Compared to 2'-deoxyguanosine (**8**) and 7-deaza-2'-deoxyguanosine (**9**), the sugar conformation of **1b** is slightly shifted towards the *N*-conformation, probably due to a stereoelectronic effect of the 5-aza-7-deazaguanine base. The conformation at the C(4')-C(5') bond of **1b** [$\gamma^{+(g)}$ 48%, γ^t 33%; $\gamma^{(g)}$ 19%] – taken from the 4',5' and 4',5'' ^{1}H,^{1}H-couplings (E. Westhof *et al.*, 1975) - is similar to that of dG and c^{7}G$_d$.

5.   X-Ray analysis of 5-aza-7-deazapurine nucleosides

Several crystal structures of 5-aza-7-deazapurine nucleosides have been reported. The crystal structure of **1a**. H_2O was published by B. Kojić-Prodić (B. Kojić-Prodić *et al.*, 1982). The nitrogen transposition compared to guanosine induced only minute steric alterations into the guanine moiety but significant changes in the electron charge distribution (S.-P. Jiang *et al.*, 1994) outing now N(1) as a hydrogen bridge acceptor. The orientation of the base relative to the sugar ring is *anti*, and the puckering of the ribose moiety is ^{2}E. The conformation at the C(4')-C(5') bond is found to be –(g).

The crystal structure of the α-*D*-anomer of 5-aza-7-deaza-2'-deoxyguanosine (**59**) has been also solved recently (Figure 1) (F. Seela *et al.*, 2002). The compound crystallizes from water as a mono hydrate in the monoclinic space group P2$_1$ which is identical to the space group in which the corresponding β-*D*-ribonucleoside crystallizes. The structure is stabilized by several hydrogen bonds. The solvent molecule forms hydrogen bonds to four symmetry equivalent nucleoside molecules acting as both, donor and acceptor. In the acceptor case the water oxygen is linked to the hydroxy groups of the sugar unit. In the donor case the water molecule forms hydrogen bonds to the nitrogen atoms N(1) and N(3) of the nucleobase.

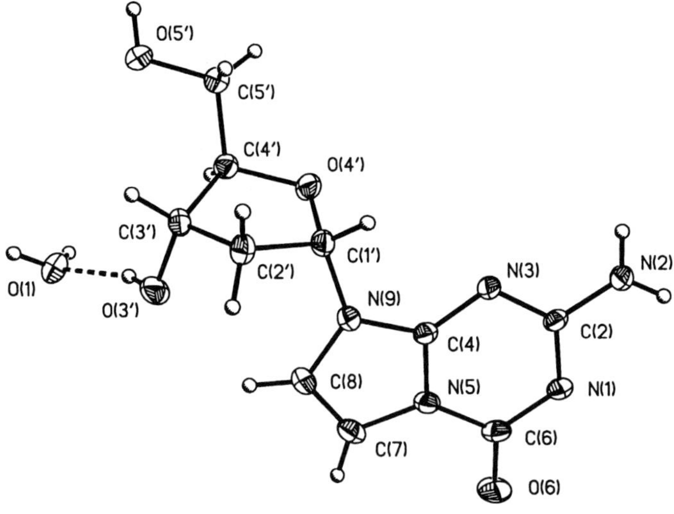

Figure 2.

The base of the nucleoside is planar; its orientation relative to the sugar lies in the high-anti range in which the C(1')-C(2') and N(9)-C(8) are nearly eclipsed. Such a conformation is quite unusual and mainly occupied by 1'-deoxy-1'-(8-aza- and 8-aza-7-deazapurine-9-yl)-2'-deoxy-β-D-ribofuranosides where it is traced back to a Coulomb repulsion between non-bonding electron pairs at O(4') and N(8) (F. Seela *et al.*, 1999). On the other hand, however, the β-D-ribonucleoside of 5-aza-7-deazaguanine (**1a**) exhibits an anti-conformation; the reason for this striking difference is still unclear. The sugar puckering of **59** in the crystalline state is $^{2'}T_{3'}$ (S-type), and the conformation about the C(4')-C(5') bond is –sc (t).

6.   Oligonucleotides containing 5-aza-7-deazapurines: base pairing properties in duplexes with parallel and antiparallel chain orientation

So far, only very few 5-aza-7-deazapurines have been incorporated into oligonucleotides. Their incorporation is interesting from the point of view to evaluate new DNA-structures and to study the recognition of unusual bases. Furthermore, new constructs for the antisense technology will be accessible in duplexes with parallel or antiparallel chain orientation. In the following the synthesis of the building blocks of 5-aza-7-deaza-2'-deoxyguanosine (**1b**) and 5-aza-7-deaza-2'-deoxyiso (**4b**) (J. J. Voegel and S. A. Benner, 1996) and the base pairing properties of the nucleosides will be discussed in some detail.

Oligonucleotides with 5-aza-7-deaza-2'-deoxyisoguanosine (**4b**). - The base pairing properties of **4b** were investigated on DNA-RNA hybrids with the ribonucleoside **80** (Py) located opposite to **4b**. Benner and co-workers (J. J. Voegel and S. A. Benner, 1996) prepared the phosphoramidite **79** from the educt **4b** as outlined in Scheme 24. The ribonucleoside **80** was incorporated into the oligoribonucleotide enzymatically with T4 RNA ligase using its 3',5'-diphosphate **81** as precursor (F. Seela and Q. H. Tran Thi, 1979; T. England and O. C. Uhlenbeck, 1978) (Scheme 25).

Incorporation of **4b** opposite to the ribonucleoside **80** within the duplex 5'-r(GAAC(**80**)AAA) • 5'-d(TTT**4b**GTTC) resulted in a DNA-RNA hybrid which was significantly destabilised (T_m-value 27°C) over the parent duplex 5'-r(GAACGAAA) • 5'-d(TTTCGTTC) carrying a dG-rC base pair (T_m = 36°C).

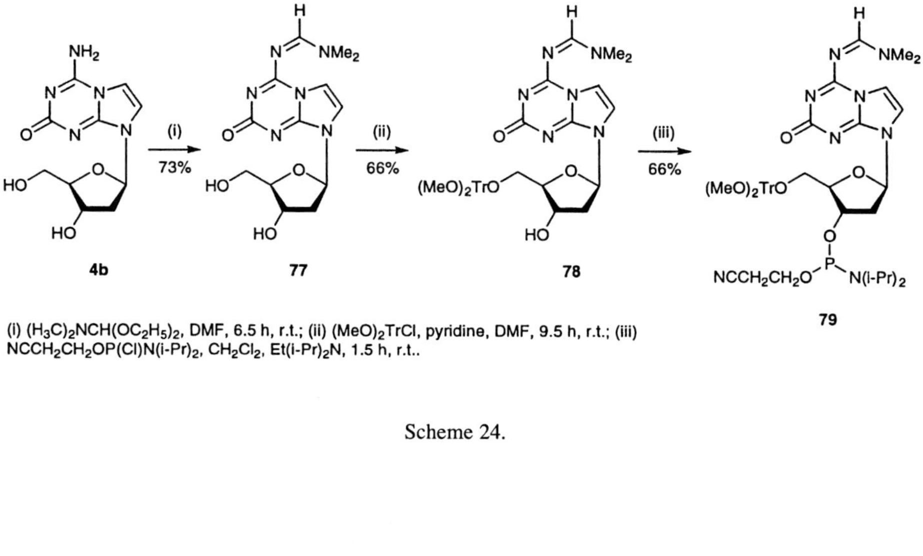

(i) (H$_3$C)$_2$NCH(OC$_2$H$_5$)$_2$, DMF, 6.5 h, r.t.; (ii) (MeO)$_2$TrCl, pyridine, DMF, 9.5 h, r.t.; (iii) NCCH$_2$CH$_2$OP(Cl)N(i-Pr)$_2$, CH$_2$Cl$_2$, Et(i-Pr)$_2$N, 1.5 h, r.t..

Scheme 24.

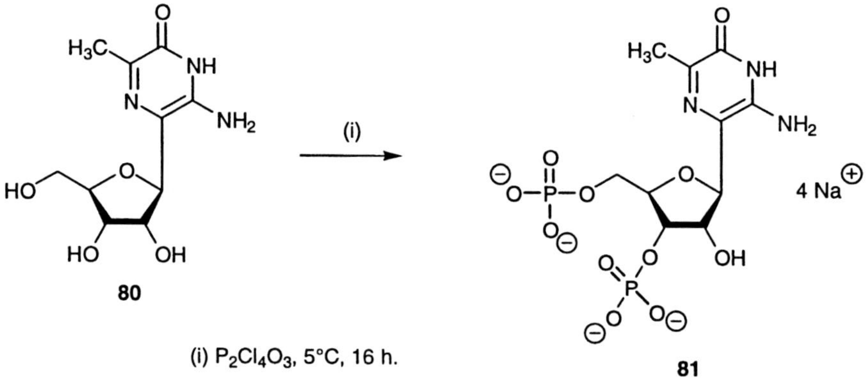

(i) P$_2$Cl$_4$O$_3$, 5°C, 16 h.

Scheme 25.

The tridentate base pair motif (I) was suggested (Scheme 26). According to results observed on C-nucleosides, the decrease of the T_m-value might not be attributed alone to the properties of the 5-aza-7-deazaguanine nucleoside. It can be also due to the lower flexibility of the ribose moiety of the C-nucleoside **80**. Reports on the lower stability of base pairs containing C-nucleoside have already been made (N. Ramzaeva *et al.*, 2000; C. Thibaudeau *et al.*, 1994).

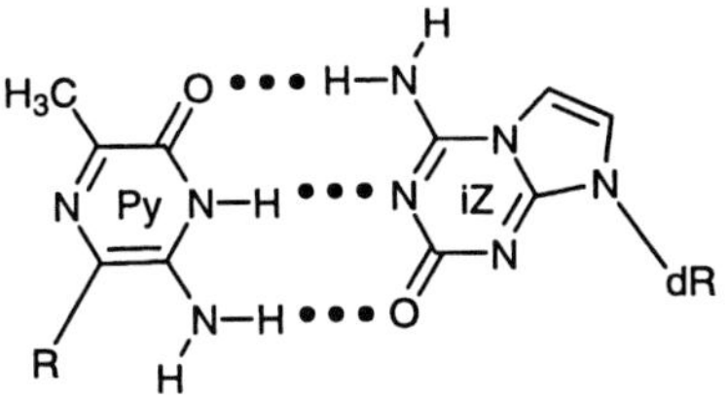

Scheme 26.

Oligonucleotides with 5-aza-7-deaza-2'-deoxyguanosine (**1b**). - The base pairing properties of 5-aza-7-deaza-2'-deoxyguanosine and of its α-*D* anomer (**1b**, ßZ$_d$; **59**, αZ$_d$) have been investigated in detail by our laboratory (F. Seela and A. Melenewski, 1999; F. Seela *et al.* 2001). For this purpose, the phosphoramidites **84** and **85** were synthesized. As educts compound **1b** and its α-*D* anomer **59** were used.

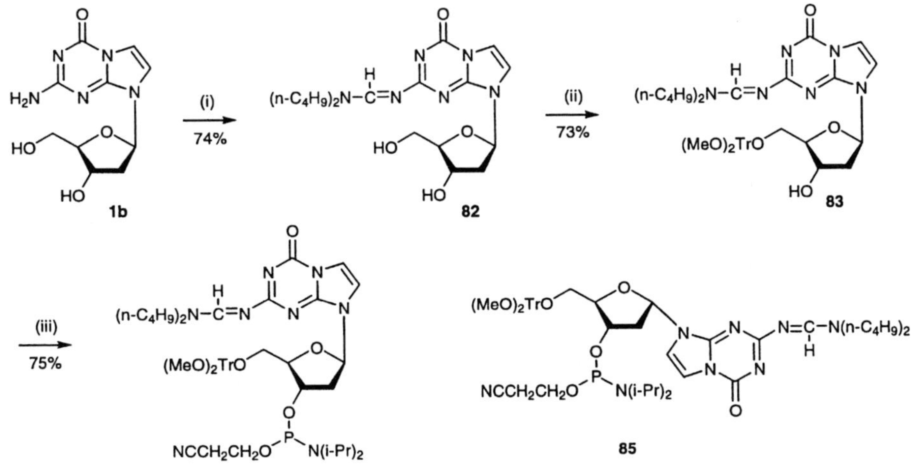

(i) (n-H$_9$C$_4$)$_2$NCH(OC$_2$H$_5$)$_2$, 2 h, 40°C; (ii) (MeO)$_2$TrCl, 4 h, r.t.; (iii) NCCH$_2$CH$_2$OP(Cl)N(i-Pr)$_2$, CH$_2$Cl$_2$, Et(i-Pr)$_2$N, 30 min, r.t..

Scheme 27.

Both nucleosides were protected at their amino groups with an N,N-di-(*n*-butyl)formamidine residue. Then, the DMT group was introduced. Phosphitylation under standard conditions furnished finally the phosphoramidites **84** and **85** (Scheme 27). These compounds were used in the solid phase synthesis of a number of oligonucleotides.

Oligonucleotides with 5-aza-7-deazaguanine opposite to cytosine. - An interesting feature of 5-aza-7-deazaguanine is its ability to act as hydrogen bond acceptor at N(1) (analogous to isocytosine) in its neutral form but as a donor (analogous to guanine) in the protonated form. Appropriate base pairs are possible with other bases providing a proton to nitrogen-1 of the 5-aza-7-deazaguanine moiety. Stable, tridentate Z_d-dC Watson-Crick (WC) base pairs are expected in acidic solution (Scheme 28, WC-base pair IIb) whereas under neutral conditions (Scheme 28, WC-base pair IIa) a lower stability is anticipated. Indeed, oligonucleotide duplexes displayed in Table 2 show that 5-aza-7-deazaguanine forms a stable base pair with cytosine in acidic solution indicated by the higher T_m-value of the duplex at pH 5.0 (compared to pH 7.0). In this case, a proton is inserted between nitrogen-1 of the 5-aza-7-deazaguanine base and nitrogen-3 of the cytosine moiety forming the third hydrogen bond (Scheme 28). The lower T_m-values of the protonated duplexes compared to a duplex containing a dG-dC pair might be due to the positive charge being present within the core of the double helix.

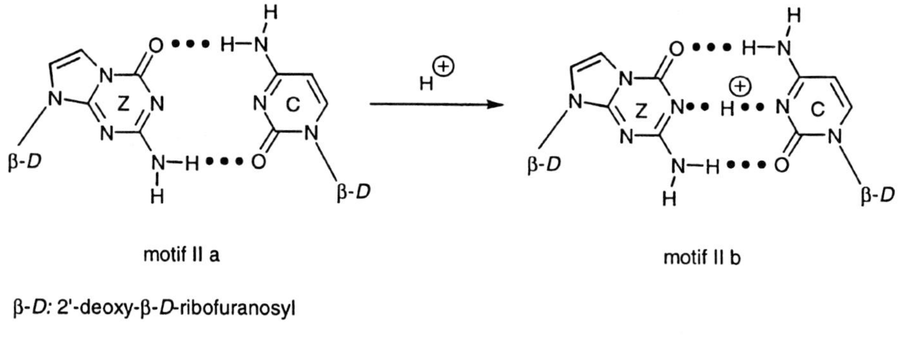

Scheme 28.

Homochiral oligonucleotides with 5-aza-7-deazaguanine opposite to guanine or isoguanine. - According to the fact that 5-aza-7-deazaguanine has the same Watson-Crick recognition site as isocytosine (Scheme 3) and isocytosine forms antiparallel duplexes with isoguanine the same was expected for compound **1b**. Furthermore, duplexes with parallel chain orientation should be accessible when 5-aza-7-deazaguanine pairs with guanine (Scheme 29).

In order to prove this hypothesis, the homochiral block oligonucleotide duplexes displayed in Table 3 were synthesized and their T_m-values have been measured UV-spectrophotometrically (F. Seela and A. Melenewski, 1999). As can be seen from Table 3, the hexamers 5'-d(GGG **1b1b1b**) (**90**) as well as the inversed oligomer

Table 2. T_m-Values of oligonucleotide duplexes containing dC-Z_d base pairs at different pH values (Z_d: 5-aza-7-deaza-2'-deoxyguanosine, **1b**).

Duplex	$T_m[°C]$ pH 5.0	$T_m[°C]$ pH 7.0
5'-d(TAGGTCAATACT) **86** 3'-d(ATCCAGTTATGA) **87**	45	46
5'-d(TA**1b1b**TCAATACT) **88** 3'-d(ATCCAGTTATGA) **87**	34	23
5'-d(TAGGTCAATACT) **86** 3'-d(ATCCA**1b**TTAT**1b**A) **89**	37	23

Measured at 260 nm in 0.1 M NaCl, 10 mM MgCl2, 10 mM Na-cacodylate, pH 7.5 at 3.5 + 3.5 μM of single strands.

5'-(**1b1b1b** GGG) (**92**) can be hybridized with the iG_d-containing complementary strands **91** and **93** to duplexes with 6 base pairs in a antiparallel manner. This results in T_m-values of 41°C and 46°C, respectively. On the other hand, hybridization of the block oligomers **92** and **90** results in the formation of a parallel-stranded duplex (T_m = 50°C), also consisting of 6 base pairs. None of the hexamers form duplexes with itself as this would imply aggregates built up from only 3 base pairs, each.

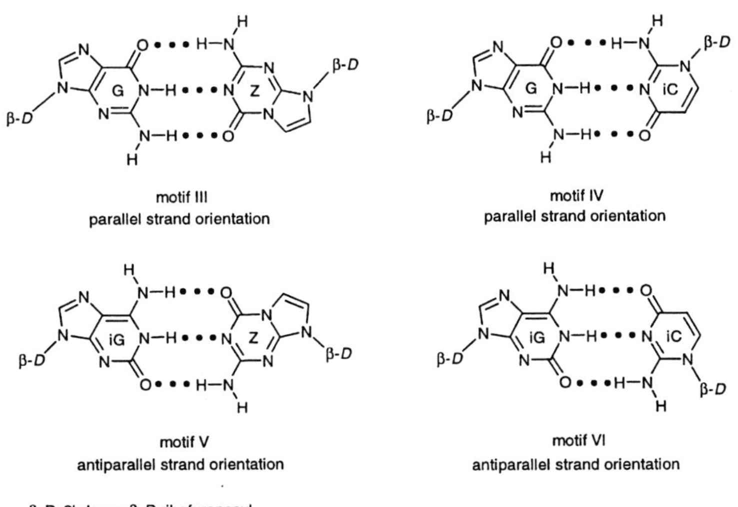

Scheme 29.

Table 3. T_m-values[a] and thermodynamic data of homochidral, antiparallel and parallel block-oligonucleotides.

Duplex	T_m [°C]	$\Delta H°$ [kcal/mol]	$\Delta S°$ [cal/mol K]	$\Delta G°_{310}$ [kcal/mol]
5'-d(GGG1b1b1b)90 3'-d(CCCiGiGiG)91	41	-41	-105	n.d.
5'-d(1b1b1bGGG)92 3'-d(iGiGiGCCC)93	46	-74	-209	-9
5'-d(1b1b1bGGG)92 5'-d(GGG1b1b1b)90	50	-50	-128	n.d.
5'-d(GGGCCC)94 3'-d(CCCGGG)94	36	-41	-112	-7

[a]   For experimental conditions see Table 2.

*Heterochiral duplexes containing the 5-aza-7-deazaguanosine α-D ribofuranoside **59**. -* From the results discussed above it is apparent that the exchange of the amino and the hydroxy group of 2'-deoxyguanosine reverses the chain orientation and generates a reverse Watson-Crick base pair instead of a Watson-Crick pair. The orientation of the chain is reversed again when one nucleoside within a base pair changes its configuration from β-*D* to α-*D*. The change of the chain orientation by changing the anomeric configuration was already postulated by U. Sequin (U. Sequin, 1973) and was studied in detail on nucleosides with canonical bases by J.-L. Imbach and co-workers (F. Morvan *et al.*, 1987). The base pair motifs and the chain orientation for the αZ_d-dG and αZ_d $-c^7G_d$ base pairs is outlined in Scheme 30. To investigate this matter, heterochiral oligonucleotides with stretches of the α-*D* and β-*D* units (block oligomers) as well as those with an alternating anomeric configuation were prepared (F. Seela *et al.*, 2001). The principle of duplex formation is shown in Table 4.

Table 5 summarizes data of duplex formation when the α-D-anomer of Z_d (**59**) forms a base pair with dG. Duplexes of high stability are formed which show antiparallel chain orientation and not an arrangement with parallel chains as it was observed in the case of Z_d (Table 3). The heterochiral "homopurine" duplexes are rather stable. The replacement of dG by c^7G_d weakens the base pair.

After the investigations of duplexes containing tracts of the modified and "purine" bases in the α-*D* or β-*D* configuration have been made, the studies were extended to self-complementary oligonucleotides in which the anomeric configuration was alternating (Table 4). As can be seen from Table 6, also in these case very stable duplexes are accessible (Table 6). This was surprising as a continuous change of the

motif VII
antiparallel strand orientation

motif VIII
antiparallel strand orientation

α–D, β-*D:* 2'-deoxy-(α,β)-*D*-ribofuranosyl

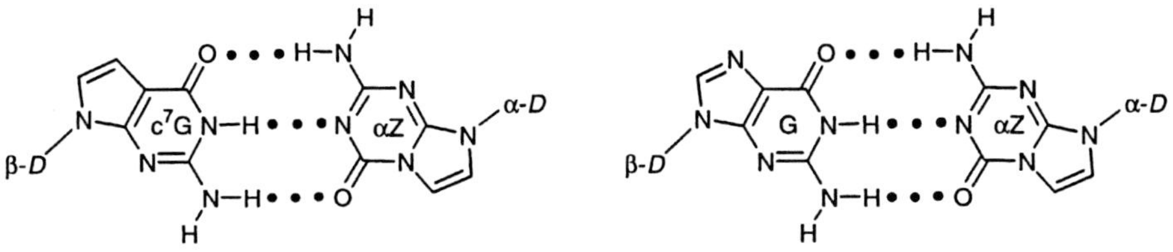

Scheme 30.

Table 4. Schematic structures of heterochiral oligonucleotides.

block-hexamers	altering hexamers
5'-d(α-α-α-β-β-β) 3'-d(β-β-β-α-α-α)	5'-d(α-β-α-β-α-β) 3'-d(β-α-β-α-β-α)
5'-d(β-β-β-α-α-α) 3'-d(α-α-α-β-β-β)	5'-d(β-α-β-α-β-α) 3'-d(α-β-α-β-α-β)

Table 5. T_m-Values[a)] and thermodynamic data of antiparallel block-oligomers.

Oligomer	T_m [°C]	ΔH° [kcal/mol]	ΔS° [cal/mol K]	$ΔG°_{310}$ [kcal/mol]
5'-d(G G G **59 59 59**) **95** 3'-d(**59 59 59** G G G) **95**	42	-33	-83	-8
5'-d(**59 59 59** G G G) **96** 3'-d(G G G **59 59 59**) **96**	54	-51	-131	-10
5'-d(c^7G c^7G c^7G **59 59 59**) **96** 5'-d(**59 59 59** c^7G c^7G c^7G) **96**	37	-22	-41	-8

a) For experimental details see Table 2.

anomeric configuration might induce conformational stress within the sugar phosphate backbone. This is obviously not the case. The replacement of the α-*D* anomer **59** by the β-*D* counterpart **1b** reverses the chain orientation again and parallel-stranded duplexes are formed. However, these duplexes are formed by only five tridentate base pairs (Table 6). Their rather high stability results partly from the nucleotide overhangs which strongly stabilize oligonucleotide duplexes by stacking interactions.

Table 6. T_m-Values[a] and thermodynamic data of heterochiral duplexes containing anomeric 5-aza-7-deaza-2'-deoxyguanosines **59** or **1b**.

Oligomer	T_m [°C]	$\Delta H°$ [kcal/mol]	$\Delta S°$ [cal/mol K]	$\Delta G°_{310}$ [kcal/mol]
5'-d(G **59** G **59** G **59**) **97** 3'-d(**59** G **59** G **59** G) **97**	55	-52	-136	-10
5'-d(**59** G **59** G **59** G) **98** 3'-d(G **59** G **59** G **59**) **98**	62	-70	-186	-12
5'-d(**59** c⁷G **59** c⁷G **59** c⁷G) **99** 3'-d(c⁷G **59** c⁷G **59** c⁷G **59**) **99**	52	-54	-141	-10
5'-d(G C G C G C) **100** 3'-d(C G C G C G) **100**	46	-55	-150	-8
5'-d(G **1b** G **1b** G **1b**) **101** 5'-d(G **1b** G **1b** G **1b**) **101**	53	-	-	-

[a]   For experimental details see Table 2.

7.   The pharmacological activity of 5-aza-7-deazapurines and their nucleosides

The biological activity of 5-aza-7-deazapurine bases and nucleosides (Scheme 31) has been investigated only scarcely. R. K. Robins and co-workers (S.-H. Kim *et al.*, 1978) studied the antiviral activity of 5-aza-7-deazaguanine (**30**) and 5-aza-7-deazaxanthine (**31**) as well as their ribonucleosides (**1a**, **2a**) against type 1 herpes virus, type 3 parainfluenza virus and type 13 rhino virus.

Inhibition of the virus-induced cytopathic effect (CPE) was used as the indicator of antiviral activity. As anti rhino virus activity was observed for compounds **1a** and **30**, the authors tested them against four additional virus species. The 5-aza-7-deazaguanine base (**30**) as well as its ribonucleoside (**1a**) showed only moderate activity against all five rhino viruses, comparable with ribavirin. On the other hand, the base **30** was

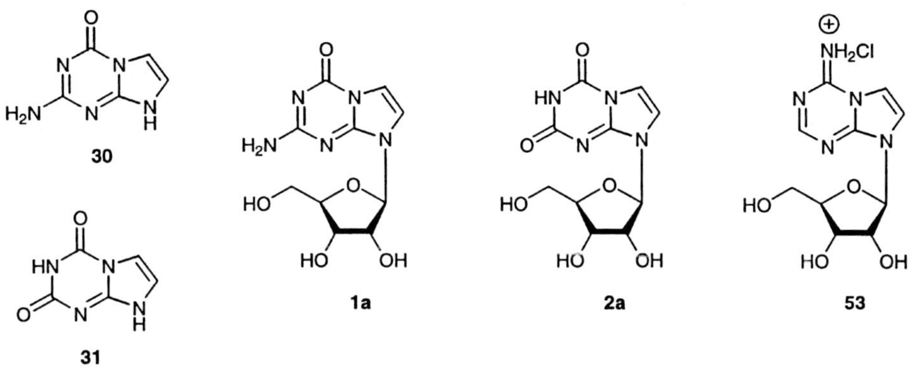

Scheme 31.

found to be a competitive inhibitor of xanthine oxidase with a K_i value of 0.55 mM (H. Rosemeyer and F. Seela, 1987).

5-Aza-7-deazapurin-6-imine-9-yl β-D-ribofuranoside (**53**) develops cytotoxicity against HeLa and L-1210 cells in culture (ED_{50} 2.6 and 0.5 μg/ml) (E. J. Prisbe *et al.*, 1978)

Assays of the acyclic nucleosides **63** and **64** (Scheme 32) against herpes simplex virus types I (HSV-I) and II (HSV-II) in cell culture brought the surprising result that the 7,8-dihydro derivative **64** exhibits a moderate antiviral activity (D. P. C. McGee *et al.*, 1985). This prompted the authors to synthesize the 7,8-dihydro nucleoside **65**. The latter, however, was inactive against both, HSV-I and HSV-II in cell culture.

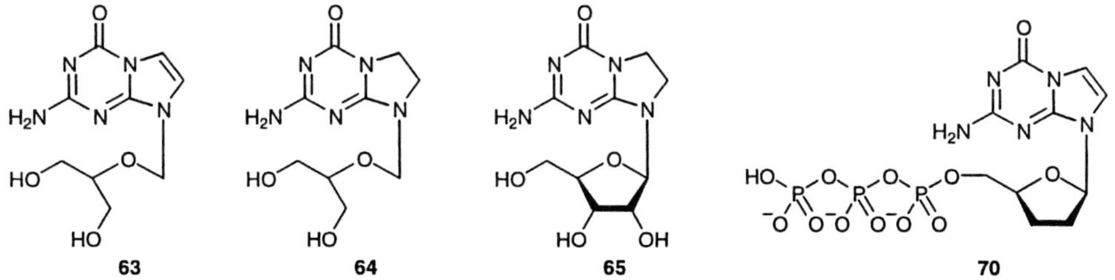

Scheme 32.

Also the 5'-triphosphate of the 2',3'-dideoxy-β-D-ribonucleoside of 5-aza-7-deaza-guanine (**70**) was prepared and tested as inhibitor of HIV reverse transcriptase (HIV-RT). However, the IC_{50} was low (> 100 μM) in comparison to AZTTP (IC_{50} of 6.6 μM) (F. Seela *et al.*, 1990).

8. Acknowledgments

We gratefully acknowledge financial support by the European Community (grant no.: OLK3-CT-2001-00506, "Flavitherapeutics").

9. References and notes

Appel R. (1975) Tertiäres Phosphan/Tetrachlormethan, ein vielseitiges Reagens zur Chlorierung, Dehydratisierung und PN-Verknüpfung. Angew. Chem.; 87: 863-874.

Balasubramanian KK, Bindumadhavan GV, Udupa MR, Krebs B. (1980) A novel thermal transformation of 2,4-di(N-aryl)amino-1,3,5-triazine-6-yl prop-2-ynyl ethers. Tetrahedron Lett.; 21: 4731-4734.

Capuano L, Schrepfer HJ. (1971) Darstellung von Pyrazolo-, Triazolo-, Oxazolo- und Thiazolo-*s*-triazinen mit Brückenkopf-Stickstoff sowie eines *N*-Äthoxycarbonyl-isopurins. Chem. Ber.; 104: 3039-3047.

Dovlatyan VV, Eliazyan KA, Agadzhanyan LG. (1977) Synthesis and thermal decomposition of haloalkoxy-*s*-triazines. I. Synthesis of chloroethoxy-*s*-triazines and their conversion to dihydroimidazo-*s*-triazines. Khim. Geterotsikl. Soedin.; 2: 262-265.

Dovlatyan VV, Pivazyan VA, Eliazyan KA, Mirzoyan RG. (1980) Synthesis and thermal decomposition of haloalkoxy-*s*-triazines. 6. Synthesis and thermolysis of 2-(2-chloroethoxy)-4-N-methyl-N-cyanoamino-6-dialkylamino-*s*-triazines. Khim. Geterotsikl. Soedin.; 11: 1558-1560.

Dovlatyan VV, Pivazyan VA, Eliazyan KA, Skhakyan SM, Mirzoyan RG. (1981) Synthesis and thermal decomposition of haloalkoxy-*s*-triazines. IX. Thermal decomposition of 2-(dialkylamino)-4-(2-chloroethoxy)-6-(cyanoamino)-*s*-triazines. Arm. Khim Zh.; 34: 576-579.

Dovlatyan VV, Pivazyan VA, Eliazyan KA, Mirzoyan RG. (1985) Synthesis and thermolysis of haloalkoxy-sym-triazines. 11. Synthesis and thermolysis of 2-(2-chloroethoxy)-4-(N-methyl-N-methoxyamino)-6-(dimethylamino)-*sym*-trazines. Khim. Geterotsikl. Soedin.; 8: 1125-1128.

England T, Uhlenbeck OC. (1978) Enzymatic oligoribonucleotide synthesis with T4 RNA ligase. Biochemistry; 17: 2069-2076.

Englisch U, Gauss DH. (1991) Chemisch modifizierte Oligonucleotide als Sonden und Agentien. Angew. Chem.; 103: 629-646.

Hinkens R, Promel R, Martin RH. (1961) Synthese de dihydro-6,7-imidazo[1,2-a]triazines-1,3,5 disubstituees. Helv. Chim. Acta; 37 : 299-309.

Jiang S.-P, Raghunathan G, Ting K.-L, Xuan JC, Jernigan RL. (1994) Geometries, charges, dipole moments and interaction energies of normal, tautomeric and novel bases. J. Biomol. Struct. Dyn.; 12: 367-382.

Katritzky AR, Rees CW, Scriven EFV, editors-in-chief; Jones G, volume editor (1996) Comprehensive Heterocyclic Chemistry II. Vol. 8; Pergamon Press.

Kim S.-H, Bartholomew DG, Allen LB, Robins RK, Revankar GR, Dea P. (1978) Imidazo[1,2-a]-*s*-triazine nucleosides. Synthesis and antiviral activity of the N-bridgehead guanine, guanosine, and guanosine monophosphate analogues of imidazo[1,2-a]-*s*-triazine. J. Med. Chem.; 21: 883-889.

Kobe J, Stanovnik B, Tišler M. (1970) Heterocyclen, 73. Mitt.: Synthesen einiger substituierter 1,3,5-Triazine und Imidazo[1,2-a][1,3,5]-triazine. Mh. Chem.; 101: 724-735.

Kojić-Prodić B, Ružić-Toroš Ž, Golič L, Brdar B, Kobe J. (1982) Conformation and structure of 2-amino-8-(β-D-ribofuranosyl)imidazo[1,2-a]-s-triazin-4-one (5-aza-7-deazaguanosine), a potent antiviral nucleoside. Biochim. Biophys. Acta; 698: 105-110.

Ludwig J. (1981) A new route to nucleoside 5'-triphosphates. Acta Biochim. Biophys. Acad. Sci. Hung.;

16: 131-133.

McGee DPC, Martin JC, Verheyden JPH. (1985) Synthesis of the 7-deaza and 5-aza-7-deaza purine analogs of the antiherpes agent 9-[(1,3-dihydroxy-2-propoxy)methyl]guanine (DHPG). J. Heterocyclic Chem.; 22: 1137-1140.

Morvan F, Rayner B, Imbach J.-L, Chang D.-K, Lown JW. (1987) α-DNA. Synthesis, characterization and base pairing properties of unnatural α-oligodeoxyribonucleotides. Nucleosides Nucleotides; 29: 1059-1062.

Nair V, Lyons AG, Purdy DF. (1991) Novel dideoxynucleoside isosteres. Tetrahedron; 47: 8949-8968.

Niedballa U, Vorbrüggen H. (1974) A general synthesis of N-glycosides. I. Synthesis of pyrimidine nucleosides. J. Org. Chem.; 39: 3654-3660.

Prisbe EJ, Verheyden JPH, Moffatt JG. (1978 a) 5-Aza-7-deazapurine nucleosides. 2. Synthesis of some 8-(D-ribofuranosyl)imidazo[1,2-a]-1,3,5-triazine derivatives. J. Org. Chem.; 43: 4784-4793.

Prisbe EJ, Verheyden JPH, Moffatt JG. (1978 b) 5-Aza-7-deazapurine nucleosides. 1. Synthesis of some 1-(ß-D-ribofuranosyl)imidazo[1,2-a]-1,3,5-triazines. J. Org. Chem.; 43: 4774-4784.

Ramzaeva N, Rosemeyer H, Leonard P, Mühlegger K, Bergmann F, von der Eltz H, Seela F. (2000) Oligonucleotides functionalized by fluorescein and rhodamin dyes: *Michael* addition of methyl acrylate to 2'-deoxypseudouridine. Helv. Chim. Acta; 83: 1108-1128.

Revankar GR, Robins RK. (1991) The synthesis and chemistry of heterocyclic analogues of purine nucleosides and nucleotides. In: Townsend LB, editor. Chemistry of Nucleosides and Nucleotides, Vol. 2; Plenum Press, New York: 161-398.

Rosemeyer H, Seela F (unpublished results, a). Compd. **34**: Anal. calcd. for $C_{11}H_{21}N_5O_5$ (303.3): calcd. C 43.56, H 6.98, N 23.09; found: C 43.72, H 6.96, N 23.17. Compd. **35**: Anal. Calcd. for $C_7H_9N_5O_3$ (211.2): calcd. C 39.81, H 4.30, N 33.16; found: C 39.69, H 4.36, N 33.22. Compd. **36**: Anal. Calcd. for $C_7H_5N_5O$ (175.2): calcd. C 48.00, N 2.88, N 39.99; found: C 47.96, H 3.18, N 39.99; ¹H-NMR (D_6DMSO): δ 7.51 (d, J = 2.3 Hz), 7.28 (d, J = 2.3 Hz); UV (MeOH): $λ_{max}$ = 276 nm (ε 14800); mp (H_2O): > 320°C.

Rosemeyer H, Seela F (unpublished results, b). *2-Amino-8-[2-deoxy-5-O-{(1,1-dimethylethyl)diphenylsilyl}-α-D-erythro-pentofuranosyl]-imidazo[1,2-a]-s-triazin-4-one and 2-amino-8-[2-deoxy-5-O-{(1,1-dimethyl-ethyl)di-phenylsilyl}-β-D-erythro-pentofuranosyl]-imidazo[1,2-a]-s-triazin-4-one:* An anomeric mixture (**1b/59**, 2:1, 500 mg, 1.87 mmol) is dissolved in anhydr. amine-free dimethyl formamide. *t*-Butyldiphenyl-silyl chloride (0.514 ml, 2 mmol) and imidazole (320 mg, 4.7 mmol) are added, and the reaction mixture is stirred for 24 h at room temperature. After evaporation of the solvent in high vacuo, the residue is dissolved in MeOH and flash-chromatographed on silica gel 60H (column: 6 × 25 cm, $CHCl_3$-MeOH, 9:1, 0.5 bar). From the faster migrating zone the α-*D* anomer (210 mg, 23%) is obtained after evaporation of the solvent as a colorless foam. From the slower migrating zone the ß-*D*-anomer is isolated (450 mg, 48%).

Rosemeyer H, Seela F. (1987) 5-Aza-7-deaza-2'-deoxyguanosine: Studies on the glycosylation of weakly nucleophilic imidazo[1,2-a]-*s*-triazinyl anions. J. Org. Chem.; 52: 5136-5143.

Rosemeyer H, Toth G, Seela F. (1989) Assignment of anomeric configuration of *D*-ribo-, arabino- 2'-deoxyribo- and 2',3'-dideoxyribonucleosides by noe difference spectroscopy. Nucleosides Nucleotides; 8: 587-597.

Rosemeyer H, Toth G, Golankiewicz B, Kazimierczuk Z, Bourgeois W, Kretschmer U, Muth H.-P, Seela F. (1990) Syn-anti conformational analysis of regular and modified nucleosides by 1D ¹H NOE difference spectroscopy: A simple graphical method based on conformationally rigid molecules. J. Org. Chem.; 55: 5784-5790.

Rosemeyer H, Seela F. (1997) Stereoelectronic effects of modified purine bases on the sugar conformation of nucleosides: pyrrolo[2,3-d]pyrimidines. J. Chem. Soc., Perkin Trans. 2; 2341-2345.

Sączewski F, Gdaniec M. (1987) Synthesis, reactions, and crystal structure of 2-(alkylthio)-7,8-dihydroimidazo[1,2-a]-1,3,5-triazine-4(6*H*)-thiones. Liebigs Ann. Chem.; 721-724.

Schaefer FC. (1955) Rearrangement reactions of 1-aziridinyl-*s*-triazines; dihydroimidazo[1,2-s]-*s*-triazines. J. Am. Chem. Soc.; 77: 5922-5928.

Seela F, Rosemeyer H, Melenewski A, Heithoff E-M, Eickmeier H, Reuter H (2000) The α-D-anomer of 5-aza-7-deaza-2'-deoxyguanosine. Acta Cryst. C58: 142-144.

Seela F, Amberg S, Melenewski A, Rosemeyer H (2001) 5-Aza-7-deazaguanine DNA: recognition and strand orientation of oligonucleotides incorporating anomeric imidazo[1,2-*a*]-*s*-triazine nucleosides. Helv. Chim. Acta; 84: 1996-2014.

Seela F, Tran Thi QH. (1979) 2'(3'),5'-Diphosphate des Nucleosides X und N³-alkylierter Uridin-Derivate. Chem. Ber.; 112: 3743-3747.

Seela F, Bourgeois W. (1989) Stereoselective glycosylation of nitrobenzimidazole anions: Synthesis of 1,3-dideaza-2'-deoxyadenosine and related 2'-deoxyribofuranosides. Synthesis; 912-918.

Seela F, Bourgeois W, Gumbiowski R, Röling A, Rosemeyer H, Mertens A, Zilch H, König B, Koch E. (1990 a) Purine analog nucleoside and nucleotide compounds. US patent 5446139; priority: 23. 04. 1990. Compound **70**: Yield 46% of a colorless foam. ³¹P-NMR [rel. to ext. H_3PO_4 (85%), D_2O/TRIS-HCl buffer, pH 8.0, 1:1, 100 mM EDTA): δ -8.30 (d, J = 19 Hz, P_γ); -10.6 (d, J = 19 Hz, P_α); -22.2 (t, J = 19 Hz, P_β)].

Seela F, Rosemeyer H, Fischer S. (1990 b) Synthesis of 3-deaza-2'-deoxyadenosine and 3-deaza-2',3'-dideoxyadenosine: glycosylation of the 4-chloroimidazo[4,5-c]pyridinyl anion. Helv. Chim. Acta; 73: 1602-1611.

Seela F, Melenewski A. (1999 a) 5-Aza-7-deaza-2'-deoxyguanosine : Oligonucleotide duplexes with novel base pairs, parallel chain orientation and protonation sites in the core of a double helix. Eur. J. Org. Chem.; 485-496.

Seela F, Becher G, Rosemeyer H, Reuter H, Kastner G, Mikhailopulo IA. (1999 b) The high-anti conformation of 7-halogenated 8-aza-7-deaza-2'-deoxyguanosines: A study of the influence of modified bases on the sugar structure of nucleosides. Helv. Chim. Acta; 82: 105-124.

Sequin U. (1973) Nucleosides and nucleotides. 5. Stereochemistry of oligonucleotides consisting of 2'-deoxy-α-D-ribosides, a study with Dreiding stereomodels. Experientia; 29: 1059-1062.

Staehle H, Koeppe H, Kummer W, Hoefke W. (1973) 2,3-Dihydro-8-phenylimidazo[1,2-a]-*s*-triazin-5-ones. Ger. Offen. 2,314,488; 23 Mar 1973.

Thibaudeau C, Plavec J, Chattopadhyaya J. (1994) Quantitation of the anomeric effect in adenosine and guanosine by comparison of the thermodynamics of the pseudorotational equilibrium of the pentofuranose moiety in *N*- and *C*-nucleosides. J. Am. Chem. Soc.; 116: 8033-8037.

Uhlmann E, Peyman A. (1990) Antisense oligonucleotides: A new therapeutic principle. Chem. Rev.; 90: 543-584.

Unishi T, Kitahama T, Shimomura Y. (1987) Synthesis of 2,4-diamino-6,7-dihydroimidazo[1,2-a][1,3,5]triazine derivatives. Nippon Kagaku Kaishi; 1: 40-44.

Unishi T, Takahashi H, Shimomura Y. (1988) Synthesis of 4-(p-toluidino)-2,6,7,8-tetrahydroimidazo[1,2-a][1,3,5]triazin-2-one. Nippon Kagaku Kaishi; 2 : 236-238.

Van Wijk J, Altona C. (1993) PSEUROT 6.2 – A program for the conformational analysis of five membered rings. University of Leiden, The Netherlands.

Veronese AC, Di Bello C, Filira F, D'Angeli F. (1971) Nonaromatic heterocycles. XI. Cycloadditions of methyl isothiocyanate onto 2-methyl- and 2-phenyl-1,3-diaza-2-cycloalkenes. Gazz. Chim. Ital.; 101: 569-580.

Voegel JJ, Altorfer MM, Benner SA. (1993) The donor-acceptor-acceptor purine analog: transformation of 5-aza-7-deaza-1-*H*-isoguanine (= 4-aminoimidazo[1,2-a]-1,3,5-triazin-2(1*H*)-one) to 2'-deoxy-5-aza-7-deaza-isoguanosine using purine nucleoside phosphorylase. Helv. Chim. Acta; 76:2061-2069.

Voegel JJ, Benner SA. (1996) Synthesis, molecular recognition, and enzymology of oligonucleotides containing the non-standard base pair between 5-aza-7-deazaisoguanine and 6-amino-3-methylpyrazin-2(1*H*)-one, a donor-acceptor-acceptor purine analog and an acceptor-donor-donor pyrimidine analog. Helv. Chim. Acta; 79: 1881-1898.

Westhof E, Röder O, Croneiss I, Lüdemann H.-D. (1975) Ribose conformation in the common purine (ß) ribosides, in some antibiotic nucleosides, and in some isopropylidene derivatives: a comparison. Z. Naturforsch.; 30c:131-140.

Wittenburg E. (1964) A new synthesis of nucleosides. Z. Chem.; 4: 303-304.

Yamakawa K, Naruse H. (1991) Novel dye forming coupler and silver halide photographic sensitive material formed by using this coupler. Japanese patent JP 3206451; publication date: 09.09.1991; 1 page.